Springer Series in Statistics

Springer Series in Statistics (SSS) is a series of monographs of general interest that discuss statistical theory and applications.

The series editors are currently Peter Bühlmann, Peter Diggle, Ursula Gather, and Scott Zeger. Peter Bickel, Ingram Olkin, and Stephen Fienberg were editors of the series for many years.

Anthony C. Atkinson · Marco Riani ·
Aldo Corbellini · Domenico Perrotta ·
Valentin Todorov

Robust Statistics Through the Monitoring Approach

Applications in Regression

Anthony C. Atkinson
Department of Statistics
London School of Economics and Political Science
London, UK

Aldo Corbellini
Department of Economics and Management
University of Parma
Parma, Italy

Valentin Todorov
Statistics Division
United Nations Industrial Development Organization
Vienna, Austria

Marco Riani
Department of Economics and Management
University of Parma
Parma, Italy

Domenico Perrotta
Joint Research Centre
European Commission
Ispra, Italy

ISSN 0172-7397   ISSN 2197-568X (electronic)
Springer Series in Statistics
ISBN 978-3-031-88364-4   ISBN 978-3-031-88365-1 (eBook)
https://doi.org/10.1007/978-3-031-88365-1

Mathematics Subject Classification: 62F35, 62Y05, 62P20

This work was supported by the European Commission's Joint Research Centre (JRC). The publication is part of the European Commission's Open Access Policy that, in line with the European Union's Horizon research and innovation programme, aims to make research outputs openly accessible.

© The Editor(s) (if applicable) and The Author(s) 2025. This book is an open access publication.

Open Access This book is licensed under the terms of the Creative Commons Attribution 4.0 International License (http://creativecommons.org/licenses/by/4.0/), which permits use, sharing, adaptation, distribution and reproduction in any medium or format, as long as you give appropriate credit to the original author(s) and the source, provide a link to the Creative Commons license and indicate if changes were made.
The images or other third party material in this book are included in the book's Creative Commons license, unless indicated otherwise in a credit line to the material. If material is not included in the book's Creative Commons license and your intended use is not permitted by statutory regulation or exceeds the permitted use, you will need to obtain permission directly from the copyright holder.
The use of general descriptive names, registered names, trademarks, service marks, etc. in this publication does not imply, even in the absence of a specific statement, that such names are exempt from the relevant protective laws and regulations and therefore free for general use.
The publisher, the authors and the editors are safe to assume that the advice and information in this book are believed to be true and accurate at the date of publication. Neither the publisher nor the authors or the editors give a warranty, expressed or implied, with respect to the material contained herein or for any errors or omissions that may have been made. The publisher remains neutral with regard to jurisdictional claims in published maps and institutional affiliations.

This Springer imprint is published by the registered company Springer Nature Switzerland AG
The registered company address is: Gewerbestrasse 11, 6330 Cham, Switzerland

If disposing of this product, please recycle the paper.

For my wife Basia and my daughters Alison and Rachel. Fellow scientists

A mia moglie Fabia e ai miei figli Jacopo e Cecilia, ai miei genitori e a mio fratello Gianfranco

A mia moglie Antonella e ai miei genitori Licia e Luigi

Alla mia famiglia, la cui forza si estende dalle coste spazzate dal vento dell'estremo sud alle maestose vette dell'estremo nord

For my wife Bistra, my children Marta and Diman and my grandson Nickolas

Preface

Why We Wrote This Book

Our book is about statistics, computing, and graphical methods in the service of robust data analysis. The recent development of the computational technique we call monitoring provides an envelopment of many robust methods, allowing the data to choose important parameters of the analysis. The procedures we describe yield highly informative graphical output that allow comparison of a variety of robust methods for each set of data. Robustness is required because data may be contaminated and models may be wrong. Successful data analysis requires methods that reveal any divergence between the data and the fitted model and provide indications of how both may be reconciled. When the data contain outliers, that is observations in disagreement with the model, the methods of traditional robust statistics are intended to give such observations reduced importance in the analysis, either by reducing the weight of the outliers, or by deleting them altogether.

Traditional robust statistics was first formalized in *The Princeton Robustness Study* of 1973 (see Sect. 1.1). From that time the robust analysis of data has had something of a black box about it. The box needs some inputs apart from the data. The most important is the proportion of outliers that the data analyst believes is present in the data. Given the choice of robust procedure, the box then delivers a robust estimate of the parameters, by downweighting or deleting observations during the fitting process. If too few observations are deleted, the parameter estimates will be biased by the outliers. If too many observations are deleted, information is lost and the parameter estimates will be unnecessarily imprecise. We call this the static approach.

Over the last 25 years, we have been developing an alternative dynamic approach to robust statistics, which we call the Forward Search. The search starts from a small, robustly chosen, subset of the data that excludes outliers. We then move forward through the data, adding observations to the subset used for parameter estimation and removing any that have become outlying due to the changing content of the subset. As we move forward, we monitor statistical quantities, such as parameter estimates, residuals, and test statistics. One importance of this approach is that sequential testing for outliers can be used to determine the proportion of outliers in the data and so to provide an empirical estimate of how many observations should be deleted in the robust estimation of the parameters.

These ideas were presented in the 2000 book *Robust Diagnostic Regression Analysis* by Anthony Atkinson and Marco Riani. The companion 2004 book on the analysis of multivariate data, *Exploring Multivariate Data with the Forward Search*, co-authored with Andrea Cerioli, extended the Forward Search to the analysis of multivariate data.

Subsequent developments in widely available computer graphics and our exploitation of them led to the idea of linking plots. A typical sequence could begin by noticing that towards the end of the search a group of outliers was identified. What are the properties and effects of these observations? For example, where do they lie in the space of the response and explanatory variables? Highlighting the observations in a plot of residuals against subset size leads to their identification in any linked plot, not just of their locations, but also, for example, what effect do these observations have on test statistics, such as those that indicate the transformation of the response in a regression model?

All of these ideas were developed for the Forward Search. What motivated the writing of this book was the idea of dynamic monitoring of the performance of a range of static robust estimators. The theoretical performance of these estimators is often derived from asymptotic results. The realization was that procedures similar to those that were used for the Forward Search could be used to provide non-asymptotic results for other estimators by enumerating their behaviour over a range of values of the expected proportion of outliers in the data. It is thus possible to compare the finite sample performance of many robust estimators, either through simulation or by comparative analyses of the same set of data. One particularly informative plot is that of monitoring residuals. For many robust estimators, there is a dramatic change in the pattern of residuals as the expected proportion of outliers in the data is reduced. Such a break provides an empirical breakdown point for the estimator, which may be far from the asymptotic one.

This is in stark contrast to typical R packages with robustness functions such as **robustbase** (Maechler et al., 2024) and **rrcov** (Todorov and Filzmoser, 2009). These contain implementations of robust routines, generally with a prefixed value of the required robustness or efficiency. In such R packages, there is no straightforward way to check the stability of the results to the prespecified value of robustness or efficiency. Further, there are no explicit methods to compute the various functions we describe in Chap. 2, that make possible insightful comparisons of the various formulations of robust statistics. A severe limitation is that these R packages do not contain any routines for interactive graphics. Functions for robust data transformation, the subject of our Chap. 6, were added only recently (Raymaekers and Rousseeuw, 2021). Non-parametric transformations, the subject of our Chap. 7, are not included.

The graphics-rich approach, combined with monitoring, that we have pioneered makes heavy demands on programming. Typically we require 50 robust regression fits per analysis, a computational burden only made possible by the efficiency of the **FSDA** (Flexible Statistics and data analysis) robust library, first references to which are in Riani et al. (2012, 2015). By now there are many examples where the computational requirement is at least as important and intellectually challenging as the methodological one (Riani et al., 2022a, c; Rousseeuw et al., 2019, are some cases). For this reason, we feel that the success of this book will in part depend on the care with which we have treated every computational detail, even some that today are relegated to the history of computation (Azzini et al., 2023). As evidence of this effort, and for the reader to test, Appendix A.1 contains a list of functions contained in our software.

What Is in Our Book

There are three main themes in our book:

1. Both theoretical and small-sample properties of static robust estimators. Establishing the latter depends on simulation;
2. The Forward Search and monitoring of static robust estimators for regression data and
3. The analysis of numerous sets of data and the provision of software for all methods described in the book.

Chapter 1 of the book introduces robust estimation for a simple sample and contrasts analyses from the traditional robust approach with the monitoring approach. Chapter 2 presents several methods for robust estimation of the mean and variance of a univariate sample. A thorough study is made of M-estimation and its developments and the properties of the ρ functions that are a part of these methods. Standard and robust regression are described in Chap. 3. Robust methods discussed include the Forward Search, S and MM-estimation and Least Trimmed Squares. Chapter 4, the Monitoring Approach in Multiple Regression, exhibits the behaviour of these methods in the analysis of six datasets of size up to 1949. We also introduce a general method of outlier detection based on the Bayesian Information Criterion (BIC) which enables us to select the explicit level of trimming. Chapter 5 uses extensive simulations to compare the size and power of outlier tests and introduces a general framework for comparing robust regression estimators as the distance between the main population and the outliers changes. The robust transformation of the response in a regression model is covered in detail in Chap. 6. Examples include the Box-Cox transformation and its extensions to responses that can be negative. All the models discussed so far have been fully parametric. Chapter 7 provides robust methods for the non-parametric transformation of the response and of the explanatory variables. The algorithm is a robustified and extended version of the AVAS (Additivity and VAriance Stabilization) algorithm of Tibshirani. Several extensions of the robust regression model are treated more briefly in Chap. 8: Bayesian regression; the analysis of heteroskedastic data; regression clustering; the robust monitoring of a specific time series model with level shifts, trend and time-varying seasonality and the analysis of compositional data and, very briefly, of censored regression data. Chapter 9 investigates several approaches to robust model selection with Chap. 10 presenting five robust analyses of regression data that illustrate the use of techniques introduced earlier in the book. The analysis of data on customer loyalty in Sect. 10.4 concludes in Sect. 10.4.6 with recommendations for a structured approach to modern robust regression analysis.

Our book is about robust statistical methods and procedures and mostly about their application to linear regression. Our 2000 book was more limited as the focus was on the Forward Search. The Preface promised a second book to cover multivariate analysis, which duly appeared in 2004. This time we have no such intention, but see Cerioli et al. (2018b).

For Whom We Wrote Our Book and How to Approach It

We have written our book to be of use and interest both to professional statisticians and to other scientists concerned with insightful data analysis, as well as to postgraduate students. Statistics is heavily used by governmental organisations in the study, assessment and implementation of their policies. Two of us, coming from this professional background in the European Commission and the United Nations Industrial Development Organization, have seen at first-hand the importance of creating awareness of the cautious use of real data if there are important decisions at stake. We hope that this book will help in this way in such environments.

More generally, our book is intended to serve as a resource for those who have data to analyse, especially if they are, as they should be, curious about the structure of the data and want to use the monitoring approach to get further into their data than they can using the standard statistical analyses, either robust or non-robust. The book is also intended as a text for a course on modern interactive robust regression. There are both theoretical exercises, which we use to deepen and extend the coverage of the text, and those that require use of the packages mentioned above. Both kinds of exercise have detailed solutions. We give references to the statistical and computing literatures but believe our book is reasonably self-contained.

The idea of monitoring conveyed by the book is simple to grasp but cannot be fully appreciated without playing and *interacting* with the data. We strongly encourage the reader to do so. To this end, we have created the GitHub repository

https://github.com/UniprJRC/MonitoringBook/

Inside this repository the user can find a link to a YouTube video describing the overall purpose and structure of the book and links to individual chapters. Inside each chapter can be found:

Preface

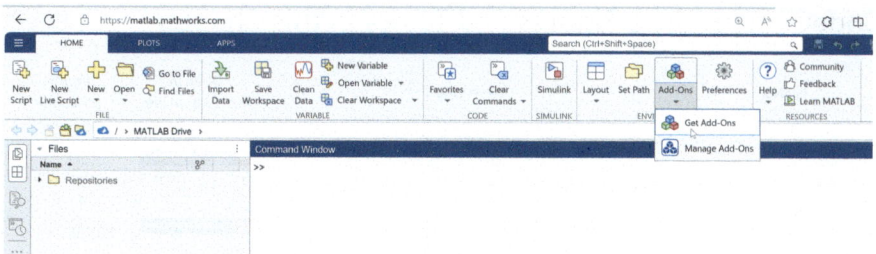

Fig. 1 In order to install **FSDA** from MATLAB desktop or online it is enough to click on Get Add-Ons (or Explore Add-Ons in MATLAB 2025) and in the Add-On Explorer window which appears, to search for FSDA. Published with permission from MathWorks, 2024, all rights reserved

- the abstract;
- a series of links to YouTube videos;
- files to reproduce the figures and tables.

The codes make extensive use of the **FSDA** toolbox, which the reader can find (free of course) in GitHub or in the space that MathWorks (the **MATLAB** company) assigns to relevant third party contributions (the **MATLAB** Central File Exchange).

In greater detail, in order to install **FSDA** as an AddOns directly in MATLAB desktop or online (see Fig. 1) it is enough to click on Get Add-Ons and in the Add-On Explorer window which appears, to search for FSDA. Note that after registration from the MathWorks website each user has free access to the **MATLAB** online environment (currently 20 h per month).

Because of the importance of Python in the wider data science community, each .m file in the GitHub web site of the book has a corresponding .ipynb file which contains the code in Jupiter format and the output which is produced. Specifically, for each Chapter, the README.md contains a table in which the second column indicates the figures the file can produce or the task which can be performed. The button in the third column enables the user to run the file in MATLAB on line and the fourth column provides the content of the file in Jupiter notebook format with attached output (see for example Fig. 1 for the README.md file of Chap. 2).

In the same spirit of openness towards the programming habits of data analysts, we managed to call a limited set of functions from the R language through the package **fsdaR** (Todorov and Sordini, 2023), downloadable, as is customary, from CRAN.

Readers who are more software-oriented and want to have an overview of the robust routines (systematic naming and brushing or linking capabilities) should refer to Appendix A.1. The complete list of datasets is in Appendix A.2.

We encourage readers to signal every error they find, both in the GitHub website of the book and in the code, through the Issue section of the GitHub repository. The same encouragement applies to errors found in the book. Everything will be detailed in the errata page of the GitHub repository.

Code to reproduce Figures and Tables in this Chapter

FileName	Description	Open in MATLAB on line	Jupiter notebook
areNormT.m	Compute empirical and theoretical ARE(Me, Mean) under the normal distribution and the Student t. This file creates Figure 2.3.	Open in MATLAB Online	[ipynb]
areVarComparison.m	Variance comparison under the contamination model. This file creates Figure 2.4.	Open in MATLAB Online	[ipynb]
bdpeffRhoFuncCompare5.m	Comparison of rho functions in terms of bdp. This file creates Figures 2.14 and 2.15.	Open in MATLAB Online	[ipynb]
bdpeffRhoFuncPD.m	Breakdown point and efficiency for PD link. This file creates Figure 2.13.	Open in MATLAB Online	[ipynb]
compareIFlocation.m	Compare IF of different location estimators (use bdp=0.5 or eff=0.95). This file creates Figures 2.16 and 2.17.	Open in MATLAB Online	[ipynb]

Fig. 2 Example of an extract of README.md file for each Chapter of the book, describing the purpose of each file. The "Open in MATLAB Online" buttons enable free running of the analysis described in the corresponding file, while the ipynb link shows the content of the file in Jupiter notebook format, together with the output of the file. Published with permission from MathWorks, 2024, all rights reserved

Thanks

Anthony Atkinson and Marco Riani have been working together on the Forward Search since 1996. Anthony Atkinson's statistical career however stretches back almost 60 years. He would like to take this opportunity to mention a few of the many people who have helped and inspired him: W. G. ("Bill") Hunter of the University of Wisconsin, Art Bobis at American Cyanamid, David Cox and the wonderful company of staff and visitors that he gathered together at Imperial College and, at the London School of Economics, Neil Shepherd and Martin Knott. This project has led to many very enjoyable and profitable visits to the University of Parma. Conversations with Andrea Cerioli about monitoring multivariate data and other statistical topics are much appreciated. It has been very enjoyable working on the book, not only with the co-authors from Parma but with Valentin Todorov and Domenico Perrotta. Visits to Ispra have been a pleasure.

Marco Riani's career started in 1990 where his advisors, at the University of Parma, were Profs. Sergio Zani and Giovanni M. Marchetti. Professor Sergio Zani inspired his research for at least two decades. He also would like to thank his Ph.D. supervisors (Prof. Giampiero M. Gallo from the University of Florence and Prof. Andrew C. Harvey from the London School of Economics). Their guidance enabled him to win the prize for the best Italian Ph.D. thesis in statistics.

Marco Riani and Aldo Corbellini would like to mention the colleagues and friends from the University of Parma with whom they continue to interact: Andrea Cerioli, Fabrizio Laurini, Gianluca Morelli, Luigi Grossi, and Piero Ganugi. The development of the monitoring approach and of the forward search in multivariate analysis has been joint work with Andrea Cerioli. The first applications of the Forward Search to international trade data were by Marco Riani together with his Ph.D. student Francesca Torti, who also received the prize for the best Italian Ph.D. thesis in statistics.

Domenico Perrotta studied Computer Science at the University of Milan and the computational theory of automatic learning at the École Normale Supérieure de Lyon, where he obtained his Ph.D. His interests have always been at the interface between computer science and statistics. When he met Marco Riani and his colleagues in Parma, he was taken by the relevance of robust statistical methods for his work at the European Commission in the field of anti-fraud. Since then, the passion for these themes has only increased.

Valentin Todorov received a doctoral degree in statistics from Vienna University of Technology, Austria and holds a master's degree in mathematics and statistics from the University of Sofia, Bulgaria. His main research interest is in computational aspects of robust statistics. He developed the first R package for robust statistics on CRAN as early as 2004 and since then has authored and maintains several packages on CRAN, of course with the support and cooperation of many people to whom his thanks go, especially to Profs. Peter Rousseeuw, Mia Hubert, Peter Filzmoser and Martin Maechler.

There are many researchers with whom we have collaborated over many years. To mention just a few: Profs. Alessio Farcomeni, Anna Clara Monti, Luca Greco, Francesca Greselin, Silvia Salini, Nadia Solaro, Matilde Bini, Paolo Mariani, Corrado Crocetta, Paolo Giudici and Emanuela Raffinetti.

Finally, all authors would like to mention the continuous collaboration with the research group of the University of Valladolid (Profs. Alfonso Gordaliza, Agustin Mayo Iscar and Luis Angel Garcia Escudero). The part of the book on robust regression clustering has strongly benefitted from their research work.

For more than 15 years, Domenico Perrotta and Marco Riani have been contributing—from their specific perspectives—to the implementation of the EU anti-fraud and Customs Union policies. They would like to mention those in the European Commission who have continuously believed in this direction of research and have cultivated our solid partnership, intentionally or indirectly through their collaborative spirit: Spyros Arsenis, Delilah al Khudairy, Charles MacMillan, Juergen Marke, Eric Yperman, Winfried Kleinegris, Sara Piller, Patrick Wallez and Dimitra Triantafyllidou.

Finally, Domenico Perrotta wishes to express gratitude to his colleagues at the JRC, who have worked with him on these issues with dedication for so many years, almost 20 in the case of Francesca Torti, and also Giuseppe Sgarlata and Emmanuele Sordini.

Berwick-upon-Tweed, UK	Anthony C. Atkinson
Parma, Italy	Marco Riani
Parma, Italy	Aldo Corbellini
Ispra, Italy	Domenico Perrotta
Vienna, Austria	Valentin Todorov
February 2025	

Acknowledgements As a general principle, the European Commission supports open access publishing. Its Joint Research Centre (JRC), which is funded under Horizon Europe and the Euratom Research and Training Programme, has ensured the financial support to provide free of charge on-line access to readers of our book.

Many datasets and problems discussed in the book have been studied in relation to activities with financial support from different sources. The authors acknowledge in particular:

- The Hercule Anti-fraud Programme of the European Union.
- The National Recovery and Resilience Plan (NRRP), Mission 4 Component 2 Investment 1.5—Call for Tender No. 3277 of 30/12/2021 of the Italian Ministry of University and Research funded by the European Union—NextGenerationEU. Award Number: Project code ECS00000033, Concession Decree No. 1052 of 23/06/2022 adopted by the Italian Ministry of University and Research, "Ecosystem for Sustainable Transition in Emilia-Romagna" (Ecosister).
- The University of Parma project "Robust statistical methods for the detection of frauds and anomalies in complex and heterogeneous data".
- The Ministry of Education, University and Research project "Innovative statistical tools for the analysis of large and heterogeneous customs data" (2022LANNKC).

Our research has benefited from the High Performance Computing (HPC) facilities of both the University of Parma and the JRC. In Parma, we wish to mention Prof. Roberto De Renzi and Roberto Alfieri and Doctors Fabrizio Russo, Fausto Pagani, Fabio Spataro, Federico Prost and Giampietro Giudice for their continuous support. At the JRC, we thank Dr. Pierre Soille for having extended the Big Data Analytics Platform to the use of **MATLAB**.

We also wish to thank two anonymous referees for their helpful comments, as well as Dr. Alessandro Sellerio, Dr. Marika Palme and Dr. Lorena Marcaletti for their careful reading of the text and for providing suggestions for detailed improvements.

Berwick-upon-Tweed, UK	Anthony C. Atkinson
Parma, Italy	Marco Riani
Parma, Italy	Aldo Corbellini
Ispra, Italy	Domenico Perrotta
Vienna, Austria	Valentin Todorov

Competing Interests The authors have no competing interests to declare that are relevant to the content of this manuscript.

Contents

1	**Introduction and the Grand Plan Samples**		1
	1.1	The Grand Plan	1
	1.2	Estimates of Location	3
		1.2.1 Consistency	4
		1.2.2 Breakdown Point, *bdp*	4
		1.2.3 Asymptotic Relative Efficiency, *eff*	5
		1.2.4 Asymptotics and Simulation	5
		1.2.5 Rate of Convergence	6
	1.3	Estimates of Scale	6
	1.4	Univariate Data Analysis Using the Traditional Robust Approach	8
	1.5	Univariate Data Analysis Using the Monitoring Approach	12
	1.6	Further Reading	15
	1.7	Some Pre-history of Robustness	16
2	**Introduction to M-Estimation for Univariate Samples**		21
	2.1	Properties of Estimators	21
		2.1.1 The Sensitivity Curve and the Influence Function	21
		2.1.2 The Asymptotic Variance	25
		2.1.3 Rates of Convergence	27
	2.2	One Univariate Sample: Introductory Concepts	28
		2.2.1 M-Estimation: Downweighting or Soft Trimming	29
		2.2.2 Properties of Location M-Estimators	31
	2.3	M-Estimation of Scale	32
		2.3.1 Properties of M-Estimators of Scale	34
		2.3.2 Analysis of the ρ, ψ, ψ', and Weight Functions	35
		2.3.3 Simultaneous Estimation of Location and Dispersion	44

	2.4	Comparison of ρ Functions	45
		2.4.1 Asymptotic Comparisons	45
		2.4.2 Comparison Using the IF	48
		2.4.3 Comparison Using the Change-of-Value Curve	51
		2.4.4 Small-Sample Comparisons	53
		2.4.5 The Issue of Multiple Solutions	59
	2.5	Background on Convergence and Asymptotics	60
3	**Robust Estimators in Multiple Regression**		**65**
	3.1	Multiple Regression: Basic Concepts	65
		3.1.1 Parameter Estimates from Least Squares	65
		3.1.2 Formal Tests	67
	3.2	Deletion Diagnostics	68
		3.2.1 Deletion Residuals	69
		3.2.2 A List of Residuals	70
		3.2.3 Mean Shift Outlier Model	70
	3.3	Outlier Detection	71
	3.4	Added Variables	72
	3.5	Examples of Residual Plots for the AR Regression Data	73
	3.6	Three Classes of Estimators for Robust Regression	77
	3.7	M-Estimators for Multiple Regression	78
		3.7.1 The Iterative Reweighted Least Squares Algorithm	78
		3.7.2 Theoretical Aspects	79
	3.8	Introduction to S-Estimators	81
	3.9	Computation of the Tuning Constants	82
	3.10	S-Estimation: Computational Details	86
	3.11	Introduction to LMS and LTS	87
		3.11.1 LMS—Least Median of Squares	88
		3.11.2 LTS—Least Trimmed Squares and the Consistency Correction for Variances	89
	3.12	MM-Estimation, τ-Estimators, and Reweighted LMS and LTS Estimators	90
		3.12.1 MM-Estimators	90
		3.12.2 τ-Estimators	91
		3.12.3 LMS and LTS Reweighted	92
	3.13	Traditional Robust Analysis for the AR Regression Data	92
4	**The Monitoring Approach in Multiple Regression**		**97**
	4.1	The Forward Search	97
	4.2	Testing for Outliers	99
	4.3	The Forward Search for Regression Outlier Detection	100
	4.4	Graphical Presentation	101
	4.5	Two Examples	102
		4.5.1 Hertzsprung–Russell Diagram of Star Cluster CYG OB1 (Stars Data)	102

		4.5.2	Wool Data	104
	4.6	Monitoring Plots, Linking Them, and Brushing		106
	4.7	The Empirical Breakdown Point and Efficiency		109
	4.8	Monitoring Tests for Regression Coefficients: The Extended Added Variable Plot		110
	4.9	The Monitoring Approach in Practice		112
		4.9.1	Correlation	112
		4.9.2	The Stars Data	113
		4.9.3	The AR Regression Data	116
		4.9.4	The Hawkins Data	128
		4.9.5	Surgical Unit Data	134
	4.10	The Bank Data		137
	4.11	Generalized BIC for Outlier Detection		142
		4.11.1	Information Criteria	143
		4.11.2	Mean Shift Outlier Model, FS, LMS, and LTS	145
		4.11.3	Extended BIC for Outlier Detection	146
		4.11.4	Finite Sample Extended BIC for Outlier Detection	147
		4.11.5	Example: Mental Illness Data	147
		4.11.6	Further Examples and Developments	148
5	**Practical Comparison of the Different Estimators**			151
	5.1	False Signals and Empirical Size: Outliers from High Breakdown Regression		151
		5.1.1	The Sizes of Many Individual and Simultaneous Tests for Outliers	152
		5.1.2	Simulation Results	153
		5.1.3	Distributional Issues	157
		5.1.4	The Chi-Squared Approximation to the Distribution of Scaled Residuals	159
	5.2	Size Comparisons		163
	5.3	The Power of the Tests		164
	5.4	A Parametric Framework for Comparing Robust Regression Estimators		169
		5.4.1	Introduction	169
		5.4.2	Bias and Maxbias	170
	5.5	A Parameterized Family of Departures		171
		5.5.1	Background	171
		5.5.2	The Theoretical Overlapping Index	173
		5.5.3	The Numerical Effect of Overlap: Normal Contamination	174
		5.5.4	The Numerical Effect of Overlap: Point Contamination	181
	5.6	Comparisons Using the Extended BIC Criterion for Outlier Detection		184

5.7	Discussion		189
5.8	Two or More Populations		189
6	**Transformations**		**195**
6.1	Introduction: The Box–Cox Transformation		195
	6.1.1	Maximum Likelihood Estimation for the Box–Cox Transformation	197
	6.1.2	The Score Test	200
	6.1.3	The Fan Plot	201
	6.1.4	Surgical Unit Data	202
	6.1.5	Loyalty Cards Data	203
	6.1.6	The Logarithmic Transformation and Gamma Generalized Linear Models	206
	6.1.7	The Shifted Power Transformation	208
6.2	The Yeo–Johnson Transformation and Its Extension		210
	6.2.1	The Yeo–Johnson Transformation	210
	6.2.2	Constructed Variables	211
	6.2.3	The Extended Yeo–Johnson Transformation	211
	6.2.4	The Extended Fan Plot: Establishing Two Transformations	214
	6.2.5	Investment Funds Data	214
	6.2.6	Balance Sheet Data	219
6.3	An Automatic Procedure for Determining Transformations		223
	6.3.1	The Extended BIC and Response Transformation	223
	6.3.2	The Agreement Index AGI	224
	6.3.3	The Coefficient of Determination R^2	225
	6.3.4	The Automatic Procedure for the Box–Cox Transformation	225
	6.3.5	Automatic Analysis of the Loyalty Cards Data	226
	6.3.6	The Automatic Procedure for the Extended Yeo–Johnson Transformation	228
6.4	Further Response Transformations		229
	6.4.1	The General Robust Procedure	229
	6.4.2	The Folded Power Transformation for Proportions	230
	6.4.3	The Guerrero–Johnson Transformation	231
	6.4.4	The Aranda-Ordaz Transformation	232
6.5	Transformations Including Explanatory Variables		232
	6.5.1	The Box–Tidwell Transformation for Explanatory Variables	233
	6.5.2	Transformation of Both Sides of a Regression Model	234
6.6	Further Reading		239

7 Non-parametric Regression ... 241
- 7.1 Introduction ... 241
- 7.2 Background ... 242
  - 7.2.1 The Generalized Additive Model (GAM) ... 242
  - 7.2.2 Definition of Backfitting ... 243
  - 7.2.3 Properties of Backfitting ... 244
- 7.3 Introduction to AVAS ... 244
  - 7.3.1 The AVAS Algorithm ... 244
  - 7.3.2 The Numerical Variance-Stabilizing Transformation ... 245
- 7.4 RAVAS: Five Extensions to AVAS ... 245
  - 7.4.1 Initial Calculations ... 246
  - 7.4.2 Response Transformation: Option `tyinitial` ... 246
  - 7.4.3 Ordering Explanatory Variables in Backfitting: Option `scail` ... 246
  - 7.4.4 Robust Outlier Detection: Option `rob` ... 246
  - 7.4.5 Ordering Predictor Variables: Option `orderR2` ... 247
  - 7.4.6 Flowchart for Initialization of RAVAS ... 247
  - 7.4.7 Flowchart for the Main (Outer) Loop of RAVAS ... 248
  - 7.4.8 Numerical Variance-Stabilizing Transformation: Option `trapezoid` ... 249
  - 7.4.9 Simplified Notation for the Variance-Stabilizing Transformation ... 252
  - 7.4.10 Flowscheme for Calculation of the Numerical Variance-Stabilizing Transformation ... 255
- 7.5 Simulated Examples ... 256
  - 7.5.1 Example 1—Robustness ... 256
  - 7.5.2 Example of Effects from Options in RAVAS ... 257
- 7.6 A Graphical Representation of the Systematic Evaluation of Options ... 261
- 7.7 Prediction of the Weight of Fish ... 265
- 7.8 Internet Marketing Data ... 270
- 7.9 A Statistical Comparison of AVAS and RAVAS ... 273
- 7.10 Discussion ... 274
- 7.11 Further Reading ... 275

8 Extensions of the Multiple Regression Model ... 281
- 8.1 Bayesian Analysis and Prior Information ... 281
  - 8.1.1 Introduction ... 281
  - 8.1.2 The Normal Inverse-Gamma Prior Distribution ... 282
  - 8.1.3 Prior Distribution from Fictitious Observations ... 283
  - 8.1.4 Posterior Distributions ... 283
- 8.2 The Bayesian Search ... 284
  - 8.2.1 Parameter Estimation ... 284
  - 8.2.2 Forward Highest Posterior Density Intervals ... 285

		8.2.3	Outlier Detection	285
	8.3		Data Analyses with Prior Information	286
		8.3.1	Windsor House Price Data	286
		8.3.2	The Importance of International Trade Data	290
		8.3.3	Lobster Data 2002: Least Squares Analysis	293
		8.3.4	Lobster Data 2003: Analysis with the Bayesian Forward Search	295
		8.3.5	Lobster Data 2004: Analysis with the Bayesian Forward Search	296
		8.3.6	Timeliness and Power	297
	8.4		Heteroskedasticity	298
		8.4.1	Homoskedasticity and Heteroskedasticity	298
		8.4.2	Models for Non-constant Variance and the Forward Search	299
		8.4.3	Example: International Trade Data 1	300
		8.4.4	Example: International Trade Data 2	303
	8.5		Robust Regression Clustering	304
		8.5.1	Face Masks: A Trade Dataset with Several Groups	307
		8.5.2	Analysing the Face Masks Data with Random Start Forward Searches	308
		8.5.3	Clusterwise Linear Regression: Basic Concepts	312
		8.5.4	Monitoring Clusterwise Linear Regression	316
		8.5.5	Analysing the Face Masks Data Using The Traditional Information Criterion	321
		8.5.6	Analysing the Face Masks Data with the Monitoring Approach	322
	8.6		Robust Monitoring of Time Series	325
		8.6.1	A Non-linear Time Series Model	326
		8.6.2	LTSts Estimation	327
		8.6.3	FSRts for an Adaptive Data-Driven Time Series Monitoring	327
		8.6.4	Contaminated Airline Data	328
	8.7		Regression with Compositional Data	330
		8.7.1	Coordinate Representation of Compositional Data	332
		8.7.2	Regression with Compositional Covariates	335
		8.7.3	Monitoring Compositional Regression	339
	8.8		Censored Regression	343
9	**Model Selection**			349
	9.1		Introduction	349
	9.2		Significance of the Explanatory Variables	350
	9.3		Arbitrary Numerical Rules for Model Selection and Hald's Cement Data	352
		9.3.1	Mallow's C_p and the Generalized Candlestick Plot	353

		9.3.2	The Generalized Candlestick Plot and Two Analyses of the Ozone Data	355
	9.4	Regularization Methods		357
		9.4.1	Least Absolute Shrinkage and Selection Operator (LASSO)	358
		9.4.2	Sparse LTS Regression	363
		9.4.3	Monitoring Sparse LTS	366
10	**Some Robust Data Analyses**			**371**
	10.1	Introduction		371
	10.2	Income Data 1: United States Census Bureau		372
	10.3	Income Data 2: An Italian Municipality		381
	10.4	Customer Loyalty		389
		10.4.1	Data and Regression	389
		10.4.2	Robust Box-Cox Transformation	395
		10.4.3	Non-robust Analysis of the Loyalty Data with AVAS	399
		10.4.4	Robust Non-parametric Regression with Response and Explanatory Variable Transformations: RAVAS	400
		10.4.5	Interpretation	402
		10.4.6	A Procedure for a Structured Approach to Modern Robust Regression Analysis	406
	10.5	Modified Customer Loyalty Data		408
		10.5.1	Least Squares Analysis	408
		10.5.2	AVAS and RAVAS	413
	10.6	NCI60 Cancer Cell Panel Data		416
		10.6.1	Monitoring and Stability	416
		10.6.2	Least Squares	417
		10.6.3	Response Transformation	419
		10.6.4	The Generalized Candlestick Plot	424
		10.6.5	AVAS	425
		10.6.6	RAVAS	425

Appendix: Software and Datasets ... 433

Solutions ... 445

References ... 525

Author Index ... 537

Subject Index ... 541

Acronyms

$[\cdots]$	Integer part; also the outer set of brackets in the hierarchy $[\{(\cdots)\}]$
α	Tuning parameter in the PD estimator; also denotes the trimming proportion in the trimmed mean
ϵ	Error term in the regression model
ε	Fraction of contamination
λ	Transformation parameter in the Box-Cox or Yeo-Johnson families or, in Chap. 9, tuning parameter controlling the amount of regularization
μ	Location parameter
$\hat{\mu}$	Estimator (estimate) of the location parameter
$\hat{\mu}_0$	Initial estimate of location in the iterative loop
ρ	ρ function
σ	Scale parameter
$\hat{\sigma}$	Estimator (estimate) of the scale parameter
$\hat{\sigma}_0$	Initial estimate of the scale in the iterative loop
Φ	Cumulative distribution function (cdf) of the standard normal distribution. Φ_{μ,σ^2} is the cdf of the $\mathcal{N}(\mu, \sigma^2)$
ϕ	Density of the standard normal distribution
ψ	ψ function (first derivative of the ρ function)
ψ'	ψ' function (derivative of the ψ function)
ACE	Alternating Conditional Expectation algorithm
AIC	Akaike Information Criterion
AS	Andrews' sine estimator
AVAS	Additivity and VAriance Stabilization transformation
$\mathcal{B}(a, b)$	Beta distribution with parameters a and b
BIC	Bayesian Information Criterion
bdp	Breakdown point
c	Tuning constant or consistency factor for a nominal value of *bdp* or *eff*. In the context of regression clustering (see 8.5.4) c is the optimal restriction factor among the variances of the error components
CV	Change of value curve
CVC	Change of variance curve

CVS	Change of value sensitivity
det(·)	Determinant
E(·)	Expectation
e_i	Raw residual for unit i: $e_i = y_i - x_i^T \hat{\beta}$. Note that $\hat{\beta}$ can be the output of any possibly robust estimator including least squares at step m of FS
eff	Asymptotic relative efficiency
F_ε	Contamination model (2.4)
F_θ	Null model without contamination; e.g. $F_\theta = \mathcal{N}(\mu, \sigma^2)$
$F_\theta(y)$	Parametric model indexed by a parameter θ
$\mathcal{F}(\nu_1, \nu_2)$	Fisher–Snedecor distribution with parameters ν_1 and ν_2
$\mathcal{G}(a, b)$	Gamma distribution with parameters a and b
GAM	Generalized additive model
GES	Gross error sensitivity, sup(IF)
HA	Hampel estimator
HYP	Hyperbolic tangent estimator
$1_{(\cdot)}$	Indicator function
IC	Information criterion
IF	Influence function
$\mathcal{IG}(a, b)$	Univariate inverse gamma distribution with parameters a and b
IQR	Interquartile range, $y_{[n-[n/4]+1]} - y_{[n/4]}$
IQRN	Normalized Interquartile range, IQR /1.349
K	E(ρ) in the estimating equation for the scale estimate
k	Tuning constant in the hyperbolic tangent estimator which controls the sup of CVC. Note that in Chap. 2, k is also the index of the iteration number when finding the robust estimates of location and scale. In Chap. 8 k is the number of groups in regression clustering
log	Natural logarithm
$L(\theta)$	Likelihood
$\mathcal{L}(\theta)$	Loglikelihood
LS	Ordinary least squares
LMS	Least median of squares
LR	Likelihood ratio test
LTS	Least trimmed squares
MAD	Minimum absolute deviation from the median
MADN	Normalized Minimum absolute deviation from the median MAD ×1.4826
MB	Maximum bias
m	Subset size in the forward search
med	Sample median
$\mathcal{N}(\mu, \sigma^2)$	Univariate normal distribution with mean μ and variance σ^2
n	Sample size
$n\,\text{var}(T)$	Asymptotic variance of estimator T
OPT	Optimal estimator
PD	Power divergence estimator
Pr	Probability

$\mathbb{R}$	Set of real numbers
$\mathbb{R}^p$	Set of real numbers in p dimensions
R^2	Coefficient of determination
R^2_{adj}	Adjusted coefficient of determination
RAVAS	Robust Additivity and VAriance Stabilization transformation
RE	Relative efficiency
r_i	Scaled residual (e_i/s) for unit i
$\tilde{r}_i$	Studentized residual for unit i
r_i^*	Deletion residual for unit i
SC	Sensitivity curve
s	Standard deviation
sgn	Sign function
TB	Tukey biweight estimator
t_ν	Student's t random variable on ν degrees of freedom
t_k	t test for variable k in regression
T_n	$T_n = T(y_1, y_2, \ldots, y_n)$ an estimator based on n observations
$\mathcal{U}(a, b)$	Uniform distribution with limits a and b
var($\cdot$)	Variance
Y_n	A sample of n observations $y_1, \cdots, y_n$
y_i	i-th observation from a sample of size n
$y_{[i]}$	i-th order statistic. The symbol [] is also used to denote the integer part
$\bar{y}$	Sample mean
$\bar{y}_\alpha$	Trimmed mean of order α. We discard a proportion α of the largest and smallest observations
Z	Random variable $\sim \mathcal{N}(0, 1)$.

Chapter 1
Introduction and the Grand Plan Samples

Abstract Robust statistical methods provide understanding of suitable models for data that are contaminated by systematic or random departures. A discussion of the "The Grand Plan" for robust statistics provides a historical context. We start robust data analysis with an introduction to estimates of location, introducing the ideas of breakdown point *bdp*, the proportion of outlying observations that a particular robust analysis can be expected to adjust for, and of the asymptotic relative efficiency *eff*, how much information is lost by the robust procedure if the data are not contaminated. As one measure increases, the other decreases. We also introduce robust estimates of scale. In the traditional robust approach, analyses are usually made for one value of *bdp* or *eff*. Our book focuses on the monitoring approach to robust statistics in which analyses are performed over a range of values of *bdp* and *eff*. For some simple examples, including the transformation of income data, we compare the information obtained from the traditional static and the dynamic monitoring approaches, the latter providing extra insights into the data. The Grand Plan for the wider application of robust procedures can be realized through the monitoring approach, leading to appreciably more effective data analyses.

1.1 The Grand Plan

Data rarely follow the simple models of mathematical statistics. Often, there will be distinct subsets of observations so that more than one model may be appropriate. Further, parameters may gradually change over time. In addition, there are often dispersed or grouped outliers. This distance between mathematical theory and data reality has led, over the last 70 years, to the development of a large body of work on robust statistics. An early use of the term is due to Box (1953). Twenty years after Box, the outlines of the theory of robust statistics were becoming clearly established. At that point, according to Stigler (2010), it was expected that in the near future "any author of an applied article who did *not* use the robust alternative would be asked by the referee for an explanation". Now, a further 50 years on, there is a large literature on robust statistics, some references to which are given in 1.6. Unfortunately, this activity seems largely to be statisticians talking to other statisticians; there does

© The Author(s) 2025
A. C. Atkinson et al., *Robust Statistics Through the Monitoring Approach*,
Springer Series in Statistics, https://doi.org/10.1007/978-3-031-88365-1_1

not seem to have been the breakthrough into the wider scientific universe that was foreseen 50 years ago. The purpose of our book is to provide a system of graphically rich interactive procedures for robust data analysis, which finds full realization in the MATLAB toolbox **FSDA** and some representative instances in the R package **fsdaR** (Todorov & Sordini, 2023). Details are in Appendix A.1.

The implementations contained in **FSDA** not only present a wide variety of robust methods (parametric and non-parametric) but, most importantly, provide a systematic framework for the comparative use of the methods in data analyses (allowing robust data transformation and interactivity through brushing and logical linking); different methods may reveal various distinct aspects of the data. The fundamental idea is to monitor the change in fitted models and inferences as we move from very robust fits to those that are highly efficient. This monitoring procedure is accompanied by interactive graphics that enable discovery of the effect of individual observations on inferences being drawn from the data. Introductory examples are in Sect. 1.5, the simplest of which is estimation of the location of a simple sample. The median is the most robust estimator (allowing for 50% of contamination in the data). But this estimate of location has higher variance than that of the arithmetical mean, which is to be preferred if there are no outliers. What estimate is best if there is an unspecified proportion of outliers? How is the best estimator to be determined; that is, how can the data be used to find the optimum trimming level?

Although the term "robust" was first popularized by Box, the idea of considering the distribution of statistics under departures from the assumption of normality goes back at least to E.S. Pearson's review (Pearson, 1929) of the second edition of Fisher's *Statistical Methods for Research Workers*, an interest that was to stay with Pearson until the end of his scientific life (Pearson & Please, 1975). The current understanding of "robust" was much more the creation of Tukey, starting with Tukey (1960), and of Huber (1964). It is instructive to quote at greater length the passage from Stigler.

"... by 1972 a number of the early workers in robust statistics expected that from the 1970s to 2000 we would see the same development with robust methods—extensions to linear models, time series, and multivariate models, and widespread adoption to the point where every statistical package would take the robust method as the default.... This was, and I [Stigler] will call it, a Grand Plan. But that plainly is not what has occurred". Stigler then presents a lively and warm discussion of the early history of robust statistics, suggesting that the first signs of trouble with the Grand Plan were already evident in 1972 at the time of the publication of the Princeton Robustness Study (Andrews et al., 1972). To quote him again: "From the full set of 10,465 estimates of a location parameter they had considered, they reported in detail on the accuracy of 68 estimates that had received extensive study, focusing upon small samples and an inventively wide selection of 32 distributions, nearly all of which were symmetric scale mixtures of normal distributions".

A powerful reason for the lack of opening to the scientific world is that robust statistics, as often understood and practised, has led to a new mathematical statistics, more complicated than the old, in which ever more refined solutions are presented to a few well-defined problems. An impressive example is Field and Hampel (1982). There is not the hoped for "robust method" mentioned above, but a plethora of competing methods with a wide variety of specific properties, often only asymptotic. Suggestions for further reading are in Sect. 1.6.

This "grand plan" failed, in part, not only because of the plethora of robust alternatives, but also because the properties of the methods are, mostly, only proven asymptotically. Asymptotics may provide a poor guide to the analysis of small samples of data. Here "small" may sometimes be several thousand. See the comparisons of the sizes of asymptotically equivalent tests for outliers in Sect. 5.2. Now, a further 50 years on, a consensus seems to be emerging as to the correct route to an appropriate robust data analysis.

An important aspect of our book is that we are able to compare the behaviour of competing robust methods, not just for very large samples but, through efficient simulations and data analyses, for the data configuration of interest. An example is in the comparison of various ρ functions for M-estimation of Sect. 2.3.2. Provided the same value of efficiency or breakdown point (*bdp*) is chosen, it makes little difference which ρ function is used. Our monitoring of robust procedures has at least two consequences. One is that we produce insightful data analyses. Another is that, by considering a variety of procedures for robust regression, we are able to determine which are the critical parameters, distinguishing them from those that are only of secondary importance.

1.2 Estimates of Location

We start by considering robust methods for the simple problem of estimation of a location parameter in the presence of outliers. These methods and principles provide some tools for the more complicated problems arising in regression, which are described in Chap. 3.

We have a set of observations $y_1, y_2, \ldots, y_n$ with order statistics $y_{[1]}, y_{[2]}, \ldots, y_{[n]}$. The parametric model is $F_\theta(y)$ with $\theta \in R^p$ and $T_n = T(y_1, y_2, \ldots, y_n)$ is an estimator based on the n observations. In classical statistical theory, the observations are distributed according to F_θ. As an example, for the normal distribution $F_\theta = \mathcal{N}(\mu, \sigma^2)$ and so $\theta = (\mu, \sigma^2)$. In robust estimation, the goal is to provide statistical procedures that behave well under deviations from the assumed model. The simplest deviations are either that some observations have been contaminated by the addition of arbitrary quantities, or that some observations have a greater variance than the majority. However, robust procedures are not confined only to the detection of such anomalies.

The classic estimator of location is the sample mean:

$$\hat{\mu} = \bar{y}_n = \bar{y} = \frac{1}{n} \sum_{i=1}^{n} y_i. \tag{1.1}$$

A well-known robust estimator is the sample median

$$\text{med} = \begin{cases} y_{[(n+1)/2]} & n \text{ odd} \\ 0.5\left(y_{[n/2]} + y_{[n/2+1]}\right) & n \text{ even.} \end{cases} \quad (1.2)$$

Between these two lies the α trimmed mean in which we discard a proportion α of the largest and smallest observations. More precisely, let $\alpha \in [0, 0.5)$ and $m = [(n-1)\alpha]$ where $[\cdot]$ is the symbol which denotes the integer part. The α–trimmed mean ($\bar{y}_\alpha$) is defined as

$$\bar{y}_\alpha = \frac{1}{n - 2m} \sum_{i=m+1}^{n-m} y_{[i]}. \quad (1.3)$$

For $m = 0$ this is the sample mean and as $\alpha \to 0.5$, $m = [(n-1)\alpha] \to n/2$, the asymptotic definition of the median.

1.2.1 Consistency

This is the first of some important general properties of estimators.

If there are no outliers, the sample mean is a consistent estimator of μ. For any n, the expectation of $\bar{y}_n$, $E(\bar{y}_n)$ is μ and the variance of $\bar{y}_n$ decreases as $1/n$. The same is true for the sample median.

1.2.2 Breakdown Point, bdp

The comparison of trimmed means with various trimming proportions shows that the greater the trimming, the larger can be the number of outliers before the estimator becomes inconsistent. The breakdown point is the proportion of observations that can be replaced by *any* outliers with the estimator remaining in a bounded set. It does not remain in the set if the proportion is greater.

Let $\hat{\theta}_n(y)$ be an estimator for a sample of size n. The replacement finite sample breakdown point of an estimator $\hat{\theta}_n$ at y is the largest proportion of observations which can be arbitrarily replaced by outliers without $\hat{\theta}_n(y)$ leaving a set which is bounded and is also bounded away from the boundary of the parameter space Θ. For example, the finite sample breakdown point of the sample mean is zero, since making one of the y_i arbitrarily large makes $\bar{y}$ arbitrarily large. For the sample median when n is odd, $(n-1)/2$ observations can be given arbitrary values and the median remains unchanged (we require the smallest set of outliers for which this holds, so ignore the possibility of outliers in both tails of the distribution). The proportion of deleted observations is therefore $(n-1)/2n$.

1.2 Estimates of Location

For the general estimator $\hat{\theta}_n(y)$ as $n \to \infty$ we obtain the asymptotic breakdown point *bdp*. For the sample median the limit, as n increases, of $(n-1)/2$ gives a *bdp* = 0.5. The *bdp* for each estimator has to be treated separately. A high value is desirable, but other properties are also necessary. For example an "estimator" $\hat{\theta}_n = \theta_0, \theta_0 \in \Theta$, has *bdp* =1, but it is not consistent.

Values of *bdp* are available for many estimators. However, data analysts are provided with finite, often small, samples for which the theoretical results may provide a poor guide. An important feature of the monitoring approach is that it allows estimation of *bdp* for a specific finite sample across sets of estimators. The calculations of Sect. 5.4.1 compare the behaviour of several robust estimators as the contamination moves from being near to being far from the model for the majority of the data.

1.2.3 Asymptotic Relative Efficiency, eff

If there are, in fact, no outliers, the trimming of observations leads to parameter estimates with increased variance. For a fixed underlying distribution F, the relative efficiency of an estimator $\tilde{\theta}_n$ relative to that of the maximum likelihood estimator $\hat{\theta}_n$ is

$$\mathrm{RE}(\tilde{\theta}_n; \hat{\theta}_n) = \frac{\mathrm{variance}(\hat{\theta}_n)}{\mathrm{variance}(\tilde{\theta}_n)}. \tag{1.4}$$

The interpretation is that $\hat{\theta}_n$ requires only RE times as many observations as $\tilde{\theta}_n$ for the same variance. For the comparison of biased estimators the mean squared error (MSE) is used. The asymptotic relative efficiency, *eff*, is the limit of RE as $n \to \infty$:

$$\mathit{eff} = \mathit{eff}(\tilde{\theta}_n; \hat{\theta}_n) = \lim_{n \to \infty} \frac{\mathrm{variance}(\hat{\theta}_n)}{\mathrm{variance}(\tilde{\theta}_n)}. \tag{1.5}$$

1.2.4 Asymptotics and Simulation

Even in the simple problem of the estimation of the location parameter we have had to refer to asymptotic results for the variance of the median and for the asymptotic relative efficiency *eff*. Simulation is often used to obtain small-sample results. Asymptotics enter more directly into the definition of the *bdp*, which is defined by considering what happens as observations become arbitrarily distant from the main body of the data (Sect. 1.2.2). However, practical interest is often in the detection of outliers that are only moderately remote. It is possible, as in the analysis of the AR data in Sect. 4.9.3, for two estimators to have similar values of *bdp*, but to vary markedly in the detection of moderate outliers. Such a distinction is often most easily detected by simulation.

An important general point is the inverse relationship between *eff* and *bdp*; as one increases, the other decreases. A discussion of this for M-estimation is in Sect. 2.3.2.

1.2.5 Rate of Convergence

Rates of convergence quantify the (stochastic) order of magnitude of an estimation error as a function of the sample size n. The rate of convergence is an important criterion for selecting the best possible estimator for a given problem. For most parametric problems, it is well known that the optimal (i.e. fastest possible) convergence rate is $n^{-1/2}$. In non-parametric regression or density estimation, the optimal convergence rate is only $n^{-2/5}$, if the underlying function is twice continuously differentiable. Note that in general maximum likelihood estimators of an unknown parameter usually possess a rate of convergence of $n^{-1/2}$. In the next chapters, we will give the rates of convergence of the different robust estimators which are to be introduced.

1.3 Estimates of Scale

We now move to the robust estimation of the scale parameter and include a consistency factor to ensure that the estimates of σ or σ^2 converge to unbiased estimators as the sample size increases.

The standard, non-robust, consistent estimator of the population standard deviation is the root mean square estimator $s = \sqrt{\{\sum_{i=1}^{n}(y_i - \bar{y})^2/(n-1)\}}$. A robust alternative is the interquartile range:

$$\text{IQR} = \text{IQR}(Y_n) = y_{[n-[n/4]+1]} - y_{[n/4]}, \tag{1.6}$$

the difference of two-order statistics.

However, at $F = \mathcal{N}(\mu, \sigma^2)$, the asymptotic expectation of IQR$= 2\Phi^{-1}(.75)\sigma \neq \sigma$ where Φ is the cdf of the standard normal distribution (Exercise 1.2). The normalized interquartile range is

$$\text{IQRN} = \text{IQRN}(Y_n) = \frac{1}{2\Phi^{-1}(0.75)} y_{[n-[n/4]+1]} - y_{[n/4]}. \tag{1.7}$$

The quantity $1/2\Phi^{-1}(0.75) = 0.7413$ is called a consistency factor. Although it provides a consistent estimate of σ, for small-sample sizes a further correction, the small-sample correction factor may be required to provide a sufficiently accurate estimate of σ. It is clear from the definition of the IQR that the breakdown point is 25%.

1.3 Estimates of Scale

An estimator with 50% *bdp* uses medians both to estimate the centre of location and then to estimate σ from the residuals around med(y). The resulting median absolute deviation is

$$\text{MAD} = \text{MAD}(Y_n) = \text{med}\{|y_i - \text{med}(Y_n)|\}. \tag{1.8}$$

At the normal distribution, we introduce a consistency factor and use the normalized version (Exercise 1.3):

$$\text{MADN} = \text{MADN}(Y_n) = \text{MAD}(Y_n)/\{\Phi^{-1}(0.75)\} = 1.4826\text{MAD}(Y_n). \tag{1.9}$$

Comparison of the consistency factors for the IQR and the MAD shows that, for the normal (and indeed for any symmetrical distribution) IQR = 2MAD.

For the MAD the consistency factor $1/\Phi^{-1}(0.75) \approx 1.4826$ is again asymptotic. Figure 1.1 shows the results of a simulation study based on 100000 samples generated from the normal distribution with size $n = 5, 6, \ldots, 100$. For each sample we calculate the MAD. Finally the mean of the empirical MADs is taken across the 100000 simulations. This figure shows the reciprocal of the mean of the empirical MADs as a function of the sample size. The horizontal line is the asymptotic consistency factor. The figure shows that, as the sample size increases, the two lines become closer. However, the empirical consistency factor when $n = 100$ is 1.4942, still slightly greater than the asymptotic value used in (1.9).

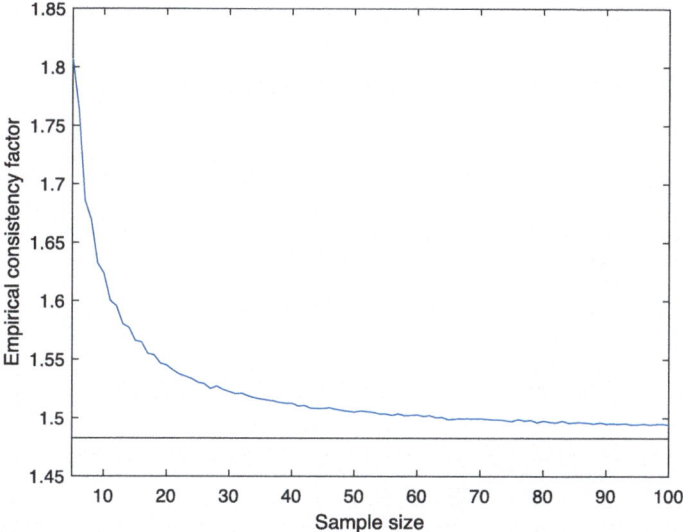

Fig. 1.1 MAD: Empirical and asymptotic consistency factor under the hypothesis of normality. The decreasing curve represents the reciprocal of the average value of the MAD across 100000 simulations

1.4 Univariate Data Analysis Using the Traditional Robust Approach

As an illustration of univariate data analysis we start with the analysis of the variable HTOTVAL in the dataset `Income1`. This variable is the total annual household income in dollars of a random sample of 200 adults from a survey of the United States Census Bureau, available in their Annual Social and Economic Supplements. The full data survey can be downloaded from https://www2.census.gov/programs-surveys/cps/datasets/2021/march/asecpub21csv.zip.

Figure 1.2 shows the histogram and the boxplot of the income values. It is customary to transform income data to approximate normality before statistical analysis. Clearly, in order to reach approximate normality, these highly skewed data have to be transformed.

Figure 1.3 shows boxplots of the distribution of income for the untransformed data and for the data after three power transformations: the square root, logarithmic, and reciprocal square root. To avoid numerical problems, the data were scaled, before transformation, to have a maximum value of one. This scaling does not affect the information about transformation. The calculations were performed using the normalized Box–Cox power transformation, discussed in Sect. 6.1.1. The transformation depends on a parameter λ. The data are untransformed when $\lambda = 1$, with the

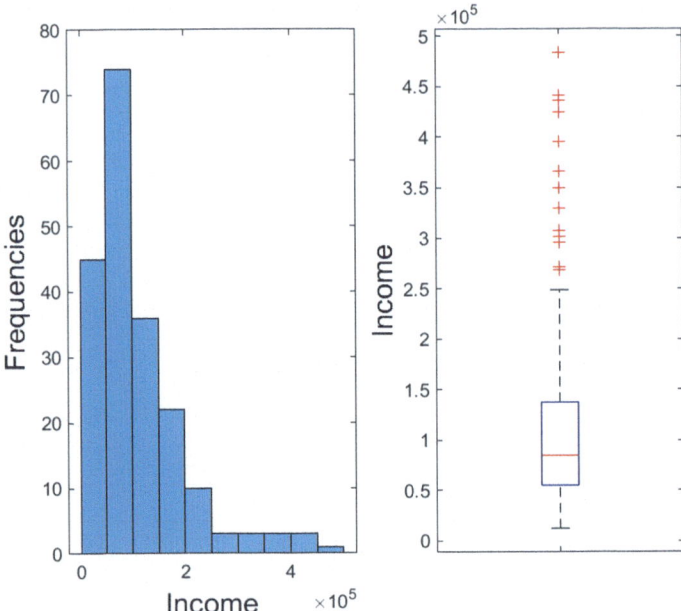

Fig. 1.2 Univariate income data from the United States Census Bureau: histogram and boxplot; positive skewness is evident in both panels

1.4 Univariate Data Analysis Using the Traditional Robust Approach

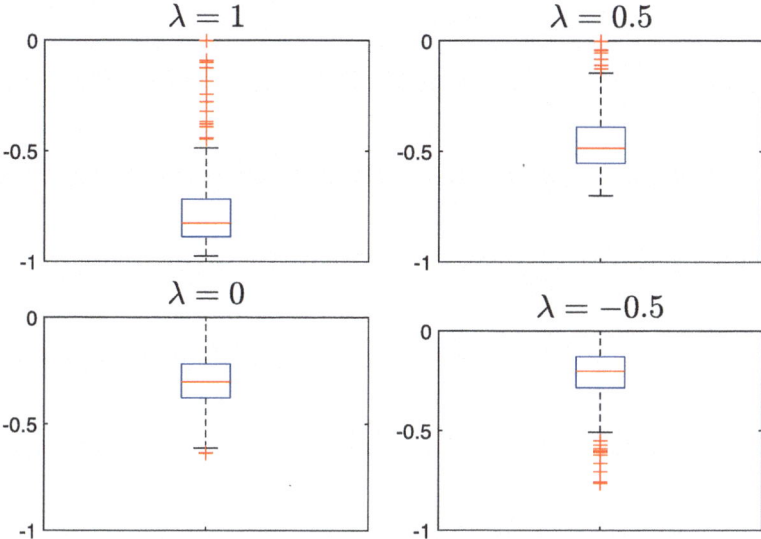

Fig. 1.3 Univariate income data from the United States Census Bureau: boxplots for four values of λ using the normalized Box–Cox power transformation after preliminary rescaling of the data to a maximum value of one

log transformation corresponding to $\lambda = 0$. The four values of λ used in Fig. 1.3 are 1, 0.5, 0, and -0.5 corresponding to the cases of no transformation, square root, log, and inverse square-root transformation. It seems that the most appropriate transformation in this case is the log.

This qualitative analysis, visually comparing boxplots, can be made more quantitative through an analysis of the score tests, that is the t tests which come from the addition of a constructed variable (Sect. 6.1.2) and serve as a useful approximation to the likelihood ratio test of specific values of λ. The smaller the absolute value of the t-statistic from including this constructed variable in the analysis, the greater is the indication that we are testing a value of λ which is close to the maximum likelihood estimate (Exercise 1.5). Table 1.1 shows the value of the score tests for the five most common values of λ, values which will be repeatedly used in our data analyses.

This table is the traditional way of analysing and presenting a summary of a data analysis. The traditional robust way simply presents the values of the same statistics computed using a robust estimator based on a specific value of breakdown point or efficiency, as well as such parameters as tuning constants, tolerances for terminating iterative loops in numerical calculations, small-sample consistency factors, and so forth. Such a tabulation of results might be contrasted with a robust analysis, for example, the procedure of Marazzi et al. (2009) which is based on the use of MM-estimators (Sect. 3.12.1) with a *bdp* equal to 0.5 and a value of *eff*= 0.95. The results of looking at another simple table, this time with this robust analysis, are unilluminating. If the analysis starts from a grid of λ values 0, 0.01, ..., 1, the estimate of λ is 0.18. If, alternatively, the larger grid of values $-1, -0.99, ..., 1$, is used, the robust estimate

Table 1.1 Univariate income data from the United States Census Bureau: score tests from the five most common values of λ. The smaller the absolute value of the score test, the greater is the indication that we are testing a value of λ close to the maximum likelihood estimate

Transformation	λ	Score test
Inverse	-1	25.68
Reciprocal square root	-0.5	12.91
Logarithmic	0	1.61
Square root	0.5	-9.46
None	1	-21.68

of λ is now -0.97. This indicates that the function being maximized, unlike the Box–Cox likelihood, does not have a unique maximum but is multimodal. It tells us little else.

In general, from a table such as Table 1.1, even if it were based on robust score tests, we would not know whether the small value of the score test for the log transformation is due to the presence of atypical observations nor what is the percentage of observations which is in agreement with the simple location model for each value of λ. Moreover, from this static table, we do not know what is the effect that each of the 200 units of the sample has on the value of the score test and, similarly, we do not know whether there is a single population or whether there are subgroups of observations. If there are subgroups, how similar are such units to each other?

Once a value of λ has been found, the traditional robust approach implies the comparison of the non-robust estimators of location and scale (e.g. the mean and the standard deviation) with the robust ones (e.g. the median and rescaled MAD) or with the use of the trimmed mean with a prefixed value of trimming.

Table 1.2 is a typical output from this approach. This table shows that, while the values of the mean and the median (and of the standard deviation and MADN) seem to be substantially different in the original scale, in the transformed scale these differences are reduced, due to the transformation. Also the ratios of these estimates are closer to one. There is little else to be said from these tabulated results.

Table 1.2 Univariate income data from the United States Census Bureau: robust and non-robust estimates of location and scale

Transformation	Original data	Logged data
Mean	109557.52	11.335
Trimmed mean $\bar{y}_{0.10}$	95172.62	11.345
Median	84881.00	11.349
Standard deviation	86394.86	0.752
MADN	63084.72	0.701

1.4 Univariate Data Analysis Using the Traditional Robust Approach

Table 1.3 Univariate income data from a municipality in the north of Italy: score tests for four values of λ. The smaller the value of the score test, the greater is the indication that we are testing a value of λ close to the maximum likelihood estimate

Transformation	λ	Score test
	−2.0	3.20
	−1.5	−1.07
Inverse	−1	−5.47
Reciprocal square root	−0.5	−10.12
Logarithmic	0	−15.17
No transformation	1	−27.15

Table 1.4 Univariate income data from a municipality in the north of Italy: robust and non-robust estimates of location and scale

Transformation	Original data	Inverse transformation (×100000)	$\lambda = -1.5$ (×10000000)
Mean	18395.17	6.00	4.76
Trimmed mean $\bar{y}_{0.10}$	16883.81	6.10	4.80
Median	16093.50	6.21	4.90
Standard deviation	7247.04	1.55	1.72
MADN	2771.72	1.17	1.42

Proceeding in a similar way we can also, with a traditional exploratory approach, analyse an income dataset coming from a municipality in the north of Italy. Problem 1.4 suggests calculation of the boxplots for the data with $\lambda = -2, -1.5, -1, -0.5, 0$ and 1. This unconventional set of values for λ is chosen to ensure a change of sign in the medians of the boxplots, so providing a first indication of a good parameter value for transformation of the data. The score tests for these values of λ are presented in Table 1.3.

Among these values the best transformation seems to be the implausible value $\lambda = -1.5$. It is hard to find an explanation for such a transformation.

As for the previous dataset, we now show, in Table 1.4, robust and non-robust measures of location and scale. These are in the original scale, in the transformed reciprocal scale, for which there might be an explanation, and for $\lambda = -1.5$.

The results in this table are more structured than those in Table 1.2, the values of robust and non-robust estimates changing in a structured way with λ. For $\lambda = 1$, the original data, the mean is higher than the location estimates from the two robust procedures, which are relatively close together. For $\lambda = -1$, the mean and trimmed mean are as far apart as the trimmed mean and the median. Finally, for $\lambda = -1.5$, the mean and the trimmed mean are closest together. There is clearly a trend with λ. The transformation also has a systematic effect on the ratio of the non-robust standard deviation to the MADN. Moving across the table as λ becomes more negative, the values of the ratios are 2.61, 1.32, and 1.21. By this measure, the transformation appears to be achieving normality. However, the value -1.5 for the estimate of λ suggests further exploration of the data is advisable.

1.5 Univariate Data Analysis Using the Monitoring Approach

The purpose of this book is not to present the results of analyses based on a predetermined value of the breakdown point or of the efficiency, chosen by relying on asymptotic properties. Instead, we seek to understand how and why the results are modified as we change these parameters. We give particular attention to methods which start with a subset of the data (eventually robustly chosen) and learn how particular statistics and calculated quantities, such as residuals, fitted values, goodness-of-fit measures, and t-statistics, change as we add more observations to this subset. Similarly, if a particular estimator depends on the choice of a particular value of *bdp* or *eff*, our aim is to understand how the results change with these input parameters. Plotting methods allowing linking of plots are crucial to this undertaking of bringing to light the hidden structure of the data.

Figure 1.4 shows, for the Univariate income data from the United States Census Bureau, the values of the trimmed mean for different values of the trimming propor-

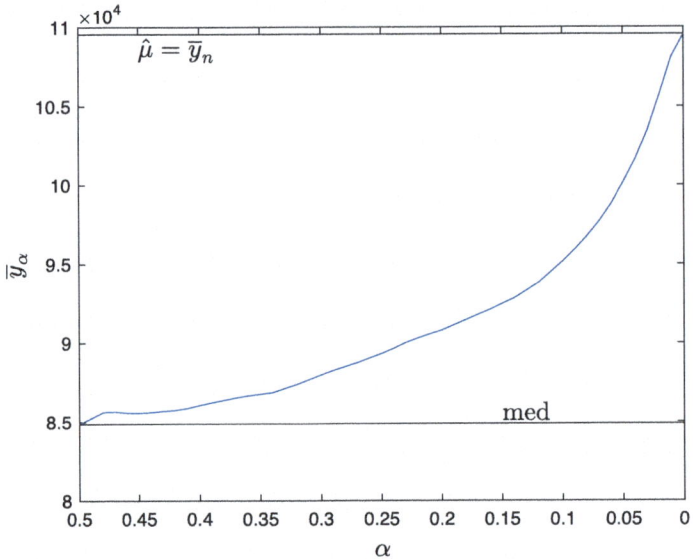

Fig. 1.4 Univariate income data from the United States Census Bureau: monitoring the truncated mean as a function of the trimming proportion α

1.5 Univariate Data Analysis Using the Monitoring Approach

tion α. In this figure, the x-axis is reversed so that $\alpha = 0$ is on the right. The purpose, as in all plots where this is relevant, is to plot the least robust fit (no trimming or downweighting) on the right.

It is well known that income data have positive skewness, so that the mean is greater than the median. This plot monitors estimates from the truncated mean. The figure clearly shows the passage from the very robust estimator of location given by the median (84881 dollars) to the completely non-robust location estimator given by the mean (109557 dollars). The beauty and importance of the monitoring approach is that we do not have arbitrarily to select a value of α in advance of the analysis, but can let the data determine the most informative value. Here, there is no clearly best value. The rationale of the trimmed mean is that the data have an underlying symmetrical distribution contaminated by some outliers. The implication is that this is not a useful model for these data.

Instead of presenting a table with a static value of a particular statistic (like those of the score test in Tables 1.2 and 1.4), we monitor the value of the statistic as a function of the breakdown point or of the size m of the subset used to fit the data. Given a subset of size m it is possible to find an estimate of θ based on the m units and hence to estimate the residuals for all n observations. A new larger subset is formed by the $m + 1$ units with the smallest absolute residuals, providing a new parameter estimate and so on. This is the basis of the forward search, described more fully in Sect. 4.3. Figure 1.5 shows the monitoring of the score test statistic for transformation of the income data from the United States Census Bureau as a function of the subset size. In this figure, we base our analyses on five separate searches with values of $\lambda : -1, -0.5, 0, 0.5$ and 1. The data are first transformed for a particular value of λ and a small subset m_0 found robustly to serve as a starting point. The searches

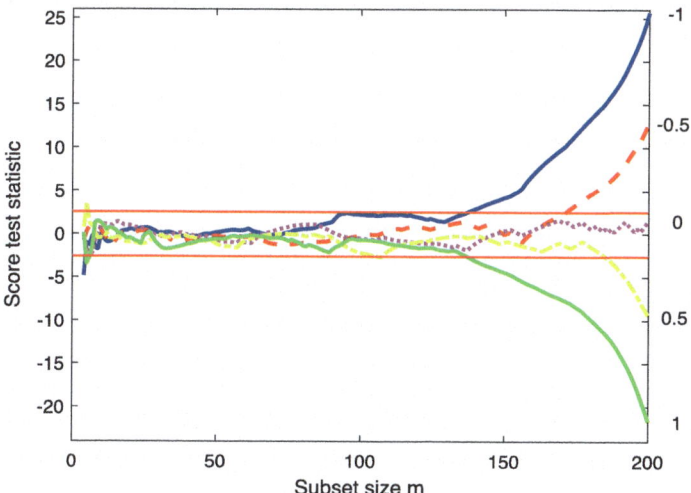

Fig. 1.5 Univariate income data from the United States Census Bureau: monitoring plot of the score test statistics for five values of λ as a function of the subset size m

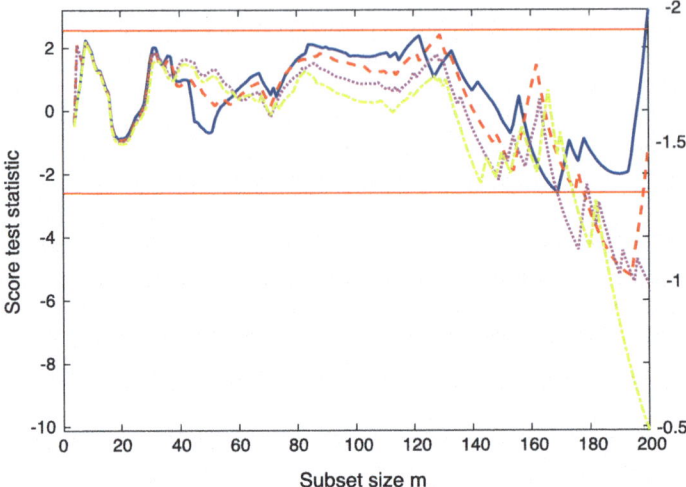

Fig. 1.6 Univariate income data from a municipality in north of Italy: monitoring plot of the score test statistics for four values of λ as a function of the subset size m

yielding each trajectory proceed independently. The central horizontal bands in the figure are at ±2.58, containing 99% of a standard normal distribution. For obvious reasons, we refer to this kind of forward plot as a fan plot.

Initially, apart from the very beginning when results may be unstable, there is no evidence against any transformation. When the subset size is close to $m = 140$ (70% of the data), $\lambda = -1$ and $\lambda = 1$ are rejected. The next rejections are $\lambda = -0.5$ at 86% and 0.5 at 93%. The value of $\lambda = 0$ is supported not only by all the data ($m = n$), but also by the complete sequence of subsets in the figure. Note that, as a bonus of the monitoring approach, we can easily understand what is the percentage of observations which are in agreement with each value of the transformation parameter.

When we monitor the score test statistic for the second set of univariate income data (Fig. 1.6), we extract completely different information from that provided by the first set. In this case, all the four trajectories ($\lambda = -2, -1.5, -1$ and -0.5) are inside the bands up to m around 170. The trajectory for $\lambda = -1$ is the first to exceed the lower band. The effect of about 30 observations which have a crucial effect on the value of the score test statistics is evident. In particular, the plot shows how misleading the results of Table 1.3 are in suggesting the transformation $\lambda = -1.5$. It is true that the value of the score statistic for this value of λ lies within the 99% band at the end of the search, that is, when all observations are used in fitting. But the figure shows that this value was rejected from $m = 178$ and only returned inside the bound for $m = 199$, that is, very close to the end of the search.

This simple example serves as an illustration of the power of monitoring in understanding the effect that a subset of observations can exert on a particular statistic. At the time of the grand plan, vast computational resources were needed to obtain robust estimates based on iterative loops. Despite this effort, the goal was simply to compare, by use of static tables, the results from traditional estimators with those based on a prefixed value of breakdown point. In the 1970s John Tukey (Tukey, 1979) said that "which robust/resistant methods you use is not important – what is important is that you use some. It is perfectly proper to use both classical and robust/resistant methods routinely, and only worry when they differ enough to matter. But when they differ, you should think hard". In the case of the example of the second univariate income dataset, we have seen that it is not possible to claim that all the features in the data have been caught, even from a series of static analyses, that is, those from the use of a prefixed value of *bdp* or of *eff*. Such analyses prevent us from understanding what is the effect that individual observations or subsets of observations exert on the model fitted to our data.

In the next chapter, we continue and extend the analysis of data from a simple sample to introduce several static methods of robust data analyses and state their, often asymptotic, properties. Most of the rest of the book is concerned with linear regression, which is described in detail in Chap. 3. The extension to generalized additive models and non-parametric inference is in Chap. 7. A main theme is the comparison of the traditional approach to robust statistics, based on a single "snapshot" of the data with the monitoring approach where we can watch a "film" or "movie" of the data, ordering the observations from those most in agreement with a model to those least in agreement with it.

1.6 Further Reading

The identification of outliers and the immunization of data analysis against both outliers and failures of modelling are important aspects of modern statistics. The detection of outliers has a long history. Over the centuries scientists have used vigorous common sense to discard data that look suspicious. For example, the data on the boiling point of water presented in Forbes (1857) contain one outlier from the regression model. According to Weisberg (2005a, p. 38), Forbes noted on his copy of the paper that this pair of values was "evidently a mistake". With larger and more complicated datasets more formal methods of outlier detection have had to be developed. There is an appreciable history of the ideas of outlier detection and robustness, which is summarized in the opening chapters of several books. Hawkins (1980) is concerned with tests for outlier detection. The emphasis in the books of Cook and Weisberg (1982) and Atkinson (1985) is on diagnostic procedures, especially in regression, which illuminate the effect of individual, or very small groups of observations on parameter estimates and inferences. These methods make use of the explicit algebraic formulae for the effect of deletion of observations, summarized in Sect. 3.2.

There are several well-cited books on robust statistics. The Princeton Robustness Study (Andrews et al., 1972), mentioned in Sect. 1.1, is now only of historical interest. This was followed, after some years, by the more mathematical approach to more general problems of Huber (1981) and its lightly revised second edition (Huber and

Ronchetti, 2009) as well as Hampel et al. (1986). Hoaglin et al. (1983) is more focused on data analysis. The systematic use of computing makes its appearance in Rousseeuw and Leroy (1987), although this was well before the development of current graphical methods for visualizing data analyses. The focus of Rousseeuw and Leroy (1987) is high breakdown regression. A succinct historical survey of robust regression opens Rousseeuw (1984). A recent exposition of the theory and methods of robust statistics is Maronna et al. (2019) which will be much referred to in what follows. However, despite the title, the book is not concerned with computer-based data analysis, although there is an accompanying R package called **RobStatTM**. Olive (2020) is an online text of which, the author says, "revisions are ongoing". It presents interesting ideas about graphics and data analysis, as well as some trenchant criticisms of algorithms for robust fitting.

The history of the Forward Search is much shorter. The idea of fitting a model to subsets of an increasing size in such a way as to provide robustness was introduced by Hadi (1992) and Atkinson (1994) for multivariate data and by Hadi and Simonoff (1993) for regression. The ensuing development of the forward search, combining flexible downweighting with a focus on data analysis, is covered in two books, Atkinson and Riani (2000) on regression and Atkinson et al. (2004), which is concerned with multivariate procedures. Atkinson et al. (2010), which includes interesting discussions, provides a survey of developments since the publication of our two books. Our more recent work on the forward search will be much cited in the following chapters.

Apart from Chap. 7, we are concerned with parametric models. The standard work on robust non-parametric statistical methods is Hettmansperger and McKean (2011). The most recent of Jana Jurečková's books on robust statistics is Jurečková et al. (2019). A survey of robust method for non-parametric regression models is given by Salibian-Barrera (2023) who cautions that non-robust estimation methods for these models are very sensitive to the presence of outliers, a phenomenon we observe in Chap. 7.

1.7 Some Pre-history of Robustness

The following extract from Thucydides indicates that awareness of outliers and a method to cope with them has existed for at least 2000 years.

In his third book about the Peloponnesian War (III 20, 3–4), Thucydides describes how in 428 BCE the Plataeans used concepts of robust statistics in order to estimate the height of the ladder which was needed to overcome the fortifications built by the Peloponnesians and the Boeotians who were besieging their city.

"The same winter the Plataeans, who were still being besieged by the Peloponnesians and Boeotians, distressed by the failure of their provisions, and seeing no hope of relief from Athens, nor any other means of safety, formed a scheme with the Athenians besieged with them for escaping, if possible, by forcing their way over the enemy's walls. ... At first all were to join: afterwards, half hung back, thinking the risk to be great; about two hundred and twenty, however, voluntarily persevered in the attempt, which was carried out in the following way. Ladders were made to match the height of the enemy's wall, which they measured by the layers of bricks, the side turned towards them not being thoroughly whitewashed. These were counted by many persons at once; and though some might miss the right calculation, most would hit upon it, particularly as they counted over and over again, and were no great way from the wall, but could see it easily enough for their purpose. The length required for the ladders was thus obtained, being calculated from the height of the brick".

The implication is that the mode, if not the median or a sort of trimmed mean, was thought to be a more reliable estimate than the sample mean. This simple example, brought to our attention by Dr. Spyros Arsenis (European Commission, Joint Research Centre), whose English translation has been used above, shows that concerns about the robustness of statistical methods in the sense of insensitivity to grossly wrong measurements may be as old as the experimental approach to science.

Problems

1.1 Trimmed means
Using the definition of the trimmed mean given in (1.3) and the first 15 observations of variable HTOTVAL in dataset `Income1`, compute the 0.05 and 0.10 trimmed means.

1.2 Consistency factor for IQR
Show that under the normal model

$$E(IQR) = 2\Phi^{-1}(0.75)\sigma.$$

1.3 Consistency factor for MAD
Show that under the normal model

$$E(MAD) = \Phi^{-1}(0.75)\sigma.$$

1.4 Univariate analysis
Using variable income in dataset `Income2`:

(a) Compute the trimmed mean using the following values of α, $0, 0.01, \ldots, 0.50$ and plot these values vs. α using a reverse scale for α. Add two horizontal lines with the values of the mean and the median. Comment on the plot you obtain.

(b) Compute the boxplots using the normalized Box–Cox power transformation for the following six values of λ : $1, 0, -0.5, -1, -1.5$ and -2, having first rescaled the data to have a maximum value of one.

1.5 Likelihood concepts

The likelihood $L(\theta; y)$ is a function of θ and y that gives the probability (or density) of observing a sample under a postulated distribution, treating the observations as fixed. From a pure optimization perspective, the likelihood is a particular choice of objective function that reflects the probability of the observed outcome.

In this book, we assume that observations are independent and so the joint probability of the sample values is the product of the probability of the individual observations:

$$L(\theta; y) = \prod_{i=1}^{n} f(y_i, \theta).$$

Since the logarithm is a strictly increasing function, maximizing the natural logarithm of the likelihood (denoted $\mathcal{L}(\theta) = \log\{L(\theta)\}$) leads to the same solution as maximizing $L(\theta)$. Note that in our notation we suppress the random variables $y = (y_1, \ldots, y_n)$ unless this could cause confusion. Working with the loglikelihood is preferable because the product over n likelihood contributions becomes a sum which facilitates numerical calculation and the derivation of analytical properties of the maximum likelihood estimator. The maximum likelihood estimator $\hat{\theta}$, maximizing the likelihood, is the value under which the random sample is the most likely to have been generated. Scientific reasoning may lead to the choice of a physically meaningful parameter value close to $\hat{\theta}$. See Chap. 6 for instances of this in choosing the estimate of λ for the Box–Cox transformation. Because a maximum of the likelihood has been obtained does not guarantee that a good summary of the data has been found; the model may be quite wrong. The robust procedures of this book provide powerful and transparent methods of checking agreement between data and a fitted model.

Consider a null hypothesis H_0 that imposes restrictions on the possible values of θ, relative to an unconstrained alternative H_1. For example, the restricted model could be $H_0 : \lambda = 0$ against $\lambda \in \mathbb{R}$. The likelihood testing procedure involves fitting the two models and obtaining the maximum likelihood estimators $\hat{\theta}$ and $\hat{\theta}_0$ of the parameters under H_1 and H_0, respectively. The null hypothesis H_0 states that "the reduced model is an adequate simplification of the full model". Likelihood theory provides three main classes of statistics for testing this hypothesis: these are likelihood ratio test statistics, denoted LR, which measure the decrease in loglikelihood (vertical distance) from $\mathcal{L}(\hat{\theta})$ to $\mathcal{L}(\hat{\theta}_0)$; Wald test statistics, denoted by W, which consider the standardized horizontal distance between $\hat{\theta}$ and $\hat{\theta}_0$, and score tests denoted by T_{SC} which are the scaled gradient (derivative of) $\mathcal{L}(\theta)$ evaluated only at θ_0.

(a) Under the hypothesis of normal likelihood, show graphically the difference between these three tests.

(b) What is the value of T_{SC} when $\theta_0 = \hat{\theta}$?

(c) Further suppose that interest is in the normal location model with known variance. What is the geometric form of $\mathcal{L}(\theta)$ for a single observation and for a sample of n observations? What is the precise relationship between the values of the three test statistics?

Open Access This chapter is licensed under the terms of the Creative Commons Attribution 4.0 International License (http://creativecommons.org/licenses/by/4.0/), which permits use, sharing, adaptation, distribution and reproduction in any medium or format, as long as you give appropriate credit to the original author(s) and the source, provide a link to the Creative Commons license and indicate if changes were made.

The images or other third party material in this chapter are included in the chapter's Creative Commons license, unless indicated otherwise in a credit line to the material. If material is not included in the chapter's Creative Commons license and your intended use is not permitted by statutory regulation or exceeds the permitted use, you will need to obtain permission directly from the copyright holder.

Chapter 2
Introduction to M-Estimation for Univariate Samples

Abstract The purpose of this chapter is to introduce and discuss the tools, such as the influence function, used to evaluate the properties of the various estimators of location and scale in univariate samples. We focus on M-estimators, that is, estimators of maximum likelihood type, in which the estimating equations for least squares are modified by a ρ function that downweights large residuals. We consider the estimation of location and scale and the simultaneous estimation of both parameters and provide algorithms. In Sect. 2.3.2, we show the properties of the ρ functions and their derivatives, which are important in the algorithms for estimation; all functions depend on parameters which specify the resulting values of *bdp* and *eff*. In Sect. 2.4.1.1, we compare the estimators in terms of *bdp* and *eff*, varying the parameters. These asymptotic calculations of *bdp* and *eff* for the estimators show that, because of the relationship between the two properties, there is surprisingly little difference between the performance of the ρ functions. The chapter concludes with some small-sample comparisons of the estimators and consideration of multiple solutions to numerical algorithms for parameter estimation.

2.1 Properties of Estimators

2.1.1 The Sensitivity Curve and the Influence Function

Let $T_n = T_n(y_1, \ldots, y_n)$ be an estimator of a quantity of interest based on a sample of n observations from F_θ. In robust procedures F_θ is considered as a mathematical abstraction, which is an idealized approximation to reality; the goal is to produce statistical procedures which still behave fairly well under deviations from the assumed model. We now illustrate what happens to an estimator when we slightly change the distribution of the data. For this purpose, we assume that we have $n - 1$ fixed observations $Y_{n-1} = (y_1, \ldots, y_{n-1})$ and that we add observation n at y, which can be any real number. The standardized sensitivity curve, $\text{SC}(y, \hat{T}_n, Y_{n-1})$ of the estimator $\hat{T}_n$ for the sample Y_{n-1} after adding observation y, is defined as

$$\text{SC} = \text{SC}(y, \hat{T}_n, Y_{n-1}) = n\{\hat{T}_n(Y_{n-1}, y) - \hat{T}_{n-1}(Y_{n-1})\}. \tag{2.1}$$

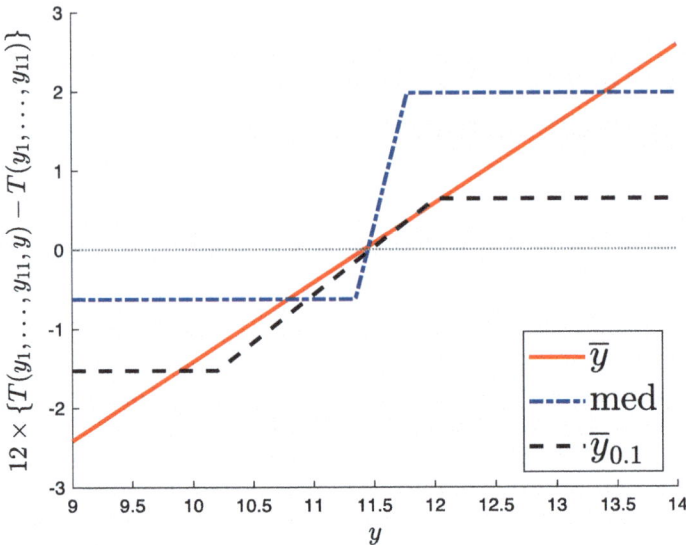

Fig. 2.1 Logged univariate income data from the United States Census Bureau (first 11 observations): sensitivity curve for the arithmetic mean, the median, and the truncated mean with $\alpha = 0.10$

The effect of the addition of a further observation to the sample, the $n-1$ observations remaining fixed, will decrease as n increases. The value of n before the parenthesis in (2.1) serves to neutralize the effect of sample size. For the arithmetic mean, it is easy to show that (Exercise 2.1)

$$\text{SC}(y, \hat{T}_n, Y_{n-1}) = y - \bar{y}_{n-1}. \tag{2.2}$$

On the other hand, for the median, supposing that $n = 2m+1$ is odd, Exercise 2.2 shows that

$$\text{SC}(y, \hat{T}_n, Y_{n-1}) = \begin{cases} n\{y_{[m]} - \frac{1}{2}(y_{[m]} + y_{[m+1]})\} = \frac{n}{2}(y_{[m]} - y_{[m+1]}) & \text{if } y < y_{[m]} \\ n\{y - \frac{1}{2}(y_{[m]} + y_{[m+1]})\} & \text{if } y_{[m]} \leq y \leq y_{[m+1]} \\ n\{y_{[m+1]} - \frac{1}{2}(y_{[m]} + y_{[m+1]})\} = \frac{n}{2}(y_{[m+1]} - y_{[m]}) & \text{if } y > y_{[m+1]} \end{cases} \tag{2.3}$$

To illustrate these concepts, we use the first 11 observations of the income data taken from the United States Census Bureau. The logarithms of the first 11 values are 10.84 11.35 11.82 10.20 11.05 11.99 11.45 12.01 11.78 11.01 11.99. Figure 2.1 shows the value of the sensitivity curves of the mean, median, and trimmed mean with $\alpha = 0.10$ when y is a value belonging to the sequence 9, 9.01, 9.02, ..., 13.99, 14.

We see that all curves are bounded, except that corresponding to the arithmetic mean, which grows linearly without bound with y (see Eq. 2.2). The curve for the trimmed mean shows that it does not reject large and small observations, but just

2.1 Properties of Estimators

limits their influence. The curve for the median is very steep when y is close to med$(y_1, \ldots, y_{n-1})$ (see Eq. 2.3). Note that the sensitivity curve is regarded primarily as a function of the added observation y. However, it also depends on the sample which, in this case, leads to an asymmetrical plot. To remove this dependence we can let the sample size tend to infinity. When this happens the estimator from the sample becomes the parameter of the underlying distribution $\hat{T}_\infty = \theta$ (Fisher consistency) and the sensitivity curve becomes the influence curve (or influence function IF defined below).

Throughout the book we shall be concerned with data that are either normally distributed, or can be made so by transformation (Chap. 6). Many of the robust procedures in this book have been derived for robustness against the contaminated normal model in which a proportion $1 - \varepsilon$ of the observations is generated by a normal model, with a proportion ε ($0 \leq \varepsilon < 0.5$) generated by some unknown mechanism. *Remark*: Note that for the contamination model we use the symbol ε rather than ϵ. Then, the distribution of the observations is the mixture

$$F_\varepsilon = (1 - \varepsilon) F(\theta) + \varepsilon G, \qquad (2.4)$$

with $F(\theta) = \mathcal{N}(\mu, \sigma^2)$ and G any other distribution. We do not try to estimate G. The purpose is to estimate the parameter θ of $F(\theta)$, unaffected by the proportion ε of outliers. A special case of the contaminated normal model is the, so-called, point-mass contamination model when G is a point-mass distribution and the point mass δ_{y_0} is the distribution such that $P(y = y_0) = 1$.

The IF is an approximation to the behaviour of $\hat{T}_\infty$ when the sample contains a small fraction of identical outliers and is defined as

$$\text{IF}_{\hat{T}}(y_0, F_\theta) = \lim_{\varepsilon \downarrow 0} \frac{\hat{T}_\infty \left\{ (1 - \varepsilon) F_\theta + \varepsilon \delta_{y_0} \right\} - \hat{T}_\infty (F_\theta)}{\varepsilon}. \qquad (2.5)$$

The quantity $\hat{T}_\infty \left\{ (1 - \varepsilon) F_\theta + \varepsilon \delta_{y_0} \right\}$ is the asymptotic value of the estimate in the presence of the point-mass contamination model.

Thus the IF describes the effect on the estimate, at the point we are analysing, of an infinitesimal contamination, standardized by the mass of the contamination (the asymptotic bias caused by contamination in the observations). For a robust estimator, we want a bounded influence function, that is, one which does not go to infinity as y_0 becomes arbitrarily large.

The influence function is the asymptotic version of the sensitivity curve. It is computed for an estimator T at a certain distribution F, and does not depend on a specific dataset. Instead of being a function of the observations, the estimator is a function of the distribution F. For instance $T_\infty = T(F) = \mathrm{E}_F(Y)$ is the "functional version" of the sample mean. Similarly, $T_\infty = T(F) = F^{-1}(0.5)$ is the functional version of the sample median. The influence function measures how $T(F)$ changes when the data come from F_ε given in Eq. (2.4) with G a point-mass distribution.

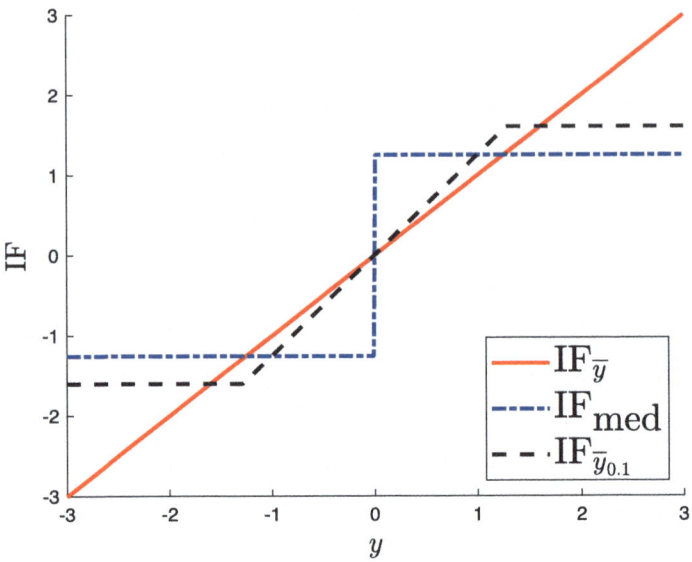

Fig. 2.2 IF, at the standard normal model, of the mean, median, and truncated mean with $\alpha = 0.1$

It is possible to show that the expressions of the IF for the mean, median, and truncated mean (under a symmetric model) are, respectively (Exercises 2.3–2.6):

$$\text{IF}_{\bar{y}}(y, F_\theta) = y - \mu \quad \text{(unbounded)}, \tag{2.6}$$

$$\text{IF}_{\text{med}}(y, F_\theta) = \frac{\text{sgn}(y - \mu)}{2f\{F^{-1}(0.5)\}} \quad \text{and} \tag{2.7}$$

$$\text{IF}_{\bar{y}_\alpha}(y, F_\theta) = \frac{\text{sgn}(y - \mu)}{1 - 2\alpha} \min\left\{|y - \mu|, F^{-1}(1 - \alpha) - \mu\right\}. \tag{2.8}$$

Figure 2.2 shows the plot of the IF of the mean, median, and truncated mean with $\alpha = 0.1$ at the standard normal model. Note that, given that the normal distribution is symmetrical, this leads to a symmetrical figure unlike Fig. 2.1.

An alternative summary of the behaviour of a robust estimator is given by the Gross Error Sensitivity (GES, often indicated by the symbol γ^*). This has the advantage of being a number rather than a function. It expresses, asymptotically, the maximum effect a fixed amount of contamination can have on the estimator T_n (hence it can be regarded as an approximate bound for the bias of the estimator). It is the maximum value of the IF.

Definition: The gross error sensitivity (Hampel, 1974) of a consistent estimator T_n of T at a distribution F is

2.1 Properties of Estimators

$$\text{GES}(T, F) = \gamma^*(T, F) = \sup_y |\text{IF}(y, F, T)|.$$

Clearly, the preference is for estimators which have a fairly small sensitivity (not just finite). It is easy to see that the maximum asymptotic bias of an estimator is simply related to the gross error sensitivity. Given the contamination model (2.4), the maximum bias (MB) from the contamination is, asymptotically:

$$\text{MB}(\varepsilon, T, F) = \sup_G |T\{(1 - \varepsilon)F(\theta) + \varepsilon G\} - T(F)|, \tag{2.9}$$

where G is the distribution of the contamination. From the definition of the IF given in Eq. 2.5, it follows that the maximum bias is approximately $\text{GES}(T, F) \times \varepsilon$.

2.1.2 The Asymptotic Variance

An important property of an estimator T_n is its asymptotic variance, defined as the limit, as $n \to \infty$ of $n \text{var}(T)$ (provided this limit exists). Clearly, our goal is to have estimators with small variance (that is with high statistical efficiency) and at the same time with low $\gamma^*(T, F)$. In the case of the sample mean, when $\text{var}(Y) = \sigma^2$, $\text{var}(\bar{y}) = \sigma^2/n$ and $\lim_{n \to \infty} n \text{var}(\bar{y}) = \sigma^2$. Therefore the asymptotic variance of the sample mean is σ^2 regardless of the distribution of Y. Huber (1981, Chap. 3) shows that, under certain regularity conditions, the asymptotic variance of a consistent estimator T_n of θ is

$$n \, \text{var}(T) = \int_{-\infty}^{\infty} \{\text{IF}_{\hat{T}}(y_0, F_\theta)\}^2 f_\theta(y) dy. \tag{2.10}$$

This result shows that the asymptotic variance is proportional to the IF, which implies a specific efficiency. Using this result gives the expression for the asymptotic variance of the sample median. Combining Eqs. (2.7) and (2.10) shows that the asymptotic variance of the median is

$$n \, \text{var}(\text{med}) = \int_{-\infty}^{\infty} \left\{\text{IF}_{\hat{T}}(y_0, F_\theta)\right\}^2 f_\theta(y) dy, \tag{2.11}$$

$$= \int_{-\infty}^{\infty} \left[\frac{\text{sgn}(y - \mu)}{2f\{F^{-1}(0.5)\}}\right]^2 f_\theta(y) dy \quad \text{and} \tag{2.12}$$

$$= \frac{1}{4\left[f\{F^{-1}(0.5)\}\right]^2}. \tag{2.13}$$

Using the above expression we obtain (see Exercise 2.7) the asymptotic relative efficiency (introduced in Eq. 1.5) of the sample median relative to the sample mean.

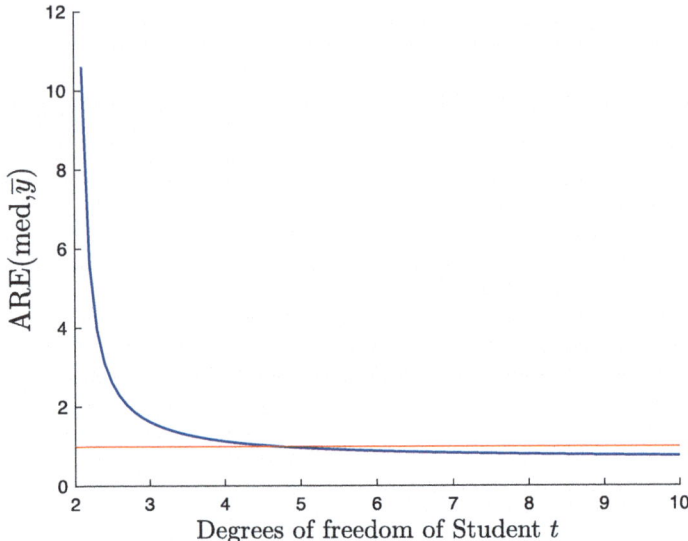

Fig. 2.3 Analysis of ARE(med; $\bar{y}$) as a function of the degrees of freedom of the Student's t distribution. When the degrees of freedom are smaller than 4.6785 the variance of the sample mean is greater than the variance of the sample median. The magnitude of the difference between these two variances increases as the degrees of freedom decrease

ARE(med; $\bar{y}$) = $2/\pi \approx 64\%$ when the data come from the normal distribution.

For the Student's t distribution (see Exercise 2.7):

$$t_5 \text{ ARE(med; } \bar{y}) \approx 96.07\%$$
$$t_4 \text{ ARE(med; } \bar{y}) \approx 112.50\%$$
$$t_3 \text{ ARE(med; } \bar{y}) \approx 162.11\%$$
$$t_2 \text{ ARE(med; } \bar{y}) = \infty.$$

Figure 2.3 plots the ARE as a function of the degrees of freedom of the t distribution. This plot suggests that the heavier are the tails of the distribution from which we sample, the greater is the desirability of using a robust estimate of location.

We may also wonder what happens to the variance of the sample mean (or median) when all measurements are not equally precise. Suppose that the data come from the special case of the contamination model given in Eq. 2.4:

$$F_\varepsilon = (1 - \varepsilon)\mathcal{N}(\mu, \sigma^2) + \varepsilon\mathcal{N}\{\mu, (\tau\sigma)^2\}. \tag{2.14}$$

In this case (Exercise 2.8)

2.1 Properties of Estimators

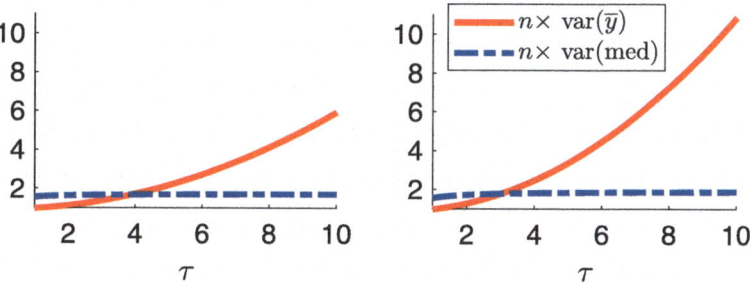

Fig. 2.4 Analysis of $n\,\text{var}(\bar{y})$ and $n\,\text{var}(\text{med})$ for the contamination model $F_\varepsilon = (1-\varepsilon)\mathcal{N}(\mu,\sigma^2) + \varepsilon\mathcal{N}\{\mu,(\tau\sigma)^2\}$. Left-hand panel: $\varepsilon = 0.05$. Right-hand panel: $\varepsilon = 0.10$

$$\text{var}(\bar{y}) = \sigma^2 \frac{(1-\varepsilon) + \varepsilon\tau^2}{n}$$

$$\text{var}(\text{med}) = \frac{1}{n} \cdot \sigma^2 \frac{\pi}{2(1-\varepsilon+\varepsilon/\tau)^2}.$$

Figure 2.4 shows the evolution of these two variances as function of τ for two values of ε. As τ increases, the variance of the sample mean tends to become much greater than that of the median. The crossing of the two curves takes place for lower values of τ for larger ε; with $\varepsilon = 0.05$ ARE(med; $\bar{y}$) > 1 when $\tau > 3.857$. With $\varepsilon > 0.10$ ARE(med; $\bar{y}$) > 1 when $\tau > 3.005$.

The moral of the story of this section is the following: under the ideal conditions that the observations all have equal precision and that the data are from the normal distribution and that there are no outliers, the variance of the sample mean is roughly 64 percent of that of the sample median. However, when the data come from distributions with fat tails, or there are outliers, or the observations are not measured with the same precision, the variance of the sample mean can be much greater than that of the sample median.

2.1.3 Rates of Convergence

Rates of convergence quantify the (stochastic) order of magnitude of an estimation error as a function of the sample size n. This order of magnitude is usually represented using the symbols: O_P and o_P. Let $\{Y_n\}_{n=1,2,3,\ldots}$ be a sequence of random variables, and let $\{c_n\}_{n=1,2,3,\ldots}$ be a sequence of positive (deterministic) numbers.

- We will write $Y_n = O_p(c_n)$ if, for any $\epsilon > 0$, there exist numbers $0 < M < \infty$ and m such that
$$P(|Y_n| \geq M \cdot c_n) \leq \epsilon \quad \text{for all} \quad n \geq m.$$

- We will write $Y_n = o_p(c_n)$ if

$$\lim_{n \to \infty} P\left(|Y_n| \geq \epsilon \cdot c_n\right) = 0 \quad \text{for all} \quad \epsilon > 0.$$

- With $c_n = 1$ for all n, $Y_n = O_p(1)$ means that the sequence $\{Y_n\}$ is stochastically bounded, i.e. for any $\epsilon > 0$ there exist numbers $0 < M < \infty$ and m such that

$$P\left(|Y_n| \geq M\right) \leq \epsilon \quad \text{for all} \quad n \geq m.$$

- With $c_n = 1$ for all n, $Y_n = o_P(1)$ is equivalent to $Y_n \xrightarrow{p} 0$, i.e. Y_n converges in probability to zero.

- $Y_n = O_p(c_n)$ is equivalent to $Y_n/c_n = O_p(1)$.

- $Y_n = o_p(c_n)$ is equivalent to $Y_n/c_n = o_p(1)$.

Definition: An estimator $\hat{\theta} = \hat{\theta}_n$ of a parameter θ possesses the rate of convergence n^{-r} if and only if r is the largest positive number with the property that

$$\left|\hat{\theta}_n - \theta\right| = O_P\left(n^{-r}\right).$$

The rate of convergence quantifies how fast the estimation error $\left|\hat{\theta}_n - \theta\right|$ decreases when increasing the sample size n.

It follows that, if $\hat{\theta}_n$ is an unbiased estimator of an unknown parameter θ satisfying $\text{var}\left(\hat{\theta}_n\right) = Cn^{-1}$ for some $0 < C < \infty$, then $\hat{\theta}_n$ possesses the rate of convergence $n^{-1/2}$ (Exercise 2.9). Therefore, given the expressions for the variances of the mean (σ^2/n), or of the median (2.13), it is clear that they both have the rate of convergence $n^{-1/2}$.

Section 2.5 contains further material on convergence and also gives references to work on asymptotic statistical procedures, in particular, those with applications to robust estimation.

2.2 One Univariate Sample: Introductory Concepts

The location parameter of a probability distribution is a scalar- or vector-valued parameter μ which determines the location or shift of the distribution. If μ is a location parameter, the cumulative distribution function (cdf) can be written as $F_\mu(y) = F(y - \mu)$ and the density as $f_\mu(y) = f(y - \mu)$.

The scale parameter σ of a probability distribution determines the statistical dispersion of the probability distribution. If σ is large, then the distribution will be more spread out; if σ is small then it will be more concentrated. More rigorously, σ is a scale parameter if the associated cdf can be written as

2.2 One Univariate Sample: Introductory Concepts

$$F(y, \sigma) = F(y/\sigma),$$

or the density as

$$f(y, \sigma) = \frac{1}{\sigma} f(y/\sigma).$$

A location estimator $\hat{\mu}$ is called location equivariant and scale equivariant iff

$$\hat{\mu}(aY_n + b) = a\hat{\mu}(Y_n) + b$$

for all samples Y_n and all $a \neq 0$ and $b \in \mathbb{R}$. A scale estimator $\hat{\sigma}$ is called location invariant and scale equivariant iff

$$\hat{\sigma}(aY_n + b) = |a|\hat{\sigma}(Y_n).$$

2.2.1 M-Estimation: Downweighting or Soft Trimming

In this section, we introduce the class of M-estimators of location and scale which contains the mean and the median as special cases and give algorithms to compute them. They are called M-estimators because they are estimators of a maximum likelihood type.

In the previous chapter, we introduced the concept of breakdown point in loose terms. We now provide a more formal definition using the contamination model.

Definition: The asymptotic breakdown point of the estimator $\hat{\theta}$ at F, abbreviated *bdp*, is the largest $\varepsilon^* \in (0, 1)$ such that for $\varepsilon < \varepsilon^*$, $\hat{\theta}\{(1-\varepsilon)F(\theta) + \varepsilon G\}$ remains bounded away from the boundary of Θ for all G. Note that, in the case of a scale estimator, the boundary of Θ is $0, +\infty$.

As we have seen in Chap. 1, the *bdp* quantifies the behaviour of the M-estimator in the presence of contamination. For equivariant location estimators, the breakdown point can be at most 50%:

$$\varepsilon_n^*(\hat{\mu}, X_n) \leq \frac{1}{n}\left[\frac{n+1}{2}\right] \approx 50\%.$$

Intuitively: With more than 50% of outliers, the estimator may treat the outliers as the regular observations.

The breakdown value of a scale estimator is defined as the minimum of the explosion breakdown value ($\varepsilon_{\text{expl}}^*$), and the implosion breakdown value ($\varepsilon_{\text{impl}}^*$). Explosion is when the scale estimate is inflated ($\hat{\sigma} \to \infty$). Implosion is when the scale estimate becomes arbitrarily small ($\hat{\sigma} \to 0$). Note that this becomes a problem, because scale estimates often occur in the denominator of a statistic (such as in the computation of scaled residuals).

In the next sections, we also discuss the properties of M-estimators in terms of IF, *bdp*, asymptotic variance, and *eff*.

2.2.1.1 Estimation of Location

In this section, we initially assume that μ is a scalar and that this is the only unknown parameter which characterizes the underlying distribution.

In the location model, the maximum likelihood estimator of μ maximizes the loglikelihood:

$$\hat{\mu} = \arg\max_{\mu} \sum_{i=1}^{n} \log f(y_i - \mu) = \arg\min_{\mu} \sum_{i=1}^{n} -\log f(y_i - \mu). \tag{2.15}$$

The M-estimator of location is then defined as

$$\hat{\mu} = \arg\min_{\mu} \sum_{i=1}^{n} \rho(y_i - \mu). \tag{2.16}$$

If ρ is differentiable, with $\psi(u) = \rho'(\psi) = d\rho(u)/du$, $\hat{\mu}$ is the solution of the equation:

$$\sum_{i=1}^{n} \psi(y_i - \hat{\mu}) = 0. \tag{2.17}$$

If $\psi(.)$ is discontinuous, $\hat{\mu}$ is the point at which the sign of the sum changes.

The maximum likelihood estimator is an M-estimator, with $\rho(y) = -\log f(y)$, so that $\psi(y) = -f'(y)/f(y)$. Then (2.17) is the score equation. For the standard normal distribution $\rho(y) = y^2/2$ so that $\psi(y) = y$ and $\hat{\mu} = \bar{y}$, as above. If $\rho(y) = |y|$, $\psi(y) = \text{sgn}(y)$ and $\hat{\mu} = \text{med}$, that is the median of y. This is the maximum likelihood estimator for the double exponential distribution. For most of the functions $\rho(\cdot)$ used in robust estimation, there is no corresponding distribution f and no explicit form for $\hat{\mu}$.

M-estimation of θ in (2.4) with $\rho(y) = |y|$ provides the data median as the estimate; that is all but one or two observations are downweighted to zero. With $\rho(y) = y^2/2$ no observations are downweighted. There are numerous other ρ functions which, depending on the parameters in ρ, downweight an intermediate number of observations. We now elucidate some of their properties.

The estimating Eq. (2.17) shows the importance of $\psi(y)$. The equation can be rewritten to provide an algorithm for estimation of μ as

$$\sum_{i=1}^{n} w_i(y_i - \hat{\mu}) = 0 \quad \text{where} \quad w_i = \psi(y_i - \hat{\mu})/(y_i - \hat{\mu}). \tag{2.18}$$

2.2 One Univariate Sample: Introductory Concepts

The iterative procedure starts with some $\hat{\mu}_0$, for example, the sample median. Given the estimate $\hat{\mu}_k$ at stage k, compute

$$w_{i,k} = w(y_i - \hat{\mu}_k), \quad i = 1, \ldots, n \tag{2.19}$$

and then find

$$\hat{\mu}_{k+1} = \sum_{i=1}^{n} w_{i,k} y_i / \sum_{i=1}^{n} w_{i,k}. \tag{2.20}$$

Since each step is a weighted mean, which is a special case of weighted least squares, this algorithm is called iteratively reweighted least squares (IRLS). This algorithm stops when $|\hat{\mu}_{k+1} - \hat{\mu}_{k+1}| < \delta$, an appropriate small tolerance (say 1e-07).

So far we have considered just the estimation of the location parameter and ignored the existence of a scale parameter which may characterize the underlying distribution. In the unlikely case that the scale parameter σ_0 is known, u in $\rho(u)$ and $\psi(u)$ are replaced by $(y - \mu)/\sigma_0$. Then the M-estimate of location is found as

$$\hat{\mu} = \arg\min_{\mu} \sum_{i=1}^{n} \rho\left(\frac{y - \mu}{\sigma_0}\right), \tag{2.21}$$

or equivalently as

$$\sum_{i=1}^{n} \psi\left(\frac{y - \hat{\mu}}{\sigma_0}\right) = 0. \tag{2.22}$$

If the scale parameter is not known it is necessary to replace σ_0 with a robust consistent estimate $\hat{\sigma}_0$ such as the MADN (1.9). Note that, in such cases, the estimate of σ is kept fixed; it does not change between iterations. Whether σ is known or unknown, the algorithm stops when $|\hat{\mu}_{k+1} - \hat{\mu}_k| < \hat{\sigma}_0 \delta$ or $< \sigma_0 \delta$.

At the end of this chapter, we will see that for monotone M-estimators (that is, estimators associated with a ψ function which is non-decreasing), this algorithm is guaranteed to converge to the (unique) solution of the estimating equation; the starting point μ_0 just affects the number of iterations. If, on the other hand, ψ is non-monotonic, then $\hat{\mu}_0$ must be robust in order to ensure convergence to a good solution.

2.2.2 Properties of Location M-Estimators

The properties of location M-estimators (under suitable regularity conditions) can be summarized as follows:

- They are Fisher consistent if and only if

$$\int \psi(x) dF(x) = 0.$$

- Their Influence function (IF) is

$$\text{IF}(x, T, F) = \frac{\psi(x)}{\int \psi'(y) dF(y)}. \tag{2.23}$$

  The influence function of an M-estimator is proportional to its ψ-function. A bounded ψ-function thus leads to a bounded IF.

- Their asymptotic distribution is normal with asymptotic variance (Maronna et al., 2019), Chap. 10:

$$V(T, F) = \int \text{IF}(x, T, F)^2 dF(x) = \frac{\int \psi^2(x) dF(x)}{\left(\int \psi'(y) dF(y)\right)^2} = \frac{E_{F_0}\{\psi(y)^2\}}{E_{F_0}\{\psi'(y)\}^2},$$

  where $\psi'(y) = d\psi(y)/dy$. By the information inequality (or Cramer–Rao lower bound), the asymptotic variance satisfies

$$V(T, F) \geq 1/I(F),$$

  where $I(F) = \int \{-f'(x)/f(x)\}^2 dF(x)$ is the Fisher information of the model.

2.3 M-Estimation of Scale

This section initially assumes that μ is known. The maximum likelihood estimate (mle) of σ for the scale family $\left(\frac{1}{\sigma}\right) f\left(\frac{y_i - \mu}{\sigma}\right)$ is

$$\arg\min_{\sigma} \prod_{i=1}^{n} \left(\frac{1}{\sigma}\right) f\left(\frac{y_i - \mu}{\sigma}\right).$$

Moving to the loglikelihood and differentiating with respect to σ yields the estimating equation:

$$\frac{1}{n} \sum_{i=1}^{n} \left\{ -\frac{f'\left(\frac{y_i - \mu}{\sigma}\right)}{f\left(\frac{y_i - \mu}{\sigma}\right)} \right\} = 1. \tag{2.24}$$

2.3 M-Estimation of Scale

The estimate of σ^2 produced using this equation is not robust. To bound the effect of large values of $(y_i - \mu)/\sigma$, note that what is in the braces { } in (2.24) is a generic function of $\dfrac{y_i - \mu}{\sigma}$ and can be replaced by a ρ function to give an estimating equation:

$$\frac{1}{n}\sum_{i=1}^{n} \rho\left(\frac{y_i - \mu}{\sigma}\right) = K. \tag{2.25}$$

In order to have consistency, $K = E\rho(u) = \int \rho(t) dF(t)$. Here u has the standard normal distribution. The quantity K adjusts for the downweighting of observations due to the ρ function and leads, under the normal distribution, to an asymptotically consistent estimate of σ.

As we did in (2.18), we can rewrite this estimating equation in a weighted form to provide an iterative algorithm for estimation, here of σ. Let $w(u) = \rho(u)/u^2$. Then (2.25) can be rewritten as

$$\frac{1}{n}\sum_{i=1}^{n} w\left(\frac{y_i - \mu}{\sigma}\right)\left(\frac{y_i - \mu}{\sigma}\right)^2 = K, \tag{2.26}$$

or equivalently as

$$\sigma^2 = \frac{1}{nK}\sum_{i=1}^{n} w\left(\frac{y_i - \mu}{\sigma}\right)(y_i - \mu)^2, \tag{2.27}$$

a weighted mean square estimate. Remember, μ is taken as known.

The iterative procedure starts with some initial estimate $\hat{\sigma}_0$, for example, the MADN (1.9). In general, given the estimate $\hat{\sigma}_k$ at step k, the weights are calculated as $w\{(y_i - \mu)/\hat{\sigma}_k\}$. Then (2.27) can be expressed as

$$\hat{\sigma}_{k+1}^2 = \frac{1}{nK}\sum_{i=1}^{n} w\left(\frac{y_i - \mu}{\hat{\sigma}_k}\right)(y_i - \mu)^2, \tag{2.28}$$

or equivalently

$$\hat{\sigma}_{k+1}^2 = \hat{\sigma}_k^2 \frac{1}{nK}\sum_{i=1}^{n} w\left(\frac{y_i - \mu}{\hat{\sigma}_k}\right)\frac{(y_i - \mu)^2}{\hat{\sigma}_k^2}. \tag{2.29}$$

Then, from the definition of w,

$$\hat{\sigma}_{k+1}^2 = \hat{\sigma}_k^2 \frac{1}{nK}\sum_{i=1}^{n} \rho\left(\frac{y_i - \mu}{\hat{\sigma}_k}\right) = \hat{\sigma}_k^2 \frac{1}{K}\bar{\rho}\left(\frac{y_i - \mu}{\hat{\sigma}_k}\right), \tag{2.30}$$

where
$$\bar{\rho} = \frac{1}{n}\sum_{i=1}^{n} \rho\left(\frac{y_i - \mu}{\hat{\sigma}_k}\right).$$

2.3.1 Properties of M-Estimators of Scale

- In general it is possible to prove that, for equivariant scale estimators, the breakdown value is at most 50%:
$$bdp(\hat{\sigma}, Y_n) \leq \frac{1}{n}\left[\frac{n}{2}\right] \approx 0.5.$$

- The breakdown point of an M-estimator of scale is
$$\varepsilon^*(\hat{\sigma}_M) = \min\left(\varepsilon^*_{\text{expl}}, \varepsilon^*_{\text{impl}}\right) = \min\left(\frac{\delta}{\rho(\infty)}, 1 - \frac{\delta}{\rho(\infty)}\right),$$

which is 0% for unbounded ρ and 50% for a bounded ρ with $\delta = \rho(\infty)/2$.
Remark recall Sect. 2.2.1 for "explosion" and "implosion".

- At the model distribution F we have $\hat{\sigma} = 1$ by Fisher consistency, and
$$\text{IF}(y, T, F) = \frac{\rho(y) - K}{\int x\rho'(x)dF(x)}.$$

As a consequence, the influence function of an M-estimator is proportional to $\rho(y) - K$. Therefore, as for location estimators, a bounded ρ-function leads to a bounded IF.

- The estimators are asymptotically normal with asymptotic variance:
$$V(T, F) = \int \text{IF}(y, T, F)^2 dF(y).$$

By the Cramer–Rao lower bound (information inequality), the asymptotic variance satisfies
$$V(T, F) \geq 1/I(F),$$

where $I(F) = \int \left\{-1 - \frac{xf'(x)}{f(x)}\right\}^2 dF(x)$ is the Fisher information of the scale model. For $F = \Phi$ we find $I(F) = 2$ and IF$(y; \text{mle}, \Phi) = (y^2 - 1)/2$.

2.3.2 Analysis of the ρ, ψ, ψ', and Weight Functions

As we have seen in previous sections, the characteristics of the estimates of location and scale depend on the properties of the ρ, ψ, ψ', and weight functions. Plots of these quantities together with the weight functions can be informative about the properties of specific M-estimators. We now present six ρ functions in a standard notation with $u = (y - \mu)/\sigma$. Depending on the problem formulation, one or both of μ and σ may have to be estimated.

Huber. The first ρ-function was proposed in Huber (1964):

$$\rho(u) = \begin{cases} (u^2/2) & |u/c| \leq 1 \\ c|u| - c^2/2 & |u/c| > 1, \end{cases} \quad (2.31)$$

when

$$\psi(u) = \begin{cases} u & |u/c| \leq 1 \\ c \, \text{sgn}(u) & |u/c| > 1. \end{cases} \quad (2.32)$$

The limiting cases, $c \to \infty$ and $c = 0$, are the mean and median provided we define $\psi(0) = \text{sgn}(u)$. In fact, when $c = 0$, the objective function defined in Eq. (2.31) is only linear, whereas as $c \to \infty$, this equation becomes quadratic throughout.

This ρ function is unbounded. Although it only increases as the absolute value of c, rather than quadratically as does the ρ function for the normal distribution, both functions go to infinity as $|u|$ does (see Fig. 2.5).

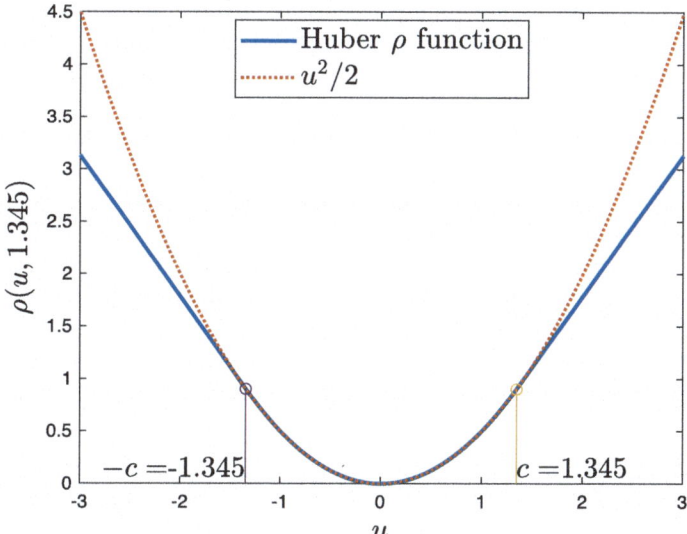

Fig. 2.5 Comparison between Huber's ρ function with $c = 1.345$, and the quadratic ρ function for the normal distribution

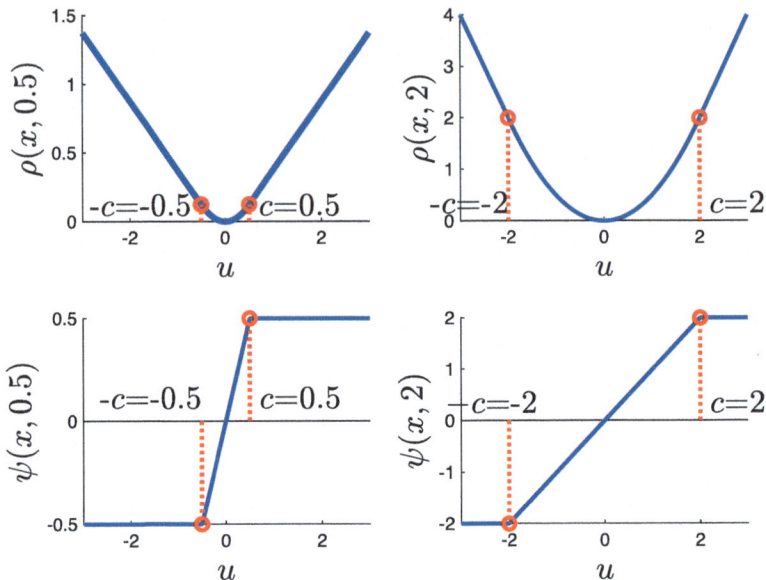

Fig. 2.6 Comparison between Huber ρ and ψ functions with $c = 0.5$ (left-hand panels) and $c = 2$ (right-hand panels)

The resulting estimator therefore has a breakdown point of zero apart from the degenerate limit as $c \to 0$. The corresponding ψ function, as is shown in Fig. 2.6, is linear in the centre and constant in the tails for $|u| > c$.

Given that the IF is proportional to the ψ function, see Eq. (2.23), this implies that the IF for Huber's estimator is constant for all the observations beyond a certain point.

In general parameter c (also called the tuning constant) inside the ρ function is linked to *bdp* and we further explore this point in Sect. 3.9.

M-estimators can be improved by having the ψ function (and hence the IF) return to zero as $|u|$ becomes large. Increasingly remote outliers have a diminishing effect on such redescending estimators. The rejection point is defined as the least distance from the location estimate beyond which observations do not contribute to the value of the estimate (for a given auxiliary scale estimate).

Hampel's ρ function. The first redescending estimator was the Hampel (or the three part redescending) estimator (Hampel et al., 1986), p. 150, defined as

$$\rho(u) = \begin{cases} \frac{1}{2}u^2 & \text{if } |u/c| \leq c_1 \\ c_1|u| - \frac{1}{2}c_1^2 & \text{if } c_1 < |u/c| \leq c_2 \\ c_1 c_2 - 0.5 c_1^2 + 0.5(c_3 - c_2)c_1 \left\{ 1 - \left(\frac{c_3 - |u|}{c_3 - c_2}\right)^2 \right\} & \text{if } c_2 < |u/c| \leq c_3 \\ c_1 c_2 - 0.5 c_1^2 + 0.5(c_3 - c_2)c_1 & \text{if } |u/c| > c_3, \end{cases}$$

(2.33)

2.3 M-Estimation of Scale

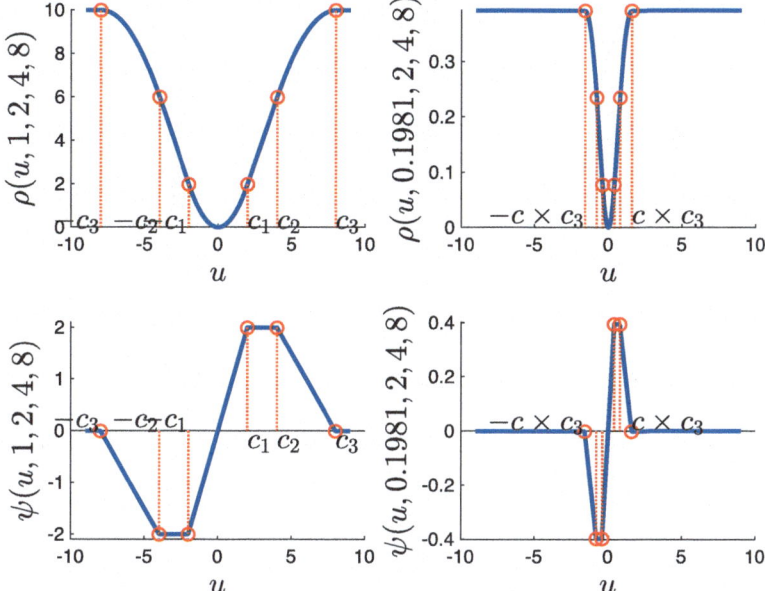

Fig. 2.7 Hampel estimator: ρ and ψ functions. The left-hand panels use $c = 1$, $c_1 = 2$, $c_2 = 4$, and $c_3 = 8$. The right-hand panels use $c = 0.1981$, $c_1 = 2$, $c_2 = 4$, and $c_3 = 8$

with

$$\psi(u) = \begin{cases} u & |u/c| \leq c_1 \\ c_1 \operatorname{sgn}(u) & c_1 \leq |u/c| < c_2 \\ c_1 \frac{c_3 - |u|}{c_3 - c_2} \operatorname{sgn}(u) & c_2 \leq |u/c| < c_3 \\ 0 & |u/c| >= c_3. \end{cases}$$

Figure 2.7 shows ρ and ψ for the most used values of c_1, c_2, and c_3, namely, 2, 4, and 8 and two values of the tuning constant c. The first derivative is piecewise linear and vanishes outside the interval $[-c \times c_3 + c \times c_3]$. The crucial tuning constant is c_3. (Huber and Ronchetti (2009), p.101) suggest that the slope of the tail of the ψ function joining c_2 and c_3 should not be too steep. The value of c equal to 1 (left panels) corresponds to a nominal efficiency of 0.9897 and $bdp = 0.0494$ (see Sect. 3.9) while the value of $c = 0.1981$ corresponds to $bdp = 0.5$ and $eff = 0.2923$.

In the case of the Hampel estimator, for a given estimate of scale $\hat{\sigma}_0$, $u = (y - \hat{\mu}_k)/\hat{\sigma}_0$, where $\hat{\mu}_k$ is the location estimate at iteration k; the rejection point is $c \times c_3 \hat{\sigma}_0$. Observations beyond the rejection point do not contribute to the value of the estimate (except possibly through the scale estimate $\hat{\sigma}_0$). The chosen value of the tuning constant c reflects beliefs in the proportion of outliers in the data. When the sample is thought to be more outlier prone, a smaller value of c is more appropriate. Then the rejection point $c \times c_3$ decreases and more observations will tend to be trimmed (downweighted).

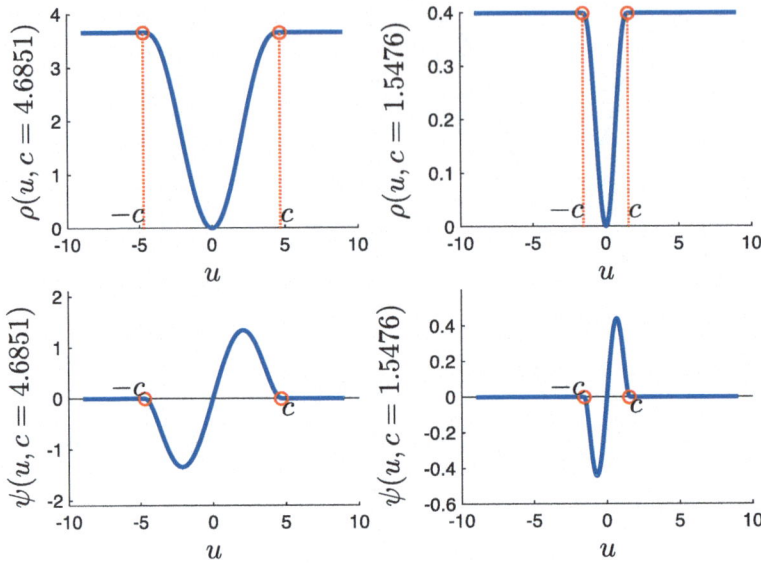

Fig. 2.8 Tukey estimator: ρ and ψ functions. The left-hand panels use $c = 4.6851$ (*eff* = 0.95). The right-hand panels use $c = 1.5476$ (*bdp* = 0.5)

The most widely used of such redescending ψ functions is **Tukey's biweight (or bisquare) function** (Beaton and Tukey, 1974):

$$\rho(u) = \begin{cases} \frac{u^2}{2} - \frac{u^4}{2c^2} + \frac{u^6}{6c^4} & \text{if } |u| \leq c \\ \frac{c^2}{6} & \text{if } |u| > c, \end{cases} \quad (2.34)$$

when

$$\psi(u) = \begin{cases} (c^2/6)u\{1 - (u/c)^2\}^2 & |u/c| \leq 1 \\ 0 & |u/c| > 1, \end{cases}$$

the first derivative of which vanishes outside the interval $[-c + c]$ (the rejection point is therefore c). Again, in this case, the tuning constant c in the biweight can be chosen to balance breakdown point and efficiency. Figure 2.8 shows the ρ and ψ functions for two typical values of c. The value of c equal to 4.6851 (left-hand panels) corresponds to a nominal efficiency of 0.95 and *bdp* = 0.1194 while the value of $c = 1.5476$ corresponds to *bdp* = 0.5 and *eff* = 0.2868. The complete table of the tuning constants associated with the most used values of *bdp* or *eff* for this ρ function is given in Table 3.3.

(Hampel et al. (1986), p.328) considered a different optimization problem leading to a new form of ρ function. This minimizes the asymptotic variance of the M-estimate, subject to a bound on the supremum of the Change of Variance Curve (CVC) of the estimate. The CVC describes the infinitesimal increment of the logarithm of

2.3 M-Estimation of Scale

the variance of the M-estimator in the vicinity of the null normal model, in the same way that the influence function reflects the infinitesimal asymptotic bias. This leads to the **Hyperbolic Tangent ρ function**, which, for suitable constants c, k, A, B, and d, is defined as

$$\rho(u) = \begin{cases} \frac{1}{2}u^2 & \text{if } |u| \leq d \\ \frac{d^2}{2} - 2\frac{A}{B}\log\cosh\{\frac{1}{2}\sqrt{\frac{(k-1)B^2}{A}}(c-|u|)\}+ \\ +2\frac{A}{B}\log\cosh\{\frac{1}{2}\sqrt{\frac{(k-1)B^2}{A}}(c-d)\} & \text{if } d \leq |u| \leq c \\ \frac{d^2}{2} + 2\frac{A}{B}\log\cosh\{\frac{1}{2}\sqrt{\frac{(k-1)B^2}{A}}(c-d)\} & \text{if } |u| > c, \end{cases} \quad (2.35)$$

$$\psi(u) = \begin{cases} u & |u| \leq d \\ \sqrt{A(k-1)}\tanh\left\{\sqrt{(k-1)B^2/A}(c-|u|)/2\right\}\text{sgn}(u) & d \leq |u| < c \\ 0 & |u| \geq c, \end{cases} \quad (2.36)$$

where $0 < d < c$ is such that

$$d = \sqrt{A(k-1)}\tanh\left\{\frac{1}{2}\sqrt{\frac{(k-1)B^2}{A}}(c-d)\right\}. \quad (2.37)$$

The parameters A and B are found as

$$A = E\{\psi^2(x)\} \quad \text{and} \quad B = E\{\psi'(x)\}.$$

The value of d is found by applying the Newton–Raphson method to (2.37). New values of A and B are obtained (through numerical integration) and the procedure is iterated to convergence. For additional details see Hampel et al. (1985). The parameter k is defined as

$$k = \sup_x\{CVC(\psi, x)\}.$$

Figure 2.9 shows the hyperbolic ρ and ψ functions for two typical values of c and two values of k. The value of c equal to 6 (left-hand panels) when $k = 5$ corresponds to $bdp = 0.0596$ and to a nominal efficiency of 0.9861 and when $k = 3$ to $bdp = 0.0933$ and $eff = 0.9052$. Clearly the smaller is k, the steeper is the ρ (ψ) function in the central part, the greater is the bdp but the smaller is the efficiency. The value of c equal to 2.3563 (right-hand panels) when $k = 5$ corresponds to a nominal efficiency of 0.646 and $bdp = 0.30$ and when $k = 3$ to $eff = 0.3324$ with $bdp = 0.4767$.

Yohai and Zamar (1997) introduced a ρ function which minimizes the asymptotic variance of the regression M-estimate, subject to a bound on a robustness measure called contamination sensitivity. Therefore, this function is called the **optimal ρ function**.

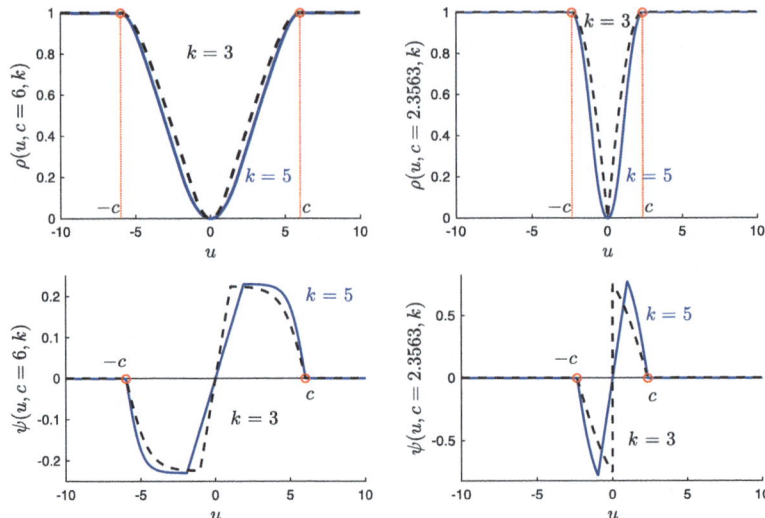

Fig. 2.9 Hyperbolic estimator: ρ and ψ functions. The left-hand panels use $c = 6$. The right-hand panels use $c = 2.3563$. In all panels we show two values of change of variance sensitivity of the M-estimator, k, namely, 3 and 5. In the upper panels, the curves have been rescaled in order to have $\sup(\rho) = 1$

$$\rho(u) = \begin{cases} 1.3846 \left(\frac{u}{c}\right)^2 & \text{if } |u| \leq \frac{2}{3}c \\ 0.5514 - 2.6917 \left(\frac{u}{c}\right)^2 + 10.7668 \left(\frac{u}{c}\right)^4 - 11.6640 \left(\frac{u}{c}\right)^6 + \\ +4.0375 \left(\frac{u}{c}\right)^8 & \text{if } \frac{2}{3}c < |u| \leq c \\ 1 & \text{if } |u| > c. \end{cases} \quad (2.38)$$

Now the first derivative vanishes outside the interval $[-c + c]$. The resulting M-estimate minimizes the maximum bias under contamination distributions (locally for a small fraction of contamination), subject to achieving a desired nominal asymptotic efficiency when the data are normally distributed. The plot of the ρ (ψ) is not shown but it is very similar to the previous ones with the same interpretation for c although the tuning constants to obtain a nominal level of breakdown point or efficiency are different (see Sect. 3.9).

Power Divergence. Riani et al. (2020) used ideas from minimum power density divergence estimation to obtain a single parameter ρ function, which is a simple function of the pdf of a normal distribution:

$$\rho_\alpha(u) = 1 - \exp\{-\alpha u^2/2\}, \quad (2.39)$$

$$\psi_\alpha(u) = \alpha u \exp\{-\alpha(u^2/2)\}. \quad (2.40)$$

2.3 M-Estimation of Scale

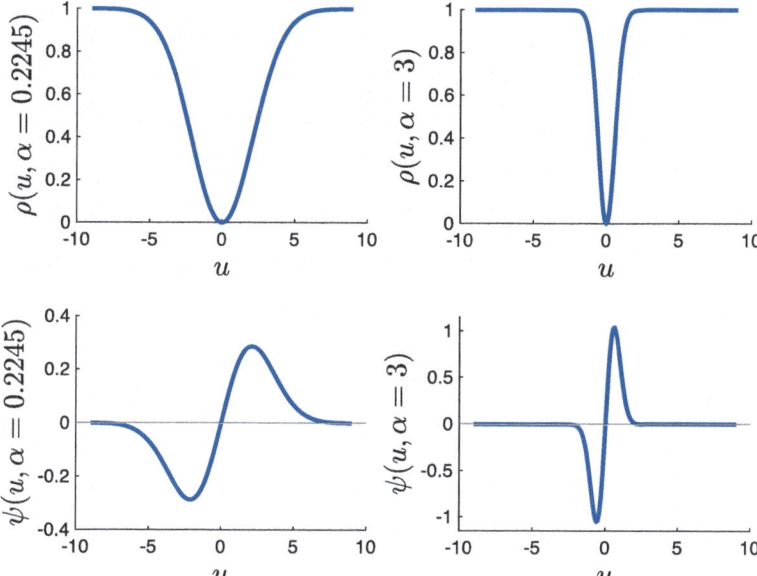

Fig. 2.10 Power divergence estimator: ρ and ψ functions. The left-hand panels use $\alpha = 0.2245$. The right-hand panels use $\alpha = 3$

This is a slight reparameterization of an otherwise unreferenced ρ function attributed to Welsh. It is straightforward to show that $\psi_\alpha(u) = \alpha u \exp(-\alpha u^2/2)$ and $w(u) = \alpha \exp\{-\alpha u^2/2\}$, a scaling of the pdf of the standard normal distribution. Riani et al. (2020) also obtained explicit expressions for *bdp* and *eff*:

$$\mathit{eff} = \frac{\sqrt{(1+2\alpha)^3}}{(1+\alpha)^3} \quad \text{and} \quad \mathit{bdp} = 1 - \frac{1}{\sqrt{1+\alpha}}. \tag{2.41}$$

Figure 2.10 shows the ρ and ψ for two typical values of α. The value of α equal to 0.2245 (left-hand panels), as can be seen from (2.41), corresponds to $bdp = 0.0963$ and to a nominal efficiency of 0.95. The value of α equal to 3 (right-hand panels) corresponds to a nominal efficiency of 0.2894 and $bdp = 0.5$. Equation 2.40 shows that the rejection point for this ρ function is ∞. So, at first sight, one might think that it does not provide sufficient protection against large outliers. The actual rejection point in the estimation procedure will depend upon the details of the numerical algorithm. When $u = \pm 7$ the value of the corresponding ψ function is of order 10^{-31}, when $u = \pm 9$ is of order 10^{-52}. Therefore, the magnitude of the outliers is completely counterbalanced by the exponential speed of convergence of $\exp(-u^2)$ to zero. Given that, as we have seen in Eq. 2.23, the IF is proportional to the ψ function, in the case of power divergence estimation, the effect of gross outliers is practically zero.

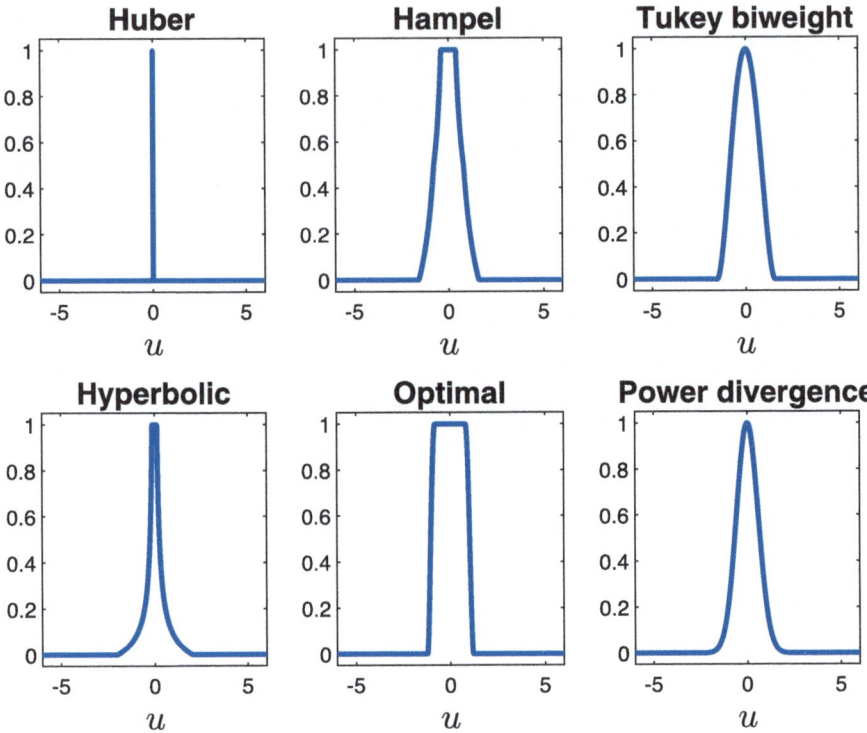

Fig. 2.11 The weight function $\psi(x)/x$ for six estimators with $bdp = 0.5$

An important feature is that the relationships in (2.39) and (2.40) can be inverted to give the value of α for a specified *bdp* or *eff*. This single parameter α in the ρ function makes explicit the relationship between *bdp* and *eff*. For other ρ functions, the relationship is implicit although, as we shall repeatedly see later, as one increases the other decreases. In later chapters, we shall see through monitoring data analyses that methods that try to escape from this relationship are likely to fail.

Figures 2.11 and 2.12 provide plots of the weights $w(u) = \psi(u)/u$ for the six ρ functions introduced in this section. In all cases in Fig. 2.11, we have used the tuning constant associated with $bdp = 0.5$. Note that for the Huber ρ function in this case the tuning constant c is set equal to 0; for all other values of c, $bdp = 0$. All are shaped like brimmed hats; high in the "crown" and mostly decreasing smoothly towards the "brim". Despite the similarity in general shape, there are important differences. All except the biweight and power divergence weights are constant in the centre, corresponding to the ρ function for the normal distribution. In addition all of them give zero weight to extreme observations. For the Huber, the extreme observations are all those above or below the median. For all the others, the extreme observations with zero weights are those beyond the rejection point. Note that for the PD ρ function the weight is not exactly zero but tends to 0 exponentially fast. For example,

2.3 M-Estimation of Scale

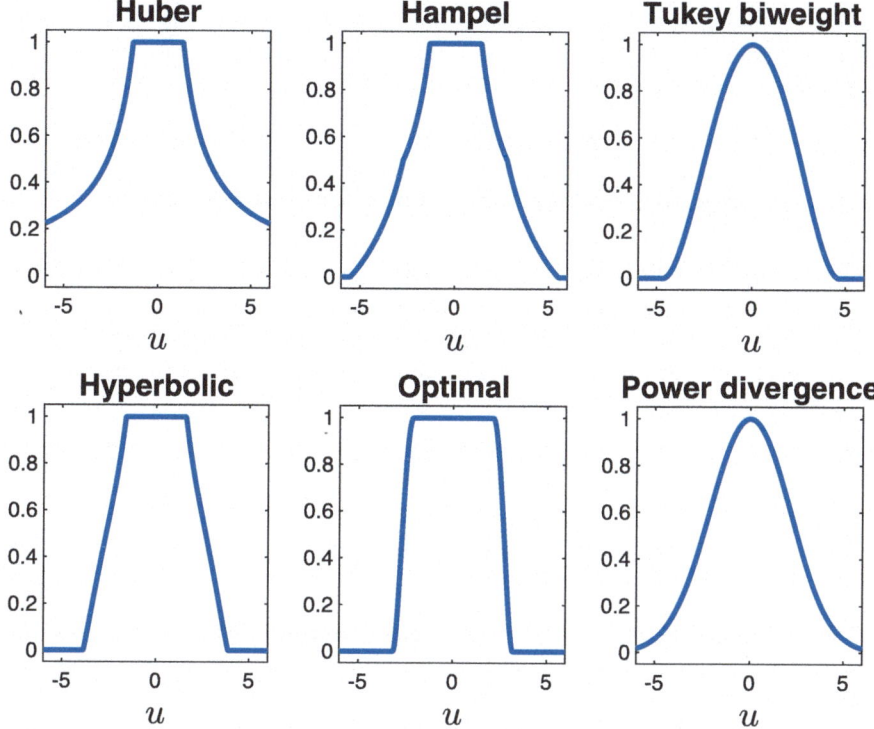

Fig. 2.12 The weight function $\psi(x)/x$ for six estimators with *eff* = 0.95

when $u = \pm 4$ the PD weight is already of order 10^{-11}. Figure 2.12, on the other hand, shows the weights when the nominal efficiency is set to 0.95. It is important to remark that, if we increase efficiency, we have reduced protection against outliers, by giving non-negligible weights to intermediate observations. In this case, all weights redescend to zero for large $|u|$ and are zero beyond the rejection point, or virtually zero in the case of PD; the Huber weights converge more slowly than those for power divergence.

Andrews' Sine Function. Five of the six ρ functions presented above have a similar shape, for example, that of the power divergence ρ function of Fig. 2.10. Andrews (1974) suggested the use in multiple regression of a sine-based ρ function, rather than the normal distribution of the PD estimator, so that

$$\rho(u) = \begin{cases} c\{1 - \cos(u/c)\} & \text{if } |u| \leq c\pi \\ 2c & \text{if } |u| > c\pi. \end{cases} \tag{2.42}$$

Like the PD function and several other ρ functions, this function also has a single adjustable parameter. We leave the exploration of the properties of this proposal for robust estimation to the exercises (see Exercise 4.1).

2.3.3 Simultaneous Estimation of Location and Dispersion

In the comparison of the previous section, we have assumed that the true σ (or that a fixed consistent robust estimate of σ) has been used.

In most data analyses, it is necessary to estimate both location and scale. We combine the results of Subsects. 2.2.1.1 and 2.3 and solve the system of two equations:

$$\sum_{i=1}^{n} \psi_{\text{loc}}\left(\frac{y_i - \hat{\mu}}{\hat{\sigma}}\right) = 0$$

$$\sum_{i=1}^{n} \rho_{\text{scale}}\left(\frac{y_i - \hat{\mu}}{\hat{\sigma}}\right) = K. \qquad (2.43)$$

Here we distinguish between ρ_{scale} used for scale estimation and ψ_{loc} and its derivatives used for location. The two need not be different and are, indeed, often the same.

Given starting values $\hat{\mu}_0$ and $\hat{\sigma}_0$ the pair of reweighting equations moves forward from stage k using the calculations:

1. Find the location weights $w_{i,k} = w_{\text{loc}}\{(y_i - \hat{\mu}_k)/\hat{\sigma}_k\}$.
2. Calculate the new location estimate as

$$\hat{\mu}_{k+1} = \sum_{i=1}^{n} w_{i,k} y_i / \sum_{i=1}^{n} w_{i,k}.$$

3. The new scale estimate is

$$\hat{\sigma}_{k+1} = \hat{\sigma}_k \left[\frac{1}{nK} \sum_{i=1}^{n} \rho_{\text{scale}}\{(y_i - \hat{\mu}_{k+1})/\hat{\sigma}_k\}\right]^{0.5}.$$

4. Return to Step 1 until the standardized change in the estimates

$$\left|\frac{\hat{\sigma}_{k+1}}{\hat{\sigma}_k} - 1\right|$$

is less than a prespecified tolerance (say $1e - 7$).

This alternating algorithm converges to a point with zero derivatives, which may be a minimum, a maximum, or a saddle point. As with many such algorithms, it should be run from a variety of starting points. In the presence of multiple solutions, we have two strategies. The first is to take the solution that gives the absolute minimum of $\sum_{i=1}^{n} \rho(u_i)$. The second approach is to take the solution closest to the median (Hoaglin et al., 1983), p. 387.

The breakdown point of the simultaneous estimates of location and scale requires the solution of a nonlinear system of equations (Huber, 1981), p. 141 and is less than 0.5 even if we use tuning constants associated with a nominal *bdp* of 0.5. Therefore, the standard approach in the robust statistical literature is first to estimate the scale and to use the value of the robust scale in the iterative loop for location. This leads to MM-estimation which will be analysed in Sect. 3.12.1.

2.4 Comparison of ρ Functions

After introducing all these ρ functions, the natural reaction is to compare them and to provide guidelines to situations of data analysis in which each provides the most appropriate estimators. In this section, we start with an asymptotic comparison, first in terms of *bdp* and *eff* and subsequently in terms of the change-of-value curve. We then move to a small-sample comparison with data contaminated by a variety of mechanisms.

2.4.1 Asymptotic Comparisons

2.4.1.1 Comparison in Terms of *bdp* and *eff*

We start our comparisons of the asymptotic properties of ρ functions with those for the power divergence ρ function (2.39). The values of the asymptotic breakdown point *bdp* and the asymptotic efficiency *eff* are given in (2.41) as a function of the single parameter α and so are readily calculated. Figure 2.13 shows these two properties as functions of α over the range $0 \leq \alpha \leq 3$. As *bdp* increases from zero towards 0.5, *eff* decreases from 1 to 0.2894. These are generic shapes for robust estimators, quantifying the trade-off between robustness and efficiency. Figure 2.14 shows plots of efficiency against breakdown point for the power divergence ρ function and four of the other ρ functions of Fig. 2.12 (the Huber function being excluded because it has a zero breakdown point). In order to generate these curves, we fix a particular value of breakdown point and find the associated tuning constant α for PD or c for the other estimators (the details are in Sect. 3.9). In the case of the Hampel ρ function, the three extra parameters c_1, c_2, and c_3 have been set equal to 2, 4, and 8. For the hyperbolic tangent estimator, the extra parameter k which reflects the log of the change of variance sensitivity of the M-estimator has been set equal to 4.5. Given the value of the tuning constant we found the corresponding value of the efficiency.

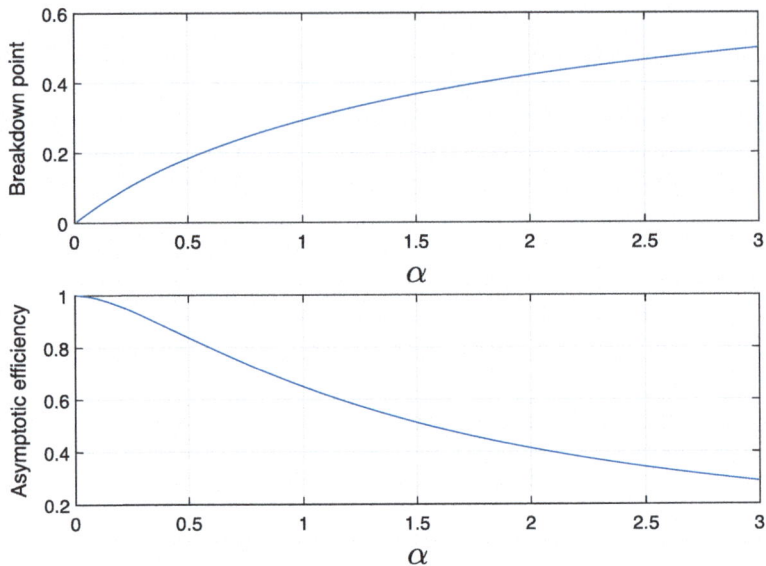

Fig. 2.13 Power divergence: breakdown point and efficiency as functions of α

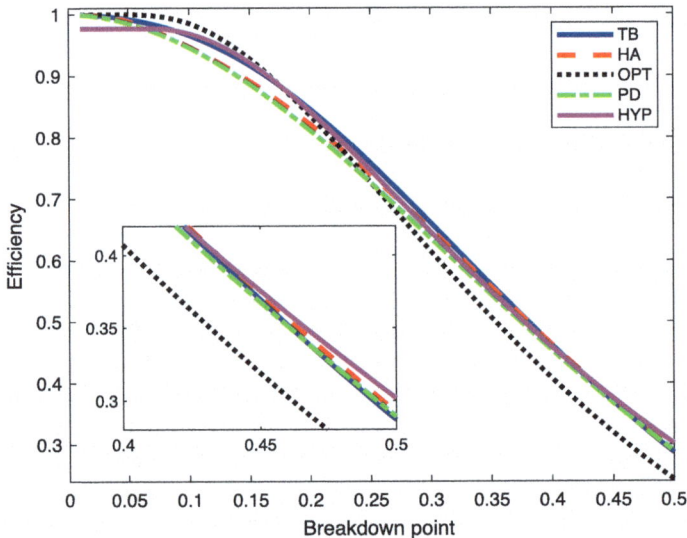

Fig. 2.14 Breakdown point and efficiency as parameters vary for five rho functions: TB: Tukey biweight; HA: Hampel; OPT: optimal; PD: power divergence; and HYP: hyperbolic. The inset is a zoom of the main figure for high breakdown point

2.4 Comparison of ρ Functions

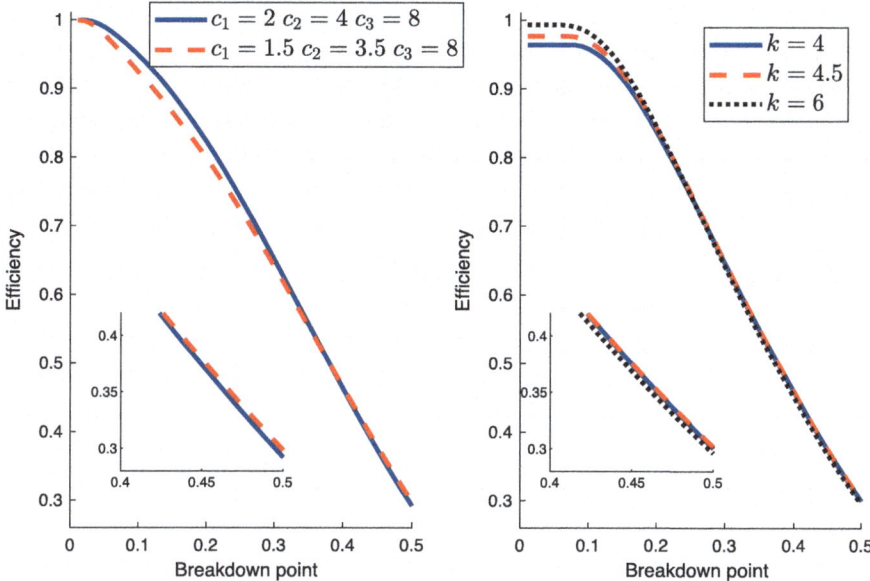

Fig. 2.15 Breakdown point and efficiency as parameters vary for the Hampel and Hyperbolic ρ functions

It is clear from the figure that the general asymptotic performance of the five methods is similar. The optimal function is best for small *bdp*, but worst for values slightly larger than 0.25. The situation for Hampel is the reverse, being worst for small *bdp* and best for *bdp* values above approximately 0.4. For small *bdp* the power divergence is the second worst, but behaves much like the hyperbolic and biweight functions for larger values of *bdp*. For 50% *bdp* (as the inset in the figure shows) the ordering is (in parenthesis we give the exact numbers) hyperbolic (0.3019), Hampel (0.2924), power divergence (0.2894), biweight (0.2868), and lastly the optimal (0.2428). Hössjer (1992) proves that, for normal theory linear models, the maximum efficiency when *bdp* = 0.5 is 0.329.

Some further insight into the balance between breakdown point and efficiency comes from varying the parameters of the Hampel and hyperbolic functions. In Fig. 2.14, the parameters for the Hampel were $c_1 = 2$, $c_2 = 4$, and $c_3 = 8$. The left-hand panel of Fig. 2.15 compares the breakdown point and efficiency of Hampel's ρ function with these values to those when $c_1 = 1.5, c_2 = 3.5$, and $c_3 = 8$. The original procedure is better for *bdp* less than 0.3, with the modified version being slightly better for larger values. For the hyperbolic ρ function in the right-hand panel, the freely variable parameter, other than *c*, is *k*. The curves for three values of *k* are shown in the right-hand panel of Fig. 2.15. The difference is largest for small values of *bdp*, when $k = 6$ has the highest efficiency. In other words, imposing a looser constraint through the change of variance parameter produces higher efficiency for small values of *bdp*. For breakdown points near 0.5, the order is reversed, with $k = 6$

being the least efficient, although, in this region, the differences are less than for low *bdp*. The conclusion from this figure reinforces that from Fig. 2.14 that no one ρ function has the highest breakdown point and efficiency over the whole range of *bdp* from 0 to 0.5. These results also implicitly show that the choice of the ρ function is not a crucial aspect since all (provided they are bounded) have similar behaviour in terms of breakdown point and efficiency. These theoretical results are in line with the empirical findings in Salini et al. (2015) where it is demonstrated that the size of the test for outlier detection is much more affected by the choice of the requested level of efficiency or breakdown point than by the choice of the ρ function.

2.4.2 Comparison Using the IF

In this section, we compare the different ρ functions using the IF and their associated GES (gross error sensitivities).

2.4.2.1 Estimate of Location

The comparisons of IF for six estimators with $bdp = 0.5$ are in Fig. 2.16, with those for $eff = 0.95$ in Fig. 2.17. The two figures show that the ρ function with the highest GES for location is OPT, both with $bdp = 0.5$ and $eff = 0.95$. The hyperbolic ρ function in Fig. 2.16 is the steepest and so is the closest to the median.

When $eff = 0.95$, Fig. 2.17 shows that the IF for the different estimators is very similar in the central part. The difference is mainly concerned with their redescending to zero.

Table 2.1 summarizes the results on GES extracted from these two figures. The major conclusion is that, for the estimation of location, the optimal ρ-function is worst.

2.4.2.2 Estimate of Scale

Companion plots of the IF for estimation of scale are in Figs. 2.18 and 2.19. It is possible to show that MADN is the M-estimate of scale with the smallest GES at the normal model (Rousseeuw and Croux, 1993). In the scale case, the ρ function with the largest GES is the hyperbolic when $bdp = 0.5$. On the other hand, when $eff = 0.95$ the largest GES is associated with the PD ρ function. Numerical results are in the right-hand half of Table 2.1.

2.4 Comparison of ρ Functions

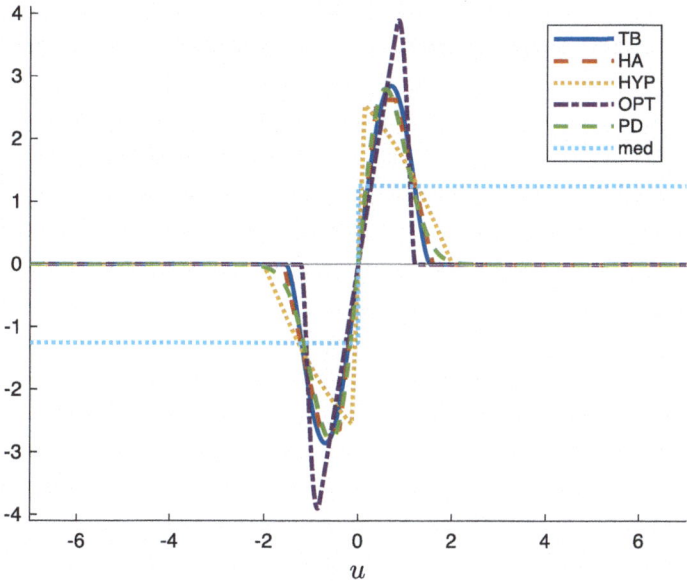

Fig. 2.16 Estimate of location. Analysis of IF for six estimators with $bdp = 0.5$: TB: Tukey biweight; HA: Hampel; OPT: optimal; PD: power divergence; HYP: hyperbolic and median

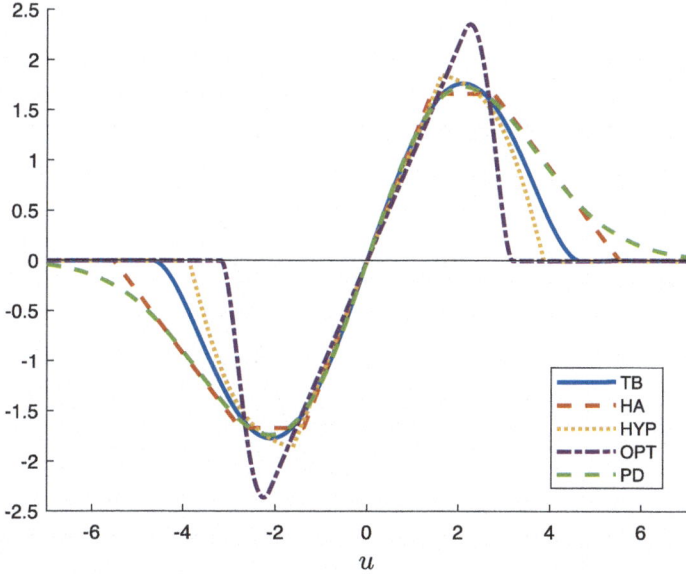

Fig. 2.17 Estimate of location. Analysis of IF for five estimators with $eff = 0.95$: TB: Tukey biweight; HA: Hampel; OPT: optimal; PD: power divergence; and HYP: hyperbolic

50 2 Introduction to M-Estimation for Univariate Samples

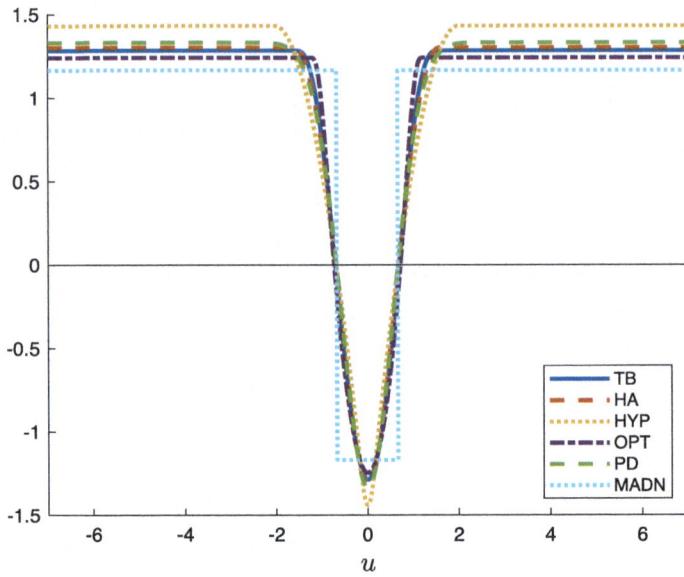

Fig. 2.18 Estimate of scale. Analysis of IF for six estimators with $bdp = 0.5$: TB: Tukey biweight, HA: Hampel, OPT: optimal, PD: power divergence, HYP: hyperbolic, and MADN

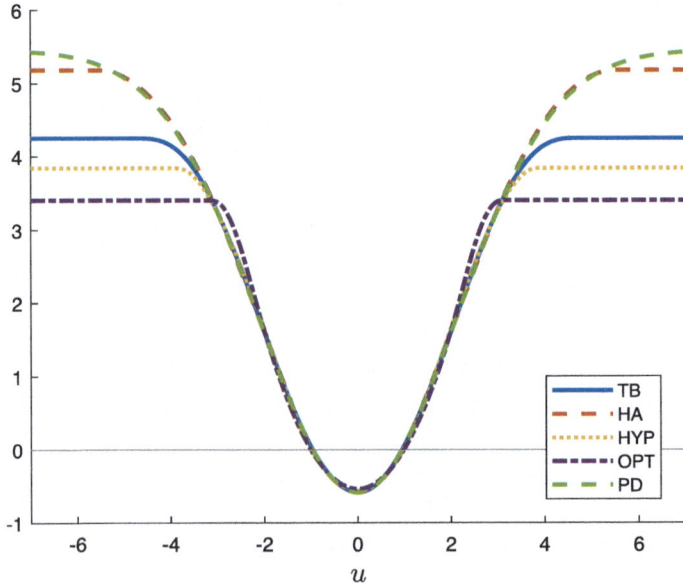

Fig. 2.19 Estimate of scale. Analysis of IF for five estimators with $eff = 0.95$: TB: Tukey biweight, HA: Hampel, OPT: optimal, PD: power divergence, and HYP: hyperbolic

2.4 Comparison of ρ Functions

Table 2.1 Analysis of GES under the normal model for 5 ρ functions, after fixing $bdp = 0.5$ and $eff = 0.95$. Columns 2 and 3 refer to the estimate of location. Columns 4 and 5 to the estimates of scale

Estimator	Location		Scale	
	$bdp = 0.5$	$eff = 0.95$	$bdp = 0.5$	$eff = 0.95$
Median	1.2533			
MADn			1.1670	
Tukey biweight	2.8499	1.7696	1.2842	4.2512
Hampel	2.6323	1.6647	1.3038	5.1866
Hyperbolic	2.5420	1.8576	1.4343	3.8478
Optimal	3.8962	2.3575	1.2422	3.4089
Power divergence	2.8014	1.7345	1.3333	5.4294

2.4.3 Comparison Using the Change-of-Value Curve

The sensitivity curve shows the effect on an estimate of adding or deleting an observation. The influence curve extends this notion from sample to distribution. The change-of-value curve, which we now discuss, shows the effect of varying the value of an observation.

Definition: The change-of-value curve of an estimator T_n, with sample $y_1, \ldots, y_{n-1}$ and n-th observation at y, is the rate of change of $T_n(y_1, \ldots, y_{n-1}, y)$ with y. Namely:

$$CV(T_n) = CV(y_1, \ldots, y_{n-1}, y, T_n) = \frac{\partial}{\partial y} T_n(y_1, \ldots, y_{n-1}, y). \qquad (2.44)$$

From this definition, it is clear that the change-of-value curve corresponds to the slope of the sensitivity curve. For M-estimation of location, we take the implicit equation for T_n,

$$\sum_{i=1}^{n} \psi\left(\frac{y_i - T_n}{\hat{\sigma}}\right) = 0. \qquad (2.45)$$

If we assume that extreme values of y have no effect on the robust estimate of scale, then we can take $\partial \hat{\sigma}/\partial y = 0$. It then follows that $CV(T_n)$ is proportional to $\psi'(u)$ (Hoaglin et al., 1983), p. 380.

The CV is a function of u. In order to summarize the CV with a unique value, we introduce the concept of change-of-value sensitivity (CVS) as

Definition: The change-of-value sensitivity of an M-estimator is

$$\text{CVS} = E_F[\{CV(y, F)\}^2] = \int \psi'^2 dF.$$

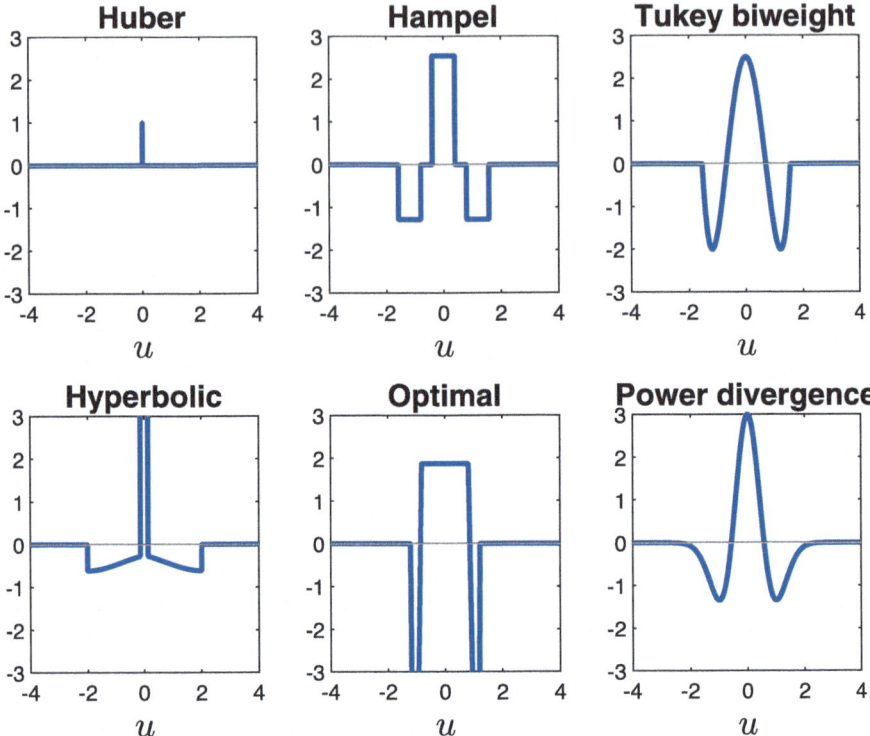

Fig. 2.20 Analysis of $\psi'(u)$ for six estimators associated to $bdp = 0.5$: TB: Tukey biweight, HA: Hampel, OPT: optimal, PD: power divergence, and HYP: hyperbolic

Note that the variance of an M-estimator is $E_F[\{IF(y, F)\}^2]$. Therefore, the CVS augments the analysis of the variance of the M-estimator to give a better understanding of the properties of the different ρ functions for particular values of *bdp* and *eff*.

Figure 2.20 provides, when $bdp = 0.5$, plots of $\psi'(u)$ for the six ρ functions introduced in the previous section. In order to provide a fair comparison of the different ρ functions they have all (with the exception of that of the Huber estimator) been normalized to have $\sup(\rho) = 1$. It is immediate that, when $bdp = 0.5$, the optimal estimator has the largest fluctuations in $\psi'(u)$. Table 2.2 shows that the smallest CVS comes from the hyperbolic function. Figure 2.21 shows the related plot with *eff* taken as 0.95.

Table 2.2 shows the CVS under the normal model for the two cases $bdp = 0.5$ and $eff = 0.95$. When $bdp = 0.5$, the estimator which is characterized by the smallest value of CVS is the hyperbolic tangent, followed by TB. On the other hand, the estimator which indeed shows the highest value of CVS is the optimal estimator. When $eff = 0.95$ the estimator with the smallest CVS is PD, with optimal larger and all other estimators yielding much higher values: 20 to 30 times those for PD.

2.4 Comparison of ρ Functions

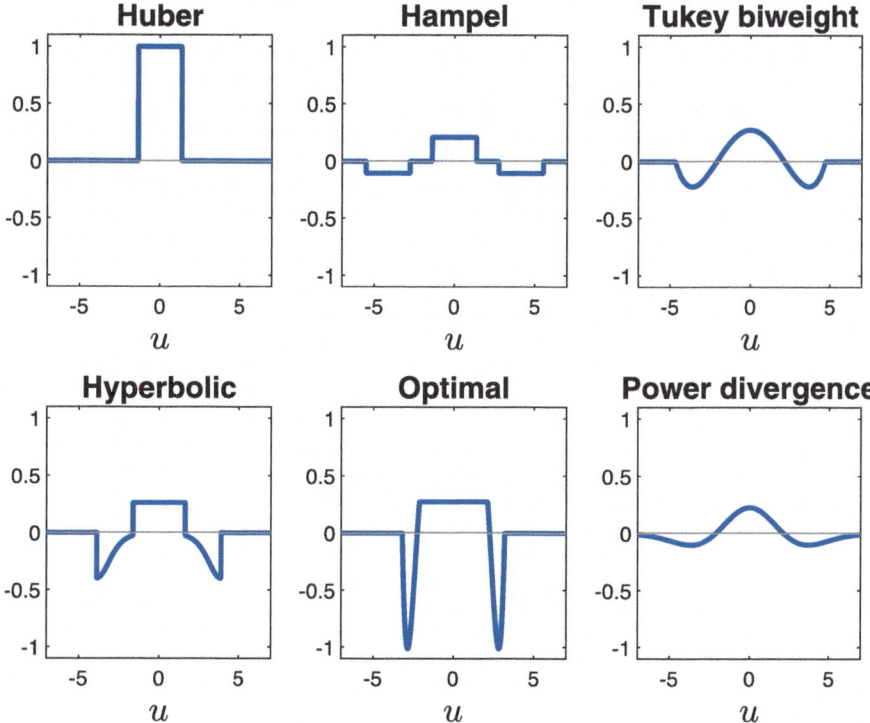

Fig. 2.21 Analysis of $\psi'(u)$ for six estimators with $\mathit{eff}=0.5$: TB: Tukey biweight, HA: Hampel, OPT: optimal, PD: power divergence, and HYP: hyperbolic

Table 2.2 Analysis of Change of Variance Sensitivity (CVS) under the normal model for five ρ functions, with $bdp = 0.5$ and $\mathit{eff} = 0.95$

Estimator	$bdp = 0.5$	$\mathit{eff} = 0.95$
Hampel	0.3869	0.8345
Tukey biweight	0.3833	0.6659
Hyperbolic	0.1085	0.9031
Optimal	5.9631	0.0813
Power divergence	2.3603	0.0319

2.4.4 Small-Sample Comparisons

The conclusion from the asymptotic comparisons of the five ρ functions in Sect. 2.4.1.1 was that there was little difference between their properties. We now extend these comparisons using simulations with small normal samples, some of which are contaminated. For such samples, the comparisons are dependent on whether location or scale is being estimated. They are affected by the presence

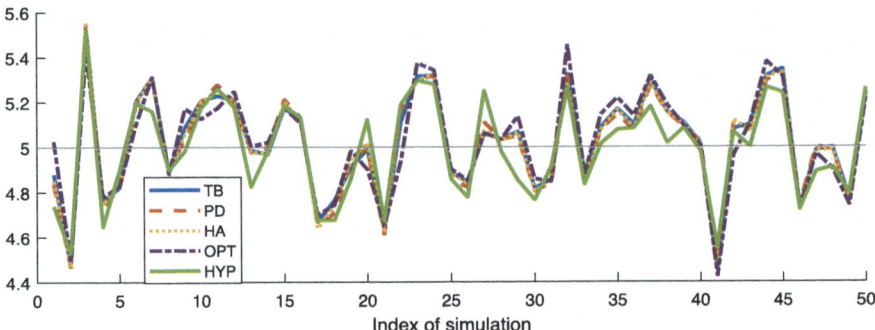
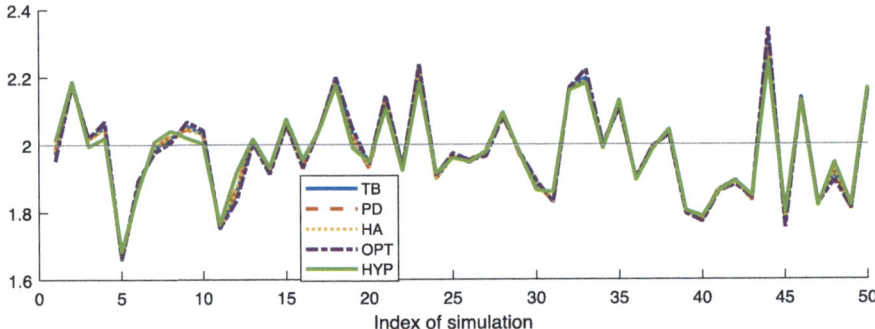

Fig. 2.22 50 samples of 200 observations generated from $\mathcal{N}(5, 4)$; no outliers. Parameter estimation with $bdp = 0.5$. Upper panel: location estimate using the iterative procedure of (2.20) with σ estimated using MADN. Lower panel: scale estimate using the iterative procedure of Eq. (2.30) with $\mu = 5$

or otherwise of outliers and on the value of *bdp* used. Surprisingly, the effects of assuming an over high value differ between estimation of location and scale. We remain with univariate samples.

Figure 2.22 shows results from estimation on 50 independent samples of 200 observations generated from the normal distribution $\mathcal{N}(5, 4)$. No outliers were introduced. The upper panel shows the results from estimating location using the iterative algorithm of (2.20) with scale estimated by the MADN (1.9); the lower panel uses the iterative algorithm for scale (2.30), with the location taken as 5. Both estimates seem randomly scattered around the parameter values used in the simulation. In general, the lines on the figures for the different ρ functions are close together, increasing or decreasing in a similar way for the different samples. In the estimate of scale, the results for the five ρ functions are virtually indistinguishable, whereas there are slight differences for the estimation of location.

Figure 2.23 shows the effect of 30% contamination on the estimate of location. Now there are 200 simulated $\mathcal{N}(5, 4)$ samples with 200 observations with location estimated as in the upper panel of Fig. 2.22. In both panels of the plot, the samples

2.4 Comparison of ρ Functions

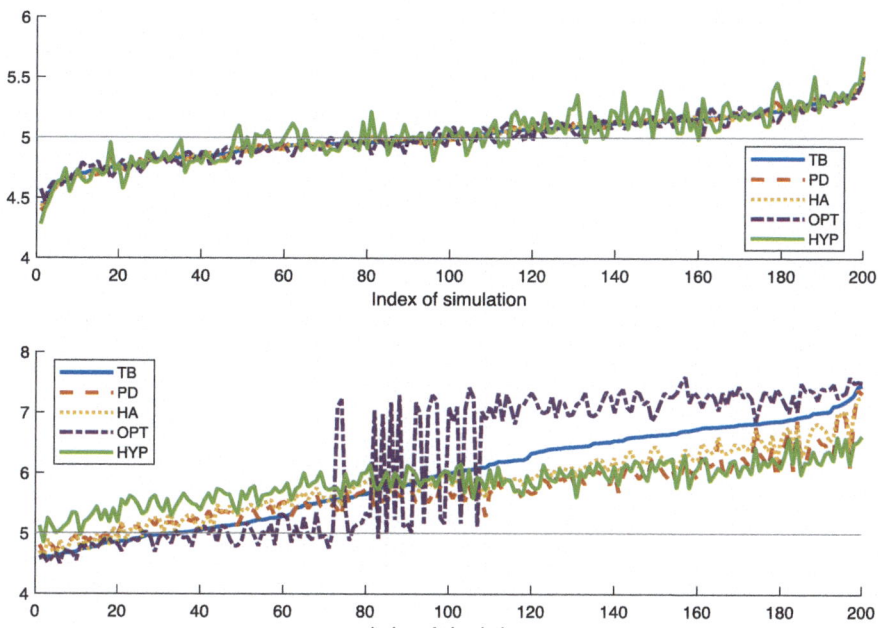

Fig. 2.23 200 samples of 200 observations generated from $\mathcal{N}(5, 4)$; 30 percent contamination. Estimates of location as in the upper panel of Fig. 2.22. Ordering based on the estimates of location using Tukey's biweight. Upper panel: shift contamination with a value of 10. Lower panel: point-mass contamination of 10

have been ordered by location estimates from Tukey's biweight. The lines joining the results of distinct simulations in this and the previous plot have been introduced to aid visual clarity. In the upper panel, the contamination was in the form of shift outliers; each contaminated observation had y increased by ten. All ρ functions gave similar estimates centred around the parameter value of 5, with HYP having a slightly higher variability.

The very different structure in the lower panel of the figure is caused by point-mass contamination in which the 30% of contaminated units all had the value of y changed to 10. The estimates now mostly have positive bias. For the lowest one-third of the values, as ranked by the TB, all procedures behave in an approximately similar manner, including the power divergence function. The most striking performance is that of the optimal ρ, which initially provides the lowest valued estimates, close to five. There is then a period of oscillation before the estimates settle down close to seven. This is quite distinct from the behaviour of the TB, which increases steadily, with higher values than those of the other three functions. The estimates from these other three functions also increase steadily. Rocke and Woodruff (1996) consider why point-mass contamination is the most disruptive form of outlier contamination.

We now turn from general comparison of the ρ functions to illustrating the effect of the estimation of σ on the comparative performance of the ρ functions. We start

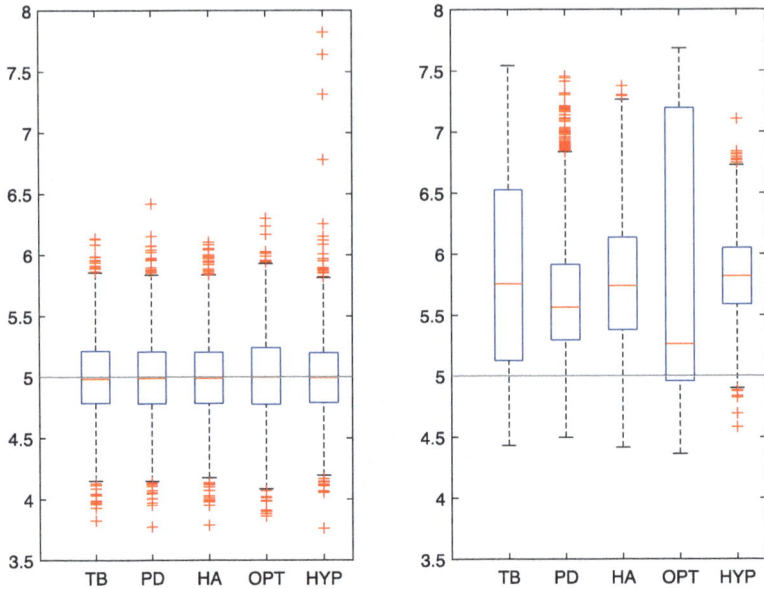

Fig. 2.24 200 samples of 200 observations generated from $\mathcal{N}(5, 4)$; 30 percent shift contamination. Boxplots of location estimates as in Fig. 2.22. Left-hand panel: $\sigma = 2$. Right-hand panel: σ estimated with MADN

in Fig. 2.24 with the strongest contrast in these estimates. In the left-hand panel, the true value of σ has been used whereas, in the right-hand panel, σ was estimated by the MADN. The effect is dramatic. The boxplots in the left-hand panel all have medians close to 5. The spreads of the boxplots are also similar, apart from the hyperbolic function, which gives rise to four large estimates. The estimates all also appear to be normally distributed which illustrates the result on asymptotic normality in Sect. 3.7.2. The median estimates of location in the right-hand panel, on the other hand, show positive bias. The two ρ functions with the least biased median estimates are the power divergence and optimal. However, the optimal is the most skewed.

The next two figures present plots of the effect of specified *bdp* on the simultaneous estimation of μ and σ^2 of Sect. 2.3.3. In both, the upper panel presents estimates of location and the lower panel estimates of scale. In Fig. 2.25, the calculations used a *bdp* of 0.5, although only ten percent of the observations had been contaminated. In Fig. 2.26 a *bdp* of 0.2 was used, closer to the 0.1 fraction of contaminated observations. The type of contamination used is point-mass. However, the result do not change if we use shift contamination. Comparison of the upper panels of the two figures shows that there are only negligible differences in the median estimates of location provided by the five robust estimators. Both sets of boxplots seem approximately normal, although those with a *bdp* of 0.5 all have a very few low values. However, the important difference is in the spread of the estimates. Those for *bdp* = 0.5 all have a greater spread than those for *bdp* = 0.2. This is to be expected, since

2.4 Comparison of ρ Functions

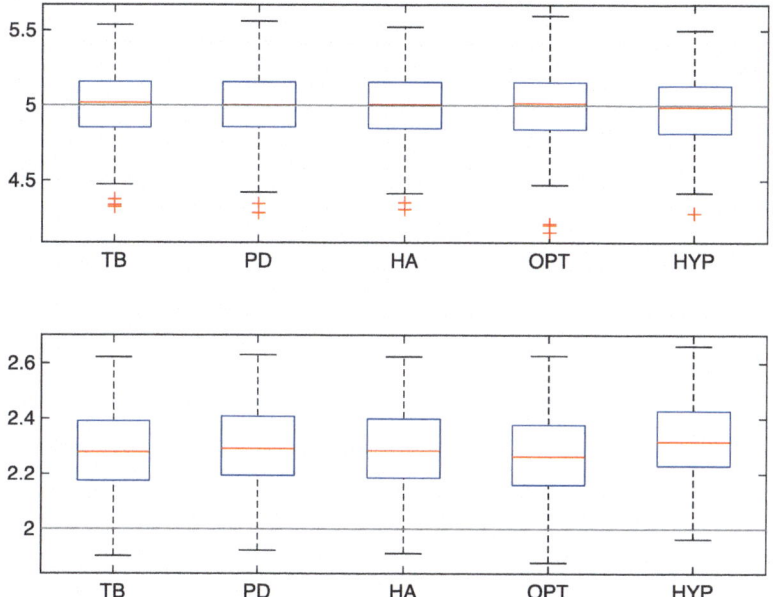

Fig. 2.25 200 samples of 200 observations generated from $\mathcal{N}(5, 4)$; 10 percent point-mass contamination, $bdp = 0.5$. Simultaneous estimation of μ and σ given in Sect. 2.3.3. Upper panel: boxplots of the estimates of location. Lower panel: boxplots of the estimates of scale

the higher value of *bdp* leads to an increased loss of information due to heavier downweighting and so to an increased variance for the parameter estimate.

The boxplots for the estimate of the scale σ in the lower panels of the two figures show some pattern in the median of the scale estimates, the pattern being more evident in Fig. 2.26, that is, for $bdp = 0.2$. The lowest median comes from use of OPT. A fainter version of this structure is seen in Fig. 2.25. A more important aspect of all estimates is that they are biased. In these simulations $\sigma = 2$. This value is plotted as a line for $bdp = 0.5$. For all estimators, the value is in the lower tail of the distribution; the estimators are biased. As the lower panel of Fig. 2.26 shows the bias is appreciably stronger for $bdp = 0.2$; the parameter value of 2 is well off the bottom of the plot. For $bdp = 0.5$, the median scale estimate is around 2.3, whereas when $bdp = 0.2$, it is around 2.8.

The conclusion from these simulations is that, given a good estimate of scale, estimation of location is well behaved. Only the crude MADN estimate of scale in the right-hand panel of Fig. 2.24 gave consistently poor estimates of location. A further conclusion is that, when we simultaneously estimate μ and σ, estimates of scale calculated with a high *bdp* value, such as 0.5, appear to have better properties. However, the estimate of location should have as low a *bdp* as is compatible with removing, or, at least, strongly downweighting, all outliers from the estimation of location. Figure 2.27 gives boxplots for estimates of location when 15% of the observations

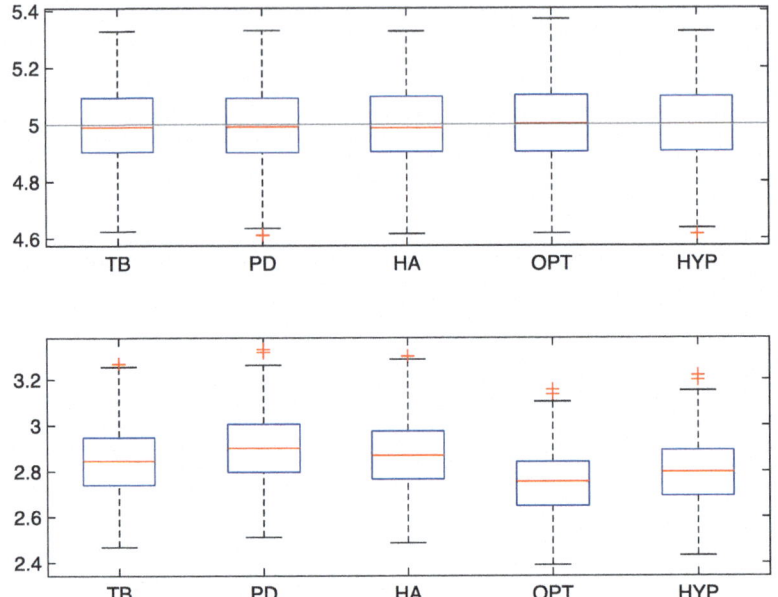

Fig. 2.26 200 samples of 200 observations generated from $\mathcal{N}(5, 4)$; 10 percent point-mass contamination, $bdp = 0.2$. Simultaneous estimation of μ and σ as in Fig. 2.25. Upper panel: boxplots of the estimates of location. Lower panel: boxplots of the estimates of scale

have shift contamination. In estimating location, the true value of σ was used. In the left-hand panel, the value of *bdp* in the iterative calculation of location was taken as 0.5. In the right-hand panel, the value was 0.15, the proportion of contamination in the observations. The improvement from reducing the value of *bdp* is striking; the scatter of the estimates is greatly reduced and the slight biases in the medians just visible in the left-hand panel are not apparent when the lower value of *bdp* is used.

These results, for M-estimation, led to MM-estimation (Yohai, 1987) and Sect. 3.12.1 in which a very robust estimate of scale is calculated, not simultaneously with estimation of location, and then used as a fixed value for the estimation of location (which may include regression parameters), with a *bdp* appropriate for the data being analysed. Of course, it is rare that it is known what the proportion of outliers will be in any given dataset. The monitoring methods of Chap. 4 provide a mechanism for detecting such an empirical breakdown point. Alternatively, the forward search of Sect. 4.1 provides an empirical breakdown point avoiding the use of a fixed estimate of scale.

2.4 Comparison of ρ Functions

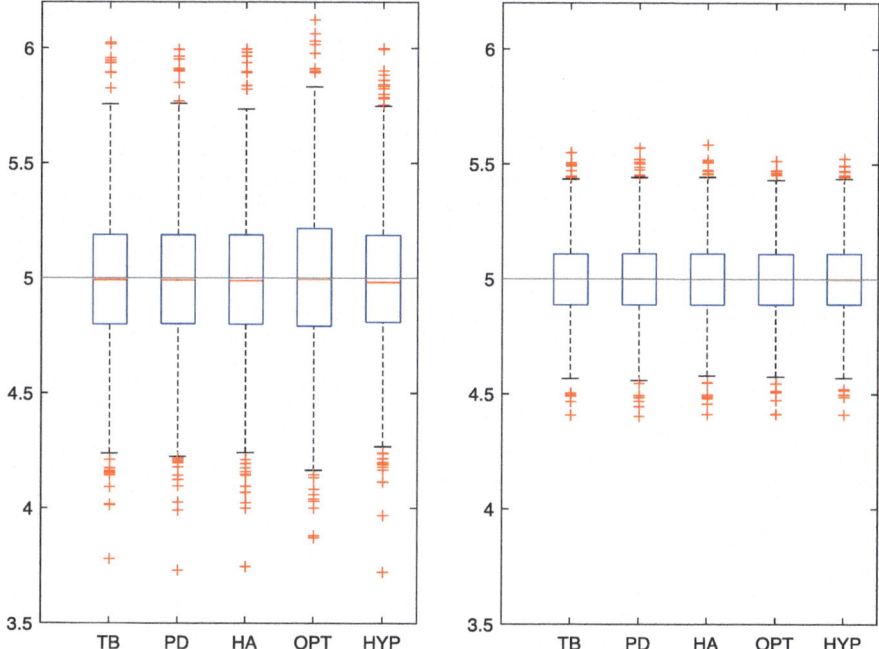

Fig. 2.27 2000 simulations of 200 observations generated from $\mathcal{N}(5, 4)$; 15 percent shift contamination. Boxplots of the distribution of the estimates of location when the true value of σ is used in the iterations of (2.20). Left-hand panel: $bdp = 0.5$. Right-hand panel $bdp = 0.15$

2.4.5 The Issue of Multiple Solutions

Finally, we return to M-estimation of the location parameter μ with a given estimate of σ. Equation (2.22) provides a condition on the estimate $\hat{\mu}$ which requires a function of ψ to be zero. If ψ is increasing, the iterative solution of this equation leads to a unique estimate of location. Alternatively, with redescending ψ, there can be multiple solutions. In order to exemplify this concept we take the first 20 observations from the dataset on income from the United States Census Bureau introduced in Chap. 1. To these observations, we add 3 fictitious incomes equal, respectively, to 600000, 575000, and 590000. Figure 2.28 shows, for the six ρ functions and a range of values of μ, the plot of

$$\overline{\psi}_{\mathrm{loc}}(u) = \overline{\psi}_{\mathrm{loc}}\left(\frac{y - \mu}{\hat{\sigma}}\right) = \frac{1}{n} \sum_{i=1}^{n} \psi_{\mathrm{loc}}\left(\frac{y_i - \mu}{\hat{\sigma}}\right), \qquad (2.46)$$

where $\hat{\sigma}$ is the MADN (1.9) for all six ρ functions.

If there are multiple solutions to the estimating equation, there will be more than one value of μ for which $\overline{\psi}_{\mathrm{loc}}(u) = 0$. For the Huber function, ψ is not decreasing

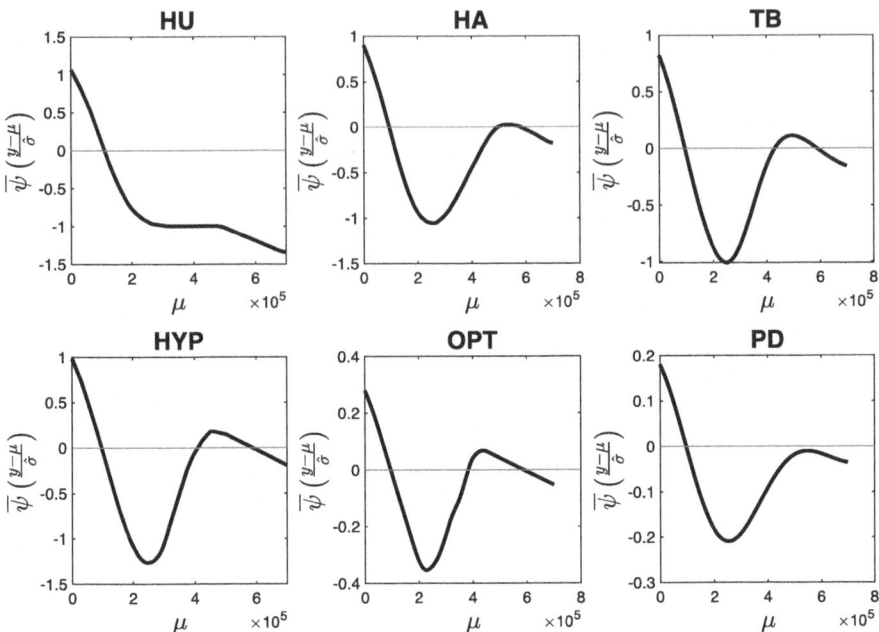

Fig. 2.28 Illustration of multiple solutions to an estimating equation. Average of $\psi(x-\mu)/\hat{\sigma}$ (2.46) as a function of μ

and there is a unique solution. In all the other cases, the curve is non-monotonic. It is interesting to notice that while for hyperbolic, optimal and Tukey ρ functions, there are 3 clear solutions, in the case of Hampel there are just 2 solutions. On the other hand, in the case of power divergence, there is just one solution. This could suggest that with exponentially redescending tails, as in the PD function, the problem of multiple solutions becomes less severe. Be that as it may, we note the similarity in shape of the curves for all ρ functions except for Huber.

2.5 Background on Convergence and Asymptotics

Stochastic convergence. Let $\{Y_n\}_{n=1,2,3,\ldots}$ be a sequence of random variables. Mathematically, there are different kinds of convergence of $\{Y_n\}$ to a fixed value c. The three most important are

- Convergence in probability (abbreviated $Y_n \xrightarrow{p} c$):

$$\lim_{n\to\infty} P\left(|Y_n - c| > \epsilon\right) = 0 \quad \text{for all} \quad \epsilon > 0.$$

- Almost sure convergence (abbreviated $Y_n \xrightarrow{a.s.} c$):

2.5 Background on Convergence and Asymptotics

$$P\left(\lim_{n\to\infty} Y_n = c\right) = 1.$$

- Convergence in quadratic mean (abbreviated $Y_n \xrightarrow{q.m.} c$):

$$\lim_{n\to\infty} E\left\{(Y_n - c)^2\right\} = 0.$$

Note that:

- $Y_n \xrightarrow{a.s.} c$ implies $Y_n \xrightarrow{p} c$;
- $Y_n \xrightarrow{q.m.} c$ implies $Y_n \xrightarrow{p} c$.

Consistency of estimators. Based on a sample $y_1, \ldots, y_n$, let $\hat{\theta}_n = \theta_n(y_1, \ldots, y_n)$ be an estimator of an unknown parameter θ. Then $\hat{\theta}_n$ is called "weakly consistent" if

$$\hat{\theta}_n \xrightarrow{p} \theta \quad \text{as} \quad n \to \infty;$$

$\hat{\theta}_n$ is called "strongly consistent" if

$$\hat{\theta}_n \xrightarrow{a.s.} \theta \quad \text{as} \quad n \to \infty.$$

Remark: For most statistical estimation problems, it is usually possible to define many different estimators. The real problem is to find a good estimator which approximates the true parameter θ with the maximum possible accuracy. Consistency is generally seen as a necessary condition which has to be satisfied by any reasonable estimator.

In statistical practice, usually only weak consistency is derived, which generally follows from weak laws of large numbers.

Example: Assume again an i.i.d. sample $y_1, \ldots, y_n$ with mean $\mu = E(y_i)$ and variance $\sigma^2 = \text{var}(y_i) < \infty$. As stated above we then have

$$E\left\{(\bar{y} - \mu)^2\right\} = \text{var}(\bar{y}) = \sigma^2/n \to 0 \quad \text{as } n \to \infty.$$

Therefore, $\bar{y} \xrightarrow{q.m.} \mu$. The latter implies that $\bar{y} \xrightarrow{p} \mu$, i.e. $\bar{y}$ is a (weakly) consistent estimator of μ.

Convergence properties are described in detail in any medium-level book on the theory of probability, such as Gnedenko (1962). Chapter 9 of Cox and Hinkley (1974) provides a statistically motivated introduction to asymptotic theory; Sect. 9.4 illustrates calculation of the distribution of an M-estimator of location through an asymptotic expansion. These calculations are based on a first-order Taylor expansion of the statistic. Higher accuracy may be obtained by use of higher order expansions in powers of $n^{-1/2}$. An example is the Edgeworth expansion, which is good for the centre of the distribution, but may give negative values in the tails. Field and Hampel (1982) derive several higher order approaches to the distribution of M-estimators of location which are amazingly accurate for very small samples, for example, $n = 4$. The algebra involved is, however, complicated. A review of such "small-sample asymptotics" is given by Field and Ronchetti (1990) who devote their Sect. 7.3 to problems in robust statistics. In all these references, the estimation of scale is ignored. Cox and Hinkley (1974, p. 351) mention results from a bivariate Taylor series expansion which show that, asymptotically, the estimate of location is independent of the estimate of scale when this satisfies (2.25). However, this conclusion says little about estimation in small samples. A careful introduction to higher order asymptotics is Barndorff-Nielsen and Cox (1989) with more advanced material in Barndorff-Nielsen and Cox (1994). It is interesting that, despite its theoretical bent, Maronna et al. (2019) does not include any references to the work cited in this paragraph.

In robustness two kinds of asymptotics are invoked. One is that customary in statistics, described in this section, of considering the properties of parameter estimates and tests as the sample size $n \to \infty$. The other is what happens to the estimator of a parameter θ as the outliers in G move from being remote from the main population F to being much closer to it? In Sect. 5.5, we present a systematic approach to the solution of this problem.

Problems

2.1 SC for the arithmetic mean

Show that the sensitivity curve of the arithmetic mean is

$$SC(y, \hat{T}_n, Y_{n-1}) = y - \overline{y}_{n-1}.$$

2.2 SC for the median

Suppose $n = 2m + 1$ is odd. Show that the sensitivity curve of the median is

$$SC(y, \hat{T}_n, Y_{n-1}) = \begin{cases} n\{y_{[m]} - \frac{1}{2}(y_{[m]} + y_{[m+1]})\} = \frac{n}{2}(y_{[m]} - y_{[m+1]}) & \text{if } \quad y < y_{[m]} \\ n\{y - \frac{1}{2}(y_{[m]} + y_{[m+1]})\} & \text{if } y_{[m]} \leq y \leq y_{[m+1]} \\ n\{y_{[m+1]} - \frac{1}{2}(y_{[m]} + y_{[m+1]})\} = \frac{n}{2}(y_{[m+1]} - y_{[m]}) & \text{if } \quad y > y_{[m+1]} \end{cases}$$

and comment on this expression.

2.3 IF for the mean

Show that the IF for the mean is equal to the expression given in (2.6).

2.4 IF for the median

Show that the IF for the median is equal to the expression given in (2.7). You can assume that $f(Y)$ is continuous and non-zero at the population median (well-behaved distribution).
(a) Derive the result starting from the sensitivity curve.
(b) Derive the result starting from the definition of IF given in (2.5).

2.5 IF for the quantile $y_s = F^{-1}(s)$

(a) Write the functional version of the s-th quantile.

(b) Find IF$_{y_s}(y, F_\theta)$.
(c) Show that when $s = 1/2$ we obtain the IF for the median of Eq. (2.7).

Remark In these exercises, you can assume that F has a non-zero finite derivative f, at $F^{-1}(s)$, so that it is possible to take the derivative under the integral sign, it is legitimate to integrate by parts and that corner points $F^{-1}(\alpha)$ and $F^{-1}(1-\alpha)$ are uniquely determined (i.e. F^{-1} does not have jumps there).

2.6 IF for the truncated mean
(a) Write the functional version of the truncated mean.
(b) Show the IF for the truncated mean for a general asymmetric model.
(c) Show that if F is a continuous unimodal symmetric density, the IF is equal to the expression given in (2.8).

2.7 ARE when the underlying distribution is normal or Student's t
Show that ARE(med; $\bar{y}$) $= 2/\pi \approx 64\%$ when the data come from the normal distribution. Compute ARE(med; $\bar{y}$) when the data come from the Student's t distribution with 2, 3, 4, and 5 degrees of freedom.

2.8 ARE when not all measurements are equally precise
Using the contamination model in Eq. 2.14, show that the asymptotic variances of the sample mean and of the sample median are

$$(a) \, n\,\text{var}(\bar{y}) = \sigma^2\{(1-\varepsilon) + \varepsilon\tau^2\},$$
$$(b) \, n\,\text{var}(\text{med}) = \sigma^2 \frac{\pi}{2(1-\varepsilon+\varepsilon/\tau)^2}.$$

2.9 Rate of convergence
Show that if $\hat{\theta}_n$ is an unbiased estimator of an unknown parameter θ satisfying $\text{var}(\hat{\theta}_n) = Cn^{-1}$ for some $0 < C < \infty$, then $\hat{\theta}_n$ possesses the rate of convergence $n^{-1/2}$. Prove that the $n^{-1/2}$ is the rate of convergence of $\bar{y}$.

Open Access This chapter is licensed under the terms of the Creative Commons Attribution 4.0 International License (http://creativecommons.org/licenses/by/4.0/), which permits use, sharing, adaptation, distribution and reproduction in any medium or format, as long as you give appropriate credit to the original author(s) and the source, provide a link to the Creative Commons license and indicate if changes were made.

The images or other third party material in this chapter are included in the chapter's Creative Commons license, unless indicated otherwise in a credit line to the material. If material is not included in the chapter's Creative Commons license and your intended use is not permitted by statutory regulation or exceeds the permitted use, you will need to obtain permission directly from the copyright holder.

Chapter 3
Robust Estimators in Multiple Regression

Abstract The first five sections describe multiple regression with least squares, including methods for outlier detection. Section 3.6 introduces three distinct approaches to robust regression: (i) soft trimming or downweighting, which extends the M-estimation of Chap. 2 to regression, principally S-estimation (Sect. 3.8); (ii) hard trimming, Least Trimmed Squares (LTS, Sect. 3.11) in which a specified proportion of the observations is trimmed; and (iii) adaptive hard trimming, the Forward Search (FS, Sect. 4.1). This monitoring method is explored more thoroughly in Chap. 4. Unlike the FS, the methods in (i) and (ii) provide a single robust analysis under chosen specified conditions. That for LTS depends on the chosen trimming proportion, which should, hopefully, trim all outliers and fit the model to all the uncontaminated data. For the downweighting methods, the severity of downweighting is determined by the choice of tuning constants to give desired robustness properties. The calculations are described in Sect. 3.9. The algorithm for S-estimation is in Sect. 3.10 with that for LTS in Sect. 3.11.2. Further developments of M-estimation (MM- and τ-estimators) are described in Sects. 3.12.1 and 3.12.2. Reweighted LTS estimators are introduced in Sect. 3.12.3. Section 3.13 concludes the chapter with comparisons of traditional robust data analyses for a single specified target of robustness.

3.1 Multiple Regression: Basic Concepts

3.1.1 Parameter Estimates from Least Squares

We start with an introduction to least squares regression. This is often said to be the workhorse of applied statistics; that is, the most used and useful set of procedures for the analysis of data. The methods described in this book were built on least squares.

The linear regression model in matrix form is written as $E(Y) = X\beta$, where there are n observations on a continuous response and X is an $n \times p$ matrix of the constant term and the values of the $p - 1$ explanatory variables with ith row x_i^T. The elements of X are known constants. The model for the ith of the n observations can be written in several ways as, for example,

$$y_i = \eta(x_i, \beta) + \epsilon_i = x_i^T \beta + \epsilon_i = \beta_0 + \sum_{j=1}^{p-1} \beta_j x_{ij} + \epsilon_i. \tag{3.1}$$

The "second-order" assumptions are that the errors ϵ_i have zero mean, constant variance σ^2, and are uncorrelated. In non-robust inference, it is standard to assume that, in addition, the errors are normally distributed. This is often called Ordinary Least Squares (OLS) regression. In this book OLS is abbreviated to LS.

Remark Note that we use ϵ_i for the error of observation i while in the previous chapter and elsewhere we use the symbol ε to denote the fraction of contamination.

Under the second-order assumptions and normality, the least squares estimates $\hat{\beta}$ of the parameters β are the maximum likelihood estimators. They minimize the sum of squares $S(\beta) = (y - X\beta)^T(y - X\beta)$ and so are solutions of the estimating equation

$$X^T(y - X\hat{\beta}) = 0, \tag{3.2}$$

the non-robust regression version of the score equation (2.17) for the M-estimate of location. For the linear model considered here, the estimating equation is usually written in the form of the normal equations $X^T X \hat{\beta} = X^T y$, with the solution $\hat{\beta} = (X^T X)^{-1} X^T y$. The vector of n predictions from the fitted model is $\hat{y} = X\hat{\beta} = X(X^T X)^{-1} X^T y = Hy$, where the projection matrix H is called the hat matrix because it "puts the hats on y". It has an important role in the algebra of least squares. For example, the vector of *least squares residuals* is

$$e = y - \hat{y} = y - X\hat{\beta} = (I - H)y \tag{3.3}$$

and the residual sum of squares

$$S(\hat{\beta}) = \sum_{i=1}^{n} e_i^2 = y^T(I - H)y. \tag{3.4}$$

The residual mean square estimate of the variance σ^2 is $s^2 = S(\hat{\beta})/(n-p)$.

The least squares residuals have variance

$$\text{var}(e) = (I - H)\sigma^2. \tag{3.5}$$

With h_i the ith diagonal element of H, given by

$$h_i = x_i^T (X^T X)^{-1} x_i, \tag{3.6}$$

$\text{var}(e_i) = (1 - h_i)\sigma^2$. Division of the least squares residuals by s gives the scaled residuals. The values of the *scaled residuals* $r_i = e_i/s$ do not depend on the value of σ^2. But, like the least squares residuals, they do not in general have the same

3.1 Multiple Regression: Basic Concepts

variance. The *studentized residuals*, which do have equal variance, are then defined as

$$\tilde{r}_i = \frac{e_i}{s(1-h_i)^{0.5}} = \frac{y_i - \hat{y}_i}{s(1-h_i)^{0.5}}. \tag{3.7}$$

The studentized residuals are widely used in model checking. Although they all have the same variance, they are not independent, nor do they follow a Student's t distribution. Cook and Weisberg (1982, p. 19) show that $\tilde{r}_i^2/(n-p)$ has a beta distribution $\mathcal{B}\{1/2, (n-p-1)/2\}$. Since the first parameter of the beta distribution is 0.5, the distribution is unbounded at the origin. In Sect. 5.1.4, we discuss the effect on outlier detection of approximating this scaled beta distribution by the χ_1^2 distribution, which is also unbounded at the origin.

The quantity h_i is called the leverage of the ith observation. The sum of the h_i is p, so that the average value is p/n, with $0 \leq h_i \leq 1$. The variance of a fitted value $\hat{y}_i$ is $\text{var}(\hat{y}_i) = \sigma^2 h_i$. A large value of h_i indicates a "leverage point". For such points the variance of $\hat{y}_i$ will be close to σ^2, indicating that the fit is mostly determined by the value of y_i. Likewise $\text{var}(e_i) = (1-h_i)\sigma^2$ will be small. An outlier at a leverage point causes the fitted least squares line to pass close to the observed value, an effect clearly shown for the Stars data in Fig. 4.9. Inspection of plots of least squares, or even studentized, residuals may not indicate how influential such observations are for the fitted model.

3.1.2 Formal Tests

To test the terms of the model we use F tests or, equivalently, t tests if interest is in a single parameter. The overall effect of the model can be assessed by the coefficient of determination R^2 and the results summarized in an analysis of variance table.

If the total corrected sum of squares of the observations is

$$S_0 = \sum_{i=1}^{n}(y_i - \bar{y})^2,$$

where $\bar{y} = \sum y_i/n$, the coefficient of determination is defined as

$$R^2 = \{S_0 - S(\hat{\beta})\}/S_0. \tag{3.8}$$

A value near one indicates that a large proportion of the total sum of squares of the observations has been explained by the regression. The value of R^2 can be modified to allow for the degrees of freedom of the two sums of squares, to give the adjusted R^2

$$R_{\text{adj}}^2 = 1 - \frac{S(\hat{\beta})/(n-p)}{S_0/(n-1)}. \tag{3.9}$$

However, for both measures, a large value, while encouraging, says nothing about the contribution of particular groups of observations to various aspects of the fit, such as their importance, if any, in the estimation of specific parameters.

Calculation of the t tests requires the variances of the elements of $\hat{\beta}$. Since $\hat{\beta}$ is a linear function of the observations, $\text{var}(\hat{\beta}) = \sigma^2 (X^T X)^{-1}$. Let the kth diagonal element of $(X^T X)^{-1}$ be v_k, when the t test for testing that $\beta_k = 0$ is

$$t_k = \hat{\beta}_k / (s_v^2 v_k)^{0.5}, \qquad k = 1, \ldots, p, \tag{3.10}$$

where $s_v^2 = S(\hat{\beta})/(n - p)$ is an estimate of σ^2 on $\nu = n - p$ degrees of freedom. If $\beta_k = 0$, t_k has a t distribution on ν degrees of freedom.

These individual t tests are the signed square roots of the F tests from the difference in the sums of squares when β_k is and is not included in the model. If the explanatory variables are correlated, dropping x_k from the model may cause appreciable change in the significance and even the signs of the remaining t statistics.

3.2 Deletion Diagnostics

If there are I outliers, all of which are deleted, and the model refitted, the outlying observations will be revealed by the use of suitable tests and plots. Exact formulae for linear regression models provide expressions for the effect of deletion on the residuals, residual sum of squares, and parameter estimates, so that the explicit deletion of individual observations and repeated refitting of the model are not necessary. The methods, collectively known as deletion diagnostics, are described in the books of Belsley et al. (1980), Cook and Weisberg (1982) and of Atkinson (1985). They use these exact formulae to examine the importance of individual observations to inferences about the fitted model. If there is a single outlier it is revealed by calculating the effect of deletion of each observation in turn. Historically the methods were a step in understanding the effect of individual observations on fitted models and the ensuing inferences. However, the methods work backwards from the full dataset. They are accordingly subject to *masking*, in which the presence of many outliers renders them undetectable from a fit to all the data, and also to *swamping*, in which "good" observations appear, under the same conditions, to be outlying. There is also a combinatorial explosion in the number of sets of observations to be deleted (single observations, pairs, triples, ...). Atkinson and Riani (2000), especially Chap. 4, present examples in which diagnostic methods fail to reveal the structure of outliers revealed by the FS.

3.2 Deletion Diagnostics

3.2.1 Deletion Residuals

These residuals quantify the effect of deletion of the ith observation on prediction at x_i. In particular, does the observed y_i agree with the prediction $\hat{y}_{(i)}$ obtained when the ith observation is excluded from estimation of β? Let the parameter estimate after deletion of observation i be $\hat{\beta}_{(i)}$. Since y_i and $\hat{y}_{(i)}$ are independent, the difference $y_i - \hat{y}_{(i)} = y_i - x_i^T \hat{\beta}_{(i)}$ has variance

$$\sigma^2 \{1 + x_i^T (X_{(i)}^T X_{(i)})^{-1} x_i\}. \tag{3.11}$$

To estimate σ^2 we use the deletion estimate

$$s_{(i)}^2 = \{(n-p)s^2 - e_i^2/(1-h_i)\}/(n-p-1), \tag{3.12}$$

which is also independent of y_i (Exercise 3.2). The test for agreement of the observed and predicted values is

$$r_i^* = \frac{y_i - x_i^T \hat{\beta}_{(i)}}{s_{(i)}\{1 + x_i^T (X_{(i)}^T X_{(i)})^{-1} x_i\}^{0.5}}, \tag{3.13}$$

which, when the ith observation comes from the same population as the other observations, has a t distribution on $(n-p-1)$ degrees of freedom. Deletion results make it possible to simplify (3.13) to obtain

$$r_i^* = e_i/\{s_{(i)}(1-h_i)^{0.5}\} = (y_i - \hat{y}_i)/\{s_{(i)}(1-h_i)^{0.5}\}. \tag{3.14}$$

We call r_i^* the *deletion residual*. Comparison with (3.7) shows that the deletion residual differs from the studentized residual only in the estimate of σ employed. This is enough to ensure that the square of r_i^* has the F distribution as opposed to the beta distribution of $\tilde{r}_i^2$ (Sect. 3.1.1). The difference between the two residuals is most acute if there is a single outlier at a point of high leverage (Exercise 3.7).

In the forward search, we consider the effect of adding observations. Given a parameter estimate $\hat{\beta}$, a design matrix X and an estimate s^2 of σ^2, we add an observation y_i at x_i. Then the test for agreement between observation and prediction (3.13) becomes

$$r_i^* = \frac{y_i - x_i^T \hat{\beta}}{s\{1 + x_i^T (X^T X)^{-1} x_i\}^{0.5}} = \frac{y_i - x_i^T \hat{\beta}}{s(1 + h_i)^{0.5}}, \tag{3.15}$$

which is the usual t test for the agreement of a new data point with a previous set of observations. Since, apart from replications, x_i is not a row of X, it may be that here h_i is greater than one. The power of the test is improved by using s^2 as the estimate of error variance, rather than updating for the new observation.

3.2.2 A List of Residuals

Here is a list of the residuals defined in this chapter. See (3.6) for h_i and (3.12) for $s_{(i)}^2$.

Raw	$e_i = y_i - \hat{\beta}^T x_i$
Scaled	$r_i = e_i/s$
Studentized	$\tilde{r}_i = e_i/\{s(1-h_i)^{0.5}\} = r_i/(1-h_i)^{0.5}$
Deletion	$r_i^* = e_i/\{s_{(i)}(1-h_i)^{0.5}\} = (y_i - \hat{y}_i)/\{s_{(i)}(1-h_i)^{0.5}\}$.

In the raw residuals $\hat{\beta}$ can come from LS or any robust estimator. The nomenclature for residuals is fluid. In the **R** environment, including Maronna et al. (2019), the scaled residuals r_i are known as "standardized' residuals". In these residuals, the choice of scale estimator s is again catholic. The definitions of studentized and deletion residuals refer to least squares estimation, including the Bayesian version. We leave to the exercises' derivation of the relationships, for LS, between $\hat{\beta}_{(i)}$ and $\hat{\beta}$ (Exercise 3.1), $s_{(i)}^2$ and s^2 (Exercise 3.2), and between the deletion residual r_i^* and the studentized residual $\tilde{r}_i$ (Exercise 3.3).

3.2.3 Mean Shift Outlier Model

In the mean shift outlier model, the deletion of observation i is modelled by extending the regression model to include an extra parameter ϕ_i with the explanatory variable d_i which is all zeroes apart from a single one in position i. Least squares estimation then gives a zero residual for observation i. More generally, let there be m observations remaining in the fitted model. Then $n - m$ observations will have been deleted. This can be expressed by writing the regression model as

$$y = X\beta + D\phi + \epsilon. \tag{3.16}$$

Here D is an $n \times (n - m)$ matrix with a single one in each of its columns and in $n - m$ rows, all other entries being zero. These entries specify the observations that are to have individual parameters or, equivalently, are to be deleted (Cook and Weisberg, 1982, p. 20; Insolia et al., 2021).

For deletion of the single observation i, straightforward algebra applied to (3.16) shows, for example, that the squared deletion residual is the F test for $\phi_i = 0$ on one and $n - p - 1$ degrees of freedom (Exercise 3.5).

The standard use of the mean shift outlier model is for inferences about the observations in a sample of size n. In Sect. 4.11, we apply this method to the comparison of fitted models with differing numbers of observations. We render outlier detection and deletion compatible with the information criterion BIC Schwarz (1978) by

3.3 Outlier Detection

For LS regression the most powerful test that observation i is outlying is to compare the deletion residual r_i^* with the quantiles of the t distribution on $n - p - 1$ degrees of freedom. This has the advantage that, if observation i is outlying, the value of e_i does not contribute to inflation of the estimate of σ^2, thus avoiding the consequent reduction in the power of the test using the other residuals described in Sect. 3.2.2. However, in robust regression, provided the breakdown point is sufficiently large, the robust estimate of σ will be unaffected by outliers. It is therefore standard to use the scaled residuals $r_i = e_i/s$ to test for outliers. Particularly in view of the uncertainties in the statistical literature on robustness about the correct form of tests for robust regression with M-estimators, mentioned in Sect. 3.7.2, values of robust r_i are referred to the quantiles of the normal distribution.

If interest is in a specific value of i, observation y_i is declared outlying if

$$|r_i| = |e_i/s| > \Phi^{-1}(1 - \alpha), \qquad (3.17)$$

for some appropriate α such as 1% or 0.1%.

This provides an individual testing procedure. Under the null hypothesis $H_{0,i}$ that $y_i \sim \mathcal{N}(x_i^T \beta, \sigma^2)$, the expected number of false outliers will be $n\alpha$ for any uncontaminated dataset. Atkinson and Riani (2006) adapt and extend a sophisticated simulation method of Buja and Rolke (2003) to investigate the distribution of the surprisingly high number of false outliers that can be detected.

Simultaneous inference, on the other hand, tests the hypothesis that simultaneously none of the observations is outlying. With the null hypothesis for the individual test written as $H_{0,i}$, the hypothesis that there is no contamination in the data is

$$H_{0,\text{all}} : H_{0,1} \cap \ldots \cap H_{0,n}.$$

Under this hypothesis the intention is that (at least) one outlier will be detected in a proportion α of the datasets. The distribution of the test statistic for this hypothesis is approximated by using the Bonferroni correction for simultaneity with level $\alpha^* = \alpha/n$, so taking the $1 - \alpha^*$ cut-off value of the reference standard normal distribution. Then the individual outlier rule (3.17) is replaced by

$$|r_i| = |e_i/s| > \Phi^{-1}(1 - \alpha^*). \qquad (3.18)$$

An index plot of residuals including these boundaries is frequently used to detect patterns in any outlying observations. See, for example, Fig. 3.2.

3.4 Added Variables

We now introduce a less-standard aspect of regression, developments of which are of importance in the design and monitoring of robust procedures for data analysis. Further algebraic details of the material of this section are in Atkinson and Riani (2000, cap. 2). The starting point is to fit a model including all variables except the one of interest, the "added" variable. The added variable can be replaced by a "constructed" variable derived from the data. We use this form for the score test for transformations derived in Chap. 6. For the moment we concentrate on regression variables.

We extend the regression model to include an extra explanatory variable, the added variable w, so that the regression model becomes

$$E(Y) = X\beta + w\gamma, \qquad (3.19)$$

where γ is a scalar. The least squares estimate $\hat{\gamma}$ can be found explicitly from the normal equations for this partitioned model. If the model without γ can be fitted, $(X^T X)^{-1}$ exists and

$$\hat{\gamma} = w^T(I - H)y/w^T(I - H)w = w^T A y / w^T A w. \qquad (3.20)$$

We now write (3.20) in terms of residuals. Since $A = (I - H)$ is idempotent, $\hat{\gamma}$ can be expressed in terms of the two sets of residuals $e = \tilde{y} = (I - H)y = Ay$ and the residuals of the added variables $\tilde{w} = (I - H)w = Aw$. Then

$$\hat{\gamma} = w^T(I - H)y/w^T(I - H)w = w^T A y / w^T A w = \tilde{w}^T e / (\tilde{w}^T \tilde{w}). \qquad (3.21)$$

Thus $\hat{\gamma}$ is the coefficient of linear regression through the origin of the residuals e on the residuals $\tilde{w}$ of the new variable w, both sets of residuals coming from regression on the variables in X.

Because the slope of this regression is $\hat{\gamma}$, a plot of e against $\tilde{w}$ can be used as a visual assessment of the evidence for a regression and for the assessment of the contribution of individual observations to the relationship. Such a plot is called an added variable plot or a constructed variable plot if w is not a straightforward explanatory variable. However, the plot is one of residuals against residuals. Since points of high leverage tend to have small residuals, if something important to the regression happens at a leverage point, it will often not be revealed by the plot. Cook and Weisberg (1994) provide a rich collection of informative plots for regression residuals and related diagnostics.

To calculate the t statistic requires the variance of $\hat{\gamma}$. Since, like any least squares estimate in a linear model, $\hat{\gamma}$ is a linear combination of the observations, it follows from (3.20) that

$$\text{var}(\hat{\gamma}) = \sigma^2 w^T A^T A w / (w^T A w)^2 = \sigma^2 / (w^T A w) = \sigma^2 / (\tilde{w}^T \tilde{w}). \qquad (3.22)$$

Calculation of the test statistic also requires s_w^2, the residual mean square estimate of σ^2 from regression on X and w, given by (Exercise 3.6)

$$(n - p - 1)s_w^2 = y^T y - \hat{\beta}^T X^T y - \hat{\gamma} w^T y = y^T A y - (y^T A w)^2/(w^T A w). \quad (3.23)$$

The t statistic for testing that $\gamma = 0$ is then

$$t_w = \hat{\gamma}/\{s_w^2/(w^T A w)\}^{0.5}. \quad (3.24)$$

If w is the explanatory variable x_k, (3.24) is an alternative way of writing the usual t test (3.10).

In Sect. 4.8, we use an extension of the added variable method, combined with the FS, to provide an informative method of monitoring the evolution of t tests for the individual parameters in the model.

3.5 Examples of Residual Plots for the AR Regression Data

In these data, introduced by Atkinson and Riani (2000, Sect. 1.2.2), there are three explanatory variables and 60 observations, with a structure of six masked outliers. In this section, we show that the plots of residuals from LS fail to reveal this structure.

Figure 3.1 is the yX plot of the data, that is, a side-by-side presentation of the scatter plots of y against each of the three explanatory variables. There is no obvious evidence of outliers. The plot suggests linear relationships between y and x_2 and x_3. The relationship with x_1 is less certain.

We now turn to the identification of outliers. Figure 3.2 shows index plots of two forms of residual. Those in the upper panel are the studentized residuals $\tilde{r}_i$ (3.7), with the lower panel showing the deletion residuals r_i^* (3.15). Also included are the 95% and 99% pointwise and 95% simultaneous bands for the normal distribution, the latter found from the Bonferroni inequality. Both plots indicate observation 43 as an outlier, although not a very remote one. Observation 43 is not one of the six masked outliers included in the data.

The QQ-plot of studentized residuals against normal order statistics is another plot for checking outliers as well as more general aspects of the distribution of residuals. Interpretation is enhanced by the inclusion of simulation envelopes giving empirical 90% confidence intervals for the distribution of the residuals. The QQ-plot in the left-hand panel of Fig. 3.3 shows that the studentized residual for observation 43 is large, but not larger than would be expected for an extreme residual; the bands for central residuals are appreciably tighter. The right-hand panel shows a plot of studentized residuals against fitted values. Observation 43 appears outlying in this plot, but there is no indications of a systematic lack of fit, nor of the dependence of the variance of the residuals on the fitted values. In the latter case, a response transformation should be considered. See Chap. 6.

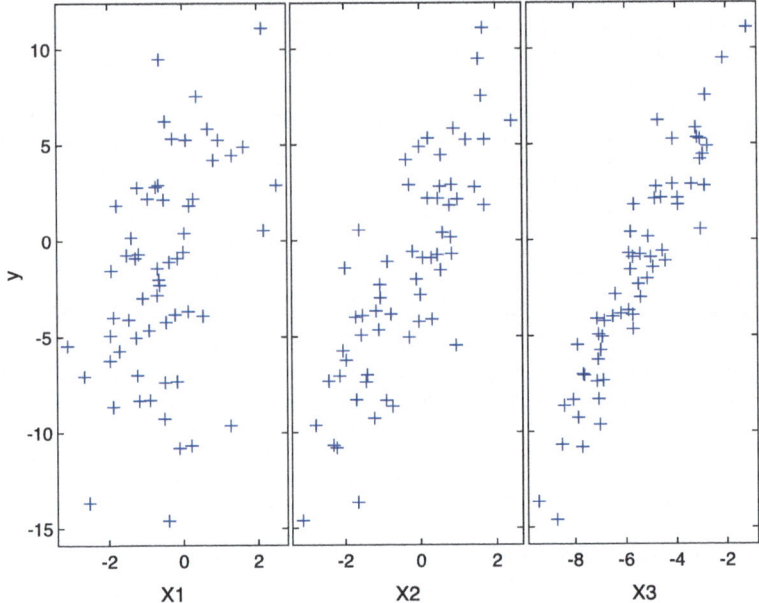

Fig. 3.1 AR regression data: yX plot of y against each of the three explanatory variables

Since these are regression data, we should also check further aspects of the relationship between the response and the explanatory variables. For this we use the added variable plot of Sect. 3.4. In Fig. 3.1 regression on x_1 is most in question so we treat x_1 as the added variable with X containing the values of x_2 and x_3 (as well as a column of ones). As we know the identity of the six masked outliers, in Fig. 3.4, we compare the added variable regression with and without the outliers. The added variable plot shows that the outliers are extreme. The regression on all units has a negative slope with a t value of -1.26. If the subset of outliers is excluded from the fit, the slope becomes positive with a t value of 2.25. Note that Fig. 3.4 is the added variable plot for all 60 observations. If the masked outliers were deleted, all points on the plot would change, as well as six disappearing. The slope of the line in the figure was calculated without the excluded units, but has here been plotted on the added variable plot for all observations. Added variable regression lines all pass through the origin.

We repeat this analysis in Fig. 3.5, but now focus on the effect of observation 43. We see that this lies on the opposite side of the regression line from the six outliers. Its exclusion from the fit causes the t value to become more negative, changing from -1.26 to -1.93. Deletion of observation 43 results in a more negative slope whereas deletion of the six outliers leads to a positive relationship between y and x_1.

Least squares led to the identification of observation 43 as mildly outlying. However, the informative analysis in Fig. 3.4 requires the identification of the masked outliers. The results of the analysis of these data in Sect. 4.9.3 show that both

3.5 Examples of Residual Plots for the AR Regression Data

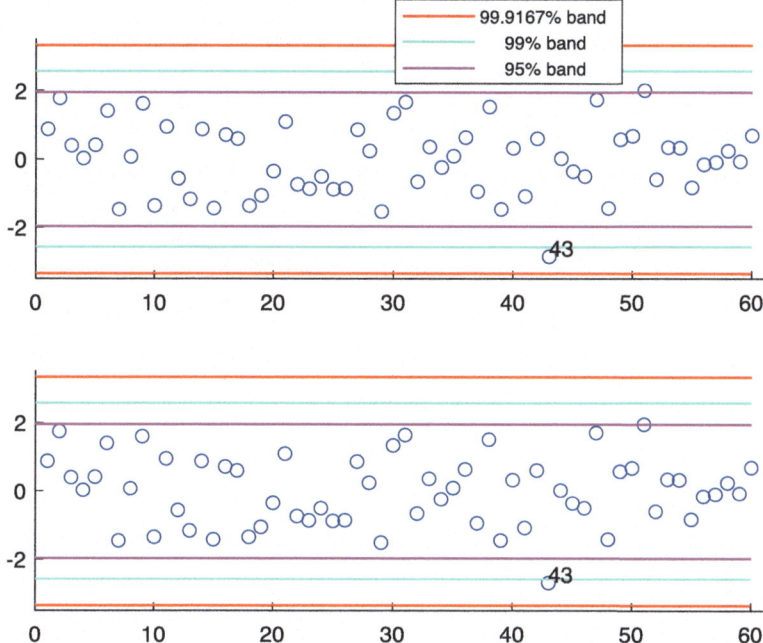

Fig. 3.2 AR regression data: index plot of residuals. Upper panel studentized residuals; lower panel deletion residuals, with 95% and 99% pointwise and 95% simultaneous bands from the normal distribution

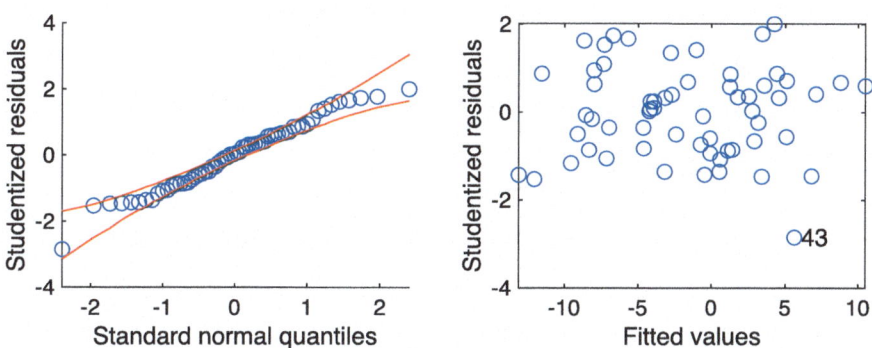

Fig. 3.3 AR regression data: left-hand panel, QQ-plot of studentized residual with 90 percent simulation envelopes; right-hand panel, studentized residuals against fitted values

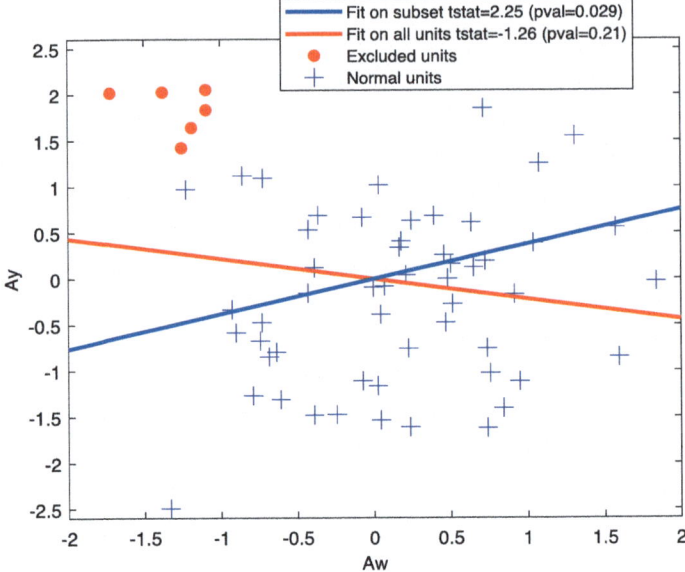

Fig. 3.4 AR regression data: added variable plot for all observations with x_1 as the added variable. Slope of regression with and without the six outliers

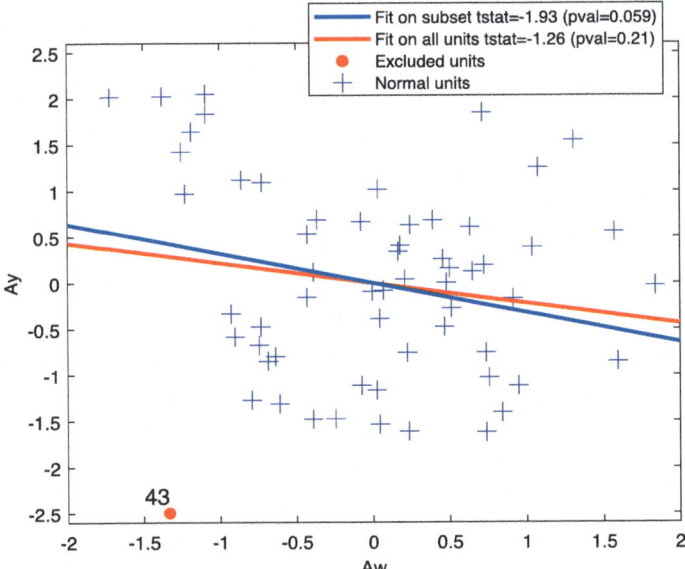

Fig. 3.5 AR regression data: added variable plot for all observations with x_1 as the added variable. Slope of regression with and without observation 43

the FS and monitored M-estimation, among other methods of robust regression, readily identify the masked outliers, and are informative about the properties of observation 43.

3.6 Three Classes of Estimators for Robust Regression

The least squares regression of the preceding sections is not robust. The parameter estimation method has zero *bdp*. In the remainder of this chapter, we extend several of the robust methods of Chap. 2 to regression. Some methods use developments of M-estimation and so downweight observations. Others are based directly on trimming. Since the trimming is applied to least squares residuals from subsets of the data, the computational and inferential problems are much greater than those that arise from trimming a simple sample, such as those in Sect. 1.4.

It is helpful to divide methods of robust regression into three classes Hampel et al. (1986); Atkinson et al. (2004); Farcomeni and Greco (2015).

1. Soft trimming (downweighting). M-estimation for univariate samples was discussed in detail in Chap. 2. The small-sample comparisons of Sect. 2.4.4 illustrated the importance of the estimation of σ^2. In this chapter, M-estimation is extended in Sect. 3.7 to regression models. We also include several derived methods, which depend on the way in which the residual variance σ^2 is estimated. S-estimation is in Sect. 3.8 with MM- and τ-estimation in Sect. 3.12.
   In all methods, the intention is that observations near the regression plane retain their value, but the ρ function (Sect. 2.2.1.1) ensures that increasingly remote observations have a weight that decreases with distance from the regression plane which contains the bulk of the data. The desired value of either the asymptotic breakdown point *bdp* or of the efficiency *eff* has to be specified. Since the algorithms perform calculations using residuals, this efficiency is that of estimation when the method is applied to a sample from the normal distribution.
2. Hard {0,1} Trimming. Very robust regression was introduced by Rousseeuw (1984) who developed suggestions of Hampel (1975) that led to the Least Median of Squares (LMS) and Least Trimmed Squares (LTS) algorithms. In their original form both find estimates of β with 50% *bdp*. The LMS estimator minimizes the hth-ordered squared residual $e_{[h]}^2(\beta)$ with respect to β, where $h = [0.5(n + p + 1)]$ with the symbol [·] denoting integer part. In least trimmed squares, the amount of trimming of n observations is again determined by the same parameter h which is asymptotically $n/2$.
   Lower values of *bdp* are obtained by taking larger values of h, which are specified in advance of the analysis. For LS $h = n$, we introduce LMS and LTS in Sect. 3.11.
3. Adaptive Hard Trimming. The methods of robust regression listed above are conventional and static, in the sense that the constants in the methods are chosen in advance and used to provide a single data analysis. In the Forward Search (FS),

described in Chap. 4, the observations are again hard trimmed, but the value of h is determined by the data, being found adaptively by the search Riani et al. (2014a).

3.7 M-Estimators for Multiple Regression

3.7.1 The Iterative Reweighted Least Squares Algorithm

In least squares, the estimate of β is independent of the estimate of the scale σ. Section 2.3.3 gave an outline of an algorithm for the robust simultaneous estimation of μ and σ in the location-scale model. However, we mentioned the result of Huber (1981) that the *bdp* of estimators using this algorithm is lower than the nominal values for individual estimation of the two parameters. In the extensions of M-estimation introduced in Sect. 3.6, the estimation of σ is invoked in attempts to break away from the relationship between *eff* and *bdp* explored in Sect. 2.4. We start by considering M-estimation in regression with a given estimate of scale σ_0.

For the extension of the model to regression, the residuals for some parameter estimate $\tilde{\beta}$ are $e_i(\tilde{\beta}) = y_i - \tilde{\beta}^T x_i$. For least squares regression, the estimating Eq. (3.2) can be written as

$$X^T(y - X\hat{\beta}) = X^T e(\hat{\beta}) = \sum_{i=1}^{n} x_i e_i(\hat{\beta}) = 0_{p \times 1}, \qquad (3.25)$$

independently of the estimate of σ.

For a given estimate σ_0 of σ, the M-estimator of the parameters of the linear model is the solution $\hat{\beta}$ of

$$\sum_{i=1}^{n} x_i \psi\{e_i(\hat{\beta})/\sigma_0\} = 0_{p \times 1}, \qquad (3.26)$$

the extension of (3.25) with $e_i(\hat{\beta}) = y_i - x_i^T \hat{\beta}$.

For M-estimation, the estimating equation can be written in an extension of the weighted least squares form of (2.18). The scaled residuals $r_i = e_i/\sigma_0$, when

$$\sum_{i=1}^{n} x_i \psi\{r_i(\hat{\beta})\} = \sum_{i=1}^{n} x_i r_i(\hat{\beta}) \frac{\psi\{r_i(\hat{\beta})\}}{r_i(\hat{\beta})} = \sum_{i=1}^{n} x_i r_i(\hat{\beta}) w_i = 0, \qquad (3.27)$$

where $w_i = \dfrac{\psi\{r_i(\hat{\beta})\}}{r_i(\hat{\beta})} = \psi(r_i)/r_i$.

3.7.2 Theoretical Aspects

This section summarizes results on the robust analogues of normal theory t tests (Wald tests) for the parameters in a linear model. Under the normal theory assumptions, the LS estimate of location, the mean, is normally distributed. Under the second-order assumptions the central limit theorem ensures that the estimate is again normally distributed, converging at rate $n^{-0.5}$. Asymptotic results for regression models also require conditions on X, to ensure the stability of the information matrix. Let $V_x = X^T X$. Then as n tends to infinity, $V_x/n \xrightarrow{p} M_X$. The requirement is that $\det(M_X) \neq 0$. For random X the condition is that $E(xx^T)$ be finite. The purpose is to ensure increasing information on all parameters as n increases. Then under suitable regularity conditions, such as the continuity of the ρ function, the M-estimates $\hat{\beta}_M$, are asymptotically normal and converge as $n^{-0.5}$:

$$\sqrt{n}(\hat{\beta}_M - \beta) \to \mathcal{N}_p(0, \sigma^2 \gamma V_x^{-1}), \tag{3.28}$$

provided the estimate of σ also converges. Here γ is a scalar depending on ρ. For details see Maronna et al. (2019, Sect. 10.10). Asymptotic normality is an important aspect allowing for Wald-type tests and derived confidence intervals. We explore these tests in this section. In Sect. 3.11.1, we discuss the slower rate of convergence of LMS.

For LS $\text{var}(\hat{\beta}) = \sigma^2 V_x^{-1}$. The correction factor γ in (3.28) is given by

$$\gamma = \frac{A}{B^2} = \frac{E\psi(r_i^2)}{\{E\psi'(r_i)\}^2}, \tag{3.29}$$

which tends to one for the normal distribution (Exercise 3.8). It is estimated (Maronna et al., 2019, p. 100) by

$$\hat{\gamma} = \frac{1}{n-p} \sum_{i=1}^{n} \psi\left(\frac{e_i}{\hat{\sigma}}\right)^2 \Big/ \left\{\frac{1}{n}\sum_{i=1}^{n} \psi'\left(\frac{e_i}{\hat{\sigma}}\right)\right\}^2. \tag{3.30}$$

The factor $(n - p)$ is used instead of n in order to obtain the classical formula when $\psi(x) = x$, that is, LS.

Huber and Ronchetti (2009, Sect. 7.6) suggest further expressions to estimate V_X which are included in Table 3.1, where the version using (3.30) is called *covrob*.

These further robust estimates require the additional tuning constants

$$K_c = 1 + \{CV(\psi')\}^2 p/n \quad \text{and}$$

$$CV(\psi') = \sqrt{\frac{\text{var}\{\psi'(e_i/\hat{\sigma})\}}{\{\sum_{i=1}^{n} \psi'(e_i/\hat{\sigma})/n\}^2}}.$$

Table 3.1 A summary of different strategies to estimate $\text{cov}(\hat{\beta})$ and associated formulae

Acronym	Formula
covrob	$\sigma^2 \hat{\gamma} (X^T X)^{-1}$
covrob1	$\sigma^2 K_c^2 \hat{\gamma} (X^T W X)^{-1}$
covrob2	$\sigma^2 K_c^2 \hat{\gamma} (X^T X)^{-1}$
covrob3	$\sigma^2 K_c \frac{n}{n-p} \frac{\sum_{i=1}^n \psi(e_i/\hat{\sigma})^2}{\sum_{i=1}^n \psi'(e_i/\hat{\sigma})} (X^T W X)^{-1}$
covrob4	$\sigma^2 K_c^{-1} \frac{1}{n-p} \sum_{i=1}^n (\psi(e_i/\hat{\sigma}))^2 (X^T W X)^{-1} X^T X (X^T W X)^{-1}$

In these expressions, the notation $CV(\psi')$ stands for the coefficient of variation of ψ', where ψ and ψ' are, respectively, the first and second derivatives of the ρ function discussed in Sect. 2.3.2; *covrob1* is equal to *covrob2* with $X^T X$ replaced by $X^T \hat{W} X$.

Under the assumptions of a symmetric error distribution, a symmetric continuous ρ function and a matrix X with all leverages equal to p/n, (Huber, 1973) showed that $\hat{\gamma}(X^T X)^{-1}$ contains a bias of order $O(p/n)$ and derived a correction K_c^2 that makes $\hat{\gamma}(X^T X)^{-1}$ unbiased up to terms of order $O(p^2/n^2)$. However, all leverages will only be equal to p/n for experiments satisfying D-optimality Kiefer (1959); Atkinson et al. (2007). Even if we have such data, we will be looking at subsets of observations and the condition will not hold. Huber and Ronchetti (2009, p. 161) admit that there is no general agreement on the "desirability of safeguarding against leverage points in an automatic fashion".

Moreover, when the sample size is small, it sometimes happens that the scaled residuals r_i may be very close to zero. This affects the ψ' function, making $CV(\psi')$ unexpectedly large and also making the associated correction factor K_c large (see Table 3.1). Note that this has strong effects on the corresponding associated hypothesis tests (or confidence intervals). This effect is overwhelming in *covrob1* and *covrob2* because, in these cases, the value of K_c is raised to the power of two. Under these conditions, M-estimators may become unreliable. As n becomes large, the distribution of the r_i is better approximated by their asymptotic normal distribution, and the effect decreases.

All the expressions *covrobj* given in Table 3.1 are asymptotically equivalent. Salini et al. (2022) find the threshold value of CV above which *covrob* is a better estimator of the covariance matrix based than *covrob1* or *covrob2*. As we demonstrate in Fig. 4.22, through use of the monitoring approach, the differences in conclusions which come from the use of the various proposals may be slight.

This discussion is concerned with robustness against outlying observations. Croux et al. (2004) extend the discussion to t-statistics which are also robust against correlation and heteroskedasticity of the additive errors in the simple regression model.

3.8 Introduction to S-Estimators

The definition of M-estimation in Sect. 3.7.1 does not depend on how s, the estimate of σ is obtained. Clearly, if we want to keep the M-estimate robust, s should also be a robust estimate. Here we assume that the same ρ is used in the estimation of β and σ, which is customary in practice. The generalization of Eq. (2.25) to the regression case leads to defining the M-estimator of scale, say s, for any particular β, as the solution to the equation:

$$\frac{1}{n}\sum_{i=1}^{n}\rho\left(\frac{e_i}{s}\right) = \frac{1}{n}\sum_{i=1}^{n}\rho\left(\frac{y_i - \beta' x_i}{s}\right) = K. \tag{3.31}$$

The solution s of this equation is sometimes called the dispersion. In order to have a consistent scale estimate for normally distributed observations we require that, for normal data,

$$K = \mathrm{E}_{\Phi_{0,1}}\left\{\rho\left(\frac{e_i}{s}\right)\right\}, \tag{3.32}$$

where $\Phi_{0,1}$ is the cdf of the standard normal distribution. To see consistency notice that $\mathrm{E}_{\Phi_{0,1}}(\rho) = K$ implies

$$\frac{\mathrm{E}_{\Phi_{0,\sigma^2}}(\rho)}{K} = \frac{K\sigma^2}{K} = \sigma^2.$$

Therefore K has the same interpretation as the K in univariate samples in (2.25).

Equation (3.31) is solved, at least in principle, among all $(\beta, s) \in \mathbb{R}^p \times (0, \infty)$, where $0 < K < \sup \rho$. Unlike estimation of σ^2 in LS, there is here no unique analytical solution. Rousseeuw and Yohai (1984) defined S-estimators by minimization of the dispersion s of the residuals

$$\hat{\beta}_S = \min_{\beta \in \mathbb{R}^p} s\{e_1(\beta), \ldots, e_n(\beta)\}, \tag{3.33}$$

with final scale estimate

$$\hat{\sigma}_S = s\{e_1(\hat{\beta}_S), \ldots, e_n(\hat{\beta}_S)\}.$$

The S-estimates can therefore be thought of as self-scaled M-estimates whose scale is estimated simultaneously with the regression parameters. When the scale and the regression parameters are simultaneously estimated, S-estimators for regression also satisfy, for example Maronna et al. (2019, p. 125)

$$\hat{\beta}_S = \min_{\beta \in \mathbb{R}^p} \sum_{i=1}^{n} \rho\left(\frac{e_i}{s}\right). \tag{3.34}$$

The estimator of β in (3.33) is called an S-estimator because it is derived from a scale statistic in an implicit way.

The function ρ, extensively discussed in Chap. 2, is the key to many important properties of M- and S-estimates. Rousseeuw and Leroy (1987, p. 139) show that if the function ρ satisfies the following conditions:

1. It is symmetric and continuously differentiable, and $\rho(0) = 0$.
2. There exists a $c > 0$ such that ρ is strictly increasing on $[0, c]$ and constant on $[c, \infty)$.
3. It is such that
$$K/\rho(c) = bdp \quad \text{with} \quad 0 < bdp \leq 0.5, \tag{3.35}$$

then the asymptotic breakdown point of the S-estimator tends to bdp when $n \to \infty$. If $\rho(c)$ is normalized in such a way that $\sup \rho(c) = 1$, the constant K becomes exactly equal to the breakdown point of the S-estimator. However, the Huber ρ function is unbounded when $c > 0$, so that the bdp associated with this function is zero. The other ρ functions introduced in Chap. 2 are bounded and therefore can have the desired value of bdp. In Chap. 2, we compared the bdp (and the corresponding eff) for various ρ functions. In the next section, we provide the details of how the tuning constants can be calculated and elucidate their inter-relationships.

3.9 Computation of the Tuning Constants

For Tukey's biweight, the parameter c is also the rejection point. Under the Gaussian model, conditions (3.35) and (3.32) imply that c for this ρ-function must satisfy

$$\int_{-c}^{c} \left(\frac{x^2}{2} - \frac{x^4}{2c^2} + \frac{x^6}{6c^4} \right) d\Phi_{0,1}(x) + \frac{c^2}{6} \Pr(|X| > c) = bdp \frac{c^2}{6}. \tag{3.36}$$

Therefore, given a value of c, the value of bdp is automatically determined and vice versa. Similarly for the asymptotic efficiency eff at the Gaussian model

$$\mathit{eff} = \frac{\left\{ \int_{-c}^{c} \psi'(x) d\Phi(x) \right\}^2}{\int_{-c}^{c} \{\psi(x)\}^2 d\Phi(x)}. \tag{3.37}$$

For Tukey's biweight ρ function

$$\psi^2(x) = \begin{cases} \frac{1}{c^8} x^{10} - \frac{4}{c^6} x^8 + \frac{6}{c^4} x^6 - \frac{4}{c^2} x^4 + x^2 & \text{if } |x| \leq c \\ 0 & \text{if } |x| > c, \end{cases} \tag{3.38}$$

$$\psi'(x) = \begin{cases} \frac{5}{c^4} x^4 - \frac{6}{c^2} x^2 + 1 & \text{if } |x| \leq c \\ 0 & \text{if } |x| > c. \end{cases} \tag{3.39}$$

3.9 Computation of the Tuning Constants

Riani et al. (2014b) show that if X follows a standard normal distribution, then the k-th-order truncated central moment (where k is an even number) is given by

$$\int_{-c}^{c} x^k d\Phi(x) = (k-1)!! F_{\chi^2_{k+1}}(c^2), \tag{3.40}$$

where the double factorial $j!!$ is defined as follows:

$$j!! = \begin{cases} j \cdot (j-2) \cdot \ldots \cdot 5 \cdot 3 \cdot 1, & j > 0, \text{ odd} \\ j \cdot (j-2) \cdot \ldots \cdot 6 \cdot 4 \cdot 2, & j > 0, \text{ even} \\ 1, & j = -1, 0, \end{cases}$$

and $F_{\chi^2_{k+1}}(\cdot)$ if the cdf of the χ^2 distribution with $k+1$ degrees of freedom.

Therefore, for fixed *eff*, we obtain a value of c and a corresponding value of *bdp*. Similarly, given *bdp*, we can obtain a value of c and a corresponding value of *eff*.

The upper panel of Fig. 3.6 shows the parameter c and the lower panel the value of *eff* as a function of *bdp* for S-estimation with Tukey's biweight. The computations use (3.59) (further details are in Exercise 3.9). Breakdown values close to 0.5 correspond to an efficiency smaller than 0.3. The numerical results, which are in Table 3.2, agree with those of Hössjer (1992), who proved that an S-estimate with breakdown point equal to 0.5 has an asymptotic efficiency under normally distributed errors that is not larger than 0.33.

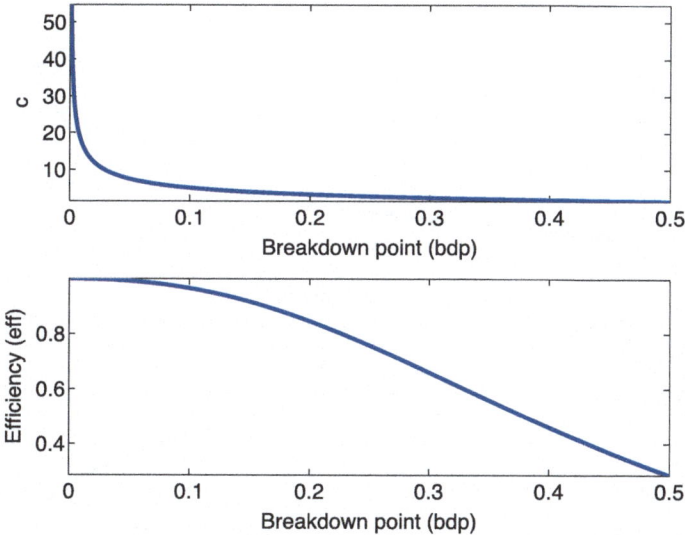

Fig. 3.6 Parameter c (upper panel) and *eff* (lower panel) as a function of *bdp* for Tukey's biweight ρ function

Results equal to traditional least squares are obtained as c tends to infinity when $bdp \to 0$, as shown in the lower panel of the figure. The shape of this curve and the results of Table 3.2 show that the transition of bdp from 0 to 0.1 does not cause a significant decrease of eff which is above 0.95 in this range. On the other hand, an increase of bdp from 0.3 to 0.4 decreases the eff from 0.661 to 0.462. The figure shows that such increases in higher values of bdp cause similar decreases in eff, although the curve is not quite straight.

Figure 3.7 shows the parameter c (top panel) and the breakdown point (bottom panel) as a function of the efficiency. This plot and Table 3.3 clearly show that values of the efficiency greater than 0.9 lead to a considerable decrease in the breakdown point.

Table 3.2 Asymptotic breakdown point, parameter c and asymptotic efficiency at the normal model for Tukey's biweight ρ function

Breakdown point bdp	Parameter c	Efficiency eff
0.05	7.5453	0.9924
0.10	5.1824	0.9662
0.20	3.4207	0.8467
0.25	2.9370	0.7590
0.30	2.5608	0.6613
0.40	1.9880	0.4619
0.50	1.5476	0.2868

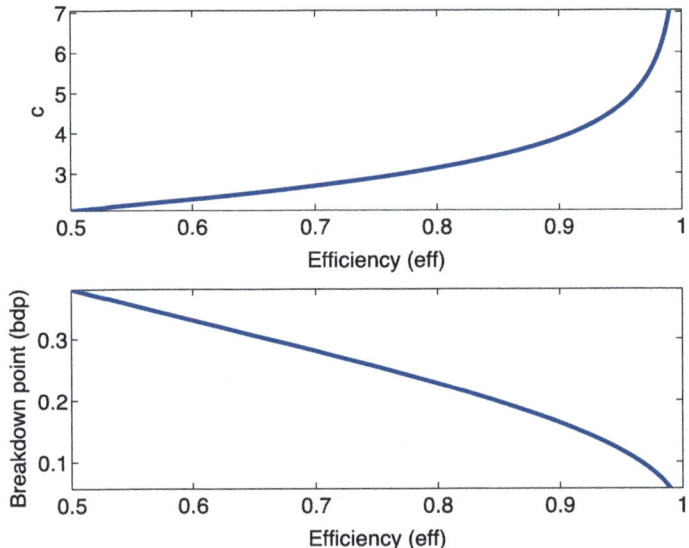

Fig. 3.7 Parameter c (top panel) and breakdown point (bottom panel) as a function of the efficiency (eff) for Tukey's biweight ρ function

3.9 Computation of the Tuning Constants

Table 3.3 Efficiency, parameter c, and breakdown point at the normal model for Tukey's biweight ρ function

Asymptotic efficiency at the normal model (*eff*)	Parameter c	Breakdown point (*bdp*)
0.50	2.0871	0.3804
0.60	2.3666	0.3304
0.70	2.6972	0.2806
0.75	2.8972	0.2548
0.80	3.1369	0.2276
0.85	3.4437	0.1980
0.90	3.8827	0.1638
0.95	4.6851	0.1194
0.99	7.0414	0.0570

For the other ρ functions, the computation of the tuning constants follows similar lines. In the case of the Hampel function, we must also fix in advance the constants c_1, c_2, and c_3, to compute c, while in the case of the hyperbolic tangent estimator we must also select the value of k in advance.

Unlike Tukey's biweight and the optimal ρ functions, Hampel's (2.35) and the hyperbolic ρ functions also include absolute moments of random variables. In general, the odd truncated absolute moments of order k, where truncation is outside the interval $[a\ b]$, are equal to Riani et al. (2014a)

$$\int_{a<|x|<b} |x|^k d\Phi(x) = \sqrt{\frac{2}{\pi}}(k-1)!! \left\{ F_{\chi^2_{k+1}}(b^2) - F_{\chi^2_{k+1}}(a^2) \right\}. \tag{3.41}$$

The tuning constant c can readily be found using the iterative procedure for Tukey's biweight (see Exercise 3.10). For the Hampel, given *bdp* or *eff* and the additional parameters c_1, c_2, and c_3, the value of c can be automatically determined; they remain fixed during the estimation procedure.

Similarly, given k for the hyperbolic, after fixing *bdp* or *eff*, c can be automatically determined. However, for this estimator, it may happen that, for certain combinations of k and c, the constants A, B, and d which characterize the ρ and ψ functions, see (2.35) and (2.36), may either not exist or only be calculable up to a coarse tolerance. For these ρ functions care may therefore be needed in the choice of k and c. For the PD, on the other hand, no iterative procedure is needed because *bdp* and *eff* are both simple functions of the parameter α (2.41).

Detailed names of the functions inside **FSDA** which compute *bdp* or *eff* are given in Appendix A.1.

3.10 S-Estimation: Computational Details

The results of Sect. 2.4.5 demonstrate that the use of redescending ψ functions implies multiple solutions to the estimating equations. In this section, we present an outline of an algorithm Salibian-Barrera and Yohai (2006) which is designed for S-estimation in the presence of multiple solutions. The difficulty in the optimization problem is that the objective function has many local minima and saddle points; which is found depending upon the starting point of the algorithm. The standard numerical procedure in such cases is to start the algorithm from many initial parameter values and, in this case, after further calculations, to select as the solution that providing the minimum value of $\hat{\sigma}$. The sets of initial parameter values are found by taking, at random, many subsets of the observations of size p, the fitted model to each providing a set of initial parameter values. In S-estimation, the algorithm uses a bound that avoids the iteration to convergence of poor starting points. Even so, the calculation of S-estimates involves many weighted least squares fits to subsets of observations.

In the algorithm below, Steps 1–3 describe the calculation of preliminary parameter estimates from each subset. In Steps 4–6, the estimates from Step 3, for a small randomly chosen set S_1 of the subsets, are iterated further to give an estimate of σ for each member of S_1. The largest such estimate provides an upper bound on the estimate of σ. In Steps 7–8, the estimates from Step 3 for the many subsets not in S_1 are compared with the bound. If the estimate is below the bound, it is iterated further and enters S_1, the bound is sharpened, and the set with the previous maximum estimate of σ leaves S_1. In Step 9, the estimates from the subsets in S_1 are iterated to convergence. The minimum estimate of σ in S_1 then provides the S-estimate of scale.

1. Initialization. We start with a set of n observations from which we extract *nsamp* subsets each of size p (generally *nsamp* is the minimum of $\binom{n}{p}$ and 1000).
2. Compute $\hat{\beta}_{LS}$ for each subset. Since these are elemental subsets, there are p parameters and p observations in the subset; the fit is exact.
3. Refining steps for each subset. For subset j, use the value of $\hat{\beta}_{LS}$ together with the MADN of the associated residuals $y - X\hat{\beta}_{LS}$ as the starting point of the iterative procedure described in Sect. 2.3.3, adapted to regression, using a small number of iterations (refining steps). Generally the maximum number of iterations (*refsteps*) is set to 3. The final estimates of β, σ, and the residuals for subset j are called $\hat{\beta}_{rw}(j)$, $\hat{\sigma}_{rw}(j)$, and $r_{rw}(j)$.
4. Subset selection and initial calibration. We take an arbitrary set of size *bestr* of the results from the *nsamp* analyses of Step 3. Typically *bestr* = 5. We call this set S_1 with S_2 the set of the remaining *nsamp—bestr* results.
5. To the elements of S_1 we apply the routine of Sect. 2.3 to estimate the scale. That is, for each $j \in S_1$, we keep the estimate of β fixed and take $r_{rw}(j)$ and $\hat{\sigma}_{rw}(j)$ as the initial values for the iterative procedure. We store the final estimate of the scale, $\hat{\sigma}_{rwf}(j)$.

6. Among the set S_1 of *bestr* subsets we find the largest estimate of the scale (we call it $\hat{\sigma}_{\text{worst}}$):
$$\hat{\sigma}_{\text{worst}} = \max \hat{\sigma}_{rwf}(j), \ j \in S_1.$$

7. The remaining part of the algorithm searches through the subsets in S_2, first using a simple bound to check whether each new subset might produce an estimate of σ which is smaller than $\hat{\sigma}_{\text{worst}}$. If so, σ is estimated for this subset.

8. For all $j \in S_2$, it is easy to show (Exercise 3.12) that if
$$\overline{\rho}(r_{rw}/(c\hat{\sigma}_{\text{worst}})) \geq K \qquad (3.42)$$
we do not have to consider subset $j \in S_2$ because $\hat{\sigma}_{rw}(j) > \hat{\sigma}_{\text{worst}}$.
Therefore, if:
$$\overline{\rho}(r_{rw}/(c\hat{\sigma}_{\text{worst}})) < K \qquad (3.43)$$
we find the index associated with $\hat{\sigma}_{\text{worst}}$ and replace its properties with the corresponding estimates of $\hat{\beta}_{rw}$, $\hat{\sigma}_{rw}$, and r_{rw} coming from $j \in S_2$. We also apply the routine to estimate the scale described in Step 5 using as initial value of the iterative procedure $\hat{\sigma}_{rw}(j)$ and we store the final estimate of the scale $\hat{\sigma}_{rwf}(j)$. This estimate cannot be greater than $\hat{\sigma}_{\text{worst}}$, so no further check is needed for this j.

9. Final Iterations to Convergence. For each of the *bestr* tentative solutions in S_1, we again apply the iterative procedure in Point 3, with the maximum number of iterations a very large number (often *refsteps* = 50), Finally, we select the solution with the smallest value of the estimate of the scale, which we call $\hat{\sigma}_S$, the associated $\hat{\beta}$ which we call $\hat{\beta}_S$ and the associated residuals $y - X\hat{\beta}_S$.

3.11 Introduction to LMS and LTS

We now consider the two hard trimming methods, Least Median of Squares (LMS) and Least Trimmed Squares (LTS), introduced in Sect. 3.6. The original versions of these very robust regression estimators have a *bdp* of 50%. That is, they share the property that the parameter estimates are not affected when up to half of the data are outliers. As in the case of S-estimation, the calculation of both estimators involves many least squares fits to subsets of observations.

3.11.1 LMS—Least Median of Squares

The LMS estimator for general *bdp* minimizes the hth-ordered squared residual $e_{[h]}^2(\beta)$ with respect to β (see Sect. 3.6):

$$SS_{\text{LMS}} = \min_{\hat{\beta}} \ SS\{\hat{\beta}(h)\} = e_{[h]}^2\{\hat{\beta}(h)\}. \tag{3.44}$$

The LMS estimate may be written as that value minimizing the scale $\hat{\sigma}$ given by (3.31) with discontinuous ρ function $\rho(t) = I(|t| < 1)$ and $K = 0.5$. For a general K, the solution to (3.31) is order statistic h of the e_i with $h = n - [nK]$. The corresponding solution is called a least $1 - \alpha$ quantile estimate, where $1 - \alpha = h/n$. In general, therefore, the LMS estimator can be considered as an S-estimator and the algorithmic machinery of Sect. 3.10 could be used. Stromberg (1993a) and Stromberg (1993b) give an exact algorithm for computing the LMS estimate, but the number of operations required is of order $\binom{n}{p+1}$ and therefore the algorithm is only feasible for very small values of n and p. For other numerical approaches, see Agulló (1997) and Insolia et al. (2021).

Current algorithms for LMS find an approximation to the estimator, selecting the best fit, that is, the one giving the smallest h-th-ordered squared residual out of all n residuals using the parameter estimates from fits to randomly chosen subsets of p observations. It is customary to define h as $[(n + p + 1)/2]$. In the PROGRESS algorithm of Rousseeuw and Leroy (1987, Sect. 4.4) sampling of observations is at random. Samples containing subsets that are not of full rank are rejected, although the same subset may be selected more than once.

An advantage of LMS is that an estimate of σ^2 is not required for the estimation of β. Let $\hat{\beta}_{\text{LMS}}$ be the LMS estimate of β. Rousseeuw (1984) bases the estimate of σ on the value of the median squared residual $e_{\text{med}}^2(\hat{\beta}_{\text{LMS}})$. Rousseeuw and Leroy (1987, p. 202) define

$$\tilde{\sigma}_{\text{LMS}} = 1.4826\{1 + 5/(n - p)\} \left\{e_{\text{med}}^2(\hat{\beta}_{\text{LMS}})\right\}^{0.5}, \tag{3.45}$$

where $1.4826\{1 + 5/(n - p)\}$ is a generalization of the correction factor for MADN (1.9). The additional factor $1 + 5/(n - p)$ was found by simulation by Rousseeuw and Leroy (1987, p. 202) to take into account sample size.

The best value of $e_{[h]}^2\{\hat{\beta}(h)\}$ in (3.44) in our implementation is found as

$$\tilde{\sigma}_{\text{LMS}} = \frac{1}{\Phi^{-1}(0.5 + h/(2n))}\{1 + 5/(n - p)\} \left\{e_{[h]}^2(\hat{\beta}_{\text{LMS}})\right\}^{0.5}. \tag{3.46}$$

We declare as an outlier any observation i for which the absolute scaled residual

$$|e_{\text{LMS},i}^S| = |e_i(\hat{\beta}_{\text{LMS}})|/\hat{\sigma}_{\text{LMS}} > \Phi^{-1}(1 - \alpha^*) \quad (i = 1, \ldots, n). \tag{3.47}$$

There are several disadvantages to least median of squares. The original proposal has a *bdp* of 50% so can be expected to be extremely inefficient if there are few, or no, outliers. It displays a marked sensitivity to central data Shertzer and Prager (2002), that is, it is not locally stable. Moreover, Davies (1990) shows that $\hat{\beta}_{\text{LMS}}$ converges to β at the slow rate of $n^{-1/3}$. We mentioned in Sect. 3.7.2 that estimates based on a smooth ρ function have the usual convergence rate of $n^{-1/2}$. Some discussion of the asymptotic properties of LMS is in Rousseeuw and Leroy (1987, Sect. 4.4). Pronzato and Pázman (2013, Sect. 3.6) provide further references on convergence at the rate of $n^{-1/3}$. Rousseeuw and Van Driessen (2006) comment that "Nowadays we think that the LMS estimator should be replaced by the ... LTS estimator". The LMS estimator can however be used as a starting point for more efficient estimators. We do, indeed, use the LMS estimator to provide a starting point for our FS algorithm. Monitoring results in Fig. 4.20 show how unstable LMS estimates can be compared to those from procedures that converge as $n^{-1/2}$.

3.11.2 LTS—Least Trimmed Squares and the Consistency Correction for Variances

As opposed to minimizing the median squared residual, in LTS we find $\hat{\beta}_{\text{LTS}}$ to

$$SS_{\text{LTS}} = \min_{\hat{\beta}} \ SS\{\hat{\beta}(h)\} = \sum_{i=1}^{h} e_{[i]}^2 \{\hat{\beta}(h)\}, \ i \in S_1, \quad (3.48)$$

where S_1 is the set of the h smallest squared residuals. For any subset j of size h the parameter estimates $\hat{\beta}(h)$ are straightforwardly estimated by least squares. Let the minimum value of (3.48) be $SS_T(\hat{\beta}_{\text{LTS}})$. We base the estimator of σ^2 on this residual sum of squares. However, since the sum of squares contains only the central h observations from a normal sample, the estimate needs scaling. The variance of the truncated normal distribution containing the central h/n portion of the full distribution is

$$\sigma_T^2(h) = 1 - \frac{2n}{h} \Phi^{-1}\left(\frac{n+h}{2n}\right) \phi \left\{ \Phi^{-1}\left(\frac{n+h}{2n}\right) \right\}, \quad (3.49)$$

where $\phi(.)$ and $\Phi(.)$ are, respectively, the standard normal density and cdf (see Croux and Rousseeuw, 1992, Eq. (6.5), or the results of Tallis, 1963 on elliptical truncation). To estimate σ^2 we accordingly take

$$\hat{\sigma}_{\text{LTS}}^2 = SS_T(\hat{\beta}_{\text{LTS}})/\{h \times \sigma_T^2(h)\}. \quad (3.50)$$

Outliers are found as in (3.47) but with LTS parameter estimates.

The preceding consistency factor is not sufficient to make the LTS estimate of scale unbiased for small samples. Pison et al. (2002) use 82 simulation and curve fitting to obtain improved correction factors for LTS.

A simple LTS algorithm has the same structure as that for LMS in Sect. 3.11.1. That is, parameter estimates are obtained from elemental subsets. These estimates provide n residuals from which $nsamp$ values of $SS_{LTS}\{\hat{\beta}(h)\}$ are calculated. The minimizing subset provides the parameter estimates in (3.48). This algorithm can be slow as the number of observations increases. Rousseeuw and Van Driessen (2006) introduced an improved "fast" algorithm with increased speed and several refinements. The most important of these is the concentration step, which is similar to the step used in the FS in moving from a subset of size m to one of $m + 1$; parameter estimates from a subset of observations provide residuals for all n observations. The next subset is chosen based on properties of these residuals. Here the new subset is of size h.

The algorithm has some structure in common with that of Sect. 3.10, but is simpler, since iterations are not needed to estimate σ^2.

1. Initialization. We start with a set of n observations from which we extract $nsamp$ subsets each of size p (generally $nsamp$ is the minimum of $\binom{n}{p}$ and 1000).
2. Compute $\hat{\beta}_{LS}$ for each subset.
3. Initial concentration step for each subset. For subset j, use the value of $\hat{\beta}_{LS}$ to calculate n residuals $y - X\hat{\beta}_{LS}$ as the starting point of the iterative refinement procedure.
4. Refining steps for each subset. From the n residuals find the subset S_1 of h smallest squared residuals, estimate β, and calculate a new set of n residuals.
   Repeat this a small number of times; typically *refsteps* is set to 3.
5. Final iterations to convergence. Out of the *nsamp* values of refined $SS_{\text{LTS}}(\hat{\beta}_j)$, take the five smallest and iterate to convergence (*refsteps* is set to 50).
   The j giving the smallest value of $SS_{\text{LTS}}(\hat{\beta}_j)$ provides the subset of h observations from which $\hat{\beta}_{\text{LTS}}$ and $\hat{\sigma}^2_{\text{LTS}}$ are calculated using LS.

3.12 MM-Estimation, τ-Estimators, and Reweighted LMS and LTS Estimators

3.12.1 MM-Estimators

We have repeatedly emphasized (e.g. Sect. 2.4.1, Sect. 3.9) the asymptotic relationship between *bdp* and *eff* for M-estimators; as one increases, the other decreases. In an attempt to break this relationship, (Yohai, 1987) introduced MM-estimation, which extends S-estimation. In the first stage, the breakdown point of the scale estimate is set at 0.5, thus providing a high breakdown point. This fixed estimate is then used in the estimation of β for which a new value of the tuning constant

3.12 MM-Estimation, τ-Estimators, and Reweighted LMS and LTS Estimators

c can be chosen to provide an estimator of β with a high efficiency. Maronna et al. (2019, p. 130), recommend a value of 0.85 for this efficiency. We show results of monitoring MM-estimates in Sect. 4.9.2 for a range of values of *eff*.

3.12.2 τ-Estimators

The final estimator of the M-class that we consider is an extension of S- and MM-estimation introduced by (Yohai and Zamar, 1988). In τ-estimation, unlike MM-estimation, there is no global precalculated estimate of scale. In the general procedure, the function ρ_0 used to estimate scale is chosen to give the maximum breakdown point for regression estimates. On the other hand, the function ρ_1 used for estimation of β is chosen to give high efficiency. More precisely, if $\hat{\sigma}(\beta)$ solves the usual scale equation

$$\frac{1}{n}\sum_{i=1}^{n}\rho_{c_0}\left(\frac{e_i(\hat{\beta})}{\hat{\sigma}(\hat{\beta})}\right) = K_{c_0}$$

for a given bounded function ρ_{c_0}, where we write ρ_{c_0} to stress that this ρ function depends on a tuning parameter c_0 which controls the *bdp*, the scale τ is defined as

$$\tau(e^2) = [\hat{\sigma}\{e(\hat{\beta})\}]^2 \frac{1}{n}\sum_{i=1}^{n}\rho_{c_1}\left(\frac{e_i(\hat{\beta})}{\hat{\sigma}(\hat{\beta})}\right),$$

where ρ_{c_1} is another bounded ρ function which depends on a tuning constant linked to the efficiency. A regression τ-estimate is defined by

$$\hat{\beta}_\tau = \min_{\beta} \tau\{e^2(\beta)\},$$

where $e(\beta) = (e_1(\beta), \ldots, e_n(\beta))^T$. Both β and σ are iteratively estimated in an alternating manner as described in Sect. 3.10. The resulting weights in the iterative reweighted least squares algorithm are a linear combination of the ρ_{c_0} and ρ_{c_1}. The hope is that, by an adequate choice of the two ρ functions, the estimate can have the maximum possible *bdp* and can be made arbitrarily close to the LS estimate and therefore arbitrarily efficient at the normal distribution. However, the justification relies again on an asymptotic argument with very remote outliers. Sometimes a value as high as 0.95 is suggested. It is important to stress that these are asymptotic values for extremely well-separated data; less fortunate forms of data can give rise, for example, to biased estimates. In Table 4.2, our analysis of the AR regression data reveals that analyses with *eff*= 0.95 all collapse to least squares.

3.12.3 LMS and LTS Reweighted

Versions of the fast algorithm of Sect. 3.11.2 all include fitting to subsets of asymptotic size $n/2$. They thus provide consistent estimates of the parameters and overcome the objections to reliance on elemental subsets voiced by Hawkins and Olive (2002). But, since the estimates are based on half the data, they will have an unnecessarily low value of efficiency when there are fewer outliers; a *bdp* of 50% will be too protective.

To increase efficiency, reweighted versions of the LMS and LTS estimators can be computed. These reweighted estimators are found by giving weight zero to observations which conditions (3.17) or (3.18) suggest are outliers. We then obtain a sample of reduced size $n - k \geq h$, possibly outlier free, to which LS is applied.

More precisely, in the reweighted version of the LTS or LMS estimator (LTSr, LMSr) we first set $h = [(n + p + 1)/2]$, for which preliminary and very robust estimates of β and σ, say $\hat{\beta}^*$ and $\hat{\sigma}^*$, are computed, respectively, through (3.44) and (3.48). Note also that $\hat{\sigma}^*$ is the estimate of σ after applying both the consistency factor and the small-sample correction.

In the second stage, we obtain the final parameter estimates by applying a weighted LS approach, with weights:

$$w_i = \begin{cases} 1 & \text{if } \left(\frac{e_i^*}{\hat{\sigma}^*}\right)^2 \leq \chi^2_{1,1-\alpha} \\ 0 & \text{if } \left(\frac{e_i^*}{\hat{\sigma}^*}\right)^2 > \chi^2_{1,1-\alpha}, \end{cases} \quad (3.51)$$

where $e_i^* = y_i - \hat{y}_i^*$, $\hat{y}_i^* = x_i^T \hat{\beta}^*$ and $\chi^2_{1,\gamma}$ denotes the γ-th quantile of the χ^2_1 distribution. It is customary to use $1 - \alpha$ set to 0.975. Clearly this weighting, depending on the dataset, may lead to the exclusion of good observations and the inclusion of outliers in the subset of supposedly clean data, especially when the two groups are close or partially overlapping.

3.13 Traditional Robust Analysis for the AR Regression Data

The methods of robust regression described in this chapter require the specification of the amount of robustness required, either through the value of *bdp*, or through the value of *eff*. The purpose of the analysis of the AR regression data in this section is to illustrate how important this specification is.

We start in Fig. 3.8 with S-estimation combined with Tukey's biweight ρ function. In the upper panel *bdp* = 0.25. Here only observation 43 lies outside the 99% band. This was the one observation that was shown as mildly outlying in the plots based on LS of Sect. 3.5. The lower panel of Fig. 3.8 is for a *bdp* of 0.5, that is, very robust regression. Now there are six outliers well above the 99% confidence band. These are

3.13 Traditional Robust Analysis for the AR Regression Data 93

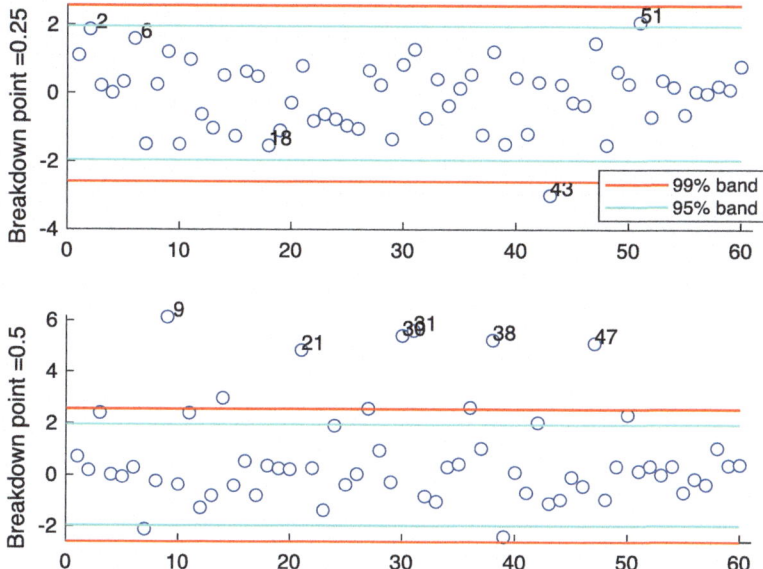

Fig. 3.8 AR regression data: S-estimation. Index plot of residuals for two values of *bdp*, Tukey's biweight ρ function. Upper panel, *bdp* = 0.25; lower panel, *bdp* = 0.5

the six masked outliers highlighted in the added variable plot of Fig. 3.4. There are also two marginally outlying observations. The effect of the increase in the specified value of *bdp* has been to produce a very different analysis, which is no longer like that from LS. An important question is whether there is a value of *bdp* less than 0.5 which will identify just the six outliers. If so, this robust regression will produce parameter estimates with improved efficiency compared to those with *bdp* = 0.5.

Figure 3.9 provides index plots of residual plots from MM-estimation, again using Tukey's biweight. In the upper panel *eff* = 0.9. There are six outliers well above the upper 99% confidence band, all of which are the six masked outliers. In the lower panel, *eff* is increased to 0.95. The effect is again to identify observation 43 as mildly outlying, but now with observation 51 just significantly outlying.

The conclusion from these analyses is that a too mild choice of downweighting parameter leads to the LS analysis. The question then is whether there is another value, not as extreme as very robust regression, that does not collapse to least squares, but which gives efficient parameter estimates while identifying all outliers. Monitoring plots in Sect. 4.9.3 for several methods of robust regression clearly show a transition from the identification of the six outliers to LS and the identification of just observation 43 as outlying. A value of *bdp* or *eff* close to the robust side of the transition point provides the best robust analysis for the particular combination of estimation method and ρ function. Table 4.2 shows that some combinations may fail to provide a robust analysis.

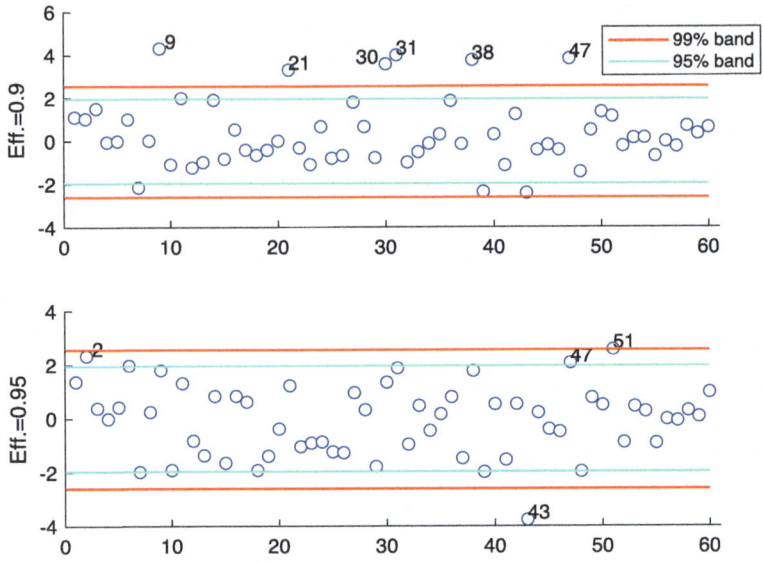

Fig. 3.9 AR regression data: MM-estimation. Index plot of residuals for two values of *eff*, Tukey's biweight ρ function. Upper panel, *eff* = 0.9; lower panel, *eff* = 0.95

Problems

3.1 Relationship between $\hat{\beta}_{(i)}$ and $\hat{\beta}$
Show that:
$$\hat{\beta}_{(i)} - \hat{\beta} = -(X^T X)^{-1} x_i e_i / (1 - h_i). \tag{3.52}$$

3.2 Relationship between $s_{(i)}^2$ and s^2
(a) Show that the change in residual sum of squares on deletion of observation i is $e_i^2/(1 - h_i)$. In other words, show that

$$(n - p - 1)s_{(i)}^2 = (n - p)s^2 - e_i^2/(1 - h_i). \tag{3.53}$$

(b) Explain why $\hat{\beta}_{(i)}$ and $s_{(i)}^2$ are independent of y_i.

3.3 Relationship between deletion residual r_i^* and studentized residual $\tilde{r}_i$
Show that:
$$s_{(i)} r_i^* = \frac{y_i - x_i^T \hat{\beta}_{(i)}}{\sqrt{1 + x_i^T (X_{(i)}^T X_{(i)})^{-1} x_i}} = \frac{y_i - \hat{y}_i}{\sqrt{1 - h_i}} = s\tilde{r}_i. \tag{3.54}$$

3.4 Interpretation of $h_i/(1 - h_i)$
Show that the quantity $h_i/(1 - h_i)$, the ratio of variance of the ith predicted value

Problems

$(\text{var}(\hat{y}_i) = \sigma^2 h_i)$ to the variance of the ith ordinary residual $\{\text{var}(e_i) = \sigma^2(1 - h_i)\}$, can be interpreted as the ratio of the part of $\hat{y}_i$ due to y_i to the part due to the predicted value $x_i^T \hat{\beta}_{(i)}$; that is, show that

$$\hat{y}_i = (1 - h_i)x_i^T \hat{\beta}_{(i)} + h_i y_i.$$

3.5 r_i^* and mean shift outlier model
Show that r_i^* can be interpreted as the t-statistic for testing the significance of the ith unit vector d_i in the following model (mean shift outlier model) Sect. 3.2.3:

$$E(Y) = X\beta + d\theta. \tag{3.55}$$

Comment on the distribution of r_i^*.

3.6 Find s_w^2 in the added variable test
Prove Eq. (3.23) (Sect. 3.4).

$$(n - p - 1)s_w^2 = y^T y - \hat{\beta}^T X^T y - \hat{\gamma} w^T y = y^T A y - (y^T A w)^2 / (w^T A w). \tag{3.56}$$

3.7 Comparison of the distributions of squared deletion residuals and squared studentized residuals
Compare graphically the density of the squared deletion residual and the squared studentized residual in the interval (0 0.6] for $n = 50$ and $p = 5$ under the normal regression model.

Show that the difference between the two residuals is most acute if there is a single outlier at a point of high leverage. See Eq. (3.12).

3.8 Value of γ in Sect. 3.7.2 when the data come from the normal distribution
Show that asymptotically:

$$\gamma = \frac{A}{B^2} = \frac{E\psi(r_i^2)}{\{E\psi'(r_i)\}^2} \to 1, \tag{3.57}$$

when the data come from the normal distribution (under the assumption of using a consistent estimator of σ).

3.9 Computation of the tuning constant c for the Tukey biweight
Show that the constant c in the Tukey biweight ρ function for the location estimator which produces a given breakdown point (bdp) must satisfy the following equality:

$$bdp = \frac{3}{c^2}\left\{{}_c\Psi_3 - 3\frac{{}_c\Psi_5}{c^2} + 5\frac{{}_c\Psi_7}{c^4} + \frac{c^2}{3}(1 - {}_c\Psi_1)\right\}, \tag{3.58}$$

while the tuning constant c necessary to obtain a fixed efficiency *eff* must satisfy the equality

$$\textit{eff} = \frac{\left(15\frac{c\Psi_5}{c^4} - 6\frac{c\Psi_3}{c^2} + {}_c\Psi_1\right)^2}{945\frac{c\Psi_{11}}{c^8} - 420\frac{c\Psi_9}{c^6} + 90\frac{c\Psi_7}{c^4} - 12\frac{c\Psi_5}{c^2} + {}_c\Psi_3}, \qquad (3.59)$$

where ${}_{[a,b]}\Psi_k = \Pr(\chi_k^2 < b^2) - \Pr(\chi_k^2 < a^2)$ and ${}_{[0,b]}\Psi_k = {}_b\Psi_k = \Pr(\chi_k^2 < b^2)$.

3.10 First truncated moment in folded normal distribution

Starting from (3.41) show that

$$E(|X|) = \int_{a<|x|<b} |x| d\Phi(x) = 2\left(\phi(a) - \phi(b)\right),$$

where $\phi(b)$ is the density function of the standard normal evaluated at b. If $a = 0$ and $b = \infty$ what is the value of $E(|X|)$?

3.11 Equality of the truncated mean in a folded normal distribution to the MAD in a standard normal distribution

Find the tuning constant c for which the truncated mean defined in the interval $[0\ c]$ in a folded normal distribution is equal to the MAD in a $\mathcal{N}(0, 1)$.

3.12 Test whether the current solution in Sect. 3.10 is an admissible candidate to find the minimum scale

Suppose that during the subsampling algorithm we have already examined $j - 1$ subsets and that $\hat{\sigma}_{\min}$ is the current minimum. When we draw the j-th subset which, after the iterative algorithm described in Sect. 3.7.1, yields the candidate solution $\hat{\beta}_{rw}(j)$, and the corresponding residuals $r_{rw}(j) = y - X\hat{\beta}_{rw}(j)$ show that if

$$\overline{\rho}\{r_{rw}(j)/(c\hat{\sigma}_{\min})\} \geq K, \qquad (3.60)$$

we may discard $\hat{\beta}_{rw}(j)$ because $\hat{\sigma}_j \geq \hat{\sigma}_{\min}$.

Open Access This chapter is licensed under the terms of the Creative Commons Attribution 4.0 International License (http://creativecommons.org/licenses/by/4.0/), which permits use, sharing, adaptation, distribution and reproduction in any medium or format, as long as you give appropriate credit to the original author(s) and the source, provide a link to the Creative Commons license and indicate if changes were made.

The images or other third party material in this chapter are included in the chapter's Creative Commons license, unless indicated otherwise in a credit line to the material. If material is not included in the chapter's Creative Commons license and your intended use is not permitted by statutory regulation or exceeds the permitted use, you will need to obtain permission directly from the copyright holder.

Chapter 4
The Monitoring Approach in Multiple Regression

Abstract This chapter introduces algorithms for the monitoring approach to robust statistical analysis and compares their performance in the analyses of several sets of data. The Forward Search (FS) is introduced in Sect. 4.1. This algorithm fits subsets of the data of increasing size in such a way that the most outlying observations are included towards the end of the search. To monitor other forms of robust regression, we estimate the parameters using, typically, a grid of 50 values of *bdp* or *eff*. Monitoring plots of residuals, parameter estimates, and t-tests from very robust regression to LS are introduced in Sect. 4.6; these plots are enriched by brushing and linking to other plots. Frequently, the plots from monitoring residuals show an abrupt change from robust to LS analyses. In Sect. 4.7, we introduce the empirical *bdp* defining the most efficient robust estimator for each dataset, thus overcoming the arbitrariness of the conventional approach to robust statistics. Many examples are given in Sect. 4.9 for a variety of estimators. The chapter concludes in Sect. 4.11 with a generalized Bayesian Information Criterion (BIC) for model choice which, via the mean shift outlier model, makes it possible to compare models in which different numbers of observations have been deleted.

4.1 The Forward Search

The Forward Search (FS) is a robust adaptive trimming method based on least squares regression. It has already been briefly mentioned in Sects. 1.5 and 3.6. The purpose of adaptive trimming methods is to select the largest outlier free subset of observations, which then provides unbiased parameter estimates of minimum variance. Outliers, if any, enter the subset towards the end of the search, which provides an ordering of the observations from those closest to the fitted model to those most remote. This important ordering can be used to monitor how properties of the fitted model, such as parameter estimates, residuals, and significance tests, change as the parameters of the model are estimated from an increasing number of observations. Particular interest is often in the effect of outliers on quantities of inferential interest.

The robust methods detailed in Chap. 3 nearly all provide a single analysis of the data for a specified value of *bdp*, *eff* or, equivalently, of h. The exceptions are the reweighted procedures LMSr and LTSr, which can provide two analyses. At the end of Sect. 1.5, we contrasted this traditional approach to robust statistics, based on a single "snapshot" of the data, to the monitoring approach where we can watch a "film" or "movie" of the data as the robustness of the fit changes. Although the idea of monitoring was initially developed with the FS, in this chapter, we follow Riani et al. (2014a) and also apply it to the robust procedures, such as S-estimation, of Chap. 3. Instead of just one robust analysis, we fit 50 robust regressions plus LS, going from most robust ($bdp = 50\%$) to LS ($bdp = 0$), or with *eff* going from 0.5 to 1. In Sect. 4.6, the results of monitoring are presented as linked and brushed plots.

The forward search fits subsets of observations of size m to the data, with $m_0 \leq m \leq n$. Let $S^*(m)$ be the subset of size m found by the forward search, for which the matrix of regressors is $X(m)$. Least squares on this subset of observations yields parameter estimates $\hat{\beta}(m)$ and $s^2(m)$, the mean square estimate of σ^2 on $m - p$ degrees of freedom. Residuals can be calculated for all observations including those not in $S^*(m)$. The n resulting least squares residuals are $e_i(m) = y_i - x_i^T \hat{\beta}(m)$. The search moves forward with the augmented subset $S^*(m + 1)$ consisting of the observations with the $m + 1$ smallest absolute values of $e_i(m)$.

To start we take $m_0 = p$ and search over subsets of p observations to find the subset that yields the LMS or LTS estimate of β. However, this initial estimator is not important, provided masking is broken. Our computational experience is that randomly selected starting subsets also yield indistinguishable results over the last one-third of the search, provided there is a single unambiguous model generating the data. Figure 4.12 shows an example, including outliers. This result does not hold if there is an appreciable number of structured outliers. Section 8.5.2 describes a specific form of Random Start FS in which the forward searches from a large number of starting points, perhaps 1,000, are run to completion, that is, to $m = n$. Then, for example, if the data come from a mixture of two distinct regression models, the random starts lead, in the middle of the search to at least two distinct sets of trajectories, which must then converge by the end of the search.

To test for outliers the deletion residual (3.15) is calculated for the $n - m$ observations not in $S^*(m)$. These residuals, which form the maximum likelihood tests for the outlyingness of individual observations, are

$$r_i^*(m) = \frac{y_i - x_i^T \hat{\beta}(m)}{\sqrt{s^2(m)\{1 + h_i(m)\}}} = \frac{e_i(m)}{\sqrt{s^2(m)\{1 + h_i(m)\}}}, \quad i \notin S^*(m),$$

where the leverage $h_i(m) = x_i^T \{X(m)^T X(m)\}^{-1} x_i$. Let the observation not in $S^*(m)$ nearest to those forming $S^*(m)$ be $i_{\min}$ where

$$i_{\min} = \arg \min_{i \notin S^*(m)} |r_i^*(m)|.$$

To test whether observation $i_{\min}$ is an outlier we use the absolute value of the minimum deletion residual

4.2 Testing for Outliers

$$r_{\min}(m) = \frac{e_{i\min}(m)}{\sqrt{s^2(m)\{1+h_{i\min}(m)\}}}, \qquad (4.1)$$

as a test statistic. If the absolute value of (4.1) is too large, the observation $i_{\min}$ is considered to be an outlier, as well as all the other, more remote, observations not in $S^*(m)$.

4.2 Testing for Outliers

The test statistic (4.1) is the $(m+1)$st-ordered value of the absolute deletion residuals. We can therefore use distributional results to obtain envelopes for our plots. The argument parallels that of Riani et al. (2009) where envelopes were required for the Mahalanobis distances arising in applying the FS to multivariate data.

Let $Y_{[m+1]}$ be the $(m+1)$st-order statistic from a sample of size n from a univariate distribution with cdf $G(y)$. Then the c.d.f of $Y_{[m+1]}$ is given exactly by

$$P\{Y_{[m+1]} \leq y\} = \sum_{j=m+1}^{n} \binom{n}{j} \{G(y)\}^j \{1-G(y)\}^{n-j}. \qquad (4.2)$$

See, for example, Lehmann (1991, p. 353). We then apply properties of the beta distribution to the RHS of (4.2) to obtain

$$P\{Y_{[m+1]} \leq y\} = F_B(G(y), m+1, n-m), \qquad (4.3)$$

where $F_B(y; \nu_1, \nu_2)$ is the cdf of the beta distribution with parameters ν_1 and ν_2. From the relationship between the F and the beta distributions (4.3) becomes

$$P\{Y_{[m+1]} \leq y\} = P\left\{\mathcal{F}_{2(n-m),2(m+1)} > \frac{1-G(y)}{G(y)} \frac{m+1}{n-m}\right\},$$

where $F_{2(n-m),2(m+1)}$ is the F distribution with $2(n-m)$ and $2(m+1)$ degrees of freedom (Guenther, 1977). Thus, the required quantile of order γ of the distribution of $Y_{[m+1]}$ say $y_{m+1,n;\gamma}$ is obtained as

$$y_{m+1,n;\gamma} = G^{-1}(q) = G^{-1}\left(\frac{m+1}{m+1+(n-m)x_{2(n-m),2(m+1);1-\gamma}}\right), \qquad (4.4)$$

where $x_{2(n-m),2(m+1);1-\gamma}$ is the quantile of order $1-\gamma$ of the F distribution with $2(n-m)$ and $2(m+1)$ degrees of freedom.

In our case, we are considering the absolute values of the deletion residuals. If the c.d.f of the t distribution on ν degrees of freedom is written as $T_\nu(y)$, the absolute value has the cdf

$$G(y) = 2T_\nu(y) - 1, \qquad 0 \le y < \infty.$$

The required quantile of $Y_{[m+1]}$ is given by

$$y_{m+1,n;\gamma} = T^{-1}_{m-p}\{0.5(1+q)\},$$

where q is defined in (4.4). To obtain the required quantile, we calculate an inverse of the F and then an inverse of the t distribution.

Because the estimate of σ^2 is based on the central m/n proportion of the observations we need to apply the consistency correction given in (3.49). Since the outlier tests we are monitoring are divided by an estimate of σ^2 that is too small, we need to scale up the values of the order statistics to obtain the envelopes

$$y^*_{m+1,n;\gamma} = y_{m+1,n;\gamma}/\sigma_T(m).$$

To be specific, in the case of the 99% envelope, $\gamma = 0.99$ corresponds to a nominal pointwise size $\alpha = 1 - \gamma$ which is equal to 1%. We expect, for the particular step m which is considered, to find exceedances of the quantile in a fraction 1% of the samples under the null normal distribution. We however require a sample-wise probability of 1% of the false detection of outliers, that is, over all values of m considered in the search. The algorithm in the next section is accordingly designed to have a size of 1%.

4.3 The Forward Search for Regression Outlier Detection

We require appropriate bounds for the outlier test (4.1). Because we are testing for the existence of an outlier at each step of the search, we have to allow for the effect of simultaneous testing. If there are a few large outliers they will enter at the end of the search, as in Sect. 4.5.1, and their detection is not a problem. However, even relatively small numbers of outliers can be difficult to identify and may cause peaks well before the end of the search. Masking may then cause the plot to return inside the envelopes at the end of the search, as in Fig. 4.30. Methods of using the forward search for the formal detection of outliers have to be sensitive to these two patterns—a few "obvious" outliers at the end and a peak earlier in the search caused by a cluster of outliers.

To use the envelopes in the forward search for outlier detection we accordingly propose a two-stage process. In the first stage, we run a search on the data, monitoring the bounds for all n observations until we obtain a "signal" indicating that observation $m^\dagger$, and therefore succeeding observations, may be outliers, since the value of the statistic lies beyond our threshold. In the second part, we superimpose envelopes for values of n from this point until the first time we introduce an observation we recognize as an outlier. The conventional envelopes shown, for example, in Figs. 4.2 and 4.24 consist roughly of two parts; a flat "central" part and a steeply curving "final" part. Our procedure FS for the detection of a "signal" takes account of these two parts and is similar to the rule used by Riani et al. (2009) for the detection of multivariate outliers. In our definition of the detection rule, we use the nomenclature $r_{\min}(m, n^*)$ to denote that we are comparing the value of $r_{\min}(m)$ with envelopes from a sample of size n^*.

1. Detection of a Signal

There are four conditions, the fulfilment of any one of which leads to the detection of a signal.

- In the central part of the search, we require three consecutive values of $r_{\min}(m, n)$ above the 99.99% envelope or 1 above 99.999%.
- In the final part of the search, we need two consecutive values of $r_{\min}(m, n)$ above 99.9% and 1 above 99%.
- $r_{\min}(n - 2, n) > 99.9\%$ envelope.
- $r_{\min}(n - 1, n) > 99\%$ envelope. In this case, a single outlier is detected and the procedure terminates.

The final part of the search is defined as

$$m \geq n - \left[13\,(n/200)^{0.5}\right],$$

where here [] stands for rounded integer. For $n = 200$ the value is slightly greater than 6% of the observations.

2. Confirmation of a Signal

The purpose of, in particular, the first point is to distinguish informative peaks from random fluctuations in the centre of the search. Once a signal takes place (at $m = m^\dagger$) we check whether the signal is informative about the structure of the data. If $r_{\min}(m^\dagger, m^\dagger)$ is less than the 1% envelope, we decide the signal is not informative, increment m and return to Step 1.

3. Identification of Outliers

With an informative signal, we start superimposing 99% envelopes taking $n^* = m^\dagger - 1, m^\dagger, m^\dagger + 1, \ldots$ until the final, penultimate, or antepenultimate value is above the 99% threshold or, alternatively, we have a value of $r_{\min}(m, n^*)$ for any $m > m^\dagger$ which is greater than the 99.9% threshold. Examples are in Figs. 4.4 and 4.6.

4.4 Graphical Presentation

Although we have adjusted our decision rule to give the desired samplewise error rate, the check of whether the value of $r_{\min}$ is above or below a threshold is pointwise for each m. We can therefore transform the vertical scale at each m without changing the rule.

We have already mentioned the curved shape of the conventional envelopes obtained by Riani and Atkinson (2007). For ease of reading the graphs produced by the FS algorithm, we can use the normal probability transformation to straighten the envelopes. This is a natural choice, although of course not unique, since values around 2 and 2.6 have a familiar interpretation for those accustomed to statistical analyses.

Let the level of some envelope be γ^*. Then the normal probability transformation yields the value $\Phi^{-1}(\gamma^*)$. To calculate the confidence level for the observed value of $r_{\min}(m)$ we invert (4.4) and (4.5) to obtain the confidence level γ as

$$\gamma = 1 - F_{2(n-m),2(m+1)} \left\{ (m+1) \left[\frac{1}{2T_{m-p}\{r_{\min}(m)\sigma_T(m)\}} - 1 \right] \frac{1}{n-m} \right\}, \quad (4.5)$$

for $m = m_0, m_0 + 1, \ldots, n - 1$. Here F and T are the cdfs of the F and t distributions. The plot in normal coordinates uses $\Phi^{-1}(\gamma)$, so improving the clarity of the figure.

4.5 Two Examples

4.5.1 Hertzsprung–Russell Diagram of Star Cluster CYG OB1 (Stars Data)

As a first illustration of our procedure, we use a standard example in which the outliers are easily found. The Hertzsprung–Russell star data were introduced into the statistical literature on robustness by Rousseeuw and Leroy (1987, p. 27). There are 47 observations with a single explanatory variable, the logarithm (to base 10) of the effective temperature at the surface of the stars in kelvins. The response is the logarithm of the light intensity. The data are plotted in Fig. 4.1. The figure shows that there are four outliers clearly separated from the body of the data. The conclusion of the comparisons of Riani et al. (2014d) is that this canonical form of data is one for which many methods of very robust regression behave well. We return to further analysis in Sect. 4.9.2.

Figure 4.2 shows the forward plot of minimum deletion residuals. This is the conventional plot in the scale of residuals. Figure 4.3 repeats the plot with the normal-scores transformation (4.5). Both plots, of course, give the same statistical information: there is a signal at $m = m^{\dagger} = 43$ since $r_{\min}(43, 47)$ is greater than the 99.999% point of the envelope. Although the statistical information is the same, we find the graphical representation in Fig. 4.3 to be clearer than that of Fig. 4.2. We now investigate this first indication of the presence of outliers by looking at the superimposition plots in Fig. 4.4.

Since there is a signal at $m = 43$, we start the process of resuperimposing envelopes at $m = 42$. Figure 4.4 shows the resuperimposed envelopes for $n^* =$

4.5 Two Examples

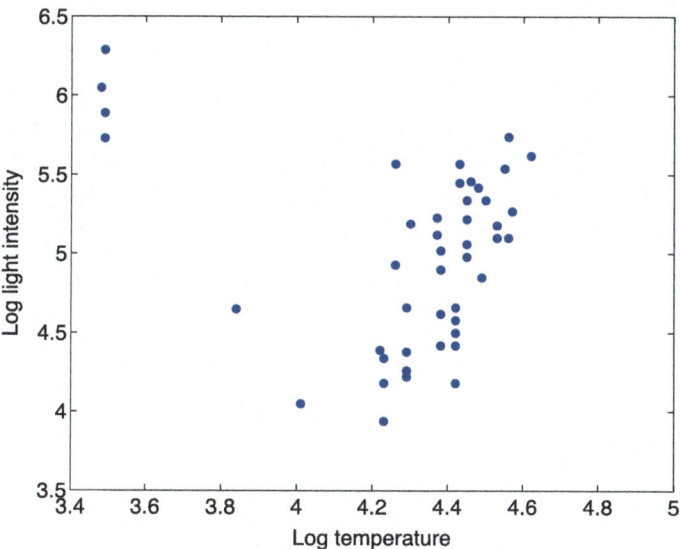

Fig. 4.1 Stars data: scatter plot of the 47 observations

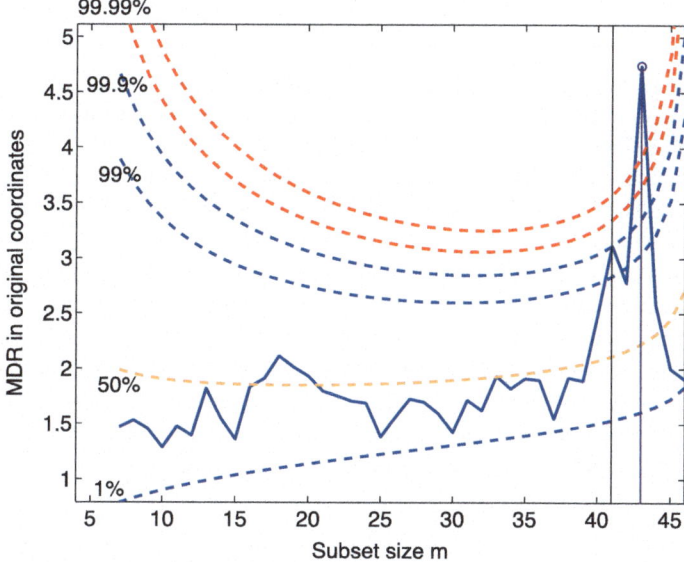

Fig. 4.2 Stars data: conventional forward plot of minimum deletion residuals with envelopes in the scale of residuals. The signal is in step $m = 43$. The vertical line at $m = 41$ is the division between the central and final parts of the search as defined in the algorithm

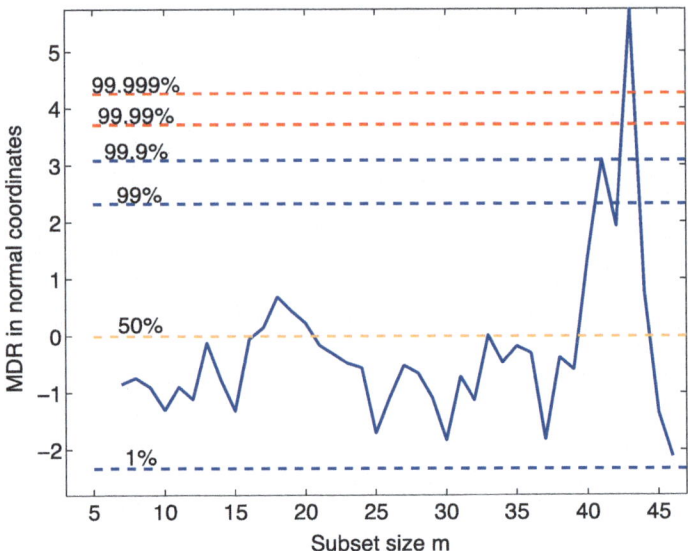

Fig. 4.3 Stars data: forward plot of minimum deletion residuals with envelopes. Normal-scores transformation. The signal is in step $m = 43$

42, 43, and 44. We stop the process at $n^* = 44$ since $r_{\min}(44, 44)$ lies above the 99% envelope and so is an outlier. There are therefore 43 good observations available for fitting the model and for statistical inference and four outliers.

The four identified outliers are those in the upper left-hand part of Fig. 4.1. There was no difficulty in identifying them and, in this straightforward example, the initial signal at $m = 43$ correctly indicated that the last four observations to enter the search are outlying. Note however how the plot in Fig. 4.3 returns within the central envelopes as the outliers enter the subset used in fitting; the parameter estimates are so distorted that the outliers no longer appear so. When $m = 46$ the last observation to enter has a small deletion residual. The least squares residuals when all observations are used in fitting will have an even smaller maximum and there will be no indications of any outliers. The zero breakdown point of least squares leads to complete masking of these clustered outliers.

4.5.2 Wool Data

We now analyse a small set of 27 observations taken from Box and Cox (1964) that require transformation. We show that, despite the small value of n, the forward regression procedure identifies many outliers in the untransformed data—the good properties of the procedure are not solely for large samples. When the data are correctly transformed, we do not identify any outliers and all observations are used

4.5 Two Examples

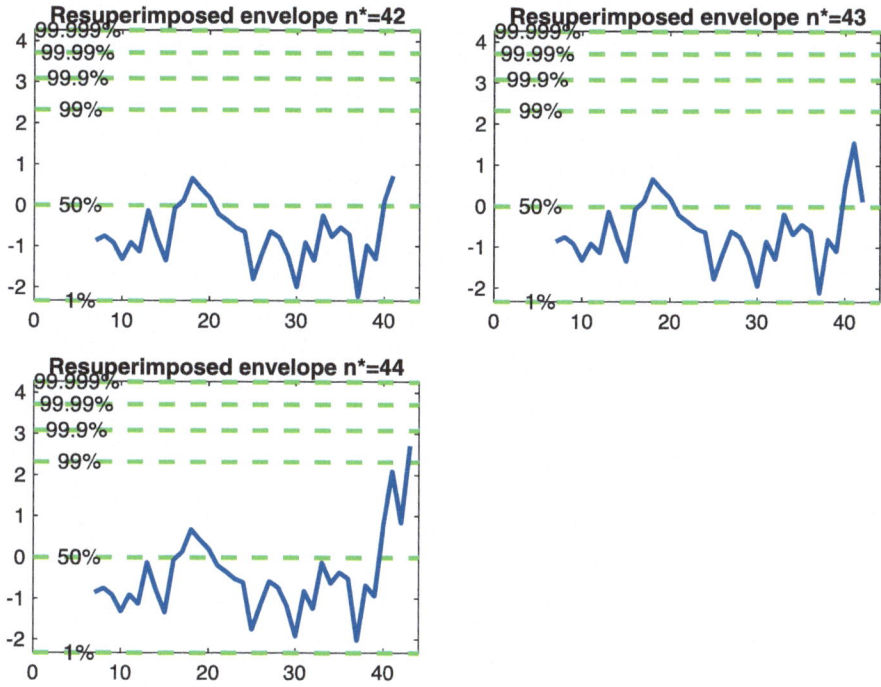

Fig. 4.4 Stars data: resuperimposed envelopes for steps 42, 43, and 44. Normal-scores transformation. In this process, the first outlier is detected at $n^* = 44$ so there are 43 observations available for estimation and 4 outliers

in fitting. The outliers found in the analysis of the untransformed data are not artefacts of our method of analysis.

The data are the number of cycles to failure of a worsted (woolen) yarn under cycles of repeated loading. The results are from a single 3^3 factorial experiment. Box and Cox (1964) recommend analysis on the logarithmic scale for y. The forward analysis of Atkinson and Riani (2000, Sect. 4.3) using the forward search supports this conclusion.

Figure 4.5 shows the forward plot of the minimum deletion residuals in normal coordinates. A signal is received at $m = m^\dagger = 17$ because $r_{\min}(17, 27)$ is greater than the 99.999% envelope. The superimpositions of envelopes are shown in Fig. 4.6. We start with $n^* = 16$. The first outlier occurs at $n^* = 19$ so that there are 18 good observations for parameter estimation and 7 outliers.

Figure 4.7 shows scatter plots of y against the three explanatory variables (a yX plot), with the outliers numbered and plotted as circles. The forward search regression has identified the nine largest observations as outliers. Because the data need transforming these observations all seem remote from the main body of the data.

The structure of these forward plots, with a higher proportion of outliers, is more complicated than those for the stars data. Here in Fig. 4.5 the plot has a second peak

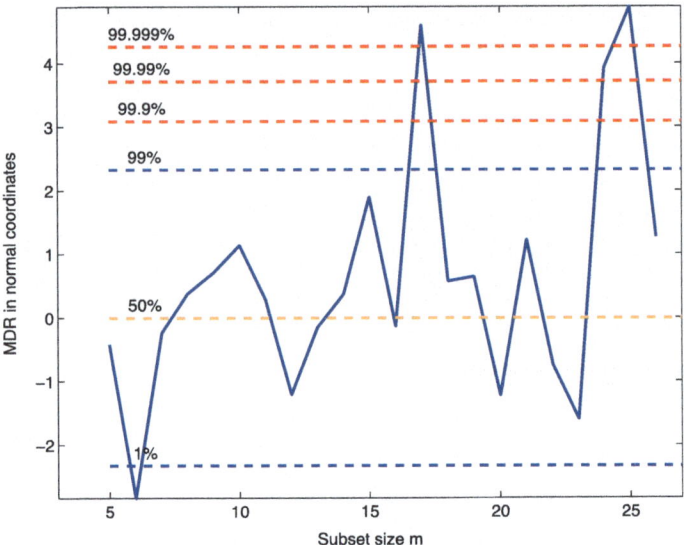

Fig. 4.5 Untransformed wool data: details of the procedure showing 99% and 99.99% envelopes. Normal-scores transformation. A signal is obtained at $m = 17$ because $r_{\min}(17, 27) > 99.999\%$ envelope

of outlying observations close to the end. However, the series of superimpositions in the panels of Fig. 4.6 is clear. There are no outliers when $n^* = 18$, but the outlying observation when $n^* = 19$ indicates that there are only 18 good observations; masking has caused the 19th observation to appear consistent with the rest of the data.

To conclude we analyse the data after log transformation as suggested by Box and Cox. Figure 4.8 shows the forward plot of minimum deletion residuals; there are no signals, no outliers are detected, and all 27 observations are used in fitting. The forward procedure has not produced any untoward artefacts in this now well-behaved set of data.

4.6 Monitoring Plots, Linking Them, and Brushing

It is often suggested that a very robust fit, asymptotically resistant to 50% of aberrant observations, be compared with a non-robust fit. For example, Rousseeuw and Leroy (1987, p. 111) present comparative index plots of least squares (LS) and least median of squares (LMS) residuals. That is a comparison of two extreme forms of regression: the most and least robust. If these are thought of as providing two snapshots of the data, our monitoring can be thought of as providing a film of which the two snapshots are stills from the beginning and end of the film.

4.6 Monitoring Plots, Linking Them, and Brushing

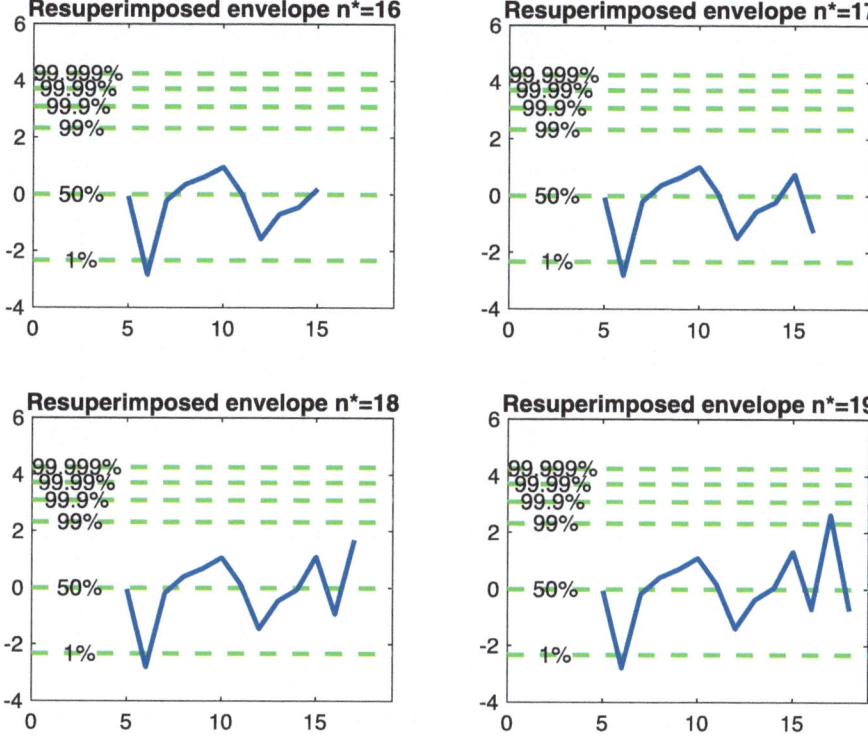

Fig. 4.6 Untransformed wool data: resuperimposed envelopes for steps 16, 17, 18, and 19. Normal-scores transformation. In this process the first outlier is detected at $n^* = 19$ so there are 18 observations available for estimation and 9 outliers

In Sect. 4.5, we used forward plots of deletion residuals, combined with distributional envelopes, to detect outliers. The idea of monitoring as m increases can be extended to any quantities of interest in regression modelling. An obvious candidate is to look at a forward plot of all residuals, not just the individual minimum deletion residuals. Such plots are often highly informative about departures from the model. They can also provide confirmation of stability of the model if that is the case. It is also interesting to plot parameter estimates, to illustrate the effect of any outliers. In Sect. 6.1.3, we monitor the score test from a series of transformation parameters.

In monitoring the FS, we move from a fit using a basic subset to LS with zero *bdp*, although we usually plot from a maximum *bdp* of 0.5. We use a similar idea for monitoring other forms of robust regression. In monitoring S-estimators, we customarily vary the *bdp* from 0.5 to 0.01. For MM-estimators, it is more convenient to monitor changes as the efficiency goes from 0.5 to 0.99. In monitoring the FS, we monitor as each observation is added to the subset $S^*(m)$, unless n is too large. In that case Torti et al. (2021a) suggest adding the observations in small batches of, for example, five observations, omitting fitting to the intermediate subset sizes. For

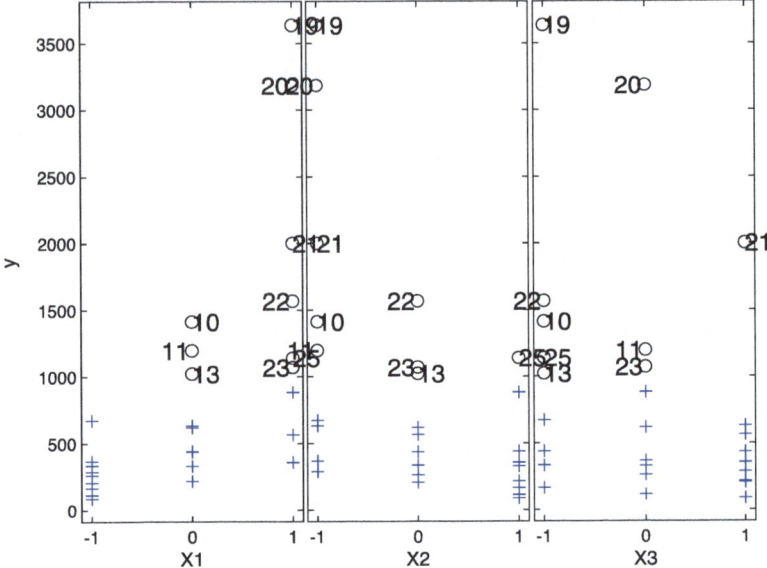

Fig. 4.7 Untransformed wool data: yX plot. Outliers marked with o and numbered

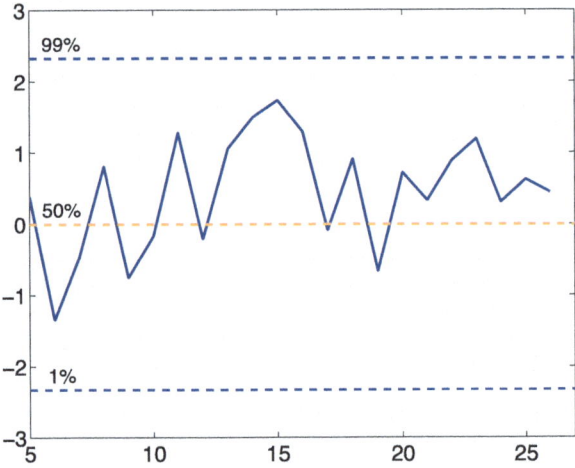

Fig. 4.8 Log transformed wool data: forward plot of minimum deletion residuals. Normal-scores transformation. An uneventful plot

other robust estimators we take 51 values, including LS. We also apply this grid to generalized forms of LMS and LTS in which the value of h/n goes from 0.5 to 1, with rounding to give an integer number of observations. Examples are in Figs. 4.19 and 4.20.

Further, it is often informative to link plots via brushing. This methodology is typically used to link features in a forward plot to another feature of the data. Often, observations that are shown as outlying in the forward plot of minimum deletion

4.7 The Empirical Breakdown Point and Efficiency

One consequence of our monitoring of robust procedures is that we produce insightful data analyses. Another is methodological. By considering a variety of procedures for robust regression, we are able to determine which are the critical parameters in determining the properties of the robust fit, distinguishing them from those that are only of secondary importance. We believe it is thus possible to provide comparatively simple prescriptions for robust regression.

The most difficult problem in a conventional robust data analysis is often specification of the desired efficiency or, equivalently, breakdown point. More generally, this is asking what proportion of outliers are expected in the particular set of data being analysed. A second important question is the nature of robust estimator that is required. In the regression case, the choice is between various forms of M-estimation, such as MM- and S-estimation and fixed or adaptive hard trimming. The third choice is that of the ρ function, which we have already discussed in Sect. 2.4. We show how monitoring makes some of these decisions redundant and illuminates which remaining ones are important.

As a result of monitoring, we observe an *empirical breakdown point*, the point at which the fit switches from being robust to being non-robust least squares. This important property depends both on the nominal properties of the estimator and on the particular dataset being analysed. Using too low a value of *bdp* leads to estimates contaminated by outliers. Conversely, using too high a value of *bdp* leads to a low efficiency of estimation. Table 3.2 showed the relationship between *eff* and *bdp* for Tukey's biweight. The downweighting or trimming in robust estimators can lead to estimates with an appreciably reduced efficiency compared to least squares. For finite samples and outliers that are not remote, the situation may be much worse than these numbers suggest. The sense of security in using an estimator with a $bdp = 0.5$ may be security bought at too high a price. It may also be false security, unless the empirical breakdown point of the method is checked by monitoring the specific combination of data and robust estimator.

A final advantage of monitoring is in overcoming some problems of robust statistical inference. An example is the loss, through the use of robust methods, of the simplicity of distributional theory associated with least squares estimation and related tests. Table 3.1 listed five suggestions of Huber and Ronchetti (2009) that lead to robust estimates of the variances of the parameter estimates and so to t-like tests of the parameter values. The simplest treats the estimates as if they came from LS. Trajectories of the t statistics from the FS analysis of the AR data are in Fig. 4.15. Monitoring plots of t-like tests based on S-estimates for the same data are in Fig. 4.22. The upper panel of this figure does indeed treat the estimates as if they came from LS. The lower panel uses the first of the approximations in Table 3.1, that is, *covrob1*. The shapes of all three sets of monitored trajectories are sufficiently similar to indicate that, for these data, inferences about the parameter values do not much depend on which form of statistic is used.

4.8 Monitoring Tests for Regression Coefficients: The Extended Added Variable Plot

Figure 4.15 shows forward plots of the t statistics for the three explanatory variables of the AR regression data. Due to the ordering of the observations introduced by the FS, the null distribution of the statistics is not Student's t. In this section, we use a development of the added variable plot due to Atkinson and Riani (2002a) to provide a forward plot of statistics which, under the normal theory assumptions, do exactly have t distributions, provided a suitable estimate of σ^2 is available.

In order to obtain forward plots of t tests for all variables, write the regression model for all n observations in the added variable form of Sect. 3.4 as

$$y = Q\theta + \epsilon = X\beta + w\gamma + \epsilon,$$

where Q is $n \times p$ and γ is a scalar. We in turn take each of the columns of Q as the vector w (except the column corresponding to the constant term in the model). The t statistic for testing that $\gamma = 0$ is thus

$$t_\gamma = \hat{\gamma}/\{s_w^2/(w^T A w)\}^{1/2}, \tag{4.6}$$

with $\hat{\gamma}$ given by (3.21) and s_w^2 by (3.23). There are $p-1$ such statistics at each step of the search.

In the forward search, the quantities in (4.6) are calculated for a subset of size m. We now derive the effect on the values of $\hat{\gamma}$ and of t_γ of adding observation $m+1$.

From (3.23) the residual sum of squares of regression of m observations only on X can be written as

$$R(y, y) = y^T A y.$$

Let the new observation be y_+, with explanatory variables x_+ and w_+. The leverage of the new observation is

$$h_+ = x_+^T (X^T X)^{-1} x_+,$$

which is ≥ 0, but, unlike leverages for deletion, need not be ≤ 1; see the discussion of (3.23). The residual for the new observation is

$$e_+ = \tilde{y}_+ = y_+ - x_+^T \hat{\beta}. \tag{4.7}$$

4.8 Monitoring Tests for Regression Coefficients: The Extended Added Variable Plot

We rewrite e_+ as $\tilde{y}_+$ in order to exhibit the change in (4.6) on addition of a new observation as a function of squares and products of observations and the excluded explanatory variable.

Let $R_+(y, y)$ be the residual sum of squares of the $m+1$ observations after regression on X and x_+. Then

$$R_+(y, y) = R(y, y) + e_+^2/(1 + h_+) = y^T A y + \tilde{y}_+^2/(1 + h_+). \tag{4.8}$$

The expressions for $\hat{\gamma}$, s_w^2 and t_γ are all functions of the form $R(a, b)$ where the vectors a and b are either w or y. It then follows from (4.8) that these residual sums and products after the addition of one more observation become

$$R_+(a, b) = R(a, b) + \tilde{a}_+ \tilde{b}_+/(1 + h_+) = a^T A b + \tilde{a}_+ \tilde{b}_+/(1 + h_+), \tag{4.9}$$

where $\tilde{a}_+$ and $\tilde{b}_+$ are the residuals of a_+ and b_+, as in (3.21), after regression on X.

As a result of these relationships, we can, for example, write the t test (4.6) for $m + 1$ observations as

$$t_\gamma^+ = \frac{(m + 1 - p)^{1/2} R_+(w, y)}{\{R_+(y, y) R_+(w, w) - R_+^2(w, y)\}^{1/2}}, \tag{4.10}$$

with the quadratic forms given by (4.9).

The argument that the null distribution of the added variable test is not affected by the ordering of the data depends on the properties of residuals. In calculating the added variable test, we fit the reduced model $E(Y) = X\beta$, the residuals from which are used to determine the progress of the search. We do not include w in the model. The choice of observations to include in the model thus depends solely on y and X. But the results of Sect. 3.4 show that the added variable test (4.10) is a function solely of the residuals $\tilde{w}$ and $\tilde{y}$ which, by definition, are in a space orthogonal to X. The ordering of the observations using X does not affect the null distribution of the test statistic. Since, for normally distributed errors, the estimates $\hat{\gamma}$ and s^2 are independent, it would follow that the null distribution of the statistic is Student's t if we have an unbiased estimate of σ^2.

Although the null distribution of the test statistic is unaffected by the FS, the value of the non-centrality parameter is dependent on the search, since the values of the w_i in the subset will depend upon the ordering of the observations by the search. Atkinson and Riani (2002a) explore the dependence of the non-centrality parameter on X and discuss the effect of correlated explanatory variables on the procedure.

As a consequence of the results of this section, there are three competing methods for monitoring t statistics for regression coefficients:

1. Monitor the values created during the search.
2. Monitor the values created during the search, but apply the consistency correction factor to the estimate σ^2 for each value of m (see 3.49).
3. Use the extension to the added variable plot described in this section.

The disadvantage of 1 above is that, at the beginning of the search, the estimate of σ^2 can be very small, giving large values of t statistics. The resulting trajectories accordingly tend to decline during the search. Use of asymptotic correction factors provides some adjustment to the estimate of σ^2, but not sufficient for small values of m when the small-sample correction factor mentioned in Sect. 3.11.2 may be required. In the added variable plot, the values of the statistics increase with evidence for the significance of the explanatory variables.

The analysis of the surgical unit data in Sect. 4.9.5 includes a comparison of the three monitoring plots in Figs. 4.31, 4.32, and 4.33. The analysis of the bank data in Sect. 4.10 presents monitoring plots from point 2 above for both FS and S-estimations. Monitoring the added variable plot from the FS, point 3 is included for comparison as Fig. 4.41. It provides the sharpest information about the effect of outliers on inferences for the importance of the explanatory variables. Giudici et al. (2024) use monitoring of the added variable plot in the analysis of the time series dynamics of Bitcoin prices.

4.9 The Monitoring Approach in Practice

4.9.1 Correlation

We illustrate the use of monitoring with three examples of increasing complexity. For effective monitoring, we require a method that moves from a very robust fit to least squares. Although all methods have such properties asymptotically, the comparisons in Chap. 5 and of Riani et al. (2014d) show that the finite sample properties of the various methods vary depending on the distance between the main body of the data and the contamination. In particular, these results suggest we need to avoid methods which are tuned to have a very high efficiency for the parameters of the linear model but which are liable to failure unless the contamination is extremely remote.

Many of the plots in this section show the monitoring of all n residuals. The inclusion of outliers is usually signalled by a sudden change in residuals and a more gradual change in parameter estimates. For simple structures, like our first example, there is a clear division of the solutions into a robust fit and a non-robust one, with a sharp break between them. For more complicated examples the point of transition is not so clearly visible. But in all cases we find that the structure of the plot is well summarized by measures of the correlations between the residuals at adjacent monitoring values. We show three standard measures of correlation:

1. Spearman. The correlations between the ranks of the two sets of observations (Spearman, 1904).
2. Kendall. Concordance of the pairs of ranks (Kendall, 1938).
3. Pearson. Product-moment correlation coefficient (Pearson, 1895).

4.9.2 The Stars Data

We start with a return to the easily understood example of the stars data, already analysed in Sect. 4.5.1, where the FS identified four outliers. Our further analysis of these data provides a clear illustration of some of the properties of the monitoring approach. Figure 4.9 is again a plot of the data, augmented by fitted lines and identification of observations.

Included in the figure are six linear fits: least squares (LS), which is attracted towards the cluster of four outliers, and five robust lines that fit predominantly to the main group. The steepest line is LTS, followed by S and then MM with *eff* = 0.85 and 0.95. The FS line is the least steep. As the analysis in Sect. 4.5.1 shows, the FS identifies four outliers. The higher the desired efficiency, the closer is the fit to least squares. There are thus two very different groups of fitted lines, those from high breakdown estimators and that from a procedure with zero breakdown. Our monitoring plots each consider a single estimator as changes are made in the coefficients determining robustness. Virtually all show a robust and a non-robust fit as the two extremes. The interesting comparison is at what empirical breakdown point this transition occurs. For hard trimming, if we treat observations 7 and 9 as outliers, there are six outliers out of 47 observations, so we would hope to achieve the minimum breakdown point, for these data, of 12.8%, with correspondingly high efficiency. Methods that monitoring shows have a higher empirical breakdown point require a more robust fit to reveal the data structure. We can expect that, as data contamination increases, they may fail to reveal any contamination whatsoever.

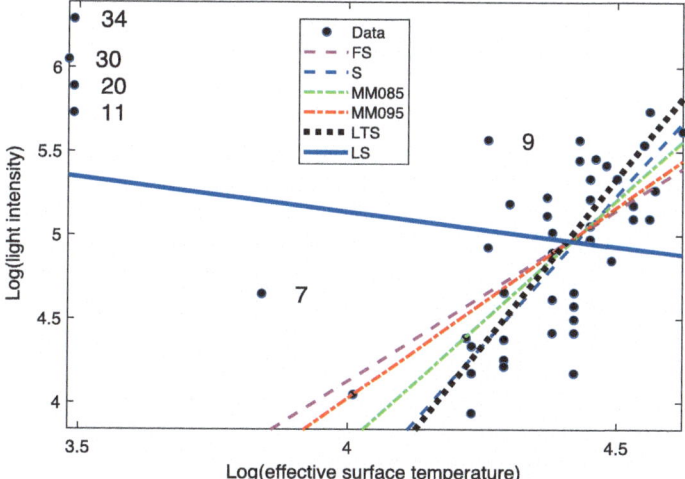

Fig. 4.9 Stars data: scatter plot and fitted regression lines: LS: least squares, which is attracted towards the cluster of outliers and five robust fits. The steepest line is LTS, followed by S and then MM with efficiencies of 0.85 and 0.95; FS is least steep. The higher the desired efficiency, the closer is the fit to least squares

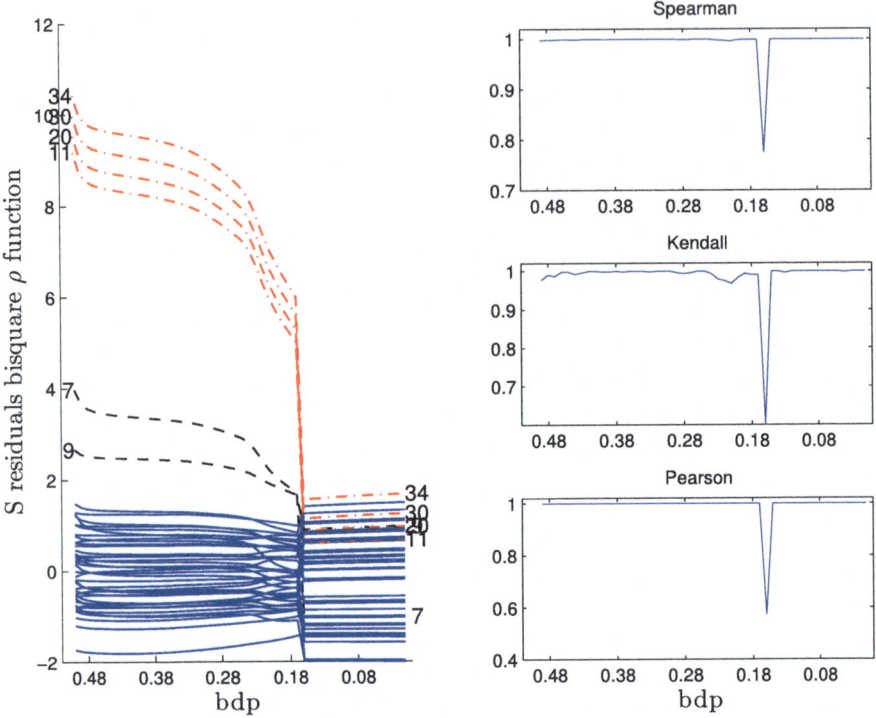

Fig. 4.10 Stars data: S-estimation, Tukey's biweight ρ function. Left-hand panel, plot of scaled residuals. Three sets of residuals (the cluster of outliers, the two intermediate outliers, and the main cluster of the data) are clearly seen for *bdp* values down to around 0.2. Right-hand panel, three measures of the correlations of adjacent residuals. The abrupt switch to LS at *bdp* = 0.16 is evident

Figure 4.10 shows a typical monitoring plot, in this case for S-estimation. This is generated by a series of robust fits, starting from *bdp* = 0.5 and decrementing the value by 0.01 up to 0.01. There are therefore 50 robust fits leading to the plot of scaled residuals in the figure, for which, in this case, we used Tukey's biweight. The plot of scaled residuals for high *bdp* fits clearly shows three sets of residuals: four very large (units 11, 20, 30, and 34), two intermediate (units 7 and 9), and the cluster of the remainder. There is some slight decrease in the values around a *bdp* of 0.2 and then an abrupt change at 0.16 when the fit switches to least squares. The exact point of this change shows clearly in the right-hand panel, the plot of correlation values.

To monitor MM-estimation, we first find σ^2, with a breakdown point of 0.5, and then increase the nominal efficiency of estimation of β from 0.5 to 0.99 with an increment of 0.01, the estimate of σ^2 remaining fixed. The results in Fig. 4.11 show that the very robust fit predominates, the change to least squares occurring at *eff* = 0.99. The comparison of this monitoring plot with that of Fig. 4.10 is clarified by the results of Table 3.3. These show that an efficiency of 0.99 corresponds to, for Tukey's biweight, a breakdown point of 0.0570. The minimum value in the plot of

4.9 The Monitoring Approach in Practice

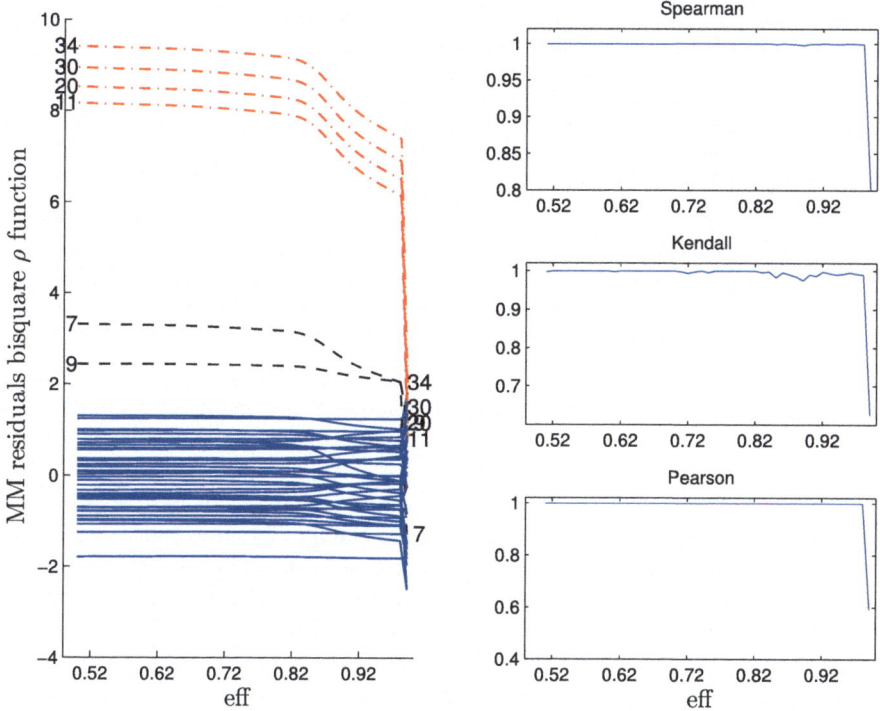

Fig. 4.11 Stars data: MM-estimation, Tukey's biweight ρ function. Left-hand panel, plot of scaled residuals. The three sets of residuals are clearly seen for virtually all efficiency values. Right-hand panel, three measures of the correlations of adjacent residuals. The abrupt switch to LS is at *eff* = 0.99, corresponding to a *bdp* of 0.057

correlations is in the last step when we compute the correlation between the residuals with *eff* = 0.98 and those for *eff* = 0.99.

We explored the τ residuals for three somewhat high values of nominal efficiency of estimation of β: 0.85, 0.9, and 0.95. The plots are similar to those already given, except that the value of the breakdown point associated with the step immediately before the change from very robust to least squares fits increases from 0.14 to 0.16 and then 0.26. As the required efficiency of estimation of β is increased, the empirical breakdown point of the procedure also increases. This means that if the ρ-function associated with nominal efficiency is higher, it is necessary to set the ρ-function associated with nominal *bdp* to a high value, because the empirical *bdp* increases.

All of these robust procedures were also used with the other three ρ functions: hyperbolic, optimal, and Hampel. For this example the choice is mostly not critical, apart from τ-estimation. For S-estimation with all four ρ functions, the change in the plots is at a breakdown point of 0.17. For MM-estimation with the Hampel function, the first change to non-robust estimation is at an efficiency of 0.97 and at 0.98 for the hyperbolic. Although for the optimal function there is a change at 0.93, causing

Table 4.1 Stars data: empirical breakdown point (*bdp*) or efficiency (*eff*) for MM-estimation: five estimators and five ρ functions. The values are for the step before the switch to a non-robust fit

Estimator		Biweight	Optimal	Hyperbolic	Hampel	PD
S	*bdp*	0.17	0.17	0.17	0.17	0.17
$\tau = 0.85$	*bdp*	0.14	0.14	0.14	0.16	0.17
$\tau = 0.90$	*bdp*	0.17	0.16	0.16	0.20	0.21
$\tau = 0.95$	*bdp*	0.26	0.21	0.24	–[a]	0.21
MM	*eff*	0.98	0.99	0.97	0.96	0.95

[a] For this combination of τ and ρ-function, only non-robust solutions were obtained during monitoring

a rescaling of the residuals, the three groups found by robust estimation remain over the whole range up to an efficiency of 0.99. Only τ-estimation is somewhat sensitive to the form of the function.

For the hyperbolic and optimal ρ functions, the behaviour of the τ-estimate is similar to that for the biweight, as it is with the Hampel function and $\tau = 0.85$ or 0.9. However, with Hampel's ρ function, the method completely breaks down when the efficiency is set to 0.95, producing a uniform plot of least squares residuals. These results are summarized in Table 4.1. The performance of PD is closest to that of Hampel. However, for τ-estimation, there is no problem with $\tau = 0.95$, the performance agreeing with that of the optimal ρ-function.

It is also interesting to look at the plots of residuals that arise from LTS, LMS, and the FS, but we leave this until we consider the more complicated structure of Example 2.

4.9.3 The AR Regression Data

In these data, introduced in Sect. 3.5, there are three explanatory variables and 60 observations, with a structure of six masked outliers. The plots of LS residuals failed to reveal the outliers (see Sect. 3.5). The traditional ANOVA table based on all the observations or after deleting unit 43 shows that variable 1 becomes almost significant (part (a) of Exercise 4.1). The analysis based on MM-estimation using Tukey's biweight and a nominal efficiency of 0.95 (part (b) of Exercise 4.1) shows that using a confidence level of 0.975, 3 units are declared as outliers and that then variable 1 becomes strongly significant.

We now compare a variety of robust methods for their ability to identify the structure of the data. The minimum breakdown we can expect from hard trimming is 6/60, i.e. 0.1.

We start with an FS analysis, first looking at the structure of residuals during the search, but augment this with plots of the t statistics for the three explanatory variables. We amplify this analysis through the use of brushing linked plots. We then

4.9 The Monitoring Approach in Practice

move to the comparative analysis of monitored robust residuals in which emphasis is, in part, on the empirical breakdown point of the different procedures and their relationship to nominal asymptotic values.

The monitoring plots of scaled residuals in the two panels of Fig. 4.12 clearly reveal that there are six trajectories whose values are distinct. From this pair of plots, it is not clear what effect this group of six units exerts on the identification of outliers and on such other statistics as the t-tests for regression. Further, it is not clear whether the entry into the subset of the six potential outliers causes some units already in the subset to leave it. This would be the case if there were masking of outliers towards the end of the search.

The two panels of the plot come from different searches. Careful inspection shows that the trajectories in the two panels are not identical, although they become increasingly so as the FS progresses. This is a reflection of the stochastic nature of the start of the search. In the random start FS (Sect. 8.5.2) when the data come from several models, the intention is that different starts will give rise to sets of distinct plots. Here, with data generated from a single model, there is no effect on conclusions drawn from the plots arising from different starting points.

We explore the structure of the data by brushing the set of units in the subset at particular steps in the FS. When option *databrush* is activated, it is possible to click on any value of the x-axis (here values of m) and show the trajectories of the residuals for the units belonging to the subset for that subset size. For example, clicking the x-axis at $m = 30$ (upper panel of Fig. 4.12) shows the units belonging to subset $S^*(m)$ at step $m = 30$ highlighted in red. We notice six clear outliers and that a neighbouring group of less outlying observations are separate from the remaining residuals. As the search progresses, the six outliers become less distinct and the scatter of values of residuals in the centre of the plot increases. The lower panel of the plot (which is obtained after clicking the x-axis for $m = 53$) shows that only the six outliers, together with observation 43, are separate from the main group. As m increases further, the necessary inclusion of the outliers in $S^*(m)$ leads to masking; only observation 43, which was not in the subset when $m = 30$ or 53, initially becomes more outlying.

We now illustrate the use of spinning point clouds to obtain insight into the location of observations in $p + 1$-dimensional space. Unfortunately, we are limited to $p = 2$, that is, to the response and subsets of two explanatory variables. We start in the upper panel of Fig. 4.13 with a representation of all 60 values of y, x_1, and x_2. The points for the six outliers and observation 43 are highlighted using linking of plots. This upper panel is unrevealing about the outlying nature of the observations. However, after some experimental rotations, the lower panel of Fig. 4.13 reveals that the six outliers fall in a distinct region of the space. In neither projection is the outlying nature of observation 43 apparent. As a final illustration of the information readily obtainable from linking plots, Fig. 4.14 shows the location of the six outliers in the scatter plots of y against each explanatory variable. An advantage of this plot over the spun point clouds of Fig. 4.13 is that we can include any number of variables. However, as might be expected from the upper panel of Fig. 4.13, the outlying nature

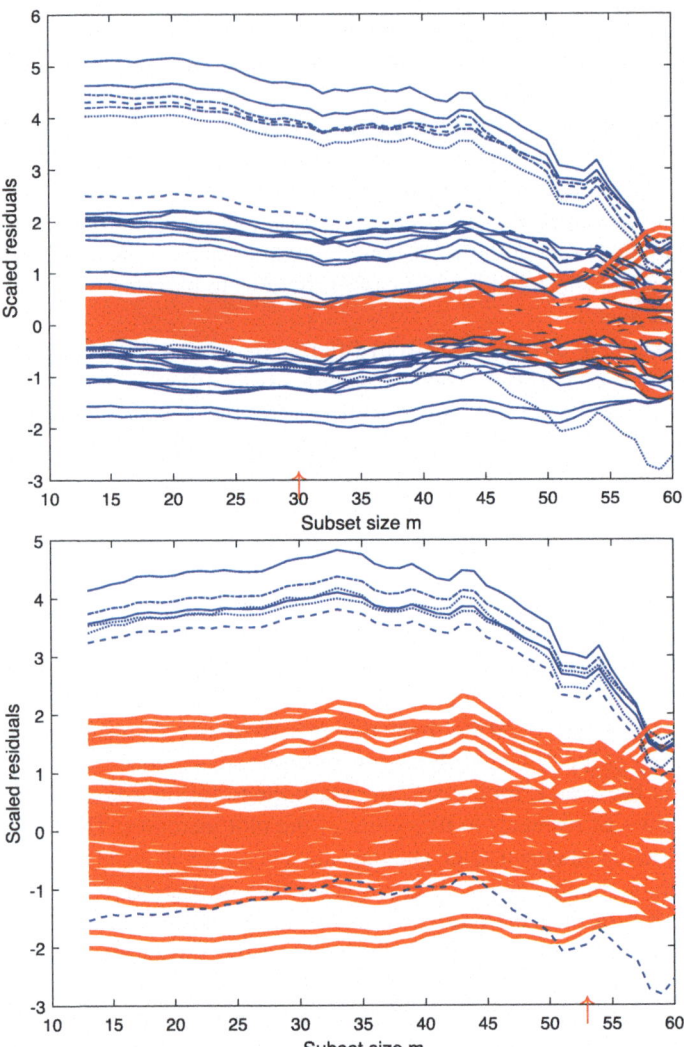

Fig. 4.12 AR regression data: FS analysis. Monitoring the units belonging to two subsets $S^*(m)$ selected by use of option *databrush*. Upper panel, $m = 30$ and, lower panel, $m = 53$. Results from two different forward searches; the final part of the search is unaffected by the slight differences in the initial parts. The red vertical arrows together with the highlighted trajectories are generated by clicking on points on the x-axis

4.9 The Monitoring Approach in Practice

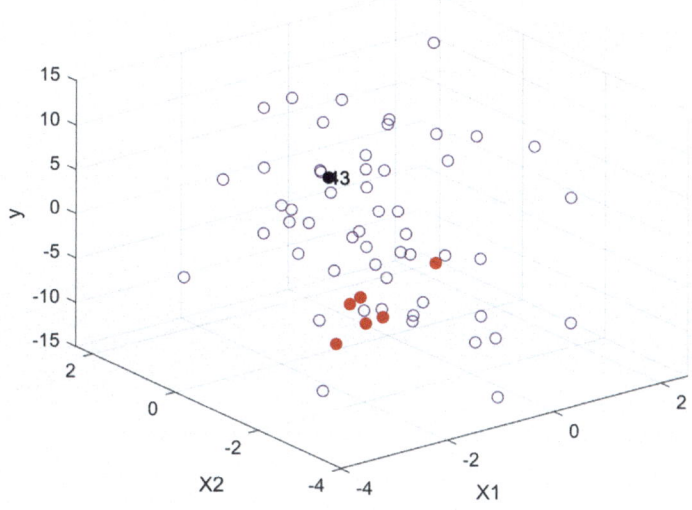

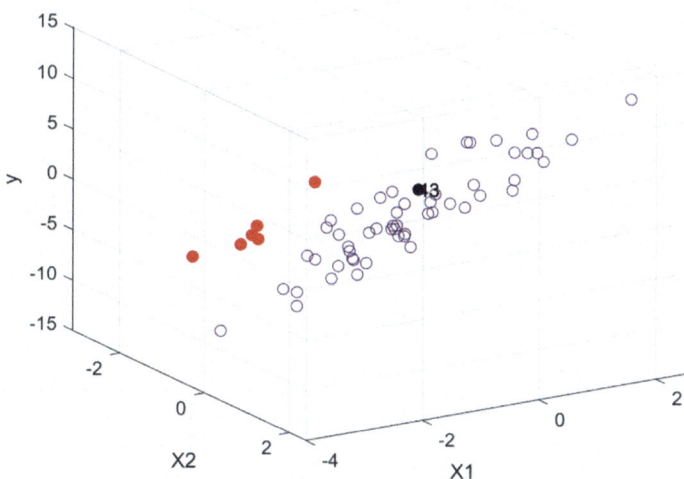

Fig. 4.13 AR regression data: xyz plot before and after rotation. Despite the labelling, the three new axes in the lower panel are linear combinations of y, x_1 and x_2

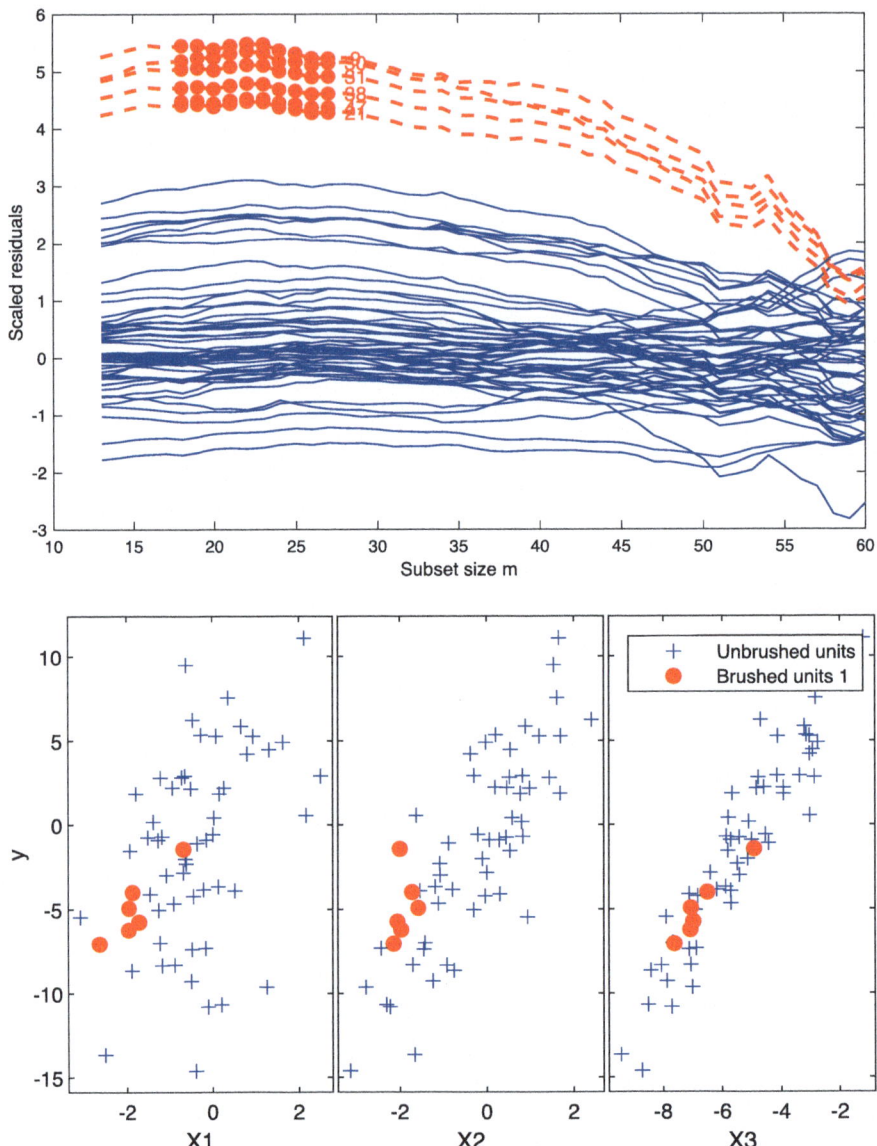

Fig. 4.14 AR regression data: FS analysis. Upper panel: forward plot of scaled residuals with the initial trajectories of the six outliers highlighted. Lower panels: scatter plots of y against the three explanatory variables; six outliers highlighted by linking plots

4.9 The Monitoring Approach in Practice 121

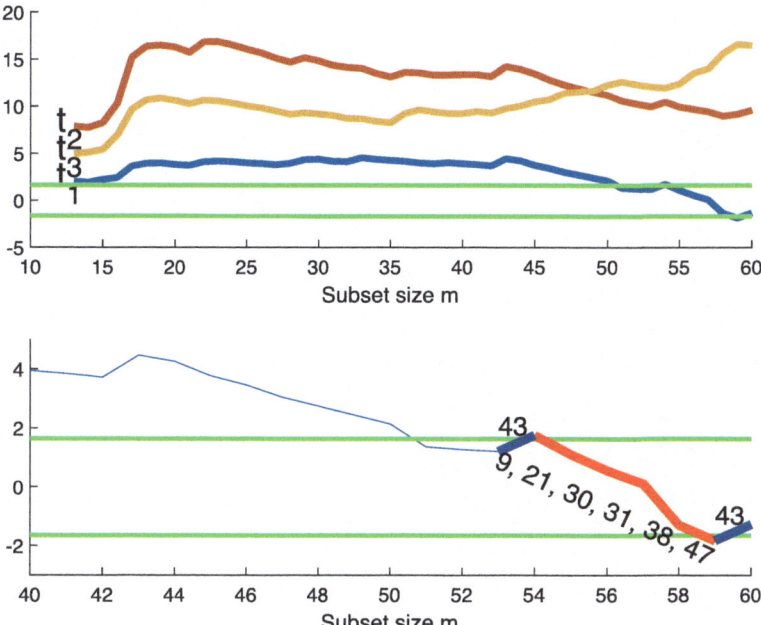

Fig. 4.15 AR regression data: FS analysis. Forward plots of t statistics for the three explanatory variables with 90% confidence intervals. Upper panel: virtually the whole search: lower panel: monitoring the effect of the observations that enter in the last seven steps of the search. In step $m = 58$, observations 31 and 47 enter the subset and observation 43 leaves it to re-enter in the final step

of the six observations is not obvious, although they form similar, not quite extreme, point clouds in the first two panels.

Since these are regression data, we now turn to the important question as to how the parameter estimates and t statistics for the variables evolve during the FS. The upper panel of Fig. 4.15 shows the t statistics for virtually the whole search calculated including the consistency correction (3.49). The parameters for x_2 and x_3 are significant throughout. However, the lower panel shows, in an interesting way, that the significance of x_1 decreases towards the end of the search. Initially the significance starts to decrease around $m = 43$; by the time the six outliers start to be included in the subset, the coefficient of the variable is still positive, becoming significant when observation 43 enters the subset. However introduction of the six outliers leads to a, barely significant, negative estimate. In the final stage of the search, observation 43 is reintroduced to the subset and the significance of the negative estimate is reduced.

At the end of this section, we compare Fig. 4.15, which shows the forward plots of t statistics as a function of the subset size, with Fig. 4.22, which shows plots of robust t statistics from monitoring S-estimation with the optimal ρ function.

We start our comparisons with robust estimators in Fig. 4.16 with the forward plot of S residuals using the optimal ρ function. Again there is a clear division of the plot

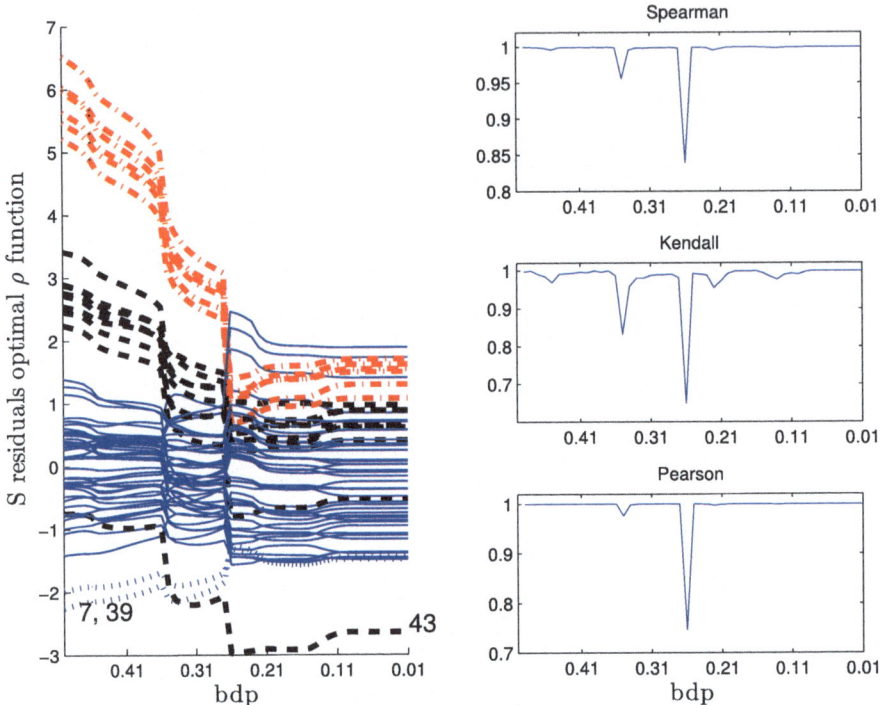

Fig. 4.16 AR regression data: S-estimation, optimal ρ function. Left-hand panel, plot of scaled residuals. Four groups of observations are evident for a *bdp* greater than 0.34 and three until the *bdp* is 0.27. Thereafter the plot shows the LS fit with one outlier. Right-hand panel, three measures of the correlation. Kendall's measure, in particular, shows the two transitions during monitoring

into an initial, very robust, region which at a *bdp* of 0.27 becomes close to the least squares fit. In the initial part of the plot, the six remote outliers are clearly visible. However, there is also interesting fine detail in the plot.

For high values of *bdp*, that is, greater than 0.34, there are four groups of observations, with observations 7 and 39 having the most negative residuals. At a *bdp* of 0.34 the two central groups coalesce and observation 43 becomes as outlying as observations 7 and 39. At the next transition, the outliers with the largest positive and negative residuals form a single group, with observation 43 outlying. This structure remains stable for lower values of *bdp*.

So far we have not reported results for the hyperbolic and Hampel's ρ functions in any detail. Figure 4.17 shows monitoring plots of S residuals from these functions. The two plots are virtually identical. Compared with the plot for the optimal ρ function in Fig. 4.16, they only show one abrupt transition, that is, from four groups of observations to one plus a mild outlier, although observation 43 does become increasingly remote before this transition. However, either plot would serve to signal the difference between the very robust and non-robust fits.

4.9 The Monitoring Approach in Practice 123

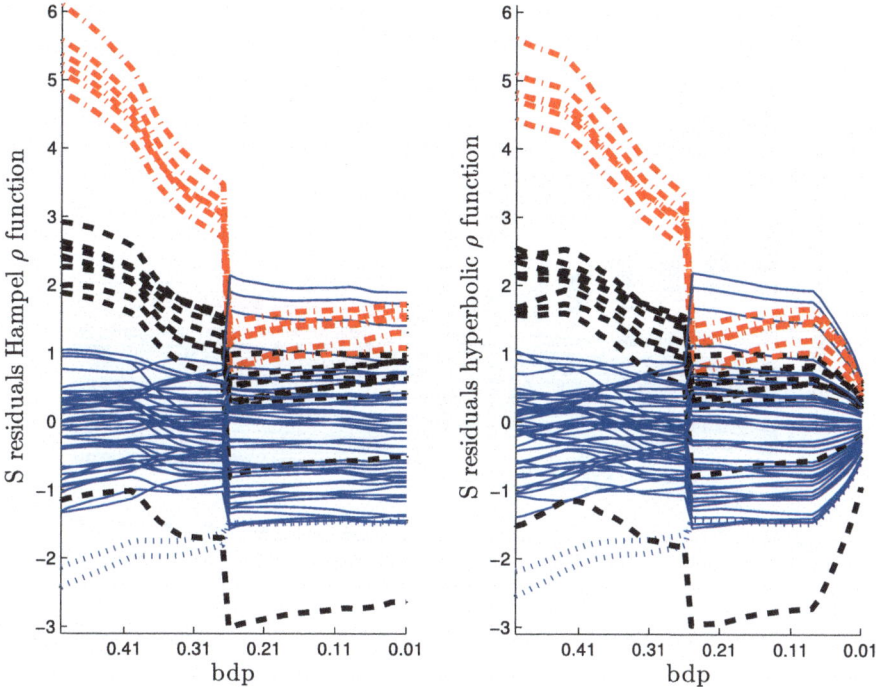

Fig. 4.17 AR regression data: S-estimation, plots of scaled residuals. Left-hand panel, Hampel's ρ function. Right-hand panel, hyperbolic ρ function. To be compared with Fig. 4.16

Table 4.2 AR regression data: empirical breakdown point *bdp* (or efficiency *eff* for MM), during monitoring for the transition between very robust and least squares regression: six estimators and five ρ functions. The values are for the step before the switch to a non-robust fit

Estimator		Biweight	Optimal	Hyperbolic	Hampel	PD
S	*bdp*	0.27	0.27	0.26	0.27	0.27
$\tau = 0.80$	*bdp*	0.32	0.36	0.33	0.34	0.36
$\tau = 0.85$	*bdp*	0.38	0.40	0.41	0.41	_a
$\tau = 0.90$	*bdp*	0.45	0.48	_a	0.50	_a
$\tau = 0.95$	*bdp*	_a	_a	_a	_a	_a
MM	*eff*	0.91	0.97	0.90	0.87	0.86

[a] For these values of τ and ρ function, only non-robust solutions were obtained during monitoring

Plots for the MM-estimator and all ρ functions are also of this kind, with the optimal ρ function again showing two transitions. As with Fig. 4.11, for MM-estimation in the stars data, the transition to a non-robust fit occurs at a high nominal efficiency, although not so high as in that figure. In the case of many τ-estimators, the plots only show the least squares fit. We accordingly summarize these results in Table 4.2. The main conclusion of these comparisons, and of those of Chap. 2, is that the form of ρ

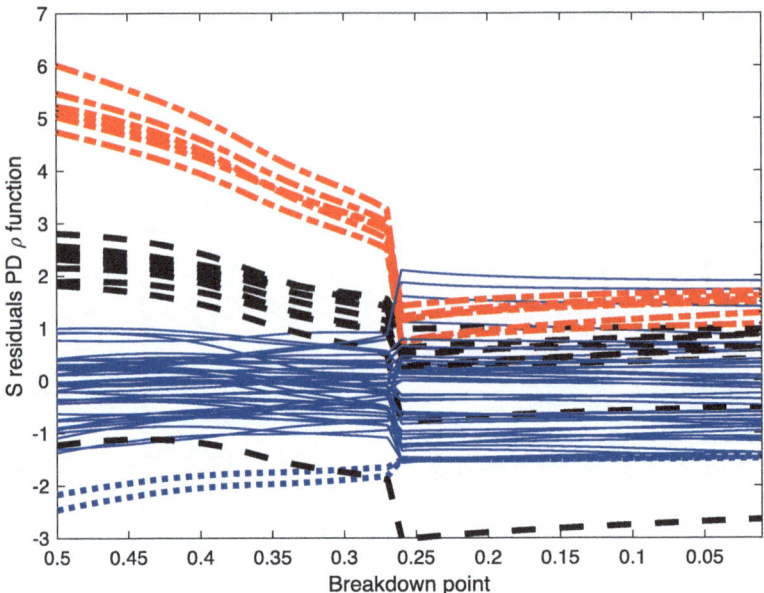

Fig. 4.18 AR regression data: S-estimation with power divergence ρ function. Plot of scaled residuals

function is unimportant for the S-estimator. But the choice of ρ in this example does seem to have some effect on the performance of the MM-estimator. However, if the purpose of the analysis is to establish different possible structures for the data, which are then to be further examined, the choice of ρ-function is not crucial for these data when MM-estimation is used. It is the results for the τ-estimator that cause concern. For an *eff* of 0.85, the breakdown point is around 0.4. For higher values of efficiency, it is either closer to 0.5, or so high that we only see a non-robust fit to the data. Use of the value of 0.95 for τ would lead to monitoring which gave no indication of any departure from the least squares model.

We next use the power divergence ρ function combined with S-estimation. The results are shown in Fig. 4.18. This plot is similar to those in Fig. 4.17 for Hampel's and the hyperbolic ρ functions. In all three plots, the same groups of outliers are visible, all collapsing to least squares with a single outlier (observation 43) at a *bdp* of around 0.26. The most interesting individual feature of this plot is that it is more stable than any of the other monitoring plots of residuals in this section. This is in line with the results of Sect. 2.4.

We now turn to hard trimming procedures. The residuals for LTS are in Fig. 4.19. As h/n increases from just above 0.5 to 0.99 and the *bdp* correspondingly reduces, the residuals gradually decrease in magnitude as the number of observations used in fitting increases. This contrasts with the plots for the robust analysis of the stars data, when the plots are sensibly constant within sectors. For the AR regression data, Figs. 4.16 and 4.17, the plots are also constant in the last region of monitoring. The

4.9 The Monitoring Approach in Practice

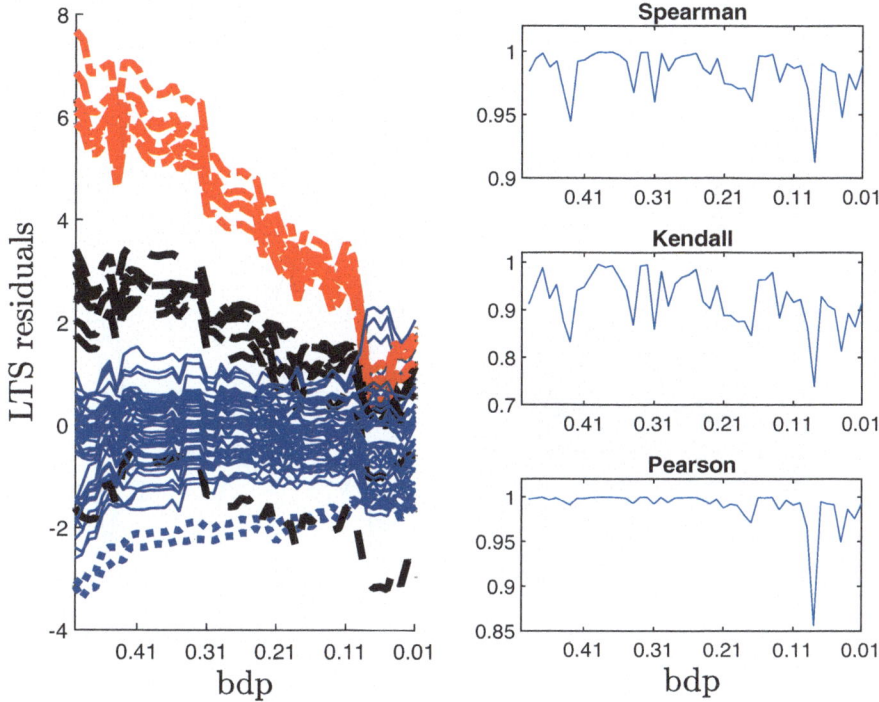

Fig. 4.19 AR regression data: LTS estimation. Left-hand panel, plot of scaled residuals. The four groups of observations are evident for a *bdp* greater than 0.23, with the main group further split close to *bdp* equal to 0.5. The groups then gradually move closer together, with the change to one group plus one outlier for a *bdp* of 0.09. Right-hand panel, three measures of the correlation. All show the final transition during monitoring

plot of residuals in Fig. 4.19 shows all the detail of groups and their coalescence evident in Fig. 4.16, together with a division of the main group for a *bdp* close to 0.5. The six large residuals are clearly visible throughout, becoming included in the fit from a *bdp* of 0.09. This feature is clearly shown in the plots of correlation functions. (Recall that there are six major outliers in the 60 observations.)

In LMS, the median of the squared residuals is used to provide an estimate of σ^2, rather than the trimmed sum of squares of the residuals as in LTS. To provide a generalization of LMS suitable for monitoring we vary the efficiency of estimation by minimizing the $100h/n$ percentage point of the distribution of the squared residuals. The general structure of the plot in Fig. 4.20 is similar to that of Fig. 4.19 but with much greater variability, reflecting the slow $n^{-1/3}$ rate of convergence of LMS (Rousseeuw and Leroy, 1987, p. 178). The transition to a non-robust fit is hardly evident in the plot of residuals, although it is indicated by the plots of correlation.

Note that all these algorithms for robust regression are stochastic in the sense that they depend on the number of subsets which are extracted. This is particularly

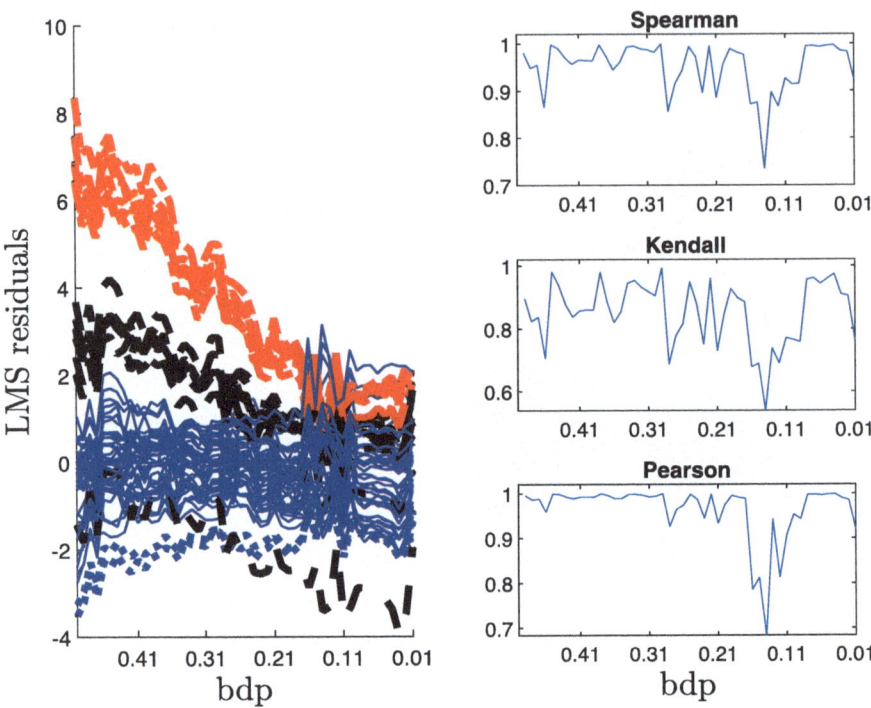

Fig. 4.20 AR regression data: LMS estimation. Left-hand panel, plot of scaled residuals. The four groups of observations are just evident for the highest values of bdp but gradually converge. The curves are much rougher than in the other plots. Right-hand panel, three measures of the correlation. All show some transition around a *bdp* of 0.12

evident for the figures generated using LTS and LMS (Figs. 4.19 and 4.20). The user who tries to reproduce these two figures, using the code associated with the book, will notice slight differences from run to run.

Finally we consider FS, a forward plot of the residuals for which is given in Fig. 4.21. Now the abscissa is the number of observations m in the subset used for fitting. The scaled residuals in this plot tell a similar story to those from monitoring LTS for a *bdp* below 0.42—chiefly that there are four groups, that observations 7, 39, and 43 behave differently and that there are six appreciable outliers from robust fits. In both LTS and FS, the six outliers remain evident until near the end of monitoring, giving estimates with higher efficiency than those for the empirical *bdp* with S-estimation in Figs. 4.16, 4.17, and 4.18.

These three plots (Figs. 4.19, 4.20, and 4.21) are different from those for the S-estimates and variations shown in the earlier plots. One difference is that they are less stable, reflecting the effect of individual observations that have weights zero or one. In the other robust procedures, the observations are smoothly downweighted. The second difference is that these three plots decline as efficiency increases. This effect is less marked in the case of FS. Since the residuals are scaled by the square

4.9 The Monitoring Approach in Practice 127

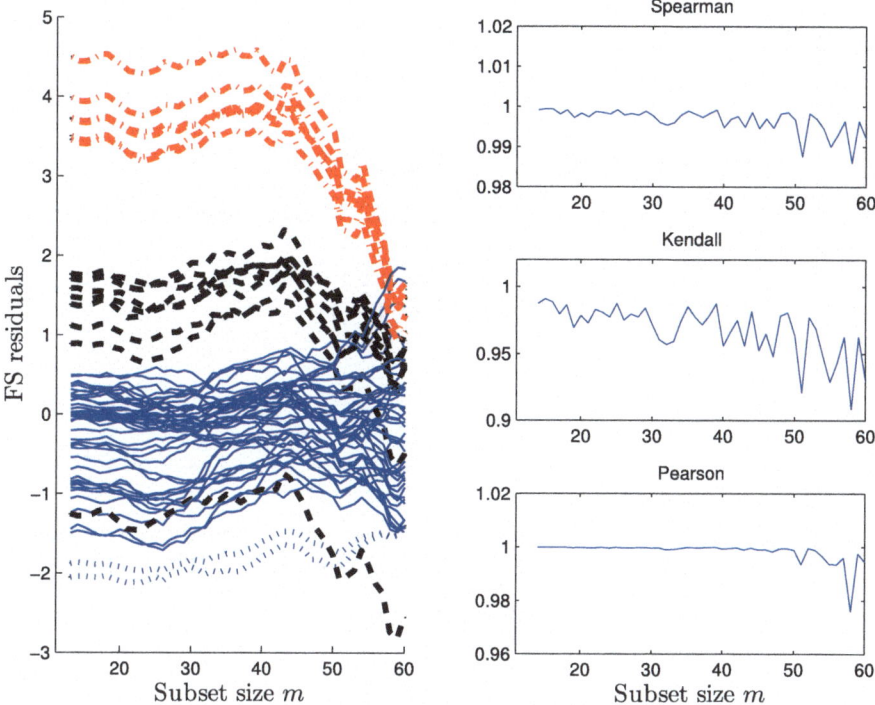

Fig. 4.21 AR regression data: FS. Left-hand panel, plot of scaled residuals. The four groups of observations are initially evident, and particularly so around $m = 40$. The six outliers remain distinct until $m = 55$. Observation 43 becomes increasingly outlying from $m = 43$ and observations 7 and 39 are distinct until $m = 58$. Right-hand panel, three measures of the correlation. Kendall's measure, in particular, shows the gradual change in the outlyingness of the various units

root of the final estimate of σ^2 the differences in the plots reflect the changes in the estimates of β as monitoring evolves. These remain constant over long periods for the estimators with flexible trimming as the weights from the ρ function remain constant. However for LMS, LTS, and the FS, the estimates change for each new observation included in the fit.

Although the plots of the residuals for the groups of methods can be very different, the plots of t statistics from monitoring S-estimation and the FS are similar. Figure 4.22 shows that the very robust fit, in this case S-estimation with the optimal ρ function, finds significant effects, in size order, for x_2, x_3, and x_1. However, the fits with low *bdp* show that x_1 is hardly, or not at all, significant, the values lying around -1.96, the lower limit of the 95% asymptotic confidence interval. Now x_3 is more significant than x_2, as it is in Fig. 4.15 for the FS using normal theory statistics. The two panels of the plot give two versions of the robust t statistics from Table 3.1. The upper panel is for *covrob* in which the robust nature of the parameter estimates is ignored by taking the estimated covariance matrix as $\sigma^2 \hat{\gamma}(X^T X)^{-1}$. The lower

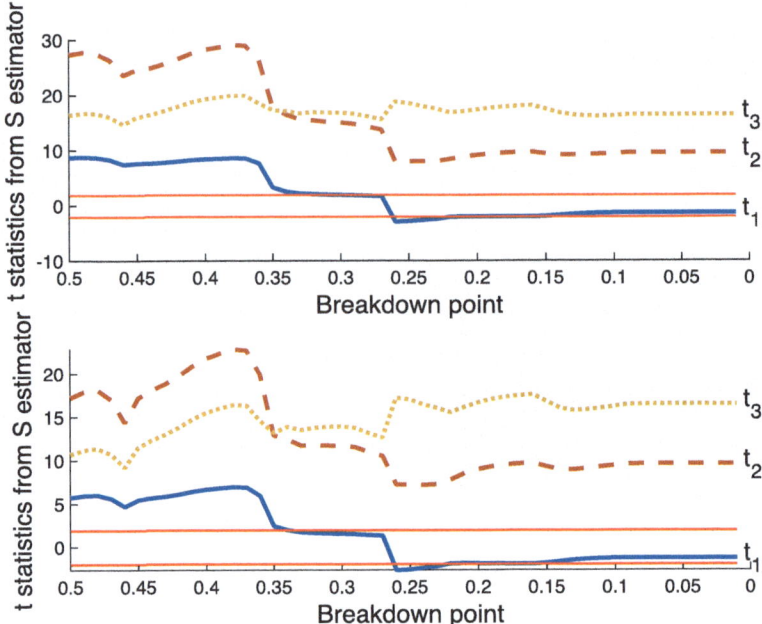

Fig. 4.22 AR regression data: S-estimation with the optimal ρ function. Monitoring plots of two versions of t statistics for the three parameters of the linear model. Upper panel, "traditional" version using *covrob* from Table 3.1; lower panel, using *covrob1*

panel shows the effect of approximation *covrob1* on the evolution of the t statistics. The conclusions to be drawn from the two plots about the significance of the three variables are virtually identical. The plots of t statistics from the forward search in Fig. 4.15 are less smooth than those from S-estimation, but lead to very similar conclusions about the significance of the variables. A difference in the plots is that the S-estimate of β_1, which is significant for all plotted results for high breakdown point estimators, ceases to be significant well before the value of 0.1 for the FS in Fig. 4.15. In general, the changes in the parameter estimates for the FS occur at lower value of *bdp* than similar changes from S-estimation.

Additional analysis of this dataset is in Exercise 4.1 which is focused on the traditional static approaches (robust and non-robust).

4.9.4 The Hawkins Data

In the analyses of the AR regression data, we were able to identify two groups of outliers, a main group of observations and one observation (unit 43) that was apparently an outlier for estimation with a low *bdp*. We now analyse a set of data

4.9 The Monitoring Approach in Practice 129

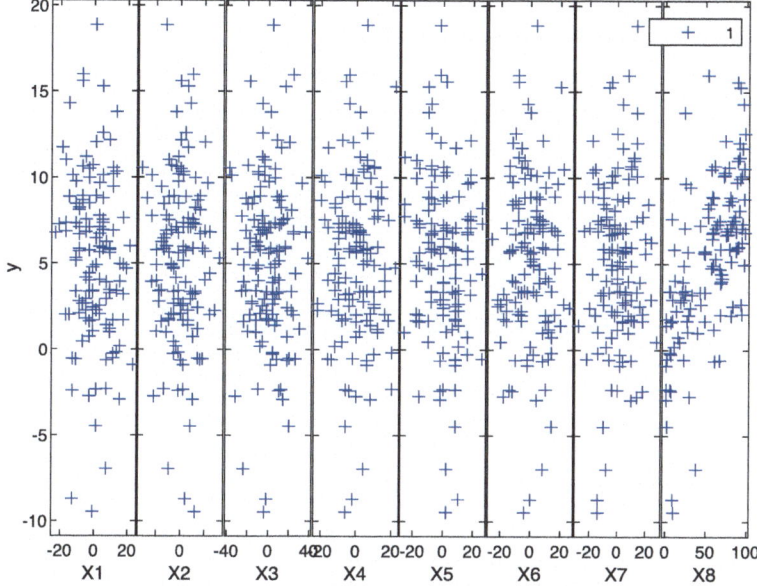

Fig. 4.23 Hawkins data: yX plot

constructed by Hawkins that was intended to be misleading for standard regression methods. An analysis is given by Atkinson and Riani (2000, Sect. 3.1), with the data in their Table A.4. There are 128 observations and 8 explanatory variables. The yX plot is given in Fig. 4.23.

We start by applying the FS with brushing to these data. The upper panel of Fig. 4.24 shows the forward plot of scaled residuals. The plot is symmetrical, with a central group that, in the initial stages of the search, gives rise to very small residuals. This is not like anything we have seen before. The clearly separated bands, their horizontal nature, and, especially, the symmetric nature of the plot are good indicators of the constructed nature of the data. The lower panel of the figure shows the forward plot of minimum deletion residuals with confidence bands. There are three large peaks of similar shape, but different scale, each corresponding to the inclusion of a group of outliers. In the upper panel, we have brushed on the most extreme residuals which, as the highlighting in the lower panel shows, are the last observations to enter the subset. This peak nicely illustrates the effect of masking. Initially the observations from this group entering the subset have large residuals. But, after several have entered, the parameter estimates are corrupted and the new outliers appear less extreme. In the final stages of the search, there is no indication of outliers. The value of the minimum deletion residual in the final step is inside the confidence bands.

Figure 4.25 repeats the process, but brushing the two most extreme groups of outliers, which are highlighted in the upper panel of the plot. The less extreme outliers are shown to give rise to the second peak in the plot of minimum deletion residuals. Figure 4.26 shows the effect of brushing all three groups of outliers; now

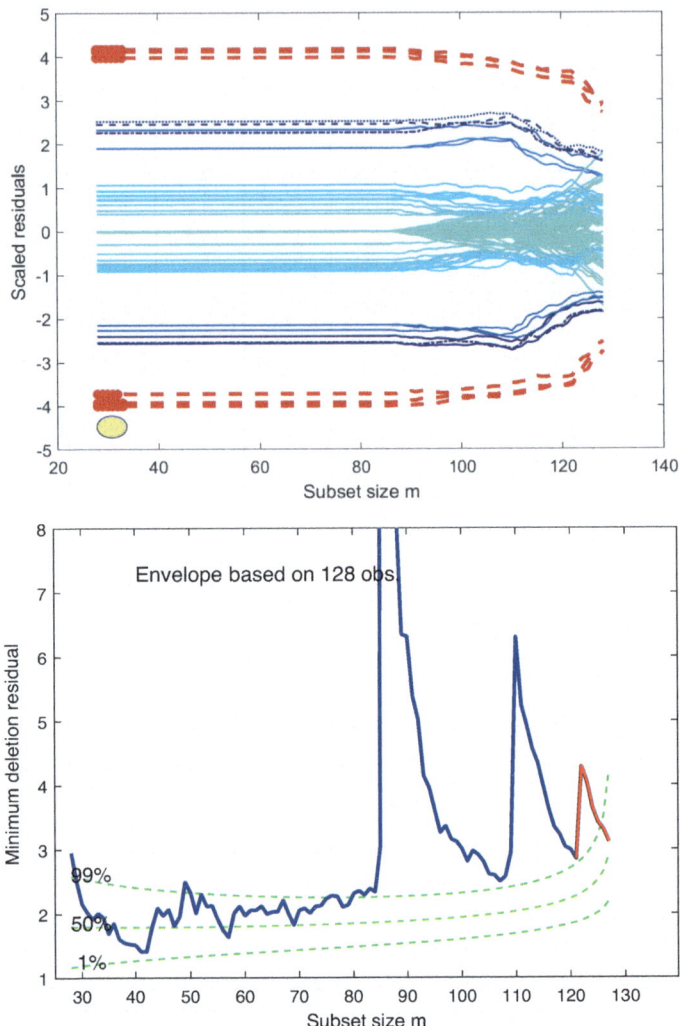

Fig. 4.24 Hawkins data: Forward Search. Upper panel, forward plot of scaled residuals with brushing on the most extreme group of outliers. Lower panel, forward plot of minimum deletion residuals. The brushing in the upper panel identifies the final peak of outliers

all three peaks in the forward plot of deletion residuals are explained. Note that the initial part of the trajectory of the minimum deletion residual comes from a search with a different starting point. However, the final part is unaffected by the starting point.

The effect on the regression analysis is equally clear. Figure 4.27 shows the yX plot for all variables, after all brushed units have been removed. Brushing has removed 43 observations corresponding to a trimming of 0.34. There is now a clear linear

4.9 The Monitoring Approach in Practice

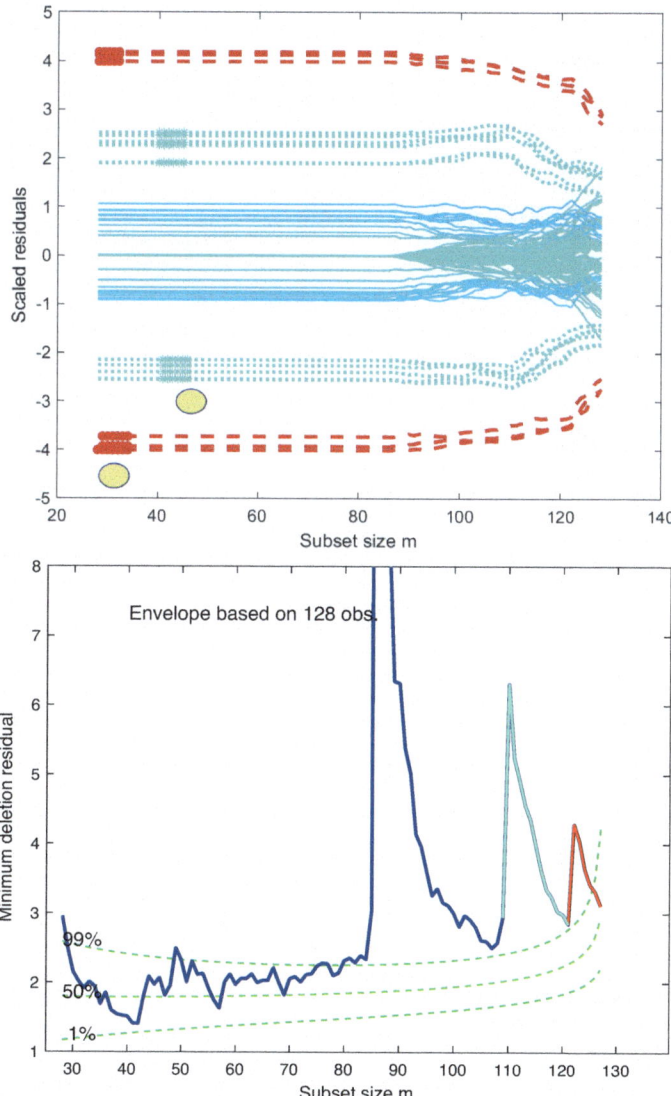

Fig. 4.25 Hawkins data: Forward Search. Upper panel, forward plot of scaled residuals with brushing on the two most extreme groups of outliers. Lower panel, forward plot of minimum deletion residuals. The brushing in the upper panel identifies the two final peaks of outliers

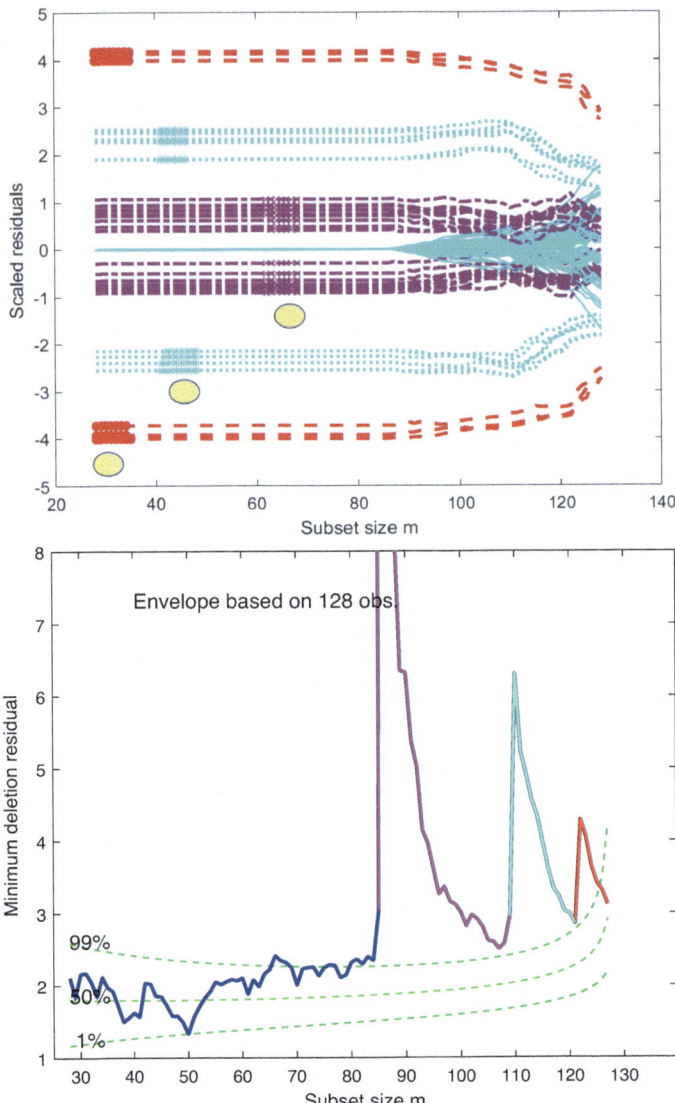

Fig. 4.26 Hawkins data: Forward Search. Upper panel, forward plot of scaled residuals with brushing on the three groups of outliers. Lower panel, forward plot of minimum deletion residuals. The brushing in the upper panel identifies the three peaks of outliers

4.9 The Monitoring Approach in Practice

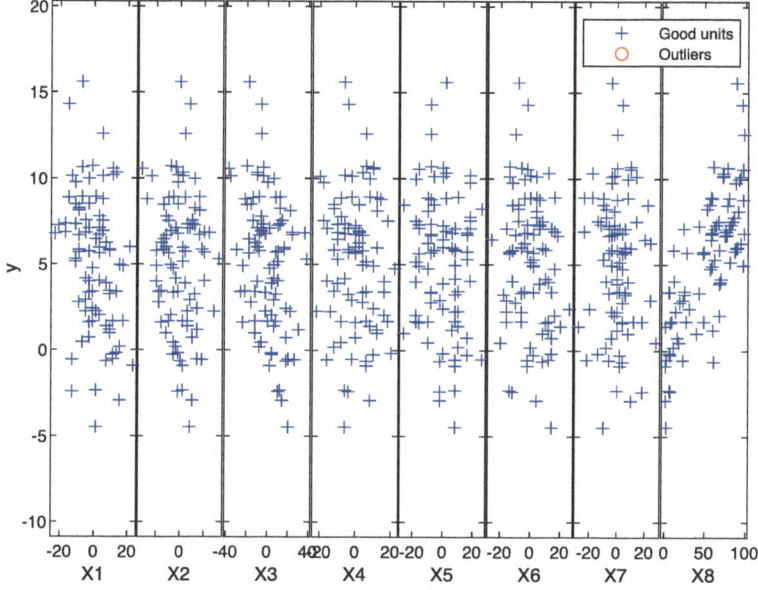

Fig. 4.27 Hawkins data: yX plot for the central group (units not declared as outliers)

relationship between y and x_8, with no obvious relationship between the response and any other variable. In fact, it looks as if the relationship with x_8 may consist of at least four parallel lines.

We now turn to two forms of S-estimation, the first using the Tukey biweight ρ function and the second that for power divergence. The two panels of Fig. 4.28 show the monitoring plots of robust residuals using the two ρ functions. We again see the bands of residuals that are such a feature of Figs. 4.24, 4.25, and 4.26, although these large residuals reduce in magnitude as *bdp* increases. There is little difference between the analyses using different ρ functions. The panels of Fig. 4.29 show a zoom of the earlier plots, focusing on the transition at a *bdp* of 0.31. The residuals of the inner group of observations shrink to virtually zero and then, as the next group of observations enters the subset, they start to increase and merge with the residuals from the next group. When *bdp* is very small there is no group of very small residuals. The upper panel of Fig. 4.24 shows how the FS residuals gradually merge into one group. The panels of Fig. 4.29 show a similar process at work for Tukey biweight and PD-residuals.

The conclusion of this analysis is that the FS is able to exhibit the structure of the data through a forward plot of residuals, the consequences of which on the plot of minimum deletion residuals are illustrated through brushing. The two monitoring plots of residuals from S-estimation show a similar pattern of strips of residuals at increasing distance from the central band of small residuals. In all three cases, the change to initial inclusion of outliers happens from a *bdp* slightly greater than 0.3.

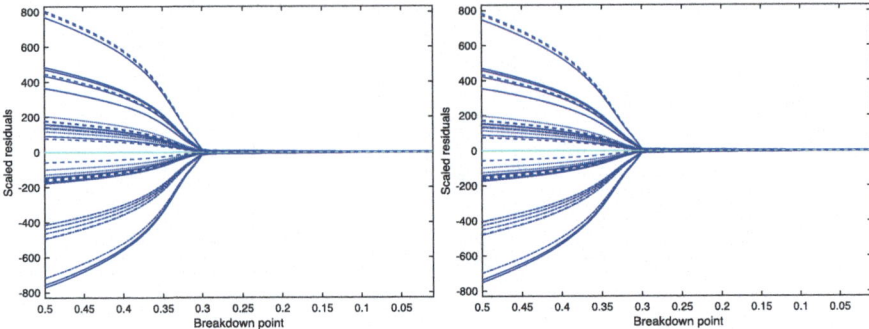

Fig. 4.28 Hawkins data: monitoring residuals using, left-hand panel, Tukey biweight and, right-hand panel, power divergence ρ functions. The residuals of the inner group are shown in pale blue

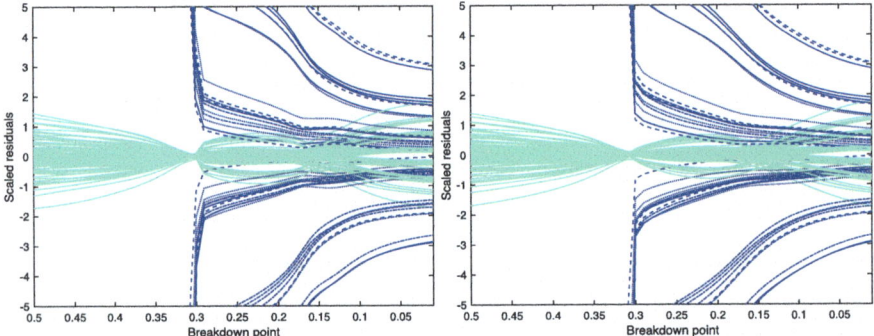

Fig. 4.29 Hawkins data: monitoring residuals using, left-hand panel, Tukey biweight and, right-hand panel, power divergence ρ functions. Zoom of Fig. 4.28. The residuals of the inner group are shown in pale blue

4.9.5 Surgical Unit Data

Neter et al. (1996, p. 334) introduced data on the logged survival time of 54 patients undergoing liver surgery, together with four potential explanatory variables. On their p. 437 another 54 observations are introduced to provide an out-of-sample check of the model fitted to the first 54. Their comparison suggests there is no systematic difference between the two sets.

Figure 4.30 is a forward plot of the minimum deletion residuals for the combined sample of 108 observations. For ease of interpretation this is plotted using the normal-scores transformation. The figure has the surprising feature of an extreme value of the statistic at the centre of the plot. This occurs at $m = 55$. At the end of the search the plot of minimum deletion residuals is well within the distributional bounds; outliers, if any, are well masked. The peak in the plot suggests that the two groups are different, being caused by the initial entry of observations from one group, followed

4.9 The Monitoring Approach in Practice 135

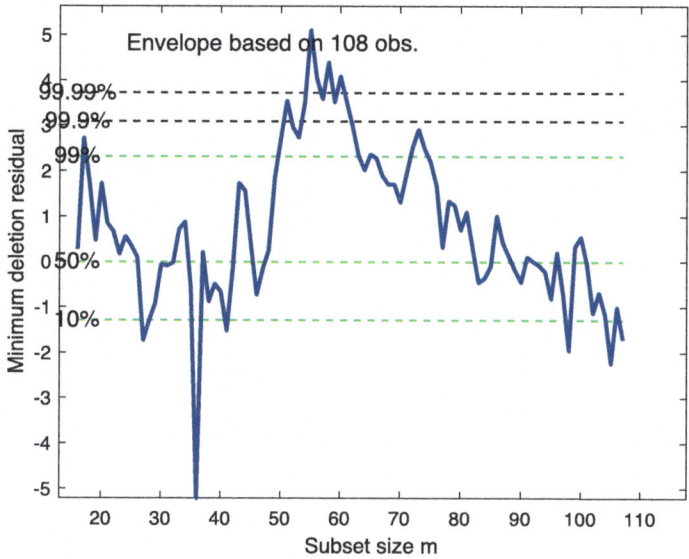

Fig. 4.30 Surgical unit data: plot of minimum deletion residuals using the normal sores transformation. All 108 observations

by observations from the other group. If the two groups were indistinguishable, entry would be random for observations from the groups.

In the left-hand panel of Fig. 4.31, we use the ordering from the FS to monitor the proportion of observations in the subset that come from Unit 1. This rises to a series of peaks around $m = 50$ and then decreases from $m = 60$ as observations from Unit 2 perforce enter the subset. The right-hand panel of the figure shows forward plots of the standard t-test for the four variables with the consistency corrected estimate of variance. The two plots are very similar in shape, even to the initial dip in which several observations from Unit 2 enter before those from Unit 1 start to form a majority.

Section 4.8 described three plots for monitoring the values of t statistics for regression during the search. The right-hand panel of Fig. 4.31 is an example of the second option: monitoring the values created during the search, but applying the consistency correction factor (3.49) to the estimate of σ^2. The importance of this correction is apparent from Fig. 4.32, which shows the same monitoring, but without rescaling the estimate of σ^2. With very small values for the estimate of σ^2 at the beginning of the search, very large values of the statistics are produced (the upper limit in the plot has been truncated to 300). Even so, as the search proceeds and better estimates of σ^2 are obtained, the plotted values continue to decrease. The non-monotone structure is caused by the effect of the two groups.

Figure 4.33 shows the t statistics from the four searches using the method of added variables introduced in Sect. 4.8. The evidence for regression on all variables except x_4 increases strongly when observations from Unit 2 start to enter the subsets of the

136 4 The Monitoring Approach in Multiple Regression

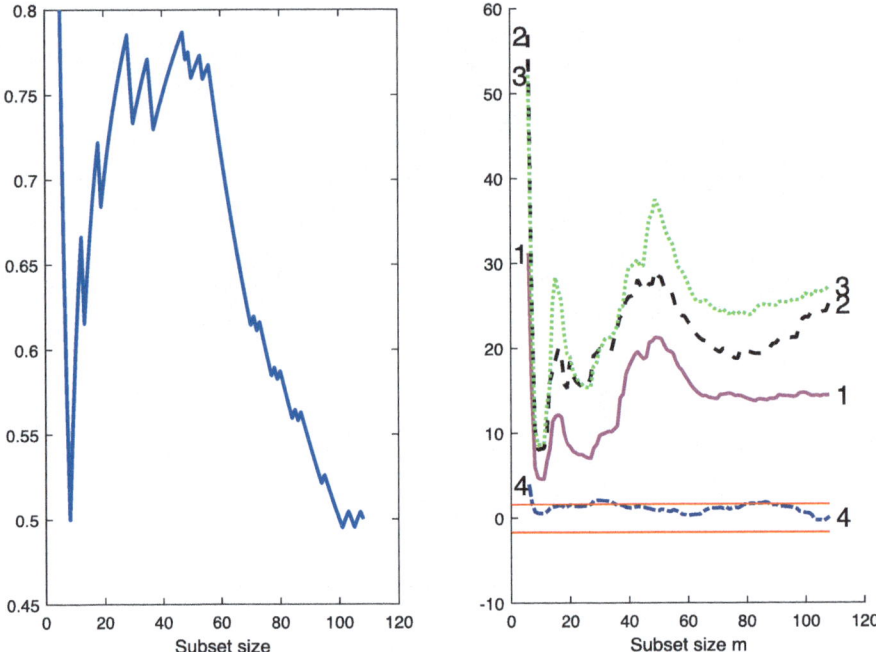

Fig. 4.31 Surgical unit data: forward Search. Left-hand panel, proportion of observations in the subset $S^*(m)$ that come from Unit 1. Right-hand panel, monitoring plot of t statistics for the parameters, with consistency corrected estimates of variance

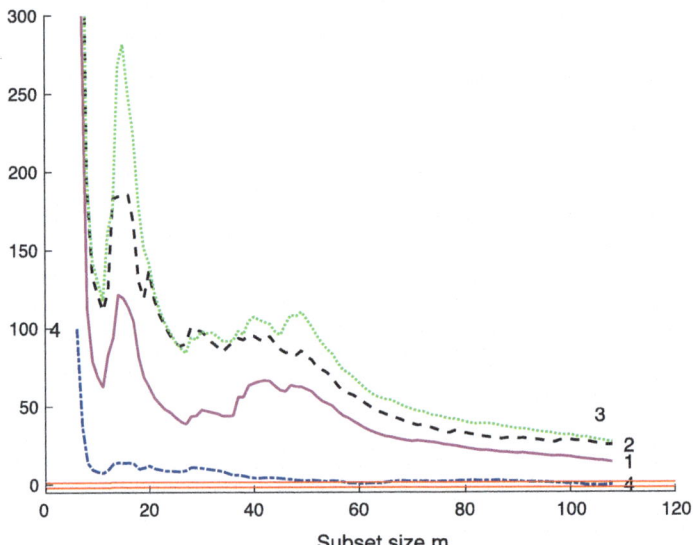

Fig. 4.32 Surgical unit data: monitoring plot of t statistics for the parameters when the variance estimates are not corrected

4.10 The Bank Data

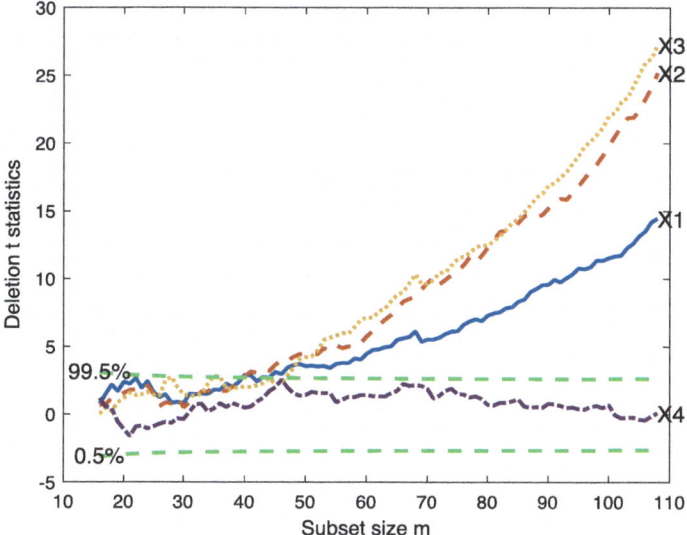

Fig. 4.33 Surgical unit data: monitoring plot of the four added variable t statistics showing the growth of significance of t statistics for the parameters. The evidence for regression on all variables except x_4 increases strongly when observations from Unit 2 start to enter the subsets

searches. These statistics all have exactly a t-distribution if the error distribution in the data is independently normal with constant variance. At $m = n$ these statistics have the same values as those in the right-hand panel of Fig. 4.31 and those in Fig. 4.32.

The way in which the two sets of data differ can be explored by further FS analyses. The dummy variable for a shift in mean between the two sets is highly significant, with a t value of -7.83 at the end of the search (see Exercise 4.3). However, the forward plot of minimum deletion residuals still has its only significant peak in the centre of the search. Some remaining structure is indicated that is more than a simple shift. Analysis of the two groups separately shows that the second group of observations is homogeneous. However, the first group shows evidence of further subgroups, with two peaks in the centre of the forward plot of minimum deletion residuals. The plot, however, returns to non-significant values by the end of the search. The standard static analysis of this dataset is considered in Exercise 4.4.

4.10 The Bank Data

We conclude the detailed data analyses of this chapter with a more complicated example, in which there is no simple model for all the data. Furthermore, the aberrant observations do not form a simple cluster.

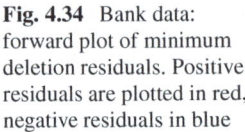

Fig. 4.34 Bank data: forward plot of minimum deletion residuals. Positive residuals are plotted in red, negative residuals in blue

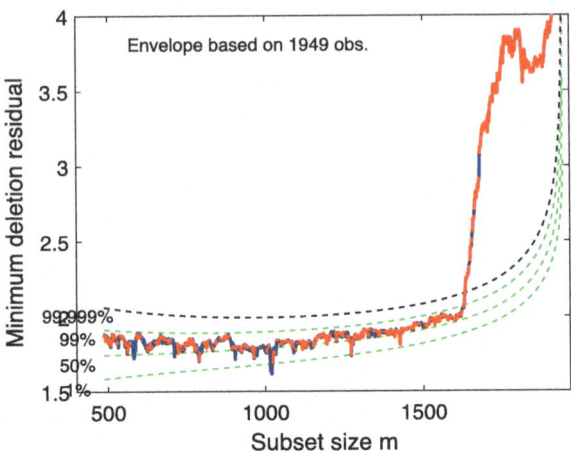

The data were introduced by Riani et al. (2014a). There are 1,949 observations, taken from a larger dataset, on the amount of money made from personal banking customers over a year. There are 13 potential explanatory variables, listed by Riani et al. in their Appendix C, describing the services used by the customers, all of which are discrete, one being binary. The prime interest in the analysis is to discover which activities are particularly profitable.

We begin with the FS. Figure 4.34 shows the plot of minimum deletion residuals. The automatic outlier detection procedure involving the superimposition of envelopes leads to the identification of 255 outliers, corresponding to $bdp = 0.13$. In the plot, positive residuals are plotted in red and negative residuals in blue. Not only is there a large number of residuals, but the pattern of residuals changes during the search. In the major part of the search, until m is around 1700, the positive and negative residuals enter in a mixed manner but, thereafter, the majority of the residuals entering are positive. The outliers are outlying in a particular way.

Figure 4.35 shows the monitoring plot of FS residuals, with the six largest residuals highlighted. The plot is surprisingly stable; the six largest residuals are free of all others for most of the search. Even after the outliers begin to enter the search at $m = 1649$, the pattern of residuals changes only gradually.

The six highlighted residuals are shown in the linked yX plots of Fig. 4.36. The observations corresponding to these residuals, indicated by the monitoring plot of Fig. 4.35, are outlying in all 13 scatter plots.

Before continuing with the FS analysis of the data, we compare this analysis with that using S-estimation combined with the biweight ρ function. The left-hand panel of Fig. 4.37 shows a monitoring plot of the scaled S residuals, with the six most extreme residuals highlighted in red. Allowing for the change in horizontal scale, this has a similar structure to the plot of scaled FS residuals in Fig. 4.35. The analysis based on the PD ρ function in Fig. 4.38 is virtually indistinguishable.

4.10 The Bank Data

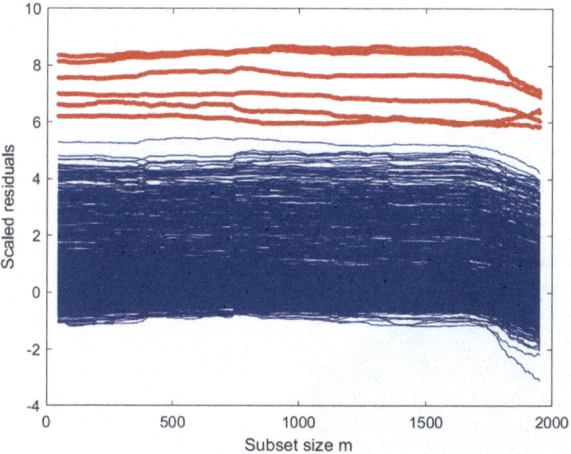

Fig. 4.35 Bank data: monitoring plot of scaled FS residuals, with the six largest residuals highlighted

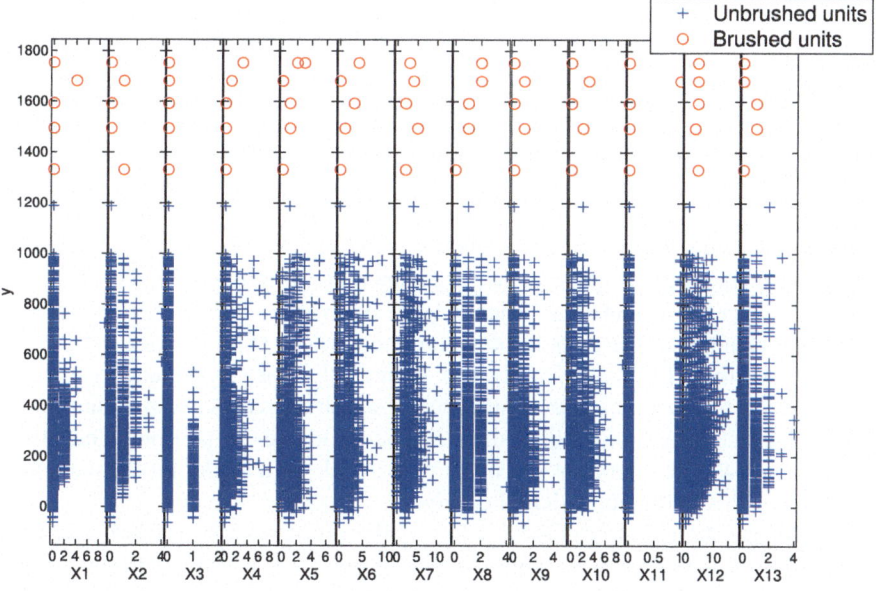

Fig. 4.36 Bank data: linked scatter plots (yX plot) showing the residuals brushed in Fig. 4.35

All three plots of correlation coefficients in the right-hand panels of Figs. 4.37 and 4.38 show clear dips at a *bdp* of 0.14 that is only fractionally larger than that of 0.13 from the FS. The similarity of the monitoring plots of scaled S and FS residuals carries over to other analyses.

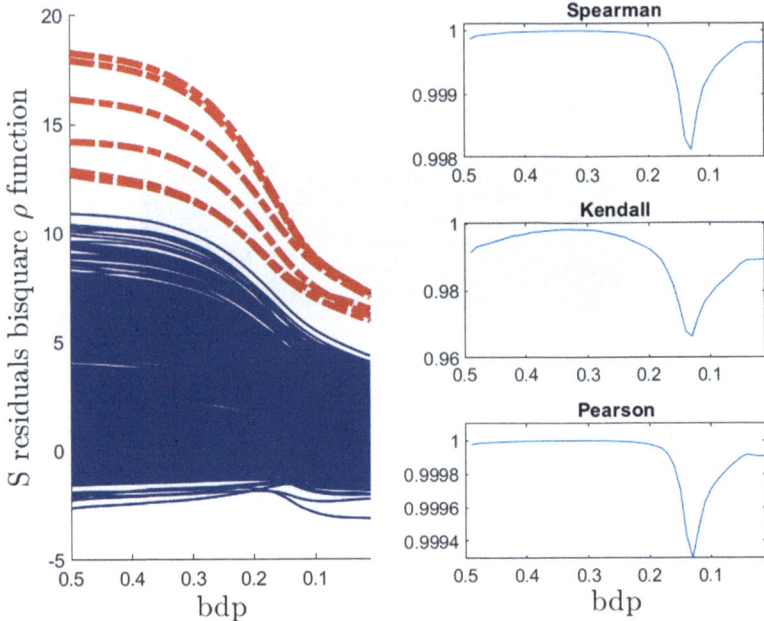

Fig. 4.37 Bank data: S-estimation with biweight ρ function. Left-hand panel, monitoring plot of scaled residuals. This distribution is highly skewed, with a few large outliers. Right-hand panel, three measures of the correlation. All show a clear transition at a *bdp* of 0.14

We now consider the behaviour of the t statistics from two analyses. The t statistics from S-estimation based on the biweight ρ function are in Fig. 4.39. There are 13 panels, one for each explanatory variable. If the data are not homogenous we would expect any changes in the plots to occur around a *bdp* of 0.13. The most dramatic changes seem to be for x_4, which become appreciably more significant, x_5 and x_{12} which lose significance, x_{10}, the significance of which decreases and x_9, which goes from having a slightly significant positive coefficient to one that is strongly negative. Such changes will be important when trying to decide on a model for the data with non-zero weights in the fits with higher *bdp*.

The plot of t statistics from the FS in Fig. 4.40, starting from $m = 1000$, is similar in general shape to that for S-estimation. Although the vertical scales of the panels are not identical, the changes in importance of x_4, x_5, x_9, x_{10}, and x_{12} are the same. The statistics for x_1 and x_2 change in a similar way in the latter halves of the panels in the two figures. However, the changes in Fig. 4.40 are more concentrated at lower values of *bdp*, a phenomenon also observed in the comparison of Figs. 4.15 and 4.22. Since the S-estimator is only monitored for 50 values of *bdp*, the plots for the S-estimator are smoother than those for the FS.

We now look at the interesting and informative monitoring plot of the added variables for the t statistics in Fig. 4.41. The statistics only start to grow in value after $m = 900$. The growth in values stops around $m = 1650$, after which the values of the

4.10 The Bank Data

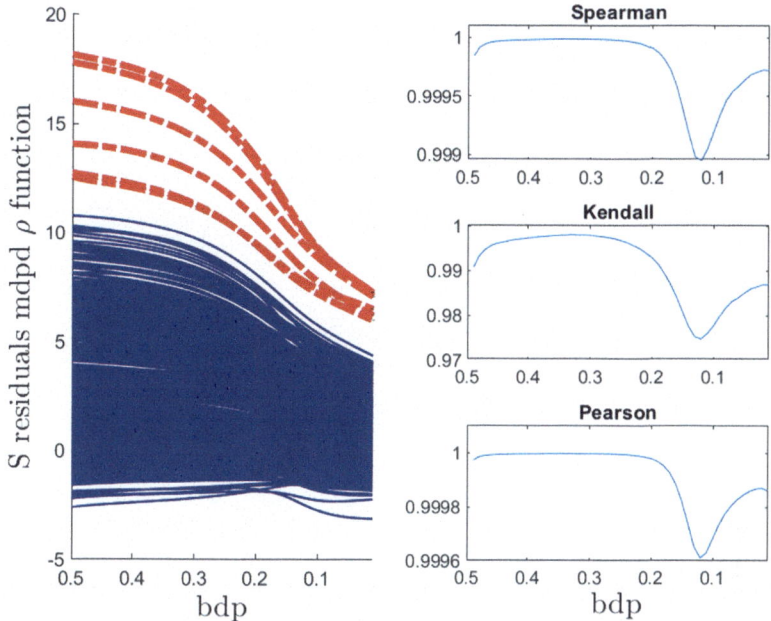

Fig. 4.38 Bank data: S-estimation with PD ρ function. Left-hand panel, plot of scaled residuals. These are highly skewed, with a few large outliers. Right-hand panel, three measures of the correlation. All show a clear transition at a *bdp* of 0.14

statistics tend to decline in significance. The first 900 observations to enter the subset are not informative about regression. The next 750 observations are informative; the peaks for the trajectories do not all occur at the same value of m because the statistics are calculated from distinct searches. Although the subsets will only differ by a single explanatory variable, the ordering of entering of units may differ between searches.

Finally, we look at scatter plots of the non-outlying and outlying observations determined by monitoring or by the FS. The left-hand panel of Fig. 4.42 shows the scatter plots of y against each x for the larger portion into which the FS has divided the data by automatically identifying 255 outliers. The right-hand panel shows the same plot for these outlying observations. The two sets are clearly different. It is not just that the values of y in the right-hand panel are, in general, higher than those in the left-hand panel; they also have a different structure. For example, as the plots in Fig. 4.40 of the t statistics for x_1 (personal loans) and x_2 (financing and hire purchase) show, there is positive regression on these variables for the data in the left-hand panel. However, as the customers from the right-hand panel are included towards the end of the search, the regression decreases, as it does for x_{10} (credit cards). On the other hand, the plot suggests that the second group has a higher uptake of life insurance (x_4). A striking feature of the scatter plots in the right-hand panel of Fig. 4.42 are the six large outliers, which are indeed visible in Fig. 4.37.

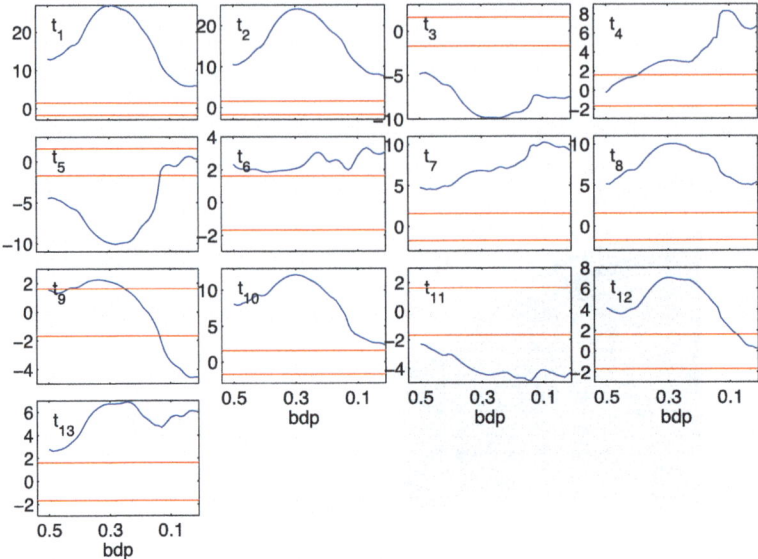

Fig. 4.39 Bank data: S-estimation with the biweight ρ function. Monitoring plots of modified t statistics for the 13 parameters of the linear model, with option *covrob1* of Table 3.1. The confidence level of the bands is 0.95

The methods of robust regression have divided these data into two groups with rather different properties. The added variable plots of t statistics of Fig. 4.41 confirm that there are two distinct groups of customers. A further analysis of these data might build two regression models since the order of importance of the variables varies considerably between groups. It may be that the groups differ in affluence and in the use they make of the bank's facilities.

4.11 Generalized BIC for Outlier Detection

In order to detect outliers in the forward plot of minimum deletion residuals, we have used the somewhat complicated and arbitrary rule introduced in Sect. 4.3 to allow for the effect of the simultaneous testing for outliers during the search. In Sect. 4.5.1, we illustrated the use of this outlier detection procedure on the stars data; the four most extreme outliers were detected by the FS. We returned to the analysis of these data in Sect. 4.9.2. Figure 4.9 contains several robust lines, including LTS, which excludes six observations: the four extreme outliers and also the less extreme observations 7 and 9. There is clearly some difference in the power of the tests from the two methods. As a flexible alternative with a grounding in asymptotic theory, Riani et al. (2022b) use a method based on the Bayesian Information Criterion (BIC Schwarz 1978) for the comparison of different models on a fixed set of data of given size.

4.11 Generalized BIC for Outlier Detection

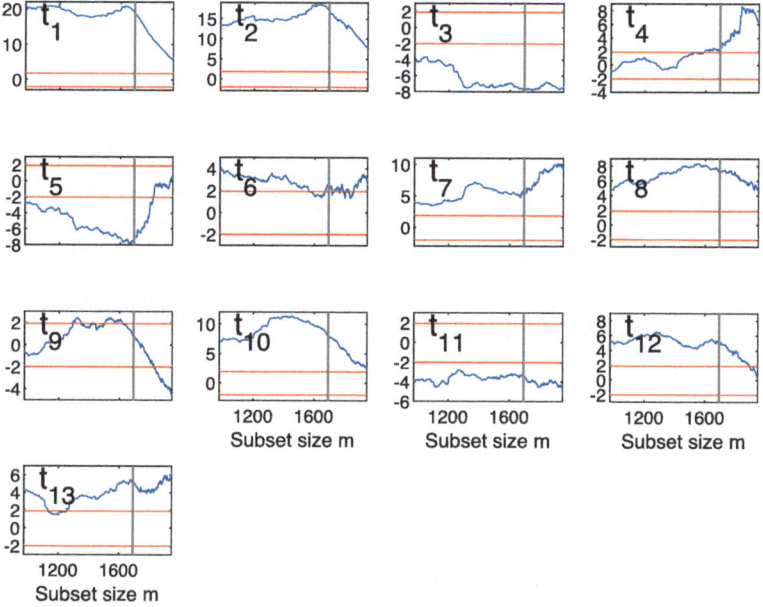

Fig. 4.40 Bank data: FS estimation. Forward plots of t statistics with consistency corrected estimates of variance for the 13 parameters of the linear model, from $m = 1000$. Vertical lines at $m = 1694$

This criterion is extended, through use of the mean shift outlier model of Sect. 3.2.3, to allow comparisons between models based on different numbers of observations. The extended BIC can then be monitored during the search, providing a general form for outlier detection, which has no explicit dependence on sample size or arbitrary constants, such as the significance levels, which come from hypothesis testing.

The following brief summary omits technical details, which are to be found in Riani et al. (2022b).

4.11.1 Information Criteria

There is a large literature on the use of a variety of information criteria in choosing the best model for a set of data. Claeskens and Hjort (2008) provide a treatment with a nice combination of mathematics and data analysis.

Let $\mathcal{L}(\theta)$ be the loglikelihood of the n observations y_i, with the parameter vector θ of length p. With $\hat{\theta}$ the maximum likelihood estimate of θ, a general form of information criterion is $IC = 2\mathcal{L}(\hat{\theta}) - k(p, n)$, where $k(p, n)$ is a function that penalizes more complicated models. For BIC, $k(p, n) = p \log n$, so that the penalty increases with sample size. That model is selected for which BIC is largest.

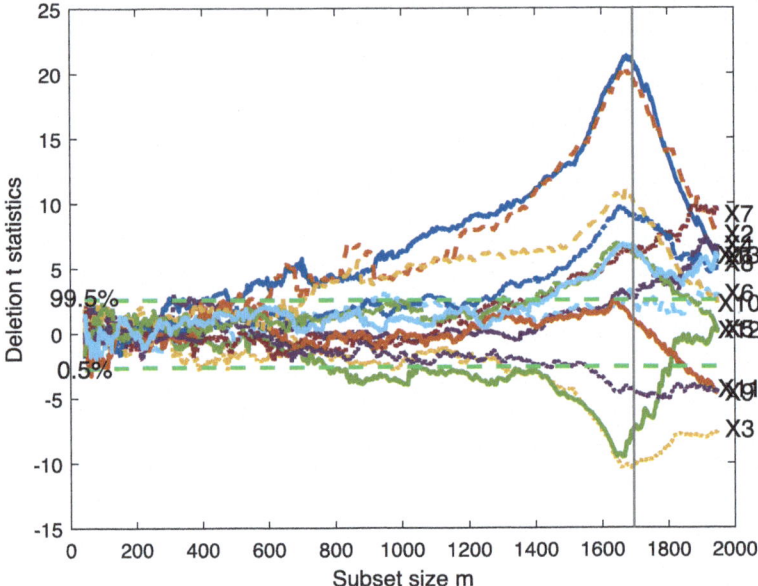

Fig. 4.41 Bank data: monitoring plot of the 13 added variable t statistics of Sect. 4.8 showing the growth of significance of t statistics for the parameters until m is just less than 1694

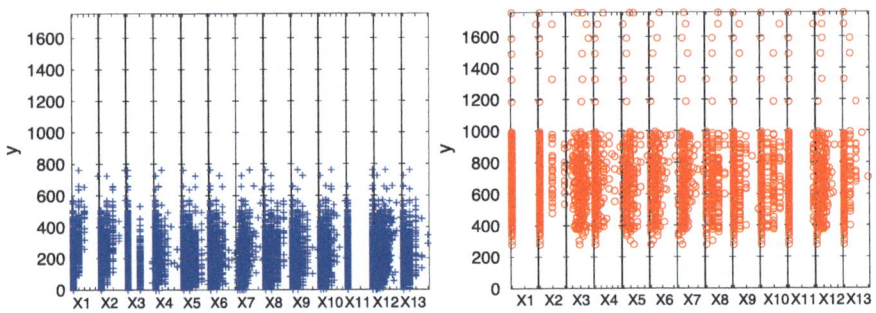

Fig. 4.42 Bank data: scatter plots of y against individual explanatory variables for the two parts of the data. Left-hand panel, the main body of the data. Right-hand panel, the remaining, somewhat different, 255 observations

For the linear regression model with univariate response and independent normal errors of constant variance σ^2, where $\hat{\beta}$ is the least squares estimate of the p parameters β of the linear model and $R(\hat{\beta})$ is the residual sum of squares of the y_i,

$$BIC = -n \log\{R(\hat{\beta})/n\} - p \log n, \qquad (4.11)$$

after constants irrelevant to the comparison of models are ignored.

4.11 Generalized BIC for Outlier Detection

We recall that use of BIC provides consistent selection of the true model, if that is included in the set of models under consideration. Justifications of the word "Bayesian" in the name BIC are given, among others by Claeskens and Hjort (2008, p. 78) and Bhat and Kumar (2010), expanding the original cryptic presentation of Schwarz (1978).

In the use of BIC in the choice of a regression model, the comparison is between models with different terms included or removed. As a preliminary to the results of Sect. 4.11.3, we consider BIC for nested regression models. Let the true model be the linear model with $p \times 1$ parameter β_p and $n \times p$ matrix of explanatory variables X_p. A model with $q \times 1$ parameter β_q, $q < p$ will be called *false* and a model with r parameters, $r > p$ is called *correct* provided $\beta_q \in \beta_r$, but is not minimal. For asymptotic results, we require the stability of the information matrix, namely, that $X_p^T X_p / n \xrightarrow{p} M_X$ with $\det(M_X) \neq 0$.

We first test a false model. The likelihood ratio test for β_q against β_p has an asymptotic non-central χ^2_{p-q} distribution with non-centrality parameter λ, which increases as n. The BIC penalty for the comparison of these two models is $(p-q)\log n$, increasing more slowly with n, so that the true model will be chosen as n increases.

For a correct model, the likelihood ratio test for β_p against β_r has asymptotically a central χ^2_{r-p} distribution and the BIC penalty is $(r-p)\log n$. Thus, for large n the true model will be preferred. Putting these two together demonstrates the consistency of the BIC.

4.11.2 Mean Shift Outlier Model, FS, LMS, and LTS

Use of the forward search or least trimmed squares to provide robustness against outliers leads to the comparison of fitted models with differing numbers of observations. We render outlier detection and deletion compatible with BIC through use of the mean shift outlier model of Sect. 3.2.3 in which deleted observations are each fitted with an individual parameter, so having a zero residual.

To stress that the extended BIC can be used to monitor LMS and LTS, as well as the FS, in this section, we change notation so that the FS fits subsets of observations of size h to the data, with $h_0 \leq h \leq n$. The subset used in fitting is now called $S^*(h)$. Then $n - h$ observations will have been deleted. In (3.16), that is, in the mean shift outlier model

$$y = X\beta + D\phi + \epsilon,$$

D will now be an $n \times (n - h)$-dimensional matrix.

Conditions under which application of the extended BIC to the FS gives a correct indication of the proportion of outliers in the data require the consistency of the FS estimator when the data contain no outliers, proved by Cerioli et al. (2014) for multivariate data and by Johansen and Nielsen (2016) for regression. We now allow outliers in the data generating distribution.

We consider the distribution of residuals. In the absence of outliers in $S^*(h)$, the distribution of $e_i^2(h) \xrightarrow{p} \sigma^2 \chi_1^2$. The residuals for individual outliers have asymptotically the non-central chi-squared distribution $\sigma^2 \bar{\chi}_1^2(\lambda_{hi})$. Let the outliers belong to $\mathcal{H}^0$. Asymptotic results about robustness require a separation condition. We assume that, for all $i \in \mathcal{H}^0$, $\lambda_{hi} = o(n)$. This rather strong, but standard, separation condition requires outliers to be increasingly far from the clean observations as n grows, in such a way that the non-centrality parameter of the resulting chi-squared distribution grows faster than n.

Let the uncontaminated observations belong to the set $\mathcal{H}$ of cardinality h^*. We call a *correct* ordering of the observations one in which $S^*(h^*) = \mathcal{H}$; that is, that at step h^* there are no outlying observations in $S^*(h^*)$. Of course, there are then no uncontaminated observations left in the remaining $n - h^*$ observations.

4.11.3 Extended BIC for Outlier Detection

To incorporate deletion of observations in BIC (4.11), let the residual sum of squares for some parameter estimate b, when $n - h$ observations are deleted, be $R_h(b)$. To allow for the additional parameters in (3.16), BIC (4.11) is accordingly replaced by

$$\text{BICH} = -n \log\{R_h(\hat{\beta}_h)/h\} - (p + n - h) \log n. \qquad (4.12)$$

The rationale in (4.12) is that, under the model, all observations are still included in the estimation set (hence the use of n), but only h are used for computation of the residual sum of squares. Finally, $(p + n - h)$ is the number of parameters. Results in Riani et al. (2022b) prove that the maximum value of BICH leads to a correct ordering of the observations, which is the result required for outlier detection.

In (4.12) σ^2 is estimated by $\{R_h(\hat{\beta}_h)/h\}$. For the FS we improve this estimate by use of the consistency factor (3.49) and scale up the value of $R_h(\hat{\beta}_h)$ to obtain the corrected BIC

$$\text{BICW} = -n \log[R_h(\hat{\beta}_h)/\{h\sigma^2(h)\}] - (p + n - h) \log n. \qquad (4.13)$$

For LTS we use two factors: the consistency factor (3.49) and, in addition, the small-sample correction factor (see Sect. 3.11.2).

4.11.4 Finite Sample Extended BIC for Outlier Detection

We note that the results above also apply to monitored LTS, in which the best subset $S^*(h)$ is found for a range of values of h, rather than for a single specified trimming value as in the original proposal (Rousseeuw, 1984). The difference is then in the algorithms used for calculating $S^*(h)$, $h_l \leq h < n$, where h_l is the lower limit of subset size that is of interest. If both algorithms result in the same value of $S^*(h)$ over the range of h, the residual sums of squares will be identical as will be the values of BICH.

Let $h^\dagger$ be the size of the subset for which BICW, the rescaled value of BIC, (4.13) is maximized. The theorem mentioned in Sect. 4.11.3 proves that asymptotically $S^*(h^\dagger)$ will not include any outliers. However, the asymptotic conditions may not be satisfied for finite sample sizes and small separation of the outliers. Then $S^*(h^\dagger)$ may contain some outliers and miss some non-outlying observations. Furthermore, in the analysis of data the outliers may not be very large and outlying observations will appear less so as the FS progresses and the parameter estimates are corrupted once some outliers are included in $S^*(h)$. Although the value of $R_h(\hat{\beta}_h)$ will continue to increase with h it may not do so sufficiently fast to outweigh the decrease in the penalty of $-h \log n$. As a consequence, the value of BICW may increase for $h > h^\dagger + 1$.

For hard trimming (4.13) can be rewritten with weights $w_i = 0, i \in n - h$ and one otherwise as

$$\text{BICW} = -n \log \left[R_h(\hat{\beta}_h) / \{\sigma^2(h) \sum_{i=1}^{n} w_i\} \right] - \left\{ p + \sum_{i=1}^{n} (1 - w_i) \right\} \log n, \quad (4.14)$$

where $\sum_n w_i = h$.

4.11.5 Example: Mental Illness Data

We now illustrate the finite sample properties of the procedure through the comparison of analyses using LTS and the FS. We use monitoring plots of BICW to illuminate the procedure. But, for automatic outlier detection, we are not initially interested in monitoring over a series of grids, but in selecting a single value for further investigation.

Kleinbaum and Kupper (1978, p. 148) describe 137 observational data on the assessment of mental illness of 53 patients. The data come from a psychiatrist's assessment of mental retardation and degree of distrust of doctors in newly hospitalized patients. After 6 months of treatment, a value is assigned for the degree of illness of each patient. Atkinson et al. (2021) showed that when degree of illness is regressed on the two initial assessments, there is strong evidence for transformation of the response. The Box–Cox transformation indicates the log transformation. After transformation the data are well behaved. We study the effect of outliers by modifying three of the smallest observations (17, 30, and 53), setting them equal to one. This contamination causes the log transformation to be rejected.

We consider least squares analysis of the logged contaminated data. The left-hand panel of Fig. 4.43 shows the plot of BICW for LTS estimation and the right-hand panel that for FS. Both show the almost linear increase as more observations are included in the fit until a peak for h near n, after which there is a sharp decline. For the FS the peak is at $h = 50$. For LTS the bdp at the maximum is 0.07, which indeed corresponds to $h = 50$. Here both LTS and the FS have correctly identified the three outliers without the need for a complicated outlier detection rule. The stepping of LTS arises because, for breakdown point d, h is found as $\lceil n(1 - d) \rceil$.

4.11.6 Further Examples and Developments

Riani et al. (2022b) explore the properties of the extended BIC through simulations and data analyses. They also provide versions of the criterion for two soft trimming methods. The simulations focus on the comparison between the FS and monitored LTS estimation. In part because of the restriction to 50 values of h when monitoring LTS, the FS has superior performance. However, the LTS completely fails to identify outliers in an example where the outliers are at leverage points.

Two further data analyses are of the stars data, introduced in Sect. 4.5.1 and of the Balance Sheet data, analysed in Sect. 6.2.6. For the stars data, both methods detect six outliers, as in Sect. 4.9.2. The application of the extended BIC to the Hawkins data is in Exercise 4.5.

Theoretical developments, again illustrated by data analyses, are to soft trimming methods, specifically S- and MM-estimation. Application of the extended BIC to soft trimming requires the identification of outliers, which are defined as having iterative weights less than 0.01. Interpretation of the peak value of BIC is provided

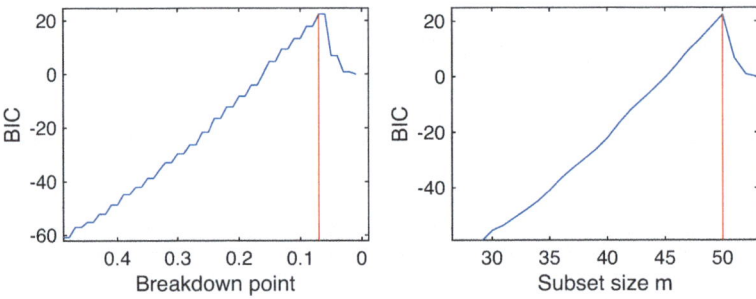

Fig. 4.43 Logged mental illness data: monitoring plots of BICW. Left-hand panel, Least trimmed squares (LTS); right-hand panel, Forward Search (FS). The stepping for LTS arises from a search over 50 values of h

by monitoring plots of the weights. These methods are compared with a development of a criterion for robust model selection of Maronna et al. (2019), who use Akaike's AIC (Akaike, 1974) to penalize more complex models. Replacement of AIC by BIC should lead, asymptotically, to consistent selection of the correct model.

Problems

4.1 AR data: standard static way of data analysis (non-robust and robust)
(a) Show the ANOVA table based on all observations and that after deleting unit 43.
(b) Use MM-estimation with a nominal efficiency of 0.95 and the hyperbolic ρ function to perform outlier detection using a confidence level of 0.975. Show the plot of the weight for each observation and the scaled residuals. Show the ANOVA table based on the units not declared as outliers. Compare the scaled residuals from the hyperbolic ρ function with those from the Tukey biweight in Fig. 3.9.
(c) Compute the scaled residuals coming from MM-estimation based on a nominal efficiency of 0.95 using Andrews' sine and the power divergence ρ functions. Compare them with those based on the Tukey biweight and the hyperbolic ρ functions.

4.2 Hawkins data: brushing residuals from monitoring S residuals
Figure 4.29 shows the monitoring of residuals as a function of *bdp* using, left-hand panel, Tukey's biweight and, right-hand panel, power divergence ρ functions. In one of the two panels:
(a) Brush the residuals shown in pale blue and show their entry order in the monitoring plot of the minimum deletion residuals.
(b) Brush the six largest residuals (in absolute value) and show their position in the monitoring plot of the minimum deletion residuals.

4.3 Surgical Unit data: analysis to test the difference between the two groups
Test the difference between the two groups by adding a dummy variable for the observations which come from Unit 1.
(a) Calculate the ANOVA table and comment on the importance of this new variable.
(b) Plot the monitoring of the five added variable t statistics.

4.4 Surgical Unit data: traditional robust analysis
Show the residuals coming from S regression using a *bdp* = 0.5 and a *bdp* = 0.25 using the ρ function for TB. Repeat the analysis using the PD ρ function. Comment on the difference between these two sets of residuals.

4.5 Hawkins data: BIC monitoring
Show and comment on the plots of BICW for LTS and FS estimation.

Open Access This chapter is licensed under the terms of the Creative Commons Attribution 4.0 International License (http://creativecommons.org/licenses/by/4.0/), which permits use, sharing, adaptation, distribution and reproduction in any medium or format, as long as you give appropriate credit to the original author(s) and the source, provide a link to the Creative Commons license and indicate if changes were made.

The images or other third party material in this chapter are included in the chapter's Creative Commons license, unless indicated otherwise in a credit line to the material. If material is not included in the chapter's Creative Commons license and your intended use is not permitted by statutory regulation or exceeds the permitted use, you will need to obtain permission directly from the copyright holder.

Chapter 5
Practical Comparison of the Different Estimators

Abstract Results for small n and moderate p show that the size and power of robust tests of regression models can be far from the nominal asymptotic values. This chapter uses simulation to investigate the properties of outlier tests for moderate sample sizes. Section 5.1 compares the size of outlier tests for 30 very robust estimators, which leads to the selection of five estimators for comparison in the next sections: S, MM, LTS, and LTSr, the reweighted LTS estimator, together with FS. Section 5.2 compares the sizes of the five test for n from 100 to 1,000 and Sect. 5.3 compares the average power (the proportion of outliers correctly identified) over a range of values of the shift in the contamination. Section 5.4 describes a parametric framework for comparing robust regression estimators. In this a set of outliers is moved along some trajectory in the space of y and X and the effect on inferences calculated; typically bias and variance of parameter estimates and average power. In all these comparisons, all estimators, apart from FS, have fixed, very robust, settings. Monitoring the comparisons is provided in Sect. 5.6 by using the extended BIC (Sect. 4.11) to determine the robustness level at which the properties of the various estimators are assessed.

5.1 False Signals and Empirical Size: Outliers from High Breakdown Regression

In Chap. 4, we compared robust estimators for regression by monitoring their behaviour in the analysis of various sets of data. In this chapter, we use simulation to compare the size and power of tests in robust regression. Earlier results suggest that for small n and moderate p the sizes of the tests can be far from the nominal values. For example, Rousseeuw and Van Zomeren (1990) describe a very robust method ($bdp = 50\%$) for the detection of outliers in multivariate data using algorithms based on elemental subsets. In a discussion of this paper, Cook and Hawkins (1990) examined fits from large numbers of elemental sets. They titled the contribution summarizing their experience "Outliers Everywhere". We see similar results here for small-sample sizes. There are several related questions: which robust methods lead to outlier tests with good size and power; does the ordering of the methods depend on the sample size and, furthermore, if the size is incorrect, can it be adjusted; if so, what is the effect on power?

Section 5.1.1 extends the simulation results of Salini et al. (2015) on the size of outlier tests for many estimators to include the PD estimator. The test statistic is a chi-squared approximation to the distribution of the scaled squared residuals, that is, the square of the normal statistic in (3.17). The properties of this approximation are investigated, in the absence of outliers, in Sect. 5.1.4. The remainder of the chapter considers five estimators: S, MM, LTS, LTSr, and FS. The sizes of outlier tests are compared in Sect. 5.2 and their power in Sect. 5.3. The theory of a general numerical procedure for finding the non-asymptotic bias, variance, and power of outlier tests is in Sect. 5.4, with numerical examples in Sect. 5.5. We conclude in Sect. 5.6 by comparing the results from using the extended BIC of Sect. 4.11 in several situations, including point-mass contamination at leverage points, for which LTS fails.

5.1.1 The Sizes of Many Individual and Simultaneous Tests for Outliers

Our analysis compares robust procedures using outlier tests based on the scaled residuals $r_i = e_i/s$, defined in Sect. 3.1.1 for LS. Here the parameters β and the error variance σ^2 are estimated robustly. For individual outliers we test using the value of r_i or equivalently its absolute value (3.17) or its squared value as in (3.51), in which case the reference distribution is χ_1^2. We are thus assuming that the scaled residuals can be taken as being normally distributed, all with variance one. For LS this becomes true as $n \to \infty$ and under the conditions on $X^T X$ given in Sect. 3.7.2. Comments on the approximate normality of the r_i for robust estimation are similar to those in Sect. 3.7.2 on the tests of hypotheses about robustly estimated parameters of the linear model.

The threshold to test the n individual hypotheses is given by $\chi^2_{1,0.99}$ for all procedures. The empirical size of the test for individual outliers is the proportion of *observations* identified as being outlying. For simultaneous tests (Exercise 5.1) we use the Bonferroni threshold $\chi^2_{1,1-0.01/n}$, given in (3.18). The empirical size of the test is the proportion of *samples* identified as containing one or more outlier.

The situation for the FS is slightly more complicated since it is designed to have an approximate simultaneous size of 1%. The individual test sizes are found by using the parameter estimates from the FS to calculate the n individual residuals and then subjecting each to the χ_1^2 test of (3.51).

For estimation of the size of outlier tests, calculations using r_i, $|r_i|$, or r_i^2 all give identical results, given that comparisons are with the appropriate reference distribution. However, the signed values of the r_i are to be preferred in the analysis of data when, for example, sets of adjacent residuals of the same sign may be an indication of lack of fit of the model.

5.1.2 Simulation Results

In the simulations of this section, the focus is on comparing the behaviour of an appreciable number of robust procedures in a few situations. For flexible trimming, we evaluate the behaviour of S-estimation using all of the ρ functions of Sect. 2.3.2 except the Huber function. Extending the results of Salini et al. (2015), we also include the PD ρ function. For MM-estimation, we give results on the five ρ functions, for three values of *eff*: 0.95, 0.9, and 0.85. Two hard trimming methods are included in our results: LTS and LTSr as well as the adaptively trimming FS. Since most extreme conditions are often the most informative, we consider two values of *bdp*: 50 and 25%, where this is an appropriate adjustable parameter. The abbreviations used for these estimators are listed in Table 5.1. We took two values of n: 50 and 200 and also two values of p: 2 and 10. For each set of values of n and p we simulated 10, 000 replicates from a first-order model with standard normal errors; the values of $X_1, \ldots, X_p$, sampled independently from standard normal distributions, were fixed for each simulation. Note that, due to the equivariance of the estimators (see Sect. 2.2), we can for simplicity generate data from the standard normal distribution both for the errors and for the explanatory variables. The outlier tests were performed with 0.01 level both individual and simultaneous.

The results of the simulations are presented in four figures. In order we work from the smallest to the largest. That is $n = 50$, $p = 2$; $n = 200$, $p = 2$; $n = 50$, $p = 10$; and, finally, $n = 200$, $p = 10$. The most severe conditions, that is, those most likely to show departures from a 1% size are $n = 50$, $p = 10$, and the least severe, $n = 200$, $p = 2$.

Figure 5.1 shows the results for $n = 50$ and $p = 2$. As in all figures of this kind, the size of the independent tests is in the upper panel, with the size for simultaneous tests in the lower panel. This pairing emphasizes, for a given n and p, the remarkable similarity of the shapes of the profiles, over the various robust procedures, of the sizes of the tests for individual and simultaneous outliers. In all figures the sizes of the simultaneous tests are appreciably greater than those of the individual tests. There is also a more general structure in all eight panels of the four figures.

The red horizontal line represents the nominal size of 1%. The exceptions to the large empirical sizes are results for LS and the FS, represented by the black dots in the lower corners of the plots. These often seem to even lie below the nominal 1%, a point we investigate analytically and numerically in Sect. 5.1.4.

The structure of Fig. 5.1 is readily described. In both panels it does not much matter which ρ function is used for S-estimation. However, the improvement in size when a *bdp* of 25% is used is pronounced, with a reduced difference in the sizes for the various ρ functions. The size of the tests for the MM-estimators lies between the two values for S-estimation and does not much depend on the value of *eff* or on the ρ function, although OPT is worst, especially for the individual tests. The worst performance is that of very robust LTS, although moving to a *bdp* of 0.25 leads to sizes similar to those for MM-estimation. The size of LTSr for both values of *bdp* are

Table 5.1 Nomenclature for (mostly robust) regression procedures, for different values of *bdp*, *eff* and five ρ functions

Acronym	Description		
LS	Classical (Ordinary) Least squares estimator		
Sbdp050TB	S-estimator	With Tukey's ρ function	and $bdp = 0.5$
Sbdp050HA	S-estimator	With Hampel's ρ function	and $bdp = 0.5$
Sbdp050OP	S-estimator	With Optimal ρ function	and $bdp = 0.5$
Sbdp050HY	S-estimator	With Hyperbolic tangent ρ function	and $bdp = 0.5$
Sbdp050PD	S-estimator	With Power divergence ρ function	and $bdp = 0.5$
Sbdp025TB	S-estimator	With Tukey's ρ function	and $bdp = 0.25$
Sbdp025HA	S-estimator	With Hampel's ρ function	and $bdp = 0.25$
Sbdp025OP	S-estimator	With Optimal ρ function	and $bdp = 0.25$
Sbdp025HY	S-estimator	With Hyperbolic tangent ρ function	and $bdp = 0.25$
Sbdp025PD	S-estimator	With Power divergence ρ function	and $bdp = 0.25$
MMeff085TB	MM-estimator	With Tukey's ρ function	and $eff = 0.85$
MMeff085HA	MM-estimator	With Hampel's ρ function	and $eff = 0.85$
MMeff085OP	MM-estimator	With Optimal ρ function	and $eff = 0.85$
MMeff085HY	MM-estimator	With Hyperbolic Tangent ρ function	and $eff = 0.85$
MMeff085PD	MM-estimator	With Power Divergence ρ function	and $eff = 0.85$
MMeff090TB	MM-estimator	With Tukey's ρ function	and $eff = 0.90$
MMeff090HA	MM-estimator	With Hampel's ρ function	and $eff = 0.90$
MMeff090OP	MM-estimator	With Optimal ρ function	and $eff = 0.90$
MMeff090HY	MM-estimator	With Hyperbolic Tangent ρ function	and $eff = 0.90$
MMeff090PD	MM-estimator	With Power Divergence ρ function	and $eff = 0.90$
MMeff095TB	MM-estimator	With Tukey's ρ function	and $eff = 0.95$
MMeff095HA	MM-estimator	With Hampel's ρ function	and $eff = 0.95$
MMeff095OP	MM-estimator	With Optimal ρ function	and $eff = 0.95$
MMeff095HY	MM-estimator	With Hyperbolic tangent ρ function	and $eff = 0.95$
MMeff095PD	MM-estimator	With Power divergence ρ function	and $eff = 0.95$
LTSbdp050	LTS estimator	With $bdp = 0.5$	
LTSbdp025	LTS estimator	With $bdp = 0.25$	
LTSrbdp050	Reweighted LTS estimator	With weights	and $bdp = 0.5$
LTSrbdp025	Reweighted LTS estimator	With weights	and $bdp = 0.25$
FS	Forward search		

5.1 False Signals and Empirical Size: Outliers from High Breakdown Regression 155

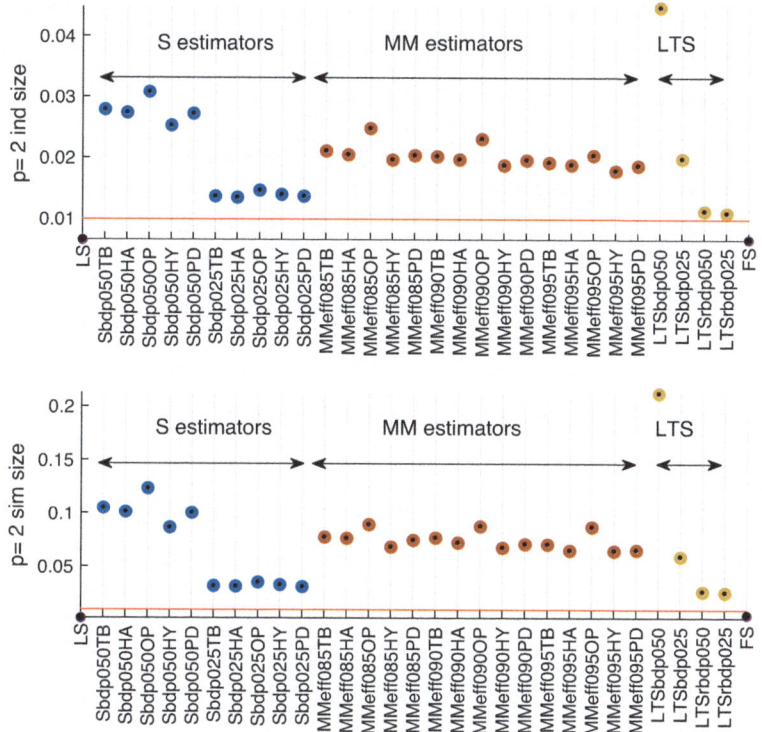

Fig. 5.1 Empirical size of outlier tests, with nominal size 1%, for the robust procedures given in Table 5.1 when $n = 50$ and $p = 2$. Upper panel, individual size; lower panel, simultaneous size

similar to S-estimation with $bdp = 0.25$. However, the results of Sect. 5.3 show that LTSr has poor power.

Figure 5.2 shows the results for the least challenging situation, that is, with $n = 200$ and $p = 2$. With this larger value of n, the values of all sizes are roughly halved compared to those in Fig. 5.1 (the sample size is quadrupled). The structure of the plot is however similar to that of Fig. 5.1, but with less relative difference between the rules within a class; in particular, the variability in performance of the MM-estimators is more uniform, with the performance of the OPT-estimator less egregious.

On the other hand, Fig. 5.3 shows the results for the most challenging set of conditions. Unsurprisingly, the sizes are all increased in comparison to those in Fig. 5.1 for $p = 2$. Except for LTS and LTSr, the structure for the individual test sizes is similar to that for $p = 2$. However, for the simultaneous sizes, there are assuredly "outliers everywhere". The majority of estimators have a size of one—every sample contained at least one apparently outlying observation. The exceptions are S-estimation and LTSr with bdp of 0.25. Only the FS and LS have sizes anywhere near the nominal value of 1%.

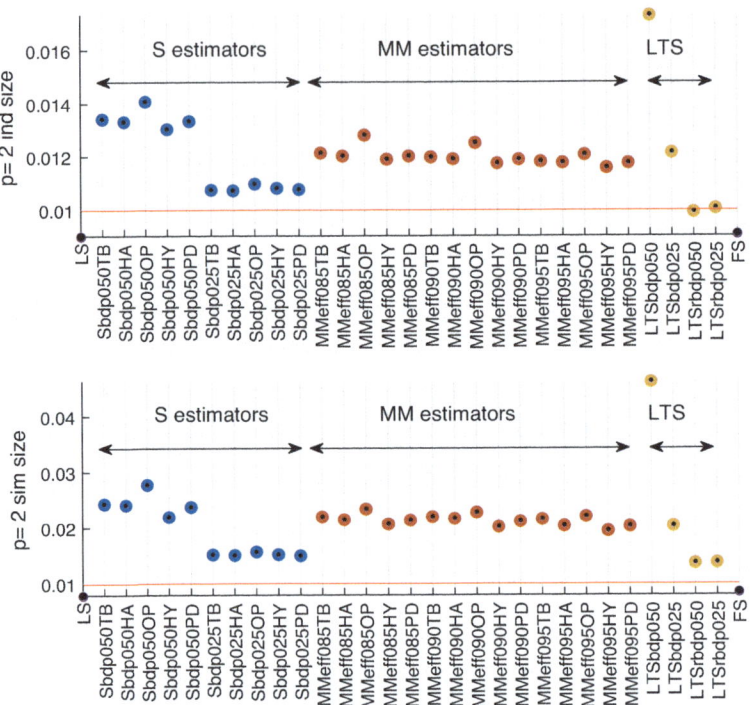

Fig. 5.2 Empirical size of outlier tests, with nominal size 1%, for the robust procedures given in Table 5.1 when $n = 200$ and $p = 2$ (the least severe case). Upper panel, individual size; lower panel, simultaneous size

Increasing the sample size to $n = 200$ in Fig. 5.4 with $p = 10$ shows a great improvement over the results in Fig. 5.3. The sizes are reduced approximately sevenfold, which is not just the effect of quadrupling n, which halved the sizes in going from Figs. 5.1 to 5.2. For the individual tests, the structure is again similar to that of Fig. 5.1, although OPT behaves relatively worse. The sizes are also close to those of Fig. 5.1, rather than to those of Fig. 5.2, which is also for $n = 200$. For the simultaneous tests in the lower panel, several of the S-estimators behave worse than LTS. Apart from LS and the FS, the sizes for LTSr are again closest to α.

In later sections of this chapter, we further study the size, as well as the power of five robust methods that these four figures suggest are typical. We combine the TB with S-estimation with $bdp = 50\%$ and with MM-estimation, with $eff = 0.85$. The most robust versions of LTS and LTSr are also included, that is, again with $bdp = 50\%$. There are thus four very robust estimators with specified asymptotic properties to which we add the FS, the only fully adaptive procedure.

5.1 False Signals and Empirical Size: Outliers from High Breakdown Regression

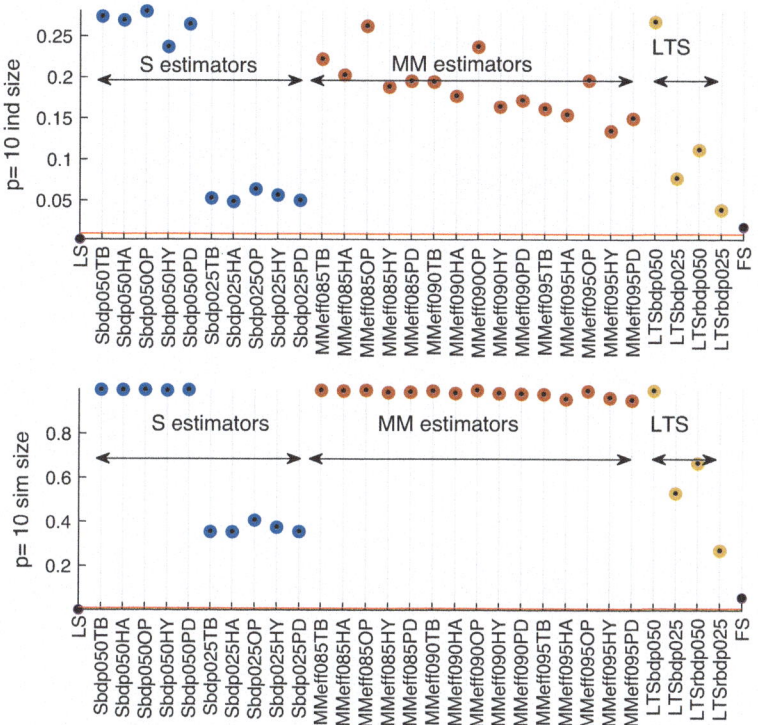

Fig. 5.3 Empirical size of outlier tests, with nominal size 1%, for the robust procedures given in Table 5.1 when $n = 50$ and $p = 10$ (the most severe case). Upper panel, individual size; lower panel, simultaneous size

5.1.3 Distributional Issues

An inferential problem with both hard and soft trimming procedures is that s^2 underestimates σ^2 when n is small or moderate, even if the Tallis consistency correction (3.49) is applied. This negative bias led Pison et al. (2002) to the simulation-based finite sample correction for LTS mentioned in Sect. 3.11.2. Maronna and Yohai (2010) discuss the same phenomenon for MM-estimation when there are relatively many explanatory variables compared to n.

Salini et al. (2015) extend the calculation of finite sample correction factors to hard trimming methods. They start with an analysis of the empirical distribution of the values of r_i^2, using the robust methods of Table 5.1, except for those with the PD ρ function. Instead of considering just the size of the test, a statistic based on a single point in the distribution of r_i^2, they use a chi-squared goodness-of-fit test for the distribution of the squared scaled residuals. The outlier detection rule assumes this to be χ_1^2. If this distribution holds, the significance levels of the outlier tests are uniformly distributed. This uniformity can be checked by use of a further chi-squared

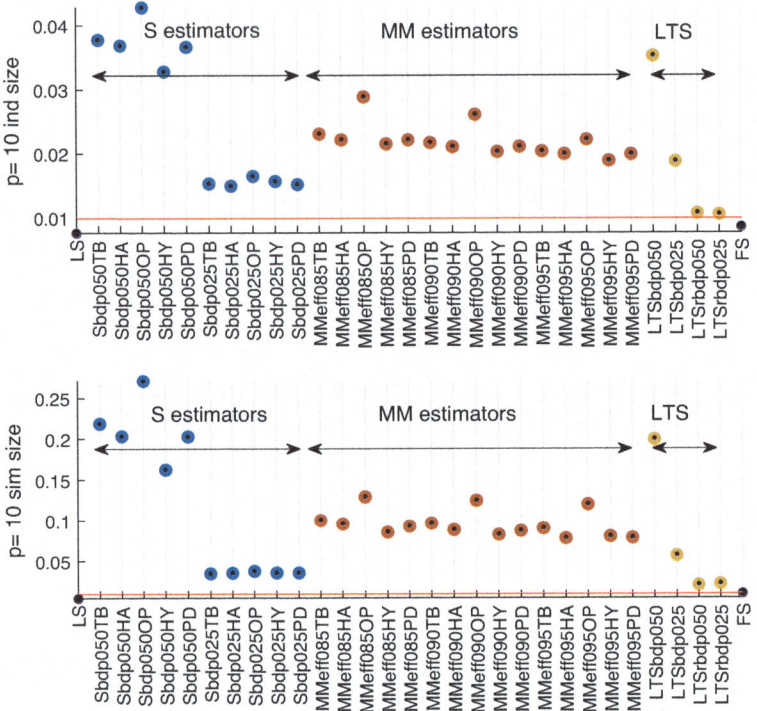

Fig. 5.4 Empirical size of outlier tests, with nominal size 1%, for the robust procedures given in Table 5.1 when $n = 200$ and $p = 10$. Upper panel, individual size; lower panel, simultaneous size

test, now of the observed significance levels; the number of categories depends on n. The procedure leads to the detection of estimation methods for which the χ_1^2 approximation does not hold. The rules that are unsatisfactory by this criterion are those which have the largest sizes for the outlier tests in Figs. 5.1, 5.2, 5.3, and 5.4.

The unsatisfactory rules all produce negatively biased estimates of σ^2. A major consequence of underestimating the residual scale is that the number of outliers flagged in each sample is typically far from the one predicted from the binomial distribution for the number of outliers under the normal model with σ^2 known. Simulations provide Salini et al. (2015) with estimates of s^2 for a grid of values of n and p, the values of s^2 varying smoothly with both quantities. Their Fig. 6 shows the curves of the estimates smoothly increasing to one from below, those for $p = 10$ converging appreciably more slowly than those for $p = 2$. From the standpoint of data analysis, a disturbing feature of this figure is the distance between the 1 and 99% quantiles of the simulated values of s^2.

Reliable diagnostics, in the sense of having a size close to nominal, can be obtained by scaling the value of s^2 from a particular data analysis by the estimate from simulation. Salini et al. (2015) provide methods of interpolation for these estimated values and show that, in the absence of outliers, they do lead to procedures with the desired size. However, the adoption of these correction factors, which downweight the estimated residuals, produces some loss in the power to detect contaminated observations. The decrease in power depends upon the form of contamination and the robust method, but, in all cases, there is some decrease. A second problem is that, once the outliers have been identified, a second application of the correction factor is required with a reduced sample size. In the FS, this problem is overcome, as described in Sect. 4.3, through the use of the resuperimposition of envelopes for outlier detection with increasing subsample sizes. Since our power comparisons in the rest of this chapter show that the FS, which does not require correction, has the highest power, we exclude from our comparisons estimators with the correction factors suggested by Salini et al. (2015).

5.1.4 The Chi-Squared Approximation to the Distribution of Scaled Residuals

There are two conclusions from the figures of Sect. 5.1.2. The first is that virtually all procedures have sizes that are too large; in the case of simultaneous tests and small n these sizes are so large that, when $n = 50$ and $p = 10$, there do indeed seem to be "outliers everywhere". The exceptions to test sizes that are appreciably too large are provided by LS, in all figures, and the FS for $p = 2$. Some of these sizes were very much smaller than nominal. We now use distributional arguments to analyse this situation for LS and present some numerical results for both LS and FS.

In the comparisons of Sects. 5.1.2 and 5.1.3, the presence of outliers was assessed by testing squared scaled residuals $r_i^2 = e_i^2/s^2$, on the assumption that the null distribution of r_i^2 is χ_1^2. We argued that this was a useful approximation for robust procedures, such as S-estimation. To investigate the small sizes observed for FS and LS we proceed by developing the remark in Sect. 3.1.1 that e_i^2/s^2 has a scaled beta distribution, rather than the chi-squared approximation used in the earlier sections of this chapter.

It follows from (3.5) that e_i is normally distributed with mean zero and variance $(1 - h_i)\sigma^2$. Then $e_i^2/\{(1 - h_i)\sigma^2\} \sim \chi_1^2$. However, when σ^2 is estimated by s^2, r_i^2 has a distribution with finite support. To see this, let $S(\hat{\beta})$ be the residual sum of squares with $n - p$ d.f. $S(\hat{\beta})$ is the sum of all n residuals including e_i^2. From Equation (3.53), the residual sum of squares can be written as

$$S(\hat{\beta}) = \frac{e_i^2}{1 - h_i} + S(\hat{\beta}_{(i)}), \tag{5.1}$$

where $S(\hat{\beta}_{(i)})$ is the residual sum of squares when observation i is deleted. Then, since $s^2 = S(\hat{\beta})/(n - p)$,

$$r_i^2 = \frac{e_i^2}{s^2} = \frac{(n-p)e_i^2}{S(\hat{\beta})} = \frac{(n-p)e_i^2}{e_i^2/(1-h_i) + S(\hat{\beta}_{(i)})}. \tag{5.2}$$

When $e_i^2 = 0$, $r_i^2 = 0$. The other extreme is when $S(\hat{\beta}_{(i)}) = 0$ (the only non-zero residual is e_i) when, from (5.2), $r_i^2 = (n-p)(1-h_i)$.

To find the distribution of r_i^2 we rewrite (5.2) as

$$\frac{r_i^2}{(n-p)(1-h_i)} = \frac{e_i^2/(1-h_i)}{e_i^2/(1-h_i) + S(\hat{\beta}_{(i)})}. \tag{5.3}$$

Here $e_i^2/(1-h_i)$ is distributed as $\sigma^2 \chi_1^2$ independently of the $\sigma^2 \chi_{n-p-1}^2$ distribution of $S(\hat{\beta}_{(i)})$. In general, let X_ν be a chi-squared random variable on ν d.f. This chi-squared distribution is a special case of the gamma distribution with parameters $1/2$ and $\nu/2$. In (5.2) interest is in the distribution of $y_B = X_1/(X_1 + X_{n-p-1})$, where X_1 and X_{n-p-1} are independent. Then $y_B \sim \mathcal{B}(1/2, (n-p-1)/2)$, a beta distributed random variable.

From (5.2), the beta distribution needs to be scaled for the range of the distribution of r_i^2 and for the variance of e_i. Then

$$\tilde{r}_i^2/(n-p) = r_i^2/\{(1-h_i)(n-p)\} \sim \mathcal{B}\{1/2, (n-p-1)/2\}. \tag{5.4}$$

See, for example, (Cook and Weisberg, 1982, p.19) who show the independence of the χ^2 distributions of e_i^2 and $S(\hat{\beta}_{(i)})$. Here we are not concerned with the closeness of the two distributions over their whole range, but with the comparison of the 99% points of the distributions. Exercise 3.7 illustrates the comparison over an extended range for specific $\mathcal{F}$ and beta distributions.

Individual outliers are declared, in general, using the threshold $x_\alpha = \chi_{1,1-\alpha}^2$, which, for probability calculations, has to be scaled for conversion to a beta variate. The scaled value is $x_\alpha/\{(1-h_i)(n-p)\}$, dependent on h_i, which was not included in the outlier detection rule. Since, in our simulations, we have generated the explanatory variables from normal distributions, the occurrence of leverage points is slight. We accordingly replace h_i by the average value p/n, to obtain the threshold for the beta distribution as

$$x_\alpha^B = \frac{n x_\alpha}{(n-p)^2}. \tag{5.5}$$

If the cdf of the Beta distribution is denoted $F_B(x; \nu_1, \nu_2)$, the probability of declaring an individual outlier is

$$p_I = 1 - F_B(x_\alpha^B; 1/2, (n-p-1)/2). \tag{5.6}$$

Simultaneous outliers, on the other hand, are detected using the Bonferroni inequality, giving the threshold $x_{\alpha^*} = \chi_{1,1-\alpha/n}^2$. Then, from (5.5), the threshold for the beta distribution for an individual observation to be declared outlying is

5.1 False Signals and Empirical Size: Outliers from High Breakdown Regression

Table 5.2 Size of tests for individual outliers: theoretical results for the χ^2 approximation calculated from (5.6). 100,000 simulations for LS, FS, and for the deletion residual (DelnRes.). Nominal size 0.01. Percentages times 10 (nominal = 10)

Number of parameters		Number of observations n			
	Procedure	50	100	200	2000
$p=2$	Theory	7.167	8.579	9.289	9.929
	LS	6.560	8.228	9.102	9.905
	FS-χ^2	6.667	8.273	9.143	9.911
	DelnRes.	10.005	9.994	9.995	9.999
$p=10$	Theory	2.776	5.956	7.884	9.781
	LS	2.581	5.775	7.735	9.760
	FS-χ^2	17.614	7.742	8.284	9.768
	DelnRes.	10.018	10.013	9.999	9.997

$$x_{\alpha^*}^{B} = \frac{n x_{\alpha^*}}{(n-p)^2}.$$

The probability that this observation is not considered an outlier is, from (5.6),

$$p^* = F_B(x_{\alpha^*}^{B}; 1/2, (n-p-1)/2).$$

The size of the simultaneous test, that is the probability that no outliers are declared, is therefore (Exercise 5.1)

$$p_S = 1 - (p^*)^n. \tag{5.7}$$

In our calculations, we continue to take $\alpha = 0.01$.

Table 5.2 summarizes the results of a simulation experiment to explore the dependence of the size of tests for individual outliers on the value of n. The values of p in the table, 2 and 10, are, respectively, situations in which the presence of explanatory variables can be expected to have a slight or an appreciable effect. This effect should decrease as n goes from 50 to 2,000. The first three rows use the χ^2 approximation to test for outliers. In the case of the FS-χ^2, the residuals from the fitted FS model were tested using the χ^2 approximation. Since these are calculations for size, the parameters β for the explanatory variables in the linear model were set to zero. The fourth line of the table, "DelnRes.", uses the deletion residual (see Sect. 3.2.1), the square of which was tested against $\mathcal{F}(1, n-p-1)$.

The results in the table are presented as ten times the nominal size as a percentage, so the target value is 10. For $p=2$ the results for LS and FS-χ^2 agree well with the predictions of the theory (5.6). The values for the test using the deletion residuals are close to ten, the value to which the sizes of the other tests tend as n increases. For $p=10$ the scaled sizes of the three empirical tests again tend to 10; LS agrees with the theory, but, when $n=50$, FS-χ^2 has a size around 18, rather than 10. For

Table 5.3 Size of tests for simultaneous outliers: theoretical results for the χ^2 approximation calculated from (5.7). 100,000 simulations for LS, FS, and for the deletion residual (DelnRes.). Nominal size 0.01. Percentage times 10 (nominal = 10)

Number of parameters		Number of observations n			
	Procedure	50	100	200	2000
$p = 2$	Theory	2.303	4.722	6.630	9.374
	LS	2.080	4.560	6.640	9.380
	FS(χ^2)	4.650	5.990	7.740	9.480
	FS	4.810	6.490	8.230	12.680
	DelnRes.	9.550	9.890	9.880	9.900
$p = 10$	Theory	0.151	1.876	4.383	8.969
	LS	0.200	1.999	4.300	9.020
	FS(χ^2)	55.24	13.60	8.830	9.400
	FS	53.89	13.99	9.520	12.79
	DelnRes.	10.17	10.36	10.10	10.06

larger n the size of FS-χ^2 is below 10. As before the deletion residual test has a size very close to 10.

Table 5.3 gives the related results for simultaneous outliers, the theoretical results being calculated from (5.7). As the results, especially, of the lower panel of Fig. 5.3 show, the size of simultaneous outlier tests is far from the nominal value when n is small and p is large.

There are now five tests to be compared. LS is the result for the Bonferronized version of the chi-squared test. The FS is designed for the testing of simultaneous outliers, so test results are not modified. FS-χ^2 uses the Bonferronized version of the χ^2 test, with the fitted FS model used to identify outliers and DelnRes. uses the Bonferronized version of the deletion residual test.

The upper panel of the table, for $p = 2$, shows more extreme values than those of the same panel in Table 5.2. For $n = 50$ and $p = 2$, the values from the theory are 2.303; LS is only slightly greater. FS and FS-χ^2 are around 4.7 and DelnRes. = 9.550. The values for this test are no longer very close to ten, due to the Bonferroni approximation. As n increases, most scaled sizes converge to ten from below. The results for $p = 10$ in the lower panel, particularly for $n = 50$, are much more extreme; the value for theory is 0.151, around a 65th of the target value. The value for LS is 0.2, whereas the values for FS and FS-χ^2 are around 54. However, when $n = 100$ the values for FS and FS-χ^2 are around 13, decreasing slightly thereafter.

The results for FS and FS-χ^2 at first sight may seem disappointing. However LS and DelnRes. refer to non-robust estimation. Figures 5.1, 5.2, 5.3, and 5.4 show that tests based on the FS have a smaller size than those of all other robust estimators, sometimes appreciably much more so. Tables 5.2 and 5.3 show how the sizes of the tests studied in this section converge towards the nominal size as n increases.

5.2 Size Comparisons

In Sect. 5.1, we reported the results of a screening experiment that compared many robust estimators for $n = 50$ and 200 and $p = 2$ and 10. In this section, the emphasis is on showing how the size of the tests depends on n for five commonly used estimators that the screening experiment showed had properties typical of many such estimators. These are LTS, LTSr, and S, with the TB ρ function, all with $bdp = 50\%$: MM, with $eff = 0.85$, again with the TB, and the LTS. There are thus four very robust estimators with specified properties and the adaptive FS.

In order to establish the size of the outlier tests, we ran simulations for sample sizes n from 100 to 1,000 for several different dimensions of problems. The results for $p = 2$ and 10 are in Figs. 5.5 and 5.6. In the simulations, the samples were allowed to grow with n, so that samples for larger values of n contained those for smaller, leading to smoother curves. Both the response and the explanatory variables were simulated from independent standard normal distributions. Since all methods are affine equivariant, the choice of the values of the coefficients of the linear model does not affect the results. For each value of n, we present the average of 10,000 simulations, in which we counted the number of observations declared as outlying, with the tests conducted at the 1% level of the upper tail of χ_1^2.

The figures show that, for three out of the five estimators, the sizes are very far from the nominal value of 1%. For $n = 100$ the sizes for MM, S, and LTS when $p = 2$ range between 0.015 for MM and 0.026 for LTS. When $n = 200$ the results agree with those in the upper panel of Fig. 5.2 which shows that LTS has the largest size. As n increases to 1000, the sizes of all three of these tests decrease towards 0.01. For $p = 10$, in Fig. 5.6, LTS and S both have strikingly similar behaviour, a feature corroborated by the upper panel of Fig. 5.4. When $n = 100$, the size of these two rules is close to 0.1, with that for MM close to 0.05. The sizes decrease with n,

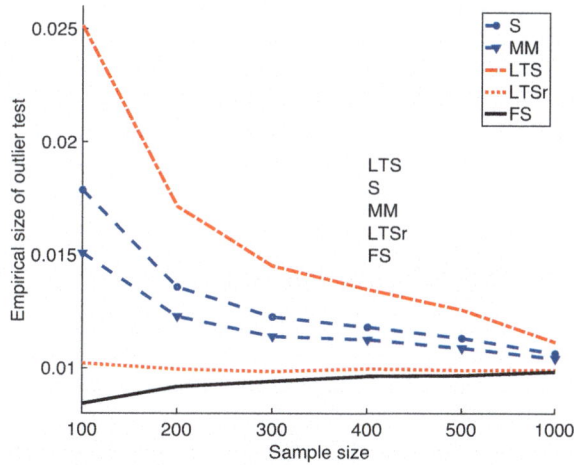

Fig. 5.5 Size of nominally 1% individual outlier tests for $p = 2$. The lettering in the centre of the panel indicates the ordering, from the largest, of the sizes of the tests. Note the contraction of the scale between $n = 500$ and $n = 1000$

Fig. 5.6 Size of nominally 1% individual outlier tests for $p = 10$. The lettering in the centre of the panel indicates the ordering, from the largest, of the sizes of the tests. Note the contraction of the scale between $n = 500$ and $n = 1000$

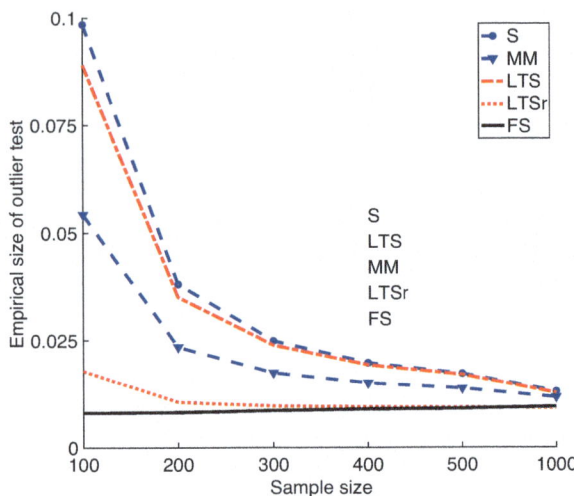

keeping the same ordering, but even so are not much less than 2% when $n = 1,000$. In both figures, the size for LTSr is closer to the nominal value; for $p = 2$, the value is just above 1% throughout. For $p = 10$, the size for LTSr decreases to 1% when n has increased to 300. Only FS has a size around 1% for both values of p and all n, particularly when the deletion residual is used for testing (see Tables 5.2 and 5.3).

These calculations show that all procedures except FS need appreciable adjustment for size, particularly for samples of 50 or less. Unfortunately, standard statistical problems rarely have sample sizes large enough for the uncorrected tests to be acceptable as indicators of the presence of outliers. A simple method of adjusting power for size is a cumulative normal, or logistic, plot of the power curves, as in Figs. 5.8 and 5.10 when the slope of the curve indicates power and the intercept size. An interesting feature of Figs. 5.5 and 5.6 is that the size of the FS test increases slightly with n, extending the results for the size of the FS outlier test in Table 5.3.

Torti et al. (2012) includes further results on the size and power of outlier tests that include both LMS and its reweighted version LMSr. Also included are several version of LTS differing in details of the algorithm employed. These comparisons exclude both S- and MM-estimation which we include.

5.3 The Power of the Tests

The comparisons of size suggest that, overall, FS and LTSr have the best performance. However, the power comparisons of this section reveal that, in many comparisons, LTSr has the lowest power of the five procedures.

In our power calculations, we shifted the mean of 5% or 30% of the observations by a shift Δ with a maximum value of 7 (the errors in our observations were standard

5.3 The Power of the Tests

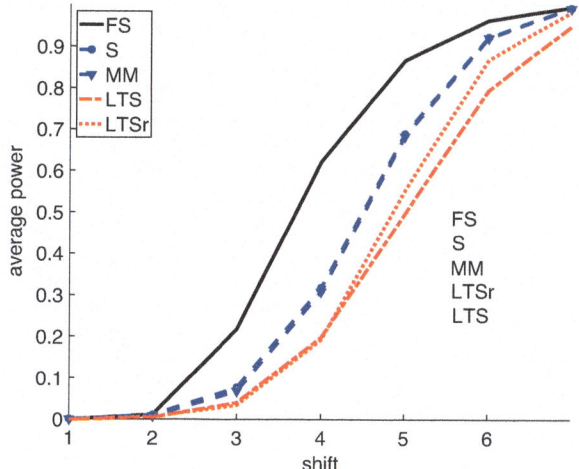

Fig. 5.7 Average power for $n = 500$, $p = 5$, and 5% contamination. The FS clearly has the best power. For all shift values Δ, S and MM have the next highest power. A third group contains LTS and LTSr, the relative performance of which improves with Δ. The labels are positioned from the top to the bottom to reflect their ranking

normal) and calculated the average power, that is, the proportion of generated outliers correctly identified. We start in Fig. 5.7, which illustrates the general shape of the curves. Here, when $\Delta < 2$, virtually no outliers are detected by any estimator. The average number of outliers detected increases with Δ, the curves approximately taking the shape of a cumulative normal distribution. When $\Delta = 7$, the average power is close to one for all estimators. Figure 5.7 is for 5% contamination when $p = 5$ and $n = 500$. FS clearly has the best power. The other methods fall into two groups; S and MM have indistinguishable performance, with LTS and LTSr close together with the lowest, indistinguishable, power for $\Delta \leq 4$. The performance of LTSr relative to LTS increases with Δ and appears close to that of S and MM for the highest level of the shift. The labels in the panels of this and other figures are positioned from top to bottom to reflect this ranking.

The conclusions from Fig. 5.7 appear clear. However, in general, there are two problems of interpretation for such straightforward plots of power. One is that they are bounded below and above by zero and one; thus the eye focuses on comparisons in the centre of the plot, that is, on powers around 50%. The other is that it is impossible to adjust by eye for the size of the different procedures. Accordingly we accompany these plots of power by logit plots.

Figure 5.8 repeats Fig. 5.7 with the power r replaced by logit$\{(r - 3/8)/(n + 1/4)\}$ Cox and Hinkley (1974, p. 470). Under this transformation procedures that plot as parallel curves will have the same power once they have been adjusted to have the same size. For an example see Atkinson (1985, Fig. 8.12). In our case, Fig. 5.8 again shows the superior performance of FS for $\Delta > 2$. The logit transformation confirms the similarity of performance of S and MM. However, it is now apparent that the seemingly superior performance of LTSr over LTS for large Δ in Fig. 5.7 does not bring it indistinguishably close to that of S and MM.

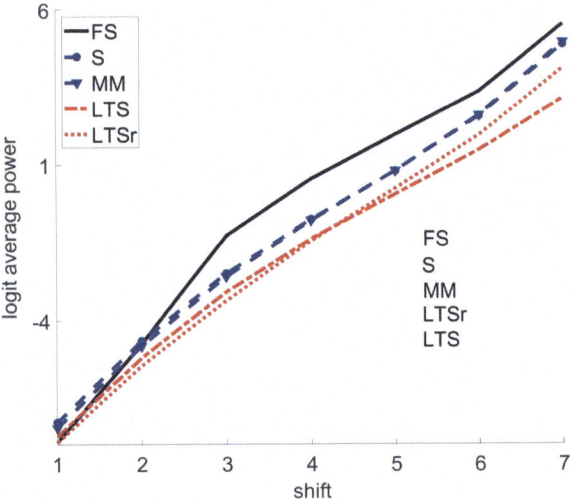

Fig. 5.8 Logit average power for $n = 500$, $p = 5$, and 5% contamination (Fig. 5.7 replotted on the logit scale). The power of FS increases more rapidly than that of the other estimators for $\Delta \leq 4$. The labels are positioned from the top to the bottom to reflect their ranking

The difference between the procedures becomes more marked as the level of contamination increases. In our latter two examples we have 30% contamination. In Figs. 5.9 and 5.10, we consider small samples with $n = 50$ when $p = 2$. The structure of Fig. 5.9 is similar to that of Fig. 5.7; FS has the greatest power, more so than before, but now, with the smaller sample, outliers are not detected for $\Delta < 3$. S and MM are close in performance, with S is slightly better than MM when Δ is not too large; the difference decreases as Δ becomes larger. LTSr again behaves worse than LTS for the smaller values of Δ, becoming more powerful than LTS for $\Delta > 6$. The highest power is for the FS, although its size was the lowest in Fig. 5.1.

An interesting feature of the logit transformed power in Fig. 5.10 is that the slope of the logits is greater for FS and LTSr than for the other three methods. The steeper slope of the curve for FS shows that the method would perform even better after adjustment for the size of the test which is $< 1\%$ (see Fig. 5.1). The size for LTSr is $> 1\%$.

The good performance of the FS becomes even more apparent for larger p and n. The results for $n = 500$, $p = 5$, and 30% contamination are in Fig. 5.11. The figure shows the outstanding performance of the FS which starts to detect outliers for $\Delta > 2$. For all other rules, the power is virtually zero until $\Delta > 5$. The logit transformation of Fig. 5.12 not only confirms the performance of the FS, but shows the effect of masking on all rules. When Δ is small, rules with a large size are more likely than are rules with a small size, to include a few genuine outliers along with observations falsely declared as outliers. As Δ increases, the undetected outliers inflate residual variance and make detection of any but large outliers increasingly difficult. The plot shows this initial decrease in power, which is less for LTS, S, and MM which all have relatively large sizes in Figs. 5.1, 5.2, 5.3, and 5.4. In interpreting Fig. 5.12 note that

5.3 The Power of the Tests 167

Fig. 5.9 Average power for $n = 50$, $p = 2$, and 30% contamination. The FS clearly has the best power for $\Delta > 3$. For all values of Δ, S and MM have the next highest power, with S more powerful than MM. The third group contains LTS and LTSr; the relative performance of LTSr improves with Δ, being better than LTS for $\Delta > 6$. Likewise, the performance of MM relative to S increases with Δ

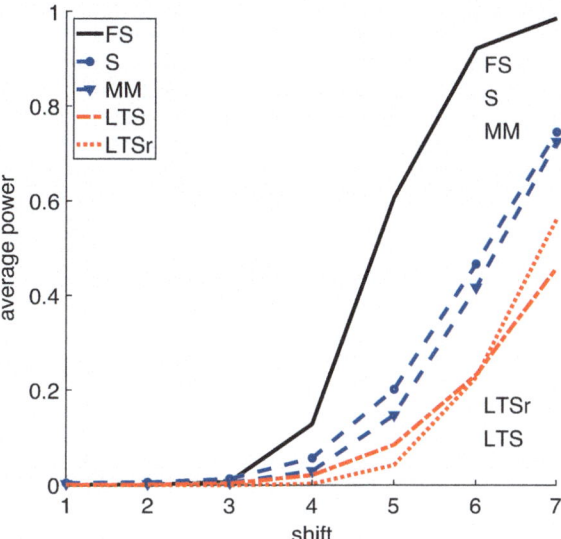

Fig. 5.10 Logit average power for $n = 50$, $p = 2$, and 30% contamination. The FS has the highest power for $\Delta > 3$. The steeper slope of the curve for FS shows that the method would perform even better after adjustment for the size of the test (see Fig. 5.1). The size of the test using LTSr is $> 1\%$

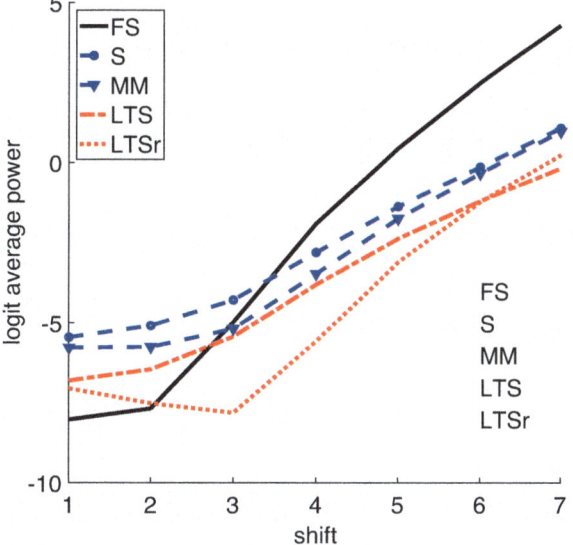

Fig. 5.11 Average power for $n = 500$, $p = 5$, and 30% contamination. The FS has by far the best power; around 0.9 before the tests using the other four estimators start to identify outliers

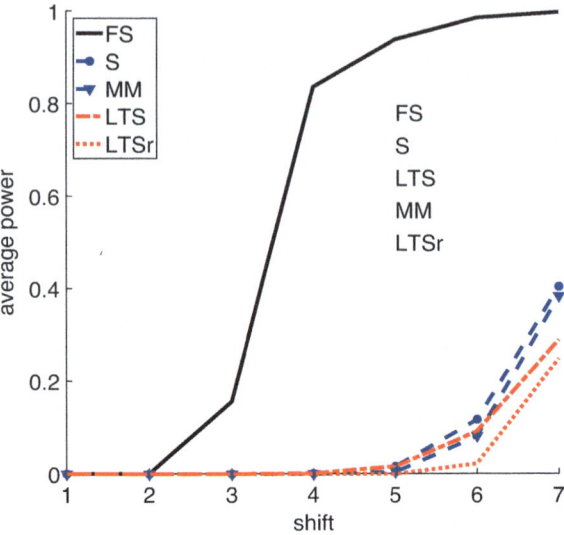

Fig. 5.12 Logit average power for $n = 500$, $p = 5$, and 30% contamination. The power of the FS increases sharply for $\Delta > 2$. The initial decrease in power for all estimators is due to masking

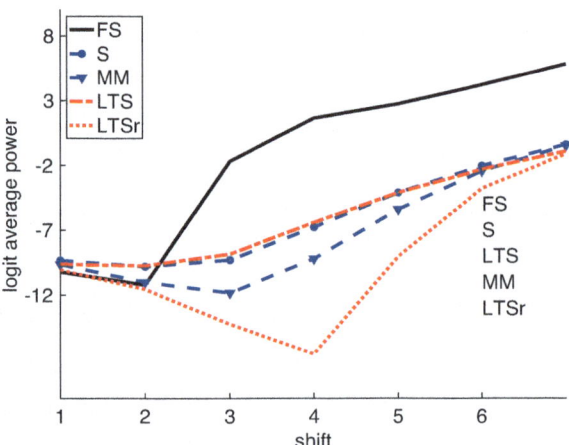

the logit of 0.01 is -4.60. The effect on data analysis of the shape of the earlier parts of the curves for any estimator except FS is negligible; all other estimators only lead to the detection of outliers for $\Delta > 5$.

The plots indicate how the relative powers of the pair S and MM and the pair LTS and LTSr change as the shift parameter increases. This is most clearly shown in the logit plot of Fig. 5.10. The slope of plot of LTSr is greater than that of LTS; LTSr provides the more powerful tests for shifts greater than 6. The plot of power for MM also has a slightly greater slope than that for S. MM is set to provide tests with greater power for the shift > 7, that is, just off the plot. Since asymptotic arguments were used to derive MM as an improvement of S and LTSr as an improvement of LTS, it is interesting to note that, for these data, large shifts are required for the improvements to become apparent.

5.4 A Parametric Framework for Comparing Robust Regression Estimators

5.4.1 Introduction

It is important to be clear that the *bdp* of a robust estimator is an asymptotic property, defined as the outlying observations become increasingly remote. In this section and Sect. 5.5, we present a systematic, parameterized framework for the non-asymptotic comparison of methods of robust regression. We continue to work with the FS and the four very robust estimators S, MM, LTS, and LTSr of the previous sections. All share the property that, asymptotically, they have a breakdown point of 50% as the main data and outliers become infinitely far apart and the sample size goes to infinity. In order to extend the size and power comparisons of Sect. 5.2 to a more general framework, we systematically study the properties of estimators as the distance between the two groups of observations decreases. In Sect. 5.5, we introduce a parameterized framework, with parameter λ, for moving outliers along a trajectory which is initially remote from the main data, but which then passes close to it before again becoming far away. We control whether, at their closest, the two populations share the same centre. We design measures of overlap to calibrate the trajectories. Importantly, the outliers in this procedure can have any specified structure.

Numerical results are in Sects. 5.5.3 and 5.5.4. In Sect. 5.5.3, we take the outliers from the regression model to have a multivariate normal distribution. This provides a very general scenario for outliers that can range from a seemingly random scatter around the regression plane to points virtually on a line. The special case of point contamination is explored in Sect. 5.5.4. Boxplots of the parameter estimates from the five methods as λ varies indeed show that, for wide separations, the methods have similar properties. However they differ markedly as the two populations converge. In order to summarize this information we look at cumulative plots, over the range of λ, of the variance and squared bias of the parameter estimates.

Another method of comparing robust estimators is by their properties for outlier detection (Cook & Hawkins, 1990). In Sect. 5.5.3, we use the number of outliers detected to calculate power curves as a function of λ for the specific structure of the outliers. In the earlier sections of this chapter, our analysis was of the effects of mean shift outliers.

5.4.2 Bias and Maxbias

In (2.4), the model for robustness was written $F_\varepsilon = (1-\varepsilon)F(\theta) + \varepsilon G$ where G is unspecified. Interest is in understanding the change in the properties of the estimators of parameters as the outliers in G become more remote. When G is close to $F(\theta)$ the effect on the estimates of θ will be small. As the distance between the two increases, but is small enough to make outlier detection difficult, the estimates of θ will become biased. As the distance increases further, detection of outliers becomes increasingly probable and the bias of the estimates eventually reduces to zero. At the same time, the variance of the parameter estimates will increase. If a robust estimator downweights or trims an excessive number of observations compared with another estimator, the parameter estimates will comparatively have an excess variance.

To compare estimators, we would ideally like to find the maximum bias (maxbias, MB, see Eq. (2.9)) over the specific G that describes the outliers habitually experienced in a particular situation. Unfortunately, this is rarely possible except numerically. The standard analysis based on the influence curve yields asymptotic results for specific G. Ideas about bias in a specific case were formalized in Sect. 2.1.1. Figure 2.1 shows the sensitivity curve for 11 data points for three estimators of location; the curve is a function of the added observation y. One conclusion from this figure is that the sample mean has maxbias $= \infty$. The influence function, Fig. 2.2, replaces the sample of observations in Fig. 2.1 by calculations at a specified distribution F, with ε infinitesimally small and G a point mass. Section 3.3 of Maronna et al. (2019) provides a fuller discussion and, in their Fig. 3.3, shows the importance of the value of ε in the calculation of the bias. Their Fig. 3.4 gives asymptotic biases for location estimation with a fixed value of ε (10%) for the Huber and Tukey biweight estimators as the location of the point contamination moves from 0 to 5. For the TB point, the bias in the estimation of location has a maximum when the point contamination is close to 2.5 and then decreases to zero, an example of the process outlined in the preceding paragraph.

Equation (2.9) shows that the maxbias depends both on the amount of contamination, through ε and on the distribution G. Figure 2.23 shows that point contamination, in which all outlying values were set equal to ten, produces higher biases and greater differences between estimators than shift outliers in which ten was added to simulated observations to produce the outliers. It is well known (Huber, 1981) that point-mass outliers present the most severe challenge to robust estimators. In the rest of this chapter, we calculate the properties of estimators numerically, with outliers generated from distributions G that are not of the simple point-mass form. We suppose that the larger part of the data, $1-\varepsilon$, where $0 < \varepsilon < 0.5$, is generated by the model $M_1(\theta_1)$ and the remaining part ε of the data is generated by the parametric model $G = M_2(\theta_2)$.

Robust methods study the properties of methods that fit $M_1(\theta_1)$ in ignorance of knowledge of the form of the outlier generating model $M_2(\theta_2)$, which can be quite general. When $M_1(.)$ is a regression model, $M_2(.)$ may be taken, for example, to distribute observations randomly over a large space, concentrate them in a cluster or to be a second regression model, a set of choices often used in comparisons of algorithms for cluster analysis (Fraley & Raftery, 2002). There is no difficulty in having $M_1(\theta) = M_2(\theta)$ but then we must have $\theta_1 \neq \theta_2$.

With $M_1(\theta_1)$ the usual regression model of Sect. 3.7.2, a general expression for the bias from fitting an incorrect model is

$$b(\hat{\beta}) = \{(\hat{\beta} - \beta)^T V_x (\hat{\beta} - \beta)\}^{0.5}. \tag{5.8}$$

The bias depends on the estimator, the distribution of y and x, the amount of contamination ε and on $M_2(\theta_2)$. For finite samples $V_x = X^T X$, the form we use in our simulations. In asymptotic analyses $V_x/n \xrightarrow{p} M_X$ with $M_2(\theta_2)$ point-mass contamination. The maxbias curve shows how the maximum bias varies with ε and G. The breakdown point of an estimator is the minimum value of ε for which $b(\hat{\beta})$ in (5.8) equals ∞. The estimators we consider all have an asymptotic breakdown point of 50%. Although 50% is customarily considered to be the maximum possible breakdown value, in the Random Start FS applied to regression clustering (Sect. 8.5.2), several of the populations may be represented by a much smaller fraction of the whole sample.

In principle, maxbias curves provide useful information for the comparison of robust estimators. Unfortunately, maxbias curves are only calculable for some estimators and distributions of x. The latter are often assumed to be elliptically symmetrical. A summary of the literature is given by Berrendero and Zamar (2001) who extend results on maxbias curves to regression models with intercepts and to regressors that have Student's t and Cauchy distributions, although without intercepts. Berrendero et al. (2007) find maxbias curves for MM-estimators, again without intercepts. Figure 5.16 of Maronna et al. (2019) plots biases of several estimators as a function of a single parameter, the slope of the regression line for the contaminating observations. Figure 3 of Garcia-Escudero et al. (2010) is more in the spirit of our numerical approach. It shows the simulated bias as a point cluster of outliers moves around a regression line. When the outliers are very close to the line, the bias is negligible, as it is when the outliers are far away and are easily downweighted by, in this case, LTS. Only for intermediate outliers is the bias appreciable.

In our numerical comparisons, we study the variance as well as the bias of the estimators. We continue to use S-, LTS-, and LTSr-estimation with *bdp* 0.5, together with the FS- and MM-estimation with *eff*= 0.85.

5.5 A Parameterized Family of Departures

5.5.1 Background

As $y_{M2} \sim M_2(\theta_2) \to \infty$ with θ_1 fixed, the observations y_{M1} and y_{M2} from the two models become increasingly well separated. Under these conditions the five estimators in our study have similar properties. We are also interested in those data configurations when the observations are not so well separated, so that estimation of each parameter θ_j will depend on observations from both y_{M1} and y_{M2}. In general, the properties of robust estimators depend on the "distance" between the two models. Configurations with moderate overlap may be highly informative about the differences in properties between robust estimators. We define a finite sample measure of the overlap of y_{M1} and y_{M2} that is designed to be informative for regression models. There is a sample S_1 of n_1 observations from $M_1(\theta_1)$ with distribution $F_1(y_i; x_i, \theta_1)$ conditional on the value of x_i. These values of x_i belong to a design region $\mathcal{X}$. The sample S_2 of n_2 observations from $M_2(\theta_2)$ has conditional expectation $E(y; x_i, \theta_2)$. Some values of x_i from S_2 may belong to $\mathcal{X}$. We define the indicator

$$1_{i,\gamma} = \begin{cases} 1 & \text{if } F_1^{-1}(\gamma/2; x_i, \theta_1) < E(y; x_i, \theta_2) < F_1^{-1}(1 - \gamma/2; x_i, \theta_1) \quad i \in S_2, x_i \in \mathcal{X} \\ 0 & \text{otherwise.} \end{cases}$$
(5.9)

The index is a function of both θ_1 and θ_2 and we examine it over sets of parameter values Θ_1 and Θ_2. For a particular pair of sets of parameter values $\theta_{1,k}$ and $\theta_{2,k}$, the overlapping index is defined as

$$O_{\gamma,k} = \sum_i 1_{i,\gamma,k} \quad i \in S_2.$$
(5.10)

With $M_1(\theta_1)$ normal theory regression, we are therefore counting the total number of observations in S_2 for which, with $x_i \in \mathcal{X}$, the conditional medians lie in a strip around the expectation of $M_1(\cdot)$. As γ decreases, the strip becomes broader in y. If also for all $i \in S_2$, $x_i \in \mathcal{X}$, then $O_{\gamma,k} \to n_2$, the number of observations in S_2.

It is informative to keep θ_1 fixed and to vary θ_2 in a smooth way with a parameter $\lambda \in \mathbb{R}$. Then we look at a set of indexes

$$O_\gamma(\lambda) = \{O_{\gamma,k}\} \quad \theta_1 \in \Theta_1 \text{ and } \theta_{2,k} \in \Theta_2(\lambda).$$
(5.11)

In particular, we vary θ_2 linearly using the combination

$$\theta_{2,k} = \lambda_k \theta_2^0 + (1 - \lambda_k)\theta_2^1 \quad (-\infty < \lambda_k \in \Lambda < \infty).$$
(5.12)

The set Λ of values considered is problem dependent. With $\theta_2^0 = \theta_1$ the centre of M_2 passes through that of M_1. Other choices of θ_2^0 can produce a trajectory in which some or all of the observations y_2 are always outlying. Our examples show how the variance and bias of the parameter estimates change in a smooth way with λ, but in different and informative ways for different estimators.

5.5.2 The Theoretical Overlapping Index

The contamination M_2 in our examples comes from a multivariate normal distribution. We now calculate the probability of intersection between this distribution and a strip around the regression plane. We call this the theoretical overlapping index.

The response and the explanatory variables lie in a space of dimension $p + 1$. Let these variables be w. Then the regression plane can be written as $b^T w - c = 0$. The equation of the normal to the plane through a point w_0 on the plane is

$$z_1 = w_0 + bd, \tag{5.13}$$

where the scalar d is the distance from the plane. The outlying observations, including the response, have a multivariate normal distribution. Let these be $W \sim \mathcal{N}(\mu, \Sigma)$. We require the probability that W lies on one side of the plane. To obtain this rotate W to a set of variables Z with z_1 (5.13) the normal to the plane. Integrating out the other p variables shows that the required probability comes from the marginal distribution of $Z_1 \sim \mathcal{N}(b^T \mu, b^T \Sigma b)$ (Exercise 5.4). Let the distance in the z_1-direction from μ to the plane be $d(c)$. Then, from (5.13), at the plane $b^T w = c = b^T \mu + b^T b d(c)$, so that

$$d(c) = (c - b^T \mu)/b^T b. \tag{5.14}$$

Since the distance $d(c)$ in the z_1-direction has been rescaled by the factor $1/b^T b$ the required probability is

$$\Pr(b^T W > c) = \Pr(Z_1 > c - b^T \mu) = \Phi\{d(c) b^T b / (b^T \Sigma b)^{0.5}\} = \Psi(c), \text{ say.} \tag{5.15}$$

We require this probability in terms of the regression model, which we now write as $y = \alpha + \beta^T x$. Then

$$b^T = (1 \quad -\beta^T), \quad w^T = (y \quad x^T) \quad \text{and} \quad c = \alpha.$$

Finally, we require the probability that W lies between two planes. For any x the required strip around this model is $y \pm 2\sigma_\epsilon$. The two planes then are defined by constants $c^+ = \alpha + 2\sigma_\epsilon$ and $c^- = \alpha - 2\sigma_\epsilon$. From (5.15) the required probability is $\Psi(c^+) - \Psi(c^-)$.

Although this theoretical overlapping index ignores $\mathcal{X}$, it does signal cases where y_2 lies close to the regression line, even if remote from $\mathcal{X}$. These observations would then be "good" leverage points, in the sense that they could improve the estimates of the regression parameters. For counting vertical outliers, we need observations that lie in $\mathcal{X}$. These are signalled by the index defined in (5.10), which has to be calculated by simulation. We therefore call this the empirical index.

5.5.3 The Numerical Effect of Overlap: Normal Contamination

Because of the flexibility of our systematic approach, we can potentially cover a wide range of possibilities. Here we look at three numerical examples with normal contamination. In the next section, we consider point contamination. We look at boxplots of the estimates over a suitable Λ and relate these plots to the overlapping indices. We separate out the variance and bias components of the estimates and compare these through cumulative plots over Λ. Finally, we compare the estimators for their power of detecting outlying observations, that is, those that come from Model 2. The detection of outliers is particularly important if we require an indication that more elaborate methods of data analysis are appropriate, such as data transformation (Chap. 6) or evidence of more than one population, when cluster analysis might be important (Chap. 8).

In our one-variable regression examples, M_1 is the regression model $y_i = \alpha + \beta x_i + \epsilon_i$, with the x_i independent and uniformly distributed in (a, b), $x_i \sim \mathcal{U}(a, b)$. These values are generated once for all observations in a particular simulation. The variance of Y is σ_ϵ and overlapping indices were calculated for a strip of width $\pm 2\sigma_\epsilon$ around $E(Y)$.

The expectation of x is $\mu_x = (a + b)/2$. The bivariate normal distribution for M_2 has mean μ and variance Σ given by

$$\mu = \begin{pmatrix} \alpha + \beta(\mu_x + d) \\ \mu_x + d \end{pmatrix} \lambda + \begin{pmatrix} \mu_2 \\ \mu_2 \end{pmatrix}(1 - \lambda) \quad \text{and} \quad \Sigma = \begin{pmatrix} \sigma_1 & \sigma_2 \\ \sigma_2 & \sigma_1 \end{pmatrix}, \quad (5.16)$$

where the first component corresponds to the response. When $\lambda = 1$ the centres of the two populations are identical when the displacement $d = 0$.

Example 5.1 We took $n_1 = 100$ with $\alpha = 10$, $\beta = 3$, $\sigma_\epsilon = 10$, $a = 0$ and $b = 10$. For the second population, $n_2 = 30$, $\sigma_1 = 20$, $\sigma_2 = 2$ and $\mu_2 = 10$. Also $d = 0$ so the centres coincide at $\lambda = 1$. There were 100 simulations for each value of λ.

Figure 5.13, to be read line by line, shows nine typical simulated datasets. As λ increases from -3 to 4, the centre of M_2 passes through that of M_1, at which point there is almost complete overlapping of the observations from the two populations. That the overlap is not complete is shown by the plots of the indices in the upper panel of Fig. 5.14, the maxima of which are less than one. The theoretical index is slightly higher than the empirical index as there is some probability of observations falling within the band of y values that are not in $\mathcal{X}$. On the other hand, the plot in the bottom right-hand panel of the squared Mahalanobis distance from the mean of M_1 to that of $M_2(\lambda)$ using the covariance matrix Σ (5.16) has a minimum of zero, showing identity of the two centres. The corresponding p-value of the distance is the lower left-hand panel.

We now consider the effect of these data configurations on the estimation of β. The left-hand panels of Fig. 5.15 show boxplots, from 100 simulations, of the

5.5 A Parameterized Family of Departures

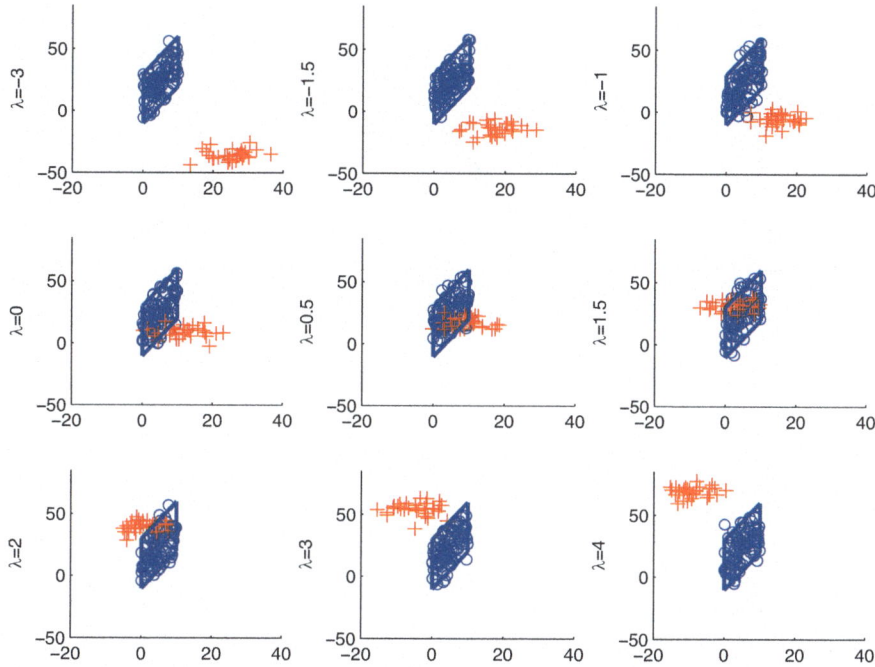

Fig. 5.13 Example 5.1: typical simulated datasets with $n_1 = 100$ and $n_2 = 30$ for nine values of λ: blue circles data from M_1. As λ increases observations from M_2 (red crosses) become close to those from M_1 and then become remote again. The parallelogram defines the region for the empirical overlapping index

values of the five estimators for a series of values of λ, together with a typical data configuration for each. For $\lambda = -3$, observations from M_2 lie below and to the right of those from M_1. If these outliers are not identified, the slope of the line is decreased. The boxplots all show some simulations where such estimates occur. LTS has the highest variance among the estimators in the main part of the boxplot, that is, for the estimates when all outlying observations are rejected, with S the second most variable. For $\lambda = -1$, LTSr and MM are most affected by the outliers. The value $\lambda = 1$ corresponds to virtually complete overlap of the two groups. All methods, on average, give estimates that are biased downwards. However, those for LTS and S are both more variable and more biased. In the last panel, for $\lambda = 3$, the outliers are not as well separated as they are in panel 1. LTSr now has appreciable negative bias, due to the inclusion of outliers in the reweighting stage. This is in agreement with the results of Sect. 5.3; the power of LTSr is greater than that of LTS only when Δ is very large.

Figure 5.16 provides a powerful summary of the results on the variance and bias of the estimates of α and β as λ varies. The left-hand panels show the partial sums

176   5 Practical Comparison of the Different Estimators

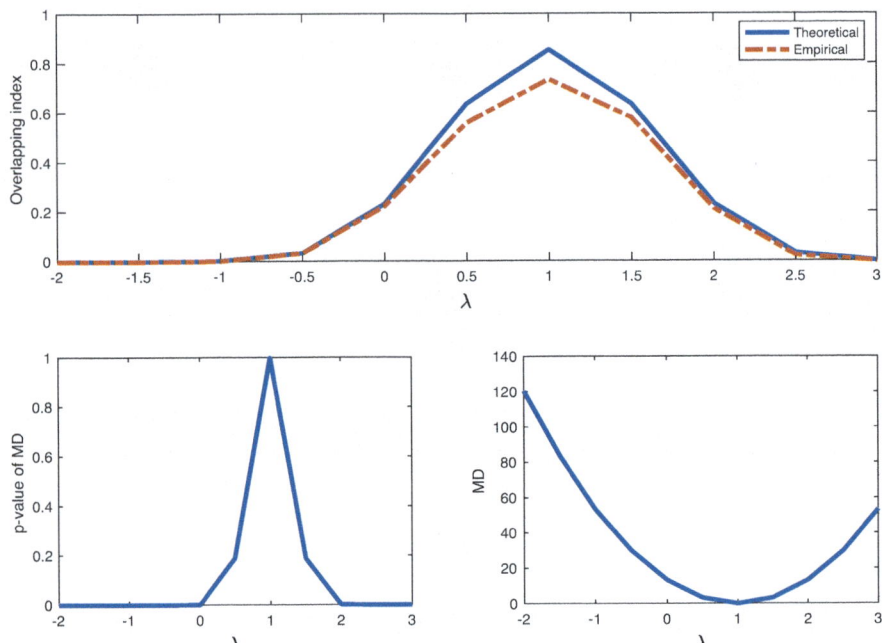

Fig. 5.14 Example 5.1: upper panel: theoretical and empirical overlapping indices for the data in Fig. 5.13, showing maxima at $\lambda = 1$. Lower panel: squared Mahalanobis distance of M_1 from M_2 (right) and corresponding p-values (left)

of the squared bias over Λ and the right-hand panels show the partial sums of the variances. The values for α are in the top row and those for β in the bottom row.

The plots illustrate the trade-off between bias and variance for some of the estimators. For values of λ up to three or so, LTS and S have the highest variances and the lowest biases and have very similar properties. Over the same range, LTSr and MM have high biases and low variances. The effect of the modification of LTS to LTSr and S to MM has, in general, been to reduce variance at the cost of an increase in bias. The bias values for FS are in between those of these two groups, but closer to the lower pair of values, especially for estimation of β. The variance of FS is close and ultimately less than the low values of LTSr and MM.

The bottom right-hand panel of Fig. 5.15 shows that for $\lambda = 3$, the outliers are becoming distinct from y_1. As λ increases further the two groups become increasingly distinct, an effect that is evident in Fig. 5.16. For the extreme values of λ, the horizontal value of the partial sums of squared bias for all estimators shows that the bias is zero at these extreme values. The two populations are sufficiently far apart that the asymptotics defining high breakdown apply. This is achieved for slightly less separation by MM than LTSr. The plots of partial sums of variances, on the other hand, increase steadily, since the estimates are always subject to the effect of the random variability in the observations. The sums of variances for S and, particularly,

5.5 A Parameterized Family of Departures

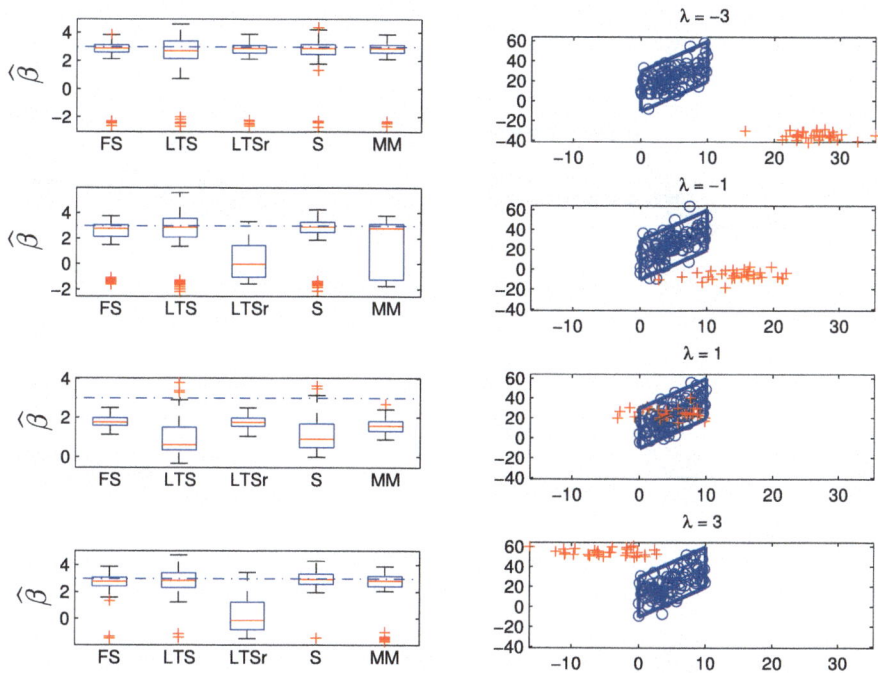

Fig. 5.15 Example 5.1: four simulated datasets for $\lambda = -3, -1, 1$ and 3. Left-hand panels: boxplots, from 100 simulations, of estimates of β (dotted and dashed line: $\beta = 3$) for FS-, LTS-, LTSr-, S-, and MM-estimators. Right-hand panels: typical simulations for these four values of λ

LTS are, however, increasing more rapidly at the ends of the region than those for the other three methods, a result in line with the rows of boxplots for $\lambda = \pm 3$ in Fig. 5.15.

These plots illustrate the differing performance of the five estimators. Since this is a book about robust statistics, we also looked at plots in which the variance of the estimators was replaced by the average median absolute deviation from the median. Those plots were close to those of the variances shown here. In addition to good parameter estimates, we would also like our estimated model to signal the presence of outliers if the model fitted to the data is incorrect. Accordingly, we calculated the average power. In testing for the presence of outliers we used a test of Bonferronized size α^*. The results are in Fig. 5.17. Outliers are not detected for central values of λ as the parameter estimates are sufficiently corrupted by observations from M_2 that no observations appear outlying. As the means of the two populations move apart, the number of outliers detected increases. Over most of the range FS has the highest power and LTSr the lowest. The other three estimates lie between these extremes with MM having lower power for values of λ near zero. These general conclusions are in line with the power comparisons of Sect. 5.3.

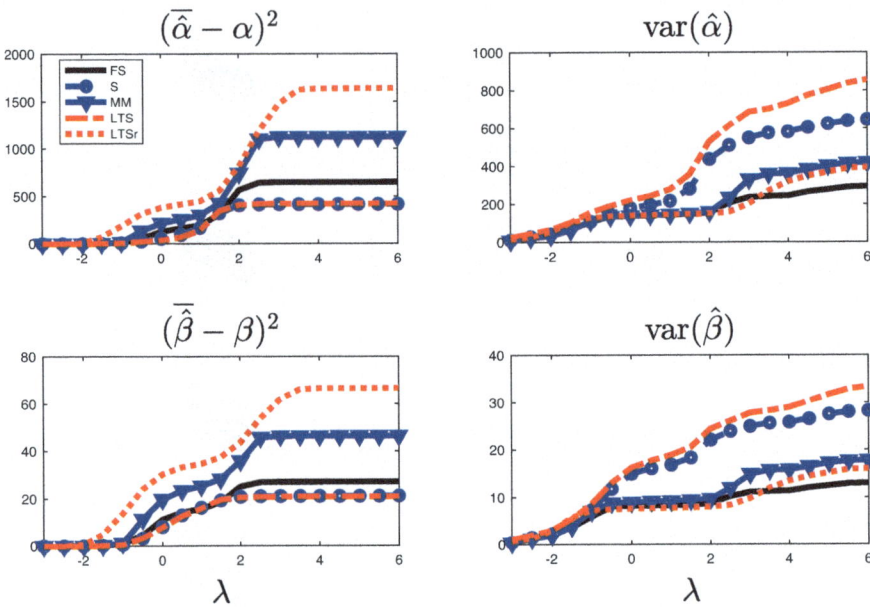

Fig. 5.16 Example 5.1: partial sums over Λ of simulated squared bias and variance of the five estimators. Left-hand panel's squared bias, right-hand panel's variance. Top line $\hat{\alpha}$, bottom line $\hat{\beta}$

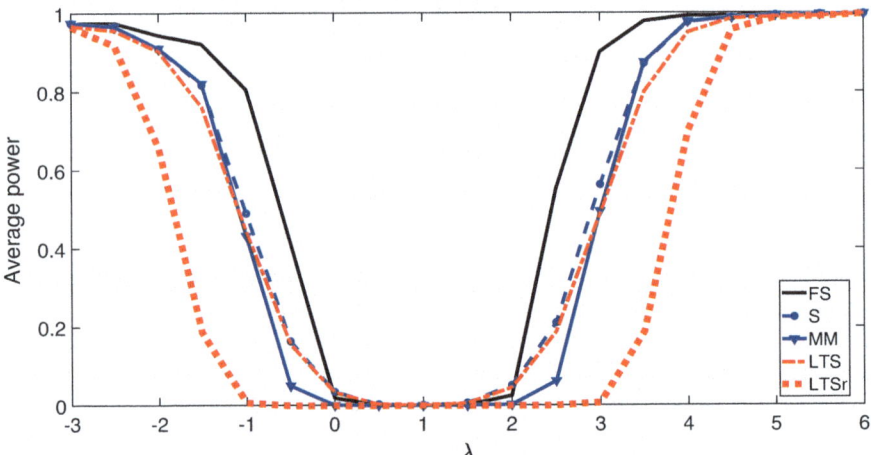

Fig. 5.17 Example 5.1: simulated average power of the five procedures over Λ

5.5 A Parameterized Family of Departures

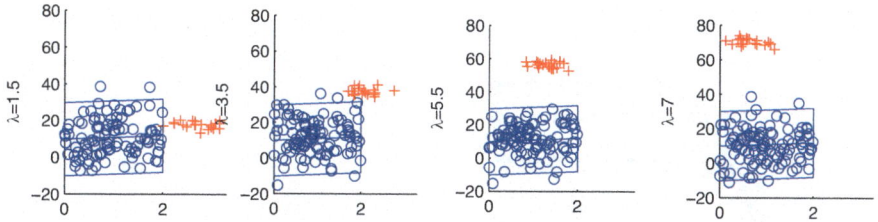

Fig. 5.18 Example 5.2: simulated datasets with $n_1 = 100$ and $n_2 = 20$ for four values of λ. As λ increases observations from M_2 become close to those from M_1 and then become remote again. The parallelogram defines the region for the empirical overlapping index

Example 5.2 We stay with a single explanatory variable but now choose a trajectory for λ such that $\theta_2^0 \neq \theta_1$, so that most of the observations y_2 are outlying. The parameter values for population 1 were $b = 2$ and $\beta = 1$. For population 2, $\Sigma = \text{diag}(4, 0.1)$, $\mu_2 = 3.4$, and $d = 2$ so that the centres no longer coincided. Also $n_2 = 20$. Figure 5.18 shows scatter plots of typical samples for four values of λ. In the first, for $\lambda = 1.5$, there is a set of horizontal outliers; these can be expected not to have an appreciable effect on the estimate of slope. As λ increases the observations from M_2 rise above those from M_1, generating increasingly remote vertical outliers.

The behaviour of the five estimators for this new situation is summarized in the partial sum plots of Fig. 5.19. The plots of variances are simply interpreted: S and LTS have high variance for both α and β over the whole range of λ with MM and LTSr having low values which are slightly less than those of FS.

The comparison of biases is less straightforward. The scatter plots of Fig. 5.18 suggest that the two populations should be adequately separated by the time $\lambda = 4$. For lower values of λ, S and LTS have similar higher biases for β. The biases for α do not show much difference for lower values of λ. In the right-hand halves of the plots in Fig. 5.19, with $\lambda > 4$, the two populations are more separated. The plots of bias are partial sum plots, so that the horizontal portions for higher λ show that S and LTS provide unbiased estimates for smaller values of λ than does MM. The LTSr estimates are not unbiased, even for the largest values of λ. The FS has excellent properties; it has the next to lowest bias for α and the lowest for β, as well as variances which are close to those from MM and LTSr.

The plot of average power for this example in Fig. 19 of Riani et al. (2014c) leads to similar conclusions to those for Example 1 in Fig. 5.17. FS has the highest power and LTSr the lowest, but now the difference between FS and the other rules is much greater. S and MM have indistinguishable performances with LTS closer to that of LTSr.

Example 5.3 The third example has five explanatory variables ($p = 6$), independently uniformly distributed on $(0, 2\sqrt{10})$ with regression parameters $\beta = 5$ for all variables and $n_1 = 200$. For population 2, $\Sigma = \text{diag}(100, I_5)$, $\mu_2 = 3, d = 2$, and $n_2 = 60$.

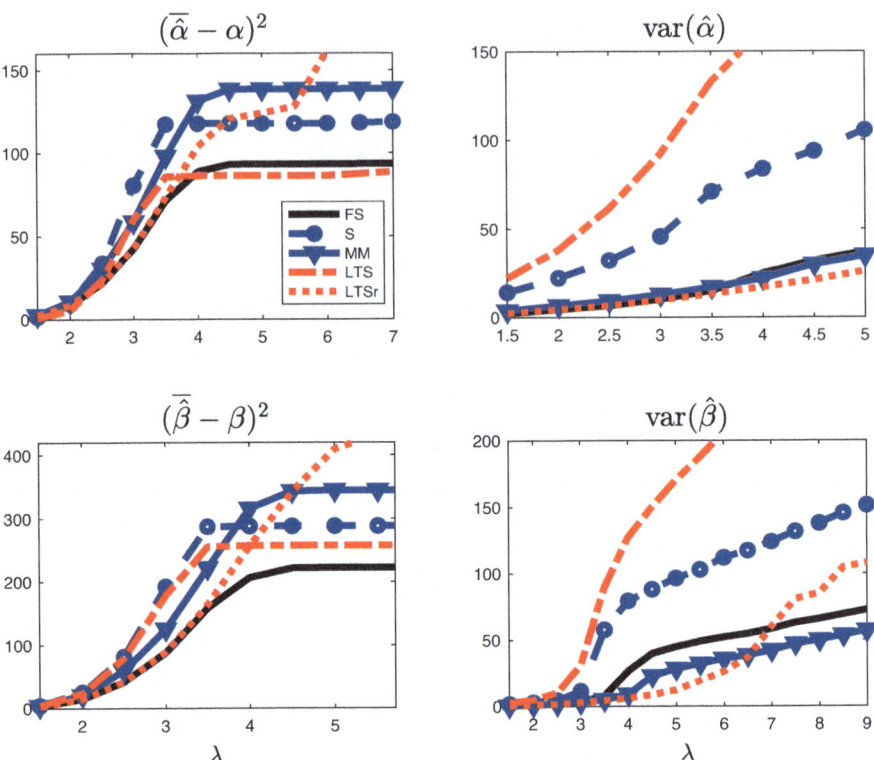

Fig. 5.19 Example 5.2: partial sums over Λ of simulated squared bias and variance of the five estimators. Left-hand panel's squared bias, right-hand panel's variance. Top line $\hat{\alpha}$, bottom line $\hat{\beta}$

This is a larger example, since $n_1 = 200$ and $n_2 = 60$. As λ increases from -1 to 2.6 the outliers "rise through" the central observations. Since $d \neq 0$, the centres of the two distributions are never identical. Unlike our other two examples, this one does not include outliers at leverage points, so that the differences in behaviour of the methods are, to some extent, reduced.

With five explanatory variables the major contribution to the mean squared error of the parameter estimates comes from β, so we only consider these values, which are plotted in Fig. 21 of Riani et al. (2014c). With independent x_j the bias and variance are the sums of those for the individual components. LTS behaves surprisingly poorly, with the uniformly highest bias and variance. LTSr and S have medium behaviour for both properties, with the order reversed for bias and variance, while MM and FS have the same, lowest values for bias and similar values for variance until $\lambda = 1$ when that for FS increases, although staying below that for S. Unlike the other two examples, the relative behaviour of the estimators is little affected by the value of λ, a reflection of the stability of the outlier pattern over Λ. Of course, the magnitude

5.5 A Parameterized Family of Departures

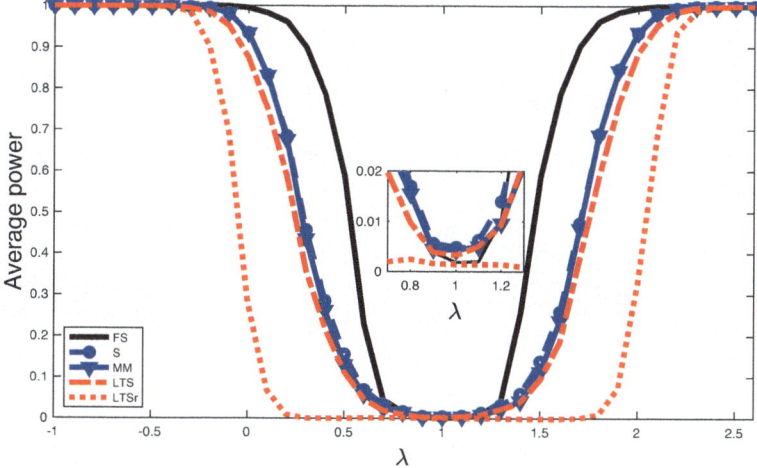

Fig. 5.20 Example 5.3: simulated average power of the five procedures over Λ with an inset zoom of the central part of the figure

of the outliers is largest for extreme values, but leverage points are not introduced or removed.

The plot of average power is in Fig. 5.20. As in the plots of average power in Sect. 5.3, FS has the highest power and LTSr the lowest. The other three estimators have very similar properties to each other. However, in assessing power, we need to be sure that we are comparing tests with similar sizes. The zoom in the centre of the plot for values of λ close to one shows that the sizes are not quite the same, with FS and LTSr having the smallest values.

For more accurate comparisons, we can scale the curves for the other three tests downwards, which will reduce the curves below the plotted values. However, even when $\lambda = 1$, the models are not identical so the displacement $d \neq 0$ and outliers are still present. We are not looking at the null distribution of the test statistics.

5.5.4 The Numerical Effect of Overlap: Point Contamination

As we saw in Sect. 5.4.2 point contamination plays an important role in the theory of robust estimation. Accordingly, we extend our simulations to such contamination. Although it is a special case of (5.16) as $\Sigma \to 0$, there are new features, both in the numerical calculation of estimates and in the results of data analysis.

The first feature is the response of the FS algorithm to several identical observations. As the search progresses observations are not only added to the subset used in estimation, but there is also the deletion of observations that have become remote due to changes in parameter estimates. If several of the identical observations are

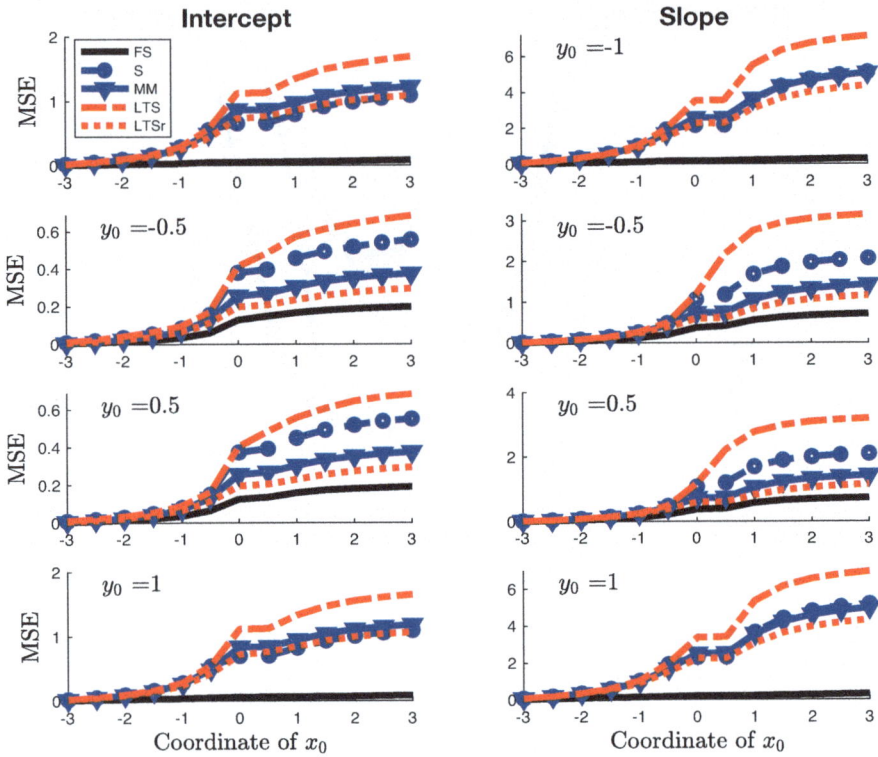

Fig. 5.21 Point contamination at (x_0, y_0): partial sums of mean squared errors of estimates of α and β for four values of y_0 as x_0 varies from –3 to 3. Symbols as in Fig. 5.20

included, those outside the point contamination will seem remote and the search will collapse, since the fitted model will be singular. If such singularity occurs, we identify all identical observations and force them to enter at the end of the search. As the figures show, in some cases, this has a powerful beneficial effect on the estimates. The second feature is that the overlapping index now has the value of either zero or one (Exercise 5.5).

We took 100 x values between 0 and 1, with the normally distributed values of y such that approximately 95% lay between –0.5 and 0.5. We added 30 identical contaminating observations at (x_0, y_0) where both the vertical and horizontal directions of contamination range from –3 to 3.

Figure 5.21 shows plots of the partial sums of the mean squared error of the estimates of the intercept and slope for four values of y_0 over a fine grid of values of x_0 from –3 to 3. The most notable features are the poor performance of LTS and the good performance of FS. This is particularly striking in the more extreme vertical contaminations, $y_0 = \pm 1$, where the FS estimates are virtually unaffected by the 30 outliers.

5.5 A Parameterized Family of Departures

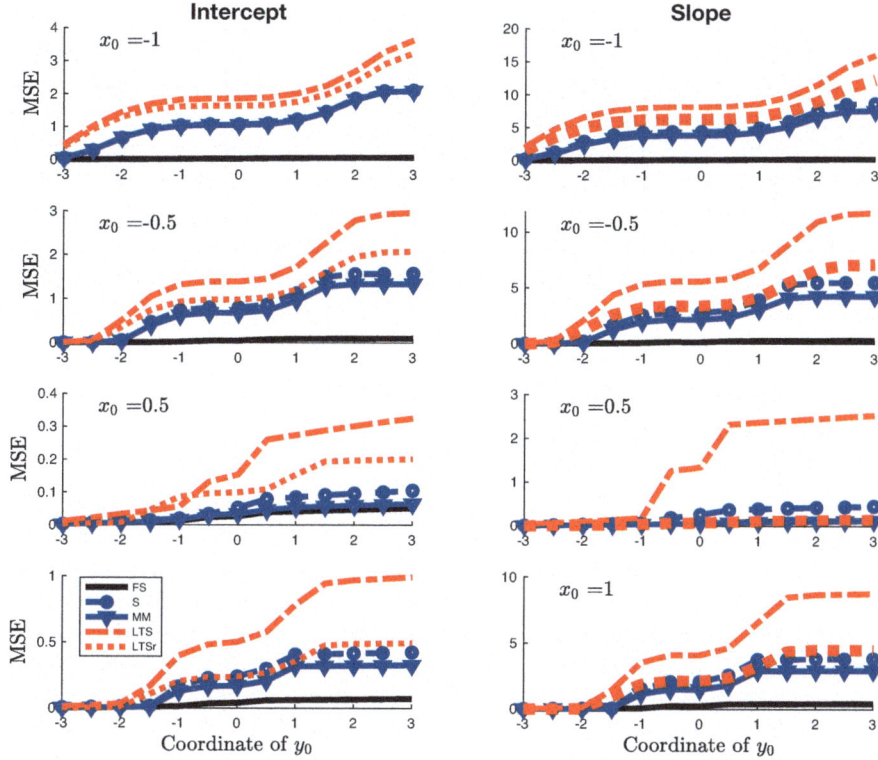

Fig. 5.22 Point contamination at (x_0, y_0): partial sums of mean squared errors of estimates of α and β for four values of x_0 as y_0 varies from –3 to 3. Symbols as in Fig. 5.20

In Fig. 5.22, we look at the same quantities as y_0 varies for four fixed values of x_0: –1, –0.5, 0.5 and 1. Recall that the values of x range from 0 to 1, so these plots are not symmetrical around $x_0 = 0$. The most striking feature is the excellent performance of the FS, which is by far the best except when the contamination passes through the centre of $\mathcal{X}$; even then it is slightly better than MM and S. LTS behaves particularly poorly when $x_0 \in \mathcal{X}$, but is uniformly poorest. S and MM are similar, and slightly better than LTSr.

The use of point contamination allows sharp comparison of the algorithms for very robust regression. In the more diffuse situations of Sect. 5.5.3, in which the contamination is scattered over a region, the plots of the power curves, such as those of Fig. 5.20, extend the comparisons to more general outlier configurations.

With this two-dimensional model for contamination it is possible to explore the properties of the estimators over a grid of values for (x_0, y_0). With higher dimensional problems, such as Example 5.3, we will again need to construct a trajectory Λ along which the point contamination moves.

Riani et al. (2014c) includes further analyses of data using our parametric framework. The second is a further motivating example; the other two are expanded versions of our analyses of Examples 5.2 and 5.3. This material is available in the supplementary material for the paper.

We have illustrated the use of our framework for comparing FS with methods designed to have a breakdown of 50%. Of course, the framework can be used for comparisons with breakdown levels more likely to be used in practice, such as the 0.3 or 0.2 that occur in the data analyses of Chap. 4 that use monitoring. The properties of FS, since they do not depend on a specified breakdown level, will not be changed.

5.6 Comparisons Using the Extended BIC Criterion for Outlier Detection

In our final set of simulations to compare the behaviour of robust estimators, we use numerical simulation combined with the extended BIC of Sect. 4.11 to provide robust parameter estimates. We are interested in behaviour as efficiency increases. For the FS we monitor performance as the value of h increases, one observation at a time. The other three methods (we exclude LTSr) were originally defined either by specifying *eff* or *bdp*, and we monitor them for 50 values of these. We compare hard trimming and downweighting methods. For hard trimming (FS and LTS), we find the best model by use of the extended BIC criterion BICW (4.14). For downweighting (S- and MM-estimation), we use the criterion $BICW_\rho$ in Table 5.4, derived from AIC-like procedures of Marazzi et al. (2009). Table 5.4 presents expressions for these two weighted forms of the BIC. The form in which BICW is written includes the consistency correction (3.49) for variance. A further small-sample correction is used for LTS.

We begin in Fig. 5.23 with the distribution of trajectories of the values of BIC for four forms of outlier detection. The simulations are for $n = 200$, with four explanatory variables ($p = 5$) simulated from standard normal distributions (the procedures are invariant to the values of the regression parameters, so these are set to zero). The observational errors are independent standard normal with a shift of $\Delta = 5$ added to 20 observations, so that there is 10% contamination. There are 200 simulations; we plot the 1, 50, and 99% quantiles of the estimated values of BICW. The top row of

Table 5.4 Two different forms of BIC

Criterion	Estimation	Formula
BICW	LS	$-n[\log R_h(\hat{\beta}_h)/\{\sigma^2(h)\sum_{i=1}^n w_i\}]$
		$-\{p + \sum_{i=1}^n (1-w_i)\}\log n$
$BICW_\rho$	MM & S	$-n\log\left\{R(\hat{\beta}_\rho)/\sum w_i\right\}$
		$-\{p + \sum_{i=1}^n (1-w_i)\}\log n$

5.6 Comparisons Using the Extended BIC Criterion for Outlier Detection

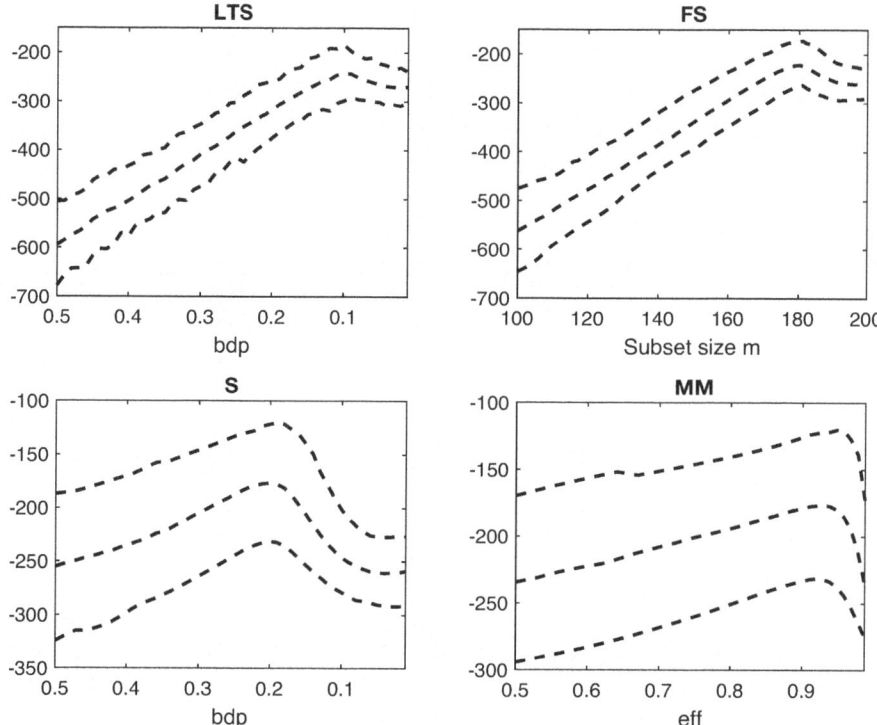

Fig. 5.23 Distribution of trajectories of BIC for outlier detection. Upper row, BICW for hard trimming; lower row, BICW$_\rho$ for soft trimming. 200 simulations, $n = 200$, $p = 5$, 10% contamination, shift $\Delta = 5$. 1, 50 and 99% points of the distribution

the figure shows the trajectories of BICW for the two hard trimming methods, monitored LTS with 50 values of h and the FS, with BICW$_\rho$ for the two soft trimming procedures, S and MM, in the lower row. All curves have a similar shape, at first increasing almost linearly to a maximum before decreasing more or less sharply.

Since we monitor from $n/2$ to n, when $n > 100$ we evaluate at fewer values of h for LTS than we do for the FS. We also start LTS estimation anew for each h. We could modify monitored LTS by evaluating at each value of $h \geq n/2$ and using the parameter estimates from $S^*(h)$ to provide starting values for estimation for $S^*(h + 1)$. These modifications would reduce some of the differences our simulations show between LTS and the FS.

The curves for hard trimming have a wider range of trajectory values than those for soft trimming, but the feature of main interest is the position of the maximum of the curves and so the indicated degree of trimming. For the hard trimming methods, the maxima are close to 0.1, correctly indicating 10% contamination. For S-estimation, the maximum is near a *bdp* of 0.2, whereas for MM-estimation the maximum is close to a high efficiency of 0.93. In all cases, the presence of some outliers is indicated.

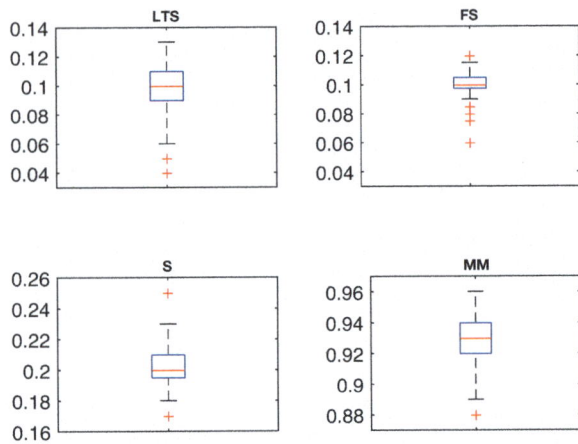

Fig. 5.24 Boxplots of position of maxima of BIC outlier detection trajectories plotted in Fig. 5.23. Upper row, BICW for hard trimming; lower row, $BICW_\rho$ for soft trimming. 200 simulations, $n = 200$, $p = 5$, 10% contamination, shift $\Delta = 5$

Simulations when there are no outliers, not shown here, yield trajectories for all four procedures that increase linearly with the maxima at non-robust least squares estimation.

Figure 5.24 shows boxplots of the position of the maxima for the trajectories of the four versions of BIC plotted in Fig. 5.23. We see that the boxplots for the two hard trimming methods in the upper row are both centred around detecting 10% of outliers, but that the variability for LTS is greater than that for the FS. The scatter for both soft trimming methods is greater than that for the FS, with the S-estimator always computed with *bdp* greater than the actual contamination level. For the remainder of this simulation section, we focus on hard trimming methods.

The outlier detection properties of the two hard trimming methods depend on the numerical details of the algorithms we have used. For each simulated set of 200 observations when using LTS we calculated BICW for 50 values of h from $0.5n$ to $0.99n$. The calculations for the different values of h are independent, using 1,000 elemental subsets with concentration steps (Rousseeuw & Van Driessen, 1999), see Sect. 3.11.2, to find the estimates of the regression parameters for each value of h. When $n = 200$ the number of observations in h changes in steps of two, so that the proportion of outliers found can only change in steps of 0.01. For the FS we take 1,000 elemental subsets of all n observations, calculate the value of the LTS criterion with bdp 50% for each subset, and take as the initial subset for the FS that yielding the minimum of the LTS criterion. At this point, use of LTS, rather than the default LMS, has no effect on the numerical results.

We now look at the proportion of observations declared as outliers as the shift Δ in the 20 outliers increases. Boxplots for these results are in Fig. 5.25. The median number of outliers for each shift is shown as a red horizontal line which may be within, or on the, boundary of each box. This shows that when $\Delta = 4.5$, the median number of outliers declared by LTS is around 0.01, whereas it is a little less than 0.1 for the FS. As Δ increases from 4.5 to 8.0, the proportion of outliers detected for both

5.6 Comparisons Using the Extended BIC Criterion for Outlier Detection

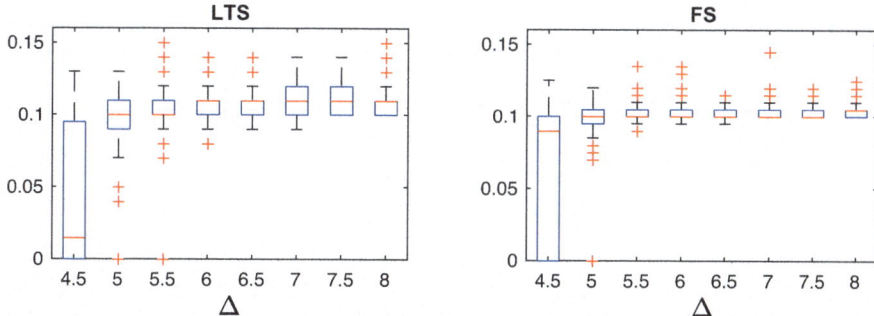

Fig. 5.25 Boxplots of the proportion of observations declared as outliers as the outlier shift Δ increases. Left-hand panel; LTS: right-hand panel FS: 200 simulations, $n = 200$, $p = 5$, 10% contamination

LTS and the FS is around 0.1. The figure illustrates in three ways the improvement in using the FS compared to monitored LTS: the detection of outliers occurs for a lower value of Δ, the mean results are more stable for larger Δ, as indicated by the width of the boxplots, and there are fewer simulations leading to false declarations of extra outliers. The plots also show the effect for LTS of increments in the proportion of outliers detected in steps of 0.01, as opposed to 0.005 for the FS.

Finally, for these simulations with $n = 200$, we look at the number of good observations declared as outliers and the number of outliers correctly detected, again as the shift in the outliers increases. The average results over 200 simulations for both $n = 200$ and 500 are in Table 5.5. As the outlier shift Δ increases from 4.5 to 8.0 both LTS and the FS detect all outliers, with the FS detecting more for lower values, especially 4.5 and 5 when $n = 500$. As Δ increases, the number of false declarations also increases, although the average number is smaller for the FS. When $n = 500$ the value of h for LTS is incremented by 5 observations. It is interesting that when $\Delta = 8$ the average number of false declarations for the FS is 0.9, whereas it is 3.4 $= 0.9 + 5/2$ for LTS, 5/2 being half of the increment size for the values of h in this simulation.

Figure 5.25, and the related Table 5.5, provides a nice illustration of the properties of outlier detection using the extended BIC. As both Δ and n increase all the outliers are identified, both by LTS and the FS. However, there is always a small number of good observations that are mistakenly declared to be outliers. These results show that the FS performs better on all measures than LTS, although the difference is not large provided $\Delta > 5$. However, with p increased to 15, further simulations showed a strong divergence in the behaviour of FS and LTS, both for $n = 200$ and 500. For $n = 500$ LTS detected a maximum of four outliers until $\Delta = 7.5$. For $\Delta = 8$ too many were detected. The unsatisfactory behaviour of LTS when $p = 15$ persisted despite the use of initial LTS subsets of 10,000, rather than 1,000, observations.

In all these simulations, the outlying responses have been within the range of the explanatory variables. As a final extension, we consider outliers at leverage points.

Table 5.5 Average number of correct ("Outliers") and incorrect ("Good") declarations of outliers for LTS and the FS as a function of outlier shift Δ and sample size; $p = 5$, 10% contamination

	$n = 200$				$n = 500$			
Δ	LTS		FS		LTS		FS	
	Outliers	Good	Outliers	Good	Outliers	Good	Outliers	Good
4.5	8.84	0.58	13.31	0.42	3.10	0.10	13.60	0.14
5	16.38	1.25	18.50	0.63	32.6	1.3	45.50	0.73
5.5	19.42	1.50	19.84	0.63	47.57	2.80	49.40	0.85
6	19.85	1.63	19.96	0.62	49.54	3.16	49.84	0.89
6.5	19.94	1.66	19.99	1.46	49.81	3.24	49.96	0.90
7	19.99	1.95	20	0.66	49.94	3.34	50.00	0.90
7.5	20	1.99	20	0.72	49.98	3.32	50	0.90
8	20	1.84	20	0.72	50	3.4	50	0.90

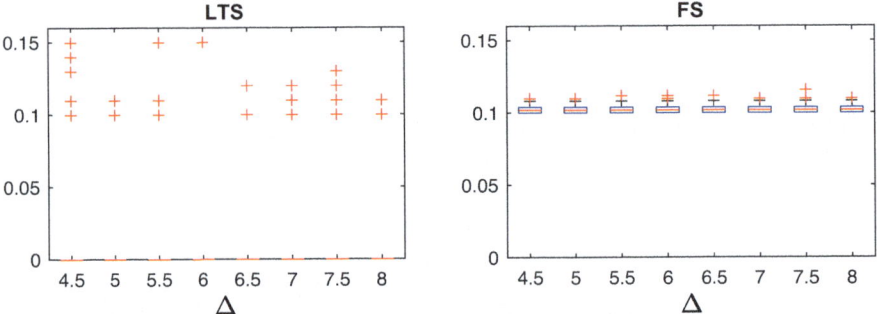

Fig. 5.26 The effect of outliers at leverage points, combined with a larger number of explanatory variables. Boxplots of the proportion of observations declared as outliers as the outlier shift and explanatory variable remoteness Δ increases when $n = 500$ and $p = 15$. Left-hand panel: LTS, initial subsets of 1,000 observations; right-hand panel FS

The 14 explanatory variables were simulated independently but were generated to have an average value of $R^2 = 0.8$ before contamination. We kept n at 500, including 50 outliers. For each value of Δ not only was this value added to the 50 outlying responses but also to the explanatory variables for each outlying response. We thus generated outliers at extreme leverage points. Boxplots for the results are in Fig. 5.26. The left-hand panel of the figure shows that LTS completely fails to detect the many outliers; the medians of the boxplots are all at zero. On the other hand, the FS reveals all the outliers, with a few extra, for all values of Δ. This figure leads to the same conclusions as those for the two simulations with $p = 15$ mentioned above, in which the outliers were not at leverage points and the initial subsets for LTS were of size either 1,000 or 10,000.

We have no explanation for the surprisingly poor behaviour of LTS when $p = 15$, although it does echo the poor performance of LTS shown in Fig. 5.21. As Olive

(2020) stresses, a complex estimator, such as many of those used in robustness, depends not only on the mathematical formulation and theoretical properties of the estimator, but also on the details of the algorithm used to provide numerical values of estimators. The comparison of analyses with different robust methods should provide security against such failures.

5.7 Discussion

The largest contrast between estimators is shown in the figures for point contamination of Sect. 5.5.4. There is also some theoretical explanation for the relative behaviour of the other estimators. In particular, the MM estimator is intended to improve the efficiency of the S-estimator and, indeed, this estimator has a lower variance in Examples 1 and 2. But this is achieved at the cost of having higher bias than the S-estimator. The same is true for the comparison of LTS and LTSr, except for those values of (x_0, y_0) in Sect. 5.5.4 for which $x_0 \in \mathcal{X}$, so that there are no leverage points to introduce serious biases and LTSr performs better than LTS. The comparisons of Sect. 5.3 indicate the LTSr performs better than LTS as the outliers become more extreme.

There are two main conclusions. The first is that the parameterized family of departures provides a cogent framework for investigating the behaviour of very robust estimators. The second is that we can clearly establish the properties of the various methods of very robust regression in terms of the bias and variance of estimators and the size and power of outlier tests. We have illustrated the use of our framework for comparing FS with methods designed to have a breakdown of 50%. Of course, the framework can be used for comparisons with breakdown levels more likely to be used in practice, such as 20 or 30%. The properties of FS, since they do not depend on a specified breakdown level, will not be changed.

5.8 Two or More Populations

In (2.4), the basic equation for the contaminated distribution of observations, it is assumed that there are two distributions of unequal importance: $F(\theta)$ to be estimated and G to provide a contamination to be guarded against, which is used to compare the performance of robust estimators. In the development of our parametric framework for comparing robust estimators in Sect. 5.4.2, we replaced G with a specified model $M_2(\theta_2)$. There, $M_2(\theta_2)$ was again used to test robust methods. But there are often situations in which data consist of observations from two or more populations. Such problems have been thoroughly studied under the heading "cluster analysis" (Fraley & Raftery, 2002) when the responses are multivariate, but lack regression structure. References to robust clustering of regression data are fewer. They include Garcia-Escudero et al. (2017) and Torti et al. (2021b).

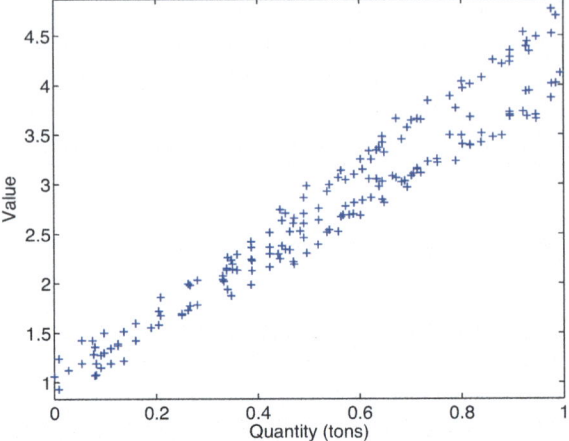

Fig. 5.27 International trade data: a mixture of two regression lines representing the values and the quantities of a given product imported by two firms. Two slopes are evident

Here we introduce the problem in the simple case of a mixture of two regression lines, with a single explanatory variable. The data, shown in Fig. 5.27, are of a kind discussed by Perrotta et al. (2009) in the detection of fraud in international trade, where false declarations of price are used in tax evasion and money laundering. An introduction to trade data is given in Sect. 8.3.2. We will analyse several datasets from international trade in Chap. 8.

The dataset is formed by 180 observations that come from two traders. The structure is of two lines that overlap for lower values; any kind of statistical separation of the data is likely to be impossible at these values. However, the two lines are clearly separate for the higher values of y and x and a robust procedure should respond to this pattern, by downweighting some of the observations in estimation and flagging them as outliers. If the outlier pattern suggests that there is a mixture of regression models, the analysis can move to clusters of regression lines, as in Garcia-Escudero et al. (2010) and in Sect. 8.5. But the first stage is the identification of outliers, for which a robust fit is required. Interest in the analysis is not in individual outliers, but whether the two lines differ.

From this perspective, it is useful to analyse the plot of the fitted lines and of the outliers identified by the five methods we have been comparing. The results are in Fig. 5.28 where the top left-hand panel repeats the plot of the data, simulated from two regression lines. The other panels show that FS, LTS, and S all provide fits to the lower of the two lines evident for the higher values of value and quantity. The other two methods, reweighted LTS and MM, provide fitted lines which lie more between the groups. Only FS indicates that there are a large number of outliers which might perhaps be modelled separately. These results are in line with the conclusions to be expected from the simulation results of earlier sections of this chapter, particularly the low power of the outlier tests for all except FS. However, the power comparisons combined with the size calculations of Sect. 5.3 show that we cannot change the level

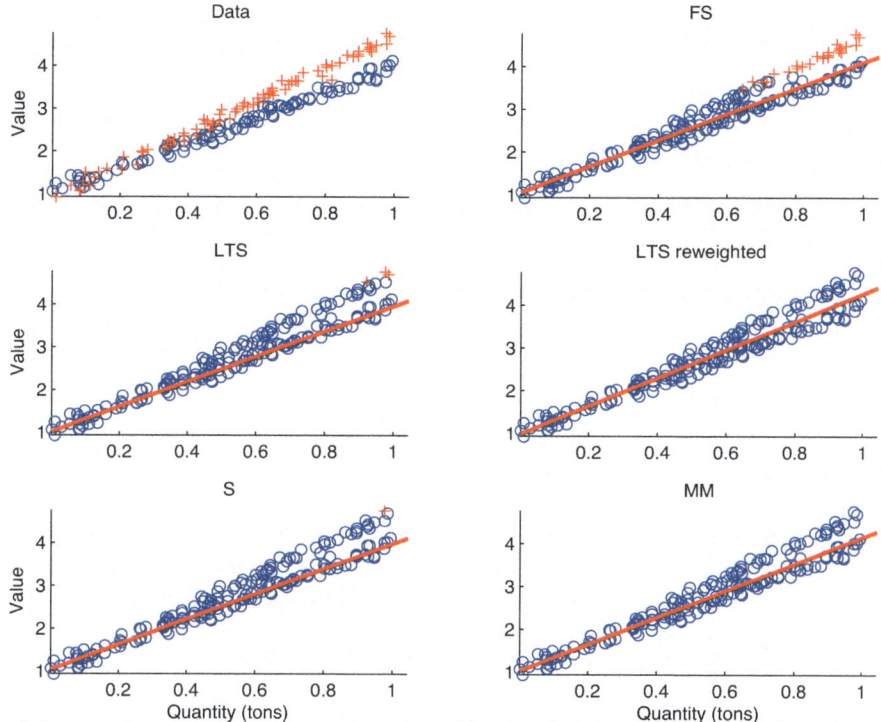

Fig. 5.28 International trade data: results of five very robust analyses. Top left-hand panel: simulated regression data for main o and secondary component +. The remaining panels contain the fitted values and the units declared as outliers +. Reading across the estimators are FS, LTS, LTS reweighted, S, and MM

of the tests without damaging the size of the test when there are no outliers and so identifying far too many outliers in the null case.

Problems

5.1 Multiple tests and the Bonferroni correction

(a) Suppose you have 20 independent hypotheses to test, and a significance level of 0.05. What is the probability of observing at least one significant result just due to chance?

(b) Plot the probability of finding at least one significant result, as a function of the number of tests in the interval [20 200] and the significance level α (consider $\alpha = 0.05$ and $\alpha = 0.01$).

(c) With the settings of point (a), compute the Bonferroni correction. Which p-value do you need to reject the simultaneous null hypothesis if you apply the Bonferroni correction?

(d) It is known that the Bonferroni correction tends to be a bit too conservative. Show this for the settings of point (a).

Remark: for this Problem and Problems 5.2 and 5.3, assume that the data are generated under the null hypothesis.

5.2 Bonferroni correction for confidence intervals

Denote by g the number of statements, comparisons or tests. The Bonferroni inequality is generally presented as

$$P\left(\bigcap_{i=1}^{g} A_i\right) \geq 1 - \sum_{i=1}^{g} P(\overline{A}_i), \tag{5.17}$$

where A_i and its complements $\overline{A}_i$ are any events. Suppose that A_i is the event "confidence interval for experiment i contains the true value". Suppose that simultaneous multiple interval estimates are desired for all g tests with an overall confidence coefficient of $1 - \alpha$. How must the individual confidence intervals be constructed to ensure that the overall confidence interval is at least $1 - \alpha$?

5.3 Theoretical individual and simultaneous size

Show, in percentages multiplied by 10, the plot of the theoretical individual and simultaneous sizes based on the χ^2 approximation calculated from (5.7) for n in the interval [30 5000]:

(a) when $p = 2$;
(b) when $p = 10$.

5.4 Probability of overlapping

Given the quantities introduced in Sect. 5.5.2, show that the probability of overlapping comes from the marginal distribution of $Z_1 \sim \mathcal{N}(b^T \mu, b^T \Sigma b)$.

5.5 Theoretical and empirical overlapping indexes with point-mass contamination

Show that both the theoretical and empirical overlapping indexes have values of either zero or one when there is point-mass contamination.

Open Access This chapter is licensed under the terms of the Creative Commons Attribution 4.0 International License (http://creativecommons.org/licenses/by/4.0/), which permits use, sharing, adaptation, distribution and reproduction in any medium or format, as long as you give appropriate credit to the original author(s) and the source, provide a link to the Creative Commons license and indicate if changes were made.

The images or other third party material in this chapter are included in the chapter's Creative Commons license, unless indicated otherwise in a credit line to the material. If material is not included in the chapter's Creative Commons license and your intended use is not permitted by statutory regulation or exceeds the permitted use, you will need to obtain permission directly from the copyright holder.

Chapter 6
Transformations

Abstract For observations from a known distribution, the variance-stabilizing transformation yields approximately normally distributed transformed variables. For data, the Box–Cox family of power transformations, indexed by a parameter λ, provides a widely applicable method for transformation of positive response data to approximate normality. The chapter describes procedures for making the transformation robust and extending the robustness to more general problems. An approximate score test for the null value λ_0 is developed in Sect. 6.1.2. A significant value leads to rejection of the value λ_0. Using the FS to monitor the score test for a set of values of λ_0 leads in Sect. 6.1.3 to the "Fan Plot". This reveals the effect of outliers on the estimated transformation and so to the deletion of outliers and selection of a value of λ for further data analysis. Section 6.2 develops related procedures (the extended Yeo–Johnson transformation) for data which can be positive or negative. The approach is visual, being based on the interpretation of fan plots. Section 6.3 provides an automatic procedure for estimating these transformations using quantities calculated from fan plots. Transformation procedures when the responses are proportions or percentages follow, as well as transformations of both sides of a statistical model.

6.1 Introduction: The Box–Cox Transformation

The transformation of response data to approximate normality was introduced in Chap. 1, Sect. 1.4, to help clarify the analysis of two sets of income data. The analyses used the Box–Cox transformation, in one example, to justify taking logarithms. This was extended in Sect. 1.5 to the analysis of regression data including demonstrations (in Figs. 1.5 and 1.6) of the information obtained by monitoring the fan plot of test statistics for transformation of positive responses. The current chapter starts with the parametric family of Box–Cox transformations for positive responses with parameter λ. No transformation corresponds to taking $\lambda = 1$, with $\lambda = 0$ being the log transformation.

Section 6.1.1 develops the maximum likelihood estimation of λ, which requires an iterative algorithm to calculate the maximum likelihood estimate $\hat\lambda$. An approximate score test for the value of λ is developed in Sect. 6.1.2, which is used in Sect. 6.1.3 to provide robust estimation of λ using the FS. Data analyses are in Sects. 6.1.4 and 6.1.5. Section 6.2 develops related procedures (the extended Yeo–Johnson transformation) for data which can be positive or negative.

The use of transformations in data analysis has a long history, as statistical histories go. References for further reading are in Sect. 6.6. The parametric family of power transformations described by Box and Cox (1964) introduced constructive methods of finding response transformations leading to approximate normality of the response in the linear model $E(Y) = X\beta$. Three stated aims of their transformation are

(i) Simplicity of structure for $E(Y)$. In a regression model, the hope is that interaction and second-order terms are not needed.
(ii) Constancy of error variance.
(iii) Approximate normality of the error distribution.

If the random variable Y has a known distribution for which $E(Y) = \mu$ and $\text{var}(Y) = V(\mu)$, the aim is to find a transformation $g(y)$ for which the variance is, at least approximately, independent of the mean. Then Taylor series expansion of $g(y)$ leads to $\text{var}\{g(Y)\} \approx V(\mu)\{g'(\mu)\}^2$. Imposition of the requirement of constant variance leads to $\text{var}\{g(Y)\} \approx V(\mu)\{g'(\mu)\}^2 = \text{constant}$. Then $g(y)$ is a solution of the differential equation $dg/d\mu = C/\sqrt{V(\mu)}$. If, without loss of generality, we have random variables standardized to have unit variance then $C = 1$ and the variance-stabilizing transformation is

$$g(t) = \int^t 1/\sqrt{V(u)}du, \tag{6.1}$$

a standard result in mathematical statistics (Bartlett, 1947).

This result, satisfying aim (ii) above, assumes that $V(\mu)$ is known. It is rare in the analysis of data that the transformation to normality is exactly known. However, all three aims are satisfied in the examples given by Box and Cox (1964), as they are in the analyses of numerous other datasets, such as those in Atkinson and Riani (2000, Chap. 4). An important consideration, which has emerged since the publication of their paper, is the large amount of statistical software available for the analysis of data with univariate or multivariate normal responses. Provided a suitable transformation can be found, statistical analysis is thereby much simplified.

Parametric and non-parametric approaches differ in the way they estimate $g(t)$. In the parametric approach of this chapter, we assume that the variance is proportional to a power of Y and that there is a parameter or vector of parameters λ that will produce a satisfactory transformation (see Exercise 6.1). In the non-parametric case (Sect. 7.3.2), we work with the reciprocal of smoothed residuals to estimate $g(t)$.

6.1.1 Maximum Likelihood Estimation for the Box–Cox Transformation

The Box–Cox transformation for non-negative responses is a function of the parameter λ. The transformed response is

$$y(\lambda) = \begin{cases} (y^\lambda - 1)/\lambda & (\lambda \neq 0) \\ \log y & (\lambda = 0). \end{cases} \quad (6.2)$$

The value $\lambda = 1$ corresponds to no transformation, $\lambda = 1/2$ to the square-root transformation, and $\lambda = -1$ to the reciprocal transformation. The use of l'Hôpital's rule as $\lambda \to 0$ shows that the logarithmic transformation is obtained for $\lambda = 0$. The transformation thus avoids the discontinuity at zero of the simple power transformation y^λ, with a great simplification in numerical procedures for the estimation of λ and for the interpretability of results.

For a given λ the model is

$$y(\lambda) = X\beta(\lambda) + \epsilon, \quad (6.3)$$

where, as before, X is $n \times p$, β is a $p \times 1$ vector of unknown parameters but now the variance of ϵ, $\sigma^2(\lambda)$, depends on λ. For given λ, the parameters are found by minimization of the sum of squares

$$S(\lambda) = \{y(\lambda) - X\beta\}^T \{y(\lambda) - X\beta\}, \quad (6.4)$$

so that the least squares estimates of the parameters are

$$\hat{\beta}(\lambda) = (X^T X)^{-1} X^T y(\lambda), \quad (6.5)$$

with the mean square estimate of σ^2 being

$$s^2(\lambda) = \{y(\lambda) - X\hat{\beta}(\lambda)\}^T \{y(\lambda) - X\hat{\beta}(\lambda)\}/(n-p) = S(\lambda)/(n-p). \quad (6.6)$$

To estimate λ it is necessary to allow for the change of scale of $y(\lambda)$ with λ. The likelihood of the transformed observations relative to the original observations y is

$$(2\pi\sigma^2)^{-n/2}[\exp\{-S(\lambda)/2\sigma^2\}]J_{BC},$$

where the Jacobian

$$J_{BC} = \prod_{i=1}^n \left| \frac{\partial y_i(\lambda)}{\partial y_i} \right| \quad (6.7)$$

allows for the change of scale due to transformation. For the power transformation (6.2),

$$\frac{\partial y_i(\lambda)}{\partial y_i} = y_i^{\lambda-1}$$

so that

$$\log J_{\text{BC}} = (\lambda - 1) \sum \log y_i = n(\lambda - 1) \log \dot{y},$$

where $\dot{y}$ is the geometric mean of the observations. A simpler form for the likelihood is found by working with the normalized transformation, defined in general as

$$z(\lambda) = y(\lambda)/J_{\text{BC}}^{1/n}. \qquad (6.8)$$

From (6.2)

$$z(\lambda) = y(\lambda)/J_{\text{BC}}^{1/n} = \begin{cases} \dfrac{y^\lambda - 1}{\lambda \dot{y}^{\lambda-1}} & (\lambda \neq 0) \\ \dot{y} \log y & (\lambda = 0). \end{cases} \qquad (6.9)$$

The likelihood can now be written as

$$(2\pi\sigma^2)^{-n/2} \exp[-\{z(\lambda) - X\beta\}^T \{z(\lambda) - X\beta\}/2\sigma^2], \qquad (6.10)$$

a standard normal theory likelihood for the response $z(\lambda)$.

For fixed λ the likelihood (6.10) is maximized by the least squares estimates

$$\hat{\beta}(\lambda) = (X^T X)^{-1} X^T z(\lambda),$$

with the residual sum of squares of the $z(\lambda)$,

$$R(\lambda) = z(\lambda)^T (I - H) z(\lambda). \qquad (6.11)$$

Division of (6.11) by n yields the maximum likelihood estimator of σ^2 as $\hat{\sigma}^2(\lambda) = R(\lambda)/n$. Replacement of this estimate by the mean square estimate $s^2(\lambda)$ in which n is replaced by $(n - p)$ does not affect the development that follows.

The loglikelihood maximized over both β and λ follows by substitution of $\hat{\beta}(\lambda)$ and $s^2(\lambda)$ into (6.10). If an additive constant is ignored, this partially maximized, or profile, loglikelihood of the observations is

$$\mathcal{L}_{\max}(\lambda) = -(n/2) \log\{R(\lambda)/(n - p)\}, \qquad (6.12)$$

so that $\hat{\lambda}$ minimizes $R(\lambda)$.

For inference about plausible values of the transformation parameter λ, Box and Cox suggest likelihood ratio tests using (6.12), that is, the statistic

$$T_{LR} = 2\{\mathcal{L}_{\max}(\hat{\lambda}) - \mathcal{L}_{\max}(\lambda_0)\} = n \log\{R(\lambda_0)/R(\hat{\lambda})\}. \qquad (6.13)$$

6.1 Introduction: The Box–Cox Transformation

A $100(1-\alpha)\%$ confidence interval for λ is given by those values of λ_0 which do not give values of T_{LR} significant at the level α.

Box and Cox do not use $\hat{\lambda}$ directly but, where possible, find a value for λ with a physical interpretation that is acceptable in the confidence interval derived from (6.13). Physical laws suggest that often such values will be ratios of small integers. Both examples in Box and Cox (1964) are from designed experiments. The first is the 48 values of survival times of poisoned animals. The value of $\hat{\lambda}$ is -0.75 with approximate 95% confidence interval from (6.13) $[-1.13 \ -0.37]$ given by Box and Cox (see Exercise 6.2). The reciprocal transformation ($\lambda = -1$) was chosen; the simple structure is therefore in death rate rather than survival time. The second example, the wool data of Sect. 4.5.2, is a 3^3 factorial ($n = 27$) with response the number of cycles to breaking of samples of worsted yarn. Multiplicative models are typical in this field and, indeed, $\hat{\lambda} = -0.06$ with surprisingly short approximate confidence interval $[-0.18 \ 0.06]$. The logarithmic transformation was chosen.

Further examples of a similar size in Atkinson and Riani (2000, Chap. 4) produce log transformations or transformations of one-third for the volume of useful timber in cherry trees, so converting the measurement to the dimension of length. However, for larger sets of data, from economics rather than the physical sciences, our experience is that such simple values for λ become less frequent, a point exemplified by the analysis of 509 observations on loyalty cards in Sect. 6.1.5.

It is important that $R(\lambda)$ in (6.12) is the residual sum of squares of the $z(\lambda)$, a normalized transformation with the approximate physical dimension of y for any λ. Comparisons of residual sums of squares $S(\lambda)$ of the power transformation $y(\lambda)$ are misleading. Suppose, for example, that the observations are of order 10^3, when the residual sum of squares $S(1)$ will be of order 10^6, whereas, when $\lambda = -1$, the reciprocal transformation, the observations will be of order 10^{-3} and $S(-1)$ will be of order 10^{-6}. However relatively well the models for $\lambda = 1$ and $\lambda = -1$ explain the data, $S(-1)$ will be very much smaller than $S(1)$. Comparison of these two residual sums of squares will therefore indicate that the reciprocal transformation is to be preferred.

An important property of the normalized Box–Cox transformation is the approximate orthogonality of λ and β (Cox & Reid, 1987); the estimates of β are little affected by the value of λ. On the other hand, the estimates of β from the unnormalized transformation are highly correlated with the value of λ. References to the confusion this has caused about appropriate inferences for the value of λ are in Atkinson et al. (2021, Sect. 2).

The practical procedure is analysis in terms of $z(\lambda)$ leading to $\hat{\lambda}$ and hence to a, hopefully, physically interpretable estimate $\tilde{\lambda}$ chosen from a grid of plausible values. Carroll (1982) argues that the grid needs to become denser as n increases. For comparisons across sets of data, parameter estimates need to be calculated using $y(\tilde{\lambda})$, rather than $z(\tilde{\lambda})$, to avoid dependence on the values of $\dot{y}$ from each set. In the light of the foregoing, we stress that our robust data analytic procedures select the value or values of λ from, often coarse, grids, the value of λ used in data analyses being guided by the data. It is not the value of $\hat{\lambda}$.

6.1.2 The Score Test

The likelihood ratio test (6.13) requires the numerical calculation of $\hat{\lambda}$, an iterative numerical procedure. In Sect. 1.4, we briefly mentioned the score test which provides an alternative to (6.13), avoiding the calculation of $\hat{\lambda}$. The score test is the t-test for the added variable w in Sect. 3.4, where the added variable is found by Taylor series expansion of $z(\lambda)$ as

$$z(\lambda) \doteq z(\lambda_0) + (\lambda - \lambda_0) \left. \frac{\partial z(\lambda)}{\partial \lambda} \right|_{\lambda = \lambda_0} = z(\lambda_0) + (\lambda - \lambda_0) w_A(\lambda_0), \quad (6.14)$$

which only requires calculations at the hypothesized value λ_0. In (6.14), $w_A(\lambda_0)$ is the constructed variable for the transformation.

The calculations for the added variable test in Sect. 3.4 only require the residual $\tilde{w}_A$ from regression on X. If, as is usually the case, X includes a constant term, constants in w_A can be ignored and the derivatives of $z(\lambda)$ simplify to

$$w_A(\lambda) = \begin{cases} \frac{y^\lambda \{\log(y/\dot{y}) - 1/\lambda\}}{\lambda \dot{y}^{\lambda-1}} & (\lambda \neq 0) \\ \dot{y} \log y (0.5 \log y - \log \dot{y}) & (\lambda = 0). \end{cases} \quad (6.15)$$

As before, the derivation when $\lambda = 0$ requires the use of l'Hôpital's rule.

The combination of (6.14) and the regression model $y = x^T \beta + \epsilon$ leads to the approximate model

$$z(\lambda_0) = x^T \beta - (\lambda - \lambda_0) w_A(\lambda_0) + \epsilon = x^T \beta + \gamma \, w_A(\lambda_0) + \epsilon, \quad (6.16)$$

where $\gamma = -(\lambda - \lambda_0)$. The approximate score statistic $T_A(\lambda_0)$ for testing the transformation is the t statistic for regression on $w_A(\lambda_0)$ in (6.16). The t test for $\gamma = 0$ in (6.16) is then the test of the hypothesis $\lambda = \lambda_0$. With $A = I - H$, the statistic can be written as

$$T_A(\lambda) = -\frac{\hat{\gamma}(\lambda)}{\sqrt{s_w^2(\lambda)/\{w_A^T(\lambda) A w_A(\lambda)\}}}. \quad (6.17)$$

The expressions for $\hat{\gamma}(\lambda)$ and $s_w^2(\lambda)$ are those in Sect. 3.4 for $\hat{\gamma}$ and s_w^2 with the substitution of $z(\lambda)$ and w_A for y and w. The negative sign arises because in (6.16) $\gamma = -(\lambda - \lambda_0)$. Since $T_A(\lambda_0)$ is the t test for regression on $-w_A(\lambda_0)$, large positive values of the statistic mean that λ_0 is too low and that a higher value should be considered.

Since w_A is a constructed variable that includes functions of the observations, $T_A(\lambda)$ cannot exactly have a t distribution. A different approximate score statistic for the Box–Cox transformation is found by Lawrance (1987), which often has an improved null distribution, although the value may not be physically meaningful (an imaginary number) if $\hat{\lambda}$ is far from λ_0. Some numerical comparisons of the two procedures are in Atkinson and Lawrance (1989).

We now use the algebra of this section, combined with the FS, to develop robust methods for data transformation that enable monitoring of the effects of individual observations on the estimated transformation parameters.

6.1.3 The Fan Plot

The final sentence of Hinkley (1975) states "Data transformation in the presence of outliers is a risky business". The risk arises particularly in robust fitting of transformations. If a symmetrical model is fitted to a skewed distribution, the observations in the longer tail may be spuriously treated as outliers and downweighted. Figure 4.7, showing the analysis of the untransformed wool data, is an example. The robust transformation of regression data is further complicated by the inter-relationship of outliers and the value of λ; any relationship between mean and variance is a strong indication of the need for response transformation, an idea explored in the development of non-parametric transformations in Chap. 7. Examples in Atkinson and Riani (2000, Chap. 4) show the effect of modifying up to four observations in the poison data of Box and Cox (1964). In the original data, the conclusion was that the inverse transformation ($\lambda = -1$) was appropriate. With four modified observations the values of 1/3 for λ were acceptable, with both no and the square-root transformation being rejected. In this section, we illustrate the use of the fan plot, glimpsed in Figs. 1.5 and 1.6, which uses the monitoring of score statistics from the FS to provide robust data transformations that are informative about the influence of individual observations on the estimated transformation.

The procedures discussed so far in this chapter, parameter estimation, likelihood ratio tests, and score tests, are all based on aggregate statistics. That is, all are calculated using all n observations. We now use the forward search to provide a robust plot of the approximate score statistic $T_A(\lambda)$ (6.17) over a grid $\mathcal{G}$ of values of λ. There is a different search for each value of λ. Observations may both leave and join the subset in going from size m to size $m + 1$. The process moves from a very robust fit to non-robust least squares. Any outliers will enter the subset towards the end of the search. We thus obtain a series of fits of the model to subsets of the data of size m, $m_0 \leq m \leq n$, for each of which we refit the model and calculate the value of the score statistics for the values of $\lambda_0 \in \mathcal{G}$. These are then plotted against the number of observations m used for estimation to give the "fan plot". The ordering of the observations in a fan plot, which reflects the presence of outliers, may depend on the value of λ_0.

We have already mentioned that the constructed variables are functions of the response so that the score statistics cannot exactly follow the t distribution. Atkinson and Riani (2002b) provide some numerical results on the distribution in the fan plot of the score statistic for the Box–Cox transformation; increasingly strong regression relationships lead to null distributions that are closer to Student's t.

6.1.4 Surgical Unit Data

Section 4.9.5 introduced data from Neter et al. (1996, pp. 334, 438) who analysed 108 observations on the times of survival of patients who had a particular kind of liver surgery. There are four explanatory variables, with response taken as the logarithm of survival time. The analysis on this logarithmic scale showed a complicated structure in which there were two groups of units, the first of which was not homogeneous. In this section, we check whether this is the best transformation in the Box–Cox family for all the data. The fan plot of Fig. 6.1 for the five most commonly used values of λ_0 ($-1, -0.5, 0, 0.5, 1$) confirms that the logarithmic transformation is suitable for these survival data, as it is for the life times to breakage of the wool data analysed by Box and Cox (1964). Here the trajectory of the score statistic for $\lambda_0 = 0$ lies within the central pointwise 99% band throughout, whereas those for the other four values of λ_0 are outside the bands by the end of the search.

This analysis shows that there are no observations that are particularly influential for the estimation of λ. However, this does not mean that there may not be outliers in the transformed scale. In general, once a transformation has been established, the FS analysis on the transformed data needs to be repeated, as in Sect. 4.9.5.

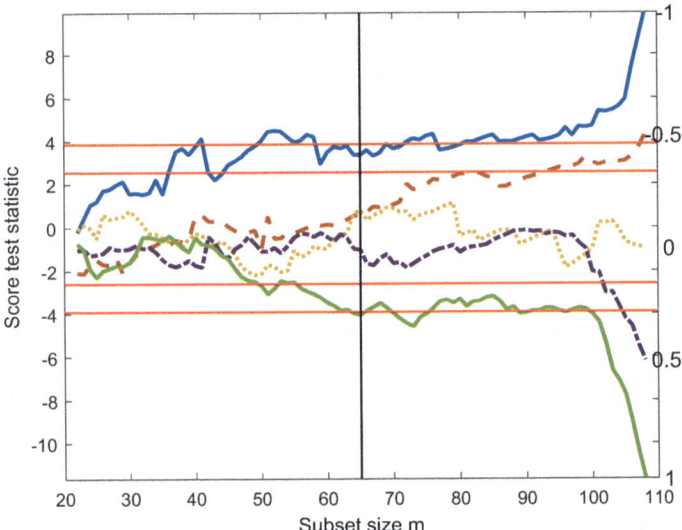

Fig. 6.1 Surgical unit data: fan plot; monitoring plot of score statistics for five values of λ_0 for the Box–Cox transformation. The horizontal bands are the 99% and 99.99% pointwise intervals for normal random variables. The vertical line at $m = 65$, the rounded value of $0.6n$, indicates the point at which monitoring starts in the automatic procedure of Sect. 6.3

6.1.5 Loyalty Cards Data

In the analysis above of the surgical unit data, the fan plot showed that there were no observations that disagreed with the overall log transformation of the data. We now analyse a larger set of data from Atkinson and Riani (2006) which demonstrates the positive insights that can be gained from monitoring, rather than just serving as a check on standard procedures.

We take 509 observations on the behaviour of customers with loyalty cards from a supermarket chain in Northern Italy. The data are themselves a random sample from a larger database. The response is the amount, in euros, spent at the shop over 6 months; the explanatory variables are: x_1, the number of visits to the supermarket in the 6-month period; x_2, the age of the customer; and x_3, the number of members of the customer's family.

The fan plot for the analysis of the surgical unit data was plotted in Fig. 6.1 for the five standard values of λ advocated by Box and Cox. Using these five values of λ for the loyalty cards data indicates that none of these values provides an appropriate transformation. The indication, for this larger dataset, is that a finer grid of values of λ is required, in line with the argument of Carroll (1982). Figure 6.2 is a fan plot for 11 values of λ (0, 0.1, ..., 1) with $\lambda = 1$ at the bottom and $\lambda = 0$ providing the top trajectory. It is clear that the trajectories for two of the five standard values of λ (0 and 0.5) steadily leave the central band in opposite directions. The trajectory for $\lambda = 0.4$ is the most central, although it changes abruptly about 20 units from the end and finishes below the lower 99.9% confidence limit. This structure may be the result of the presence of outliers. We accordingly continue our analysis with $\lambda = 0.4$, first testing for outliers.

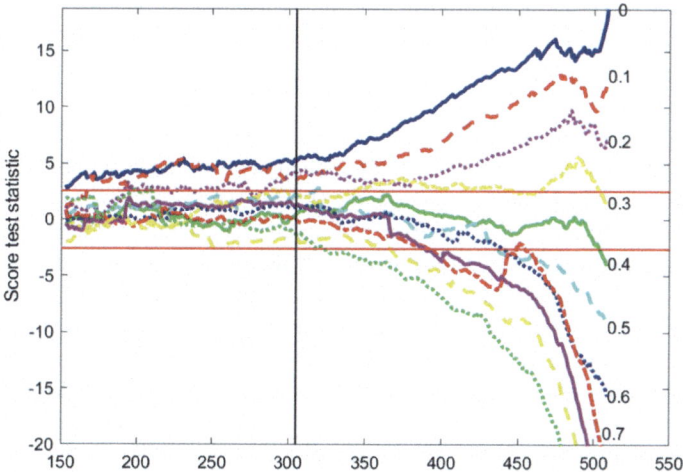

Fig. 6.2 Loyalty cards data: fan plot for 11 values of λ; (0, 0.1, ..., 1) for the Box–Cox transformation

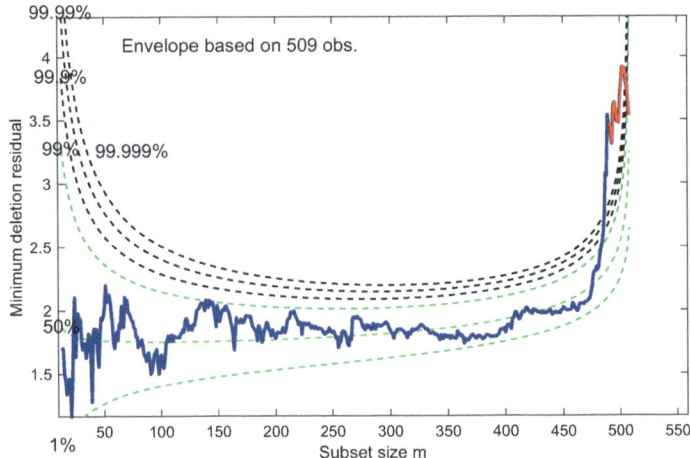

Fig. 6.3 Transformed loyalty cards data ($\lambda = 0.4$): monitoring plot of minimum deletion residuals; the 18 identified outliers are highlighted in red. The confidence bands are 0.01, 0.5, 0.99, 0.999, 0.9999, and 0.99999

Figure 6.3 shows the monitoring plot of absolute minimum deletion residuals. In addition to the residuals, the plot includes a series of pointwise percentage levels for the residuals (at 1, 50, 99. 99.9, 99.99, and 99.999%) found by the order statistic arguments of Sect. 4.2. Several large residuals occur towards the end of the search. These are identified by the procedure including the resuperimposition of envelopes illustrated in Figs. 4.4 and 4.6. In all, 18 outliers, plotted as red symbols, are identified. These form the last observations to enter the subset in the search. The figure also shows that, at the very end of the search, the trajectory of the minimum deletion residuals returns inside the envelopes, the result of masking. As a consequence, the outliers would not be detected by the deletion of single observations from the fit to all n observations.

The residuals are stable once a subset of non-outlying observations has been obtained. Figure 6.4 shows the monitoring of scaled residuals during the search with the trajectories for the 18 outliers shown in red. The outliers all have negative residuals, the values of which change little during the search until the end, when the outliers start to enter the subset $S^*(m)$. Then the residuals for the outliers decrease steadily in magnitude. At the same time, the residuals for some of the observations from the main body of the data begin to increase, an example of swamping.

The observations we have identified are outlying in an interesting way, especially as a function of x_1. Figure 6.5 shows the scatter plot of $y^{0.4}$ against x_1, with the outlying observations plotted as red crosses. The FS has identified a subset of individuals, most of whom are behaving in a strikingly different way from the majority of the population. They appear to form a group which spends less than would be expected from the frequency of their visits. On the other hand, the scatter plots of y against x_2 and x_3, given by Torti et al. (2021a), do not show any distinct pattern of outliers.

6.1 Introduction: The Box–Cox Transformation 205

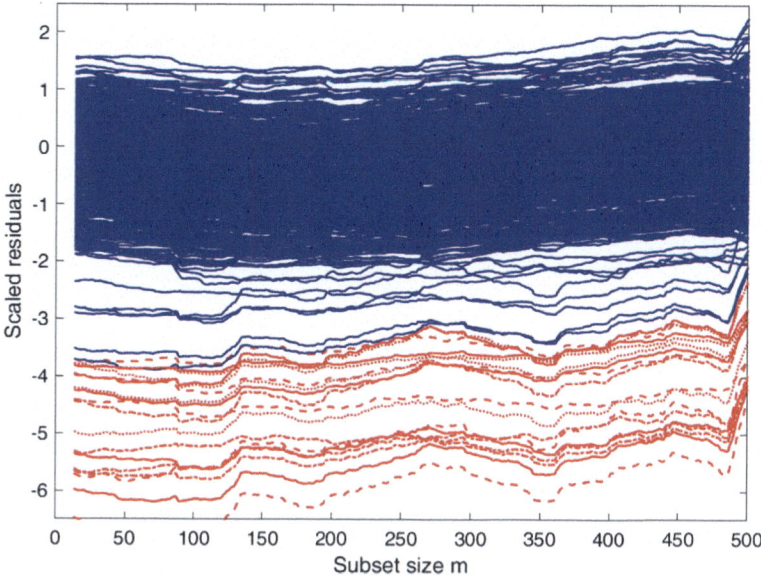

Fig. 6.4 Transformed loyalty cards data ($\lambda = 0.4$): forward plot of scaled residuals, with the 18 detected outliers highlighted in red

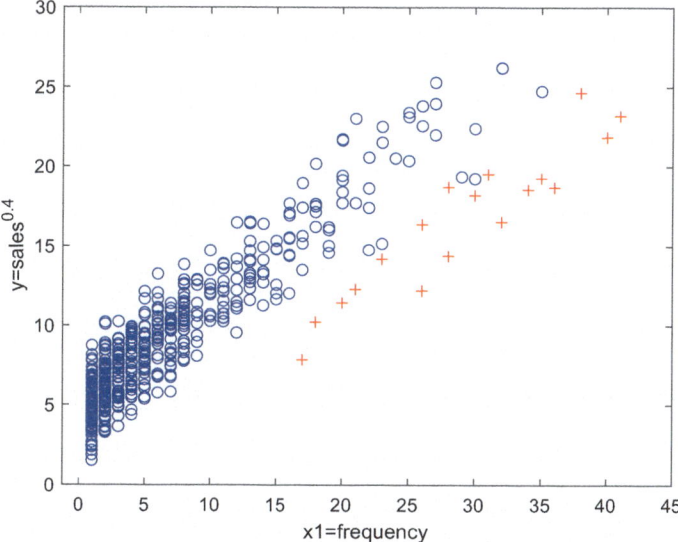

Fig. 6.5 Transformed loyalty cards data ($\lambda = 0.4$): scatter plot against x_1; the 18 outliers plotted as red crosses

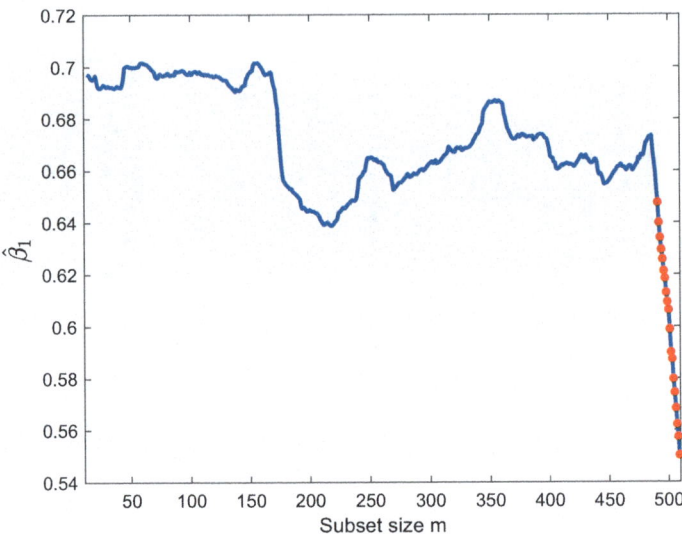

Fig. 6.6 Transformed loyalty cards data ($\lambda = 0.4$): monitoring of estimated coefficient $\hat{\beta}_1$, with the part of the trajectory corresponding to the 18 detected outliers highlighted in red

One effect of the 18 outliers on inference can be seen in Fig. 6.6 which gives the monitoring plot of the parameter estimate $\hat{\beta}_1$, again with the outlying observations plotted in red. This is by far the most dramatic of the four forward plots of the estimates of the intercept and regression coefficients given by Torti et al. (2021a). As the outliers are introduced, the estimate decreases rapidly in a seemingly linear manner. This behaviour reflects the inclusion of the outliers plotted in Fig. 6.5, all of which lie below the general linear structure; their inclusion causes $\hat{\beta}_1$ to decrease. The plots in Figs. 6.4, 6.5, and 6.6 were produced by brushing, that is, selecting the observations of interest from Figs. 6.3 or 6.4 and highlighting them in all others.

Figure 6.5 suggests that the outliers, the group that is spending less than would be expected, may be important in any further modelling and might be modelled separately. An example in which outliers are distinctly modelled is in the analysis of data on fraud in international trade introduced in Sect. 5.8.

6.1.6 The Logarithmic Transformation and Gamma Generalized Linear Models

The two most widely used transformations in the Box–Cox family are the log and the reciprocal. These are the two values that were appropriate for the examples in Box and Cox (1964). In this section, we briefly mention the relationship between the Box–Cox transformation with $\lambda = 0$ and the gamma generalized linear model (GLM) with log link.

6.1 Introduction: The Box–Cox Transformation

In the Box–Cox model, the transformed response follows a linear model. On the other hand, in generalized linear models, the linear model is transformed by the link function. For positive skew continuous data, an alternative to the Box–Cox transformation is a gamma GLM. With $E(Y) = \mu$ and the linear predictor $\eta = x^T \beta$, the link function relates the two by $\eta = g(\mu)$. We now consider the relationship between these two models. McCullagh and Nelder (1989) provide the seminal account of GLMs.

For a gamma distributed random variable, $E(Y) = \mu = \alpha/\beta$ and $\text{var}(Y) = \alpha/\beta^2 = \mu^2/\alpha$, where α is the shape parameter and β is the inverse scale or rate parameter. The variance-stabilizing transformation (Sect. 6.1 and Exercise 6.1) is $\log(Y)$. The squared coefficient of variation of a random variable Y is defined as $\tau^2 = \text{var}(Y)/\{E(Y)\}^2$. When the response has a gamma distribution, it follows that $\tau^2 = 1/\alpha$, that is, it is not a function of μ. Taylor series expansion shows that, for small τ^2, the approximate moments of $\log(Y)$ are

$$E\{\log(Y)\} = \log(\mu) - \tau^2/2 \quad \text{and} \quad \text{var}\{\log(Y)\} = \tau^2.$$

If the systematic part of the model is multiplicative on the original scale, coefficient estimates of the parameters and of their precision may be obtained by transforming to the log scale and using ordinary least squares. The intercept is then biased by approximately $-\tau^2/2$. Of course, if the exact distribution of Y is known, maximum likelihood estimation for the known distribution should be used. Firth (1988) compares the log-normal and gamma models under reciprocal mis-specification, the gamma distribution performing slightly better.

As an example, we further analyse the observational data from Kleinbaum and Kupper (1978, p. 148) on the assessment of mental illness of 53 patients, which were introduced in Sect. 4.11.5. Atkinson et al. (2021, Sect. 3) compare the logarithmic transformation with an analysis using a generalized linear model with three possible Box–Cox links in which $g(\mu) = (\mu^\lambda - 1)/\lambda$, for $\lambda = -1, 0$ and 1. The reciprocal link ($\lambda = -1$) yields the smallest deviance from fitting the gamma model, but there is no significant increase in deviance if one of the other links is used. As the right-hand panel of Fig. 6.7 shows, the plot of residuals from the GLM with $\lambda = 0$ against those from the logarithmic Box–Cox transformation is a straight line. That in the left-hand panel for the reciprocal link is distinctly curved. The insensitivity of data analyses to the exact specification of the gamma link is well established—for example, the analysis by McCullagh and Nelder (1989, p. 377) of their car insurance data. However, the relationship between response transformation and the GLM is for the log link.

The close relationship between the gamma and log-normal fits depends on τ^2 being sufficiently small. For the mental illness data, the estimated value of τ^2 is 0.4. Results of a small simulation (Atkinson, 1982) on the choice between the two models for the chimpanzee learning data of Brown and Hollander (1977) led McCullagh and Nelder to comment that discrimination between the two models may be difficult even for τ^2 as large as 0.6. The relationship also depends on the observations having a common variance. Wiens (1999) provides an example of two-group data in which

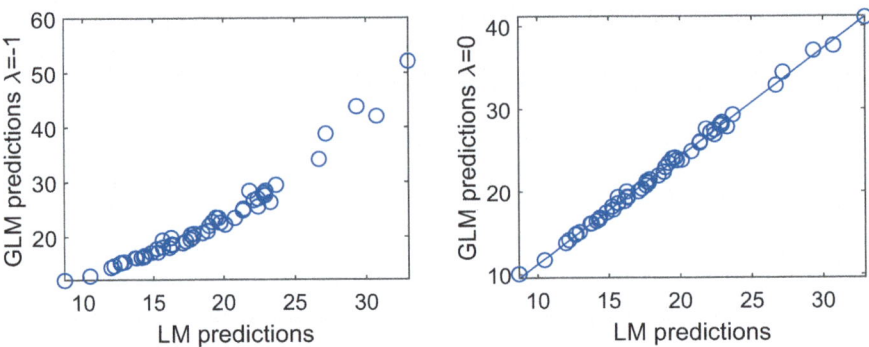

Fig. 6.7 Mental illness data: comparison of fitted values from gamma and log-normal models. Left-hand panel, reciprocal link, right-hand panel log link

the relationship fails to hold due to different variances in the two groups after log transformation.

Further discussion of the relationship between the gamma and log-normal models is in McCullagh and Nelder (1989, Chap. 8). Atkinson and Riani (2000, Chap. 6) use the goodness-of-link test of Pregibon (1980) to provide a fan plot for the parameter in the Box–Cox family of link functions.

6.1.7 The Shifted Power Transformation

To accommodate negative observations, Box and Cox (1964) introduced a shifted power transformation in which transformation of y is replaced by transformation of $y + \mu$ where $y + \mu > 0$. For a sample of observations with minimum value $y_{\min}$, it follows that $\mu > -y_{\min}$. Let $q = y + \mu$. Then the shifted version of the normalized Box–Cox transformation (6.9) is

$$z_{\text{SBC}}(\lambda, \mu) = \begin{cases} \dfrac{(y+\mu)^\lambda - 1}{\lambda G(y+\mu)^{\lambda-1}} = \dfrac{q^\lambda - 1}{\lambda \dot{q}^{\lambda-1}} & (\lambda \neq 0) \\ G(y+\mu) \log(y+\mu) = \dot{q} \log q & (\lambda = 0), \end{cases} \quad (6.18)$$

where both $G(\cdot)$ and $\dot{q}$ denote geometric mean. If the value of μ is known, the procedure is as described in Sect. 6.1.1. Estimation of both μ and λ is more problematic.

When μ is unknown, the restriction $y + \mu > 0$ makes the problem non-regular, since the range of y depends on the unknown shift parameter. The standard asymptotic results of maximum likelihood theory may not apply. Simulations in Atkinson (1985, Sect. 9.3) show that the ordinary likelihood function is unbounded at the edge of the parameter space where $\mu \to -y_{\min}$ and may, or may not, have a local maximum somewhere else. This approach to $-y_{\min}$ can be parameterized by writ-

6.1 Introduction: The Box–Cox Transformation

ing $\mu = -y_{\min}(1 - 10^\epsilon)$. Plots of the residual sum of squares of $z_{\text{SBC}}(\lambda, \mu)$ as μ decreases (to -12 in these simulations) over a range of values of λ show a characteristic hooked shape with the sum of squares decreasing as ϵ becomes more negative. For the two datasets of Box and Cox (1964), there are also local minima close $\mu = 0$, but not in some other examples.

Atkinson et al. (1991) suggest a partial resolution of this problem through the use of grouped likelihood. One consequence of the behaviour of the likelihood near the parameter boundary is that the difference of two values of μ could be statistically significant while being less than the accuracy to which the response is measured. Such an argument in favour of a grouped likelihood approach is often used for non-standard problems (Smith, 1985). We now outline the approach for the shifted power transformation. Suppose that all the observations are measured to the same accuracy Δ, so that observing y_i means that the unobserved continuous value lies in an interval $y_i \pm \Delta$. For observations rounded to the nearest integer, we take $\Delta = 0.5$. The probability that y lies within this interval is found by first transforming the interval and then applying a truncated normal distribution.

When the data are grouped, the loglikelihood is the sum of the log-probabilities of the observations lying in a particular interval, divided by the interval width. Plots of this profile loglikelihood as a function of λ and ϵ are given by Atkinson et al. (1991). These are similar in structure to those of residual sums of squares of ungrouped data in Atkinson (1985), although with an increased tendency to yield local maxima. The distinction from the ungrouped case is that the shape of the plots depends on the value of Δ. In theory, this should be the accuracy of the observations. In practice, Atkinson et al. (1991) recommend plotting the profile loglikelihood for μ for several values of Δ. In the absence of a local maximum, the chosen value of Δ should be smooth near $\mu = -y_{\min}$. Even once this inferential difficulty has been overcome, the transformation is only in the range $-y_{\min} < y \leq \infty$, with something special happening at $-y_{\min}$. The shifted power transformation, unlike the transformations of Sect. 6.2, does not provide a smooth transformation for all real y.

These problems only arise when μ has to be estimated. Some forms of data contain many zeroes and a smooth non-negative skewed distribution. An example is rainfall data when there are many dry days. It is sometimes recommended that such data be analysed by adding one to all responses and then logging the data. However, if one is of the same magnitude as the untransformed y, the distribution of $\log(y + 1)$ may be far from that of $\log(y)$ for those $y > 0$. For data values which are normally large and typically logged, such as income data, use of $\log(y + 1)$ provides a useful initial cleaning of the data which avoids the calculation of $\log(0)$. A related example is the analysis of bacterial clearance data given by Carroll and Ruppert (1988, p.158). One of the response values was 0 (to the accuracy of the experiment) and 0.05 was added to all responses before the log transformation. In this example, the largest response value was 197400. Even so, the comments on their p. 225 show that the analysis is sensitive to the exact value of the shift parameter. In Sect. 8.8, we provide a robust method for the analysis of censored data when the response may need transformation to achieve approximate normality. In applications to patterns of personal expenditure, the censoring is at zero.

6.2 The Yeo–Johnson Transformation and Its Extension

6.2.1 The Yeo–Johnson Transformation

The Box–Cox transformation requires that the observations be positive, perhaps after addition of a constant as in the shifted power transformation of Sect. 6.1.7. Yeo and Johnson (2000) generalized the Box–Cox transformation to observations that can be positive or negative. Their extension uses the same value of the transformation parameter λ for positive and negative responses. The Box–Cox transformation (6.2) has two regimes, one for $\lambda \neq 0$ and the other for $\lambda = 0$. In both cases, $y > 0$. To allow for y being either positive or negative, the Yeo–Johnson transformation of the response y requires four regimes, depending both on the transformation parameter and on the response value. The forms of the transformed response are

$$y_{\text{YJ}}(\lambda) = \begin{cases} \dfrac{(y+1)^\lambda - 1}{\lambda} & y \geq 0 \;\; \lambda \neq 0 \\ \log(y+1) & y \geq 0 \;\; \lambda = 0 \\ -\dfrac{\{(-y+1)^{2-\lambda} - 1\}}{2-\lambda} & y < 0 \;\; \lambda \neq 2 \\ -\log(-y+1) & y < 0 \;\; \lambda = 2. \end{cases} \qquad (6.19)$$

For $y \geq 0$ this is the generalized Box–Cox power transformation of $y + 1$. For negative y the transformation is of $-y + 1$ to the power $2 - \lambda$.

As for the Box–Cox transformation, analysis of data with this transformation also needs to include the Jacobian of the transformation to allow for changes of scale as λ varies. The required Jacobian, again for n observations, from Eq. (3.1) of Yeo and Johnson (2000), is

$$\log J_{\text{YJ}} = (\lambda - 1) \sum_{i=1}^{n} \text{sgn}(y_i) \log(|y_i| + 1). \qquad (6.20)$$

We continue to work with a normalized transformation $z(\lambda) = y(\lambda)/J^{1/n}$ in which the Jacobian is spread over all observations. If $\dot{y}_{\text{YJ}}$ is the nth root of J_{YJ} in (6.20),

$$\dot{y}_{\text{YJ}} = \exp\left[\sum \{\text{sgn}(y_i)\log(|y_i|+1)\}/n\right]. \qquad (6.21)$$

The normalized versions of the transformations in (6.19) then become

$$z_{\text{YJ}}(\lambda) = \begin{cases} \dfrac{(y+1)^\lambda - 1}{\lambda \dot{y}_{\text{YJ}}^{\lambda-1}} & y \geq 0 \;\; \lambda \neq 0 \\ \dot{y}_{\text{YJ}} \log(y+1) & y \geq 0 \;\; \lambda = 0 \\ -\dfrac{\{(-y+1)^{2-\lambda} - 1\}}{(2-\lambda)\dot{y}_{\text{YJ}}^{\lambda-1}} & y < 0 \;\; \lambda \neq 2 \\ -\log(-y+1)/\dot{y}_{\text{YJ}} & y < 0 \;\; \lambda = 2. \end{cases} \qquad (6.22)$$

6.2.2 Constructed Variables

For the Box–Cox transformation, there are two constructed variables in (6.15) found by series expansion of the two forms of $z(\lambda)$ in (6.2). For the Yeo–Johnson transformation, there is correspondingly a constructed variable for each of the four normalized transformations in (6.22). For $y \geq 0$ let $v_P = y + 1$ with $v_N = -y + 1$ when $y < 0$. The constructed variables are

$$w_{\text{YJ}}(\lambda) = \begin{cases} \{v_P^\lambda(\log v_P - k_P) + k_P\}/q_P, & y \geq 0 \; \lambda \neq 0 \\ \dot{y}_{\text{YJ}} \log v_P(\log v_P/2 - \log \dot{y}_{\text{YJ}}) & y \geq 0 \; \lambda = 0 \\ \{v_N^{2-\lambda}(\log v_N + k_N) - k_N\}/q_N & y < 0 \; \lambda \neq 2 \\ \{\log v_N(\log v_N/2 + \log \dot{y})\}/\dot{y}_{\text{YJ}} & y < 0 \; \lambda = 2. \end{cases} \quad (6.23)$$

In (6.23)

$$\begin{aligned} k_P &= \lambda^{-1} + \log \dot{y}_{\text{YJ}} \\ q_P &= \lambda \dot{y}_{\text{YJ}}^{\lambda-1} \\ k_N &= \log \dot{y}_{\text{YJ}} - (2-\lambda)^{-1} \\ q_N &= (2-\lambda)\dot{y}_{\text{YJ}}^{\lambda-1}. \end{aligned}$$

The approximate score test for a hypothesis about the value of λ has the same form as T_A (6.17), but with the constructed variables $w_{\text{YJ}}(\lambda)$ of (6.23).

$$T_{\text{YJ}}(\lambda) = -\frac{\hat{\gamma}(\lambda)}{\sqrt{s_w^2(\lambda)/\{w_{\text{YJ}}^T(\lambda) A w_{\text{YJ}}(\lambda)\}}}, \quad (6.24)$$

which is the test for the transformation of both the positive and negative observations. That is for $y \geq 0$, $y + 1$ has the power transformation λ whereas for $y < 0$, $-y + 1$ is transformed to the power $2 - \lambda$. In (6.24) $s_w^2(\lambda)$ again has the form given by (3.23), but is now a function of $z_{\text{YJ}}(\lambda)$ and $w_{\text{YJ}}(\lambda)$.

6.2.3 The Extended Yeo–Johnson Transformation

The transformations for negative and positive responses in Sect. 6.2.1 were determined by Yeo and Johnson (2000) by imposing the smoothness condition that the second derivative of $z_{\text{YJ}}(\lambda)$ with respect to y be smooth at $y = 0$. However some authors, for example, Weisberg (2005b), query the physical interpretability of this constraint which is indeed violated by two sets of data analysed by Atkinson et al. (2020). They show that the two classes of response may require transformation with different values of λ.

For $y \geq 0$ the Yeo–Johnson transformation is the generalized Box–Cox transformation of $y + 1$. For negative y the transformation is of $-y + 1$ to the power $2 - \lambda$. For the extended Yeo–Johnson transformation, the transformation parameter for positive y is λ_P, with that for negative y being $2 - \lambda_N$. Like the original Yeo–Johnson transformation, the extended version requires four regimes. The unnormalized transformations for these are

$$y_{\text{EYJ}}(\lambda) = \begin{cases} \dfrac{(y+1)^{\lambda_P} - 1}{\lambda_P} & y \geq 0 \;\; \lambda_P \neq 0 \\ \log(y+1) & y \geq 0 \;\; \lambda_P = 0 \\ -\dfrac{\{(-y+1)^{2-\lambda_N} - 1\}}{2 - \lambda_N} & y < 0 \;\; \lambda_N \neq 2 \\ -\log(-y+1) & y < 0 \;\; \lambda_N = 2. \end{cases} \quad (6.25)$$

Information about the values of the transformation parameters again needs to take account of the Jacobians of these transformations. The Jacobian for the original transformation is given in (6.21). For the extended transformation two Jacobians are required: one for positive observations and the other for the negative ones. For $y \geq 0$ again let $v_P = y + 1$ with $v_N = -y + 1$ when $y < 0$. For the negative observations

$$S_N = \sum_{y_i < 0} -\log(-y_i + 1) = \sum_{y_i < 0} -\log v_{i,N} \quad \text{and} \quad \dot{y}_N = \exp(S_N/n). \quad (6.26)$$

Division is by n, not n_N (the number of negative y_i), as the Jacobian is spread over all observations.

Similarly, for the non-negative observations

$$S_P = \sum_{y_i \geq 0} \log(y_i + 1) = \sum_{y_i \geq 0} \log v_{i,P} \quad \text{and} \quad \dot{y}_P = \exp(S_P/n). \quad (6.27)$$

The log of the Jacobian of the sample is then

$$\log J_{\text{EYJ}}^n = (\lambda_N - 1)S_N + (\lambda_P - 1)S_P = n\{(\lambda_N - 1)\log \dot{y}_N + (\lambda_P - 1)\log \dot{y}_P\}. \quad (6.28)$$

The normalized form of the extended Yeo–Johnson transformation is consequently

$$z_{\text{EYJ}}(\lambda_N, \lambda_P) = \begin{cases} \dfrac{v_P^{\lambda_P} - 1}{\lambda_P \dot{y}_N^{\lambda_N - 1} \dot{y}_P^{\lambda_P - 1}} & y \geq 0 \;\; \lambda_P \neq 0 \\ (\dot{y}_P / \dot{y}_N^{\lambda_N - 1}) \log v_P & y \geq 0 \;\; \lambda_P = 0 \\ -\dfrac{v_N^{2-\lambda_N} - 1}{(2 - \lambda_N)\dot{y}_N^{\lambda_N - 1} \dot{y}_P^{\lambda_P - 1}} & y < 0 \;\; \lambda_N \neq 2 \\ -\log v_N / \dot{y}_N \dot{y}_P^{\lambda_P - 1} & y < 0 \;\; \lambda_N = 2. \end{cases} \quad (6.29)$$

6.2 The Yeo–Johnson Transformation and Its Extension

This extended transformation reduces to the standard Yeo–Johnson transformation when $\lambda_N = \lambda_P$.

There are three hypotheses about the value of the pair of parameters (λ_P, λ_N) that are of interest. The associated score tests, based on constructed variables, and the alternative hypotheses are

(i) T_O, an overall test of the homogeneity of the data, that is agreement with the values of both λ_P and λ_N;
(ii) T_P, a test solely of the value of λ_P;
(ii) T_N, a test solely of the value of λ_N. The alternatives are

$$T_O : \lambda_P + \alpha, \lambda_N + \alpha; \quad T_P : \lambda_P + \alpha, \lambda_N; \quad T_N : \lambda_P, \lambda_N + \alpha, \quad (6.30)$$

tested as $\alpha \to 0$. General expressions for the constructed variables used to calculate these test statistics are in Atkinson et al. (2020).

In the calculations for the extended Yeo–Johnson transformation, we first transform the data using the null values of λ_P and λ_N and then test the hypothesis that no further transformation is needed, that is, that, for the transformed data, one or both of λ_P and $\lambda_N = 1$. There is then a simplification of the constructed variables.

For the overall test T_O:

$$w_O(1, 1) = \begin{cases} v_P(\log v_P - k_P) + k_P, & y \geq 0 \\ v_N(\log v_N + k_N) - k_N & y < 0 \end{cases}. \quad (6.31)$$

In (6.31) $k_P = 1 + \log \dot{y}_O$ and $k_N = \log \dot{y}_O - 1$ and from (6.26) and (6.27) $\dot{y}_O = \exp\{(S_P + S_N)/n\}$.

For T_P, the test of the value of λ_P:

$$w_P(1, 1) = \begin{cases} v_P(\log v_P - k_P^*) + k_P^* & y \geq 0 \\ -y \log \dot{y}_P & y < 0, \end{cases} \quad (6.32)$$

where $k_P^* = 1 + \log \dot{y}_P$. The structure is similar to that of the constructed variables w_{YJ} for general λ in (6.23). The result for $y < 0$ arises because the transformation for $y < 0$ only depends on λ_P through the Jacobian.

Similarly for T_N, the test of the value of λ_N:

$$w_N(1, 1) = \begin{cases} -y \log \dot{y}_N & y \geq 0 \\ v_N(\log v_N + k_N^*) - k_N^* & y < 0, \end{cases} \quad (6.33)$$

where $k_N^* = \log \dot{y}_N - 1$.

We now use this algebra, combined with the FS, to develop robust methods for data transformation that enable monitoring of the effects of individual observations on the estimated transformation parameters.

6.2.4 The Extended Fan Plot: Establishing Two Transformations

The extended Yeo–Johnson transformation has two parameters λ_P and λ_N. Finding robust values of these two transformation parameters is in two stages. Both depend on the numerical identification of patterns in a series of fan plots of score statistics. We use the Yeo–Johnson transformation to find an overall estimate of λ for transforming both positive and negative observations using the fan plot derived from the constructed variables (6.23). This is followed by fan plots of score tests based on the constructed variables for positive and negative observations (6.32) and (6.33); the common value λ_0 for λ_N and λ_P determines whether separate values of λ_N and λ_P are required. The general numerical procedure is to test whether $\lambda_0 = (\lambda_{P0}, \lambda_{N0})$ is the appropriate transformation by transforming the data using λ_0 and then checking an "extended" fan plot to test the hypotheses for the transformed data that no further transformations are required. The three score tests in the extended fan plot are listed in (6.30). They check the individual values of λ_{P0} and λ_{N0} as well as of the overall transformation λ_0. The second stage of determining robust values of λ_N and λ_P consists of transforming the data with various values of the two parameters until the hypothesis $\lambda_0 = 1$ for the transformed data is accepted for the first, and hopefully major, part of the search; observations indicating different transformations enter $S^*(m)$ at the end of the search. Constructed variables for these tests when $\lambda_0 = 1$ are at the end of Sect. 6.2.3.

Simulations illustrating the properties of this process are in Atkinson et al. (2020). Here we use two data analyses to exemplify the procedure, the analysis of the first of which does not reveal any outliers.

6.2.5 Investment Funds Data

The purpose is to relate the medium-term performance of 309 investment funds to two indicators. Of these funds 99 have negative performance. The data come from the Italian financial newspaper *Il Sole—24 Ore* and the variables are: y, the medium-term (36 month) performance; x_1 the short-term (12 month) performance; and x_2 the short-term volatility. Performance is measured as the percentage change in the price of units of the fund.

Scatter plots of y against the two explanatory variables are in Fig. 6.8, with the negative responses shown as circles. It is clear that there is a strong, roughly linear, relationship between the response and both explanatory variables. It is also clear that the negative responses have a different behaviour from the positive ones: the variance is less and the slope of the relationship with both explanatory variables appears to be smaller. We employ the extended Yeo–Johnson transformation to see whether we can achieve a transformation in which there is the same simple linear model for positive and negative observations, with approximately normal errors of constant variance.

6.2 The Yeo–Johnson Transformation and Its Extension 215

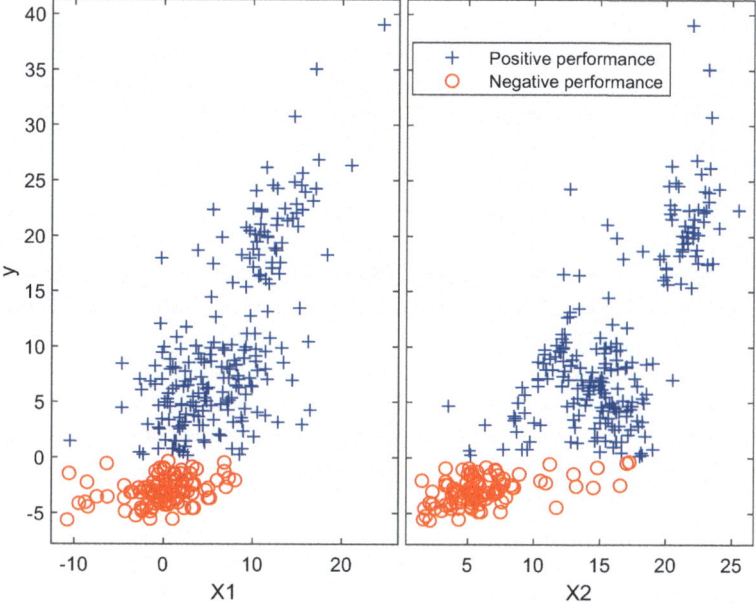

Fig. 6.8 Investment funds data: scatter plots of y against x_1 and x_2

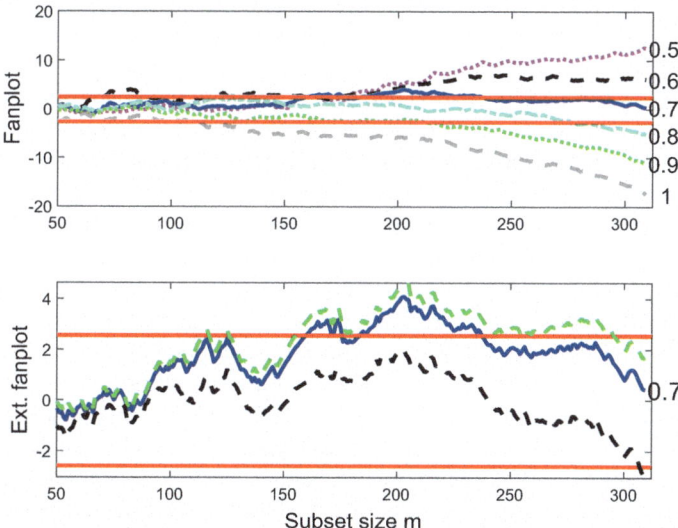

Fig. 6.9 Investment funds data: upper panel, fan plot indicating the overall transformation $\lambda = 0.7$; lower panel, extended fan plot for $\lambda_0 = 0.7$ suggesting different transformations for positive (upper green trajectory) and negative responses (lower black trajectory)

We start our analysis of the investment funds data using the original YJ transformation of Sect. 6.2.1 in which positive and negative observations are subject to the same transformation. The upper panel of Fig. 6.9 shows the fan plot for values of λ_0 in the range 0.5 to 1. All curves are relatively smooth; there are no sudden changes towards the end of the trajectories which might indicate the presence of outliers. A value of 0.7 seems to be indicated for λ, although the trajectory for this value is outside the upper 1% bound around $m = 200$, returning inside from m around 240. This behaviour is typical of the masking caused by a sizable group of observations that differ systematically from the majority of the data. Indeed, the extended fan plot for $\lambda = 0.7$ in the lower panel of Fig. 6.9 shows that the value of 0.7 is not satisfactory for all observations. The upper trajectory for the majority positive observations lies relatively close to that for an overall transformation of $\lambda = 0.7$, whereas that for the negative responses is outside the lower bound at the end of the search. This plot is a confirmation of the suggestion of a different distribution for positive and negative observations indicated by the scatter plots of Fig. 6.8.

The indication of the extended fan plot of Fig. 6.9 is that the positive observations require a transformation higher than 0.7, since, from the upper panel, higher values of λ give a lower curve. Likewise, the negative observations require a value lower. Our strategy is to try sets of pairs of values. When we have found the correct transformation, the fan plot of the transformed data will indicate that no further transformation is required; that is, we will accept the value $\lambda_0 = 1$ in this fan plot. For each tentative transformation, information from the fan plot of the transformed data should be supplemented by the scatter plots and by the full-sample regression on the two explanatory variables.

There are now two transformation parameters. We start with two values straddling 0.7 and take $\lambda_P = 1$ and $\lambda_N = 0.5$. Comparison of the scatter plots of transformed y against those for the untransformed data in Fig. 6.8 shows that the variance of the negative observations is now closer to that of the positive ones. We check this impression with the extended fan plot for $\lambda_0 = 1$ for the transformed data in the upper panel of Fig. 6.10. This plot provides a robust test of the transformation $\lambda_P = 1$ and $\lambda_N = 0.5$. Although the trajectories for the positive and negative observations are far closer together than they are in the lower panel of Fig. 6.9 for the overall transformation $\lambda = 0.7$, the correct transformations have not been found; all trajectories of score tests are well outside the lower bound at the end of the search.

The extended fan plot for $\lambda_P = 1$ and $\lambda_N = 0.25$ is in the centre panel of Fig. 6.10 with that for $\lambda_P = 1$ and $\lambda_N = 0$ in the lower panel. These plots show improving properties; that for $\lambda_N = 0$ lies in the overall bounds throughout, with the trajectories for positive and negative observations virtually identical to the overall trajectory. The conclusion is that the positive observations should not be transformed, but that the negative observations should be transformed with $\lambda = 0$. From (6.25), it follows that this is not the log transformation for negative y.

The scatter plots of this final transformation are in Fig. 6.11. The movement from the scatter plots of the original data in Fig. 6.8 to a linear model with homogeneous scatter is clear. An interesting feature of the right-hand scatter plot in Fig. 6.11 is

6.2 The Yeo–Johnson Transformation and Its Extension 217

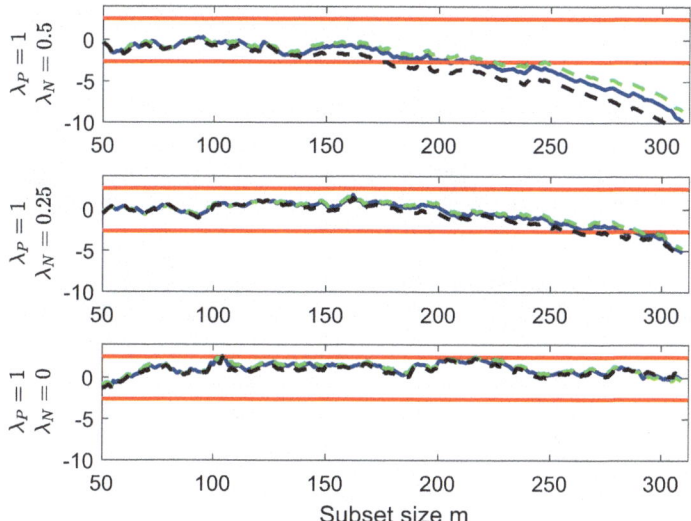

Fig. 6.10 Investment Funds data: checking the two transformation parameters. Extended fan plots for $\lambda = 1$ for the transformed data. Upper panel, $\lambda_P = 1$ and $\lambda_N = 0.5$; centre panel $\lambda_P = 1$ and $\lambda_N = 0.25$; lower panel $\lambda_P = 1$ and $\lambda_N = 0$, the preferred transformation. Upper (green) trajectory, positive observations, lower (black) trajectory, negative observations

that the observations appear to fall into three clusters, with funds with high volatility being the most profitable over this time period.

We now consider a further property of the transformed and untransformed data. Table 6.1 shows summary properties of the regression on the two variables for $\lambda_P = 1$ and four values of λ_N. The line labelled F is the value of the F statistic for testing regression on a constant, x_1 and x_2 against regression on only a constant. As λ_N goes from 1 (the untransformed data) to 0 the value of the F statistic steadily increases from 556 to 685. This improvement is in line with the results from the extended fan plots of Fig. 6.10 which show the transformation becoming increasingly acceptable over this range. Likewise, the adjusted R^2 for the regression increases from 0.783 to 0.816. Interestingly, there is little change in the two tabulated values as λ_N goes from 0.25 to 0, although the central panel of Fig. 6.10 rejects the transformation with $\lambda_N = 0.25$; the monitored robust procedure, sensitive to the effects of individual observations, detects an inadequate model, which is not detectable by the F test based on aggregate statistics.

A further assessment of the statistical properties of the transformation is given by the normal QQ-plots with envelopes in Fig. 8 of Atkinson et al. (2020). The transformation makes both tails more nearly normal and also produces a distribution which, for central values, is slightly closer to the line of expected values for a normal sample. The differences between the two panels are clarified by the envelopes, especially in the tails of the distributions. The differences may not seem large, but slightly over

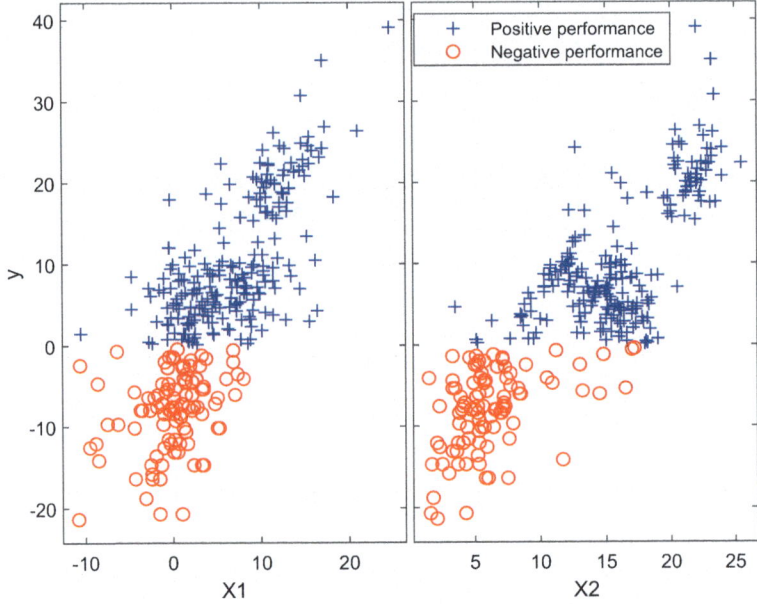

Fig. 6.11 Investment funds data: scatter plots of y against x_1 and x_2 for final transformation of the data with $\lambda_P = 1$ and $\lambda_N = 0$

Table 6.1 Investment funds data: summary properties of regression for different transformations of positive (λ_P) and negative (λ_N) observations

λ_P	1	1	1	1
λ_N	1	0.5	0.25	0
$F_{2,306}$	556	643	681	685
R^2_{adj}	0.783	0.807	0.815	0.816

two-thirds of the observations are untransformed in both plots. The scatter plots show more forcibly the effect of transformation.

These results indicate that the Yeo–Johnson transformation and its extension to differing transformations for positive and negative values has here achieved two of the goals of the Box–Cox transformation. As the scatter plots of Fig. 6.11 and the QQ-plots of Atkinson et al. (2020) show, we have achieved a homogeneous and normal distribution of errors. The results summarized in Table 6.1 quantify the increasing amount of the total variation in the data that is explained by regression using the correct transformation of the data. It is interesting that with 210 positive observations out of 309, linear interpolation in the final estimates of λ suggests an overall transformation value $\tilde{\lambda} = (99 \times 0 + 210 \times 1)/309 = 0.680$, very close to the value of 0.7 arrived at in the overall fan plot of Fig. 6.9. This satisfactory linear interpolation accords well with the Taylor series linearization used to develop

6.2 The Yeo–Johnson Transformation and Its Extension

the approximate score statistic. These results could not have been achieved without the extension of the Yeo–Johnson transformation to test for and accommodate two transformation parameters.

After transformation, the investment funds data are surprisingly well behaved. In particular, there are no obvious indications of any outliers. In Exercise 7.2, we augment the data to create a more complicated structure and ask for a comparison of the parametric analysis given here and the analysis using the non-parametric method of Chap. 7.

6.2.6 Balance Sheet Data

We now analyse a larger dataset taken from balance sheet information on limited liability companies, which, unlike the investment funds data, does include outliers.

The response is profitability of individual firms in Italy. There are 998 observations with positive response and 407 with negative response, making 1,405 observations in all. The model variables are: y, profitability, calculated as return over sales; x_1, labour share, the ratio of labour cost to value added; x_2, the ratio of tangible fixed assets to value added; x_3, the ratio of intangible assets to total assets; x_4, the ratio of industrial equipment to total assets; and x_5, the firm's interest burden, the ratio of the firm's total assets to net capital.

The aim is to explain the profitability by regression on the five explanatory variables. From scatter plots of y against the explanatory variables, it is clear that there is a negative relationship between y and x_1. The relationships with the other variables are not so obvious. However, the values of the t-tests for the coefficients in Table 6.2 show significant regression on all variables except x_4. Unlike the plot of the investment fund data in Fig. 6.8, such plots do not convey an obvious message about the need for transformation.

We check the need for a transformation using fan plots. The upper panel of Fig. 6.12 presents trajectories for the overall statistic for values of $\lambda_0 = 1.25, 1, 0.75$ and 0.5. Comparison with the fan plot for the investment funds data in Fig. 6.9 shows the increase in power consequent on moving from a sample of 309 to one of 1,405. The hypothesis of no transformation ($\lambda_0 = 1$) seems to be acceptable, although there is an abrupt increase in the value of the statistic towards the end of the search, which might indicate the presence of outliers. The extended fan plot for testing $\lambda_0 = 1$ in the lower panel of the figure clarifies this structure. This plot shows that the positive and negative observations apparently need different transformations. Unlike the extended fan plots we have already seen in this chapter, the (black) trajectory for the negative observations is uppermost, indicating the need for a value of $\lambda_N > 1$. Likewise the lowest (green) trajectory indicates that λ_P should be less than one.

Following the indication of Fig. 6.12, we continue this analysis by finding the best values of the two transformation parameters. The extended fan plot for testing the hypothesis that $\lambda_P = 0.5$ and $\lambda_N = 1.5$ is in Fig. 6.13. Although the trajectories for the three statistics do not overlap as they do for the well-behaved investment fund

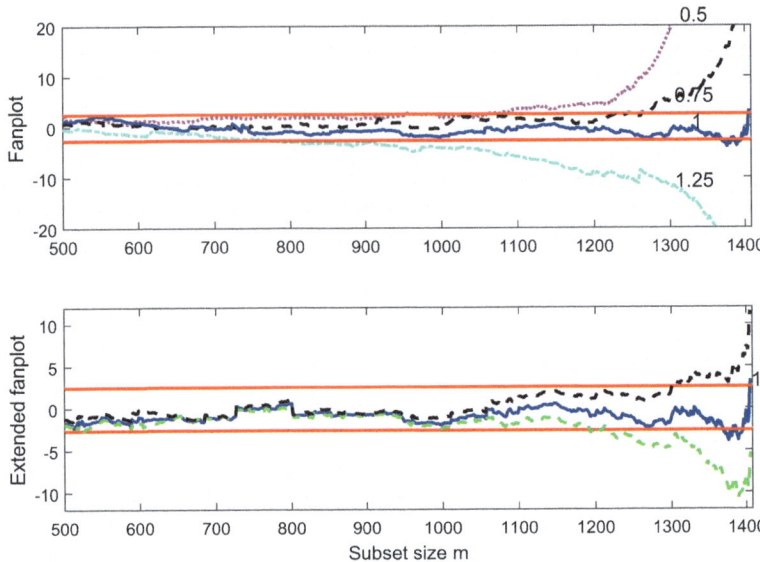

Fig. 6.12 Balance sheet data: upper panel, fan plot of overall statistic for $\lambda_0 = 0.5, 0.75, 1$ and 1.25, perhaps indicating the transformation $\lambda = 1$; lower panel, extended fan plot for $\lambda_0 = 1$; the need for different transformations for positive and negative observations is apparent, as is the effect of outliers. Upper (black) trajectory, negative observations, lower (green) trajectory, positive observations

data in Fig. 6.10, the three statistics lie virtually in the t-statistic boundaries until almost the end of the search, where the trajectories coincide when the suspected outliers enter. Other parameter values can be found for which the three trajectories are much closer, for example, $\lambda_P = 0.75$ and $\lambda_N = 1$. However, the trajectories stray outside the t-statistic boundaries and indicate around 200 outliers.

The fan plot and its extension have been central to our modelling process. The fan plot depends on the forward search which orders the observations by closeness to the model fitted for each value of the subset size m. We now finish with part of a forward search analysis of data with the recommended transformation $\lambda_P = 0.5$ and $\lambda_N = 1.5$, which leads to identification of outlying observations.

The top left-hand panel of Fig. 6.14 shows a forward plot of all 1,405 scaled residuals for values of m from 800. There is a broad upper band of residuals in pale blue, which lies between ± 2 throughout this part of the search. The slightly more extreme residuals are plotted in dark blue. The lower ones of these are slightly separated from a lower band of residuals, shown in red. What is remarkable is the stability of this pattern almost until the end of the search, indicative of a set of data lacking outliers having strong individual influence on the parameter estimates.

The upper right-hand panel of the figure shows a forward plot of minimum deletion residuals, with a 99% envelope (dotted line) for these distances for the full sample

6.2 The Yeo–Johnson Transformation and Its Extension 221

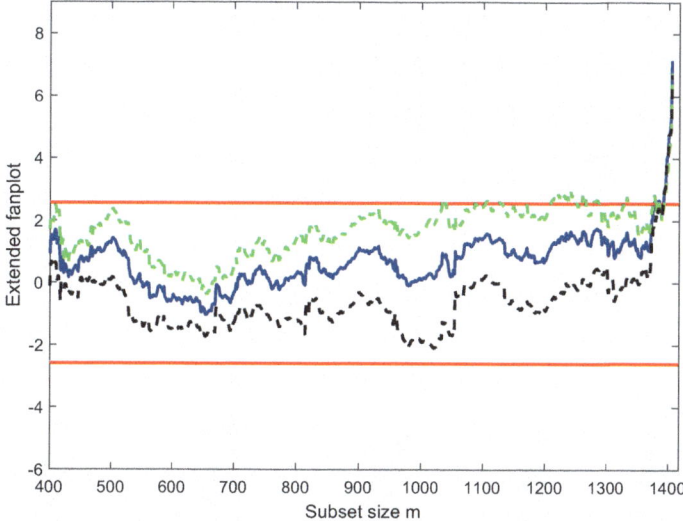

Fig. 6.13 Balance Sheet data: checking the two transformation parameters. Extended fan plots for $\lambda_0 = 1$ for the transformed data when $\lambda_P = 0.5$ and $\lambda_N = 1.5$. Upper (green) trajectory, positive observations, lower (black) trajectory, negative observations

of 1,405 observations. The lower panel of Fig. 6.14 shows scatter plots of y against the explanatory variables.

To exhibit more information, these three plots have been linked. The highlighted, red, residuals in the top left-hand panel were subjectively identified by brushing the plot. We selected the trajectories, 19 in all, that lay within the brush in the centre of the figure. The trajectory of the 19 brushed observations in the forward plot of deletion residuals in the upper right-hand panel is shown in red. Sixteen of these observations are the last observations to enter the search, the other three entering from 21 units before the end of the search. These 19 observations are plotted as filled (red) circles in the scatter plots in the lower panel. The extreme negative trajectory in the top left-hand panel is clearly caused by the outlier in the bottom left of the scatter plot for x_1. This is a gross outlier; it is impossible that x_1, the ratio of labour cost to value added, be negative. The continuous lines in the panels come from multiple regression on all observations. The panels for x_1 and x_4, in particular, show the source of the many large negative residuals.

Brushing led to the subjective choice of 19 observations with large negative residuals. However, our automatic procedure for outlier detection (Sect. 4.3) identifies a total of 42 outliers. Adding these observations to those already highlighted in the scatter plots of Fig. 6.14 reveals the structure of the outliers, which is not apparent from the plot of the untransformed data. For these plots see Atkinson et al. (2020). Further analysis is given in Exercise 6.3.

The purpose of the analysis of these data was to build a regression model relating y to the five potential explanatory variables. Table 6.2 gives summary properties of the

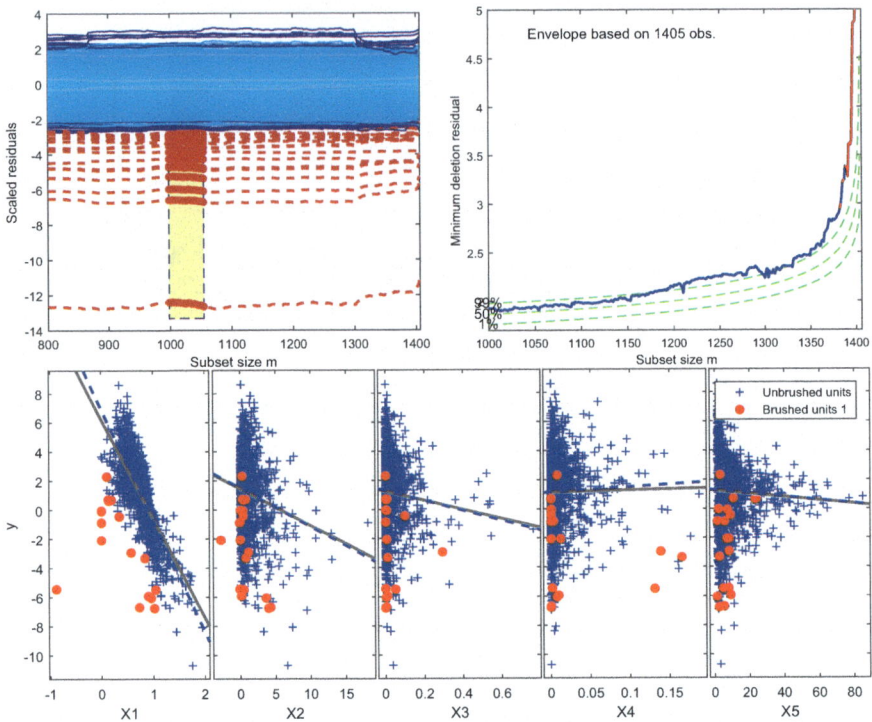

Fig. 6.14 Transformed balance sheet data: brushing linked plots from the forward search when $\lambda_P = 0.5$ and $\lambda_N = 1.5$. Upper left-hand panel, trajectories of residuals from the forward search with the residuals from 19 observations highlighted by brushing. Upper right-hand panel, linked forward plot of minimum deletion residuals during the search with the 19 brushed values shown in red. Lower panel, scatter plots of y against x_1 to x_5 showing the 19 brushed units: full line, regression with all data, dashed line regression after deletion of brushed units

regressions for the untransformed data and the transformed data with 42 observations deleted. The F statistic for the regression, as against a constant model, increases on transformation and deletion from 293 to 588, a ratio of just over 2. Thus, despite the deletion of 42 observations, the transformation has led a doubling of the amount of information available, that is, to an effective doubling of sample size; the t-test values for the intercept and all variables except x_5 show appreciable increases in significance. The strengthening of evidence for multiple regression to include x_1, from -35.8 to -51.4, is in line with the pattern of the deleted brushed units shown in the first panel of the yX plot in Fig. 6.14. However, the evidence for x_2 strengthens from -8.3 to -11.3, which is not particularly to be expected from inspection of the second panel of the figure.

The lower panels of Fig. 6.14 also include, as continuous lines, the fitted regression relationships before the outliers have been deleted. The dotted lines are for the relationships after outlier deletion. Although the significance of the fit has greatly

6.3 An Automatic Procedure for Determining Transformations

Table 6.2 Balance sheet data: summary properties of regression for original data and for data with outliers removed with different transformations of positive (λ_P) and negative (λ_N) observations

Data	All	42 deleted
λ_P	1	0.5
λ_N	1	1.5
Number of observations	1405	1363
Error d.f. ν	1399	1357
t_ν values		
Intercept	42.0	60.5
x_1	−35.8	−51.4
x_2	−8.3	−11.3
x_3	−3.6	−5.4
x_4	1.0	2.3
x_5	−3.4	−3.1
$F_{5,\nu}$ for regression	293	588
R^2_{adj}	0.509	0.683

increased due to outlier deletion, the figure shows that deletion of the outliers has a negligible effect on the estimated regression coefficients.

The negative sign for x_1 shows decreasing profitability for firms with high labour input. This is by far the most significant relationship. The negative signs for x_2 and x_3, ratios of fixed assets, and intangibles such as software are surprising. One possibility is that the investments that have been made are not yet yielding the intended advantages in, for example, labour reduction, while incurring capital costs (Bartoloni, 2013, Sect. 3). The fourth variable is not significant, whereas the negative effect of interest is to be anticipated. There is however a danger in over-interpreting the values of t-tests in multiple regression with correlated explanatory variables. A major conclusion from Table 6.2 is that the increase in the precision of estimation of the regression coefficients has been achieved by the transformation coupled with the deletion of just less than 3% of the total data.

6.3 An Automatic Procedure for Determining Transformations

6.3.1 The Extended BIC and Response Transformation

The emphasis in the robust choice of a response transformation in Sects. 6.1 and 6.2 was on the visual inspection of fan plots of score statistics for transformation parameters. The number of outliers detected in the fan plot typically depends on the value of λ. In this section, we use the extended BIC, introduced in Sect. 4.11, to provide numerical comparisons of fan plots for different values of λ, thus providing an objective alternative to human intervention in this stage of the data analyses. We supplement this new form of BIC with two further diagnostic tools, the AGreement Index, AGI, and an extended form of R^2. Section 6.3.4 describes the combination of these tools to provide the automatic procedure for the Box–Cox transformation.

The BIC, introduced in Sect. 4.11 for complete data in the absence of response transformation, is a function of the residual sum of squares $R(\hat{\beta})$. For transformed data $R(\hat{\beta})$ is replaced by $S(\lambda)$ the residual sum of squares of $y(\lambda)$. This value needs to be corrected for the Jacobian of the transformation giving

$$\text{BIC}(\lambda) = -n \log\{S(\lambda)/n\} + 2 \log J_{\text{BC}}^n - (p + n_\lambda) \log(n), \quad (6.34)$$

where there are p parameters in the linear model and, for the Box–Cox transformation, the single transformation parameter λ, so that $n_\lambda = 1$. For other transformations, a different Jacobian is required. Written in this form, large values of the index are to be preferred.

Use of the forward search to provide robustness against outliers and incorrect transformations leads to the comparison of fitted models with different numbers of observations. Let the FS terminate with h observations. Then $n - h$ observations will have been deleted. To incorporate deletion of observations in BIC (6.34) for values of λ over the grid $\mathcal{G}$, let the value of $S(\lambda)$ when $n - h(\lambda)$ observations are deleted be $S(\lambda, h)$. Then, using the mean shift outlier model of Sect. 3.2.3, BIC(λ) is replaced by

$$\text{BIC}(\lambda, h) = -n \log\{S(\lambda, h)/h\} + 2 \log J_{\text{BC}}^n - \{p + n_\lambda + n - h(\lambda)\} \log(n), \quad (6.35)$$

in which $S(\lambda, h)$ is divided by h.

Since outliers may have been deleted, the least squares estimate of error needs to be adjusted using the consistency correction due to Tallis (3.49). The value of $S(\lambda, h)$ in (6.35) is accordingly scaled up to obtain the extended BIC

$$\text{BIC}_{EXT}(\lambda, h) = -n \log[S(\lambda, h)/\{h\sigma^2(h)\}] + 2 \log J_{\text{BC}}^n - \{p + n_\lambda + n - h(\lambda)\} \log(n). \quad (6.36)$$

6.3.2 The Agreement Index AGI

To determine the best transformation, the values of $\text{BIC}_{EXT}(\lambda, h)$ are compared over $\mathcal{G}$ and the maximizing value of λ selected. Our experience suggests it is sometimes also helpful to look at a second, heuristic, quantity, the agreement index, AGI.

The value of $\text{BIC}_{EXT}(\lambda, h)$ presents the information on the transformation for the subset of size $h(\lambda)$. It can be helpful also to consider the history of the evidence for the transformation as a function of the subset size m. A correct transformation leads

6.3 An Automatic Procedure for Determining Transformations

to identification of the outliers and to a normalizing transformation of the genuine data. The most satisfactory transformation will be one which is stable over the range of m for which no outliers are detected and will lead to non-significant values of the score statistic $T_A(\lambda_0, m)$, indicating that the value of $\hat{\lambda}$ is consistent with λ_0 over a set of subset sizes.

The AGI leads to a graphical representation of this idea. The forward search procedure for testing the transformation monitors absolute values of the statistic from some lower limit M for the subset size, to $m = h(\lambda)$. The default is to take $M = 0.6n$. For ease of comparison with $\text{BIC}_{EXT}(\lambda, h)$, we take the reciprocal of the mean of the absolute values of $T_A(\lambda_0, m)$ so that large values are again desired. In order to give more weight to searches with a larger value of $h(\lambda)$, we rescale the value of the index by the variance of the truncated normal distribution. Let $\mathcal{M} = m \in [M, h(\lambda)]$. Then the agreement index is

$$\text{AGI} = \{h(\lambda) - M + 1\} / \left\{ \sigma^2(h) \sum_{m \in \mathcal{M}} |T_A(\lambda, m)| \right\}, \quad (6.37)$$

calculated for $\lambda \in \mathcal{G}$.

6.3.3 The Coefficient of Determination R^2

The main tools for determining the correct transformation are plots of $h(\lambda)$ and BIC_{EXT} with the agreement index as a further diagnostic. In addition, we calculate values of a form of R^2, extended to allow for the value of the size $h(\lambda)$ of the final cleaned sample:

$$R^2_{EXT} = R^2 / \sigma^2 \{h(\lambda)\}. \quad (6.38)$$

6.3.4 The Automatic Procedure for the Box–Cox Transformation

This section gives a sketch of the steps of the automatic procedure for robust selection of the transformation parameter and the identification of outliers for the Box–Cox transformation.

1. **Fan Plot.** In addition to the data, the FS routine requires the grid $\mathcal{G}$ of λ values to be evaluated as transformation parameters.

    a. There are two automatic choices of $\mathcal{G}$. When $n < 200$, $\lambda_0 = -1, -0.5, 0, 0.5$ and 1. For $n \geq 200$ the default grid is $\mathcal{G} = -1, -0.9, \ldots, 1$. Alternatively, specific grids of λ values can be provided.

b. Each value of λ has a separate forward search. The output structure contains numerical information to build fan plots such as Fig. 6.2.

2. **Outlier Detection, BIC, and Model Selection.**

   a. In the automatic procedure, for each $\lambda \in \mathcal{G}$, the values during the forward search of the score statistic $T_A(\lambda)$ are assessed against the standard normal distribution to estimate the maximum number $m^*(\lambda)$ of observations in agreement with that transformation. This procedure for testing the transformation monitors absolute values of the statistic from M to $m = h(\lambda)$. The default is $M = 0.6n$.
   b. The presence of any remaining outliers in the $m^*(\lambda)$ transformed observations is checked using the forward search procedure now monitoring deletion residuals. If any outliers are found in the transformed data the sample size of the cleaned data is $h(\lambda) < m^*(\lambda)$. Otherwise $h(\lambda) = m^*(\lambda)$.
   c. The extended BIC (6.36) allows comparisons of results from models with differing numbers of non-deleted observations $h(\lambda)$. The maximal value determines the preferred parameter value $\tilde{\lambda}$. The agreement index (6.37) is also calculated over $\mathcal{G}$.
   d. The results are summarized in a single plot with three panels, for example, Fig. 6.16. The main panel for decision-making is that of the extended BIC. Also included are a diagnostic plot of the agreement index and a combined plot of $h(\lambda)$ and $m^*(\lambda)$.
   e. The procedure suggests the value of λ for further data analysis and indicates a set of outliers in that scale. Further analyses using robust techniques, including model selection, could follow these suggestions (as an example see Sect. 10.6.4). Such analyses are certainly indicated if the three panels of plots like Fig. 6.16 disagree on the best transformation. Figure 6.16 shows an analysis in which all three panels agree.

6.3.5 Automatic Analysis of the Loyalty Cards Data

We now present the results of using the automatic procedure to reanalyse the loyalty cards data introduced in Sect. 6.1.5.

The correct choice of the grid of values of λ is an important aspect of the procedure. We first consider the automatic analysis with the coarse grid $\mathcal{G} = -1, -0.5, 0, 0.5$ and 1. Reference to Fig. 6.2 shows the fan plot for 11 values of λ $(0, 0.1, \ldots, 1)$ with $\lambda = 1$ at the bottom and $\lambda = 0$ providing the top trajectory. It is clear that the trajectories for two of the five values of $\lambda \in \mathcal{G}$ (0 and 0.5) steadily leave the central band in opposite directions. The automatic procedure with the coarse grid indicates in the upper panels of Fig. 6.15 that the best of the five values is $\lambda = 0.5$. However, the bottom panel of the figure shows that 37 observations do not agree with this transformation. The indication is that a finer grid of values of λ is required, in line with the argument of Carroll (1982).

6.3 An Automatic Procedure for Determining Transformations

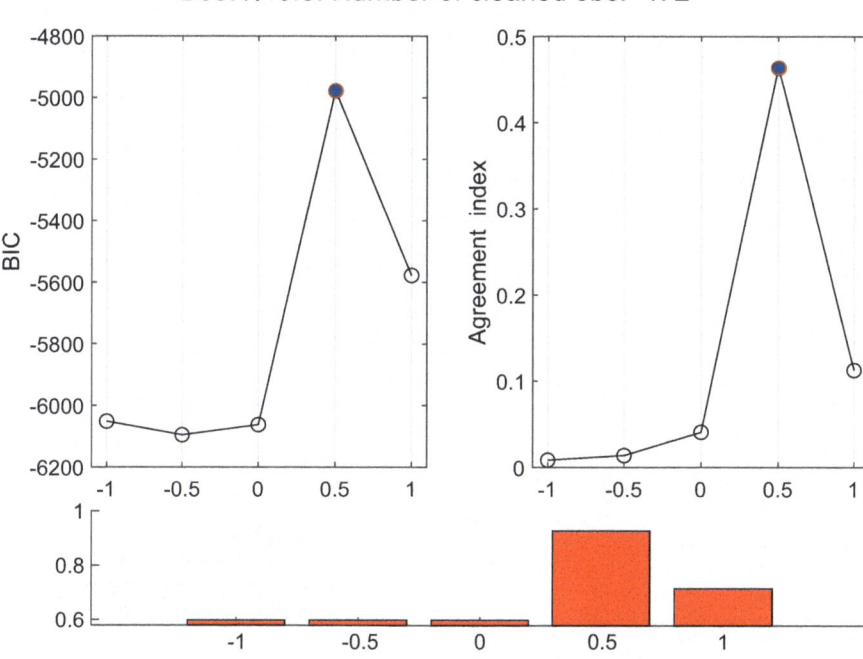

Fig. 6.15 Loyalty cards data: output from automatic analysis with coarse grid; $\mathcal{G} = -1, -0.5, 0, 0.5$ and 1. Upper left-hand panel, extended BIC; upper right-hand panel, agreement index AGI. Lower panel, proportions $h(\lambda)/n$; note that $h(0.5)/n = 0.9273$

We now use the finer grid $\mathcal{G} = -1, -0.9, \ldots, 1$. The upper left-hand panel of Fig. 6.16 shows the extended BIC plot which is complemented by the plot in the lower panel of the proportion of clean observations $h(\lambda)/n$. For λ in the range -1 to 0.2, (with the exception of $\lambda = 0.1$), $h(\lambda) = 0.6n$, the point at which monitoring starts. The plot of the extended BIC shows that $\tilde{\lambda}_G = 0.4$ with 0.3 and 0.5 giving slightly smaller values. An important feature in the lower panel is that, when $\lambda = 0.4$, checking residuals for the presence of outliers (Step e of Sect. 6.3.4), leads to the deletion of 16 observations from the $m^*(\lambda)$ found by checking the score statistic. These deleted observations are shown as the unfilled part of the bar in the plot. This automatic analysis, leading to the transformation $\tilde{\lambda} = 0.4$, agrees with the conclusions of the analysis in Sect. 6.1.5. A small difference is that the automatic analysis indicates 16 outliers, whereas the analysis in Sect. 6.1.5, using the FS outlier detection rule, indicates an extra two outliers. The overall conclusions of the analyses are not sensitive to the difference in the number of outliers found.

Riani et al. (2022a) provide further details of the automatic procedure for the Box–Cox transformation and apply the method to the analysis of the surgical unit

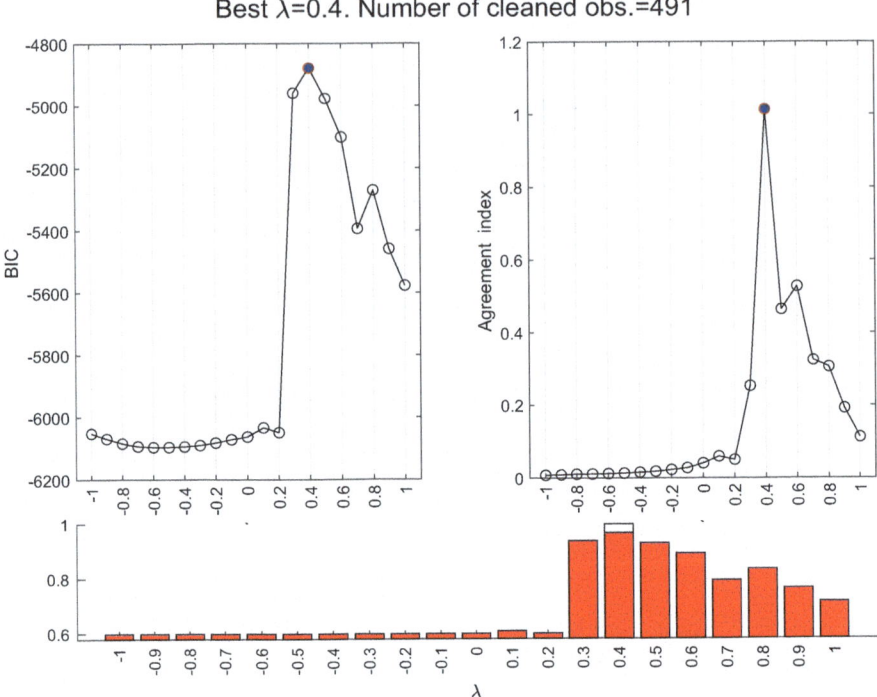

Fig. 6.16 Loyalty cards data: output from automatic analysis with finer grid; $\mathcal{G} = -1, (0.1), 1$. Upper left-hand panel, extended BIC; upper right-hand panel agreement index AGI. Lower panel $h(\lambda)/n$ as red bars; if $h(\lambda) < m^*(\lambda)$, the difference is shown as a white rectangle inside the bar

data of Sect. 6.1.4. They compare the method of this chapter with that of Marazzi et al. (2009).

6.3.6 The Automatic Procedure for the Extended Yeo–Johnson Transformation

The automatic method for the Yeo–Johnson transformation has the same structure as that for the Box–Cox transformation. Since, in both cases, the transformations depend on a single parameter, there is a single constructed variable for the Yeo–Johnson transformation. The derived score statistic is given in (6.24).

The approach to the automatic calculation for the extended Yeo–Johnson transformation is an extension to that given in Sect. 6.3.4 for the Box–Cox transformation, complicated by the need to analyse the behaviour of the fan plots for the three score tests of Sect. 6.2. As in Sect. 6.2.4, the analysis starts with the Yeo–Johnson transformation of Sect. 6.2.1 in which negative and positive observations are subject to transformation with the same value of λ. The next stage is to determine whether this transformation parameter should be used for both positive and negative observations.

The automatic numerical procedure can be divided into three parts. Fan plots for the one-parameter Yeo–Johnson transformation using the constructed variables of Sect. 6.2.2 provide information from which the BIC in the second part estimates $\tilde{\lambda}$, the best overall parameter for this transformation. The third part uses the value of $\tilde{\lambda}$ to calculate an appropriate grid $\mathcal{G}$ of parameter values, which, together with the BIC, leads to the best estimates of the two transformation parameters for the extended transformation. As before, but now with $n_\lambda = 2$, the extended BIC is calculated for the cleaned samples of observations for each combination of values of λ_P and λ_N of Sect. 6.2.3. This information is supplemented by the diagnostic use of an extended agreement index, which measures agreement between the values of the score statistics for transformation of the positive and negative observations and checks that no further overall transformation is needed. Agreement between the values of T_P and T_N is measured by the absolute value of their difference. A small value of T_O, the score statistic for the overall transformation, is also required.

Further details of the procedure for the extended Yeo–Johnson transformation are in Riani et al. (2022a, Sect. 8) followed by automatic analyses of the investment funds data and of the balance sheet data which we analysed in Sects. 6.2.5 and 6.2.6. Exercises 6.5–6.7 require the analyses of three further sets of data using the extended Yeo–Johnson transformation. A further analysis which uses the extended Yeo–Johnson transformation is in Chap. 10 (see Fig. 10.55).

6.4 Further Response Transformations

6.4.1 The General Robust Procedure

The procedure in Sect. 6.1 for developing a robust Box–Cox transformation can be generalized to the robust estimation of the parameters of several other transformations. The requirements are appropriate constructed variables and the related Jacobians of the transformations. In this section, we present the general framework and, in successive subsections, follow Atkinson (1985) in presenting the necessary algebra for the calculation of robust estimates for specific parametric transformations. Atkinson (1985) was not concerned with robustness, but used this algebra for the calculation of regression diagnostics. These include constructed variable plots, which are Fig. 3.5, but with the residual transformed response plotted against the residual constructed variable. The constructed variables can also be used to provide deletion plots of the score statistics. Both procedures illustrate the effect of deletion of a single observation on the estimate of the transformation parameter. We likewise do not here develop the robust methods based on the forward search since, given the constructed variables and Jacobians, the development is that of Sect. 6.1.3.

We assume there is some transformation of the data $u = u(\lambda)$ for which the transformed response satisfies the linear model $E(U) = X\beta$, with the errors of observation at least approximately normally distributed with constant variance. The parameter λ could, in principle, be a vector, as it was in Sect. 6.2.3. But here, for simplicity, we take λ as a scalar. The Jacobian of the transformation is

$$J = \prod_{i=1}^{n} |\partial u_i / \partial y_i|. \tag{6.39}$$

Let $\lambda v = |\partial u / \partial y|$, when an alternative form for the Jacobian (6.39) is

$$J = \prod_{i=1}^{n} \lambda v_i = \lambda^n \prod_{i=1}^{n} v_i. \tag{6.40}$$

For independent observation, we work with the normalized variables

$$z(\lambda) = u/(\lambda \dot{v}), \text{ where } \dot{v} = G(v) = \prod_{i=1}^{n} v_i^{1/n}, \tag{6.41}$$

the geometric mean of the v_i. Taking $u = y^\lambda - 1$ yields (6.9), the expression for the normalized Box–Cox transformation; differentiation of $z(\lambda)$ with respect to λ yields (6.15), that is, the constructed variable $z_A(\lambda)$.

6.4.2 The Folded Power Transformation for Proportions

For the power transformation, the limit of $y(\lambda)$ as $\lambda \to 0$ is the log transformation. For proportions, the analogue is the logit, where

$$\text{logit}(y) = \log\{y/(1-y)\} \tag{6.42}$$

used in Sect. 5.3 for the comparison of power curves. A simple transformation which yields the logit as $\lambda \to 0$ is the folded power transformation (Mosteller & Tukey, 1977, p. 92) in which

$$u_f(\lambda) = y^\lambda - (1-y)^\lambda \quad (0 \leq y \leq 1). \tag{6.43}$$

For values of y near zero the transformation behaves like the power transformation y^λ, whereas, for y near one, it behaves like $(1-y)^\lambda$.

Use of the general procedure of Sect. 6.4.1 shows that the normalized transformed response for the folded power transformation is

6.4 Further Response Transformations

$$z_f(\lambda) = \begin{cases} \dfrac{y^\lambda - (1-y)^\lambda}{\lambda G\{y^{\lambda-1} + (1-y)^{\lambda-1}\}} & (\lambda \neq 0) \\ \log\{y/(1-y)\}G\{y(1-y)\} & (\lambda = 0), \end{cases} \quad (6.44)$$

where $G(\cdot)$ is the geometric mean function of (6.41). A multiple of the desired logit is obtained for $\lambda = 0$. When $\lambda = 1$, $z(1) = y - 1/2$, that is, no transformation, provided the linear model includes a constant (Schlesselman, 1971). A disadvantage of the folded power transformation is that it does not have an explicit inverse.

Formulae for the constructed variable $w_f(\lambda)$ are in Atkinson (1985, Sect. 7.2) which also includes an analysis of gasoline data originally due to Prater. The data, which have a nested structure, are presented in Table 4.5 of Atkinson (1985). The response is gasoline yield as a percentage. Such data can be analysed by adapting the folded power transformation by the trivial replacement of $u_f(\lambda)$ with $u_{f\%}(\lambda) = y^\lambda - (100 - y)^\lambda$ $(0 \leq y \leq 100)$. Alternatively, the percentage can be divided by 100 to give proportions.

6.4.3 The Guerrero–Johnson Transformation

For an invertible family of transformations which includes the logit, Guerrero and Johnson (1982) suggested application of the Box–Cox transformation not to y but to the odds $p = y/(1-y)$. The transformation was introduced for the analysis of binary data, a task nowadays usually performed using logistic models. Here it is suggested as an alternative to the folded power family for the analysis of proportions.

The proposal is to let

$$u_{\text{GJ}}(\lambda) = p^\lambda - 1 = \{y/(1-y)\}^\lambda - 1 \quad (0 \leq y \leq 1). \quad (6.45)$$

The general procedure of Sect. 6.4.1 leads to the normalized transformed response

$$z_{\text{GJ}}(\lambda) = \begin{cases} [\{y/(1-y)\}^\lambda - 1]/\lambda G_{\text{GJ}}(\lambda) & (\lambda \neq 0) \\ \log\{y/(1-y)\}G\{y(1-y)\} & (\lambda = 0), \end{cases} \quad (6.46)$$

where $G_{\text{GJ}}(\lambda) = G\{y^{\lambda-1}/(1-y)^{\lambda+1}\}$ and $G(\cdot)$ is the geometric mean function.

The transformation thus yields the logit for $\lambda = 0$, but $z(1) \propto (2y - 1)/y$. The transformation is however readily invertible. The fitted model can thus be used to provide predictions and confidence intervals on the original scale of the observations once these have been calculated on the transformed scale. Details are in Atkinson (1985, Sect. 7.3).

6.4.4 The Aranda-Ordaz Transformation

The last transformation for proportions is due to Aranda-Ordaz (1981). It is both invertible and yields the untransformed data for $\lambda = 1$. With $p = y/(1-y)$

$$u_{AO}(\lambda) = \frac{p^\lambda - 1}{p^\lambda + 1} = \frac{y^\lambda - (1-y)^\lambda}{y^\lambda + (1-y)^\lambda}. \tag{6.47}$$

The normalized transformation is

$$z_{AO}(\lambda) = \begin{cases} \dfrac{p^\lambda - 1}{\lambda p^\lambda + 1}\{G_{AO}(\lambda)\}^{-1} & (\lambda \neq 0) \\ \log\{y/(1-y)\}G\{y(1-y)\} & (\lambda = 0), \end{cases} \tag{6.48}$$

where $G_{AO}(\lambda) = G\{v_{AO}(\lambda)\}$ with $v_{AO}(\lambda)$ coming from the application of (6.41) to (6.47).

Equation (6.48) shows that $z_{AO}(1) = y - 1/2$, confirming that the transformation does indeed combine both the logistic transformation and the untransformed data in a single family. Since the transformation is also invertible it might seem that this is the obvious transformation to use for proportions. However, the likelihood is symmetrical about $\lambda = 0$, so that the score statistic for testing $\lambda_0 = 0$ is identically zero. A procedure close to that of this chapter would be to monitor the likelihood ratio rather than the score test. This would have the disadvantage of requiring the computation of $\hat{\lambda}$ at each stage of the search, something the score test is intended to avoid. Atkinson (1985, Sect. 7.4) contains further details of the algebra for the Aranda-Ordaz transformation.

6.5 Transformations Including Explanatory Variables

So far in this chapter the focus has been on transformations of the response in a regression model. We conclude by extending our procedure to the transformation of explanatory variables. In Sect. 6.5.1, we start with the transformation of individual explanatory variables. We then extend this to include transformation of the response but, throughout, the emphasis is on the individual or joint transformation of individual terms in the model. In Sect. 6.5.2, we instead consider the transformation of both sides of the model. The response and the complete linear model are both subject to transformation with the parameter λ.

6.5.1 The Box–Tidwell Transformation for Explanatory Variables

As opposed to transformation of the response in a regression model, Box and Tidwell (1962) present a method for transformation of the explanatory variables in such models. For a single explanatory variable x_k, the parametric family of models is

$$E(Y) = \sum_{j \neq k} \beta_j x_j + \beta_k x_k^\lambda. \tag{6.49}$$

Since the response is not being transformed, there is no Jacobian included in the transformation. Taylor series expansion of (6.49) yields the linearized model

$$E(Y) \approx \sum_{j \neq k} \beta_j x_j + \beta_k x_k^{\lambda_0} + \beta_k (\lambda - \lambda_0) x_k^{\lambda_0} \log x_k.$$

The methodological novelty here is the appearance of β_k in the constructed variable. To provide an operational procedure, this is taken as the estimate from (6.49) when $\lambda = \lambda_0$, leading to the parameter estimate $\hat{\beta}_k(\lambda_0)$ and the constructed variable

$$w_{\text{BT}} = \hat{\beta}_k(\lambda_0) x_k^{\lambda_0} \log x_k, \tag{6.50}$$

effectively introduced by Box and Tidwell (1962). If the statistical focus is solely on the transformation of this one variable, the scaling factor $\hat{\beta}_k(\lambda_0)$ can be ignored, giving the constructed variable $x_k^{\lambda_0} \log x_k$.

If Box–Cox transformation of the response is also of interest, with a null value λ_1, the resulting constructed variable is

$$w_{\text{ABT}} = \hat{\beta}_k(\lambda_0) x_k^{\lambda_0} \log x_k - w_{\text{A}}(\lambda_1), \tag{6.51}$$

where $w_{\text{A}}(\cdot)$ is the constructed variable (6.15) for the Box–Cox transformation. In this case, the scaling factor $\hat{\beta}_k(\lambda_0)$ should be included as it provides the weighting for the two departures from the model which are being assessed using regression on a single constructed variable. The negative sign arises in (6.51) since the response and x_k are on different sides in the model (6.49). In addition the transformation of more than one explanatory variable can be included in the constructed variable with the further possibility of including groups of variables all transformed to the same power. Details are in Atkinson (1985, Sect. 8.3).

6.5.2 Transformation of Both Sides of a Regression Model

6.5.2.1 Introduction to the Model

There is sometimes a strong, often theoretically derived, relationship between the expected value of the response and the model $\eta(x, \beta)$, combined with variance heterogeneity. Box–Cox transformation of the response to achieve stability of variance can destroy the relationship between E(Y) and $\eta(x, \beta)$. For example, in the Michaelis–Menten model (Michaelis & Menten, 1913) for enzyme kinetics, the response goes from zero to an asymptotic value V_{max}. Transforming the response to y^λ results in a different range for the transformed response; V_{max} should also be transformed. In Sect. 6.5.2.2, we use data on mandible length in children to illustrate the difference in effect of transforming just the response as opposed to transforming the response and the model. We conclude this section with a robust analysis of data in which the volume of trees is modelled as a function of girth (perimeter of the tree at 5 ft (1.5 m) above ground level) and of height. As in the data on mandible length, both the model and the response should be transformed.

Carroll and Ruppert (1988, Chap. 4) developed a transform both sides model (TBS) for such problems, motivated by theoretical models for sockeye salmon breeding. With the Box–Cox transformation of both sides the model is

$$y(\lambda) = (y^\lambda - 1)/\lambda = \{\eta(x, \beta)^\lambda - 1\}/\lambda + \epsilon, \tag{6.52}$$

where the independent errors are normally distributed. As with the Box–Cox transformation, it is convenient to work with the normalized transformation

$$z(\lambda) = (y^\lambda - 1)/k(\lambda) = \{\eta(x, \beta)^\lambda - 1\}/k(\lambda) + \epsilon, \tag{6.53}$$

where $k(\lambda) = \lambda \dot{y}^{\lambda-1}$. Brief comments on algorithms for the estimation of the parameters β and λ are in Sect. 6.5.2.4.

6.5.2.2 Data on Mandible Length

As a first example illustrating the usefulness of the TBS model, rather than just the transformation of the response, we take data on mandible length in foetuses used by Royston and Altman (1994) to illustrate the use of fractional polynomials as explanatory variables in medical regression models. There are 158 observations on foetuses of age x less than 28 weeks. There are also nine measurements with $x > 28$, which the clinicians felt formed a different group with excessive measurement error. We work with all 167 observations.

The plot of the data in the left-hand panel of Fig. 6.17 suggests that mandible length increases linearly with gestational age and that the variance likewise increases. Royston and Altman overcome the increase in variance by use of the log transformation, but, as the right-hand panel of the figure shows, the relationship between

6.5 Transformations Including Explanatory Variables

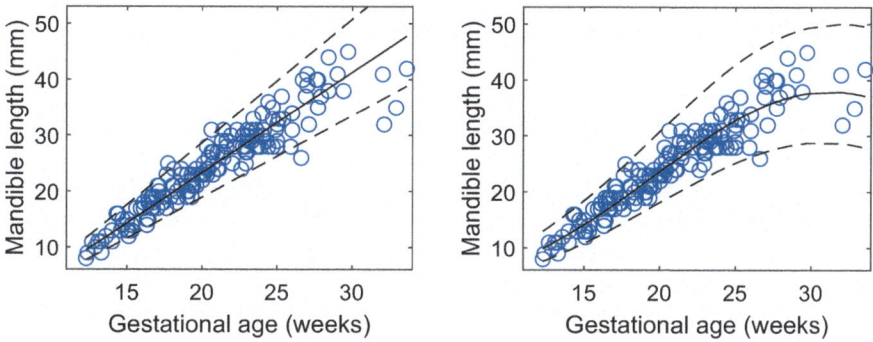

Fig. 6.17 Mandible length data: 99% prediction intervals for back-transformed data. Left-hand panel, transform both sides, $\lambda = 0$. Right-hand panel, logarithmic Box–Cox transformation with a quadratic model

log(mandible length) and age is then curved. We use the transformation of both sides to obtain a homoskedastic model in which the linear relationship is preserved.

Regression using the FS of all 167 observations on age, assuming homoskedasticity, indicates only two outliers, rather than 9. We proceed in the hope that there will be no outliers when we have transformed the data to allow for heteroskedasticity (see Sect. 8.4). This is an example of what Box and Cox (1964) call a proper attitude of sceptical optimism; we tentatively entertain the basis for analysis, rather than assuming it. The estimate for the Box–Cox transformation of the response suggests a value of 0.75 for λ—a compromise between preserving linearity ($\lambda = 1$) and a transformation to homoskedasticity ($\lambda = 0$). With a logged response there is very strong evidence for inclusion of a quadratic term. The TBS model with regression just on age also indicates the log transformation with $\hat{\lambda} = -0.08$.

The left-hand panel of Fig. 6.17 shows the 99% prediction interval for the back-transformed response from TBS. The fitted model retains the desired linearity and the prediction interval increases with gestational age in line with the heteroskedasticity of the observations. The right-hand panel shows a similar interval for the back-transformed quadratic regression with log y as the response. This panel shows that, although the quadratic model fits well to the majority of the observations, there is increasing curvature for values of $x > 28$.

It is clear that, on grounds of simplicity of interpretation, the TBS model is to be preferred to quadratic regression and a transformed response. The difference from predictions using the fractional polynomial model of Royston and Altman is not so obvious. However, TBS preserves the linear relationship between length and age and, more generally, the ability to combine theoretical models with transformation to normality.

6.5.2.3 The Volume of Shortleaf Pine Trees

The best models obtained as a result of a transformation often have some physical interpretation. An example is the reciprocal transformation of the poison data on survival times in Box and Cox (1964) which yields a simple model for the death rate of the animals. In this section, we use the transform both sides methodology to model the volume of usable timber in pine trees. The value of a tree is determined by this volume. Atkinson and Riani (2000, Table A.10) present 70 observations, taken from Bruce and Schumacher (1935), on the volume in cubic feet of shortleaf pine, together with x_1, the girth of each tree, that is, the perimeter at breast height, in inches and x_2, the height of the tree in feet. The girth and, to a lesser extent, the height are easily measured. The aim is therefore to find a formula for predicting volume y from the other two measurements. Atkinson and Riani (2000, Sect. 4.11) discuss the role of dimensional analysis in suggesting regression models for this problem. Here we view the tree trunks as cones, yielding the regression model

$$y = \alpha x_1^2 x_2 + \epsilon = \alpha v + \epsilon = \eta + \epsilon, \tag{6.54}$$

where $v = x_1^2 x_2$ is proportional to the volume of a cone and both sides have the dimension of volume. If the conical model (6.54) holds exactly, $\alpha = 1/(12^3 \pi)$ after conversion of x_1 to feet. If the data indicate a transformation of the response, the functional relationship is preserved by transforming both sides of (6.54).

In this section, we focus on constructed variables and the fan plot for the TBS model. We are therefore concerned with calculations for a given value of λ. For fixed λ, estimation of the parameters of the linear predictor η in (6.54) does not depend on whether the response is $z(\lambda)$ or the non-normalized $y(\lambda)$. Multiplication of both sides of (6.54) by $k(\lambda) = \lambda \dot{y}^{\lambda-1}$ and simplification leads to the model

$$y^\lambda = \eta^\lambda = (\beta^T x)^\lambda + \epsilon. \tag{6.55}$$

When β is a vector of parameters, minimizing the residual sum of squares in (6.55) is a non-standard nonlinear least squares problem briefly discussed in Sect. 6.5.2.4. But, when β is a scalar, as in (6.54), the simplified model (6.55) becomes

$$y^\lambda = (\alpha v)^\lambda + \epsilon = \alpha^\lambda v^\lambda + \epsilon = \delta v^\lambda + \epsilon. \tag{6.56}$$

Now let $q(\lambda) = y^\lambda$ and $u(\lambda) = v^\lambda$. Model (6.56) then reduces to the simple form

$$\begin{aligned} q(\lambda) &= \delta u(\lambda) + \epsilon \quad \lambda \neq 0 \\ q(0) = \log y &= \delta + \log u + \epsilon \quad \lambda = 0. \end{aligned} \tag{6.57}$$

For general λ this model is regression through the origin and the residual sum of squares $R(\lambda)$ is found by dividing the residual sum of squares of $q(\lambda)$ by $(\lambda \dot{y}^{\lambda-1})^{2n}$. The log model comes from (6.53) when $\lambda = 0$; there is no regression, only correction by a constant.

6.5 Transformations Including Explanatory Variables

Calculation of the score test for transformation requires the constructed variable found by Taylor series expansion of (6.53) about λ_0. Then

$$\frac{\partial z(\lambda)}{\partial \lambda} = \{y^\lambda \log y - (y^\lambda - 1)(1/\lambda + \log \dot{y})\}/k(\lambda)$$

and, likewise,

$$\frac{\partial \eta(\lambda)}{\partial \lambda} = \{\eta^\lambda \log \eta - (\eta^\lambda - 1)(1/\lambda + \log \dot{y})\}/k(\lambda).$$

The constructed variable for TBS (6.53) is found as the difference of these two, since, as in (6.51) they occur on different sides of the equation, and is

$$w_{\text{BS}}(\lambda) = \eta^\lambda \log \eta - y^\lambda \log y - (\eta^\lambda - y^\lambda)(1/\lambda + \log \dot{y}). \tag{6.58}$$

In (6.58), the multiplicative constant $k(\lambda)$ has been ignored, since scaling a regression variable does not affect the value of the t statistic for that variable. For a satisfactory model η^λ will be close to y^λ, when the values of $w_{\text{BS}}(\lambda)$ will be small, as will the value of the score statistic.

The general constructed variable (6.58) simplifies for the one-parameter model (6.57), being written in terms of $q(\lambda) = y^\lambda$ and $\delta u(\lambda) = \eta^\lambda$, provided $\lambda \neq 0$. Of course, δ is not known but is estimated from linear regression by $\hat{\delta}$, so that η^λ is replaced by $\hat{\delta}u(\lambda) = \hat{q}$ to give the constructed variable

$$w_{\text{BS}}(\lambda) = \hat{q}(\lambda) \log \hat{q}(\lambda) - q(\lambda) \log q(\lambda) - (\hat{q} - q)(1/\lambda + \log \dot{y}). \tag{6.59}$$

When $\lambda = 0$ similar reasoning leads to the variable

$$w_{BS}(0) = (\hat{q}^2 - q^2)/2 - (\hat{q} - q) \log \dot{y}.$$

We now apply this development to the construction of the fan plot of the score statistic calculated using the constructed variable (6.59). As with any regression model, the parameters of the linear model (here the scalar δ) are re-estimated at each step of the FS.

The trees are arranged in Atkinson and Riani (2000, Table A.10) from small to large, so that one indication of a systematic failure of a model would be the presence of anomalies relating to the smallest or largest observations. Here $n = 70$. To investigate transformations for these data, we use the conical model (6.54) with six transformations: the usual five values plus $\lambda = 1/3$, which has a special interpretation in the Box–Cox transformation of data such as these. If η is a linear model in x_1 and x_2, dimensional analysis suggests that the response should be $y^{1/3}$, giving a model in which all terms have the dimension length.

Figure 6.18 is the fan plot of score statistics from the constructed variable w_{BS} defined in (6.59). The plot shows that the log transformation is supported by all the

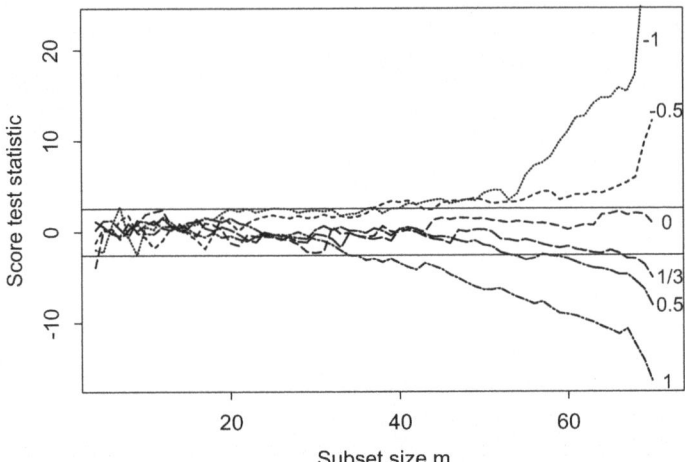

Fig. 6.18 Shortleaf Pine data: fan plot of score statistics for transforming both sides of the conical model. The logarithmic transformation is indicated. There is nothing special about $\lambda = 1/3$

data. All other values are rejected, including 1/3, which has no special dimensional significance when both sides are transformed. The monitoring plot of scaled residuals for this transformation is given in Fig. 4.39 of Atkinson and Riani (2000). The plot is very stable, with four slightly large residuals throughout, the largest belonging to observation 53, which is the last to be included in the forward search, but is far from the largest tree. The resulting model is of the form

$$\log y = \log(x_1^2 x_2) + \delta + \epsilon.$$

Their analysis shows no evidence of any departure from this model.

6.5.2.4 Numerical

The theoretical procedure is to minimize the residual sum of squares in the model (6.53), that is, with $y(\lambda)/\dot{y}^{\lambda-1}$ as the response. This is equivalent, for fixed λ, to taking $y(\lambda)/\dot{y}^{\lambda}$, as the response and the similarly transformed value of η as the model. Carroll and Ruppert (1988, p. 126) comment that, unless λ is fixed, it is not possible to use standard nonlinear regression routines for this minimization, as such routines typically do not allow the response to depend upon unknown parameters. They reformulate the problem in terms of a "pseudo-model" which they, as have we, found to have good convergence properties (the name of the routine inside **FSDA** is tBothSides). Section 4.3 of Carroll and Ruppert (1988) discusses algorithms and inference for the TBS model. In the two examples in this chapter, we were able to use the simplification to linear least squares for the single parameter model (6.57). Note, by using the fan plot we are not finding a numerical value for $\hat{\lambda}$.

6.6 Further Reading

The use of transformations in the simplification of distributions has a long history. The first four pages of Box and Cox (1964) provide a summary of developments up to the publication of their paper. Cox (1977) instances Fisher's z transformation of the correlation coefficient (Fisher, 2015). Probit analysis for binomial proportions (Bliss, 1934) is also a transformation to normality. General discussions of the history, purposes, and development of transformations are in the review article Cox (1977) and two related articles taken from the Encyclopedia of Statistics (Atkinson and Cox, 1988; Taylor, 2004). Box and Cox (1964) emphasize the effect of transformations to normality on the systematic part of the model. The transformation should provide simple, more revealing analyses that lead to sharper inferences. An extensive survey of literature from the first quarter century of the Box–Cox transformation is Sakia (1992). Hoyle (1973) lists 19 transformations, several of which are special cases of the Box–Cox transformation. The monograph of Carroll and Ruppert (1988) ranges widely over topics in the statistical transformation of data. The focus in Atkinson et al. (2021) is on the use of the FS in providing robust transformations.

Further details of the analysis of the investment funds data in Sect. 6.2.5 are given in Atkinson et al. (2020).

Problems

6.1 Variance-stabilizing transformation
Given a sample of observations for which

$$\text{var}(Y_i) \propto \{E(Y_i)\}^{2\alpha} = \mu^{2\alpha},$$

use a Taylor series expansion to find a variance-stabilizing transformation $g(y)$ such that $\text{var}\{g(Y_i)\}$ is approximately constant. What happens when $\alpha = 1$?

6.2 Confidence interval for transformation parameter λ
Compute and show the profile loglikelihood together with a 95% confidence interval for λ, both for the wool and for the poison data given in Box and Cox (1964). Comment on the stability of the estimates of λ.

6.3 Further analysis of the balance sheet data
Using the yX plot, show with 3 different symbols the normal units, the 19 brushed units, and the unbrushed outliers. Do you see any meaningful pattern?

6.4 Transformed balance sheet data and F test
Does monitoring the F test for regression for the balance sheet data show any departures from the regression models for untransformed and transformed data? The F test based on all the observations and untransformed data and that based on transformed data after deleting the outliers were given in Table 6.2. What is the third property

sought by the Box–Cox transformation? What plot would reveal whether it has been achieved? Has it?

6.5 Score test and fan plot 1

Dataset D1 contains 200 observations on 3 explanatory variables and a response.

1. Show the yX plot and provide preliminary comments about the need for transforming the response.
2. Compute the ANOVA table based on all the observations with untransformed response.
3. Compute the score test for transformation of the response and find the optimal value of λ using the grid $-1 : (0.25) : 1$.
4. Compute and comment on the fan plot. Check if positive and negative observations require a different value of λ.
5. Comment on the presence of outliers and their effect on the transformation.
6. Using the transformed data after removing any outliers, compute the ANOVA table and compare it with the one based on the original data.

6.6 Score test and fan plot 2

Same procedure as previous exercise but for dataset D2.

6.7 Score test and fan plot 3

Same procedure as previous exercise but for dataset D3.

Open Access This chapter is licensed under the terms of the Creative Commons Attribution 4.0 International License (http://creativecommons.org/licenses/by/4.0/), which permits use, sharing, adaptation, distribution and reproduction in any medium or format, as long as you give appropriate credit to the original author(s) and the source, provide a link to the Creative Commons license and indicate if changes were made.

The images or other third party material in this chapter are included in the chapter's Creative Commons license, unless indicated otherwise in a credit line to the material. If material is not included in the chapter's Creative Commons license and your intended use is not permitted by statutory regulation or exceeds the permitted use, you will need to obtain permission directly from the copyright holder.

Chapter 7
Non-parametric Regression

Abstract Unlike Chap. 6, this chapter focuses on non-parametric transformations of both the response and the explanatory variables. Transformation of the explanatory variables uses the Generalized Additive Model (GAM), described in Sect. 7.2. The transformation of the response iteratively applies the variance-stabilizing transformation to residuals from the fitted GAM. The algorithm developed, which we call RAVAS, is a robust version of Tibshirani's AVAS (Additivity and VAriance Stabilization). The developments are presented as a series of options written in pseudocode. In addition to robustness, three options use initial manipulation of the data to improve convergence of the backfitting algorithm for the GAM. Section 7.4.8 introduces an improved procedure for the variance-stabilizing transformation (option trapezoid), a detailed description of which is in Sect. 7.4.9. Section 7.5 uses simulations to illustrate the important improvement provided by this option, whether robustness is chosen or not. Section 7.6 introduces a graphical procedure, the augmented star plot, which indicates which combinations of options lead to satisfactory models. The chapter closes with further data analyses, comparisons of AVAS and RAVAS, and with references to some of the literature on non-parametric regression.

7.1 Introduction

The parametric transformations of the response studied in Chap. 6, such as that of Box and Cox, may not always be sufficiently flexible to provide satisfactory transformations for data analysis. In the Yeo-Johnson transformation and its generalization, it may be that two different response transformations are needed, but that the change between them does not occur near zero. Another possibility is that a bent stick transformation is preferable to a smooth curve. This chapter describes methods, using smoothing and robustness, that provide transformations for the best-fitting additive model in the presence of outliers. In some cases, such as the analysis of data on the weight of fish in Sect. 7.7, knowledge of the non-parametric transformations of the response and explanatory variables may lead to parsimonious parametric models capable of a physical interpretation.

Generalized Additive Models (GAMs) are a widely employed extension of linear regression models which replace the explanatory variables by non-parametrically estimated functions of those variables (Hastie and Tibshirani 1990). Typically, the response is not transformed. One of the aims of the Box-Cox transformation is the stabilization of error variance. Tibshirani (1988) extended the GAM to include response transformation using an empirical version of the variance stabilizing transformation. His algorithm AVAS (Additivity and VAriance Stabilization) provides nonparametric alternatives to the Box-Cox parametric transformation of the response and the Box-Tidwell family of power transformations of explanatory variables (see Sect. 6.5.1). Unfortunately, like these parametric methods, his procedure is not robust; the transformations of both the explanatory variables and the response can be highly distorted by outlying observations (Atkinson et al. 2021). The same problem holds for the related ACE algorithm of Breiman and Friedman (1985) and is one topic in the discussion of that paper (Fowlkes and Kettenring 1985; Buja and Kass 1985). This chapter describes a robust version of AVAS, robust both for transformations of the response and the explanatory variables, introduced by Riani et al. (2023b). A brief introduction is that of Riani et al. (2023a). For obvious reasons they call their procedure RAVAS.

The starting point for this extension of robustness to GAMs with transformed responses was the original version of Tibshirani's algorithm. This was written in "classical" Fortran, that is with few comments and uninformative variable names. There has been no further published development or comment on the details of the algorithm itself. Accordingly, we start in Sect. 7.2 with the background to AVAS, including the numerical variance-stabilizing transformation. In all, the RAVAS algorithm makes five improvements to AVAS, which are described in Sect. 7.4. These are available as options, so it is possible, as we do in Sect. 7.5.2, to compare the importance of the various modifications leading to RAVAS. One is the optional extension to robustness of fitting the GAM, with LS replaced by the choice of estimation using S or MM with Tukey's biweight or LTS or FS. Improvements to the numerical variance-stabilizing transformation are described in detail in Sect. 7.4.8.

7.2 Background

7.2.1 The Generalized Additive Model (GAM)

The generalized additive model (GAM) for observation i has the form

$$g(Y_i) = \beta_0 + \sum_{j=1}^{p} f_j(x_{ij}) + \epsilon_i. \qquad (7.1)$$

The functions f_j are unknown and are, in general, found by the use of smoothing techniques. A monotonicity constraint can be applied. If the response transforma-

7.2 Background

tion or link function g is unknown, it is restricted to be monotonic, but scaled to satisfy the technically necessary constraint that $\text{var}\{g(Y)\} = 1$ (otherwise, the zero transformation would be trivially perfect). In the fitting algorithm the transformed responses are scaled to have mean zero; the constant β_0 can therefore be ignored. The observational errors ϵ are assumed to be independent and additive with constant variance. The performance of fitted models is compared by use of the coefficient of determination R^2. Since the f_j are estimated from the data, the traditional assumption of linearity in the explanatory variables is avoided. However, the GAM retains the assumption that explanatory variable effects are additive. Buja et al. (1989) describe the background and early development of this model.

The next subsection introduces the backfitting algorithm for estimating the functions f_j when the response transformation $g(y)$ is known.

7.2.2 Definition of Backfitting

The backfitting algorithm, described in Hastie and Tibshirani (1990, p. 91), is used to fit a GAM. The algorithm proceeds iteratively using residuals when one explanatory variable in turn is dropped from the model. A discussion of convergence of iterative versions is in Schimek and Turlach (2006).

With $g(y)$ the $n \times 1$ vector of transformed responses, let $e_{(j)}$ be the vector of residuals when $f_j(x_j)$ is removed from the model without any refitting. Then

$$e_{(j)} = g(y) - \sum_{k \neq j=1}^{p} f_k(x_k). \tag{7.2}$$

The new value of $f_j(.)$ depends on ordered values of $e_{(j)}$ and x_j. Let the ordered (or sorted) values of x_j be $x_{s,j}$. The residuals $e_{(j)}$ are sorted in the same way to give the new order $e_{s,(j)}$. Within each iteration each explanatory variable is dropped in turn; $j = 1, \ldots, p$. The iterations continue until the change in the value of R^2 is less than a specified tolerance.

For iteration l the vector of sorted residuals for x_j is $e^l_{s,(j)}$. The new estimate of $f_j^{(l+1)}$ is

$$f_{s,j}^{(l+1)} = S\left\{e^l_{s,(j)}, x_{s,j}\right\}. \tag{7.3}$$

The required estimates of $f_j^{(l+1)}$ follow by restoring the elements of $f_{s,j}^{(l+1)}$ to the appropriate values of unordered x_j.

The function S depends on the constraint imposed on the transformation of variable j.

- If the transformation can be non-monotonic, S denotes a smoothing procedure. In order to compare RAVAS with Tibshirani's AVAS, both use the supersmoother

(Friedman and Stuetzle 1982), a non-parametric estimator based on local linear regression with adaptive bandwidths.
- If the transformation is monotonic, the $f_j^{(l+1)}$ come from isotonic regression (Barlow et al. 1972).
- If the transformation is linear $f_j^{(l+1)} = a + bx_j$, where a and b are the least squares coefficients.

7.2.3 Properties of Backfitting

For a linear regression model with errors that obey second-order assumptions, the least squares estimates of the parameters satisfy the normal equations. The backfitting algorithm is the Gauss–Seidel algorithm for solving this set of equations. Buja et al. (1989) extend this result for least squares to the backfitting algorithm for linear smoothers; a smoother is linear if $\hat{y} = Sy$ and S does not depend on y. This result assumes that, if necessary, the response has been transformed so that the second-order conditions are satisfied. Unfortunately, the supersmoother that is used for transformation of explanatory variables is not linear, so that convergence of the procedure is not automatically guaranteed.

The backfitting algorithm is not invariant to the permutation of order of the variables inside the matrix X, with high collinearity between the explanatory variables causing slow convergence of the algorithm: the residual sum of squares can change very little between iterations (Breiman and Friedman 1985; Hastie and Tibshirani 1988). The option `orderR2` of RAVAS, Sect. 7.4.5, attempts a solution to this problem by reordering the variables in order of importance.

7.3 Introduction to AVAS

This section first presents the structure of the AVAS algorithm of Tibshirani (1988) and then outlines the variance-stabilizing transformation used to estimate the response transformation. RAVAS has a similar structure, made more elaborate by the requirements of robustness and the presence of options.

7.3.1 The AVAS Algorithm

The algorithm has two loops. After initialization, the main (outer) loop uses the GAM fitted by the inner loop to calculate a new transformation $g(y)$. The inner loop fits the GAM for a given $g(y)$. In the descriptions of the algorithm ty and tX are transformed

7.4 RAVAS: Five Extensions to AVAS

values of the vector y and the matrix X. Thus, the fitting of the GAM in the inner loop provides new values of tX, given ty and so new values of the residuals e.

1. *Initialise Data.* Standardize response y so that $\overline{ty} = 0$ and $\text{var}(ty) = 1$, where var is the maximum likelihood biased estimator of variance. Centre each column of the X matrix so that $\overline{tX}_j = 0$, $(j = 1, \ldots, p)$.
2. *Initial call to "Inner Loop"* to find initial GAM using ty and tX; calculates initial value of R^2. Set convergence conditions on the number of iterations and value of R^2.
3. *Main (Outer) Loop.* Given values of ty and tX at each iteration, the outer loop provides updated values of the transformed response. Given the newly transformed response, updated transformed explanatory variables are found through the call to the backfitting algorithm (*inner loop*). Iterations continue until a stopping condition on R^2 is verified or until a maximum number of iterations has been reached.

7.3.2 The Numerical Variance-Stabilizing Transformation

The variance-stabilizing transformation was introduced in Sect. 6.1. The general assumption is that $\text{var}(Y)$ is a function $V(\mu)$ of the mean of Y. In the parametric approach of the Box-Cox transformation, the specific assumption is that $V(\mu) \propto \mu^{2\alpha}$, with α estimated from the data.

In the non-parametric AVAS algorithm for data, residuals are used to establish the mean-variance relationship. Specifically, $1/\sqrt{V(u)}$ in Eq. (6.1) is estimated by the vector of the reciprocals of the absolute values of the smoothed residuals sorted using the ordering based on fitted values of the model. The logged residuals in the estimation of the variance function are smoothed using the running line smoother of Hastie and Tibshirani (1986) and then exponentiated. There are n integrals, one for each observation. The range of integration for observation i, $(i = 1, 2, \ldots, n)$ goes from the smallest fitted value to its transformed value from the previous iteration. The computation of the n integrals uses the trapezoidal rule of numerical analysis. The output of this integration gives a vector of length n containing a new non-parametric estimate of transformed values for the response as in Sect. 7.4.8 (Exercise 7.1).

7.4 RAVAS: Five Extensions to AVAS

The RAVAS procedure introduces five improvements to AVAS, programmed as options. These do not have a hierarchical structure, so that there are 2^5 possible choices of the options. The augmented star plot of Sect. 7.6 provides a method for assessing these choices. We discuss the motivation and implementation for each in the order in which they are applied to the data when all are employed.

7.4.1 Initial Calculations

The structure of RAVAS is an elaboration of that of AVAS outlined in Sect. 7.3.1. Four of the five options can be invoked in the initialization of the outer loop given in Point 1 of Sect. 7.3.1. They are described here in the order in which they can be invoked. A detailed flowchart for the four options is in Sect. 7.4.6.

7.4.2 Response Transformation: Option `tyinitial`

Numerical experience shows that it is often beneficial to start from a parametric transformation of the response. This is optionally found using the automatic robust procedure for power transformations introduced in Sect. 6.3 and described in greater detail by Riani et al. (2022a). For $\min(y) > 0$ the Box-Cox transformation is used. For $\min(y) \leq 0$ the choice is the extended Yeo-Johnson transformation Sect. 6.2.3. In both cases the initial parametric transformations are only useful approximations, found by searching over a coarse grid of parameter values λ. The final non-parametric transformations sometimes suggest a generalization of the parametric ones.

7.4.3 Ordering Explanatory Variables in Backfitting: Option `scail`

In a comparison of monotone regression spline estimation with ACE (see Sect. 7.11), Ramsay (1988) observed that the fitted model obtained with ACE depends on the order of the explanatory variables (see Sect. 7.2.3). One approach is to use an initial regression to remove the effect of the order of the variables through scaling (Breiman 1988). With b_j the coefficient of $f_j(x)$ in the multiple regression of $g(y)$ on all $f_j(x)$, the option `scail` provides new transformed values for the explanatory variables: $\widehat{tX}_j = b_j f_j(x), j = 1, \ldots, p$. Option `scail` is used only in the initialization of the data.

7.4.4 Robust Outlier Detection: Option `rob`

RAVAS is robustified through the optional use of one of S, MM, LTS or FS to replace the use of all observations by a subset intended to be outlier free.

All methods provide weights for the observations, leading to the detection of outliers. As we saw in Sect. 5.2, the FS detects outliers at a simultaneous level of approximately 1% for samples of size up to around 1000. Optionally, a different level can be selected. In the other methods of robust regression, the scaled residuals for all n observations are calculated and used made of the Bonferroni inequality to give a simultaneous test for outliers with significance level for detection of individual outliers of α/n.

The algorithm works with k observations ($k = 0, 1, \ldots$) treated as outliers, providing the subset of $m = n - k$ observations used in model fitting and parameter estimation. In the backfitting algorithm, the subset S_m to calculate the transformations $f(X)$ at each iteration ignores the k outliers, which are also ignored in the calculation of the numerical variance-stabilizing transformation in Sect. 7.4.8. Since different response transformations can indicate different observations as outliers, the identification of outliers occurs repeatedly during the robust algorithm, once per iteration of the outer loop. The fan plot for response transformation of Sect. 6.1.3 shows the dependence of the observations deemed to be outlying on the value of the parameter λ. Here there is also a relationship between the transformations $f(x)$ and the declaration of outliers.

7.4.5 Ordering Predictor Variables: Option `orderR2`

To eliminate completely the dependence on the order of the variables, RAVAS includes an option that, at each iteration, provides an ordering which is based on the variable which produces the highest increment of R^2. Let $A_{j-1} = \{i_1, \ldots, i_{j-1}\}$ be the set of the $j - 1$ indexes for the columns of matrix X which have already been updated, and let i_r be the index of a column of X that has not yet been updated; then $i_r \notin A_{j-1}$. Further, let $R^2(i_r)$ be the coefficient of determination in a multiple linear regression which has $g(y)$ as response and as regressors $f_{i_1}, \ldots, f_{i_{j-1}}, f_{i_r}$. The next variable i_s to be estimated in the backfitting algorithm is that for which

$$i_s = \arg \max_{i_r \notin A_{j-1}} R^2(i_r).$$

With this criterion the most relevant features are immediately transformed and those that are perhaps irrelevant will be transformed in the final positions. For robust estimation, this procedure is applied solely to the observations in the subset S_m. Option `orderR2` is available at each call to the backfitting function.

7.4.6 Flowchart for Initialization of RAVAS

We now present the flowchart for the four preceding options in RAVAS.

(1.a) *(Option `tyinitial`). Calculation of Robust Parametric Transformation.*
(1.b) *Standardization.* If the data were not transformed in (a), standardize $y \to \widehat{ty}^{(0)}$ and centre $X \to tX^{(0)}$. If (a) was followed, $\widehat{ty}^{(0)} \leftarrow$ standardized $\widehat{ty}^{(0)}$.

(1.c) *(Option* scail*). Remove effect of order of explanatory variables.* Linear regression of $\widehat{ty}^{(0)}$ on $tX^{(0)}$ gives estimated coefficients b_j: $\widehat{tX}_j^{(0)} \leftarrow b_j X_j^{(0)}$.

(1.d) *(Option* rob*). Robust Regression.* Robust regression of $\widehat{ty}^{(0)}$ on $\widehat{tX}^{(0)}$ leads to an outlier free subset $S_m^{(1)}$ of transformed observations and the declaration of k_1 outliers.

(1.e) *(Option* orderR2*). Backfitting, with optional reordering of variables.* Given the $m_1 = n - k_1$ elements of $\widehat{ty}_i^{(0)}$ $i \in S_m^{(1)}$, use the backfitting algorithm to obtain updated values of the transformed explanatory variables $\widehat{tX}^{(1)}$ for the units belonging to $S_m^{(1)}$ and find the value of R^2.

The parameters in the backfitting procedure and the value of R^2 are calculated using only the units belonging to $S_m^{(1)}$.

7.4.7 Flowchart for the Main (Outer) Loop of RAVAS

The only option remaining to be described is the robust version of the numerical variance-stabilizing transformation of Sect. 7.4.8. This section presents the flowchart for the iterative portion of the main loop. The description of the numerical variance-stabilizing transformation follows in Sect. 7.4.8.

The looping variable is ℓ, starting from $\ell = 1$. When $\ell > 1$ robust regression in (2.b) provides outlier detection. When $\ell = 1$, this regression is not needed, since the results are available from (1.d). The numerical variance-stabilizing transformation is used in (2.c) to find updated values of the transformed response. The backfitting algorithm is used in (2.d) to find updated transformed explanatory variables. The procedure loops until a stopping condition on the change in R^2 is met or the maximum number of iterations has been reached.

(2.a) If $\ell = 1$, go to (2.c).

(2.b) *Robust Regression.* Robust regression of $\widehat{ty}^{(\ell-1)}$ on $\widehat{tX}^{(\ell-1)}$ gives outlier free subset $S_m^{(\ell)}$ of transformed observations and the declaration of k_ℓ outliers.

(2.c) *Numerical Variance-Stabilizing Transformation*, Sect. 7.4.8. Response transformation found for $n - k_\ell$ observations. Calculate $\widehat{ty}^{(\ell)}$ for all n observations.

(2.d) *Backfitting (inner loop), including option scail.* With transformed response $\widehat{ty}^{(\ell)}(m)$ obtain $\widehat{tX}^{(\ell)}(m)$, the updated values of the transformed explanatory variables using the units belonging to $S_m^{(\ell)}$. The fitted values of these m units are given by $\widehat{ty}_X^{(\ell)}(m) = \sum_j \widehat{tX}_j^{(\ell)}(m)$. Find the value of R^2 for these units. Linear interpolation or extrapolation gives $\widehat{tX}^{(\ell)}$ for all observations.

(2.e) *Check termination conditions.* If true, exit; else $\ell \leftarrow \ell + 1$ and go to (2.b).

7.4.8 Numerical Variance-Stabilizing Transformation: Option `trapezoid`

1. **Residuals and Fitted Values**
   There is a long history in regression analysis of the use of plots of residuals against fitted values to indicate the need, or otherwise, for a transformation of the response (Anscombe and Tukey 1963; Cox and Snell 1981). Further examples are in this chapter, especially in the comparison of the results of the analyses of data from the two-variable model of Sect. 7.5.2 as options are changed.
   The integral (6.1) uses fitted values, residuals, and previous estimates of transformed values of the response to estimate the non-parametric response transformation. The robust variance-stabilizing transformation is re-estimated at each iteration of the outer loop of the algorithm. At iteration ℓ the transformation is estimated from the m_ℓ residuals of the subset of observations provided by the robust regression. It is important to distinguish between vectors of length m_ℓ from the robust regression and those of length n giving quantities for all observations. At iteration ℓ the fitted values from the robust regression provide the m fitted values we denote as $\widehat{ty}^{(\ell)}(m)$. The vector of n fitted values $\widehat{ty}^{(\ell)}$ can be calculated using the parameter estimates from robust regression to predict the response at the k_ℓ outlying values detected by the regression. The membership of k_ℓ varies with the iteration. Where possible, we simplify the notation by removing dependence on ℓ.

2. **Integration and Response Transformation**
   To estimate the variance-stabilizing transformation, the fitted values $\widehat{ty}_X(m) = \sum_j \widehat{tX}_j$ have to be sorted, giving a vector of ordered values $\widehat{ty}_s(m) = (\widehat{ty}_{[1]}(m), \ldots, \widehat{ty}_{[m]}(m))'$. That is, $\widehat{ty}_{[i]}(m)$ is element i of the sorted vector $\widehat{ty}_s(m)$. The m residuals $e(m) = \widehat{ty}(m) - \widehat{ty}_X(m)$, where $\widehat{ty}(m)$ are the transformed values from the previous iteration. Arranged in the order of $\widehat{ty}_s(m)$, they are the residuals $e_s(m)$. The running line smoothing routine is applied to the logarithms of the absolute values of these sorted residuals, the result being exponentiated. The reciprocal of these smoothed residuals gives the vector $v(m)$ of estimated standard deviation values v_i to be used in the numerical integration. The details are in the flowscheme of Sect. 7.4.10 and in the Exercises.

The v_i provide estimates of $V^{-0.5}(y)$ at the m ordered points $\widehat{ty}_s(m)$. Calculation of the variance transformation (6.1) is however for all n observations and for observed, rather than fitted, transformed responses. Since the concern is robust inference, there may be several observations which fall outside the range of $\widehat{ty}_s(m)$, typically both above and below. RAVAS provides an option, perhaps confusingly called "`trapezoid`", for the choice between two methods for the extrapolation of the variance function estimate to these points. As did Tibshirani, RAVAS uses the trapezoidal rule of numerical analysis to approximate the integral with summands d_j. RAVAS uses trapezoidal d_j for the extrapolated elements, whereas AVAS leads to rectangular elements. The option `trapezoid = false` uses rectangular elements in extrapolation.

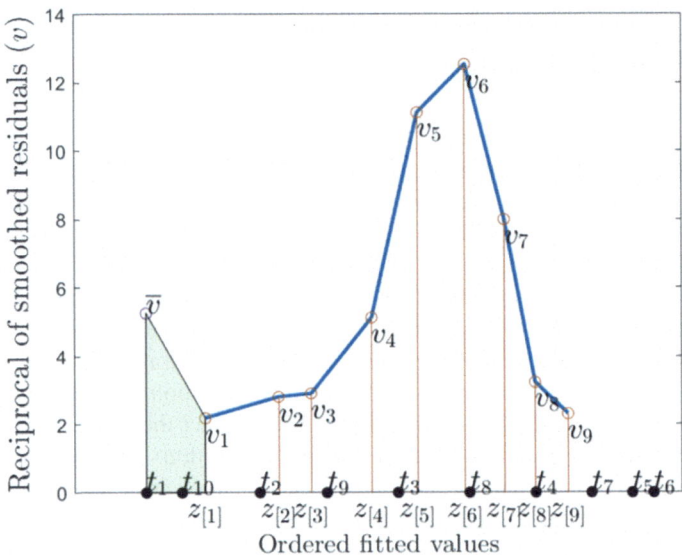

Fig. 7.1 Numerical variance-stabilizing transformation, $n = 10$ and $m = 9$. Situation $\mathcal{V}_1: t_i < z_{[1]}$. The green polygon shows the desired area (which has to be taken with a negative sign)

There are n integrals, one for each observation. The integration at iteration ℓ produces the n values of $\widehat{ty}^{(\ell)}$. In Sect. 7.4.9 we identify three separate regions, called $\mathcal{V}_l$, $(l = 1, \ldots, 3)$, determining the form of the trapezoidal approximation to the integral. For $\mathcal{V}_l$ and observation i, let the form of the summation be $D_l(i)$. Then

$$\widehat{ty}_i^{(\ell)} = \int_{\widehat{ty}_{[1]}^{(\ell)}(m)}^{\widehat{ty}_i^{(\ell-1)}} v_i \, d\hat{y} \approx D_l(i), i \in V_l. \tag{7.4}$$

The range of integration for observation i goes from $\widehat{ty}_{[1]}^{(\ell)}(m)$ (the smallest element of the vector of length m of the fitted values $\widehat{ty}_X(m)^{(\ell)}$ in iteration l) to $\widehat{ty}_i^{(\ell-1)}$, the transformed value of the response for observation i in the previous iteration. Thus the lower extreme of integration is fixed for all n integrals, with the upper limit of integration given by the elements of transformed response values in the previous iteration. Since the lower limit is the minimum of a vector of m values it may well be that $\widehat{ty}_i^{(\ell-1)}$ is smaller than $\widehat{ty}_{[1]}^{(\ell)}(m)$, which is Situation $\mathcal{V}_1$. When $\widehat{ty}_i^{(\ell-1)}$ is greater than $\widehat{ty}_{[m]}^{(\ell)}(m)$, we have Situation $\mathcal{V}_3$. Both of these involve extrapolation. For the remaining possibilities, Situation $\mathcal{V}_2$, interpolation is required. In Sect. 7.4.9 we use a simplified notation to explore the detailed structures of the summations $D_l(i)$, which are illustrated in Figs. 7.1, 7.2 and 7.3.

`trapezoid = true`. Suppose that the following condition holds:

7.4 RAVAS: Five Extensions to AVAS

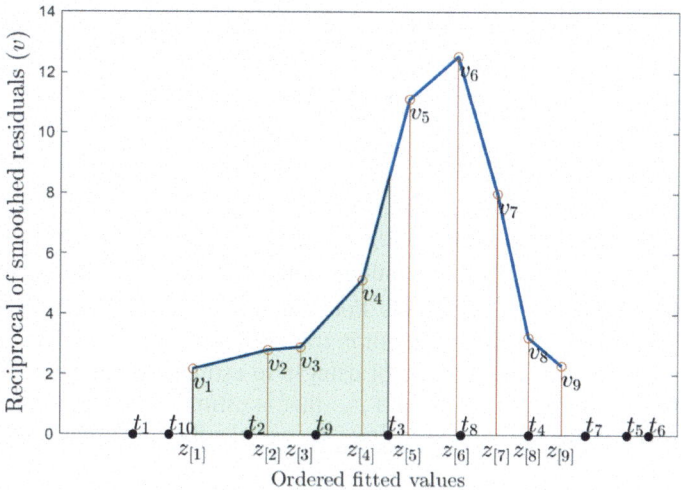

Fig. 7.2 Numerical variance-stabilizing transformation, $n = 10$ and $m = 9$. Situation $\mathcal{V}_2$: $z_{[k-1]} \leq t_i < z_{[k]}$ ($k = 2, \ldots, m$). The green polygon shows the desired area

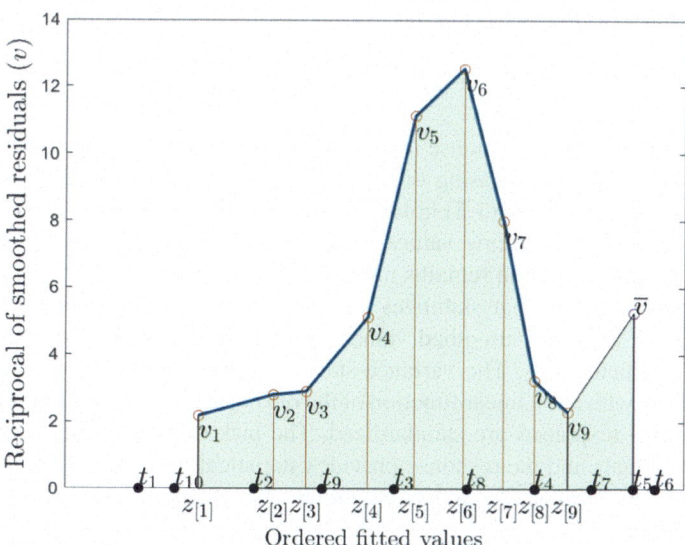

Fig. 7.3 Numerical variance-stabilizing transformation, $n = 10$ and $m = 9$. Situation $\mathcal{V}_3$: $t_i > z_{[m]}$. The green polygon shows the desired area

$$\widehat{ty}_i^{(\ell-1)} < \widehat{ty}_{[1]}^{(\ell)}(m), \qquad (7.5)$$

that is $\mathcal{V}_1$. In the absence of any information, we take the standard error component at $\widehat{ty}_i^{(\ell-1)}$ to be $\bar{v}$, the average value of the v_i, when the area of the single trapezoid is

$$D_1(i) = \{\widehat{ty}_{[1]}^{(\ell)}(m) - \widehat{ty}_i^{(\ell-1)}\}(v_1 + \bar{v})/2. \tag{7.6}$$

For any other $\widehat{ty}_{i'}^{(\ell-1)} < \widehat{ty}_{[1]}^{(\ell)}(m)$ we likewise take $v_{i'} = \bar{v}$. For

$$\widehat{ty}_{[1]}^{(\ell)}(m) < \widehat{ty}_i^{(\ell-1)} < \widehat{ty}_{[m]}^{(\ell)}(m),$$

situation $\mathcal{V}_2$, we use linear interpolation in the two values of v adjacent to $\widehat{ty}_i^{(\ell-1)}$ to give an estimate for v_i. Then depending on the value of $\widehat{ty}_i^{(\ell-1)}$, $D_2(i)$ will be a sum of between 1 and $m-1$ trapezoids. For values $\widehat{ty}_i^{(\ell-1)} > \widehat{ty}_{[m]}^{(\ell)}(m)$, $\mathcal{V}_3$, we again assume $v_i = \bar{v}$. Now $D_3(i)$ is the sum of m trapezoids.

trapezoid = false. Instead of using $\bar{v}$ to estimate the unknown standard deviations we use the values of v_1 and v_m, that is estimates based on the extreme observations. Thus, for $\widehat{ty}_i^{(\ell-1)} < \widehat{ty}_{[1]}^{(\ell)}(m)$, $\mathcal{V}_1$, we take the single rectangle

$$D_1(i) = (\widehat{ty}_{[1]}^{(\ell)}(m) - \widehat{ty}_i^{(\ell-1)})v_1 \tag{7.7}$$

with negative sign as the integral. Similarly, for i for which $\widehat{ty}_{i'}^{(\ell-1)} > \widehat{ty}_{[m]}^{(\ell)}(m)$, summand m of $D_3(i)$ is the rectangle

$$d_m = \{\widehat{ty}_{i'}^{(\ell-1)} - \widehat{ty}_{[m]}^{(\ell)}(m)\}v_m. \tag{7.8}$$

It is a requirement that response transformations be monotonic increasing. Since the values of $\widehat{ty}$ are in increasing order, the series of integrals given by (7.4) are increasing. When condition (7.5) holds, element i of the summand in (7.4) will be negative, typically for extreme values of i, and so will be the new values of ty. However, the transformation remains monotonic increasing.

For a transformation that stabilizes variance, there is no relationship between var(ty) and ty. Thus the smoothed values of the residuals will be constant, apart from random fluctuations. The variance-stabilizing transformation will produce a new value of ty which is a linear function of the old ty; that is no transformation since the transformed responses are standardized. The lack of relationship between the smoothed residuals and the response provides statistical justification for the choice of $\bar{v}$ as the stable estimate of standard deviation when the trapezoidal option is used.

7.4.9 Simplified Notation for the Variance-Stabilizing Transformation

For this section ONLY, it is helpful to adopt a simplified notation. Let

$$\begin{aligned} t_i &= \widehat{ty}_i^{(\ell-1)} & i &= 1, \ldots, n \\ z_{[i]} &= \widehat{ty}_{[i]}^{(\ell)}(m) & i &\in S_m \\ v_i &= v_i \quad \text{unchanged} & i &\in S_m. \end{aligned} \tag{7.9}$$

7.4 RAVAS: Five Extensions to AVAS

There are n values of t_i, the transformed responses at iteration $\ell - 1$. From the integral in (7.4) these form the upper limit of the integration giving the new transformed values at iteration ℓ. The variance-stabilizing transformation requires knowledge of the standard deviation at each of these values. This is found using the trapezoidal rule. The indexes of the t_i are those of the observations.

The m values of the $z_{[i]}$ are the fitted values from the robust procedure at iteration ℓ. At each, the value of v_i is known. The index of the v_i is that of $z_{[i]}$. The $z_{[i]}$ are ordered, as the notation makes clear. The v_i are not ordered.

There are three distinct situations in calculating the approximation to the integral in (7.4). These are made clear by the example in Figs. 7.1, 7.2 and 7.3 for which $n = 10$ and $m = 9$. We describe the method for `trapezoid = true`. At the end of each case, we provide the modification for `trapezoid = false`.

Situation $\mathcal{V}_1$: $t_i < z_{[1]}$. The estimate of standard deviation at t_i is $\bar{v}$, with that at $z_{[1]} = v_1$. Then the trapezoidal approximation to the integral is

$$d_0(i) = D_1(i) = (\bar{v} + v_1)(t_i - z_{[1]})/2 \qquad (7.10)$$

and

$$\widehat{ty}_i^{(\ell)} \approx d_0(i), \qquad (7.11)$$

a negative quantity.

In the example in Fig. 7.1, both t_1 and t_{10} are $< z_{[1]}$, so that $d_0(10) = (\bar{v} + v_1)(t_{10} - z_{[1]})/2$, which will be negative.

`trapezoid = false`. The estimate of standard deviation at t_i is v_1. Then for $i = 1$ or 10

$$d_0(i) = v_1(t_i - z_{[1]}). \qquad (7.12)$$

Situation $\mathcal{V}_2$: $z_{[k-1]} \le t_i < z_{[k]}$ ($k = 2, \ldots, m$). There is no extrapolation for values of v_i, so that, in this region, the choice of `trapezoid = true` or `false` is irrelevant.

First consider the case $k = 2$ and let $j = k - 1$, that is one in this case. There is then just one trapezoidal contribution to the integral. The value v_{ti} of the standard deviation at t_i is found by linear interpolation between v_j and v_{j+1}. The equality of slopes in the interpolation leads to v_{ti} being the solution of the relationship

$$\frac{v_{ti} - v_j}{t_i - z_{[j]}} = \frac{v_{j+1} - v_j}{z_{[j+1]} - z_{[j]}}. \qquad (7.13)$$

Then, for general j, the trapezoidal approximation, with interpolation, to this element of the integral is

$$d_j(i) = (v_j + v_{ti})(t_i - z_{[j]})/2. \qquad (7.14)$$

In the special case $j = 1$

$$\widehat{ty}_i^{(\ell)} \cong d_1(i).$$

In the example in Fig. 7.2, $z_{[1]} < t_2 < z_{[2]}$, so that $d_1(2) = (v_1 + v_{t2})(t_2 - z_{[1]})/2$, the only instance of $j = 1$ in this set of data.

Now consider the case when $2 < k < m$. There will then be $k - 2$ trapezoids, which do not specifically depend on t_i, of the form

$$d_j = (v_j + v_{j+1})(z_{[j+1]} - z_{[j]})/2, \quad j = 1, \ldots, k-2 \quad (7.15)$$

and one of the interpolated form (7.14) with $j = k - 1$. Putting these pieces together, the trapezoidal approximation to the integral is

$$\widehat{ty}_i^{(\ell)} \cong D_2(i) = \sum_{j=1}^{k-2} d_j + d_{k-1}(i). \quad (7.16)$$

In Fig. 7.2 $z_{[4]} < t_3 < z_{[5]}$. Then from (7.16)

$$\widehat{ty}_3^{(\ell)} \cong \sum_{j=1}^{3} d_j + d_4(3).$$

Situation $\mathcal{V}_3$: $t_i > z_{[m]}$. The estimate of standard deviation at t_i is again $\bar{v}$, with that at $z_{[m]} = v_m$. Then, for `trapezoid = true`, the contribution from this extrapolation is

$$d_m(i) = (\bar{v} + v_m)(t_i - z_{[m]})/2. \quad (7.17)$$

There are also $m - 1$ trapezoids with the form of (7.15). The approximation to this integral from all m polygons is then

$$\widehat{ty}_i^{(\ell)} \cong D_3(i) = \sum_{j=1}^{m-1} d_j + d_m(i). \quad (7.18)$$

In Fig. 7.3 $t_5 > z_{[9]}$. Now

$$\widehat{ty}_5^{(\ell)} \cong \sum_{j=1}^{8} d_j + d_9(5).$$

`trapezoid = false`. For $t_i > z_{[m]}$ the estimated standard deviation at t_i is now v_m. Then

$$d_m(i) = v_m(t_i - z_{[m]}) \quad (7.19)$$

replacing the trapezoid of (7.17) in (7.18). All other terms in the summand are unchanged.

7.4 RAVAS: Five Extensions to AVAS
255

In highly contaminated data m may be appreciably less than n. Then there may be several approximations of the form (7.11) with a single summand. In case $\mathcal{V}_2$ the number of summands increases towards $m - 1$. However, there may be several integrals with the same number of summands and there be none for specific values of j. For all t_i falling in case $\mathcal{V}_3$ there are m summands, only one of which depends directly on the value of t_i.

7.4.10 Flowscheme for Calculation of the Numerical Variance-Stabilizing Transformation

This section presents the flowchart for the calculation and ordering of the fitted values and residuals, together with their transformation for smoothing, that are required as a preliminary to the numerical integration described in Sect. 7.4.8. The notation for the various functions of residuals deliberately follows that of Tibshirani's Fortran.

1. Following the application of the robust routines which identify the m observations to be used and the consequent backfitting routine, calculate the vector of fitted values $\widehat{ty}_X^{(\ell)}(m) = \sum_{j=1}^{p} \widehat{tX}_j^{(\ell)}$. Find the residuals $e^{(\ell)}(m) = \widehat{ty}^{(\ell-1)}(m) - \widehat{ty}_X^{(\ell)}(m)$. Replace values of $|e^{(\ell)}(m)| < 10^{-10}$ with 10^{-10} (because of logarithms in the next step).

2. Create vector $e_2^{(\ell)}(m) = \log\{|e|^{(\ell)}(m)\}$. Sort the values of $\widehat{ty}_X^{(\ell)}(m)$ to give $\widehat{ty}_s^{(\ell)}(m) = \left\{\widehat{ty}_{[1]}^{(\ell)}(m), \ldots, \widehat{ty}_{[m]}^{(\ell)}(m)\right\}^T$. Sort residuals $e_2^{(\ell)}(m)$ in the order of $\widehat{ty}_s^{(\ell)}(m)$ to give $e_3^{(\ell)}(m)$.

3. Apply the running line smoother to $e_3^{(\ell)}(m)$ as a function of $\widehat{ty}_s^{(\ell)}(m)$. The smoothed residuals are called $e_4^{(\ell)}(m)$. Set $e_5^{(\ell)}(m) = \exp\{-e_4^{(\ell)}(m)\}$. Then $e_5^{(\ell)}(m) = v(m)$ are the reciprocals of the smoothed absolute values of the residuals using the ordering based on fitted values.

4. Compute the variance-stabilizing transformation by calculating the integral (7.4). The x coordinates are the sorted fitted values $\widehat{ty}_s^{(\ell)}(m)$ with y coordinates $v(m)$. The upper extremes of integration are given by the vector $\widehat{ty}^{(\ell-1)}$ of length n. When $\ell = 1$, $\widehat{ty}^{(\ell-1)}$ is simply given by the vector of standardized values $\widehat{ty}^{(0)}$. The output is the new vector of $\widehat{ty}^{(\ell)}$ (of length n). Standardize $\widehat{ty}^{(\ell)}$ using the mean and variance of $\widehat{ty}^{(\ell)} \in m_\ell$.
Return.

7.5 Simulated Examples

In this section we start with a simple example to compare the data analyses from RAVAS with the non-robust AVAS of Tibshirani when there are a few outliers. In Sect. 7.5.2 our analysis is of data without introduced outliers to demonstrate the effect of our options on the performance of the non-robust algorithm.

7.5.1 Example 1—Robustness

There are 151 observations with x equally spaced over 0, (0.1), 15. The linear model is
$$z = \sin(x) + 0.5(\mathcal{U}(0, 1) - 0.5),$$
where $\mathcal{U}(a, b)$ is the uniform distribution defined in $[a\ b]$. Outliers of value one replace the values of z in positions 100, 105, 106, ..., 110. That is there are seven outliers at $x = 9.9, 10.4, 10.5, \ldots, 10.9$. The response $y = \exp z$. Thus a logarithmic transformation of the observed responses is appropriate. We start with a non-robust analysis excluding our improvements to the algorithm. That is, there is no initial transformation of the response (tyinitial = false) and we use the rectangular extrapolation when calculating the variance-stabilizing transformation (trapezoid = false).

The top left-hand panel of Fig. 7.4 shows the transformation of y, which is roughly logarithmic, but with a slight decrease in gradient near the centre of the plot. The top right-hand panel shows the transformed values of y against fitted values. Two horizontal linear structures are apparent; a set of six values of transformed y close to one and several values close to -1.75. The bottom panel of Fig. 7.4 shows a very clear distortion of the sinusoid of $f(x)$ in the range of the outliers, that is 9.9–10.9. This "rug" plot shows the uniform distribution of the values of x.

In our robust analysis we used the initial transformation of y and set trapezoid = true. The top panels of Fig. 7.5 show plots of the transformed values of y with the 7 outliers automatically detected by the robust procedure. The plot against y shows a smoother curve than that for the non-robust analysis. However, the greatest difference is in the plot against fitted values. There is now a regression line with less scatter than in Fig. 7.4; the linear structure at the bottom of the plot is now absent and the horizontal line formed by the group of six outliers is more clearly distanced from the fitted regression. The plot also shows the isolated outlier (corresponding to $x = 9.9$). The bottom panel of Fig. 7.5 shows that the robust analysis has removed the distortion in the estimation of the sinusoidal form of $f(x)$, except where the outliers have been deleted.

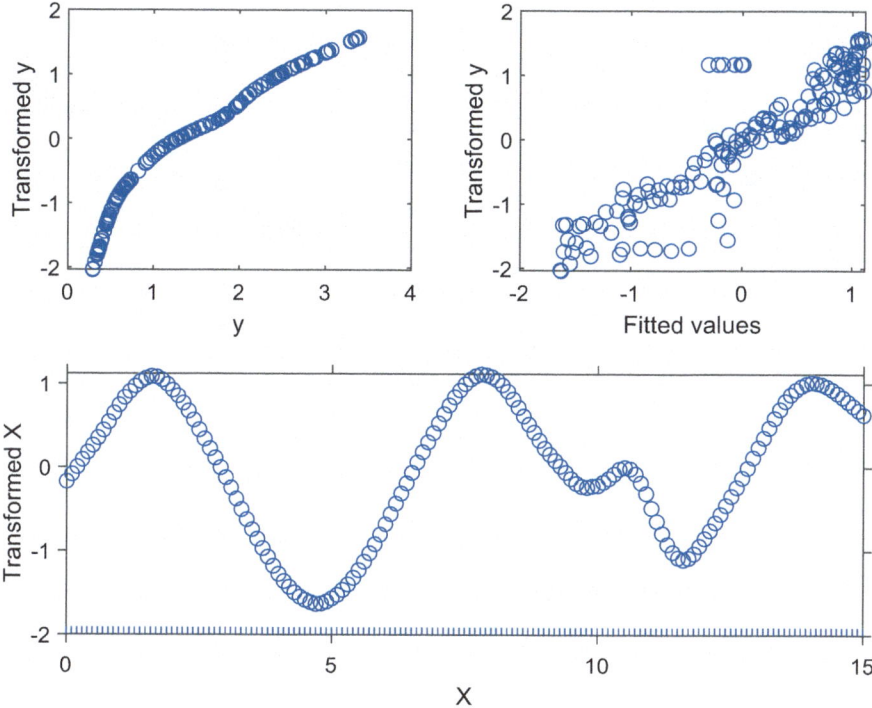

Fig. 7.4 Example 1 (seven outliers): standard non-robust analysis without options (AVAS). Top left-hand panel, transformed y against y; top right-hand panel, transformed y against fitted values; lower panel, transformed x against x with "rug" plot showing the uniform distribution of the values of x

7.5.2 Example of Effects from Options in RAVAS

The preceding example shows that the robustness of RAVAS leads to the identification of the seven outliers and to the virtual recovery of the model from which the data were simulated. In this section we use further simulations to explore how our modifications, not necessarily the robustness itself, lead to improved analyses of data.

Two-variable model. We use an example without outliers to illustrate the importance of some options including the initial transformation of the response. There are 151 observations with $x_1 \sim \mathcal{N}(0, 1)$. The linear model is

$$z = 10\{1 + \sin(x_1) + \exp(x_2)\} + \mathcal{N}(-0.5, 0.5),$$

where $x_2 = 0, (0.01), 1.5$. The observed responses are $y = z^3$, so that a 1/3rd-root transformation is expected. As well as plotting residuals and fitted values from some of the fitted models, we also calculate the value of R^2 and the significance level of

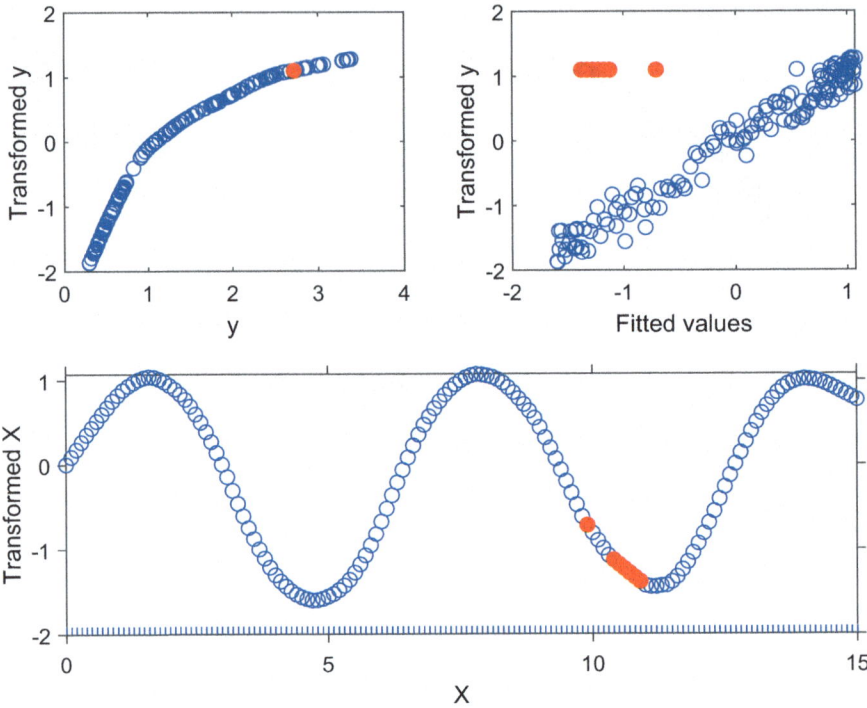

Fig. 7.5 Example 1 (seven outliers): robust analysis. Top left-hand panel, transformed y against y; top right-hand panel, transformed y against fitted values; lower panel, transformed x against x. The seven outliers are indicated by red filled symbols

the Durbin–Watson statistic (Durbin and Watson 1950) for the independence of the residuals ordered by the fitted values from the model.

Figure 7.6 shows results from using RAVAS on these data without any options. The left-hand panel shows the plot of residuals against fitted values which has a sinusoidal shape. The plot of transformed y against fitted values in the right-hand panel is logistic in shape, rather than linear. These two plots indicate that the fitted model is not satisfactory, an impression confirmed by the value of 2.3913×10^{-21} for the significance of the Durbin–Watson statistic. We do not show the plot of transformed y against y which is roughly appropriate for a transformation of 1/3.

We now proceed adding options in an ad hoc manner, starting with `scail`, that is the rescaling of the explanatory variables, Sect. 7.4.3. Figure 7.7 shows plots of the same properties as Fig. 7.6. The plot of the residuals now shows a series of rough diagonal bands with an outline for larger fitted values similar to that in Fig. 7.6. Although straighter, the plot of transformed against fitted y is again curved for higher values of y. These improvements lead to a less significant value of 2.0349×10^{-14} for the significance of the Durbin–Watson statistic. The fit is still far from satisfactory.

7.5 Simulated Examples

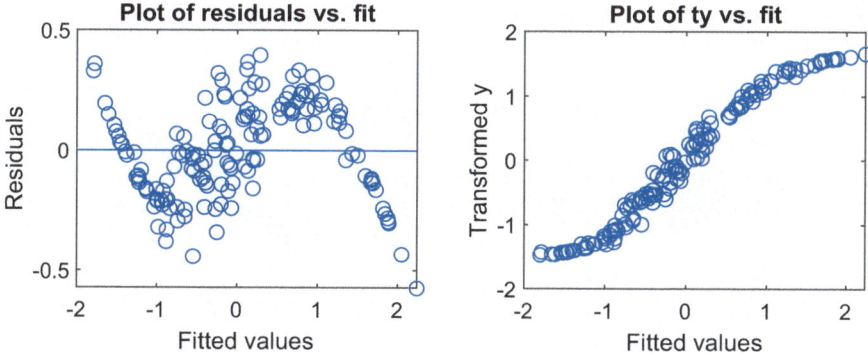

Fig. 7.6 Two-variable model: non-robust analysis without options. Left-hand panel, residuals against fitted values; right-hand panel, transformed y against fitted values

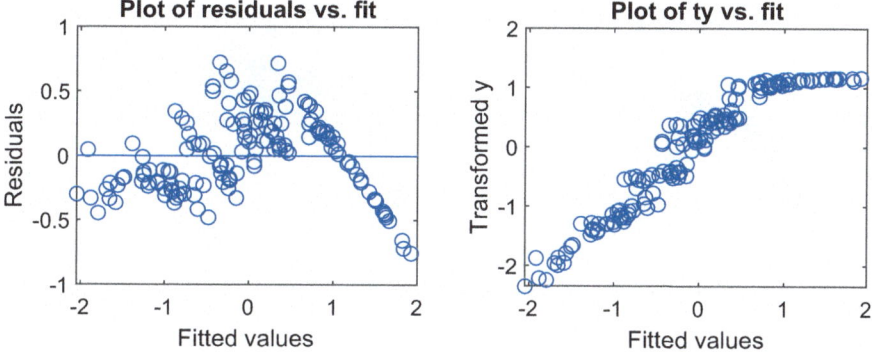

Fig. 7.7 Two-variable model: non-robust analysis with option `scail`. Left-hand panel, residuals against fitted values; right-hand panel, transformed y against fitted values

In the modelling producing the third pair of figures (Fig. 7.8), we have also included the initial transformation of the responses (`tyinitial`, Sect. 7.4.2). It is clear from scatter plots of the data that the transformation parameter λ should be positive but small. We used a grid of $\lambda = 0, 0.1, 0.2, 0.3, 0.4$ and 0.5.

The plot of residuals against fitted values in the left-hand panel of the figure is greatly improved in comparison with that of Fig. 7.7 but is still not random, showing some curvature. This is reflected in the improved, but still significantly small, value of 3.225×10^{-7} for the significance of the Durbin–Watson test. The plot of transformed y against fitted values is closer to a straight line than before.

These three different sets of options show steady improvement in the normality of the residuals. We have examined them in some detail because we observed plots with structures similar to those in the left-hand panels of Figs. 7.6 and 7.7 when we used AVAS to analyse data for response transformations in Atkinson et al. (2020). We have also failed to find any references to such phenomena. We conclude this progression with the analysis using all five options.

260 7 Non-parametric Regression

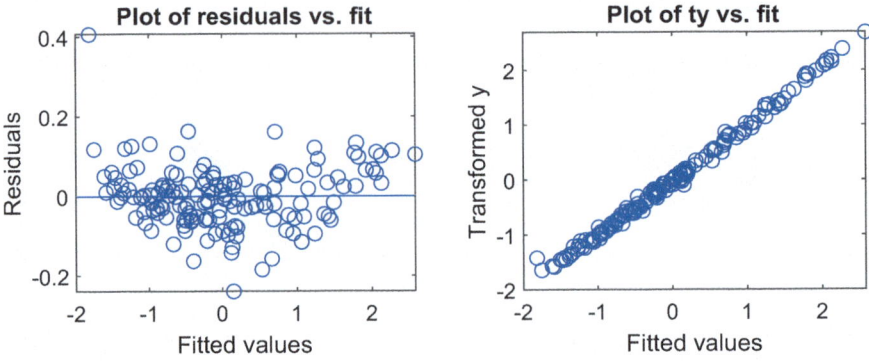

Fig. 7.8 Two-variable model: non-robust analysis with options scail and tyinitial. Left-hand panel, residuals against fitted values; right-hand panel, transformed y against fitted values

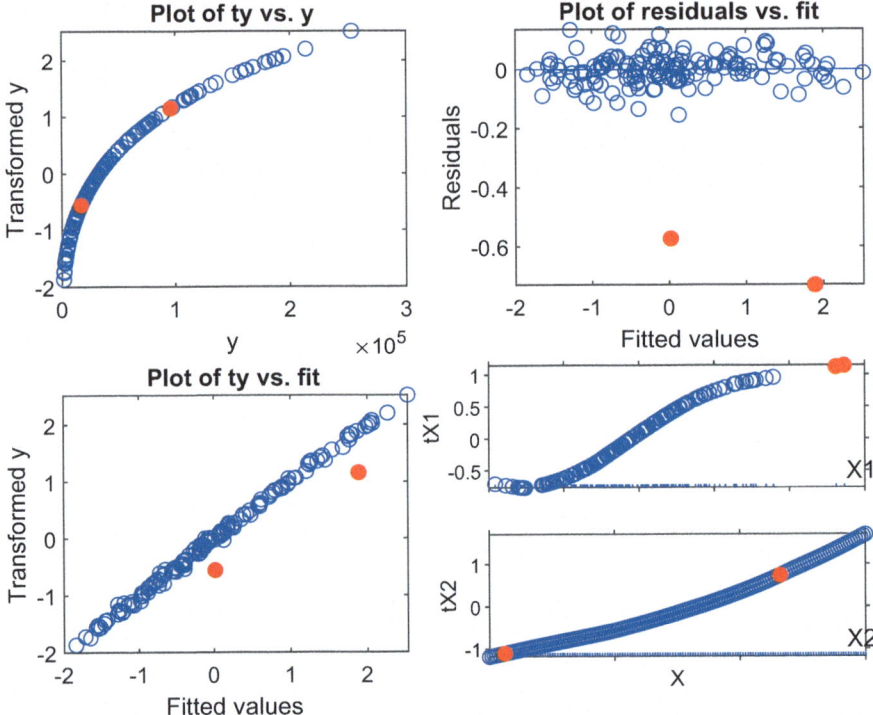

Fig. 7.9 Two-variable model: robust analysis with all options. Upper left-hand panel, transformed y against y; upper right-hand panel, residuals against fitted values; lower left-hand panel, transformed y against fitted values; lower right-hand panel, transformed explanatory variables. Two outliers are shown by red filled symbols

The upper left-hand panel of Fig. 7.9 shows the smooth curve of the transformed responses and the upper right-hand panel shows the plot of residuals against fitted values. This appears to be a random scatter with, surprisingly, two outliers at high values of x_1, an unexpected artefact of the simulation. The significance value for the Durbin–Watson statistic is 0.44, so that independence of the residuals has been achieved, once the outliers are deleted. The bottom left-hand panel shows the linear relationship between transformed response and fitted values, as is expected for a GAM, together with the two outliers. The estimated transformations of the two explanatory variables are in the bottom right-hand panel of the figure. The first explanatory variable shows the shape of a scaled and translated sine function, with the two observations for the largest values of x_1 being marked as outlying. The second explanatory variable has the desired exponential form. Plots of these nonparametric functions with superimpositions of the parametric forms can be used to check closeness to these parametric models.

7.6 A Graphical Representation of the Systematic Evaluation of Options

RAVAS adds five options to the original AVAS. There are therefore 32 combinations of options that could be chosen. It is not obvious that all will be necessary when analysing all sets of data. RAVAS provides flexibility in the assessment of these options. One possibility is a list of options ordered by, for example, the value of R^2 or of the significance of the Durbin–Watson or the normality test. In this section we describe the augmented star plot, one graphical method for visualizing interesting combinations of options in a particular data analysis. An example is Fig. 7.10. Figure 7.13 introduces a brushable version of the augmented star plot which links to residual and other plots from the fitted model.

The default setting of RAVAS does not plot analyses for which the residuals fail the Durbin–Watson and Jarque–Bera normality tests, at the 5% level (two-sided for Durbin–Watson). The Jarque–Bera (Jarque and Bera 1987) test uses a combination of estimated skewness and kurtosis to test the distributional shape of the residuals. Although there is an option for independent choice of the levels of these two tests, we used a threshold of 5% for both.

The remaining, admissible, solutions are ordered by the Durbin–Watson significance level multiplied by the value of R^2 and by the number of units not declared as outliers. Other options are available. The lengths of rays in individual panels of the plot are of equal length for those features used in an analysis; they are optionally enclosed in a polygon. All rays are in identical places in each panel of the plot; the length of the rays for each analysis is proportional to p_{DW}, the significance level of the Durbin–Watson test. Each ray has a different colour.

The ordering in which the five options are displayed in the plot depends on the frequency of their presence in the set of admissible solutions. For example, if robustness is the one which has the highest frequency, its ray is shown on the right. The remaining options are displayed counterclockwise, in order of frequency.

This is a list of the five options giving an informative text description followed by the optional input arguments.

1. Initial robust transformation of the response; tyinitial—Sect. 7.4.2.
2. Remove effect of order of explanatory variables by regression; scail—Sect. 7.4.3.
3. Robustness; rob—Sect. 7.4.4.
4. Ordering variables in the backfitting algorithm using R^2 values; orderR2—Sect. 7.4.5.
5. Trapezoidal or rectangular rule for numerical integration; trapezoid—Sect. 7.4.8.

Example 2. As an initial example we consider an extension of Example 1 simulated in Sect. 7.5.1 in which there are now four explanatory variables. There are again 151 observations with x_1 equally spaced from 0, (0.1), 15. The linear model is

$$z = \sin(x_1) + \sum_{j=2}^{4}(j-1)x_j + 0.5\mathcal{U}(-0.5, 0.5),$$

where x_{ij} ($j = 2, \ldots, 4$), are independently $\mathcal{N}(0, 1)$. Eight outliers of value one replace the values of z at $x = 8.9, 9.9, 10.4, 10.5\ldots10.9$. The response is $y = \exp z$. There are thus four explanatory variables, only one of which requires transformation, as does the response to $\log y$.

Figure 7.10 shows the augmented star plots of the two combinations of options for which the residuals pass the tests of independence and normality. The first solution uses all five options with the trapezoidal integration rule excluded from the second solution. The statistical properties of the two solutions are virtually indistinguishable (the correlation between these two solutions is 0.9999). For both $R^2 = 0.998$. For the first solution $p_{DW} = 0.32$, $p_{JB} = 0.09$, while for the second $p_{DW} = 0.18$, $p_{JB} = 0.088$. The reason the lengths of the rays are not the same is a function of the different values of p_{DW} for the two fits.

We now compare the analysis without options with that suggested in Fig. 7.10. Figure 7.11 shows the non-robust results from using AVAS. The value of $R^2 = 0.844$. Although the response has been adequately transformed the residuals are far from random (the significance level for the Durbin–Watson test is 1.64×10^{-8}), the sinusoid for x_1 is distorted and x_4 has been subjected to non-linear transformation.

The fit from RAVAS using all options identifies 9 outliers, the eight generated intentionally and one at unit 53. The upper right and lower left panels of Fig. 7.12 show, after the outliers have been deleted, that the residuals are structure free when plotted against fitted values and that there is a linear relationship between the transformed values of y and the fitted values. The lower right-hand panel plots the transformed values of x_1. The sinusoid has been recovered. We do not show the plots of the transformations of the other explanatory variables, which are straight lines, unlike some of the plots in the lower right-hand panel of Fig. 7.11.

7.6 A Graphical Representation of the Systematic Evaluation of Options

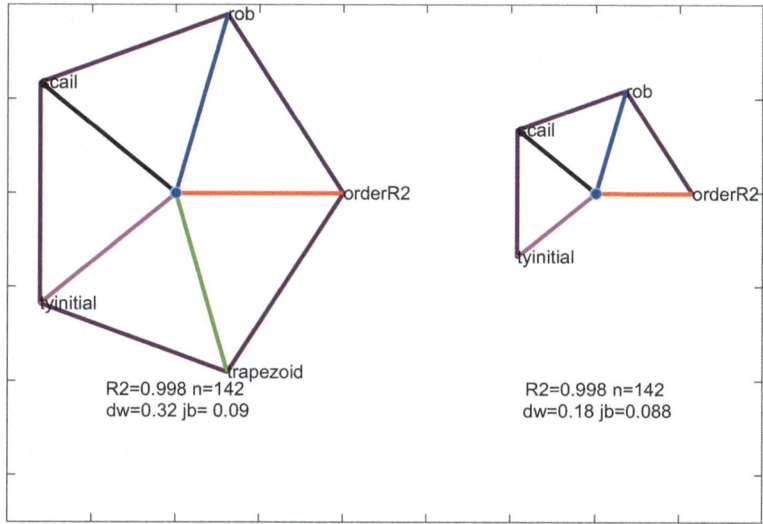

Fig. 7.10 Example 2: augmented star plot of options. The best solution has all five options while the statistically indistinguishable second-best solution excludes the trapezoidal option

Example from Wang and Murphy. Wang and Murphy (2005) used **acepack** (Spector et al. 2024) to illustrate the use of ACE and AVAS in the transformation of data. Their illustrations did not include any robust methods. We now illustrate the use of RAVAS to analyse one of their examples to which we have introduced some outliers. We also introduce an extended version of the augmented star plot which has advantages when, as here, many statistical options provide satisfactory fits to the data. The extension also allows brushing.

There are 200 observations and 4 explanatory variables all independently uniformly distributed on $[-1, 1]$. The linear model is

$$z = 4 + \sin(3x_1) + \text{abs}(x_2) + x_3^2 + x_4 + N(0, 0.1) \tag{7.20}$$

with $y = \log z$. There are 9 outliers each randomly generated as $1.9 + 0.01\mathcal{N}(9, 1)$.

Figure 7.13 shows the six combinations of options which satisfy the constraints on probabilities. As expected, these are plotted in a different order from those in Fig. 7.10. Solution 4 uses only robustness, which is included in all 6 combinations of options. The optional bars in the figure, representing the four properties of the fitted models, have been introduced as being easier to interpret, when the figure displays several solutions, than the list of numerical values, such as those in Fig. 7.10. Since the maximum number of outliers is $[n/2]$ we plot $(n - 2k)/n$, where k is the number of outliers detected when using the particular combination of options. Brushing the figure reveals plots such as those in Figs. 7.14 and 7.15. Details are in Sect. 5 of the supplementary material to Riani et al. (2023b).

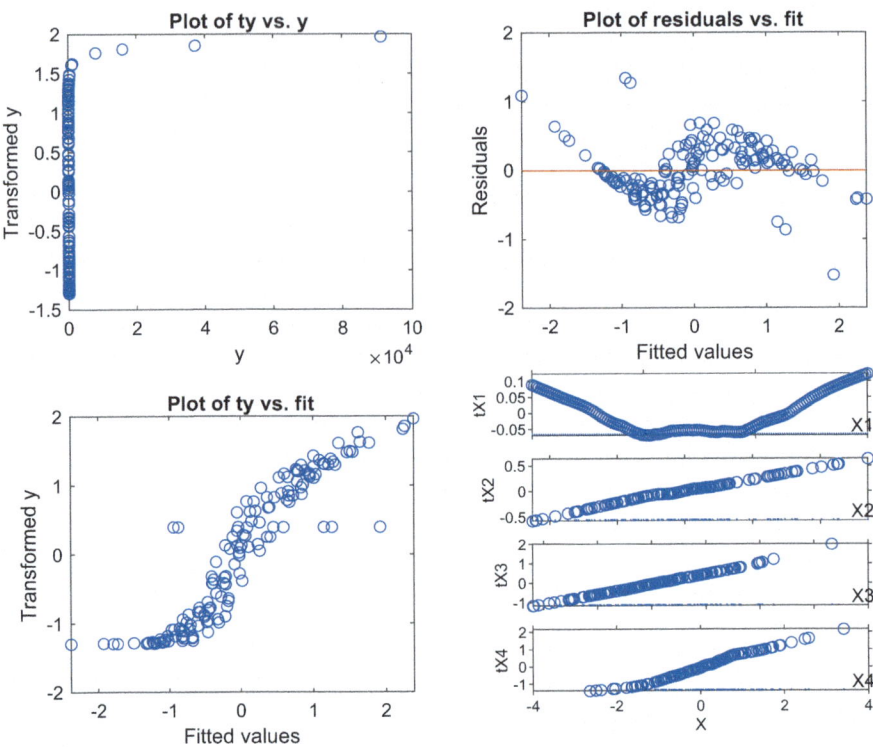

Fig. 7.11 Example 2: non-robust analysis. Upper left-hand panel, transformed y against y; upper right-hand panel, residuals against fitted values; lower left-hand panel, transformed y against fitted values; lower right-hand panel, transformed explanatory variables

The values of R^2 and of p_{JB} do not change as much as those for p_{DW}. The correlations between the pairs of fitted values for the various combinations have a minimum value of 0.996. There is virtually no difference between the fitted models, despite the four selections of options detecting 9 outliers and two detecting 11.

We look at some properties of the best solution from Fig. 7.13, which uses all options apart from the trapezoidal rule. The plot of transformed y against y in the left-hand panel of Fig. 7.14 shows the exponential relationship which is the inverse of the logarithmic transformation used to generate the data. The plot of residuals against fitted values in the right-hand panel shows the lack of relationship between the residuals and the fitted values. The nine outliers are clearly identified, although they are not evident in the scatter plots of the data. The transformations for the four explanatory variables (Fig. 7.15) recover the model used in simulation (7.20). Particularly impressive is the plot for x_2 where the sharpness of the absolute value transformation is achieved by the smoothing algorithm. Similar results are obtained by Wang and Murphy (2005) in the absence of outliers, who continue their analysis to the important step of testing the fit of parametric versions of these non-parametric forms for $f_j(x_j)$.

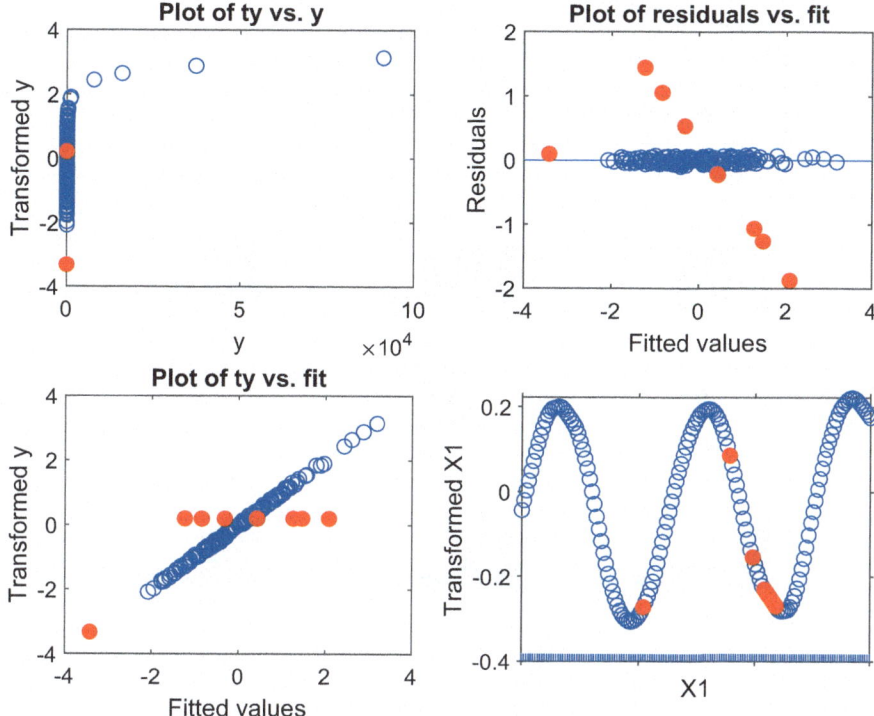

Fig. 7.12 Example 2: robust analysis with all options. Upper left-hand panel, transformed y against y; upper right-hand panel, residuals against fitted values; lower left-hand panel, transformed y against fitted values; lower right-hand panel, transformed explanatory variable x_1. Nine outliers (2 superimposed for fitted values close to 0.43) shown by red filled symbols

7.7 Prediction of the Weight of Fish

Two websites, https://jse.amstat.org/datasets/fishcatch.dat.txt and http://jse.amstat.org/datasets/fishcatch.txt present data on the weight of 159 fish caught in a lake near Tampere, Finland. Interest is in the relationship between weight and five measurements of dimensions of the fish. There are 7 species of fish including pike. These behave rather differently from the other six species so we ignore them. We use the first three lengths for which the remaining fish seem homogenous. This assumption will be tested by our robust analysis if one or more species are identified as outliers. The variables are:

- y Weight of the fish (in grams)
- x_1 Length from the nose to the beginning of the tail (in cm)
- x_2 Length from the nose to the notch of the tail (in cm)
- x_3 Length from the nose to the end of the tail (in cm).

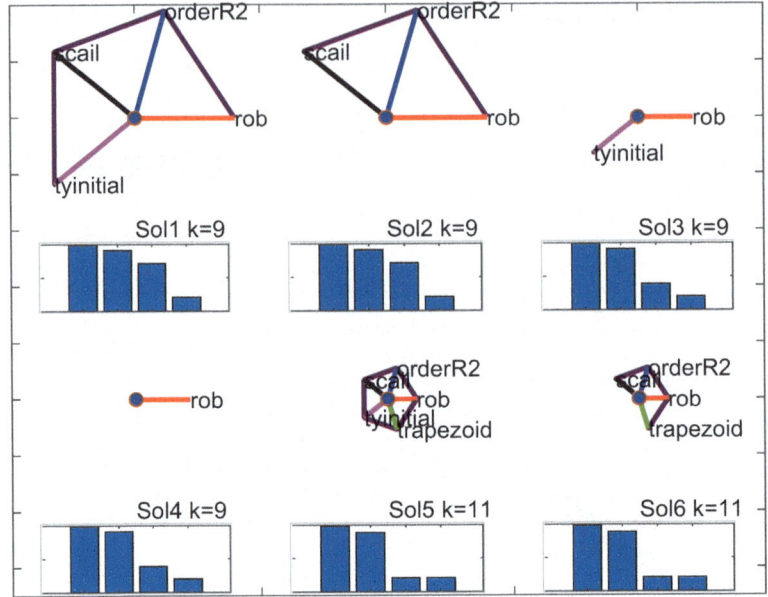

Fig. 7.13 Example from Wang and Murphy: extended augmented star plot of six options. Solution 4 employs only robustness, which is included in all options. The solutions are statistically indistinguishable. The number of outliers detected is k. From left to right the bars give values of R^2, $(n - 2k)/n$, p_{DW} and p_{JB}

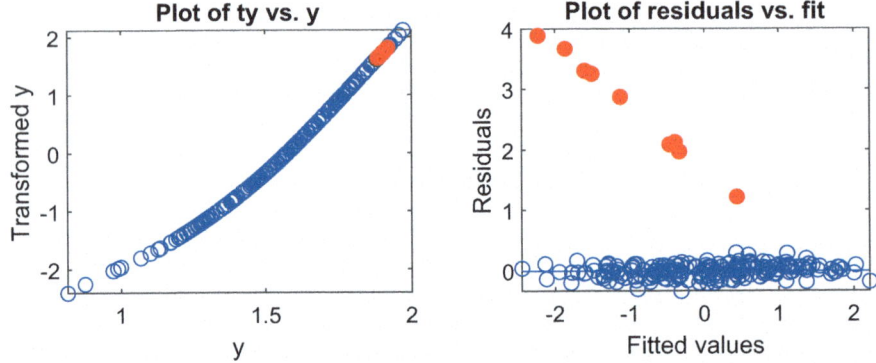

Fig. 7.14 Example from Wang and Murphy: left-hand panel, transformed y against y; right-hand panel, residuals against fitted values. Nine outliers shown by red filled symbols

After the deletion of the data on pike, 142 observations remain. Scatter plots of the response against the three explanatory variables reveal that all three lengths are highly correlated with the response, as they are with each other. It is reasonable to assume that weight increases with each of the explanatory variables. We therefore impose a monotonicity constraint on the transformations of the x_j. However, multi-

7.7 Prediction of the Weight of Fish

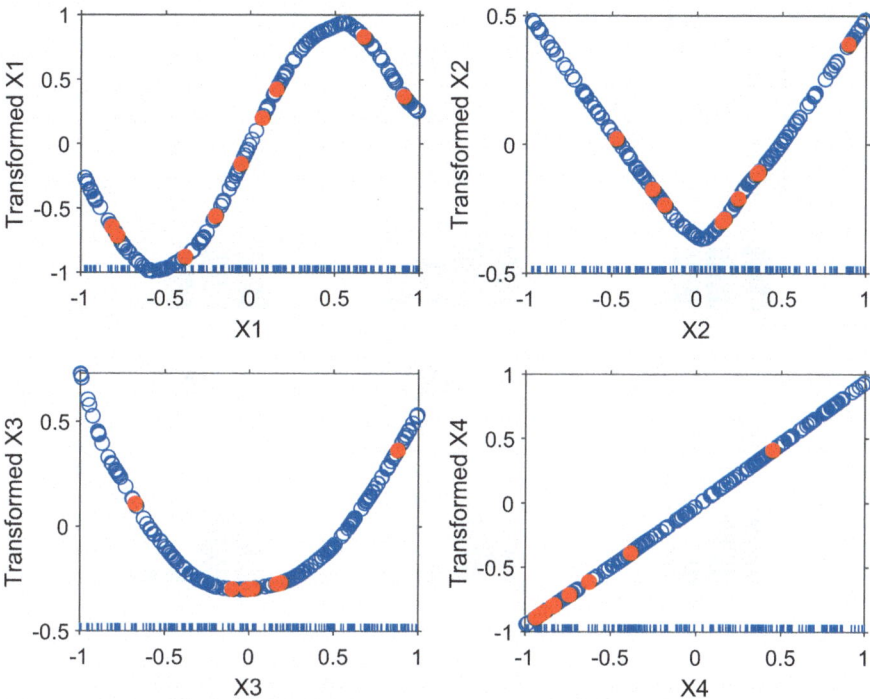

Fig. 7.15 Example from Wang and Murphy: transformations of the four explanatory variables. Nine outliers plotted as filled symbols

ple regression with highly correlated explanatory variables can lead to problems in interpretation, such as estimated effects having a physically incorrect sign.

The augmented star plot for these data is in Fig. 7.16. There are six combinations of options that satisfy the constraints on the distribution of residuals. The first solution, with an R^2 of 0.991 uses all five options except trapezoid. Robustness is used in all, succeeding selections giving R^2 values of 0.988 or 0.983.

The heatmap of the response correlations between the pairs of solutions is in Fig. 7.17. This shows that the first three solutions are strongly correlated with each other, as they are with the fifth and sixth solutions, the fifth and sixth solutions themselves having a very high correlation of 0.998. The heatmap emphasizes that solution four is appreciably different from the other five.

Due to the high correlation between the three explanatory variables, the order of the solutions from repeated analyses varies slightly, with no significant differences in the properties of the fitted models. However, the best solution in Fig. 7.16 is always obtained.

We now consider the identification of outliers using the FS. The first solution identifies three outliers. The left-hand panel of Fig. 7.18 shows that the response has been smoothly transformed. The plot of residuals against fitted values in the right-

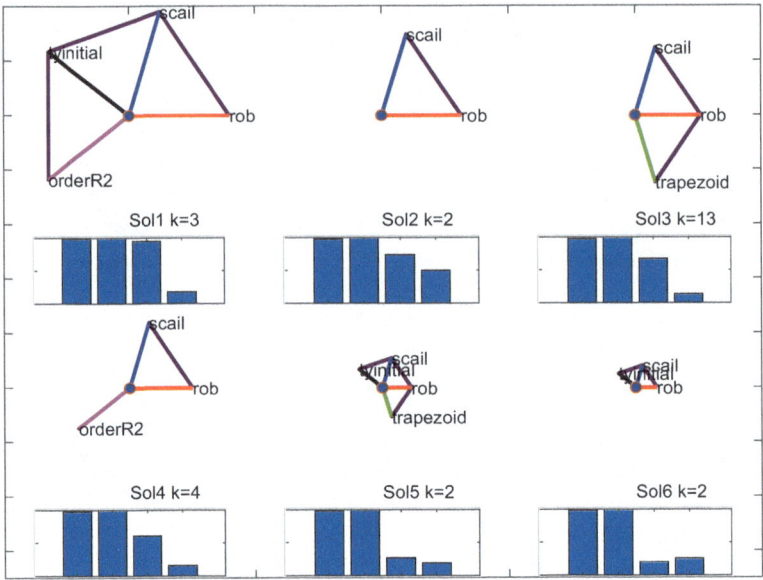

Fig. 7.16 Weight of fish: augmented star plot of six options. Option 1 excludes trapezoid

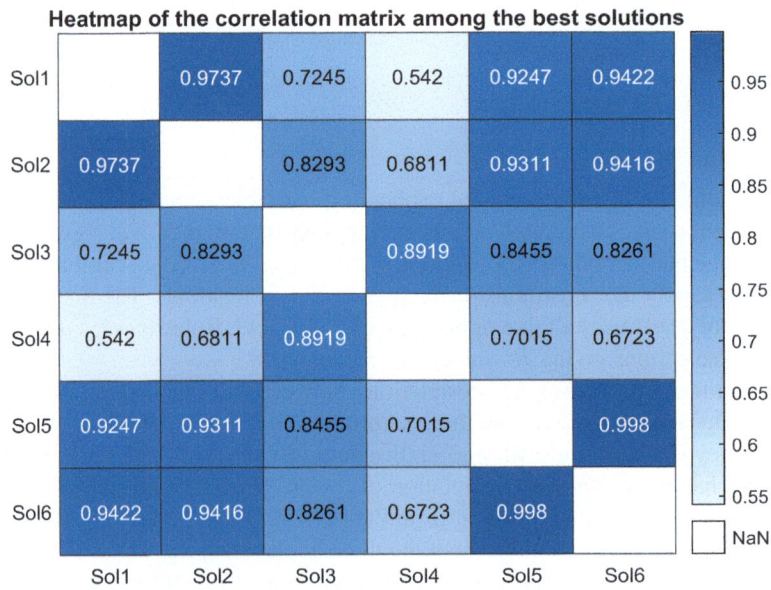

Fig. 7.17 Weight of fish: heatmap of pairwise response correlations among the six solutions

hand panel shows that there is only one remote outlier among the three and that there is no remaining structure in the residuals. The plots of transformed explanatory

7.7 Prediction of the Weight of Fish

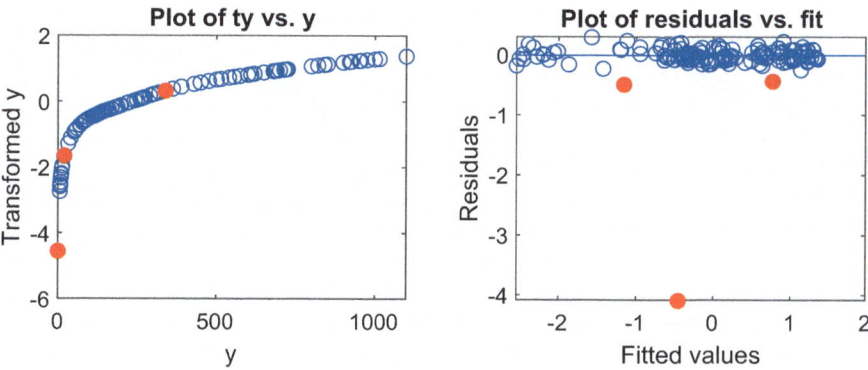

Fig. 7.18 Weight of fish: left-hand panel, transformed y against y; right-hand panel, residuals against fitted values. Three outliers are shown with red filled symbols

variables (not given here) show that $f(x_1)$ is decreasing and slightly curved. The other two functions are increasing, but only that for x_2 is almost straight, with slight curvature for the lowest values of the variable.

The interpretation of the results from fitting three explanatory variables is that the variables are too highly correlated to give individually meaningful results. In our final analysis of the data we use only x_1. The augmented star plot shows that the best selection included all options, except `orderR2`, which option is not possible with a single explanatory variable. The value of R^2 for this fit is 0.980 with the deletion of a single outlier. The three acceptable solutions had mutual fitted response correlations of 0.9994 or 1—the fitted model was stable to the choice of options.

In regressing volume on measurements of length, arguments from dimensional analysis suggest that volume should have a one-third transformation. An example is the analysis of data on the volume of shortleaf pine trees Sect. 6.5.2.3. Our final plot, Fig. 7.19, compares the transformed responses from the fits with three and one explanatory variables to $y^{1/3}$ (red continuous line), for which transformation the value of $R^2 = 0.968$. The transformed data are plotted as circles and the power transformation as a continuous line. The figure shows that both non-parametric transformations are indeed close to $y^{1/3}$ with a small systematic departure for the largest values of x. The fitted values from the single explanatory variable follow the power transformation slightly more closely than that when three variables are used. The transformation of x_1 is virtually straight with some curvature for large values. The flexibility of the non-parametric transformation provides an improved simple model compared with regression on untransformed x_1.

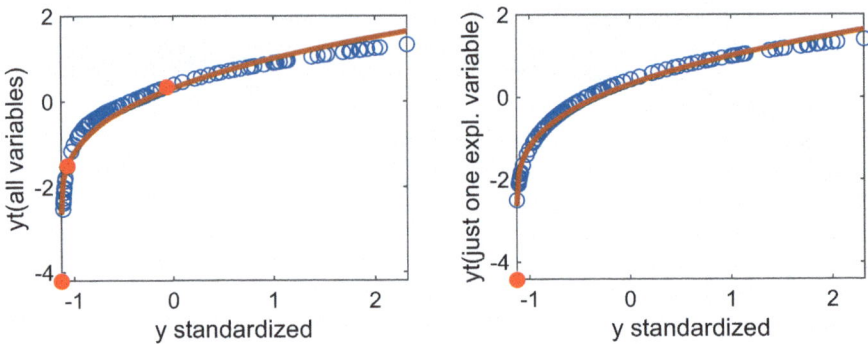

Fig. 7.19 Weight of fish: non-parametric transformation of response y compared to $y^{1/3}$. Left-hand panel, three explanatory variables: right-hand panel, only x_1. Transformed data are plotted as circles, with $y^{1/3}$ as a red continuous line. Three and one outliers are shown with red filled symbols

7.8 Internet Marketing Data

The R package **datarium** (Kassambara 2019) contains 200 results of an experiment on internet marketing. The explanatory variables $x_1 - x_3$ are the budgets for advertising on YouTube, Facebook, and newspapers and the response is sales. These data starkly show the gains in data analysis that come from using RAVAS with options rather than AVAS.

We start with linear regression on the three explanatory variables, which gives an F statistic for regression of 504 and a t statistic for x_3 of 0.658. That newspapers advertising has no effect and that there is a strong effect of advertising in the two online media is a constant pattern in all our analyses. We gain more insight into the structure of the data by using AVAS. It makes sense to assume that increasing advertising expenditures increase sales, so we include a monotonicity constraint on the transformations of the explanatory variables.

The resulting regression gives an F-value of 472 (even smaller than that for linear regression). Figure 7.20 shows that there is a strongly non-random curved pattern in the plot of residuals against fitted values, despite the smooth transformation of the response. Use of RAVAS with all options leads to the deletion of 21 observations giving an F value of 936. Despite the number of observations deleted, the residuals fail both tests—with significance levels $p_{DW} = 1.3235 \times 10^{-9}$ and $p_{JB} = 0.0045$.

Figure 7.21 shows the plot of residuals against fitted values in the RAVAS analysis with all options. This figure shows that the residuals are not yet free of structure. The deletion of the large number of outliers has reduced the curved pattern evident in Fig. 7.20. The deletion of 21 outliers has also produced relatively smooth curves in the other panels, except for the transformation of x_3 which has a bizarre stepped shape. Since this variable is not significant, the pattern can be ignored.

The curved pattern that remains in the plot of residuals against fitted values, ignoring outliers, suggests that a second-order model might be appropriate. We drop

7.8 Internet Marketing Data

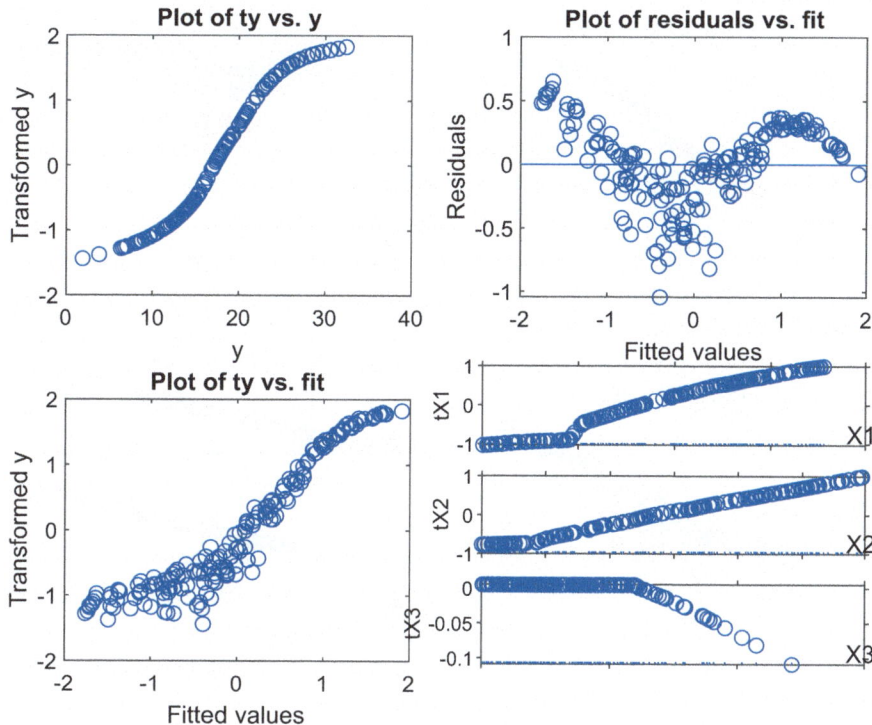

Fig. 7.20 Internet Marketing data: non-robust analysis without options (AVAS). Top left-hand panel, transformed y against y; top right-hand panel, residuals against fitted values; bottom left-hand panel, transformed y against fitted values; bottom right-hand panel, transformations of the three explanatory variables

x_3 from the model and fit a full second-order model in x_1 and x_2, including quadratic and interaction terms. The augmented star plot, Fig. 7.22, shows that the preferred model includes the options `tyinitial, rob` and `scail`. The significance levels of the two tests are $p_{DW} = 0.98$ and $p_{JB} = 0.033$. This fit detects one outlier and produces an F-statistic for regression of 5579, a more than ten-fold increase from our first fit using AVAS. The results in Fig. 7.23 show a smooth transformation of the response and a linear relationship between the transformed response and fitted values, once the outlier is ignored. The transformed explanatory variables all have smoothly changing shapes and all are significant; the least significant t-statistic is -3.01.

To conclude our analysis we look at the results of fitting (non-robust) AVAS with a quadratic model in two variables, that is, excluding x_3. The results, compared with RAVAS, which led to the deletion of one observation, are amazing. There are, of course, no deleted outliers. The F-statistic now has a value of 556, hardly better than regression on the first-order model without transformation of the response or of the explanatory variables. The plot (not given here) of tXj against Xj shows that there

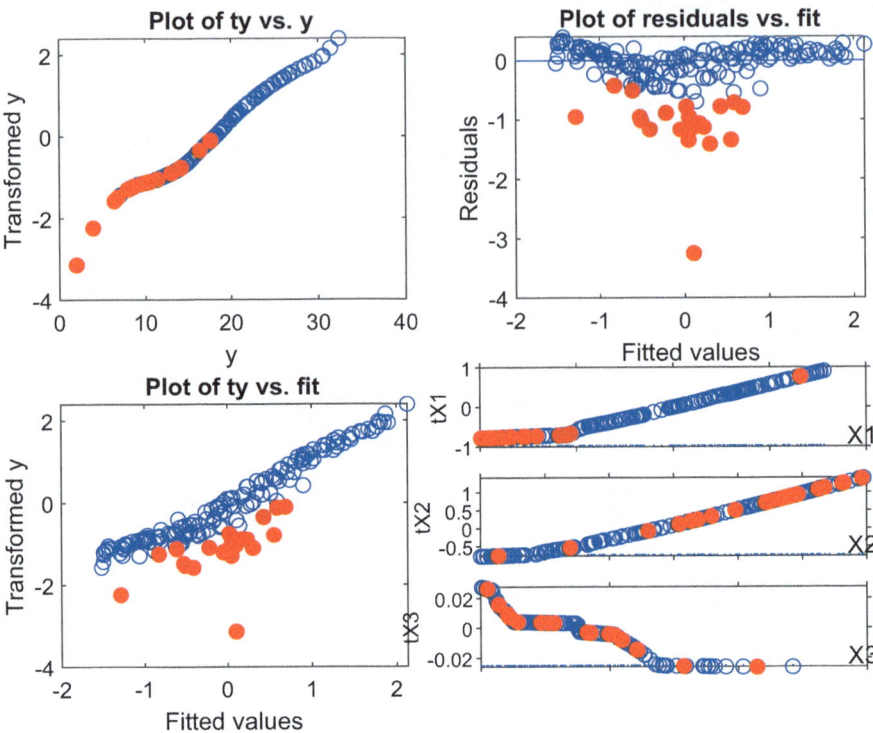

Fig. 7.21 Internet Marketing data: RAVAS analysis with all options. Top left-hand panel, transformed y against y; top right-hand panel, residuals against fitted values; bottom left-hand panel, transformed y against fitted values; bottom right-hand panel, transformations of the three explanatory variables. The 21 outliers shown by red filled symbols

Fig. 7.22 Internet Marketing data: quadratic model in x_1 and x_2. Augmented star plot with five potential options

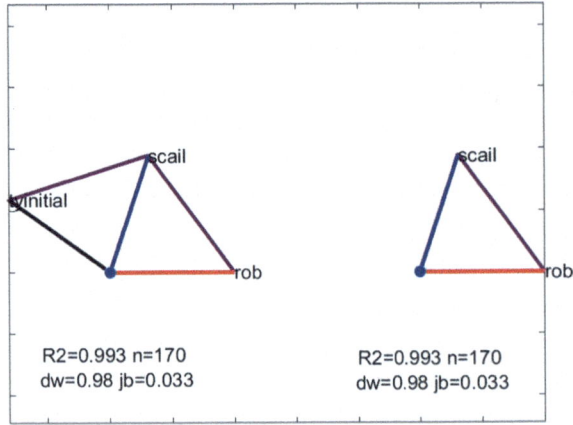

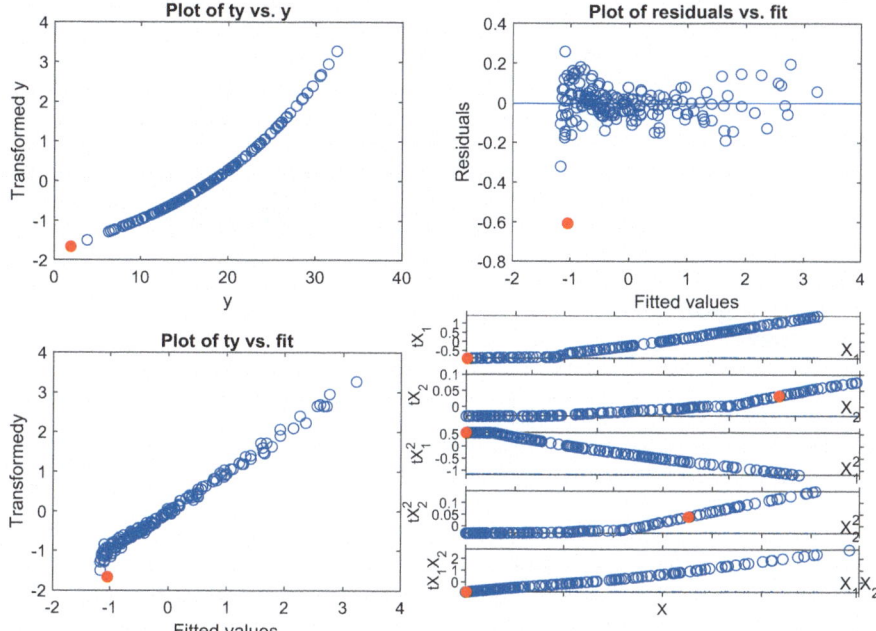

Fig. 7.23 Internet Marketing data: quadratic model. Analysis with options `tyinitial`, `rob` and `scail`. Top left-hand panel, transformed y against y; top right-hand panel, residuals against fitted values; bottom left-hand panel transformed y against fitted values; bottom right-hand panel, transformations of the five explanatory variables. The outlier is shown by a red filled symbol

are smooth transformations of all five explanatory variables. However, the plot of the transformed response against y (left-hand panel of Fig. 7.24) includes an unexpected kink. The plot of transformed y against fitted values (again not shown) is not a straight line but is lumpy for low values. Most surprisingly, the plot of residuals against fitted values (right-hand panel of Fig. 7.24), shows the inexplicable diagonal structures that we noticed in Figs. 7.7 and 7.11. Such structures were the original stimulus for our attempts to provide a reliable version of AVAS that produces interpretable data analyses.

7.9 A Statistical Comparison of AVAS and RAVAS

We now use simulations to compare some overall properties of AVAS and RAVAS. The model is linear regression with data generated to have an average value of R^2 of 0.8. The responses were standardized to have zero mean and unit variance; 10% of the observations were contaminated by a shift Δ and the responses were exponentiated. There were 1000 simulations for $n = 200$ and $n = 1000$ and 200 for $n = 10{,}000$.

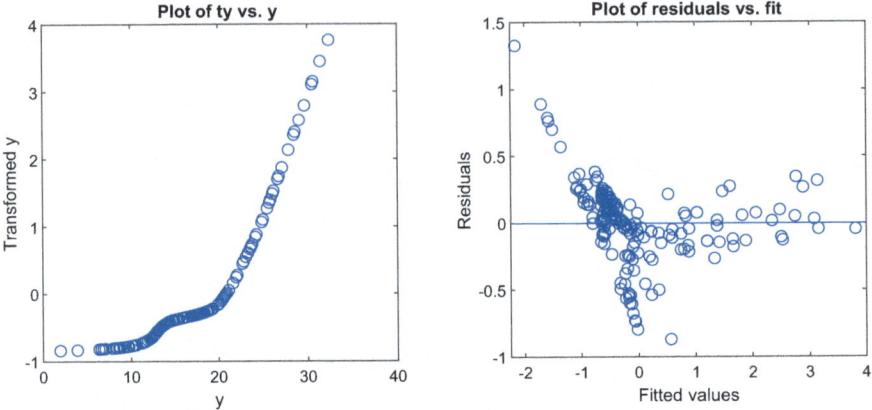

Fig. 7.24 Internet Marketing data: quadratic model. Standard non-robust analysis without options (AVAS). Left-hand panel, transformed y against y; right-hand panel, residuals against fitted values

We encountered no numerical problems in the simulations. The figures compare the performances of RAVAS (with all options) and standard AVAS (with no options). Results for RAVAS use a red dashed line.

The left-hand panels of Fig. 7.25 show the mean squared error of the parameter estimates in the linear models. For RAVAS, those for $n = 200$ and 1000 exhibit a slight increase for moderately small values of Δ which then decreases to be close to zero as Δ increases and the outliers become easier to detect. That for $n = 10,000$ is virtually constant. The results for AVAS rapidly become much larger. The right-hand column shows the average power, that is the proportion of generated outliers that are detected by RAVAS. This climbs, in all cases, steadily to one. Of course, AVAS does not detect outliers.

We also compared the number of iterations to convergence of the algorithms; the default maximum is 20. Figure 7.26 shows results for the same simulations as above. The three panels show that RAVAS converges in around 3 iterations, except for $n = 200$ when there is a peak around $\Delta = 2$, that is when the outliers are large enough to have an effect but are still difficult to detect. This behaviour is distinct from that of AVAS, where the number of iterations increases steadily both with Δ and with the sample size.

7.10 Discussion

Our examples in this chapter illustrate the excellent properties of our robust procedure. Even if robustness is not of interest, the results of Sect. 7.5.2 show how we improved on the performance of Tibshirani's algorithm. Harrell Jr (2024) claims that AVAS is unstable unless the sample size exceeds 350. We used simulation to explore

7.11 Further Reading

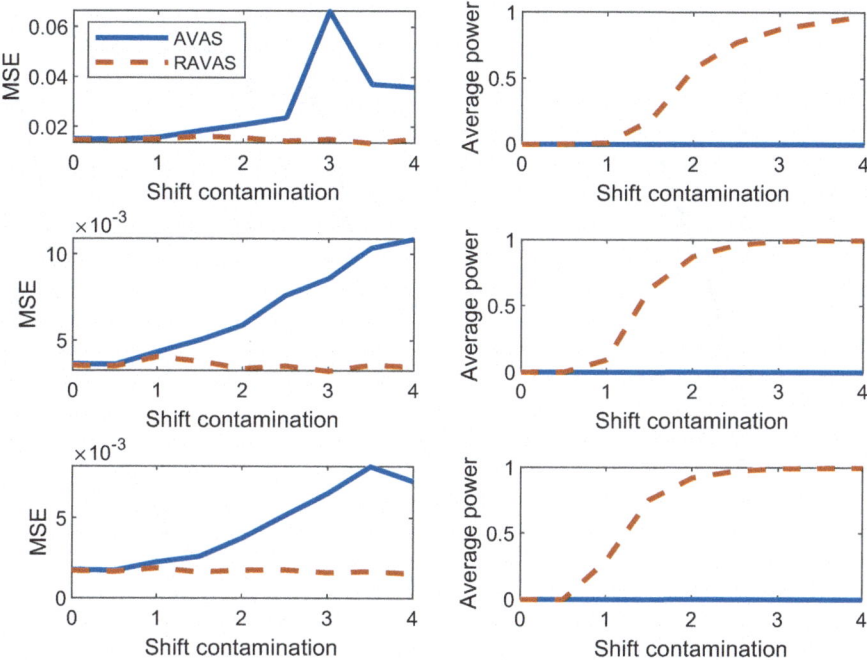

Fig. 7.25 Comparison of AVAS and RAVAS: mean squared error (MSE) and average power. Top panels $n = 200$, $p = 5$, mid panels $n = 1000$, $p = 10$, bottom panels $n = 10{,}000$, $p = 20$

the properties of our method, comparing RAVAS with traditional AVAS, that is with no options. We found no numerical instability in either procedure for sample sizes from 200 to 10,000 and values of p from 5 to 20 with 10% additive outliers. The mean squared error of parameter estimates from RAVAS remained virtually constant as the outliers became more remote, whereas that for AVAS increased steadily. The test for the detection of outliers using RAVAS tended to power one as the outliers became more remote. The average number of iterations for AVAS increased with sample size, and outlier remoteness, reaching the allowed upper limit of 20. The average number of iterations for RAVAS only rose above three for small sample sizes and moderate shift contamination, the situation in which outliers cause significant bias in parameter estimates, but are hard to detect.

7.11 Further Reading

AVAS shares with ACE (Breiman and Friedman 1985) the use of backfitting and smoothing for non-robust data analysis. Although Breiman and Friedman present ACE as a method for transformations in multiple regression and correlation, it has

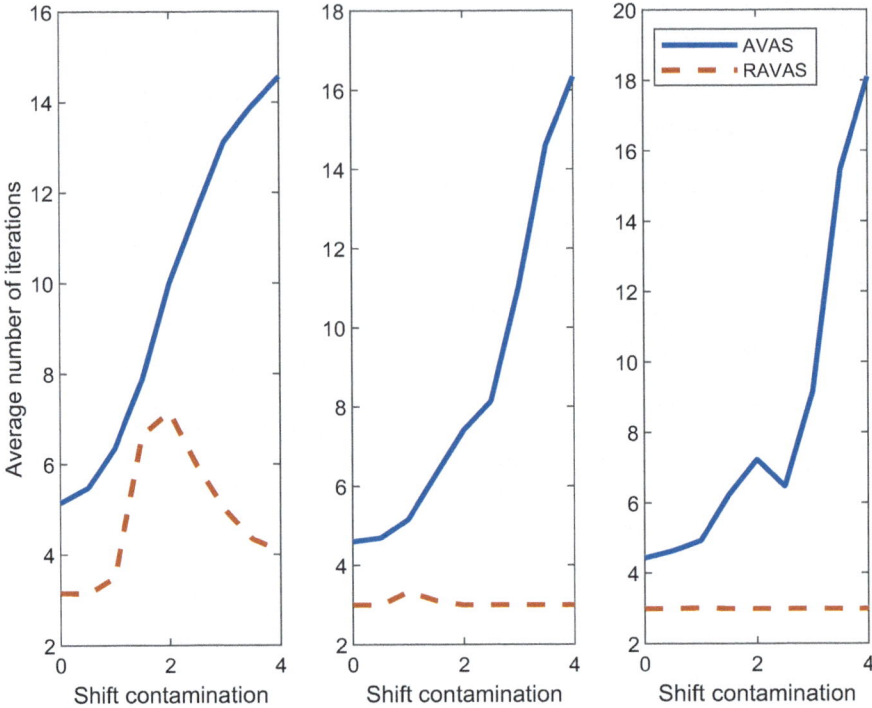

Fig. 7.26 Comparison of AVAS and RAVAS: average number of iterations to convergence. Left-hand panel $n = 200$, $p = 5$, central panel $n = 1000$, $p = 10$, right-hand panel $n = 10{,}000$, $p = 20$

some anomalous properties for estimation of response transformations, as noted by Tibshirani (1988). A discussion of the relationship of ACE and AVAS and of both to the Box-Cox transformation is in Hastie and Tibshirani (1990, Cap.7). Buja and Kass (1985) suggest a method for the robustification of ACE, but comment that "it is far easier to make this glib remark than to formulate the problem in such a way that progress [can] be made while retaining such advantages of the current ACE algorithm as the low computational cost". The same remark applies to AVAS. We believe we have provided a robust version of AVAS in the spirit of this remark, that is a procedure that retains the computational simplicity of the original AVAS. The application of our methods to robustifying ACE would follow directly from the results presented here.

Both ACE and AVAS are appreciably cited and applied and both are available, for example, in the R package **acepack** (Spector et al. 2024). For AVAS, this incorporates Tibshirani's Fortran. What is surprising is the paucity of references for improving the procedure. Foi (2008) comments that AVAS is non-asymptotic. Both he and Wang et al. (2021) use function minimization for numerical variance stabilization in an image-enhancement problem in which the errors have a Poisson-Normal distribution. Neither author nor indeed Tibshirani, mentions robustness. However, Boente et al. (2017) do provide a numerical procedure for robust backfitting in the GAM. Alternatives to AVAS include Ramsay (1988) who uses a monotone spline to estimate the response transformation, with the regression parameters estimated by least squares. An advantage is that the Jacobian of the response transformation is found by straightforward differentiation of the spline. Ramsey also uses monotone splines for transformation of the explanatory variables. Breiman (1988) queries the restriction to monotonicity for the transformation of the explanatory variables, but commends the use of splines. Subroutine *transace* of the R package **Hmisc** (Harrell Jr 2024) replaces the supersmoother in ACE and AVAS with restricted cubic smoothing splines, with a controllable number of knots.

The desirable property of AVAS, and of our robust version, is that they provide the flexibility of non-parametric models for both the response and the explanatory variables. Most of the large literature on generalized linear models (GLMs) and GAMs use smoothing or spline methods on either the response or the explanatory variables, but not both. An exception is Spiegel et al. (2019) who use spline smoothing both for the link function in a GLM and for the transformation of the explanatory variables in the GAM. Robustness is introduced to fitting the GAM by Alimadad and Salibian-Barrera (2011) who use a soft-trimming method, combined with spline smoothing. Their results extend those of Cantoni and Ronchetti (2001) for the robust fitting of GLMs. Boente et al. (2017) formulate the numerical procedure for robust backfitting in the GAM using polynomial smoothers.

Problems

7.1 New updated transformed values in AVAS

The data shown in Fig. 7.27 have been generated with ten equally spaced x coordinates 0, $(\pi/9.5)$, $9\pi/9.5$. The response is $y = \exp\{\sin(x)\}$. The first iteration in the AVAS algorithm (after calling the backfitting algorithm) produces the values for $\widehat{ty}_{[i]}^{(1)}$, $i = 1, 2, \ldots, 10$ and corresponding v_i given in Table 7.1. Compute:

(a) The initial values of $\widehat{ty}$ (namely $\widehat{ty}^{(0)}$).
(b) Show the detailed calculations of the updated value of $\widehat{ty}_1^{(1)}$ using option trapezoid = false and then with option trapezoid = true.
(c) The updated values of $\widehat{ty}$ (namely $\widehat{ty}^{(1)}$) using option trapezoid = false.
(d) The updated values of $\widehat{ty}$ using option trapezoid = true.

Hint: see function ctsub inside **FSDA**.

7.2 Augmented Investment Funds data

Dataset InvFundExt in Fig. 7.28 is an extended version of the dataset analysed in Sect. 6.2.5. It contains data on the relationship between the medium term performance of 349 investment funds and two indicators.

Fig. 7.27 Simulated data for Exercise 7.1

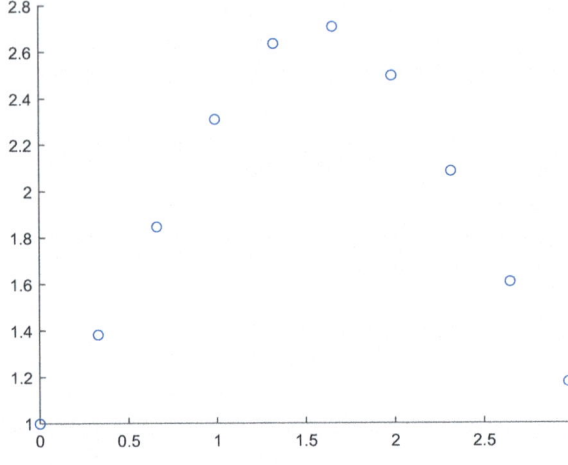

Table 7.1 Values for $\widehat{ty}_{[i]}^{(1)}$, $i = 1, 2, \ldots, 10$ and corresponding v_i in the first iteration of the AVAS algorithm using the data shown in Fig. 7.27

$\widehat{ty}_{[i]}^{(1)}$	v_i
−1.2401	2.1869
−0.8176	2.8108
−0.6309	2.9105
−0.2837	5.1341
−0.0216	11.1385
0.2502	12.5479
0.4740	8.0006
0.6507	3.2393
0.7786	2.3110
0.8404	2.3080

1. Compute the fan plot and the extended fan plot. Find the best value of the parameter λ. Discuss whether it is necessary to have two values of λ and, if this is the case, find λ_P and λ_N. Comment on the heatmap for λ_P and λ_N and also for the AGI index.
2. Perform automatic outlier detection on the transformed scale and show the yX plot in the transformed scale.
3. Compare the QQ plots, with envelopes, of the original untransformed data with those of the clean transformed data and comment on the distribution of residuals in the transformed scale.
4. Compare the values of R^2 when the untransformed data are modelled using EYJ, AVAS (without options), ACE, and ACE imposing the monotonicity constraint, with those coming from transformed and cleaned data using EYJ. These comparisons do not transform the explanatory variables.

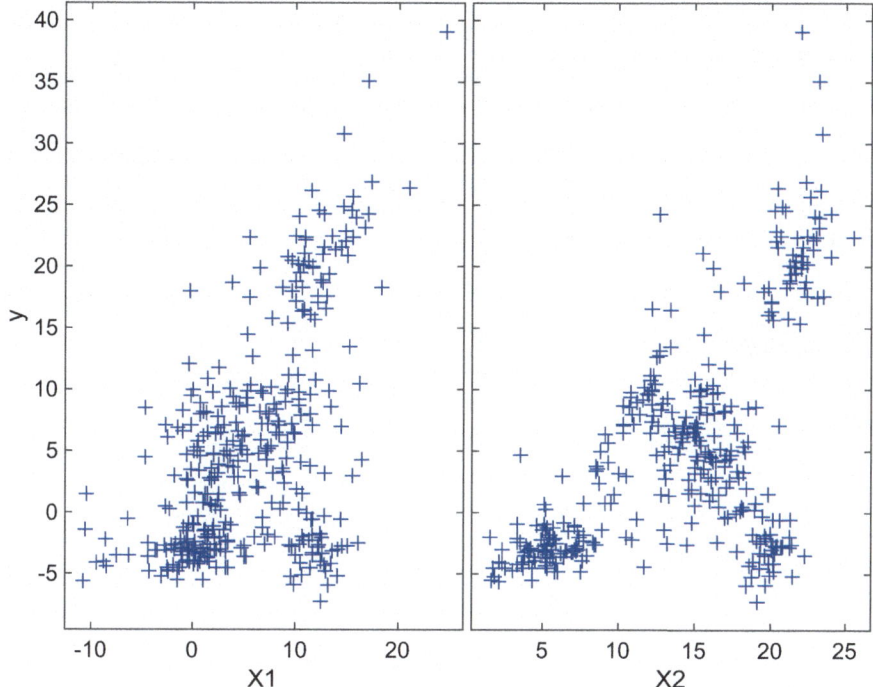

Fig. 7.28 Augmented Investment Funds data: yX plot

5. Provide plots of transformed against untransformed responses for the above four transformation methods. Are there inflection points?
6. Comment on the residuals from ACE imposing the monotonicity constraint and those from AVAS (without options) obtained in 4.
7. Perform RAVAS on the original data and compare the output of RAVAS with the previous results. Compute the ANOVA table after removing the outliers in the transformed RAVAS space and comment.

Open Access This chapter is licensed under the terms of the Creative Commons Attribution 4.0 International License (http://creativecommons.org/licenses/by/4.0/), which permits use, sharing, adaptation, distribution and reproduction in any medium or format, as long as you give appropriate credit to the original author(s) and the source, provide a link to the Creative Commons license and indicate if changes were made.

The images or other third party material in this chapter are included in the chapter's Creative Commons license, unless indicated otherwise in a credit line to the material. If material is not included in the chapter's Creative Commons license and your intended use is not permitted by statutory regulation or exceeds the permitted use, you will need to obtain permission directly from the copyright holder.

Chapter 8
Extensions of the Multiple Regression Model

Abstract The chapter explores a series of extensions to the regression model considered in earlier chapters. Section 8.1 introduces the use of prior knowledge in regression analysis, including the construction of prior distributions from fictitious observations. In Sect. 8.2 Bayesian regression is combined with the FS. Section 8.3 provides analyses of successive annual sets of trade data, in which the prior distribution is updated annually. Heteroskedastic regression is introduced in Sect. 8.4. The analysis of trade data makes clear the importance of avoiding models in which the variance goes to zero as x does. Section 8.5 extends the analysis of trade data to data from several regression hyperplanes. The number of groups is estimated in Sect. 8.5.2 from an FS which starts the search many times from random points. Regression clustering (Sect. 8.5.3) involves the choice of parameters and, for robustness, a choice of trimming level. The monitoring approach in Sect. 8.5.4 identifies solutions which do not depend on arbitrary choices of these hyper-parameters. The fourth extension, in Sect. 8.6, is to use monitoring to provide tools for modelling short-term economic time series that may have trends, time varying seasonality, and level shifts. The two final extensions, in Sects. 8.7 and 8.8 are to regression in which the explanatory variables are the components of a composition and to censored regression data.

8.1 Bayesian Analysis and Prior Information

8.1.1 Introduction

All our data analyses so far in this book have ignored any prior information and have considered each set of data in isolation. But sometimes there is prior information about parameter values, perhaps from analyses of similar sets of data, which can enrich the analysis of the data. If the prior data is correct it can sharpen inferences. We continue with linear regression models with independent normal errors and make use of the conjugate prior for this model. That is, the prior and posterior distributions of the parameters of the model have the same distribution but with differing parameter values. In our second example we make use of the helpful formulation of the prior distribution as the summary statistics of a set of n_0 fictitious observations.

Although it is straightforward to introduce prior information into the FS, an interesting technical problem arises in the estimation of the error variance σ^2. Since the sample estimate in the frequentist search comes from a set of order statistics of the residuals, the estimate of σ^2 has to be rescaled. In the Bayesian search we need to combine a prior estimate with one obtained from such a set of order statistics from the subsample of observations. This estimate has likewise to be rescaled before being combined with the prior estimate of σ^2; parameter estimation then uses weighted least squares. A similar calculation could be used to provide versions of the robust estimators described in Chap. 3 that incorporate prior information. Although our focus throughout is on linear regression, our technique of representing prior information by fictitious observations can readily be extended to more complicated models such as those based on ordinal regression described in Croux et al. (2013) or for sparse regression (Hoffmann et al. 2015).

8.1.2 The Normal Inverse-Gamma Prior Distribution

We represent prior information using the conjugate prior for the normal theory regression model leading to a normal prior distribution for β and an inverse-gamma distribution for σ^2.

If the density of the gamma distribution $\mathcal{G}(a, b)$ is written

$$f_{\mathcal{G}}(x, a, b) = \frac{b^a}{\Gamma(a)} x^{a-1} \exp(-bx), \tag{8.1}$$

the distribution has mean a/b and variance a/b^2.

If $X \sim \mathcal{G}(a, b)$, then $Y = 1/X$ has an inverse-gamma distribution $\mathcal{IG}(a, b)$ with density

$$f_{\mathcal{IG}}(x, a, b) = \frac{b^a}{\Gamma(a)} (1/x)^{a+1} \exp(-b/x) \quad (x > 0), \tag{8.2}$$

shape parameter a and scale parameter b. The mean (for $a > 1$) is $b/(a-1)$ and the variance (for $a > 2$) is $b^2/(a-1)^2(a-2)$.

Let the prior values of the parameters specifying the prior distribution be a_0, b_0, β_0 and R. Then the normal-inverse-gamma conjugate family of prior distributions for β and σ^2 has the form

$$f(\beta, \sigma^2) \propto (1/\sigma^2)^{a_0 + 1 + \frac{p}{2}} \exp\left\{-\frac{(\beta - \beta_0)^T R (\beta - \beta_0)}{2\sigma^2} - \frac{b_0}{\sigma^2}\right\}. \tag{8.3}$$

The marginal distribution of σ^2 is $\mathcal{IG}(a_0, b_0)$. Let $\tau = 1/\sigma^2$. Then $f(\tau) \propto \tau^{a_0 - 1} \exp(-b_0 \tau)$, that is $\mathcal{G}(a_0, b_0)$. The prior distribution of β conditional on τ is $\mathcal{N}\{\beta_0, (1/\tau)R^{-1}\}$.

8.1.3 Prior Distribution from Fictitious Observations

The device of fictitious prior observations provides a convenient representation of this conjugate prior information. We follow, for example, Chaloner and Brant (1988), who are interested in outlier detection, and describe the parameter values of these prior distributions in terms of n_0 fictitious observations.

We start with σ^2. Let the estimate of σ^2 from the n_0 fictitious observations be s_0^2 on $\nu_0 = n_0 - p$ degrees of freedom. Then in $f(\tau)$,

$$a_0 = \nu_0/2 = (n_0 - p)/2 \quad \text{and} \quad b_0 = \nu_0 s_0^2/2 = S_0/2, \tag{8.4}$$

where S_0 is the residual sum of squares of the fictitious observations.

Prior information for the linear model is given as the scaled information matrix $R = X_0^T X_0$ and the prior mean $\hat{\beta}_0 = R^{-1} X_0^T y_0$. Then $S_0 = y_0^T y_0 - \hat{\beta}_0^T R \hat{\beta}_0$. Thus, given n_0 prior observations the parameters for the normal inverse-gamma prior may readily be calculated.

8.1.4 Posterior Distributions

The posterior distribution of β conditional on τ is $\mathcal{N}\{\hat{\beta}_1, (1/\tau)(R + X^T X)^{-1}\}$ where

$$\begin{aligned}\hat{\beta}_1 &= (R + X^T X)^{-1}(R\beta_0 + X^T y) \\ &= (R + X^T X)^{-1}(R\beta_0 + X^T X \hat{\beta}) \\ &= (I - A)\beta_0 + A\hat{\beta},\end{aligned} \tag{8.5}$$

and $A = (R + X^T X)^{-1} X^T X$. The last expression shows that the posterior estimate $\hat{\beta}_1$ is a matrix weighted average of the prior mean β_0 and the classical LS estimate $\hat{\beta}$, with weights $I - A$ and A. If prior information is strong, the elements of R will be large, and A will be small, so that the posterior mean gives most weight to the prior mean. In the classical approach these weights are fixed, while with the forward search, as the subset size grows, the weight assigned to A increases with m; we can dynamically see how the estimate changes as the effect of the prior decreases.

The posterior distribution of τ is $\mathcal{G}(a_1, b_1)$ where

$$a_1 = a + n/2 = (n_0 + n - p)/2 \quad \text{and} \tag{8.6}$$

$$b_1 = \left\{(n_0 - p)/\tau_0 + (y - X\beta_1)^T y + (\beta_0 - \beta_1)^T R \beta_0 \right\}/2. \tag{8.7}$$

The posterior distribution of σ^2 is $\mathcal{IG}(a_1, b_1)$. The posterior mean estimates of τ and σ^2 are respectively

$$\tau_1 = a_1/b_1, \quad \text{and} \quad \tilde{\sigma}_1^2 = b_1/(a_1 - 1). \tag{8.8}$$

In our calculations we take $\hat{\sigma}_1^2 = 1/\tau_1$ as the estimate of σ^2. Unless a_1 is very small, the difference between $\hat{\sigma}_1^2$ and $\tilde{\sigma}_1^2$ is negligible.

The posterior marginal distribution of β is multivariate t with parameters

$$\hat{\beta}_1, \ (1/\tau_1)\{R + X^T X\}^{-1}, \ n_0 + n - p. \tag{8.9}$$

8.2 The Bayesian Search

8.2.1 Parameter Estimation

The posterior distributions of Sect. 8.1.4 arise from the combination of n_0 prior observations, perhaps fictitious, and the n actual observations. In the FS we combine the n_0 prior observations with a carefully selected m out of the n observations. The search proceeds from $m = 0$, when the fictitious observations provide the parameter values for all n residuals from the data. It then continues with the fictitious observations always included among those used for parameter estimation; their residuals are ignored in the selection of successive subsets.

As mentioned in Sect. 8.1, there is one complication in this procedure. The n_0 fictitious observations are treated as a sample with population variance σ^2. However, the m observations from the actual data come from a truncated distribution of m out of n observations and so asymptotically have a variance which is given by $\sigma_T^2(m)\sigma^2$ (see Eq. 3.49). An adjustment must be made before the two samples are combined. This becomes a problem in weighted least squares (for example, Rao 1973, p. 230). Let y^+ be the $(n_0 + m) \times 1$ vector of responses from the fictitious observations and the subset, with X^+ the corresponding matrix of explanatory variables. The covariance matrix of the independent observations is $\sigma^2 G$, with G a diagonal matrix; the first n_0 elements of the diagonal of G equal one, and the last m elements have the value $\sigma_T^2(m)$. The information matrix for the $n_0 + m$ observations is

$$(X^{+\mathrm{T}} W X^+)/\sigma^2 = \{X_0^T X_0 + X(m)^T X(m)/\sigma_T^2(m)\}/\sigma^2, \tag{8.10}$$

where $W = G^{-1}$. In the least squares calculations we need only to multiply the elements of the sample values $y(m)$ and $X(m)$ by $\sigma_T(m)^{-1}$. However, care is needed to obtain the correct expressions for leverages and variances of parameter estimates.

Since, during the forward search, n in (8.6) is replaced by the subset size m, X and y in (8.7) become $y(m)/\sigma_T(m)$ and $X(m)/\sigma_T(m)$, giving rise to posterior values $a_1(m)$, $b_1(m)$, $\tau_1(m)$ and $\hat{\sigma}_1^2(m)$.

The estimate of β from n_0 prior observations and m sample observations is, from (8.10),

$$\hat{\beta}_1(m) = (X^{+\mathrm{T}} W X^+)^{-1} X^{+\mathrm{T}} W y^+. \tag{8.11}$$

8.2 The Bayesian Search

In Sect. 8.1.3 $\hat{\beta}_0 = R^{-1} X_0^T y_0$, so that $X_0^T y_0 = X_0^T X_0 \hat{\beta}_0$. Then the estimate in (8.11) can be written in full as

$$\begin{aligned}\hat{\beta}_1(m) &= \{X_0^T X_0 + X(m)^T X(m)/\sigma_T^2(m)\}^{-1} \{X_0^T y_0 + X(m)^T y(m)/\sigma_T^2(m)\} \\ &= \{X_0^T X_0 + X(m)^T X(m)/\sigma_T^2(m)\}^{-1} \{X_0^T X_0 \hat{\beta}_0 + X(m)^T y(m)/\sigma_T^2(m)\}.\end{aligned}$$

8.2.2 Forward Highest Posterior Density Intervals

Inference about the parameters of the regression model comes from regions of highest posterior density. Let

$$V(m) = (X^{+T} X^+)^{-1} = \{X_0^T X_0 + X(m)^T X(m)\}^{-1}, \qquad (8.12)$$

with (j, j)th element $V_{jj}(m)$. Likewise, the jth element of $\hat{\beta}_1(m)$, $j = 1, 2, \ldots, p$ is denoted $\hat{\beta}_{1j}(m)$. Then

$$\operatorname{var} \hat{\beta}_{1j}(m) = \hat{\sigma}_1^2(m) V_{jj}(m). \qquad (8.13)$$

The $(1 - \alpha)\%$ highest posterior density (HPD) interval for β_{1j} is

$$\hat{\beta}_{1j}(m) \pm t_{\nu, 1-\alpha/2} \sqrt{\hat{\sigma}_1^2(m) V_{jj}}, \qquad (8.14)$$

with $t_{\nu,\gamma}$ the $\gamma\%$ point of the t distribution on ν degrees of freedom. Here $\nu = n_0 + m - p$.

The highest posterior density intervals for τ and σ^2 are respectively given by

$$[g_{a_1(m), b_1(m), \alpha/2}, g_{a_1(m), b_1(m), 1-\alpha/2}] \text{ and } [ig_{a_1(m), b_1(m), \alpha/2}, ig_{a_1(m), b_1(m), 1-\alpha/2}], \qquad (8.15)$$

where $g_{a,b,\gamma}$ and $ig_{a,b,\gamma}$ are the $\gamma\%$ points of the $\mathcal{G}(a, b)$ and $\mathcal{IG}(a, b)$ distributions.

8.2.3 Outlier Detection

We detect outliers using a form of deletion residual that includes the prior information. Let $S^*(m)$ be the subset of size m found by FS, for which the matrix of regressors is $X(m)$. Weighted least squares on this subset of observations yields parameter estimates $\hat{\beta}_1(m)$ and $\hat{\sigma}^2(m)$, an estimate of σ^2 on $n_0 + m - p$ degrees of freedom. The residuals for all n observations, including those not in $S^*(m)$, are

$$e_i(m) = y_i - x_i^T \hat{\beta}_1(m) \qquad i = 1, \ldots, n. \qquad (8.16)$$

The search moves forward with the augmented subset $S^*(m+1)$ consisting of the observations with the $m+1$ smallest absolute values of $e_i(m)$. To start we take $m_0 = 0$, since the prior information specifies the values of β and σ^2.

To test for outliers the deletion residuals are calculated for the $n - m$ observations not in $S^*(m)$. These residuals are

$$r_i(m) = \frac{y_i - x_i^T \hat{\beta}_1(m)}{\sqrt{\hat{\sigma}_1^2(m)\{1 + h_i(m)\}}} = \frac{e_i(m)}{\sqrt{\hat{\sigma}_1^2(m)\{1 + h_i(m)\}}}, \qquad (8.17)$$

where, from (8.12), the leverage $h_i(m) = x_i^T \{X_0^T X_0 + X(m)^T X(m)/\sigma_T^2(m)\}^{-1} x_i$. Let the observation nearest to those forming $S^*(m)$ be $i_{\min}$ where

$$i_{\min} = \arg \min_{i \notin S^*(m)} |r_i(m)|.$$

In (4.1) the estimate of σ^2 in the minimum deletion residual is the biased $\hat{\sigma}^2(m)$ and the envelopes for outlier detection were calculated for this estimate. In the Bayesian calculation (8.17) the estimator used is the unbiased $\hat{\sigma}_1^2(m)$. To recalibrate the Bayesian envelopes for testing whether observation $i_{\min}$ is outlying, we now use the absolute value of the minimum deletion residual

$$r_{\mathrm{imin}}(m) = \frac{1}{\sigma_T(m + n_0)} \frac{e_{\mathrm{imin}}(m)}{\sqrt{\hat{\sigma}_1^2(m)\{1 + h_{\mathrm{imin}}(m)\}}}, \qquad (8.18)$$

as a test statistic, where $\sigma_T(m + n_0)$ is the Tallis correction factor (3.49) for an estimate of σ^2 based on a central $m + n_0$ observations out of a total sample of size $n + n_0$. If the absolute value of (8.18) is too large, the observation $i_{\min}$ is considered to be an outlier, as well as all other observations not in $S^*(m)$.

8.3 Data Analyses with Prior Information

8.3.1 Windsor House Price Data

We now use the FS to analyse the House Price data introduced by Anglin and Gençay (1996). A Bayesian analysis of the data is given by Koop (2003, Sect. 3.9). We develop his analysis; the Forward Search shows that his prior is misspecified. We demonstrate the effect of this mis-specification on inference, for the small amount of prior information he provides and for a larger value of n_0. See Box (1980) for a consideration of agreement between prior and data which does not use the Forward Search.

The data are the sales prices of 546 houses in the city of Windsor, Ontario, Canada, during July, August, and September 1987. As well as being available in **FSDA**, they

8.3 Data Analyses with Prior Information

Table 8.1 Windsor House Price data: prior and least squares estimates of parameters

Parameter	Prior		Least squares
	Mean	V_{jj}^0	
β_0	0	2.40	-4009
β_1	10	6.0×10^{-7}	5.43
β_2	5000	0.15	2825
β_3	10,000	0.60	17,105
β_4	10,000	0.60	7635
τ	4×10^{-8}	$n_0 = 6$	3.03×10^{-9}

are also available in the R package **AER** (Kleiber and Zeileis 2008), which gives information on 12 explanatory variables. Here we follow Koop and use the following variables:

y: sales price of house i in Canadian dollars;
x_1: plot size of house i in square feet;
x_2: number of bedrooms;
x_3: number of bathrooms;
x_4: number of storeys.

The prior information used by Koop is summarized in Table 8.1. Column 2 gives the prior means of the parameters, column 3 the prior values of the diagonal elements of the inverse information matrix $V_{jj}^0 = (X_0^T X_0)_{jj}^{-1}$ with the least squares estimates of the parameters in column 4. The information matrix is taken to be diagonal, with $n_0 = 6$, so that there is slight prior information about the mean of σ^2, reflected in the value of 0.5 for a_0. The value of b_0 is 12,500,000.

The table shows differences between the prior means and the least squares estimates of the linear parameters. In addition, the value of τ is too large, providing too small a prior variance for the data. Although prior means and frequentist means for the fitted data need not be in precise agreement, the Bayesian Forward Search reveals the effects of the disagreement between prior specification and observations.

Figure 8.1 shows the forward plot of minimum deletion residuals, a well-behaved curve for the greater part of the data, with a signal for outliers first occurring at $m^\dagger = 511$. Resuperimposition of envelopes to allow for the effect of deletion of observations on the effective sample size leads to identification of 18 outliers, so that the first outlier is at $m = 529$.

Next consider parameter estimation. Figure 8.2 shows forward plots of the 95 and 99% HPD regions for the parameters (given in Sect. 8.2.2) with m going from 20 to 130; the horizontal lines give the prior values of the parameters. There is a similar pattern for all parameters of the linear model; initially the regions are wide because of the small amount of information available on the parameters. Once m is large enough the intervals decrease steadily in width. For smaller values of m than those shown here the width of the intervals is further increased by use of the t distribution

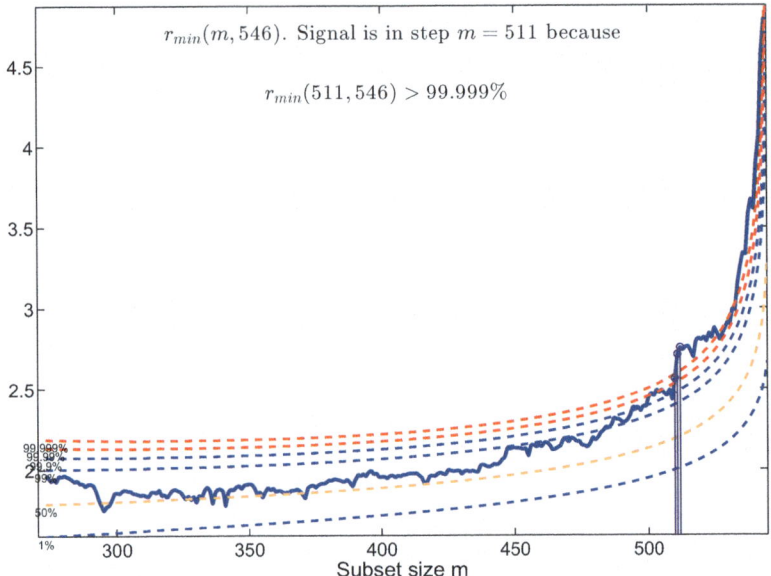

Fig. 8.1 Windsor House Price data: forward plot of absolute Bayesian minimum deletion residuals. There is a signal indicating outliers from $m = 511$. Prior information as in Table 8.1

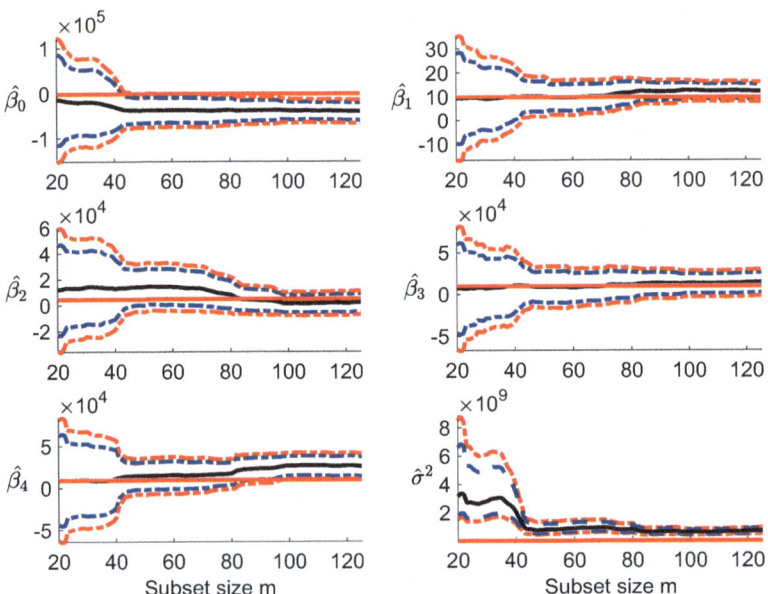

Fig. 8.2 Windsor House Price data: forward plots of 95 and 99% HPD regions for the parameters of the linear model and, bottom right-hand panel, for the estimate of σ^2. The first part of the search from $m = 20$. The horizontal red lines correspond to prior means

8.3 Data Analyses with Prior Information

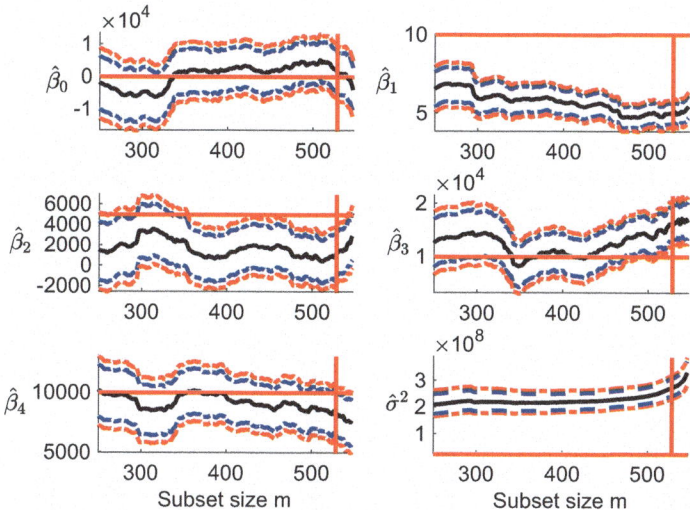

Fig. 8.3 Windsor House Price data: forward plots in the second half of the search of 95 and 99% HPD regions for the parameters of the linear model and, bottom right-hand panel, for the estimate of σ^2. The vertical red line at $m = 529$ indicates the start of the outlying observations. The horizontal red lines correspond to prior means

on few degrees of freedom. The plot of $\hat{\sigma}^2(m)$ in the bottom right-hand panel shows that the estimate remains above and far from its prior mean.

The effect of outliers is evident in Fig. 8.3 which shows the HPD regions for the parameters in the second half of the search. The vertical red line at $m = 529$ indicates the start of the outlying observations. The estimates of β_1, β_3, and β_4 appear to be little affected by the outliers, continuing the previous trend as the number of observations increases. However, the plots show that the values of $\hat{\beta}_0(m)$ and of $\hat{\beta}_2(m)$ do change, the value of $\hat{\beta}_0(m)$ decreasing with m, whereas the estimate of β_2 increases. The values of $\hat{\sigma}^2(m)$ increase sharply towards the end of the search. The horizontal red lines in the panels show the prior values. The departures of the posterior parameter estimates from the prior values is not caused by the outliers.

The yX plot in Fig. 8.4 shows the outliers detected. These observations, with few exceptions, have larger values of y than would be expected. It is clear from the plot of values against x_2 that the outliers, that is the last observations to enter the search, cause an increase in the estimate of β_2, although the correlations between the explanatory variables make further detailed comparisons difficult from these marginal plots.

Two points remain. Figure 8.5 shows the forward plot of the 95 and 99% confidence regions for the parameters from the least squares analysis. Comparison with Fig. 8.3 shows how very similar these confidence intervals are to the HPD bands. This is unsurprising with so little prior information. However, the ordering of the observations by FS is not the same for the two searches, so some differences are to be expected.

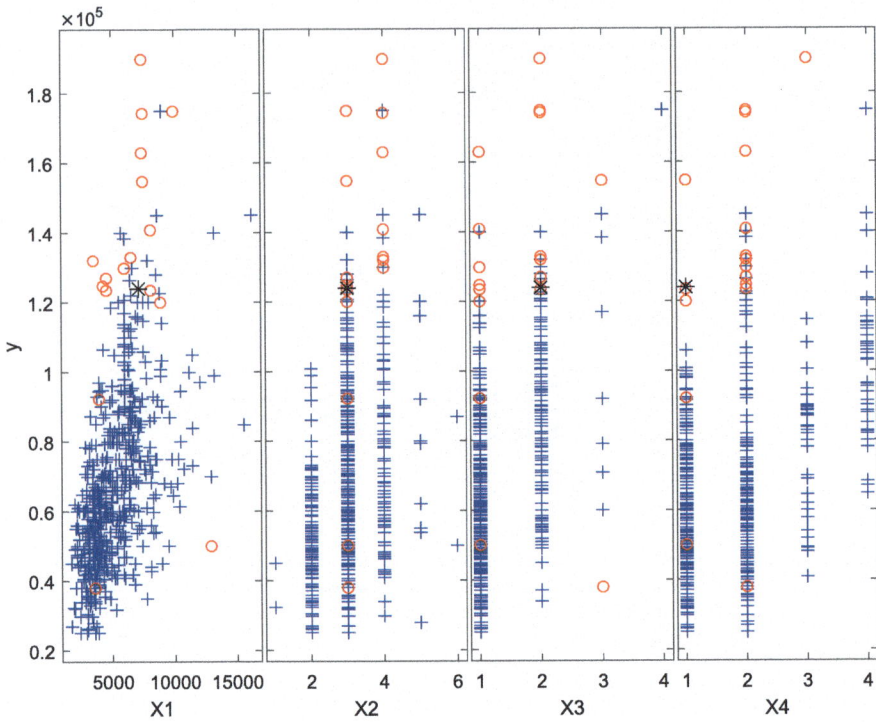

Fig. 8.4 Windsor House Price data: yX plot indicating the outliers found by the Bayesian FS. Outliers ○, other observations +. The outlier marked ∗ (most visible in the panel for x_4) was not detected by the frequentist analysis

The second point is the effect of strong prior information. We leave the prior estimates the same as before, but now take $n_0 = 250$. The result is shown in Fig. 8.6. The effect at the beginning of the search reflects disagreement between the prior estimate of σ^2 and the sample value; see (8.18), which was not evident when $n_0 = 6$. During the search the estimate of σ^2 gradually moves towards the sample value. We have already seen that the observations in the later part of the search are outlying.

Figure 8.7 shows the analysis of the corresponding plot of minimum deletion residual in the absence of any prior information. We leave to Exercise 8.1 comments about the comparison with the Bayesian analysis.

8.3.2 The Importance of International Trade Data

In Sect. 5.8 we used a set of international trade data to compare robust estimators when the data contained two distinct populations rather than a scattering of outliers. In this chapter we analyse four further distinct sets of trade data, but first we provide some background to the structure of such data.

8.3 Data Analyses with Prior Information 291

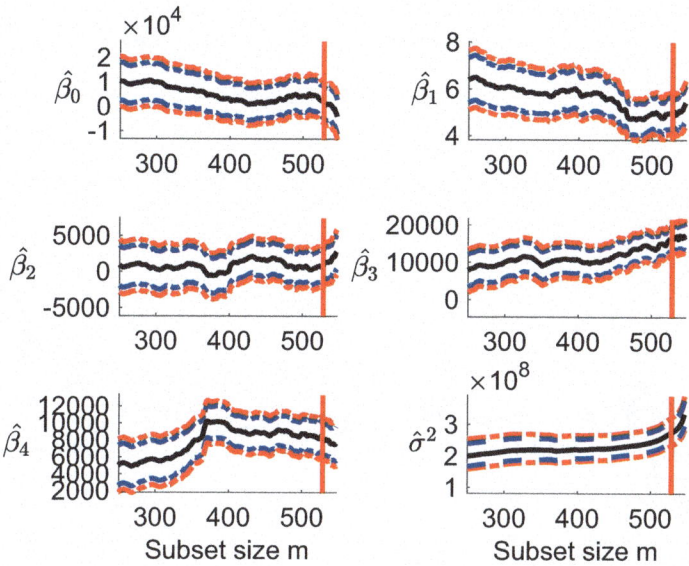

Fig. 8.5 Windsor House Price data: least squares analysis: forward plot of 95 and 99% confidence intervals for parameters of the linear model and, bottom right-hand panel, the estimate of σ^2. The vertical red line at $m = 529$ indicates the start of the outlying observations

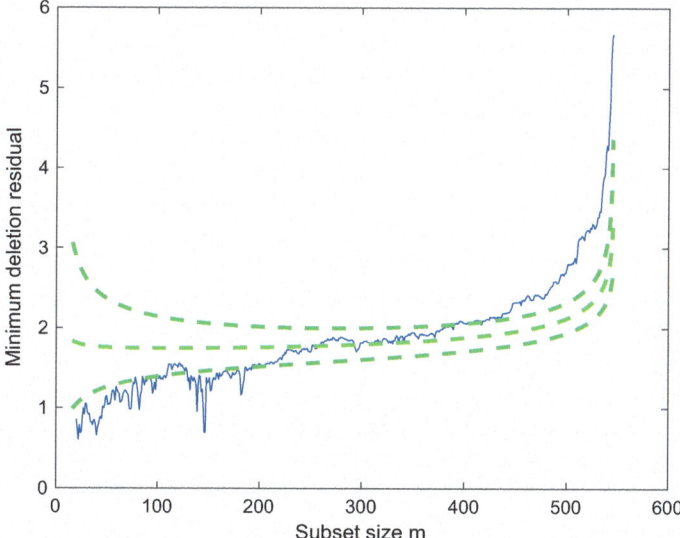

Fig. 8.6 Windsor House Price data: effect of strong prior information. Forward plot of Bayesian minimum deletion residuals when $n_0 = 250$; 1 and 99% envelopes. The effect at the beginning of the search reflects disagreement between the prior estimate of σ^2 and the sample value

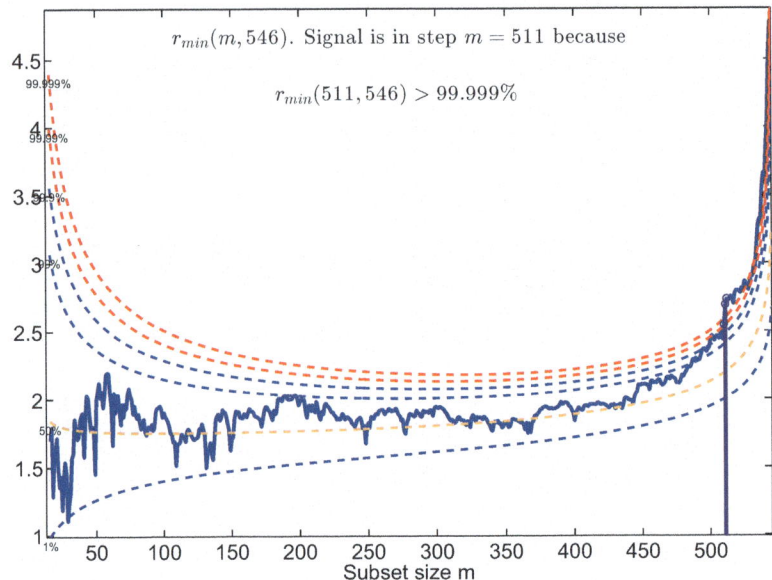

Fig. 8.7 Windsor House Price data: least squares analysis: forward plot of minimum deletion residuals

It was estimated (Economist 2014) that, in 2011, $950 billion flowed illegally out of poor countries into rich ones, mostly owing to money laundering associated with the traffic in illegal drugs and arms trading. The amount will not have decreased over time. A basic technique is misinvoicing. In our examples we look at two sets of data on imports of specific goods into the European Union from specific countries. The problem is vast, with around 220 potential source countries, monthly data, and over 1000 categories of goods (although not all countries are sources of all goods). To cope with this example of big data, robust methods are needed that function semi-automatically on relatively small individual problems (our examples have 3867 and 1302 observations), without the need for close personal intervention. The importance of statistical methods in the solution of these problems in the context of the European Customs Union is attested by the agreement of the Court of Justice of the European Union (2022) that the legal use of statistical evidence is specifically permitted, in certain cases. Such methods should be stable, such as those in this book, which are robust against specific decisions about the nominal level of trimming or a nominal efficiency.

In three out of the four examples, the observations are regression data of value against quantity. There are linear relationships followed by the majority of the data. Typically, there are also a few outliers. In economic theory, the efficient-market hypothesis asserts that, in a well organized, reasonably transparent market, the market price is generally equal to or close to a fair value. Scatter plots of annual data for a specific good sometimes show linear relationships for imports from individual countries. This structure is usually not of interest. What is important are sets of points that are systematically outlying. Systematically low observations are an indication of potential fraud; by incorrectly recording import prices, import duties and taxes such as VAT can partially be avoided. Conversely, in other sets of data, suspiciously high invoice prices allow illicit money to be laundered into legal bank accounts. Such marked departures are an indication of inefficiency in the market and again an indication of fraud. To prosecute such behaviour, it is necessary to demonstrate, as far as possible, the incontrovertible existence of outliers. This is very different from the standard intent of robust data analyses, where the purpose is to establish a single relationship between much of the data and a model; the remaining data being then either downweighted or trimmed. Since law-enforcement resources are limited, the emphasis is on detecting and controlling frauds involving large amounts of money. The "Press Release corner" of the European Anti-fraud Office (accessible at https://anti-fraud.ec.europa.eu/media-corner/news_en) contains several examples of investigations of very large economic impact.

The Lobster data in Sects. 8.3.3, 8.3.4 and 8.3.5 were used in a related analysis where the interest is in the growth over time of the evidence for fraud, which is strengthened through the use of Bayesian robust regression. Two further sets provide examples of heteroskedastic data analysis in Sect. 8.4. Depending on the purpose of the data analysis, modelling a heteroskedastic dataset as a mixture of linear homoskedastic models may be more appropriate. Such robust regression clustering is the subject of Sect. 8.5. The illustration is the classification of anti-covid face masks.

8.3.3 Lobster Data 2002: Least Squares Analysis

As a first specific example of trade data we look at data on three years importation of lobsters into the European Union from the American continent. There are 165 monthly observations in the first year and no prior information. We therefore start the analysis using the FS. The identification of outliers from this FS provides prior information on the structure of outliers for ensuing years. The observations are regression data of quantity against value. In our particular example there is a linear relationship followed by the majority of the data, a few outliers, and a second, lower, line with fewer observations. The presence of this line is an indication of potential fraud.

The left-hand panel of Fig. 8.8 shows the scatter plot of the data, in which the two lines are apparent. The right-hand panel presents the minimum deletion residuals with a Bonferroni bound to indicate outlying observations, a choice we justify later in this subsection. There is a marked increase in the value of the minimum deletion residual at $m = 150$, indicating that there are 15 outliers in the data. However, the scatter plot of the data with marked residuals in Fig. 8.9 suggests that not all of the outliers are fraudulent, let alone being sufficiently outlying to provide judicially

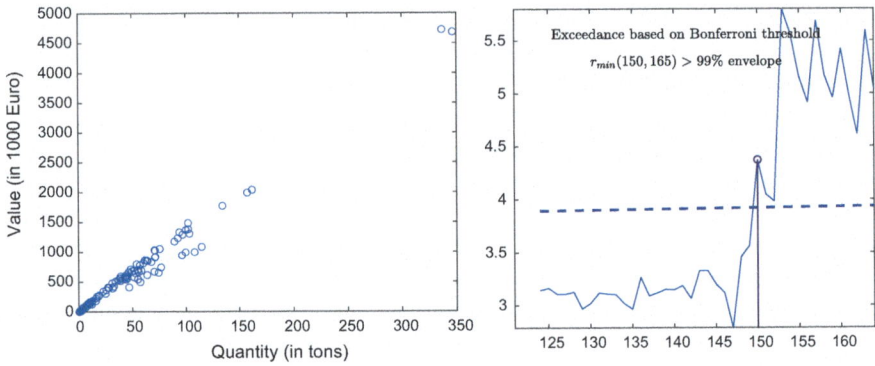

Fig. 8.8 2002 Lobster data: left-hand panel: scatter plot of value, in thousands of Euros, against quantity. Right-hand panel: monitoring plot of minimum deletion residuals; the Bonferroni bound indicates 15 outliers

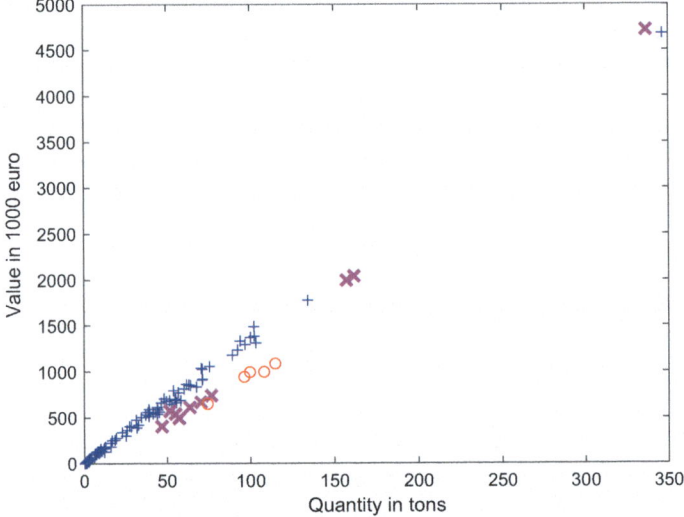

Fig. 8.9 2002 Lobster data: scatter plot of value, in thousands of Euros, against quantity with observations marked according to the size of the raw residual: + non-outlying observations, x intermediate observations within a threshold of ±300, o outliers

convincing evidence of fraud. We use the sufficient statistics from a cleaned version of the data from 2002 to provide prior information for the analysis of the data from 2003, continuing the process from year to year until overwhelming evidence of fraud has accumulated.

The marking of the residuals in Fig. 8.9 comes from the two ways we use to clean the data; one for the model for the mean and the other for the variance. In the next year, we use the non-outlying observations from the FS to determine the parameters

β of the linear model to provide an estimate of the economic fair value of the good, here lobsters. Trimming so many observations, however, indicates too many outliers to be helpful in fraud detection. Experience from those preparing legal evidence suggests that courts are most comfortable with the evidence presented in the form of raw residuals, that is differences between observed and fitted values without scaling for leverage and the estimate of error variance. Accordingly, we use a relatively generous fixed threshold around the fitted regression line to indicate which of the outliers should be excluded from the central part of the data. We use all observations within this threshold to provide the prior estimate of the error variance σ^2. Use of a fixed threshold is justified since there is little interest in detecting fraudulent small transactions. This choice of the estimated variance is motivated by analyses of many datasets, which show that the constraint that the regression line passes through the origin can give rise to a large number of very small residuals for small values of x, which can cause robust procedures to identify a large number of outliers. The desire for a procedure that is simple to explain in a legal setting also led to the use of an outlier threshold based on a Bonferroni inequality.

In Fig. 8.9 the non-outlying observations that are accepted by the FS are marked with crosses. However, there are three indicated outliers that are close to the main upper line, including one of the two observations with a quantity of around 340. We have marked with the magenta symbol x these and other intermediate observations which were identified as outliers by the FS but have raw residuals <300. Circles (red) are used to indicate outliers that have larger raw residuals. As the plot shows, the three observations close to the majority relationship are no longer suspected of being fraudulent. We also lose some outliers for small quantities, leaving the five most extreme residuals on the lower line as outliers, indicated by circles. The threshold level of 300 was found by repeated analysis of the data to provide a clear separation of appreciable outliers from those from other observations. Of course, the threshold level will depend on the goods generating the data being analysed.

As a result of the data analysis of this section, the prior information for β only uses the 150 "good" observations determined by the FS to be non-outlying. However, we have argued that the variance of this set of observations is too small. We therefore augment this set by the intermediate observations lying within the threshold, to give a set of 160 observations marked in the figure by crosses and the symbol X. These serve as prior observations for σ^2 in our analysis of the data from 2003.

8.3.4 Lobster Data 2003: Analysis with the Bayesian Forward Search

The right-hand panel of Fig. 8.10 shows that the 167 observations from 2003 have the same general pattern as the observations from 2002, that is a main line with some outliers and several observations forming a lower line. We incorporate the posterior information from the analysis of the data from 2002 into the analysis of the 2003

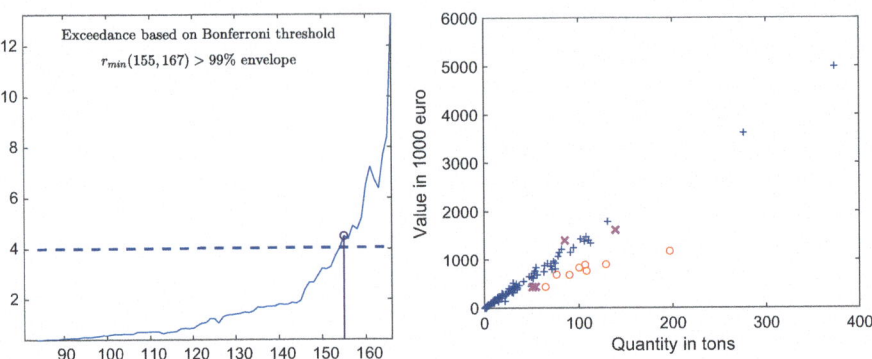

Fig. 8.10 2003 data: Bayesian search. Left-hand panel: forward plot of minimum deletion residuals. The Bonferroni bound indicates 14 outliers. Right-hand panel: scatter plot of the data and results of residual analysis: + non-outlying observations, x intermediate observations within a threshold of ±300, o outliers

data, taking $n_{0,1}$, the number of prior observations for the estimation of β, as 150 with $n_{0,2}$, for the estimation of variance as 160. The resulting forward plot of deletion residuals is in the left-hand panel of Fig. 8.10. The first outlier is at $m = 153$. The resulting classification of residuals is in the right-hand panel. The figure also shows that there are 6 observations that lie within the threshold of ±300 from the robust line. These again are marked x. The interesting observations, from the perspective of fraud detection, are the 8 observations marked o which lie on a line with a lower slope than the others.

8.3.5 Lobster Data 2004: Analysis with the Bayesian Forward Search

Finally, we consider the Bayesian analysis of data from 2004, for which there are 168 observations. We now have prior information from the non-outlying observations from both 2002 and 2003; thus $n_{0,1} = 305$ and $n_{0,2} = 319$. The right-hand panel of Fig. 8.11 shows the scatter plot of the central observations and outliers from the Bayesian FS for the 168 observations for 2004. These observations have the same general pattern as the observations from 2002 and 2003. The Bayesian FS analysis of the data, plotted in the left-hand panel, suggests there are 16 outliers. However, three of them lie within the threshold of ±300, leaving 13 outliers, one of which lies close to the upper line. The remaining 12 observations lie on an extremely clear line which warrants further investigation because it is associated with a systematic lower price. These observations came from a particular EU country. The relevance of this striking pattern was confirmed by a successful prosecution instituted independently from this analysis.

8.3 Data Analyses with Prior Information

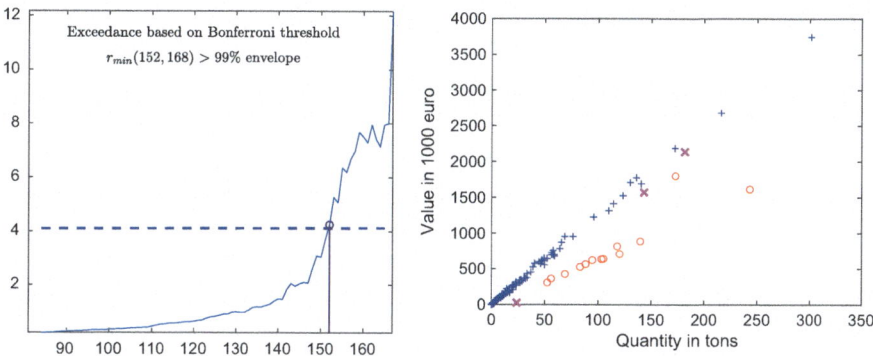

Fig. 8.11 2004 data: Bayesian search. Left-hand panel: forward plot of minimum deletion residuals. The Bonferroni bound indicates 16 outliers. Right-hand panel: scatter plot of the data and results of residual analysis: + non-outlying observations, x intermediate observations within a threshold of ±300, o outliers

8.3.6 Timeliness and Power

The purpose of the annual Bayesian analysis of these data was to monitor behaviour following initial suspicions aroused by the frequentist analysis of the 2002 data. If appropriate and possible, the analysis could be updated more frequently. The stopping point occurs when it is believed that sufficient evidence has been accumulated to lead to a successful prosecution.

Comparison of the values marked as outliers in the left-hand panels of Figs. 8.8, 8.10 and 8.11 shows that the outlying fraudulent line is appreciably clearer in the analysis of the 2004 data than in that of earlier years. This, however, is a result of the distribution of the data points, not of its analysis. The effect of the Bayesian analysis incorporating several year's data, as opposed to the analysis of a single year, is to provide parameter estimates with reduced variance and tests of increased power, particularly here, of the existence of a subset of data with a distinct relationship between price and quantity.

Atkinson et al. (2018) give further details of the analysis of the lobster data, including a comparison of the results of the Bayesian analysis with that from a frequentist analysis. They also quantify the increase in power from the use of Bayesian procedures and investigate the effect of incorrect prior information.

8.4 Heteroskedasticity

8.4.1 Homoskedasticity and Heteroskedasticity

So far we have made the assumption that the errors in the responses are either homoskedastic, that is, that they all have the same variance, or that they can be transformed to homoskedasticity, as in Chaps. 6 and 7. In both chapters the assumption was that the variance was a function of the mean response. In this section we develop the theory for, and use, the FS for robustly fitting regression models allowing for heteroskedasticity.

Of course, it is possible to ignore the heteroskedasticity. Doing so leads to unbiased but inefficient point estimates and to biased estimates of standard errors, and may result in overestimating the goodness of fit. Methods for non-robust heteroskedastic regression analysis are widely described in econometrics (Greene 2012).

M-estimation is the robust method most explored for heteroskedastic regression. An approach through weighting is in Chap. 4 of Carroll and Ruppert (1988). Further results are sketched by Welsh et al. (1994). The references to trimming are more recent. Cheng (2011) uses the model of Harvey (1976), combined with the FS and a trimmed likelihood for robust heteroskedastic regression. Neykov et al. (2012) use trimming in robust estimation of a general quasi-likelihood model, including the special case of heteroskedastic regression.

A very general method (White 1980) uses ordinary least squares (LS) combined with "heteroskedastic robust" standard errors. A major difficulty in combining down-weighting or deletion of observations with this procedure is the lack of a clear relationship between individual observations and their effect on inferences drawn from the data. We require a method of robust heteroskedastic regression to also be robust to the specification of the form of heteroskedasticity. This means that the method should include as general a model as possible for the heteroskedasticity. In what follows we provide a method for robust heteroskedastic regression which generalizes the form of heteroskedasticity described, in a non-robust context, by Harvey (1976). Atkinson et al. (2016) show how poorly the "heteroskedastic robust" procedure of White (1980) can perform when compared with a model with correctly specified heteroskedasticity and provide evidence that the parametric approach is never less efficient than LS even when the skedastic relationship is incorrectly specified. They also provide references to other work on robust heteroskedastic regression.

Our approach assumes that the conditional variance of the observations depends on a linear function of the explanatory variables in the regression, the parameters of which are to be estimated. In the next section we describe our model and comment on the relationship with Harvey's model.

8.4.2 Models for Non-constant Variance and the Forward Search

The model for heteroskedastic regression introduced by Atkinson et al. (2016) can be written
$$y_i = \beta^T x_i + \sigma_i \epsilon_i \quad i = 1, \ldots, n,$$
where the errors ϵ_i have a (homoskedastic) standard normal distribution, $\epsilon_i \sim \mathcal{N}(0, 1)$. We parameterise the variance function as
$$\sigma_i^2 = \text{Var } y_i = \sigma^2 \{1 + \exp(z_i^T \gamma)\}, \tag{8.19}$$
a form that is used, for example, in the analysis of pharmacokinetic data by Fedorov and Leonov (2014). We denote this model by ART.

In our examples the variables x and z are identical. However, all elements of the two-parameter vectors are distinct. In consequence, the information matrix is block diagonal. For given γ, estimation of β is by weighted least squares and σ^2 is estimated from the residual sum of squares. For given γ the weights are defined as
$$w_i = \sigma^2/\sigma_i^2 = \{g(z_i^T \gamma)\}^{-1}.$$

It is convenient to write
$$y_i^W = \sqrt{w_i} y_i, \quad x_i^W = \sqrt{w_i} x_i \quad \text{and} \quad W = \text{diag}\{w_i\}.$$
Then
$$\hat{\beta}^W | \gamma = (X^T W X)^{-1} X^T W Y,$$
although we shall usually notationally suppress the dependence of $\hat{\beta}^W$ on γ. Thus $\hat{\beta}^W$ is found by weighted regression of y on x or, equivalently, by unweighted regression of y^W on x^W. The FS for heteroskedastic data progresses in the same manner as that for homoskedastic data but in the space of y^W and x^W. Since $\hat{\beta}^W$ is found by least squares regression, it inherits the affine equivariance of least squares estimators. We find it informative to consider data analysis in both these spaces.

It remains to estimate γ. Numerical maximization of the likelihood is one possibility. However, since we need to estimate the parameter for each subset of the data in the FS, of which there are almost n, we use a more efficient scoring algorithm, presented in Atkinson et al. (2016).

An important property of (8.19) for the data we analyse, is that, provided $\sigma^2 > 0$, the variance does not approach zero as $\exp(z_i^T \gamma) \to 0$. A simpler model for heteroskedasticity with skedastic equation
$$\sigma_i^2 = \sigma^2 \exp(z_i^T \gamma), \tag{8.20}$$

for which the variance can go to zero, was introduced by Harvey (1976), which we denote by HAR. The properties of heteroskedastic regression with (8.20), together with a scoring algorithm, are described and illustrated by Greene (2012, pp. 554–556). A disadvantage of Harvey's model is that sometimes, as in the example of Sect. 8.4.4, it gives excessive weights to very small observations.

For numerical stability and improved convergence of algorithms, Atkinson et al. (2016) suggested reparameterizing (8.19). Let scalar $z_i = \log(x_i)$. Then, since the linear model for heteroskedasticity includes a constant γ_0, Eq. (8.19) becomes

$$\sigma_i^2 = \sigma^2(1 + \theta x_i^\alpha), \tag{8.21}$$

which follows by putting $z_i^T = (1 \ \log x_i)$, when $\gamma^T = (\log(\theta) \ \alpha)$.

8.4.3 Example: International Trade Data 1

The trade data with which we are concerned have non-negative values of the single explanatory variable x. Zero values in this case do not occur, but the data may contain many values close to zero, as is the case in Sect. 8.4.4. As a first example we analyse the 3867 observations for an import into the EU plotted in Fig. 8.12 using the skedastic equation ART (8.19). The data seem to have a simple linear structure showing heteroskedasticity, with two or three clear outliers for low values of weight, at least two outliers for intermediate values, and, certainly, some values below the

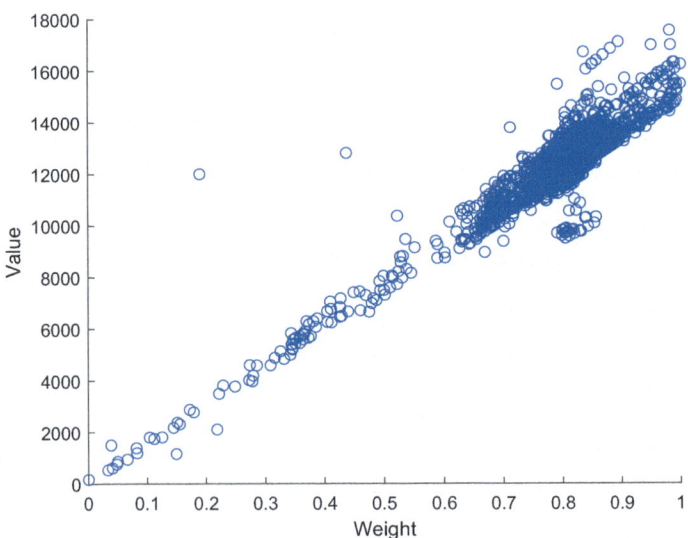

Fig. 8.12 International Trade Data 1: scatter plot of value against scaled weight

8.4 Heteroskedasticity

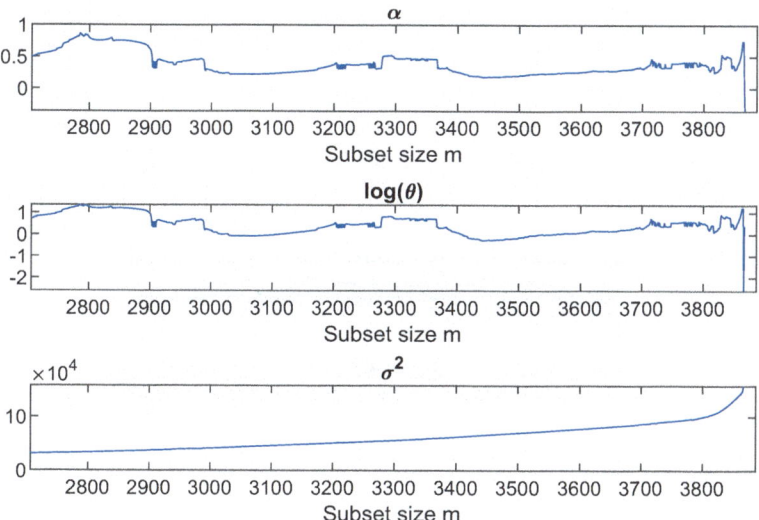

Fig. 8.13 International Trade Data 1: monitoring parameter estimates in (8.21) using skedastic equation ART. Panels reading down are estimates of α, $\log(\theta)$ and σ^2

general trend for high values of weight. The structure of the observations for high weight requires elucidation by statistical analysis.

For numerical purposes and without loss of generality, we have scaled x to lie between 0 and 1, dividing by the maximum value of x. We fit a linear regression model with intercept since even small transactions often incur fixed costs, regardless of size. In this example we allow the scoring algorithm a maximum of 100 iterations and impose maximum values of 10 on the estimates of the parameters α and $\log(\theta)$, although these bounds are not, in fact, needed.

It is clear from Fig. 8.13 that the observations that enter at the end of the search cause a sharp decrease in the estimates of the parameters α and $\log(\theta)$. The negative values of $\log(\theta)$ correspond to small values of θ so that, from (8.21), these decreases indicate that the estimates for all n observations will show less evidence of heteroskedasticity. As the bottom panel of Fig. 8.13 shows, the extra variability from fitting all observations is accommodated by the increase in the value of $\hat{\sigma}^2$.

The statistical analysis in Fig. 8.14 shows the monitoring plot of the scaled residuals in the scaled space of y^W and X^W. The plot is very stable over most of the FS, clearly showing the two extreme outliers for low values of x that are a feature of Fig. 8.12.

Figure 8.15 summarizes the results of the FS analysis by the scatter plot of the data augmented with 99% Bonferronized confidence bands. These intervals are based on the intervals in the transformed (homoskedastic) space for the robust model found by the FS. They are then transformed to heteroskedastic intervals by using the robustly estimated parameters in the heteroskedastic relationship (8.21). The confidence bands in the figure are less parallel than they might seem at a first glance; there is evidence

302   8   Extensions of the Multiple Regression Model

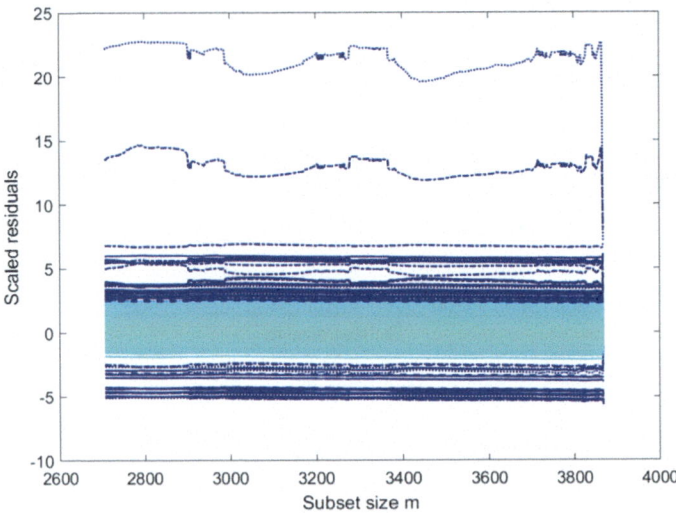

Fig. 8.14 International Trade Data 1: monitoring plot of scaled residuals, using skedastic equation ART

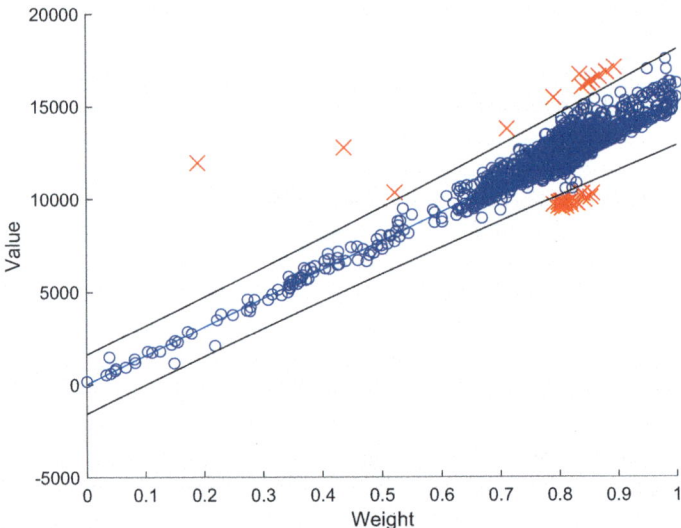

Fig. 8.15 International Trade Data 1: scatter plot (with scaled weight), showing fitted regression model and 99% Bonferronized confidence bands, using skedastic equation ART. Outliers are shown as crosses

of heteroskedasticity. In the figure the 38 outliers from FS are plotted as crosses. The resulting figure nicely complements the monitoring plot of residuals in Fig. 8.14. Apart from the two outliers mentioned above, Fig. 8.15 shows the location of the outliers forming the lower, compact, band in Fig. 8.14 and the more diffuse upper band of residuals.

Exercise 8.2 repeats the analysis using HAR (8.20), while Exercise 8.3 investigates the effect on outlier detection of ignoring the heteroskedasticity. An additional example of the effect of ignoring heteroskedasticity is in Atkinson et al. (2016).

8.4.4 Example: International Trade Data 2

We now analyse a second set of international trade data. Part of the importance of this example is to illustrate the difference in performance in data analysis of the skedastic relationship (8.19) introduced by Atkinson et al. (2016), which we called ART, and that in (8.20) suggested by Harvey (1976) (HAR) which, in the trade data since there is a single explanatory variable, goes to zero at $x = 0$, where the modelled response is also zero.

There are 1302 observations. The scatter plot of the data in Fig. 8.16 shows that the general form of this set of data is quite different from that for International Trade Data 1, shown in Fig. 8.12. There the highest density of observations was from scaled weight of 0.65 to one. Here there is a concentration of observations near the origin,

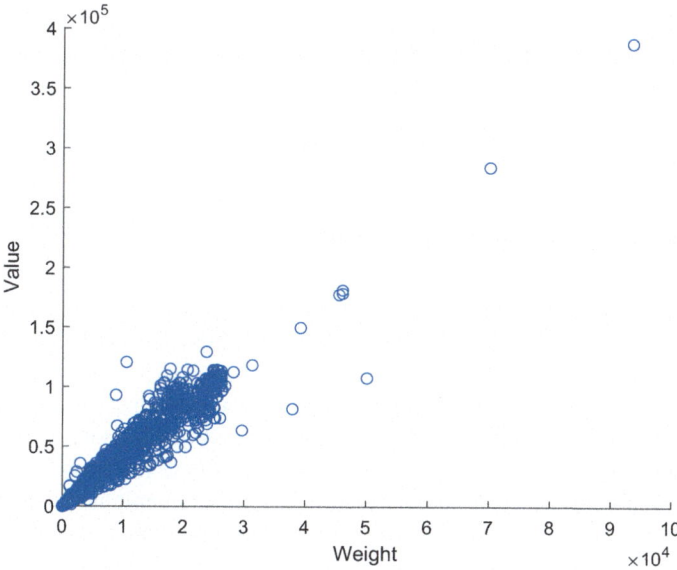

Fig. 8.16 International Trade Data 2: scatter plot of value against weight

fanning out strongly and clearly showing heteroskedasticity. There is a central line in this scatter which gives rise to leverage points and there is a line of points near to the bottom of the fan which may be outlying. There are also some seemingly too-large values at low values of x.

The upper panel of Fig. 8.17 shows the monitoring plot of scaled residuals for these data. For virtually all the search there are 7 extreme positive trajectories with an eighth less large positive trajectory. There is appreciable change in these trajectories at the end of the search, as the extreme observations enter the subset used in fitting.

The lower panel of Fig. 8.17 reveals the location of these extreme observations; five are at very low values of x and so are unlikely to be important in the detection of financial irregularity. The other 2 outliers (observations 559 and 620) are, as we have seen before, clear of the main body of the data. It appears that these 7 outliers come from a different regression line. The next stage in any investigation would be to check whether they all come from the same country. If the answer is positive, the data for preceding and subsequent years should be checked.

We conclude this introduction to the use of the FS in the analysis of heteroskedastic regression data by comparing the results from the use of the skedastic relationships HAR, Eq. (8.20) and ART (8.19). The results are in Fig. 8.18, which shows the confidence limits for smaller values of x. The upper panel shows the results for HAR. Since the variance goes to zero at $x = 0$, the analysis for outliers is very sensitive to minor perturbations for very low values of x. Indeed, the plot shows one such outlier, in fact for $x = 0.0011$. This observation is not considered an outlier in the lower panel analysed with ART. The plot using ART shows that the confidence limit still has finite width at $x = 0$. In general, the choice between the two skedastic equations will depend upon the data to be analysed and the purpose of the analysis. Here the purpose is to detect appreciable financial irregularities which are worthy of further investigation. There is a large number of datasets to be analysed, so the automatic indication of irregularities needs to be free of the identification of outliers which are irrelevant to the main purpose of the analysis.

8.5 Robust Regression Clustering

In this section we continue with the analysis of the various structures that arise in the analysis of sets of international trade data. Section 5.8 introduced data consisting of observations from two populations. In this case the best we can obtain from a single robust regression estimator is the identification of a main group of observations and a possibly large number of outliers. We now address cases in which we also model the second or additional groups separately, so that interest is in methods for *regression clustering*. That is, in (2.4), the basic equation for the contaminated distribution of observations, we are interested in the estimation of the contaminating distribution G. Here the objective is the identification of linear regression hyperplanes. There will also be an unspecified component of G to allow for the presence of outliers.

8.5 Robust Regression Clustering

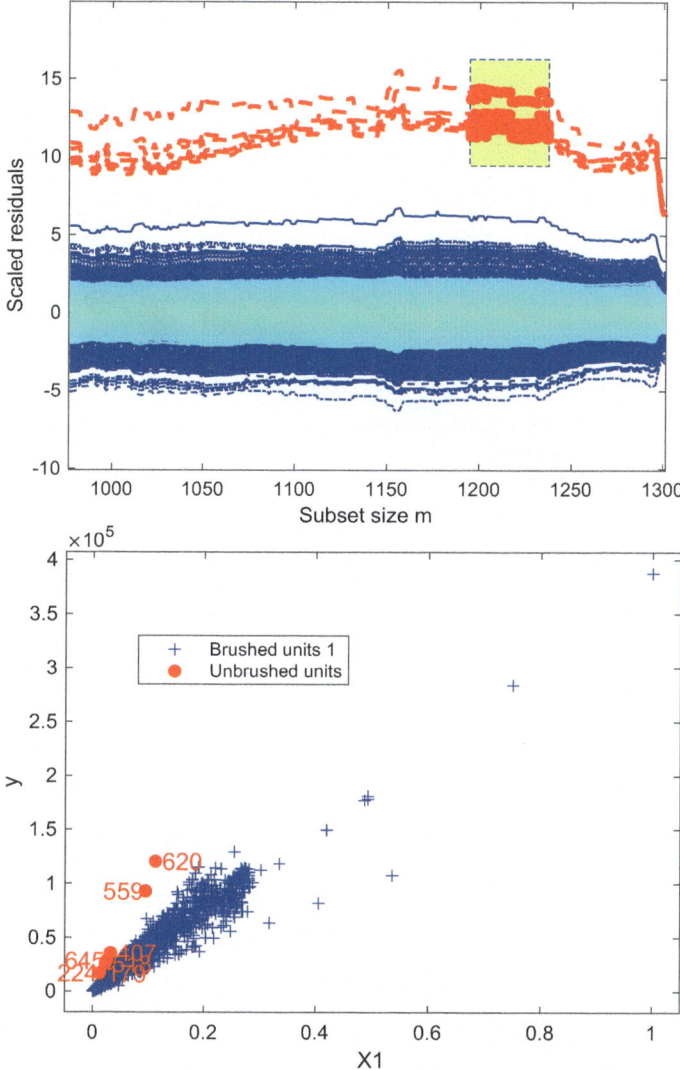

Fig. 8.17 International Trade Data 2: analysis using skedastic equation ART. Upper panel, monitoring plot of residuals with brush on extreme trajectories; lower panel, scatter plot, with scaled weight, showing the seven brushed units

A main problem in this context is the choice of the optimal number of groups k, which, in the traditional approaches, is associated with the minimization of an information criterion based on an unconstrained likelihood; for example, Fraley and Raftery (2002), which we illustrate in Sect. 8.5.5. Instead, we use the FS to analyse data, starting the search many times from random starting points. As the individual

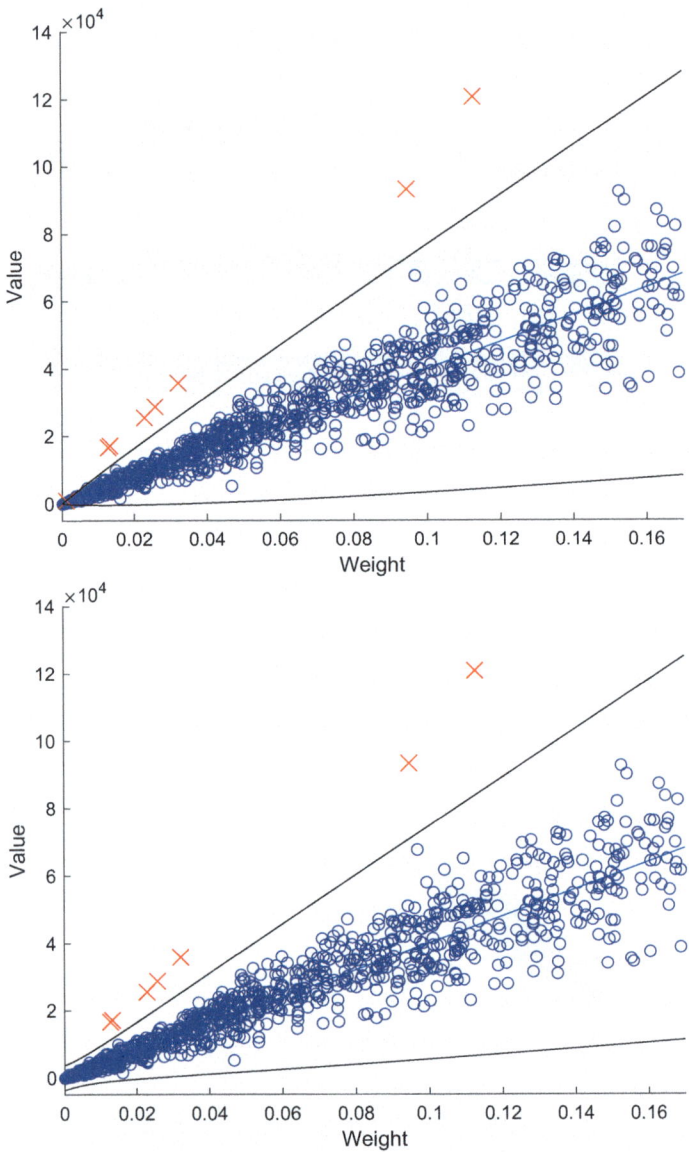

Fig. 8.18 International Trade Data 2: comparison of outliers indicated by two forms of skedastic equation for smaller values of scaled weight. Upper panel HAR, indicating an outlier at $x = 0.0011$; lower panel ART, showing one fewer outlier

trajectories of the data analyses progress, many converge to give evidence of specific groups in the form of clearly different forward trajectories of minimum deletion residuals. This is illustrated in Sect. 8.5.2.

Regression clustering, being model-based, also involves the choice of constraining parameters and an appropriate level of trimming to account for the presence of outliers. Here the monitoring approach helps to identify the so-called *optimal stable solutions,* that is solutions which do not depend on a specific choice of these hyper-parameters nor of k itself. This approach is illustrated in Sect. 8.5.4. The data that we use in the next section to illustrate these concepts come from the context of international trade, where it is essential to show that a particular classification can still be found despite minor changes of the input parameters. As we have repeatedly stressed in the earlier sections of this book, monitoring provides methods of robust analysis that avoid detailed specification, at least some, of the parameters of the particular method being employed.

8.5.1 Face Masks: A Trade Dataset with Several Groups

The dataset is taken from a study conducted by the European Commission during the COVID-19 pandemic that exploded in 2020 (Perrotta et al. 2020). The objective was to analyse the composition of the European Union trade, in order to refine the definition of the commodity codes used to monitor the import of face masks and other protective equipment from countries outside the European Union. The data plotted in Fig. 8.19 are a sample of 352 import flows from one of the newly defined codes specific for FFP2 and FFP3 face masks. The data come from one day's activity in November 2020. For each import flow, we have represented the traded value (vertical axis), weight ("W", horizontal axis left-hand panel) and number of units, technically called "Supplementary Units" ("SU", horizontal axis right-hand panel). There are therefore two explanatory variables.

In this example there are two major groups generated by the COVID crisis because of a considerable rise in the price of certain efficient protective face masks known in Europe under the name FFP. The "Filtering Face Piece" standard for mask efficiency ranges from one, the lowest efficiency, to three (the highest). The FFP2 masks, which stop more than 94% of the aerosols present in the air, possibly including the COVID-19 viruses, had suddenly been in high demand and therefore became costly. Manifestly, the upper group in the plots, which was not present before the COVID crisis, refers to these masks.

However, the fine-grained structure of the lower and more populated group is not clear, consisting of a mix of medical and community face covering masks (one-use surgical masks, cloth masks, face shields, and so on). Their price depends on the markets and also on the specific weight of the products (the weight for the FFP masks is at least $200 \, g/m^2$ while the surgical masks have lower weight). Being able to distinguish between these fine groups has contributed to the introduction, in the Integrated Tariff of the European Union, of new commodity codes specific

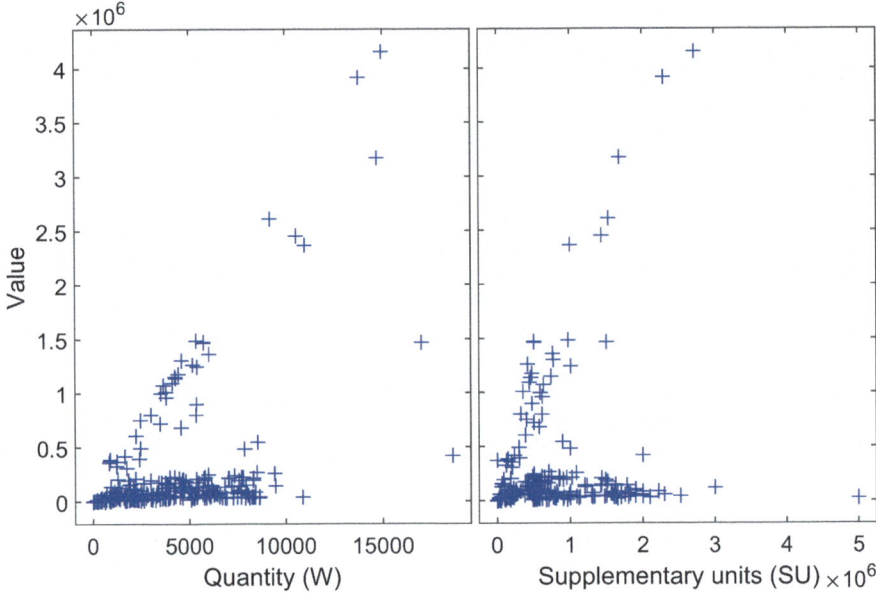

Fig. 8.19 Face Masks data: 352 imports of FFP2 and FFP3 masks into the European Union (EU) extracted on a day in November 2020 and identified in the EU Customs system with the product code 6307.90.98.10. Vertical axes: traded value in euros. Horizontal axes: left-hand panel, traded weight in Kg (W); right-hand panel, number of units (SU)

to the different types of lower-cost masks. This is important, as the commodity classification defines all measures to be undertaken by economic operators when importing or exporting goods into or from the Union.

8.5.2 Analysing the Face Masks Data with Random Start Forward Searches

We investigate the number of groups k in the dataset by running the FS many times, starting from n_R initial randomly chosen subsets, and monitoring the n_R trajectories of the minimum deletion residuals. If the dataset contains $k > 1$ groups, the n_R trajectories have the following characteristics:

- Trajectories starting from the same group will, at some point, converge to a single trajectory when the majority of observations in that particular group have entered the subset and observations from other groups have been excluded. This is a remarkable property of the forward search dynamic, which permits units to leave or join the subset during its progression, depending on their distance from the regression planes.

8.5 Robust Regression Clustering

- When the units of a group are all included in the subset and the FS progression starts to include the units of other groups, the plot of minimum deletion residuals increases forming a peak in the trajectory.
- Addition of further units may cause the original units forming the group to be seen as outliers and they may be removed by interchanges.
- As yet further units are included, the trajectories merge into a single one, typically, in the case of overlapping groups, well before all the observations have entered the subset.
- It is worth noting that the random start can produce more large values of the absolute minimum deletion residuals in the initial stages of the search than does the standard FS. This is due to interchanges of units in the subsets in the early stages of the search leading to almost collinear sets of points for fitting.

To exhibit the properties of the procedure we start by only considering regression on the weight variable W. Figure 8.20 exhibits the effects listed above in the FS analysis of the face masks data.

Given the above, trajectories consisting of units mainly coming from a single group should be separate during some part of the search, each giving rise to an extreme peak as observations from other groups are necessarily included in these subsets. The number of groups can then initially be estimated by counting the number of extreme peaks in the central part of the plot. This procedure is illustrated in the left-hand panel of Fig. 8.21, which again shows the monitoring plot of absolute minimum deletion residuals for the face masks data. In order to avoid the distraction of the extreme values of the minimum deletion residuals for small m, plotting for Fig. 8.21 starts

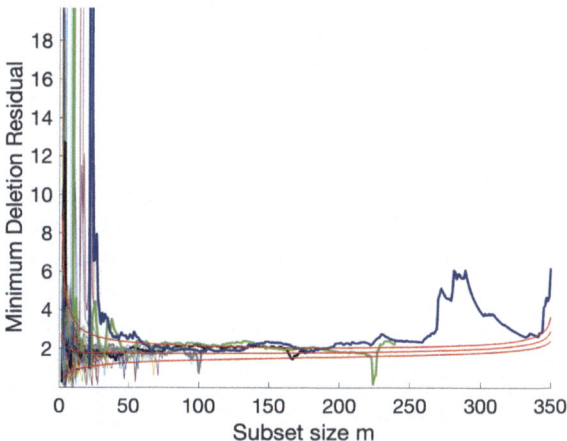

Fig. 8.20 Face Masks data: multiple start forward searches, monitoring plot of absolute minimum deletion residuals from regression on Weight. Very large values are produced in the initial stages of the search. However, the initial distinct searches rapidly converge to a relatively few trajectories with marked peaks, which provide information on the number and membership of the clusters. Near the last third of the search, all trajectories have converged

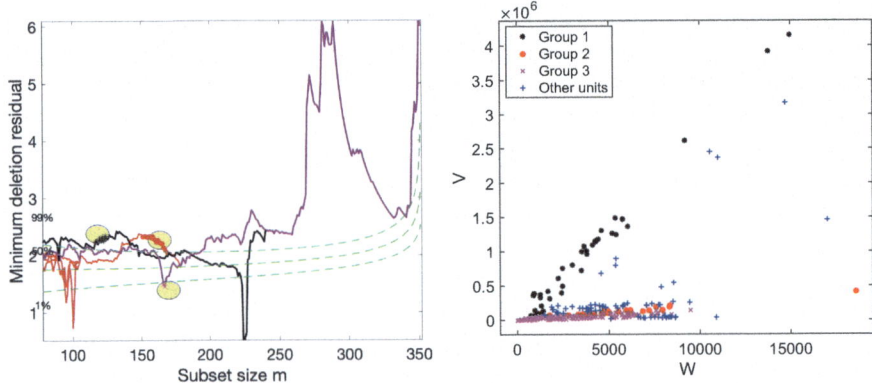

Fig. 8.21 Face Masks data: multiple start forward searches. Left-hand panel; monitoring plot of absolute minimum deletion residuals from regression on Weight. The three groups of unique trajectories selected in the central part of the forward plot correspond to distinct groups of observations in the scatter plot of the right-hand panel

around $m = 90$. From $m = 110$ there are just 3 trajectories and so a structure of three groups. To reveal the structure of the groups we select a step for each trajectory and see the units inside the subset for the associated value of m. In the figure the brushes are shown as yellow ellipses; the selected value of m is the maximum value within the ellipse. For example, selecting a trajectory from $m = 110$ to $m = 120$ reveals the units inside the subset at step $m = 120$. Scatter plots for the observations in the three groups are in the right-hand panel of Fig. 8.21. Although the choice of the step of the search to highlight is subjective (here the chosen steps are didactic) the three groups remain in the interval m [110 160].

If we force the forward search to start within one of these groups, we can expect a smooth evolution of the estimated regression coefficients, until units from a different group are included in the subset. This is shown in Fig. 8.22, for a start from each of the groups of Fig. 8.21.

After discovering the number of groups, it is of interest to determine the membership and quality of the estimated groups. To do so, we use an approach that hopefully alternates k times the identification of a homogeneous subgroup using the random start approach and its subsequent elimination, following an idea initially explored in Torti (2011) and Cerioli et al. (2019). This approach replaces the original k population (robust) estimation problem with k distinct one-population robust steps, which take advantage of the good breakdown properties of trimmed estimators when the trimming level exceeds the usual bound of 0.5. Note that so far we have assumed that the percentage of contamination ε lies in the interval [0 0.5) (see the definition of IF in Sect. 2.1.1). With several populations, observations from all except one of which are treated as outlying, it is meaningful to allow ε to be >0.5. More precisely, the approach consists in:

8.5 Robust Regression Clustering

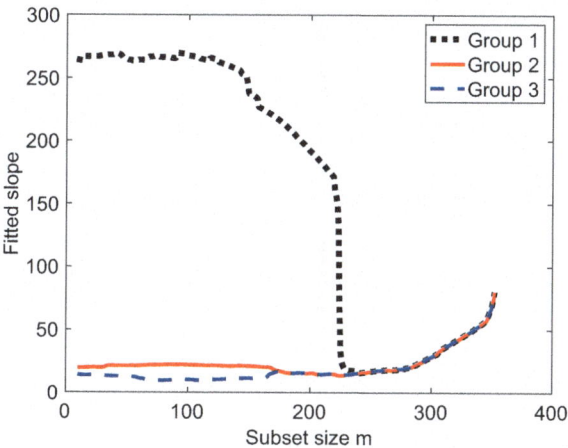

Fig. 8.22 Face Masks data: monitoring of the fitted slope (the estimated trade price), from regression on W, for three initial subsets, chosen within the groups of Fig. 8.21. When units from a different group join the subset, the fitted slopes quickly change. The merging points of the trajectories reflect the level of group overlap: trajectories initiated in groups 2 and 3 join earlier than the one in group 1

(a) At step i ($i = 1, \ldots, k$), generate the plot of the minimum deletion residuals from the *reduced* dataset cleaned of the observations belonging to the groups identified in previous steps.
(b) Brush the first peak in the trajectories lying outside the bands.
(c) The group of units associated with the brushed trajectories is automatically identified and removed from the dataset.

When $i = 1$, the procedure is applied to the plot of minimum deletion residuals for all the data.

Figure 8.23 illustrates the application of the procedure to the face masks dataset, in four consecutive iteration steps, rather than the three of Fig. 8.21. Regression remains only on W. In the plots of the minimum deletion residuals (first column panels) the colormap is proportional to the sum of the deletion residuals along the trajectory, skipping the initial 10% to avoid spurious peaks. At each step we select the trajectories with the largest sum of minimum deletion residuals and the resulting units are shown in the panels of the last column. It is remarkable that even in this case, in which the degree of overlapping between some of the groups is quite high, the approach works well in identifying the fine-grained structure of the data. The last panel of row 4 suggests that there may be some further structure to be identified, which may include allocating some of these observations to the groups already discovered. However, the procedure just discussed can provide a preliminary idea about parameter k (number of groups) to be used in the clustering methods of the next section.

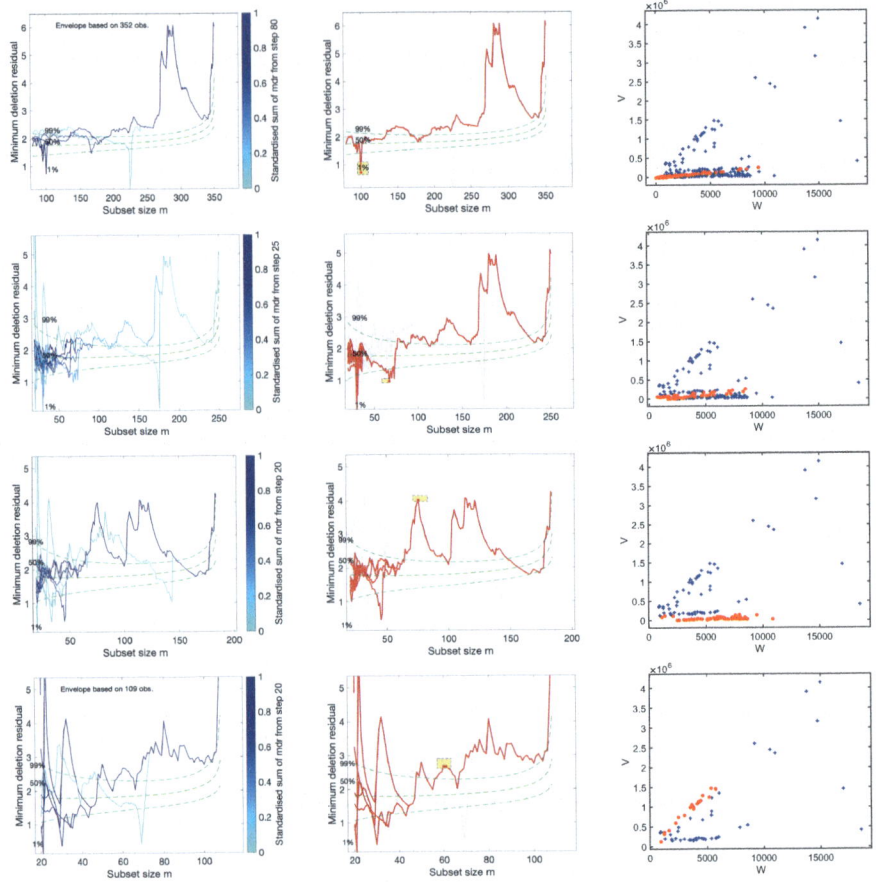

Fig. 8.23 Face Masks data: iterative FS-based random start approach; regression on W. Each row represents an iteration of the algorithm of Sect. 8.5.2; for each iteration, the three horizontal panels 1/2/3 represent respectively step *a*/*b*/*c*. As in Fig. 8.21, we skip the initial random peaks in the forward plots of minimum deletion residuals

8.5.3 Clusterwise Linear Regression: Basic Concepts

Clusterwise Linear Regression (CLR) is a general framework for clustering regression structures. We first introduce the notation, giving emphasis to the parameters that need to be monitored.

Let the vector of covariates be x (which might be univariate or multivariate) and the response variable y be defined on Ω with values in $\mathcal{X} \times \mathcal{Y} \subseteq \mathbb{R}^{p-1} \times \mathbb{R}$. Then, $\{x_i, y_i\}$, $i = 1, 2, \ldots, n$, represents an i.i.d. random sample of size n drawn from Ω. If we suppose that Ω can be partitioned into k groups, say $\Omega_1, \Omega_2, \ldots, \Omega_k$, the

8.5 Robust Regression Clustering

general formulation of the regression clustering mixture model has a density which can be written as

$$p(x, y, \theta) = \sum_{g=1}^{k} p(y|x, \theta_{y,g}) p(x, \theta_{x,g}) \pi_g, \quad (8.22)$$

where

- vector θ denotes the full set of parameters $\theta = (\theta_{y,g}^T, \theta_{x,g}^T)^T$;
- $p(y|x, \theta_{y,g})$ is the conditional density of y given x in Ω_g, which depends on the vector of parameters $\theta_{y,g}$;
- $p(x, \theta_{x,g})$ is the marginal density of x in Ω_g, which depends on the vector of parameters $\theta_{x,g}$;
- π_g reflects the importance of Ω_g in the mixture with the usual mixture constraints $\pi_g > 0$ and $\sum_{g=1}^{k} \pi_g = 1$.

It is customary to assume that in each group g the conditional relationship between Y and x, $p(y|x, \theta_{y,g})$, can be written as

$$y = \beta_{0,g} + x^T \beta_g + \epsilon_g, \quad (8.23)$$

where $\epsilon_g \sim N(0, \sigma_g^2)$, and β_g and σ_g are respectively the $(p-1) \times 1$ vector of regression parameters $(\beta_{1,g}, \beta_{2,g}, \ldots, \beta_{p-1,g})^T$ and the scale parameter for component g. This formulation was originally proposed by Gershenfeld (1997) and was developed in the context of media technology, in order to build a digital violin. With the linearity and normality assumptions, the first two conditional moments of Y given x can be written as $E(Y|x, \beta_{0g}, \beta_g, \sigma_g) = \beta_{0,g} + x^T \beta_g$, $\text{var}(Y|x, \beta_{0,g} + \beta_g, \sigma_g) = \sigma_g^2$. If, in addition, we also assume that the X distribution is multivariate normal, that is

$$p(x, \theta_{x,g}) = \phi_{p-1}(x, \mu_g, \Sigma_g), \quad (8.24)$$

where $\phi_{p-1}(x, \mu_g, \Sigma_g)$ denotes the density of a $p-1$-variate Gaussian distribution, with mean vector μ_g and covariance Σ_g, model (8.22) becomes the linear Gaussian Cluster-Weighted Model (CWM) (Gershenfeld et al. 1999) and can be written as

$$p(x, y, \theta) = \sum_{g=1}^{k} \phi(y; \beta_{0,g} + \beta_g^T x, \sigma_g^2) \phi_{p-1}(x, \mu_g, \Sigma_g) \pi_g. \quad (8.25)$$

It is interesting to notice that clustering around regression (DeSarbo and Cron 1988) can be seen as a special case of Eq. (8.25) by setting $\phi_{p-1}(x, \mu_g, \Sigma_g) = 1$, that is ignoring the distribution of X.

Equation (8.25) corresponds to a mixture of regressions with weights $\phi_{p-1}(x, \mu_g, \Sigma_g)$ depending not only on π_g, but also on the covariate distribution in each component g. This leads to defining the following loglikelihood function to be maximized, the mixture loglikelihood $\mathcal{L}_{\text{Mixt}}(\theta)$

$$\mathcal{L}_{\text{Mixt}}(\theta) = \sum_{i=1}^{n} \log \left[\sum_{g=1}^{k} \phi(y_i | b_{0,g}, x_i^T b_g, s_g^2) \phi_{p-1}(x_i, m_g, S_g) p_g \right], \quad (8.26)$$

where $\theta = (p_1, \ldots, p_k, b_{0,1}, \ldots, b_{0,k}, b_1, \ldots, b_k, s_1^2, \ldots, s_k^2, m_1, \ldots, m_k, S_1, \ldots, S_k)$ is the set of parameters satisfying $p_g \geq 0$ and $\sum_{g=1}^{k} p_g = 1$, $b_g \in R^{p-1}$, $b_{0,g} \in R^1$, $s_g^2 \in R^+$, $m_j \in R^{p-1}$ with S_j a positive semi-definite symmetric $(p-1) \times (p-1)$ matrix. The optimal set of parameter estimates from this likelihood is

$$\widehat{\theta}_{\text{Mixt}} = \arg\max_{\theta} \mathcal{L}_{\text{Mixt}}(\theta). \quad (8.27)$$

Once $\widehat{\theta}_{\text{Mixt}} = (\widehat{p}_1, \ldots, \widehat{p}_k, \widehat{b}_{0,1}, \ldots, \widehat{b}_{0,k}, \widehat{b}_1, \ldots, \widehat{b}_k, \widehat{s}_1^2, \ldots, \widehat{s}_k^2, \widehat{m}_1, \ldots, \widehat{m}_k, \widehat{S}_1, \ldots, \widehat{S}_k)$ is obtained, the observations in the sample are divided into k clusters by using posterior probabilities. That is, observation $(x_i \; y_i)$ is assigned to cluster g, if $g = \arg\max_l \phi(y_i | \widehat{b}_{0,l}, \widehat{b}_l^T x, \widehat{s}_l^2) \phi_{p-1}(x_i; \widehat{m}_l, \widehat{S}_l) \widehat{p}_l$.

In the so-called classification framework of model-based clustering, the classification loglikelihood $\mathcal{L}_{\text{Cla}}(\theta)$ to be maximized is defined as

$$\mathcal{L}_{\text{Cla}}(\theta) = \sum_{i=1}^{n} \sum_{g=1}^{k} z_{ig}(\theta) \log \phi(y_i | b_{0g}, x_i^T b_g, s_g^2) \phi_{p-1}(x_i, m_g, S_g) p_g, \quad (8.28)$$

where $\theta = (p_1, \ldots, p_k, b_{0,1}, \ldots, b_{0,k}, b_1, \ldots, b_k, s_1^2, \ldots, s_g^2, m_1, \ldots, m_k, S_1, \ldots, S_k)$ and

$$z_{ig}(\theta) = \begin{cases} 1 & \text{if } g = \arg\max_l \phi(y_i | \widehat{b}_{0,l}, \widehat{b}_l^T x, \widehat{s}_l^2) \phi_{p-1}(x_i; \widehat{m}_l, \widehat{S}_l) \widehat{p}_l, \; l = 1, 2, \ldots, k, \\ 0 & \text{otherwise.} \end{cases}$$

In this case, the optimal set of parameters is

$$\widehat{\theta}_{\text{cla}} = \arg\max_{\theta} \mathcal{L}_{\text{Cla}}(\theta) \quad (8.29)$$

and the observation $(x_i \; y_i)$ is now classified into cluster g if $z_{ig}(\widehat{\theta}_{\text{Clas},g}) = 1$.

The target functions (8.26) and (8.28) are unbounded when no constraints are imposed on the scatter parameters. It is therefore necessary to impose constraints on the maximization of the set of eigenvalues $\{\lambda_r(\widehat{S}_g)\}$, $r = 1, \ldots, (p-1)$ of the scatter matrices $\widehat{S}_g$ by imposing

$$\lambda_{l_1}(\widehat{S}_{g_1}) \leq c_X \lambda_{l_2}(\widehat{S}_{g_2}) \text{ for every } 1 \leq l_1 \neq l_2 \leq p-1 \text{ and } 1 \leq g_1 \neq g_2 \leq k$$

8.5 Robust Regression Clustering

and on the variances $\hat{s}_g^2$ of the regression error terms, by requiring

$$\hat{s}_{g_1}^2 \leq c_y \hat{s}_{g_2}^2 \quad \text{for every} \quad 1 \leq g_1 \neq g_2 \leq k.$$

The constants $c_X \geq 1$ and $c_y \geq 1$ are real numbers (not necessarily equal) which guarantee that we are avoiding the cases $|\hat{S}_g| \to 0$ and $s_g^2 \to 0$. Following Cerioli et al. (2018a), we consider the following values of the restriction parameters $2^0, 2^1, \ldots, 2^7$ which provides a sharp grid of values close to 1.

Robustness can be achieved by discarding, at each step of the maximization procedure, a proportion of units equal to α, associated with the smallest contributions to the target likelihood. More precisely, for example in the mixture modelling context, the Trimmed Cluster-Weighted Model (TCWM) parameter estimates (Garcia-Escudero et al. 2017) are based on maximization of the trimmed likelihood function

$$L_{\text{Mixt}}(\theta | \alpha, c_y, c_X) = \sum_{i=1}^{n} z^*(x_i, y_i)$$

$$\log \left\{ \sum_{g=1}^{k} \phi(y_i | b_{0,g}, b_g^T x, s_g^2) \phi_{p-1}(x_i, m_g, S_g) p_g \right\}, \quad (8.30)$$

where $z^*(\cdot, \cdot)$ is a 0–1 trimming indicator function which indicates whether observation (x_i, y_i) is trimmed ($z^*(x_i, y_i) = 0$) or not ($z^*(x_i, y_i) = 1$). A fixed fraction α of observations can be unassigned by setting $\sum_{i=1}^{n} z(x_i, y_i) = [n(1 - \alpha)]$. TCLUST-REG (Garcia-Escudero et al. 2010) can be considered as a particular case of TCWM in which the contribution to the likelihood of $\phi_{p-1}(x_i, m_g, S_g)$ is set equal to 1.

However, if the component $\phi_{p-1}(x_i, m_g, S_g)$ is discarded, α just protects against vertical outliers in Y, since these data points have small $\phi(y_i | b_{0,g}, b_g^T x, s_g^2) p_g$ values, but it has no effect in diminishing the effect of outliers in the X space. Therefore, if we adopt a TCLUST-REG approach, it is necessary to consider a second trimming step, which discards a proportion α_X of the units after taking into account their degree of remoteness in the X space, among the observations which have survived the first trimming operation. The original solution in TCLUST-REG was to fix α_X in advance (Garcia-Escudero et al. 2010), although there is no established indication of the link between this proportion and the breakdown properties of the methodology. A more recent *adaptive* solution (Torti et al. 2018) selects α_X from the data using a multivariate outlier detection procedure in the space of the explanatory variables.

The observations surviving the two trimming steps are then used for updating the regression coefficients, weights, and scatter matrices.

8.5.4 Monitoring Clusterwise Linear Regression

We have seen that CLR requires a procedure for finding:

1. The optimal number of groups k;
2. The amount of first trimming level α and
3. The optimal restriction factor c_y, among the variances of the error components and c_X among the scatter matrices of the covariates.

Here we use a two-step procedure. First we estimate one or more reasonable sets of combinations of values of c_y and k, given a large value of c_X if TCWM is used (Sect. 8.5.4.1). This choice is made using the information criteria given in the section below.

Then we find the optimal trimming level (Sect. 8.5.4.2) through a monitoring approach. The relevant literature includes Riani et al. (2014a), Cerioli et al. (2018b) and more recently Torti et al. (2021b).

8.5.4.1 Preliminary Estimate of Restriction Factor and Number of Groups

The choice of the optimal number of groups k can be made using an information criterion. Extending the results for multivariate analysis in Cerioli et al. (2018a), we have the following 3 information criteria:

$$\text{MIX-MIX}: k_{\text{opt}}(c_y, c_X|\alpha) = \arg\min_k \left\{-2\mathcal{L}_{\text{Mixt},k}(\widehat{\theta}_{\text{Mixt},k}|\alpha, c_y, c_X) + v_k^{c_y, c_X}\right\}$$
$$:= \arg\min_k F_{MM}(k, c_y, c_X|\alpha)$$

$$\text{MIX-CLA}: k_{\text{opt}}(c_y, c_X|\alpha) = \arg\min_k \left\{-2\mathcal{L}_{\text{Cla},k}(\widehat{\theta}_{\text{Mixt},k}|\alpha, c_y, c_X) + v_k^{c_y, c_X}\right\}$$
$$:= \arg\min_k F_{MC}(k, c_y, c_X|\alpha)$$

$$\text{CLA-CLA}: k_{\text{opt}}(c_y, c_X|\alpha) = \arg\min_k \left\{-2\mathcal{L}_{\text{Cla},k}(\widehat{\theta}_{\text{Cla},k}|\alpha, c_y, c_X) + v_k^{c_y, c_X}\right\}$$
$$:= \arg\min_k F_{CC}(k, c_y, c_X|\alpha),$$

where $v_k^{c_y, c_X}$ is a penalty term defined as

$$v_k^{c_y, c_X} = pk + (k-1) + (k-1)(1 - 1/c_y) + 1 + 0.5p_1(p_1 - 1)k + (p_1 k - 1)(1 - 1/c_X) + 1.$$

In our notation, "MIX-MIX" corresponds to the use of the Bayesian Information Criterion (BIC) (Fraley and Raftery 2002), while "MIX-CLA" corresponds to the use of the Integrated Complete Likelihood (ICL) method proposed by Biernacki et al. (2000). If $c_y \to \infty$ the ratio of the variances of the residuals becomes unconstrained.

8.5 Robust Regression Clustering

The same things happens to the Σ_X scatter matrices when $c_X \to \infty$. Cerioli et al. (2018a), in the context of multivariate data analysis, show in a simulation study that in presence of high overlap "MIX-MIX" seems to give the best results.

The plot which shows the values of the Information Criterion (IC) as a function of k, that we call an *elbow plot* (one example is given in the left-hand panel of Fig. 8.26), can be used to find the appropriate number of groups. In most cases, however, this trajectory is a monotonic function of k (an example of this situation is shown in Sect. 8.5.5). In the context of constrained TCWM the situation is complicated by the fact that there are different trajectories, each associated with a combination of values of the restriction factors. In some simple cases all the trajectories follow the same pattern and therefore the best solution, independent of the restriction factor values, is easy to identify. In more complex cases, different trajectories may follow different paths. In addition, the elbow plot does not provide any information about the stability of the solutions as a function of c_y (c_X) or k.

As an alternative to the traditional approach, which is based on the minimization of a particular unconstrained information criterion, we focus our attention on optimal stable solutions, that is solutions which do not depend on a specific choice of a constraining parameter. We do so by extending the graphical tool known as the car-bike plot, introduced by Cerioli et al. (2018a), to the context of regression clustering and to the case of trimmed likelihood, in order to select and visualize a ranked list of "optimal" choices for the pair (k, c_y). The procedure first detects a list with L "plausible" partitions. Such "plausible" partitions may include some solutions that are essentially the same as others already detected, because spurious clusters, consisting of a few almost collinear or very concentrated data points, have been found. In a second step, the partitions including repetitive solutions are discarded and we typically obtain a very reduced ranked list with T "optimal" (non-repetitive) partitions.

More formally, given a triple (k, c_y, c_X), let $\mathcal{P}(k, c_y, c_X)$ denote the partition into k subsets which is obtained by solving the problem (8.26) or (8.28), with the given k, c_X and c_y. In the context of TCWM, in order to prevent the presence of spurious solutions in the X, space we suggest fixing c_X at a finite large value. In what follows $\max(c_X) = 128$. Once all the other parameters have been estimated, it is possible to refine this value using the monitoring approach and the information criteria (MIX-MIX, MIX-CLA, CLA-CLA) given at the beginning of this section. If on the other hand, the data are highly non-normal (as in the case of international trade data) we suggest using TCLUST-REG with a flexible second level of trimming as mentioned in Sect. 8.5. In what follows, in order to avoid cumbersome notation, by symbol c we denote c_y (with c_X fixed at 128 in the case of TCWM). Let ARI($\mathcal{A}, \mathcal{B}$) denote the adjusted ARI index between partitions $\mathcal{A}$ and $\mathcal{B}$. We consider that two partitions $\mathcal{A}$ and $\mathcal{B}$ are "essentially the same" when ARI($\mathcal{A}, \mathcal{B}$) $\geq \varpi$, for a fixed threshold ϖ (here we take $\varpi = 0.7$). Clearly, the higher the value of the threshold the greater is the number of tentative different solutions which are considered.

By using this notation, the proposed automated procedure may be described as follows:

1. *Obtain the list of "plausible" solutions:*

    1.1 *Initialize:* Start with $K \times C$ possible (k, c) pairs to be explored. Let $\mathcal{E}_0 = \{(k, c) : k = 1, \ldots, K \text{ and } c = c_1, \ldots, c_C\}$.

    1.2 *Iterate:* If $\mathcal{E}_{l-1}$ is the set of pairs (k, c) not already explored at stage $l - 1$, then:

    1.2.1 Obtain $(k_*^l, c_*^l) = \arg\min_{(k,c) \in \mathcal{E}_{l-1}} F_m(k, c, c_X)$, where $m = MM$ (MIX-MIX), MC (MIX-CLA) or CC (CLA-CLA).

    For each "optimal" pair $(k_{\text{opt}}^t, c_{\text{opt}}^t)$, we analyse the so-called "best interval" $\mathcal{B}_t$, the set of consecutive values of c adjacent to c_{opt}^t (say c^*) for which the solution remains optimal. That is:

    $$\mathcal{B}_t = \{c^* : F_m(k_{\text{opt}}^t, c^*) \leq F_m(k_1, c_1)\}, \qquad (8.31)$$

    where $= (k_1, c_1) \in (\mathcal{E}_{l-1} \cap k_1 \neq k_{\text{opt}})$.

    and the "stable interval" is defined as

    $$\mathcal{S}_t = \{c : \text{ARI}(\mathcal{P}(k_{\text{opt}}^t, c), \mathcal{P}(k_{\text{opt}}^t, c_{\text{opt}}^t)) \geq \varpi\}. \qquad (8.32)$$

    A large interval $\mathcal{B}_t$ means that the number of clusters k_{opt}^t is "optimal" in the sense of (8.31) for a wide range of c values. A large interval $\mathcal{S}_t$ means that the solution is "stable" in the sense of (8.32), because the change when moving c in that interval is irrelevant.

    1.2.2 Remove all cluster partitions $(k, c) \in \mathcal{B}_t \cup \mathcal{S}_t$ (set of similar partitions). Take $\mathcal{E}_l$ as the set $\mathcal{E}_{l-1}$ after removing the pairs yielding "similar" partitions found at step l.

    1.3 *Finalize:* The iterative procedure ends when $\mathcal{E}_L = \emptyset$ (or when L is a positive prefixed integer number) and it returns $\{(k_*^1, c_*^1), (k_*^2, c_*^2), \ldots, (k_*^L, c_*^L)\}$ as a list with L "feasible" parameter combinations.

2. *Obtain the list of "optimal" (non-repetitive) solutions:*

    2.1 *Initialize:* Start from $\mathcal{I}_0 = \{1, \ldots, L\}$ and the $L \times L$ matrix $(d_{r,s})_{r,s=1,\ldots,L}$, where

    $$d_{r,s} = \text{ARI}(\mathcal{P}(k_*^r, c_*^r), \mathcal{P}(k_*^s, c_*^s)).$$

    2.2 *Iterate:* Given $\mathcal{I}_{t-1}$ the non discarded "plausible" solutions at stage $t - 1$:

    2.2.1 Take $(k_{\text{opt}}^t, c_{\text{opt}}^t) = (k_*^{l_t}, c_*^{l_t})$, where l_t is the tth element of $\mathcal{I}_{t-1}$ (the indexes in $\mathcal{I}_{t-1}$ are sorted from lowest to highest).

    2.2.2 Discard "repetitive" solutions (i.e. those that are similar to the already detected "optimal" ones): $\mathcal{I}_t = \mathcal{I}_{t-1} \setminus \{r : r \in \mathcal{I}_{t-1}, r > l_t \text{ and } d_{r,l_t} \geq \varpi\}$.

2.3 *Finalize:* The iterative procedure ends when $\mathcal{I}_T = \emptyset$. It returns

$$\{(k^1_{\text{opt}}, c^1_{\text{opt}}), (k^2_{\text{opt}}, c^2_{\text{opt}}), \ldots, (k^T_{\text{opt}}, c^T_{\text{opt}})\}$$

as the "optimal" set of pairs (with $T \leq L$).

The results of the procedure can be visualized in an informative plot known as the *car-bike plot*. In this plot the optimal pairs are shown by circles ("bikes"). In the circle we write two integers that rank the solution's quality. More precisely, the first integer indicates the rank of the solution among the optimal non-repetitive ones, while the second integer indicates the rank of the solution among all others. For each optimal pair, the sets $\mathcal{B}_t$ and $\mathcal{S}_t$ are shown respectively with boxes and lines ("cars"). The height of the rectangle is proportional to the goodness of the solution, in terms of the Information Criterion: the best solution has height larger than the second-best solution, which in turn has height larger than the third best and so on. This means that *a rule of thumb for choosing the best combination could be to look for the rectangle of largest area.*

Note that, in our approach, among all the possible solutions, the best ones are those which are stable along the widest interval of c values, rather than the solution which maximizes an information criterion just for a specific combination of k and c. It can also happen that rectangles can range through all values of c for different values of k. In this case, following Occam's Razor, our preferred solution would be the one associated with the rectangle with smallest k. It may also happen that the car-bike plot (as in the face masks example of Sect. 8.5.5) reveals the presence of more than one solution. In this case we suggest finding the best value of the trimming factor for each of the tentative solutions. The above procedure in the case of TCWM had kept $c_X = 128$. For each tentative solution it is possible to investigate the different values of c_X in order to monitor the stability of the results and to choose the optimal constraint among the scatter matrices in the space of the explanatory variables.

8.5.4.2 Estimate of the Optimal Level of Trimming

With the estimated k and c, we apply TCLUST-REG or TCWM on the same dataset many times, for different trimming levels α. This produces a set of plots for monitoring the change of a series of statistics among two consecutive values of α:

- The change in Adjusted Rand Index,
- The change in the regression coefficients. The formula which is used is

$$||\hat{b}_{\alpha_r} - \hat{b}_{\alpha_s}||^2 / ||\hat{b}_{\alpha_r}||^2,$$

where $\hat{b}_{\alpha_r} = \text{vec}(\hat{b}_{1,\alpha_r}, \hat{b}_{2,\alpha_r}, \ldots, \hat{b}_{k,\alpha_r})$ is the column vector of length $p \cdot k$ containing the estimates of all the regression coefficients for the k groups using a trimming level α_r, $\hat{b}_{j,\alpha_r} = (\hat{b}_{0,j,\alpha_r}, \hat{b}_{1,j,\alpha_r}, \ldots, \hat{b}_{p-1,j,\alpha_r})^T$ and symbol vec denotes the *vec* operator, while α_r and α_s denote two consecutive levels of trimming ($\alpha_r > \alpha_s$) and p is the number of explanatory variables including the intercept. Using the squared norm makes the computation easier.

Remark: given that, for each value of the trimming factor, the labels of the groups are assigned randomly, we make sure that the labels used are consistent for all values of the trimming factor. More precisely, once the labelling is fixed for the largest value of the trimming factor supplied, we change label j into label i if:

$$\sum_{q=1}^{p} \left(\frac{\hat{b}_{i,q,\alpha_r} - \hat{b}_{j,q,\alpha_s}}{\hat{b}_{i,q,\alpha_r}}\right)^2 < \min_{l \neq j} \sum_{q=1}^{p} \left(\frac{\hat{b}_{i,q,\alpha_r} - \hat{b}_{l,q,\alpha_s}}{\hat{b}_{i,q,\alpha_r}}\right)^2.$$

Groups are successively relabelled in the order of the smallest distance. Note that it may also happen that sometimes a unique relabelling is not possible in the sense that the k new groups are relabelled into $u \leq k - 2$ groups.

- The change in the error variance.

$$\|\hat{s}^2_{\alpha_r} - \hat{s}^2_{\alpha_s}\|^2 / \|\hat{s}^2_{\alpha_r}\|^2,$$

where $s^2_{\alpha_r} = (\hat{s}^2_{1,\alpha_r}, \hat{s}^2_{2,\alpha_r}, \ldots, \hat{s}^2_{k,\alpha_r})^T$, is the column vector of length k containing the estimates of the error variances for the k groups of size $n_1, \ldots, n_k$ using a trimming level α. More precisely: $\hat{s}^2_{j,\alpha} = \sum_{i=1}^{n_j}(y_i - x_i'\hat{b}_{j,\alpha})^2 / n_j$, $j = 1, 2, \ldots, k$.

In order to appreciate abrupt changes in the estimated error variance, we monitor, for each group, the values of $\hat{s}^2_j$ uncorrected and corrected for truncation. The correction for truncation takes into account that the deletion of the $n - h$ most remote observations (where $h = \sum_{j=1}^{k} n_j$) yields too small estimates of σ_j^2, because it is based on the central h observations. The variance $\sigma^2(h)$ of the truncated normal distribution containing the central h/n portion of the full distribution is given in (3.49). Therefore, assuming that the groups are subject to the same level of truncation, the (asymptotically) corrected $\hat{s}^2_c$ are computed as:

$$\hat{s}^2_{cj} = \hat{s}^2_j / \sigma^2(h).$$

In order to have an idea of the units which are on the boundaries among groups and their order of entry into the subset we monitor the posterior probabilities of each observation with respect to a reference group. Finally, in order to appreciate the units which are trimmed and the allocation we can use a series of subplots which monitor the classification for each value of α.

8.5 Robust Regression Clustering

In the next subsection we use the standard BIC combined with two clustering methods to estimate the number of groups. The two clustering methods give slightly different curves for BIC as a function of k. In Sect. 8.5.6 we use the monitoring methods of Sect. 8.5.4 to determine the clusters. The section concludes with a brief comparison of the outcomes from the two approaches.

8.5.5 Analysing the Face Masks Data Using The Traditional Information Criterion

We can obtain the traditional BIC curves for the face masks data using the Flexible Mixture Modelling R package **flexmix** (Gruen and Leisch 2007) and the Flexible Cluster-Weighted Modelling R package **flexCWM** (Mazza et al. 2018). The results, from this point, use both explanatory variables with no intercept in the model, and are in Figs. 8.24 and 8.25 respectively. Note that, unlike the random start approach of the previous section, which leads to stable trajectories, the output of these R packages strongly depends on the starting points and on the number of replicates. Figures 8.24 and 8.25 show two realizations of the approach based on the traditional BIC.

The BIC curve in the left-hand panel of Fig. 8.24 decreases monotonically; this would indicate that the best number of groups should be as large as possible. We selected as the best number of groups 4, where the curve slope starts to be smaller; despite this choice, Flexible Mixture Modelling identifies only three groups, of which the two on the bottom (red circles and blue crosses) completely overlap. The BIC of the left-hand panel of Fig. 8.25 (which is based on the R package **flexCWM**) shows a local minimum when $k = 4$ while, for a number of groups larger than 5, it decreases monotonically. This would indicate that the number of groups should

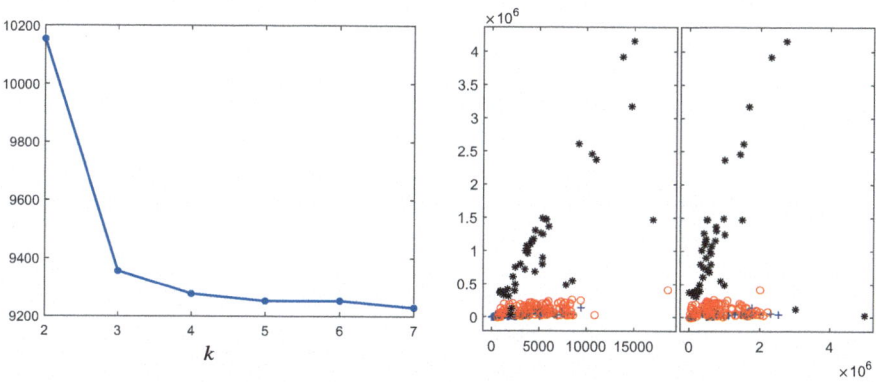

Fig. 8.24 Face Masks data: flexible mixture modelling, using both explanatory variables. Left-hand panel: BIC and number of groups k. Right-hand panel: yX plot of classification of 3 out of the 4 groups. The degree of overlapping among these groups is evident

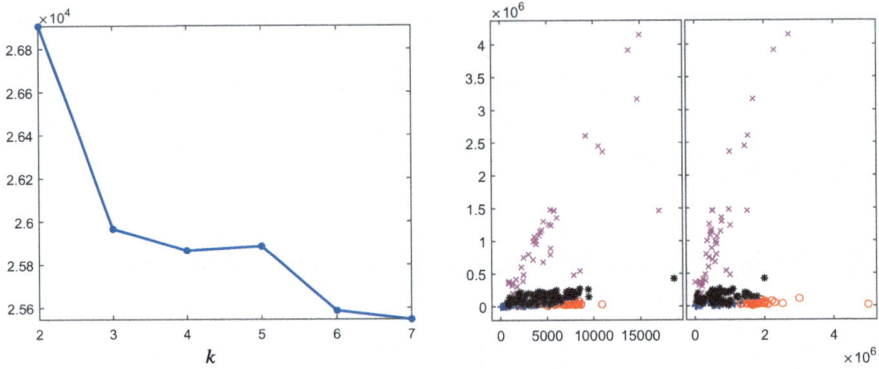

Fig. 8.25 Face Masks data: flexible cluster-weighted modelling, using both explanatory variables. Left-hand panel: BIC and number of groups k. Right-hand panel: yX plot of classification of the 4 groups. The degree of overlapping among these groups is again evident

be as large as possible. We therefore selected $k = 4$ as the best number of groups, based on the local minimum. The plot in the right-hand panel (which contains the associated classification) shows that three of the resulting groups (red circles, blue crosses and black asterisks) overlap considerably. Moreover, the group of circles is mainly associated with large values of quantity. It is clear that these complex international trade datasets cannot be analysed with standard methods, but there is a compelling need for using the monitoring tools described in this book.

8.5.6 Analysing the Face Masks Data with the Monitoring Approach

Here we apply the semi-automatic robust regression clustering tool to the face mask data, using TCLUST-REG and both explanatory variables. The non-normality of the distribution of X coupled with the fact that in this context high leverage points are highly informative about the characteristics of the different levels of price, suggest setting to zero the second-level trimming (Sect. 8.5). We use as Information Criterion the Penalized Mixture Likelihood MIX-MIX. The choice of the last information criterion is due to the considerable degree of overlap among the components (see Sect. 8.5.4.1).

The left-hand panel of Fig. 8.26 (an elbow plot) shows that the best solution suggested by the information criterion is when $c = 128$ and $k = 5$. On the other hand, the car-bike plot in the right-hand panel of Fig. 8.26, indicates that this solution is very local, being valid just for this (c, k) combination. The plot also shows two solutions for $k = 3$ and $k = 4$ which deserve particular attention.

8.5 Robust Regression Clustering 323

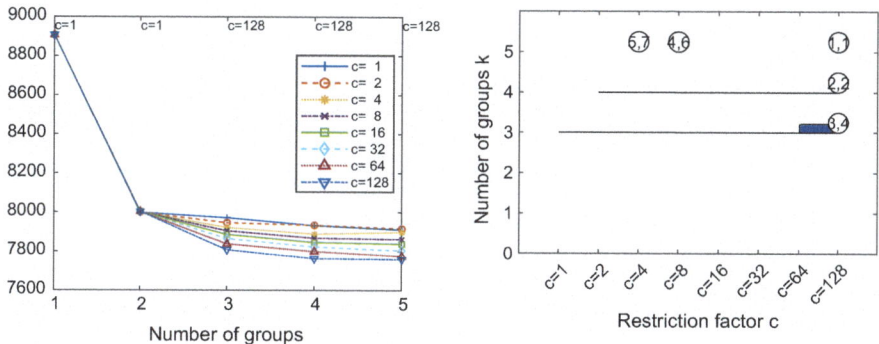

Fig. 8.26 Face Masks data: left-hand panel elbow plot (k on the horizontal axis); right-hand panel car-bike plot (k on the vertical axis). In both panels $\alpha = 0.1$

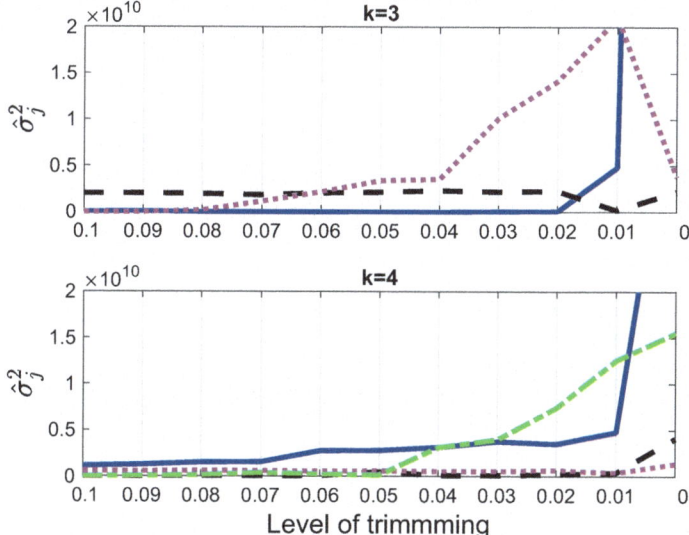

Fig. 8.27 Face Masks data: monitoring error variances when $k = 3$ (upper panel) and $k = 4$ (lower panel)

In Fig. 8.27 the monitoring of each group error variance $\hat{s}_j^2$ for $k = 3$ shows a clear increase when $\alpha = 0.03$. On the other hand, when $k = 4$, the big increase takes place when $\alpha = 0.02$. Therefore, the optimal levels of trimming are 0.04 and 0.03 respectively. Figures 8.28 and 8.29 report the final TCLUST-REG classifications of the good units (the trimmed ones are not shown) together with the estimated regression coefficients and the associated group sizes. In both cases, the data appear partitioned in very sensible groups, which capture the fine-grained structure of this (only apparently simple) dataset. The slope coefficients $\hat{b}_{1,l}$ represent the estimated prices per Kg. The classification with 3 groups splits the strip associated with the

Fig. 8.28 Face Masks data: final classification based on $k = 3$; the 4% trimmed units (denoted in the legend with symbol "+" −1 in faint grey) are not shown. Note that the legend is clickable in the sense that it is possibile to hide/show a particular group of units simply clicking on the corresponding item in the legend

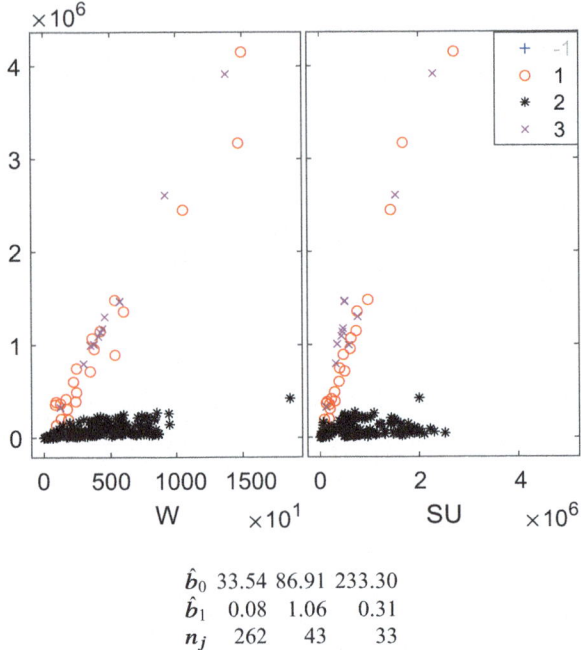

$\hat{b}_0$	33.54	86.91	233.30
$\hat{b}_1$	0.08	1.06	0.31
n_j	262	43	33

Fig. 8.29 Face Masks data: final classification based on $k = 4$; the 3% trimmed units (denoted in the legend with symbol "+" -1 in faint grey) are not shown

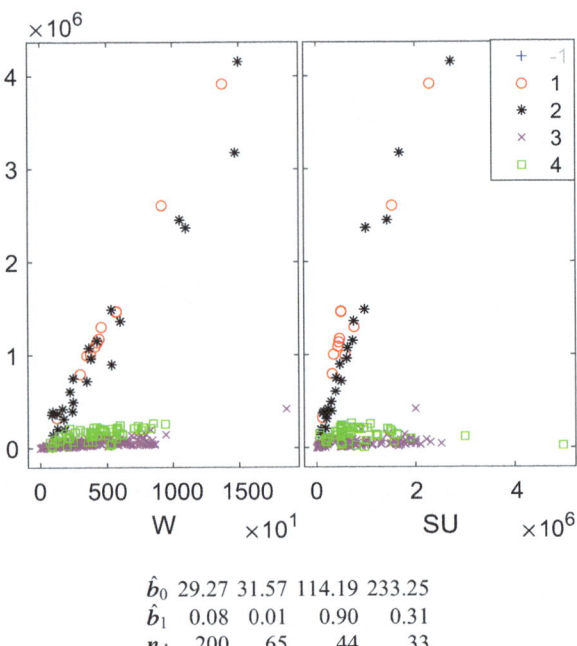

$\hat{b}_0$	29.27	31.57	114.19	233.25
$\hat{b}_1$	0.08	0.01	0.90	0.31
n_j	200	65	44	33

Table 8.2 Face Mask data: adjusted R^2 inside each group in non-decreasing order. First two columns: our final classification based on 3 and 4 groups. Third and fourth columns: classification from **flexmix** and **flexCWM**

3 groups	4 groups	**Flexmix**	**FlexCWM**
0.9999	0.9999	0.936	0.9425
0.9933	0.9909	0.7962	0.6698
0.5376	0.7849	0.3914	0.3813
	0.5074	−0.0296	

higher slope into two groups, while the strip associated with a low declared price is allocated to the same group. On the other hand the allocation with 4 groups divides both the high price and low price units into two subgroups.

We conclude the analysis of this example by comparing the degree of internal cohesion inside each group in the final classification between the suggested approach and the output from **flexmix** and **flexCWM**, Sect. 8.5.1. The values of adjusted R^2 (in non-decreasing order) inside each group for the different methods are given in Table 8.2. The different degrees of homogeneity inside each group are very evident.

8.6 Robust Monitoring of Time Series

Rousseeuw et al. (2019) extend the theory of high breakdown methods for regression and scatter matrices to the robust analysis of time series. The estimation combines ideas from the fast least trimmed squares algorithm (see Sect. 3.11.2) with alternating least squares (ALS), a technique first used with this name by (De Leeuw, 1968), probably inspired by the seminal work of Kruskal (1965). This time series approach is denoted as LTSts (*Least Trimmed Squares in time series*), as the estimation method derives from LTS.

The approach has wide applicability and is especially suitable for the modelling and forecasting of general time series presenting trends and seasonality that stay unchanged or remain reasonably stable. This is often true, for example, in relatively short economic series like those in Rousseeuw et al. (2019), where the fluctuations of official trade statistics, generated monthly by the statistical office of the European Union, repeat over a one-year period. However, the analysis of these series is complicated by the frequent presence of outliers and structural breaks. The latter are points in which a change in level takes place, following revisions of trade legislation (e.g. anti-dumping duties on imports of steel originating in certain countries), macro-economic or financial shocks (such as the one consequent to the COVID-19 pandemic), changes in the EU's political and economic relations with specific countries (for example, the introduction of sanctions on Russia during the war with Ukraine). Being able to model these patterns is of obvious interest from the economic and policy perspective and also of importance in monitoring possible cases of circumvention of trade measures.

This section argues that the monitoring approach that characterizes the Forward Search is also applicable and advantageous in this case. By analogy, the monitoring approach to time series is baptized FSRts (*Forward Search Regression in time series*).

8.6.1 A Non-linear Time Series Model

The regression model of Rousseeuw et al. (2019) for a set of observations on a variable y_t taken over time $t = 1, 2, \ldots, T$ contains trend, time varying seasonality and level shift. For the special case of monthly data:

$$y_t = \sum_{a=0}^{A} \beta_{a,0} t^a \tag{8.33a}$$

$$+ \left(1 + \sum_{g=1}^{G} \gamma_g t^g\right) \underbrace{\left[\sum_{b=1}^{B} \left\{\beta_{b,1} \cos\left(\frac{2\pi b}{12} t\right) + \beta_{b,2} \sin\left(\frac{2\pi b}{12}\right)\right\}\right]}_{S_t} \tag{8.33b}$$

$$+ \delta_1 I(t \geq \delta_2) \tag{8.33c}$$

$$+ \epsilon_t, \tag{8.33d}$$

where

Equation (8.33a): the trend is polynomial of degree A;
Equation (8.33b): the seasonal component S_n has a one-year period if $B = 1$. When $B = 2$ there are both one-year and half-year periods—and so on—with an amplitude that can vary over time in a G-degree polynomial way;
Equation (8.33c): the potential level shift of amplitude δ_1 is at an unknown time point $2 \leq \delta_2 \leq T - 1$;
Equation (8.33d): the errors are assumed i.i.d. and normally distributed, $\epsilon_t \sim N(0, \sigma^2)$.

The number of model components depends on the choice of the constants A, B, and G. Model flexibility can become progressively greater by increasing these constants and, possibly, by incorporating in the model additional covariates (to capture exogenous effects) and auto-regressive components (to absorb possible autocorrelations and account for unexplained dynamics in the response y_t). A procedure for the automatic selection of the number of terms and to detect the presence of multiple level shifts is currently under investigation.

8.6.2 LTSts Estimation

If we denote the two sets of A and B parameters by

$$\theta_B = \{\beta_{1,1}, \beta_{1,2}, \ldots, \beta_{1,B}, \beta_{2,1}, \beta_{2,1}, \ldots, \beta_{2,B}\} \text{ and } \theta_A = \{\beta_{0,0}, \beta_{1,0}, \ldots, \beta_{A,0}\},$$

then model (8.33) can be rewritten as

$$y_t = \left(1 + \sum_{g=1}^{G} \gamma_g t^g\right) S_t(\theta_B) + g_t(\theta_A) + \delta_1 I(t \geq \hat{\delta}_2) + \epsilon_t. \qquad (8.34)$$

This notation helps in showing how the parameter estimation of this non-linear model can be reduced to the repeated alternation of two linear fits, assuming a series of specific shift positions in $\hat{\delta}_2$. Each two-step iteration first solves

$$y - S_t(\hat{\theta}_B^{(k-1)}) = S_t(\hat{\theta}_B^{(k-1)}) \sum_{g=1}^{G} \gamma_g t^g + g_t(\theta_A) + \delta_1 I(t \geq \hat{\delta}_2) + \epsilon_t,$$

holding constant $\hat{\theta}_B^{(k-1)}$ from the previous step; then, with the so-obtained estimates $\hat{\theta}_A^{(k)}$, $\hat{\delta}_1^{(k)}$ and $\hat{\gamma}_g^{(k)}$ ($g = 1 \ldots G$), solves

$$y - g_t(\hat{\theta}_A^k) - \hat{\delta}_1^{(k)} I(t \geq \hat{\delta}_2) = S_t(\theta_B) \left(1 + \sum_{g=1}^{G} \hat{\gamma}_g^{(k)} t^g\right) + \epsilon_t.$$

Here we skip several technicalities, detailed in Rousseeuw et al. (2019), which contribute to the quality of the final fit, such as the initialization of the parameters, the search of $\hat{\delta}_2$ over various tentative positions and the choice of the subsets that are extracted to find the robust estimator.

8.6.3 FSRts for an Adaptive Data-Driven Time Series Monitoring

As in the standard regression context, the LTSts procedure of Sect. 8.6.2 can be used as the initial step of a Forward Search process, where we need a criterion to find the initial subset to initialize the search. For this purpose we can also use the LTSts with concentration steps or a subset supplied directly by the user. Being used only in the initialization, the selection of the number h of observations that will determine the LTSts estimator does not need to be accurate and can be set, for example, to $0.75\,T$.

The FSRts preserves all the properties of the Forward Search (even if further research is needed when there is an auto-regressive component). In particular, the outlier detection rule based on consecutive exceedances will have an overall approximate size of 1%, in the sense that for a particular step m we expect to find exceedances of the 99% quantile in a fraction of 1% of the samples under the null normal distribution. This implies that the results of the FSRts should be compared with those of the LTSts under a simultaneous size of 0.01, a comparison we make in the next section.

8.6.4 Contaminated Airline Data

We illustrate the LTSts and FSRts procedures using the classic airline data introduced in Box and Jenkins (1976). There are 144 observations of the monthly totals (in thousands) of international airline passengers between 1949 and 1960. Thus, the length of the seasonal period is 12 years.

The plot of the original series (left-hand panel of Fig. 8.30) shows a clear linear trend, suggesting $A = 1$. A seasonal component with two harmonics (4 parameters, two for the sine and two for the cosine), should be sufficient to capture the data fluctuations, therefore we can choose $B = 2$. The amplitude seems to increase linearly, which requires the order of the trend of the seasonal component to be $G = 1$. This model can be used to analyse also the contaminated airline data in the right-hand panel of Fig. 8.30, containing an introduced level shift and some outliers (as an alternative, see Exercise 8.4).

Figure 8.31 shows the application of the LTSts. The left-hand panels are obtained with a value for h set to $0.75\,T = 108$; this means that the LTSts estimator is based on 108 observations, giving a *bdp* of 25%, an intermediate choice between a high breakdown and efficient estimation. LTSts finds the contaminated observations but

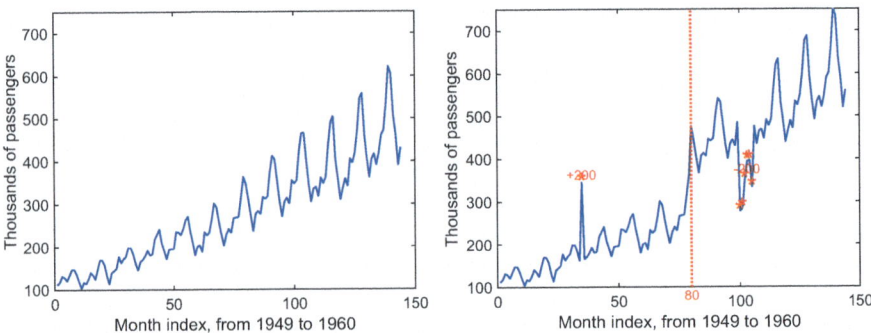

Fig. 8.30 Airline data: left-hand panel, the original data; right-hand panel, a copy contaminated at month 80 with a structural break (upward level shift of 130) and outliers at months 35, 100–105 with shifts of 200, the first upward and the other six downwards

8.6 Robust Monitoring of Time Series

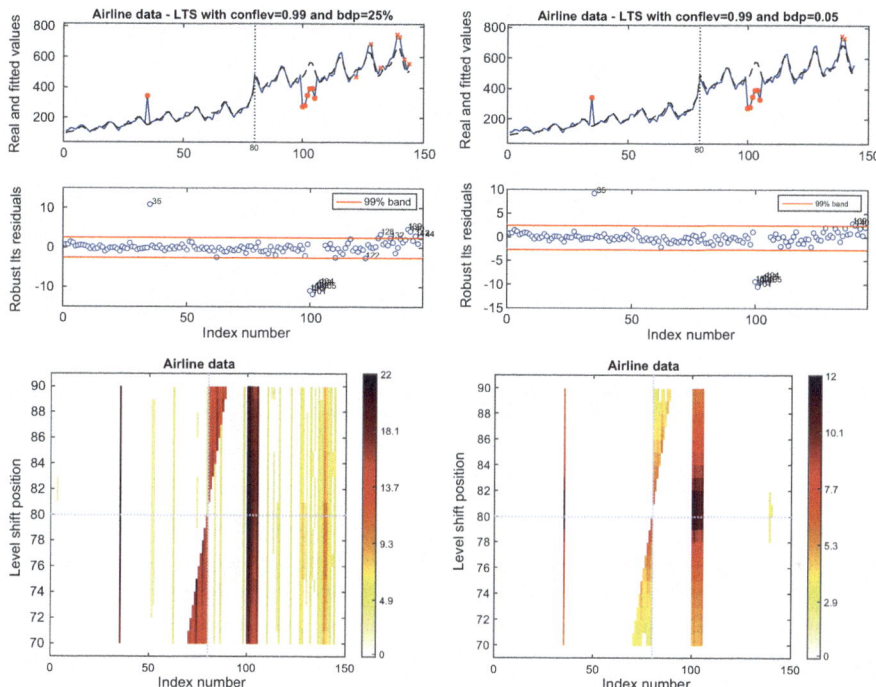

Fig. 8.31 Contaminated Airline data: analysed using LTSts with 25% (left) and 5% (right) breakdown point. Top panels: the original series in blue solid line, the fitted series in black dashed line; mid panels, the index plots of residuals (the width of the crosses depends on the size of the absolute value of the residuals). Bottom panels: the *double wedge plot* of the residuals. The y axes denote the hypothesized position of the level shift. The colour bars on the right show the magnitude of the absolute value of the residuals. The faint dashed blue lines indicate the actual position of the level shift

declares 7 additional outliers at positions 122, 128, 132, 139, 140, 142 and 144. The right-hand panels are obtained with a much lower breakdown point of 5%, leading to just a couple of additional outliers.

We close with a comment on the graphical tool in the bottom panels of Figure 8.31, named by Rousseeuw et al. (2019) *double wedge plots*. It visualizes the presence of level shifts and (groups of) outliers using the absolute scaled residuals produced by the LTSts. Actually it can be seen as a monitoring plot. The vertical axis ranges over all potential positions of a level shift. Then, for any position, it shows on the horizontal axis the absolute scaled residuals in all of the times. The colour in the plot depends on the magnitude of that absolute residual. Isolated outliers, which have a large absolute scaled residual from the robust fit, appear as dark vertical lines (observation 35). Groups of consecutive outliers generally appear as dark vertical bands (observations 100–105). Level shifts show up with two approximately triangular shapes (two wedges, in this case around $t = 73$). This last effect can be easily explained. If the true level shift is at position t^* and the algorithm is checking the

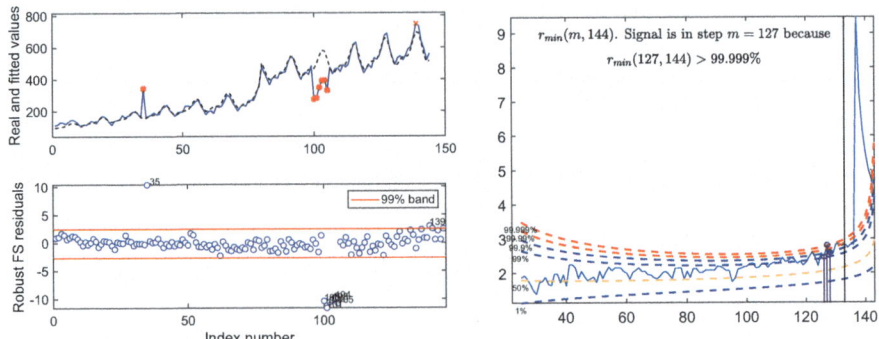

Fig. 8.32 Contaminated Airline data: analysed using FSRts. Left-hand panel: the fitted series and the index plot of residuals. Right-hand panel: the forward plot of the minimum deletion residuals. The units declared as outliers are shown with red crosses

candidate some steps earlier, those observations close to the level shift will be seen as outliers. In approaching t^* from the left, the monitored scaled residuals will form a dark upward-pointing wedge. To the right of t^* we obtain an analogous wedge pointing downward. The steps in the two wedges close to the true level shift position implicitly count the number of additional outliers which are needed to impose a level shift in that particular position.

Figure 8.32 shows the results obtained with the FSRts. As shown in the right-hand panel of the figure, the search has progressed to step 127 before detecting a signal, leading to a final outlier declaration of the 7 contaminated observations and just one additional outlier towards the end of the time series (139). Once again, it is clear that deriving the choice of observations to be used for the fitting through the monitoring approach, leads to results that are less dependent on the choice of critical hyper-parameters, such as the breakdown point.

8.7 Regression with Compositional Data

In many cases the datasets are characterized by multivariate observations (vectors) containing relative contributions of parts to a whole. Examples are geochemical composition of rocks, household budget patterns, time budgets, and ceramic compositions. A plethora of further examples can be found in Aitchison (1986; 2005) and the hundreds of papers published on this topic. Compositional data have generally been defined as a vector of proportions, a vector with strictly positive components whose sum is a constant (Aitchison 1986; Aitchison 2005; Pawlowsky-Glahn et al. 2008; Pawlowsky-Glahn and Egozcue 2006). However, this definition has changed and nowadays we refer to the term compositional to any set of multivariate observations with strictly positive components where relative rather than absolute information is relevant for the analysis (Pawlowsky-Glahn et al. 2015b; Pawlowsky-Glahn et al.

8.7 Regression with Compositional Data

2015a; Egozcue and Pawlowsky-Glahn 2019). Due to the relative character of compositional data, application of standard statistical multivariate methods, which mostly rely on Euclidean geometry, might lead to misleading results. Often this property of the data is ignored, although linear regression models are only reasonable if the covariates $x = (x_1, \ldots, x_D)^T$ carry absolute information (Hron et al. 2012). Therefore, it is usual to apply a suitable transformation. A first remedy would be a log transformation which will often reduce data skewness, but will not accommodate the compositional nature of the data. Aitchison (1986) suggested several possible transformations from the family of logratio transformations: the *additive logratio (alr)* and the *centered logratio (clr)* transformations, but these present some inconveniences for multivariate analysis (Argote-Espino et al. 2018; Filzmoser et al. 2012). These disadvantages have led to the development of the *isometric logratio (ilr)* transformation which relates the geometry on the simplex directly to the Euclidean geometry (Egozcue et al. 2003).

After a suitable transformation of compositional data we have a sample from a single population in which outliers may be present. To make the analysis insensitive to the influence of any outliers, robust methods are necessary.

Mathematically a D-part compositional dataset is a matrix X of n compositional vectors $x_i = (x_{i1}, \ldots, x_{iD})^T$ in R^D, $i = 1, \ldots, n$. A D-part simplex is defined as

$$S^D = \{x = (x_1, \ldots, x_D)^T, x_i > 0, \sum_{i=1}^{D} x_i = \kappa\}, \tag{8.35}$$

where $\kappa > 0$ is an arbitrary constant. This vector space is characterized by its own geometry called Aitchison geometry (Egozcue et al. 2011). For that vector space some basic operations are defined, corresponding to the addition of two vectors and multiplication of a vector by a real number in the Euclidean geometry and special notation is introduced. Under the operations *perturbation, powering* and *inner product* alone it can be shown (Pawlowsky-Glahn et al. 2015a) that the Aitchison geometry on the simplex forms a $(D-1)$-dimensional Euclidean vector space. The details of this geometry are described in detail in various papers (see, for example, Egozcue and Pawlowsky-Glahn 2006).

Each compositional vector x can be rescaled by a constant c when the compositions x and $y = cx$ are compositionally equivalent. This rescaling can be defined formally by the so-called *closure* operator given by

$$\mathcal{C}(x) = \frac{\kappa}{\sum_{i=1}^{D} x_i}(x_1, \ldots, x_D). \tag{8.36}$$

Even if the data are not subject to a constant sum constraint, standard statistical methods could lead to doubtful results if they were directly applied to the original data (Filzmoser et al. 2018).

Considering linear regression in the context of compositional data one is confronted with four different cases:

1. Compositional response variables and real (non-compositional) explanatory variables;
2. Real response variable(s) and compositional explanatory variables;
3. The response variables and the explanatory variables are compositions and
4. The response variable is a part of the same composition containing the explanatory variables.

Cases (1) and (3) suggest multivariate regression and thus are outside the scope of this book. In case (4), since both the response and the explanatory variables originate from one composition, it cannot be assumed that the covariates represent errorless variables as in case (2) of a real-valued response. Consequently, the use of an ordinary multiple regression model is inappropriate and can lead to biased results. Therefore, an orthogonal regression model (or, equivalently, a total least squares model) should be applied for this purpose, which is a specific type of errors-in-variable (EIV) model (Fuller 1987). This case is considered in detail by Hrusova et al. (2016).

In the remainder of this chapter we consider Case (2) with a single response variable. These regression models could be developed from scratch for the original compositions directly in the Aitchison geometry. But, in order to be able to take advantage of the standard tools for regression analysis, we use the coordinate representation of the compositions provided by the *ilr* transformation.

8.7.1 Coordinate Representation of Compositional Data

A common practice in the analysis of compositions is first to project the data onto the unconstrained real space by re-expressing the data in logarithms of ratios between parts. Several types of transformation have been proposed in the literature; the choice of the most appropriate transformation depends on the data at hand and on the type of statistical analysis envisaged. Then standard statistical methods can be applied to the transformed data, with the results back transformed to the original space.

The *additive logratio transformation (alr)* introduced by Aitchison (1986) maps a D-part composition in the simplex non-isometrically to a $D - 1$ dimensional vector in $\mathbb{R}^D$, treating one part (usually the last one) as a common denominator of the others. While very easy to interpret, the disadvantage of this transformation is that distances are not preserved; the corresponding basis on the simplex is not orthonormal with respect to the Aitchison geometry (Pawlowsky-Glahn et al. 2008).

Another transformation, also proposed by Aitchison (1986) is the *centered logratio (clr)* transformation which maps a D-part composition x from the simplex isometrically to a vector $y \in \mathbb{R}^D$, using the geometric mean as a common denominator.

8.7 Regression with Compositional Data

$$y = clr(x) = (y_1, \ldots, y_D)^T = \left(\log \frac{x_1}{\sqrt[D]{\prod_{j=1}^{D} x_j}}, \ldots, \log \frac{x_D}{\sqrt[D]{\prod_{j=1}^{D} x_j}}\right)^T. \quad (8.37)$$

While very useful in terms of interpretation, this transformation produces observations with components which sum up to zero and thus the obtained data matrix does not have full rank. The covariance matrix of the transformed data is singular which does not allow application of most multivariate robust statistical methods. The desired orthonormality is fulfilled by the *isometric logratio (ilr)* transformations from $\mathbb{R}^D$ to $\mathbb{R}^{D-1}$ which, for a given basis, can be defined as Egozcue et al. (2003) $z = (z_1, \ldots, z_{D-1})^T$ with:

$$z_i = \sqrt{\frac{D-i}{D-i+1}} \log \frac{x_i}{\sqrt[D-i]{\prod_{j=i+1}^{D} x_j}} \quad \text{for} \quad i = 1, \ldots, D-1. \quad (8.38)$$

In this representation the compositional part x_1 is contained only in the first coordinate z_1 as a scaled logratio of the part itself with the geometric mean of all the remaining parts. Thus, this coordinate can be interpreted in terms of the relative abundance of the part x_1 compared to the "average behaviour" (in terms of geometric mean) of the remaining parts in the composition. These coordinates are usually referred to as *pivot coordinates* (Hron and Fišerová 2011) because one part (e.g. x_1) is set to be a pivot and is contained just in the first coordinate. The first coordinate z_1 can be represented alternatively as a scaled sum of all pairwise logratios containing the part x_1 in the numerator, as follows:

$$z_1 = \frac{1}{\sqrt{D(D-1)}} \left(\log \frac{x_1}{x_2} + \cdots + \log \frac{x_1}{x_D}\right). \quad (8.39)$$

It should be noted that the first pivot coordinate z_1 in (8.39) is proportional to the first *clr* coefficient y_1 in (8.37):

$$z_1 = \sqrt{\frac{D}{D-1}} y_1.$$

The coordinates in (8.38) define a special role to the first part of the composition but this is not a limitation. If it is of interest to obtain interpretation for a specific part within a given composition, say $x_l, l \in \{1, D-1\}$, which is not the first part, one can simply permute the compositional parts in such a way that the part of interest is moved to the first position. Then the original composition $x = (x_1, \ldots, x_D)^T$ is replaced by the permuted composition

$$x^{(l)} = (x_l, x_1, \ldots, x_{l-1}, x_{l+1}, \ldots, x_D)^T = (x_1^{(l)}, \ldots, x_D^{(l)})^T$$

and the pivot coordinates (8.38) are constructed for the permuted composition. The coordinates $z^{(l)} = (z_1^{(l)}, \ldots, z_{D-1}^{(l)})^T$ are defined as

$$z_i^{(l)} = \sqrt{\frac{D-i}{D-i+1}} \log \frac{x_i^{(l)}}{\sqrt[D-i]{\prod_{j=i+1}^{D} x_j^{(l)}}} \text{ for } i = 1, \ldots, D-1. \quad (8.40)$$

Thus, we obtain D different orthonormal coordinate systems, which are orthogonal rotations of each other. Each of these coordinate systems emphasizes the role of the respective compositional part placed at the first position (Hron and Fišerová 2011).

As before, the relation between the first coordinate $z_1^{(l)}$ and the *clr* coefficient y_l can be generalized

$$z_1^{(l)} = \sqrt{\frac{D}{D-1}} y_l. \quad (8.41)$$

The pivot coordinates represent one-to-one mapping and it is possible to restore the original parts by the following expressions:

$$x_1 = \exp\left(\frac{\sqrt{D-1}}{\sqrt{D}} z_1^{(l)}\right),$$

$$x_j = \exp\left(-\sum_{k=1}^{j-1} \frac{1}{\sqrt{(D-k+1)(D-k)}} z_k^{(l)} \frac{\sqrt{D-j}}{\sqrt{D-j+1}} z_j^{(l)}\right), j = 2, \ldots, D-1,$$

$$x_D = \exp\left(-\sum_{k=1}^{D-1} \frac{1}{\sqrt{(D-k+1)(D-k)}} z_k^{(l)}\right). \quad (8.42)$$

8.7.1.1 Example 1: Manufacturing Value Added by Industry

Data on value added by industry were obtained from the UNIDO INDSTAT database available at https://stat.unido.org/ which comprises industrial statistics for all 22 divisions of the manufacturing sector (according to the ISIC Revision 3 classification at 2-digit level of detail: UN 2002) in 174 countries. Several variables, including the value added, are available from 1963 to 2022, but only the data for 2019 have been used for this example. To reduce the complexity of the analysis, instead of the 22 divisions, a derived classification into three technology groups *low technology*, *medium-low technology manufacturing*, and *medium-high and high-technology manufacturing*, as defined in UNIDO (2010, p. 244), was used. The final value added, aggregated for the three technology groups, was available for 101 countries. This dataset is available in **FSDA** with the name valueadded.

The left-hand panel of Fig. 8.33 presents a scatter plot of the two-dimensional *ilr*-transformed data. To better visualize the multivariate data structure we superimpose

8.7 Regression with Compositional Data

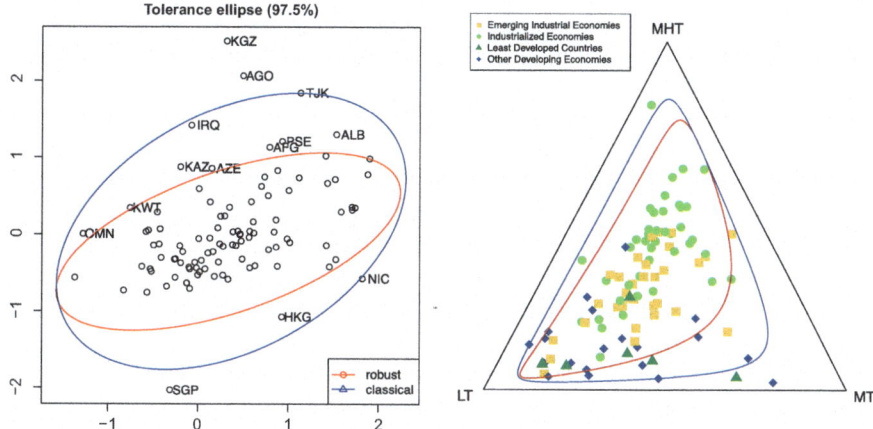

Fig. 8.33 Manufacturing Value Added data *ilr*-transformed: 0.975 tolerance ellipses based on the robust (MCD) and classical Mahalanobis distances in the left-hand panel. The outliers identified by the robust Mahalanobis distances (outside the red ellipse) are mostly covered by the classical (blue) ellipse. Right-hand panel: ternary diagram with classical and robust back-transformed Mahalanobis distance tolerance ellipses

0.975 tolerance ellipses of the Mahalanobis distances are computed by two methods. The first, non-robust, method uses the sample mean and covariance (blue). The second, robust, method, uses distances calculated using the Minimum Covariance Determinant (MCD) estimator (Rousseeuw and Van Driessen 1999) (red) which is the multivariate version of LTS regression. In the right-hand panel, the ellipses are back transformed to the original space using the inverse *ilr* transformation as proposed in Filzmoser and Hron (2008). The ellipse corresponding to the classical estimates (blue) covers all data points, while the robust one (red), based on MCD, excludes 11 points identified as potential outliers. A further three points are close to the border line and, although identified as outliers with confidence level 0.975, would not be identified if the confidence level were changed to 0.99.

A similar example is analysed using the monitoring approach of the forward search for multivariate data in Todorov (2021).

8.7.2 Regression with Compositional Covariates

In order to design a linear model between a dependent variable y and the set of independent variables x forming a composition, one can use the *ilr* transformation given by (8.38). Thus a standard multiple linear regression of y on the explanatory variables $z = (z_1, \ldots, z_{D-1})^T$ is obtained as follows:

$$E(Y|z) = \beta_0 + \beta_1 z_1 + \cdots + \beta_{D-1} z_{D-1}. \tag{8.43}$$

When n observations of the covariates are available together with n values of the response variable the model can be written in one of the D orthogonal coordinate systems given in (8.40) as

$$y_i = \beta_0^{(l)} + \beta_1^{(l)} z_1^{(l)} + \cdots + \beta_{D-1}^{(l)} z_{D-1}^{(l)} + \epsilon_i, \qquad (8.44)$$

for $i = 1, \ldots, n$ and $l = 1, \ldots, D$. Thus D different multiple regression models are obtained with $\beta^{(l)} = (\beta_1^{(l)}, \ldots, \beta_{D-1}^{(l)})^T$ the unknown regression parameters and ϵ_i the random errors in model l. It can be shown that, due to the orthogonality of the ilr bases, the least squares estimate of the intercept term $\beta_0^{(l)} = \beta_0$ is the same in all D models (Hron et al. 2012). The same is valid for the fitted values of the response variable and consequently for the residuals. The estimation of the regression parameters is usually achieved by minimizing the size of the residuals. Hron (2012) apply LS for parameter estimation but, in order to protect against outliers in the response variable or the explanatory variables, we should apply robust regression methods, such as those described in this book.

In addition to the intercept term parameter β_0, the parameters $\beta_1^{(l)}$ are of particular interest for the interpretation of the first coordinates $z_1^{(l)}$ in (8.40). The remaining coordinates $z_2^{(l)}, \ldots, z_{D-1}^{(l)}$ fully represent the remaining compositional parts $x_2^{(l)}, \ldots, x_{D-1}^{(l)}$ and thus cannot be omitted from the model, but the corresponding regression parameters are not needed for interpretation purposes. Usually the estimates of the parameters $\beta_0, \beta_1^1, \ldots, \beta_1^D$ obtained from the D different regressions are collected in a table, together with their characteristics (standard errors, t-statistics and the corresponding p-values), as if they were coming from the same model.

8.7.2.1 Example 1: Manufacturing Value Added by Industry (continued)

Life expectancy at birth is a key indicator for the Sustainable Development Goal (SDG) 3 which is one of the 17 Sustainable Development Goals established by the United Nations (2015). The official wording of SDG 3 is: "To ensure healthy lives and promote well-being for all at all ages". It is often associated with higher economic development. We continue the example in the previous section by adding life expectancy as a dependent variable to the structure of the manufacturing value added and study this extended relationship. Data on overall human development can be taken from the *Human Development Report* published by the United Nations Development Programme http://hdr.undp.org/en. For an appropriate analysis of this relation, the compositional structure of the independent covariates must be taken into consideration. Furthermore, to control and understand the influence of outlying data points, monitored robust regression estimates should be employed.

Following (8.44), D models need to be computed, where $D = 3$ is the number of compositional parts. Only the first coefficient in each model is of interest, together with the corresponding inference statistic. Parameter estimation can be carried out by

8.7 Regression with Compositional Data

Table 8.3 Life expectancy and structure of Manufacturing Value Added data: regression output of MM-estimation of *ilr*-transformed data

	Estimate	Std. error	t-value	$Pr(>|t|)$	
Intercept	77.3249	0.6211	124.4962	<2e-16	***
Low	−1.3448	0.6211	−2.1651	0.03281	*
Medium	−2.7477	0.8596	−3.1965	0.00187	**
High	4.0924	0.6251	6.5467	2.71e-09	***

Significance codes: 0 '***' 0.001 '**' 0.01 '*' 0.05 '.' 0.1 ' ' 1
Robust residual standard error: 4.221
Multiple R-squared: 0.3532 Adjusted R-squared: 0.34

the function complmrob() from the R package **complmrob** (Kepplinger 2019). This, in turn, calls $D = 3$ times, the function lmrob() from the R package **robustbase** (Maechler et al. 2024), to perform MM-estimation and collect together the parameters β_0 and β_1 from each of the regressions. Table 8.3 shows the standard way of presenting the results, using a prefixed value of efficiency (*eff*= 0.95).

The value of the parameter $\beta_1^{(l)}$ $l = 1, 2, 3$ indicates how much the response variable changes on average by a unit change of the logratio between the contribution of the sector of interest and an average of the contributions of the remaining sectors. While the relative size of the low- and medium-technology sector has a negative effect on the life expectancy, a larger high-technology sector results in a higher life expectancy. The most pronounced influence is the contribution of the high-technology industry. The estimated coefficient is high and significantly positive (4.0924; p-value is 2.71e-09), with 95% confidence interval of 2.85 and 5.33, and is therefore significantly positive. The values of $\beta_1^{(1)}$ and $\beta_1^{(2)}$ are negative, so the larger is the share of the low- or medium-technology industry in the total MVA (Manufacturing Value Added), the lower is the life expectancy. When using the model on the untransformed data (see the results presented in Table 8.4), the high-technology sector does not anymore have a significant influence on life expectancy, in contrast to the very high effect in the model with the *ilr*-transformed value added.

Understanding of these analyses can be increased through use of the monitoring plot of the added variable t-statistics introduced in Sect. 4.8 and exemplified in

Table 8.4 Life expectancy and structure of Manufacturing Value Added: regression output of MM-estimation on the original variables. Standard presentation of the results

	Estimate	Std. error	t-value	$Pr(>|t|)$	
Intercept	85.823	1.740	49.330	<2e-16	***
Low	−12.919	2.589	−4.990	2.65e-06	***
Medium	−19.036	2.843	−6.696	1.40e-09	***
High	4.310	2.506	1.720	0.0886	.

Significance codes: 0 '***' 0.001 '**' 0.01 '*' 0.05 '.' 0.1 ' ' 1
Robust residual standard error: 3.881
Multiple R-squared: 0.4158, Adjusted R-squared: 0.3978

Fig. 4.33. The application of this plot to compositional data is straightforward, apart from some slight extensions due to the nature of the explanatory variables. We use the *ilr* transformation taking each of the three columns of X in turn to form Variable 1. We then take Variable 1 as the added variable, so, in this case, using the other two variables for regression and progression in the FS. The algebra for the added variable t-test is then applied to Variable 1 and the search progresses. In this way, three trajectories of added variable t-tests are calculated.

The resulting plot, obtained by use of function FSRaddt, is in Fig. 8.34 together with a 99% significance band. It is interesting, as a start, to compare the values of the statistics in the figure for x_1, x_2 and x_3 (-2.19, -3.39 and 7.22) with those from Table 8.3 (-2.17, -3.20 and 6.55). These values from MM-estimation and from the end of the FS, that is LS, are similar except for x_3 for which the FS gives a larger statistic. Figure 8.34 shows that the positive significance of x_3 (high tech) increases steadily with m and does not depend on the presence of particular observations. The plot of the statistic for x_2 shows values significant at the 1% level for m in the range 50–92. However, for larger values of m the values are not significant at the 1% level until addition of the last two observations causes a return to significance at this level. In the centre of the search variable x_1 is not significant and remains so until the last units to be included bring it to significance at the 5% level.

If the data are homogeneous, information about each parameter will increase with m and the values of the t-statistics in such figures as Fig. 8.34 will approximately be linear functions of m. The behaviour of the statistics for x_3 in Fig. 8.34 is an example, as are the trajectories in Fig. 4.33. If outliers are present, all trajectories

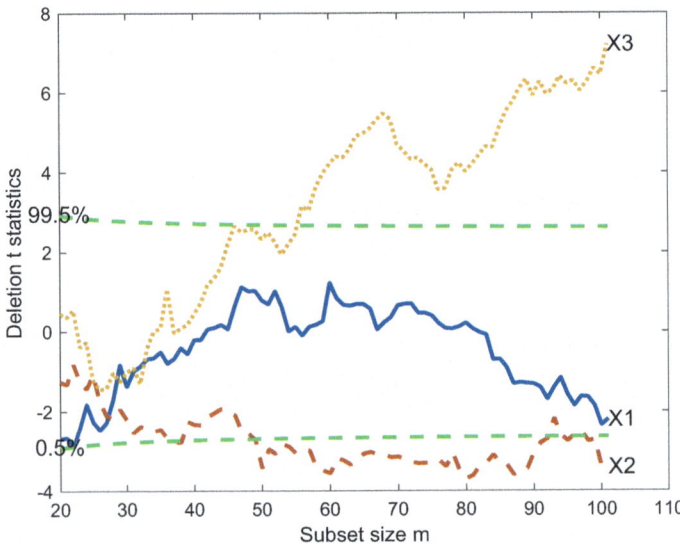

Fig. 8.34 Life expectancy and structure of Manufacturing Value Added data: monitoring plot of added variable t-statistics, *ilr* transformation

8.7 Regression with Compositional Data

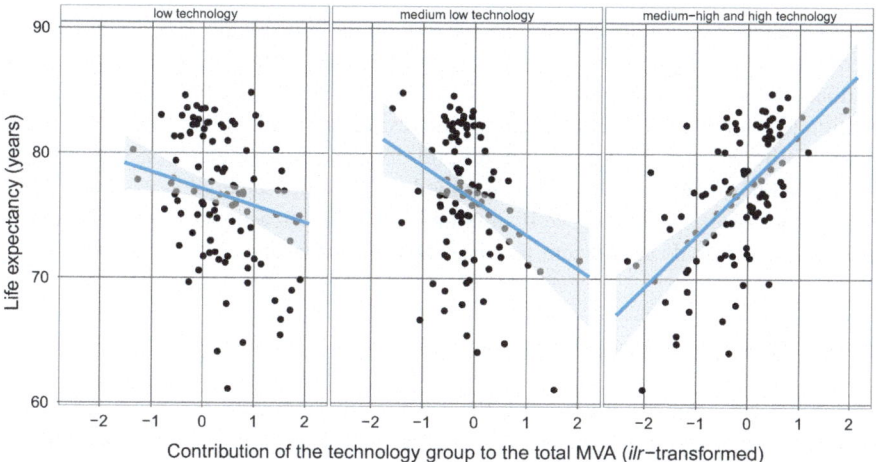

Fig. 8.35 Life expectancy and structure of Manufacturing Value Added data: life expectancy (response) and fitted values with 95% confidence intervals (grey shaded areas) versus the compositional parts of the structure of the manufacturing value added data. These are represented by the first *ilr*-coordinates $z_1^{(l)}$ of the orthonormal basis $l = 1, \ldots, D = 3$

may be affected, as in Fig. 4.41 where, from m around 1650, observations from a different population start to enter the subset; the trajectories of the added variable t-statistics change abruptly. The trajectories for x_1 and x_2 in Fig. 8.34 are of neither form, although there is evidence of outlying observations in the last few steps causing a change in the significance of both variables; the last two units to enter the FS are South Africa and Angola. However, there is an indication of more general lack of homogeneity in the data. This plot, again, shows the importance of a dynamic, rather than a static, robust analysis.

Figure 8.35 illustrates the relation of the response variable life expectancy to the compositional parts, i.e. the contribution of each technology group to the total MVA, represented by the first *ilr*-coordinates $z_1^{(l)}$ of the orthonormal basis $l = 1, \ldots, D = 3$. The fitted values, shown with a 95% confidence interval, were computed by setting the other components to the median of the values in the corresponding direction. It is clearly seen that the dependence is strongly positive only in the medium-high and high-technology categories.

8.7.3 Monitoring Compositional Regression

As already mentioned above, when using orthogonal transformations between pivot coordinate systems (8.40) and as shown by Hron et al. (2012), the estimates of the parameters $\beta_0^l = \beta_0$ remain the same for all D regressions $l = 1, \ldots, D$. The same is true for the fitted values of the response variable and consequently for the residuals.

This facilitates the monitoring approach based on the n residuals. It is not necessary to compute all D regressions for each step of the monitoring; using just one set of residuals reduces the computational burden. We could vary the breakdown point of S-estimates and observe the behaviour of the residuals in a plot or similarly, we could vary the efficiency of the MM-estimates and observe the residuals in a similar plot. The inclusion of outliers would be signalled by a sudden change in residuals with the structure of the plot being well summarized by the measures of the correlations between the residuals at adjacent monitoring values introduced in Sect. 4.9.1.

8.7.3.1 Example 2: Fish Morphology Data

As our second example illustrating the monitoring approach for regression with compositional covariates, we use the FishMorphology dataset introduced in the R package **easyCODA** (Greenacre 2019). The dataset consists of 26 morphometric measurements, in millimetres, on a sample of 75 fish of the species Arctic charr (*Salvelinus alpinus*). Additionally, to each observation, sex (male or female), habitat (littoral, close to the shore), pelagic (in deeper water far from the shore), and the body mass are recorded. The dependent variable y will be the body mass. This dataset is available in **FSDA** with the name FishMorphology. For the sake of our example we select only the former habitat (59 observations) and take the first 10 (out of the 26) morphometric measures as explanatory variables to provide a single population of moderate size. Particularly, we want to reduce the number of variables to no more than 10 since it is a rule of thumb for the methods we are considering that, for each dimension, there should be at least five observations. Of course, we could choose any other set of variables and then the results might be different from those shown here, but the principles of the illustration would not change. Thus we remain with a dataset of 59 observations on the following 10 variables: {Bg, Bd, Bcw, Jw, Jl, Bp, Bac, Bch, Fc, Fdw}. For more details on this dataset see Greenacre (2019, p. 34). It should be noted that in this dataset there is one observation which is an "obvious" outlier, the observation with $ID = 51$ which in our dataset is number 16. This "obvious" outlier can be identified by any outlier detection method and it will appear in all graphs shown below. The data are not subject to a constant sum constraint but, nevertheless, it is compositional in nature according to the more general definition provided in the introduction of the current section. Greenacre (2019), the source of our dataset, first closed the data, before starting the analysis, by applying the closure operator (8.36), but this is not necessary for our example.

After *ilr*-transforming the explanatory part of the dataset, we can carry out robust S- or MM-regression in the same way as in the previous example. The residuals will be identical in all $D = 10$ regressions, therefore, it is sufficient to carry out only the first of the regressions. Setting the *bdp* equal to 0.25, as is often recommended in the literature, produces the plot shown in the upper panel of Fig. 8.36

Only observation 16 lies well outside the 99% band with observation 57 on the border. This is quite similar to the results that we obtained with LS regression. The lower panel of Fig. 8.36 shows the results of regression with the maximal breakdown

8.7 Regression with Compositional Data

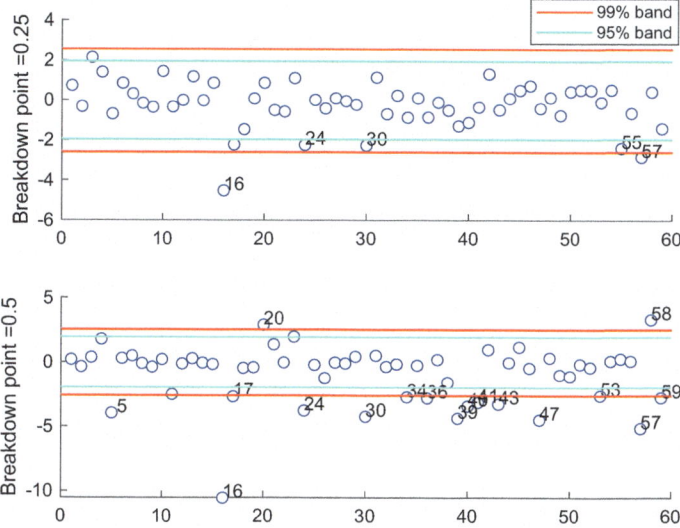

Fig. 8.36 Fish Morphology data: *ilr*-transformed, S-estimation. Index plots of residuals from Tukey's biweight ρ function; two values of *bdp*. Upper panel, *bdp* = 0.25; lower panel, *bdp* = 0.5

point, $bdp = 0.5$. Here many more units are declared as outliers lying outside the 99% bands. As in Sect. 3.13, the question arises whether there are intermediate values of *bdp* which produce results robust enough to reveal the structure of the outliers.

Figure 8.37 shows the residual plot for MM-estimation, again with Tukey's biweight ρ function. In the upper panel, with *eff* = 0.80, twelve outliers are identifiable while in the lower panel, with *eff* = 0.99, the result shown is close to the one produced by LS—only observation 16 is clearly identified and 57 is just outside the border.

Figure 8.38 shows the monitoring plot for S-estimation. This is generated by a series of robust fits on the *ilr*-transformed data, starting from $bdp = 0.5$ and decrementing the value by 0.01 down to 0.01. There are therefore 50 robust fits leading to the plot of scaled residuals in the figure. The abrupt change happens at $bdp = 0.43$ and later there is another slight change at $bdp = 0.1$. The monitoring of the MM-estimates shown in Fig. 8.39 does not show specific structure. The major change, as indicated by the monitoring plots of correlations, happens at the end of the process. The panel of residuals for S-estimation when $bdp < 0.44$ shows a similar set of residuals to MM-estimation when *eff* = 0.8. However, the monitoring plot for MM-estimation fails to show a clear change in structure and so does not provide an empirical estimate of *eff* for a conclusive robust analysis, nor even any indication that a robust analysis is needed.

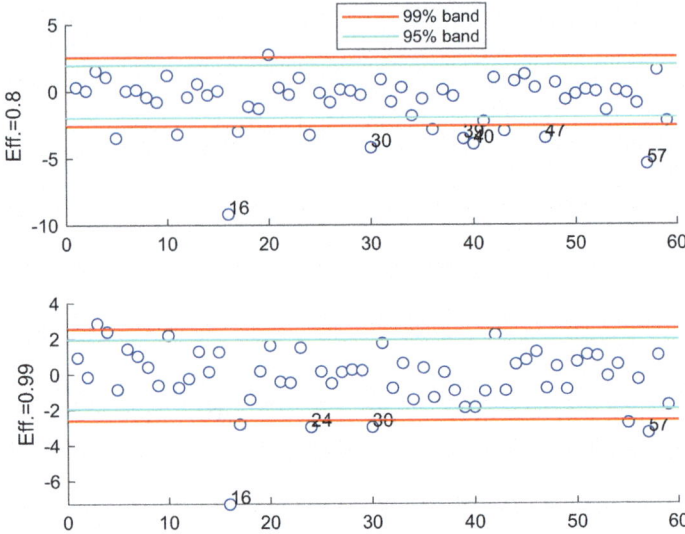

Fig. 8.37 Fish Morphology data: *ilr*-transformed, MM-estimation. Index plots of residuals from Tukey's biweight ρ function; two values of *eff*. Upper panel, *eff* = 0.80; lower panel, *eff* = 0.99

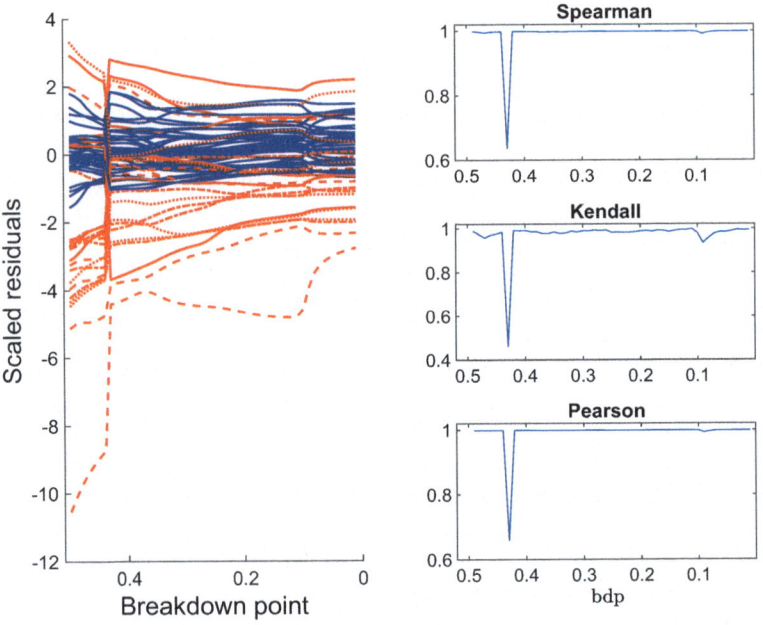

Fig. 8.38 Fish morphology data: *ilr*-transformed, S-estimation. Left-hand panel, plot of scaled residuals. Right-hand panel, three measures of the correlations of adjacent residuals. The abrupt switch at 0.43 is evident

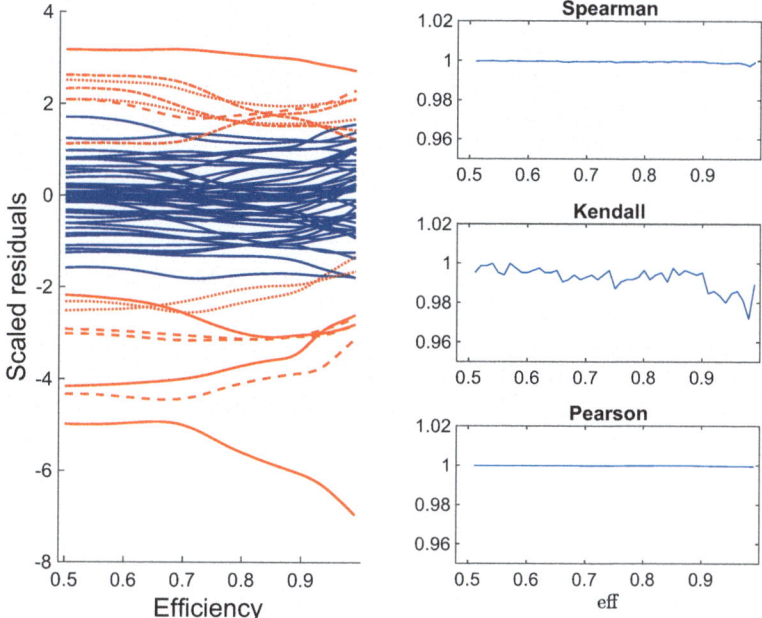

Fig. 8.39 Fish Morphology data: *ilr*-transformed, MM-estimation. Left-hand panel, plot of scaled residuals. Right-hand panel, three measures of the correlations of adjacent residuals.

8.8 Censored Regression

Censored regression data often occur in measuring consumer behaviour involving infrequent purchases. Tobin (1958) introduced a linear regression model with the response left-truncated at zero. He used it to analyse data on the purchase of durable goods; 25% of the respondents had not made such a purchase in the relevant period. The name tobit for this model was introduced by the econometrician Arthur Goldberger, presumably with reference to the rather different probit and logit models (Shiller 1999). In this section we very briefly mention work by Riani et al. (2025a) which combines robust methods including response transformation to provide robust analyses of data for which the doubly truncated regression model is appropriate.

A survey of the literature on tobit models is given by Amemiya (1984). There is potentially both left- and right-censoring in the dependent variable. The standard regression model for the partially latent response is $y_i^* = x_i'\beta + \epsilon_i$, $i = 1, \ldots, n$. Here x_i is a vector of p explanatory variables, there are p parameters β and we assume the independent errors ϵ_i are distributed $\mathcal{N}(0, \sigma^2)$. The dependence of the observed response y_i on y_i^* is given in Table 8.5, where a is the lower limit and b is the upper limit of the uncensored dependent variable and $n_a + n_u + n_b = n$. If $a = -\infty$ or $b = \infty$, the dependent variable is respectively not left-censored or right-censored.

Table 8.5 Censored ($\mathcal{A}$, $\mathcal{B}$) and uncensored ($\mathcal{U}$) observations

Set	Condition	y_i	Number
$\mathcal{A}$	$y_i^* \leq a$	a	n_a
$\mathcal{U}$	$a < y_i^* < b$	y_i^*	n_u
$\mathcal{B}$	$y_i^* \geq b$	b	n_b

This model thus assumes that there is an underlying variable y_i^* which is observed only when it is inside an interval (a b).

Censored regression models (including the standard Tobit model) are usually estimated by the maximum likelihood method. Assuming that the disturbance term ϵ follows the normal distribution with mean 0 and variance σ^2, the loglikelihood function is

$$\mathcal{L}(\beta, \sigma) = \log L(\beta, \sigma) = \sum_{i=1}^{n} \left\{ I_i^a \log \Phi \left(\frac{a - x_i' \beta}{\sigma} \right) + I_i^b \log \Phi \left(\frac{x_i' \beta - b}{\sigma} \right) \right.$$
$$\left. + (1 - I_i^a - I_i^b) \left(\log \phi \left(\frac{y_i - x_i' \beta}{\sigma} \right) - \log \sigma \right) \right\}, \quad (8.45)$$

where I_i^a and I_i^b are the functions

$$I_i^a = \begin{cases} 1 & \text{if } y_i = a \\ 0 & \text{if } y_i > a \end{cases}$$
$$I_i^b = \begin{cases} 1 & \text{if } y_i = b \\ 0 & \text{if } y_i < b. \end{cases}$$

It is clear that, in some applications, such as income distributions, the response will also need to be transformed to, at least, approximate normality for the regression model to be appropriate. Riani et al. (2025a) use the Yeo-Johnson transformation since it provides a transformed response for both positive and negative responses which is smooth at $y = 0$. Since this is the only transformation considered in this section we write the normalized transformation. $z_{\text{YJ}}(\lambda)$ of (6.22) as $z(\lambda)$. The related transformations are also required for the truncation parameters a and b, leading to unchanged membership of $z(\lambda)$ in $\mathcal{A}$ and $\mathcal{B}$. The transformation of the truncation parameters is helpful in the graphical presentation of results. The loglikelihood for the transformed observations is then

$$\mathcal{L}(\beta, \sigma, \lambda) = \sum_{i=1}^{n} \left\{ I_i^{a(\lambda)} \log \Phi \left(\frac{a(\lambda) - x_i' \beta}{\sigma} \right) + I_i^{b(\lambda)} \log \Phi \left(\frac{x_i' \beta - b(\lambda)}{\sigma} \right) \right.$$
$$\left. + (1 - I_i^{a(\lambda)} - I_i^{b(\lambda)}) \left(\log \phi \left(\frac{z_i(\lambda) - x_i' \beta}{\sigma} \right) - \log \sigma \right) \right\}. \quad (8.46)$$

8.8 Censored Regression

Note that the Jacobian of the transformation of the response in the censored regression model is only calculable from the n_c uncensored observations in C. Riani et al. (2025a) maximize this censored loglikelihood with respect to the $p+2$ dimensional parameter vector $(\beta'\ \sigma\ \lambda)$ using routine *fminunc* of the optimization toolbox of MATLAB.

For uncensored data, interpretation of the fan plot of Sect. 6.1.3 provides a robust estimate of the transformation parameter through a monitoring plot of the approximate score statistic for the transformation. The score test is found by Taylor series expansion of $z(\lambda)$ around λ_0 to yield a constructed variable. In the absence of an algebraic expression for $z(\lambda)$ for censored observations it is possible to calculate the likelihood ratio test based on (8.46), namely

$$\mathrm{LR}(\lambda_0) = -2\{\mathcal{L}(\hat{\beta}(\lambda_0), \sigma(\hat{\lambda}_0), \lambda_0) - \mathcal{L}(\hat{\beta}, \hat{\sigma}, \hat{\lambda})\}. \tag{8.47}$$

To include the information from the sign of the difference $\lambda_0 - \hat{\lambda}$ they take, as the test statistic, the signed square root of $\mathrm{LR}(\lambda_0)$ (8.47)

$$T(\lambda_0) = \mathrm{sgn}(\lambda_0 - \hat{\lambda})\{\mathrm{LR}(\lambda_0)\}^{0.5}, \tag{8.48}$$

which has, asymptotically, a standard normal distribution. In the absence of censoring, this statistic should be asymptotically equivalent to the approximate score statistic (see Exercise 1.5).

So far in this book, applications of the FS have used only the size of the residuals to include observations in the subset $S^*(m)$. With censored data the FS can be less stable due to difficulties in convergence of the optimization algorithm, particularly for values of λ_0 far from the value appropriate to the set of data being analysed and for subsets with many censored observations. For estimation of λ, Riani et al. (2025a) therefore use a balanced search in which the subset is, as far as possible, forced to have observations in sets $\mathcal{A}, \mathcal{U}$ and $\mathcal{B}$ in the ratios of n_a, n_u and n_b.

Let R_ι be the ratio of the number of units in set ι in the subset to the number n_ι in the sample. That is $R_\iota = m_\iota/n_\iota, \iota = \{\mathcal{A}, \mathcal{U}, \mathcal{B}\}$. Let ι^* indicate the set with the smallest value of R_ι and $|e_{\iota,[i]}(m)|$ be the ith ordered absolute residual in set ι. The search moves forward from $S^*(m)$ to $S^*(m+1)$ with $S^*(m+1)$ containing the units associated with the distances:

$$|e_{\iota,[1]}(m)|, \ldots |e_{\iota,[m_\iota]}(m)|, \quad \iota \neq \iota^*,$$

$$|e_{\iota,[1]}(m)|, \ldots |e_{\iota,[m_\iota+1]}(m)|, \quad \iota = \iota^*.$$

Riani et al. (2025a) use the balanced FS to establish the value of λ. Once one, or rarely a few, plausible transformations are established, they use the original (unbalanced) search to discover any effects of outliers and other observations on the estimated parameters and the residuals from the fitted model.

They initially illustrate the properties of the procedure using two simulated examples. In the first example there is upper and lower truncation. In the second there are outliers, with only a lower threshold. The final illustration is the analysis of data from 493 customer records generated by a supermarket chain in northern Italy; there are 140 zero-censored observations. Fitting the model for truncated regression together with response transformation leads to a well-fitting parsimonious model. Monitoring plots of residuals combined with brushing, clearly exhibit the effect of censoring.

Routines to generate the analyses and figures of Riani et al. (2025a) are given in Table A.1.

Problems

8.1 Frequentist analysis of the Windsor House Price data
Using the Windsor House Price data, compute and comment on the forward plot of absolute minimum deletion residuals without prior information. What is the relationship between this plot and the one based on weak prior information in Fig. 8.1?

8.2 Differences and similarities between HAR and ART heteroskedasticity
Analyse the international trade date of Sect. 8.4.3 (`inttrade1`) using Harvey's heteroskedasticity (Eq. 8.20). Show the outliers which are found and the confidence band of predicted values. Discuss the differences and similarities between the two forms of heteroskedasticity.

8.3 Analysis of an additional international trade dataset: inttrade3
Figure 8.40 shows a plot of `inttrade3` (POD_0307591000_SN_IT). Analyse this dataset using ART heteroskedasticity. Brush the units showing the largest negative residuals in the monitoring residual plot and comment on their position in the scatter plot. Compute the 99% Bonferronized confidence bands of predicted values and show, with a different symbol, the units detected as outliers.
Repeat the analysis using the homoskedastic approach and find the outliers. Comment on the dangers of neglecting heteroskedasticity.

8.4 Analysis of transformed Airline data
The amplitude of the data of Sect. 8.6.4 seems increasing over time. Apply a suitable transformation to the data and analyse them with an appropriate choice for the hyperparameters of the model (8.33). Is the logarithmic transformation sufficient? Which of the approaches of Chap. 6 would you follow?

8.5 Analysis of body measurements to predict percentage of body fat
The dataset `fat` introduced in the R package **UsingR** (Verzani 2022) contains physical measurements of 252 males. One of the variables is `body.fat` (in percentage) which we would like to predict based on (some of) the other measurements. Take as response variable the logit-transformed `body.fat` and as explanatory variables select {`neck, chest, abdomen, hip, thigh, knee, ankle, bicep, forearm, wrist`}. Why is the explanatory part of

Problems

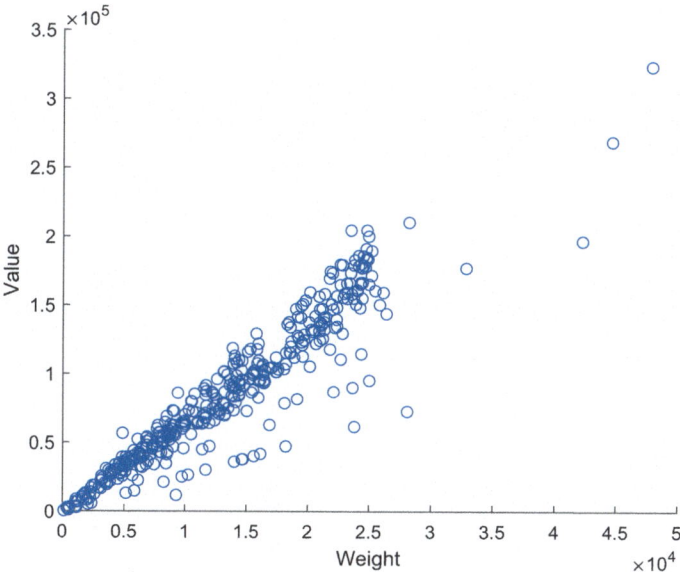

Fig. 8.40 POD_0307591000_SN_IT: scatter plot of value against weight

the dataset compositional? Carry out a monitoring analysis similar to that for Example 2 of Sect. 8.7.3.1.

Hint: For *ilr*-transforming the composition you can use the function `pivotCoord()`. This function is implemented both in the R package **robCompositions** (Filzmoser et al. 2018) and inside **FSDA**.

Open Access This chapter is licensed under the terms of the Creative Commons Attribution 4.0 International License (http://creativecommons.org/licenses/by/4.0/), which permits use, sharing, adaptation, distribution and reproduction in any medium or format, as long as you give appropriate credit to the original author(s) and the source, provide a link to the Creative Commons license and indicate if changes were made.

The images or other third party material in this chapter are included in the chapter's Creative Commons license, unless indicated otherwise in a credit line to the material. If material is not included in the chapter's Creative Commons license and your intended use is not permitted by statutory regulation or exceeds the permitted use, you will need to obtain permission directly from the copyright holder.

Chapter 9
Model Selection

Abstract This chapter considers the choice of explanatory variables to include in the linear predictor $x^T\beta$. We start with models for all of which p, the dimension of β, is $< n$. The problem arises specifically when some variables are nearly collinear when the significance of a variable in the model may depend strongly on what other variables are included. Section 9.3.1 derives Mallow's C_p from Akaike's AIC; models with more parameters are penalized. Robustness, together with the detection of influential observations, is provided by the generalized candlestick plot, illustrated by three data analyses. For the rest of the chapter we take $n < p$. Section 9.4.1 describes two regularizations: the LASSO estimates β as the minimizer of a linear combination of the L_2 norm of the residuals and the L_1 norm of the parameter estimates. The method provides model selection, the number of parameter estimates set to zero depending on a parameter λ. In ridge regression both terms use the L_2 norm, the parameter controlling the shrinkage of the parameter estimates. Neither method is robust. Section 9.4.2 describes sparse LTS which adds an L_1 penalty term with penalty parameter λ to LTS estimation. In Sect. 9.4.3 the parameter λ is estimated, for the cancer data of Sect. 9.4.1, by monitoring. Seven explanatory variables occur in many of the selected models. These are subjected to robust model selection in Sect. 10.6.

9.1 Introduction

This chapter considers the choice of explanatory variables to include in the liner predictor $x^T\beta$. The problem arises specifically when some variables are nearly collinear. Then one or the other may be significant in the regression, but the two together may not be; their regression coefficients will be poorly defined; see Sect. 9.3.

This problem has been, and continues to be, widely studied. We have already mentioned the book of Claeskens and Hjort (2008) which provides a nice mixture of mathematics and data analysis. Burnham and Anderson (2002) is one of a series of publications by these authors on the use of information criteria in model selection, with an emphasis on applications in life sciences.

It is not our purpose to review this literature. Burnham and Anderson (2002) do not mention robustness in the sense we have used in this book and much of the literature on model selection appears to ignore the possibility that the data for which a model or models are being painstakingly assessed may be corrupted. An exception is Sect. 2.10 of Claeskens and Hjort (2008), "Outlier-robust methods", which finds a model to explain an outlier in speed-skating data. However, the solution is static, in the sense that it depends on assumed values of the parameters of the robust method. In Sects. 9.3.2 and 10.6 we use monitoring of robust methods combined with a graphical display, "The generalized Candlestick Plot", to find outliers together with the detection of influential observations and so strengthen the process of model selection.

9.2 Significance of the Explanatory Variables

The t-tests for the significance of the explanatory variables in linear regression were introduced in Eq. (3.10):

$$t_k = \hat{\beta}_k / (s_v^2 v_k)^{0.5}, \qquad k = 1, \ldots, p, \tag{9.1}$$

where v_k is element k of the diagonal of $(X^T X)^{-1}$. In selecting a model we look for one in which all the included terms are significant, with no significant variables omitted. If $(X^T X)$ is orthogonal, for example from a designed experiment, the value of t_k does not depend on whether x_j is included in the model. If $(X^T X)$ is not diagonal, removing x_k from the model may have a strong effect on the significance of x_j. As the comparison of the analyses of variance in Tables 9.1 and 9.2 shows, the removal of one term from a model can make a large difference to the values of the t-statistics for all remaining variables. It is not enough to fit a model with all potential terms and then, using this fit, to delete sequentially all non-significant variables. Several models with significant variables may be missed.

Wald, or t-like statistics, were introduced in Sect. 3.7.2 to extend the use of regression t-statistics to M-estimation and related estimators. Table 3.1 lists six possible estimators of $\text{cov}(\hat{\beta})$. Monitoring plots for LS analysis of the AR regression data are in Fig. 4.15 and for S-estimation with two choices for estimation of $\text{cov}(\hat{\beta})$ are in Fig. 4.22. These are surprisingly similar when compared with the difference in the monitoring plots of residuals. Importantly, for model selection, these plots are all for statistics within one model, even if the identification of outliers due to monitoring does suggest that x_1 might be included in the model for cleaned data.

The added variable t-statistics introduced in Sect. 4.8 extend the number of models considered. Each t-statistic measures the effect of adding variable x_k to a model from which it was excluded. Equivalently the square of the statistic is the value of the F-test from the difference in the sums of squares when β_k is and is not included in the model.

9.2 Significance of the Explanatory Variables

Table 9.1 Hald's Cement data: ANOVA using all variables including intercept

	Estimate	SE	tStat	pValue
(Intercept)	62.405	70.071	0.8906	0.39913
x1	1.5511	0.74477	2.0827	0.070822
x2	0.51017	0.72379	0.70486	0.5009
x3	0.10191	0.75471	0.13503	0.89592
x4	-0.14406	0.70905	-0.20317	0.84407

Number of observations: 13, Error degrees of freedom: 8 Root Mean Squared Error: 2.45 R-squared: 0.982, Adjusted R-Squared: 0.974
F-statistic versus constant model: 111, p-value = 4.76e-07

Table 9.2 Hald's cement data: ANOVA using all variables without intercept

Estimated Coefficients:

	Estimate	SE	tStat	pValue
x1	2.193	0.18527	11.837	8.6523e-07
x2	1.1533	0.047942	24.057	1.7716e-09
x3	0.75851	0.15951	4.7551	0.0010367
x4	0.48632	0.041409	11.744	9.2484e-07

Number of observations: 13, Error degrees of freedom: 9
Root Mean Squared Error: 2.42

9.3 Arbitrary Numerical Rules for Model Selection and Hald's Cement Data

In the data analyses in this book we have not only used LS to fit first-order models but have, among other elaborations, considered transformations of response and explanatory variables and the detection of outliers. The standard methods for model selection consider first-order models in p parameters. Including and excluding each variable gives 2^{p-1} models to be compared, assuming each contains an intercept. The methods in this section can be thought of as a first attempt to reduce the number of models, without considering robustness.

Forward Selection. Starting from a model with just the intercept, the first variable to be included is that with the greatest t-statistic when it is the only variable in the model. This variable is kept in the model and the statistics for the remaining variables calculated, with again the variable with the greatest t-statistic for entry being chosen. The procedure continues until the only variables not included have non-significant t-statistics for entry.

Backwards Elimination. This starts from the full model with p parameters and successively eliminates the variables with the smallest t-statistic in the fitted model. Of the two, backwards elimination is generally preferable because it starts from a model with a better estimate of σ^2.

Stepwise Regression. There are four options at each stage: add a term, delete a term, swap a variable in the model for one not in the model or stop.

There is no theoretical basis for these methods and they often lead to the selection of different models. There is also arbitrariness in the significance levels chosen for the entry or deletion of variables and, in stepwise regression, for the precise rule for swapping variables.

This discussion is an abbreviation of that in Davison (2003, p. 400). On p. 354 he introduces Hald's cement data (Hald 1952, pp. 635–649) which we use to exemplify the behaviour of these rules for model selection.

There are 13 observations on the relationship between heat generated in the setting of cement and the concentration of four chemicals, $x_1 - x_4$. The models found by the three methods are:

Backwards: x_1 and x_2;

Forwards: x_1 and x_4;

Stepwise (MATLAB "adjrsquared" option or "stepAIC" in the R package MASS): x_1, x_2 and x_4.

Since the search for a single model has led to three different solutions for a model with only four-variable, it is sensible to consider those models that provide an adequate fit to the data. For sufficiently small problems complete enumeration is a possibility. The resulting fitted models can be evaluated by use of an information

9.3 Arbitrary Numerical Rules for Model Selection and Hald's Cement Data

criterion, such as BIC introduced in Sect. 4.11.1 or Akaike's Information Criterion (AIC) discussed in Sect. 9.3.1. These both penalize the reduction in the residual sum of squares from fitting extra terms by a penalty that increases with the number of parameters in the fitted model.

The model selection methods provide a hard selection, choosing just one model, however close in performance other models are. The information criteria, on the other hand, provide a set of models that may be satisfactory. Bayesian methods can be used to provide a weighting of the models which are then combined, for example, to give the predicted response at a particular value of x. Davison (2003, p. 593) provides an example of prediction using Bayesian model averaging for the cement data. See Raftery et al. (1997) for Bayesian model averaging in linear regression.

Instead of searching for an automatic method to provide a single model to explain this small set of data, it is interesting to look at the analysis of variance given in Table 9.1. This shows that there are no significant t statistics, neither for the intercept nor for any of the variables. However the F statistic for all variables against the constant model has a high significance level of 4.76e-07. The indication is that the model is overparameterized with near collinearity between some of the variables. Davison (2003, p. 399) observes that, if the four explanatory variables are components of a mixture, they should sum to 100%. The sums however range from 95 to 99, with the majority being either 97 or 98. It is no surprise that β_0 is poorly estimated. The methods of Sect. 8.7 for compositional data would be appropriate if the sums of the explanatory variables had a constant value. A simple first step would be to slightly change the values of the x_i so that the summation was exact.

Table 9.2 gives the analysis of variance table for the analysis of these data for a regression model without intercept, that is the first-order mixture model introduced by Scheffé (1958). Now all four variables are highly significant with the smallest t-statistic, that for x_3, equal to 4.76. Piepel and Redgate (1998) considers the use of mixture models in the analysis of the original data that Hald slightly modified for the purposes of his book. It is unfortunate that these data, including the intercept term, are so often used as an example of model selection, a procedure which makes little sense for this set of data in which the ill-conditioning is caused by improper modelling of mixture data.

9.3.1 Mallow's C_p and the Generalized Candlestick Plot

In Sect. 4.11.1 we introduced an information criterion for model selection of the general form $\text{IC} = 2\mathcal{L}(\hat{\theta}) - k(p, n)$, where $k(p, n)$ is a function that penalizes more complicated models. For BIC, $k(p, n) = p \log n$, so that the penalty increases with sample size. That model is selected for which BIC is the largest. At the end of Sect. 4.11.6 we mentioned a second information criterion, AIC (Akaike 1974), in which the penalty $k(p, n)$ equals $2p$, non-increasing with n. Special cases of AIC are widely used for the selection of regression models. In this section we use one such development, Mallows' C_p, to provide a robust form of model selection.

For the linear multiple regression model $y = X\beta + \epsilon$, let the residual sum of squares from fitting the model to all n observations be $R_p(n)$ when, for known σ^2,

$$\text{AIC}_\sigma = n \log(2\pi) + n \log \sigma^2 + R_p(n)/\sigma^2 + 2p. \tag{9.2}$$

If, as is usually the case, σ^2 is not known, the maximum likelihood estimator is

$$\hat{\sigma}^2 = R_p(n)/n. \tag{9.3}$$

With this internal estimate of σ^2 the criterion (9.2) becomes

$$\text{AIC}_I = n \log(2\pi) + n \log\{R_p(n)/n\} + n + 2p, \tag{9.4}$$

a form frequently used in the selection of non-nested time series models with normally distributed errors. Some references are in Tong (2001).

Mallows (1973) introduced the C_p statistic for the selection of a regression model, which is a special case of AIC. The application of AIC to the selection of regression variables requires an estimate of σ^2. On the assumptions of the absence of outliers and a correct model, the unbiased estimate of σ^2 comes from a large regression model with an $n \times p^+$ matrix X^+, $p^+ > p$, of which X is a submatrix. The unbiased estimator of σ^2 from regression on all p^+ columns of X^+ is $s^2 = R_{p^+}/(n - p^+)$. With this estimate the criterion (9.2) becomes

$$\text{AIC} = n \log(2\pi) + n \log\{R_p^+(n)/(n - p^+)\} + (n - p^+)R_p(n)/R_p^+(n) + 2p. \tag{9.5}$$

The variable factors are p and the regressors that are being considered. Then the choice of the model minimizing (9.5) is identical to the choice of model minimizing

$$C_p(n) = R_p(n)/s^2 - n + 2p = (n - p^+)R_p(n)/R_{p^+}(n) - n + 2p. \tag{9.6}$$

One derivation of $C_p(n)$ is that it provides an estimate of the mean squared error of prediction at the n observational points from the model with p parameters provided the full model with p^+ parameters yields an unbiased estimate of σ^2. For the visualization of model selection it is helpful to have the distribution of C_p so that the plot can provide bounds on acceptable values of the statistic. Mallows (1973) and Gilmour (1996) show that

$$C_p(n) \sim (p^+ - p)F + 2p - p^+, \quad \text{where} \quad F \sim \mathcal{F}_{p^+ - p, n - p^+}. \tag{9.7}$$

Then $\text{E}\{R_p(n)\} = (n - p)\sigma^2$, $\text{E}(s^2) = \sigma^2$ and $\text{E}\{C_p(n)\}$ is approximately p. Models with small values of C_p are preferred.

AIC and BIC differ in the penalty term $k(p, n)$. Since the BIC penalty $p \log n$ is increasing in n, the BIC asymptotically guarantees consistent selection of the true model. On the other hand, the constant penalty in AIC and $C_p(n)$, where $k(p, n) = 2p$, has the consequence that there is always a small probability of fitting too large a

9.3 Arbitrary Numerical Rules for Model Selection and Hald's Cement Data

model. However, the consistency of BIC is an asymptotic property. For finite samples this distinction may not matter and $C_p(n)$ may have a greater probability of selecting the true model (Davison 2003, Chap. 4).

Equation (9.6) provides an expression for $C_p(n)$, the whole sample value of C_p. In order to monitor the values of this statistic we look at the values over the important part of the FS. We call these statistics $C_p(m)$, obtained by replacing the value of n in (9.6) by m and using the residual sums of squares $R_p(m)$ and $R_{p^+}(m)$. Riani and Atkinson (2010, Sect. 3.4) give a fuller discussion of the distribution of $C_p(m)$ from the FS. They also show that $C_p(m)$ is the generalization of the added variable t statistic to the inclusion of more than one explanatory variable and therefore has an F distribution.

We now need to find a suitable way of presenting the results of the calculation of the $C_p(m)$. Riani and Atkinson (2010, Fig. 4) shows, for data we do not otherwise mention, the monitoring of $C_p(m)$ for six models with $p = 6$. Unlike other monitoring plots, the curves vary widely with m, revealing no clear structure. Furthermore, the plot is only for a single value of p.

Our starting point is the "candlestick" plot used to summarize such quantities as the high, low, and closing values of stocks; Google provides many examples. Riani and Atkinson (2010) introduced a generalization of this plot which provides a boxplot-like structure for each model, combining average and individual values of $C_p(m)$. Since we expect any outliers to enter the search towards the end, the last few values of $C_p(m)$ are, generally, of particular interest. The plot contrasts the average values of $C_p(m)$ over a central region with these last values, the central region extending from $m = 0.8n$ to $0.95n$, the values being rounded. Individual symbols are reserved for the influential units. The influential units are defined as the units which enter the subset in the final part of the search and bring the value of the C_p below the minimum or above the maximum value of the central part of the search. The definition of the candlesticks is:

Lowest Value: minimum in the central part of the search;

Highest Value: maximum in the central part of the search. The vertical line in the plot connects these two values;

Central Box: mean and median of the values in the central part of the search; filled if mean < median;

Red Stars: the values in the final part of the search if these lie outside the range of the C_p values in the central part of the search;

Unfilled Blue Circle: the final value.

9.3.2 The Generalized Candlestick Plot and Two Analyses of the Ozone Data

As a comparatively small example with sufficiently many potential explanatory variables to be interesting, we look at part of the data on ozone concentration used by

Breiman and Friedman (1985) when introducing ACE. These are a series of 330 daily measurements, from the beginning of the year, of ozone concentration and eight meteorological variables in California.

Atkinson and Riani (2000, Sect. 3.4) analyse the first 80 observations, so starting in January, and give further details of the data. They find that the data should be transformed by taking logs; the fan plot forms their Fig. 4.8 and that a linear time trend ("Time") should be considered as one of the explanatory variables. Together with the constant term, we therefore have $p^+ = 10$.

In this case the central part of the search is [64 76]. The final part is [77 80]. Figure 9.1 shows the candlestick plot for the reduced ozone data, for models with p going from 5 to 9. These plots are given in adjacent panels, with horizontal lines marking the 2.5% and 97.5% points of the distribution of $C_p(n)$ given in (9.7). The lines show how rapidly the distribution of $C_p(n)$ becomes tighter as p increases.

The candlestick plot provides a means of selecting a few parsimonious models for further study. In this way it is much more informative than Table 3.1 of Atkinson and Riani (2000) which used a backwards procedure to select models for a set of values of p.

As an illustration, inspection of Fig. 9.1 shows that the model with Time plus variables 4, 5, and 6, that is with $p = 5$, has a low value of C_p, with all except one property (the final value) lying within the range for this value of p. We checked the model using RAVAS, starting from the log transformation of the response and a linear time trend. There was no indication of further transformation of these two variables, three outliers were identified and the least significant variable had a t-value of 4.8 (Exercise 9.1).

In the reduced ozone data, starting from the beginning of the year, it was possible to explain the effect of time by a linear trend as the weather warmed towards summer.

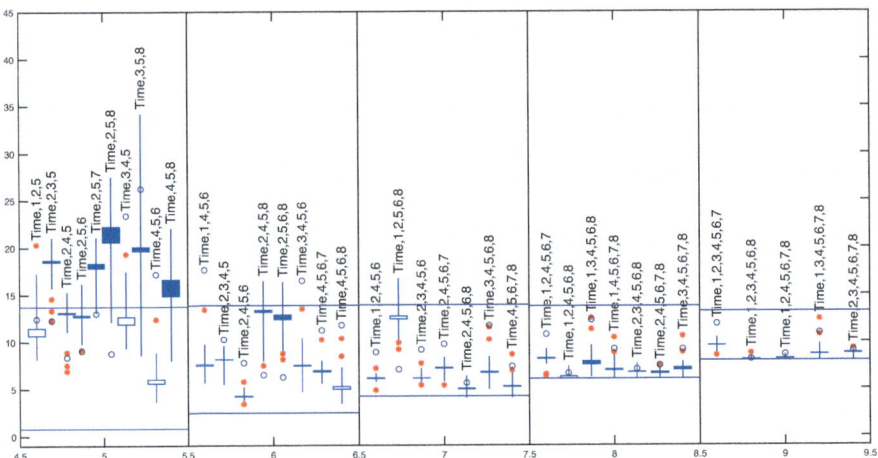

Fig. 9.1 Reduced Ozone data: generalized candlestick plot for $m = 64$–80, $p^+ = 10$. Time is a linear trend

9.4 Regularization Methods

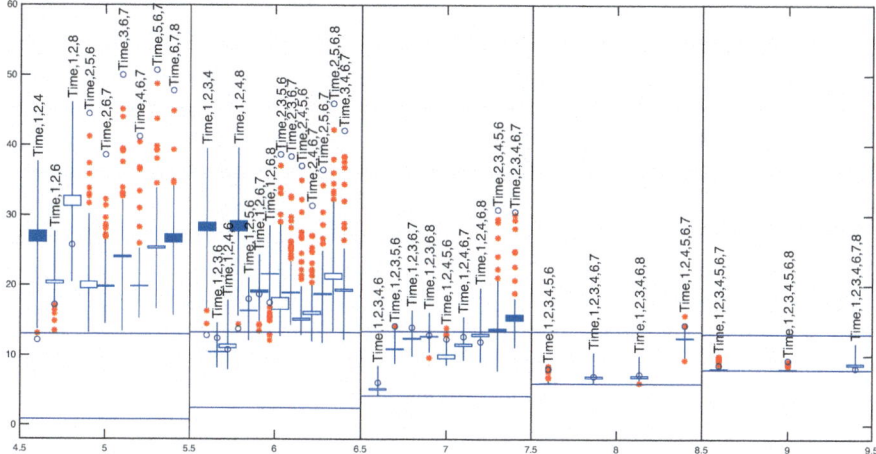

Fig. 9.2 Full Ozone data: generalized candlestick plot for $m = 264$–313, $p^+ = 10$. Time increases linearly up to $n/2$ and then decreases symmetrically as n increases further

For 330 days we take a symmetrical time trend, increasing linearly to $n/2$ and then decreasing. Figure 9.2 shows the resulting candlestick plot. Here a more complicated model is required; there are no models within the C_p bounds for $p < 6$. To check the choices made for the structure of the model, we looked briefly at the 8-parameter model with Time and variables 1, 2, 3, 4, 6, and 7. There was no evidence of any further transformation of the response, but there was a suggestion of a transformation away from the symmetrical variable we have assumed for time. The discussion of Breiman and Friedman (1985, Fig. 5) indicates that the symmetrical model for time is appropriate if the meteorological variables are excluded from the model, but not if they are included. Within the class of models we use for this example, Fig. 9.2 leads to the choice of the seven parameter model with Time and variables 1, 2, 3, 4, and 6.

9.4 Regularization Methods

The LS estimator for linear regression has good properties and can be efficiently calculated. But the estimator can be extremely unreliable if the number of predictors p nears the number of observations n, when overfitting can occur; i.e. the model fits noise in the data rather than capturing the underlying relationships. There is no unique solution if $p > n$. These problems are exacerbated by the use of robust methods which tend to use a subset of observations for fitting. Another issue in high dimensions is multicollinearity which can lead to unstable coefficient estimates and difficulty in interpreting the individual effects of variables. The analysis of the Hald cement data provides an example in low dimensions. A number of alternative strategies have been developed to address the issues of high dimensional data. The first group of such strategies is the subset selection approach (stepwise regression, best subset regression), see Sect. 9.3, which reduces the number of predictors and thus reduces the prediction error and improves interpretability. The effect of collinearity is illustrated in our analysis of Hald's cement data in Sect. 9.3. The main disadvantage of subset selection methods is that, due to the discrete process of choosing (retaining or eliminating) variables a high variance is introduced and finally the prediction error of the full model is not reduced (see Hastie et al. 2009, Sect. 3.4). Another class of approaches transforms the predictors by projecting the data onto a lower dimensional space and then fitting an LS model to the transformed variables (*dimension reduction methods*, like principal components regression and partial least squares regression). While these methods do not increase the variance, they present difficulties in the interpretation of the model, because we are no longer working with the original variables but with linear combinations of these. A completely different approach comes from regularization or shrinkage methods, which continue to work with the p original variables but constrain or *regularize* the coefficients, bringing them nearer to zero relative to the least squares estimates. These methods significantly reduce the variance at a cost of a slight increase in the bias and work well when the number of variables p is close to the number of observations n and even when $p > n$ or $p \gg n$. By putting a set of the parameter estimates equal to zero, some of these methods can perform variable selection, thus facilitating the interpretability of the model.

9.4.1 Least Absolute Shrinkage and Selection Operator (LASSO)

Tibshirani (1996) introduced the Least Absolute Shrinkage and Selection Operator (LASSO) which solves the following constrained L_1 optimization problem:

$$\underset{\beta_0, \beta}{\text{minimize}} \left\{ \sum_{i=1}^{n} \left(y_i - \beta_0 - \sum_{j=1}^{p} \beta_j x_{ij} \right)^2 \right\} \tag{9.8}$$

$$\text{subject to} \sum_{j=1}^{p} |\beta_j| \leq t.$$

In this way we can control the variance by controlling how large the coefficients can be at the price of slightly increasing the bias. This control is made through the parameter t—if $t \geq \sum_{j=1}^{p} |\hat{\beta}_j|$ where $\hat{\beta}_j, j = 1, \ldots, p$ are the LS estimates for β, the coefficients are not constrained and we obtain the LS solution. For smaller t the coefficients are shrunken towards 0 and if $t = 0$ the solution is a model with only an intercept. Note that the intercept β_0 is not included in the penalization term. While the LS regression is scale invariant (i.e. we can change the scale of the predictors and the fitted values will not change), the LASSO is not. Therefore it is important to

9.4 Regularization Methods

scale the data beforehand (x_j, $j = 1, \ldots, p$ are standardized with mean 0 and unit variance and y is centred). The centering of y is convenient, since it means that we can omit the intercept term β_0 in the optimization of (9.8). The LASSO Eq. (9.8) can be rewritten for convenience in the Lagrangian form

$$\underset{\beta}{\text{minimize}} \left\{ \sum_{i=1}^{n} \left(y_i - \sum_{j=1}^{p} \beta_j x_{ij} \right)^2 + \lambda \sum_{j=1}^{p} |\beta_j| \right\} \tag{9.9}$$

for some $\lambda \geq 0$. In matrix notation

$$\underset{\beta}{\text{minimize}} \left\{ ||y - X\beta||_2^2 + \lambda ||\beta||_1 \right\}, \tag{9.10}$$

where the L_1 norm of the coefficient vector β is $||\beta||_1 = \sum_{j=1}^{p} |\beta_j|$ and $||.||_2$ is the usual Euclidean norm of a vector. The second term in (9.9) and (9.10), which introduces a penalty on the L_1 norm of the parameter vector, often called the *shrinkage* penalty, shrinks the estimates of the parameters β_j towards 0. The parameter λ serves as a tuning parameter controlling the amount of regularization (shrinking of the estimates) and has a one-to-one correspondence to the threshold t in (9.8). If $\lambda = 0$ the penalty term has no effect and we obtain a solution without any shrinkage, i.e. we obtain the LS solution. On the other hand, when $\lambda \to \infty$ the penalty effect grows and the coefficients approach zero. In this particular form of the penalty term, the LASSO shrinks some of the parameter estimates to be exactly zero and thus performs *variable selection*. As a result, the obtained LASSO models will be much easier to interpret than the LS models or models obtained by some other forms of regularization. The LASSO yields models that contain only a subset of the p variables and we say that it generates *sparse* models. This is the main difference from the other popular regularization method, *ridge regression* (Hoerl and Kennard 1970), which also can cope with high dimensional data (i.e. can find a solution in cases when p is close to n or even when $p > n$). However, ridge regression always produces solutions including all p original variables. The ridge regression solution is obtained by solving

$$\underset{\beta}{\text{minimize}} \left\{ ||y - X\beta||_2^2 + \lambda ||\beta||_2 \right\}. \tag{9.11}$$

The only difference between (9.11) and (9.10) is the penalty term—the ridge regression is using a quadratic penalty. The ridge regression has a closed form solution

$$\beta^{\text{ridge}} = (X^T X + \lambda I)^{-1} X^T y, \tag{9.12}$$

which is another difference from the LASSO. The L_1 penalty makes the solutions non-linear in y_i and there is no closed form solution for the LASSO which is thus a quadratic programming problem. However, a fast algorithm for computing the LASSO is available through the framework of least angle regression (Efron et al. 2004) but recently a more efficient algorithm, based on cyclical coordinate descent, computed along a regularization path, is available (Friedman et al. 2010) and implemented in the R package **glmnet**. Thus the entire coefficient path can be computed with more or less the same effort as a single least squares fit. Apart from the LASSO, this package can be used to compute ridge regression models as well as other extensions in the framework of general linear models.

Example: NCI60 cancer cell panel data. As an example we analyse the well-known cancer cell panel of the National Cancer Institute which consists of data on 60 human cancer cell lines and can be downloaded via the web application *CellMiner* (http://discover.nci.nih.gov/cellminer/). It is also available in the R package **robustHD** (Alfons 2021) as a dataset nci60, suitably preprocessed since it was used for illustration of the *Sparse LTS* method in Alfons et al. (2013). The dataset contains the data frames protein—the protein expressions based on 162 antibodies, $\log_2$ transformed, and gene—the normalized gene expressions resulting in a set of $p = 22283$ predictors. In this example we will regress protein expression on the gene expression data. As a response we take the KRT18 antibody, which is the variable with the largest MAD. We screen the most correlated predictor variables, selecting the first 100. One observation had to be removed since all values were missing in the gene expression data. Thus, our example dataset contains 59 observations and 100 independent variables. The name of this dataset inside **FSDA** is nci60.

We start by estimating the LASSO for a range of the penalty parameter λ (using the function glmnet() from the R package **glmnet**). The result is shown in the left-hand panel of Fig. 9.3. The penalty parameter λ varies from 0 on the left where it does not have any effect, to some large value on the right where all coefficients β_j become 0 (λ is shown on a log scale). The upper x-axis shows the degrees of freedom (or the number of effective predictors in the model). Many of the coefficients become exactly zero for large values of λ. It is important to keep in mind that, although the number of non-zero coefficients decreases with λ, the models are not necessarily nested. That is, the variables selected for a larger value of λ may well not be a subset of those for a smaller λ value.

The right-hand panel of Fig. 9.3 shows the related coefficient paths for ridge regression (also computed with the function glmnet() from the R package **glmnet**). We see that while all coefficients are shrunk towards 0 (with λ moving from 0 to some large value) none of them becomes exactly 0. Ridge regression provides a solution, even if the number of predictors $p > n$. However, it does not allow variable selection, i.e. does not produce a sparse solution.

The results for the LASSO in the left-hand panel of Fig. 9.3 show the dependence of the number of non-zero parameter estimates on the value of the penalty parameter λ, but do not provide information as to which model is the most accurate for predicting new independent data from the same population. One traditional approach is to split the dataset randomly into two roughly equal parts, a training set on which to estimate the model and a test set from which to calculate the prediction error (mean squared error) for each value of λ (on a selected grid). This would not be possible if there were not enough observations. An alternative traditional approach is to estimate the test mean squared error by cross validation. The dataset is split randomly into k blocks (called *folds*) of roughly equal size. In turn, $k - 1$ folds are merged together, and

9.4 Regularization Methods

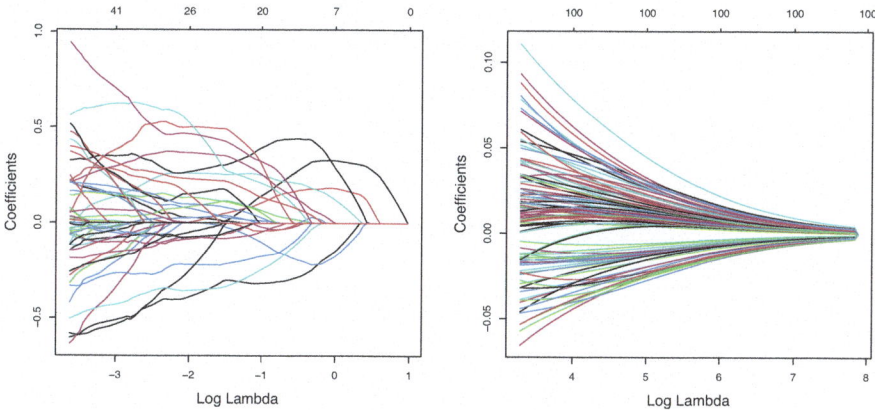

Fig. 9.3 NCI60 data: coefficient paths plotted against the penalty parameter λ (on a log scale). Upper x axes, number of non-zero coefficients. Left-hand panel: LASSO; right-hand panel: ridge regression

the model is fitted to this artificial training set and is tested (the MSE is calculated) on the remaining fold. This is repeated k times, each time leaving out one fold, and the test MSE estimate is obtained as the average of the k values. From the MSE for the different λ values we can select the most accurate model. In some cases, the number of observations might still not be sufficient for conducting cross validation as described above. Then leave-one-out cross validation could be used. In turn one observation is removed from the dataset, the model is estimated on the remaining $n - 1$ observations, and the MSE is calculated for the left out observation. Repeating this procedure n times and averaging the MSE will yield the leaving-one-out cross validation estimate of the prediction error. Such procedures are vulnerable to the presence of multiple outliers.

For our example we can perform the selection of the penalty parameter with the function `cv.glmnet()` which reports the results of the cross validation in the plot shown in Fig. 9.4. The dotted vertical lines show the value of λ which gives the minimal MSE and the largest value of λ such that error is within one standard error of the minimum. The minimal MSE is 5.152 obtained for $\lambda = 0.1559$ which selects a model with 24 effective predictors. A larger value of λ would select a model with fewer predictors. Hastie et al. (2009, p. 61) comment that "many of [these] curves are very flat over large ranges near their [minima]" and suggest a "one-standard-error" rule which picks the most parsimonious rule within one standard error of the minimum cross-validation error. Although an example of the use of this rule on their p. 244 gives a good result (a model with ten parameters is estimated to have nine), we note its seeming arbitrariness. In our case this rule selects the model with MSE equal to 6.597 for $\lambda = 0.6595$ which results in a model with 14 non-zero parameters.

The LASSO performs best when there are many predictors and we suspect that some of them are irrelevant or redundant. On the other hand, the LASSO does not

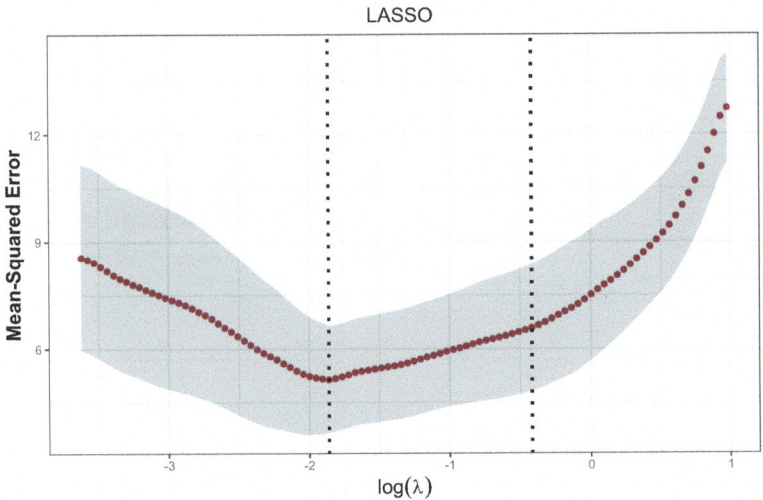

Fig. 9.4 NCI60 data: LASSO. Cross-validated estimate of mean-squared prediction error together with upper and lower standard deviation curves, as a function of the penalty parameter λ (on a log scale). The value of λ which gives the minimal MSE and the largest value of λ such that the error is within one standard error of the minimum are indicated by vertical dotted lines

perform well with highly correlated variables in which case the advantages of ridge regression shine. However, ridge regression cannot shrink any of the coefficients to exactly zero and thus improve the interpretability of the model by performing variable selection. To overcome the disadvantages of the two methods and to enjoy the advantages of both one can combine the L_1 penalty of the LASSO with the L_2 penalty of the ridge regression to obtain the *elastic net* (Zou and Hastie 2005) which solves the following optimization problem:

$$\underset{\beta}{\text{minimize}} \left\{ \sum_{i=1}^{n} \left(y_i - \sum_{j=1}^{p} \beta_j x_{ij} \right)^2 + \lambda \left[\frac{1}{2}(1-\alpha)||\beta||_2^2 + \alpha||\beta||_1 \right] \right\}, \quad (9.13)$$

with $\alpha \in [0, 1]$ a parameter controlling the compromise between the two penalties. With $\alpha = 1$ we have the LASSO penalty and, with $\alpha = 0$, the ridge penalty. In some cases of regression problems we want groups of variables to be simultaneously either included in the model or excluded from the model. A typical example is the presence of a qualitative predictor which is coded as a set of dummy variables and we want to include or exclude this group of variables together. A solution to this problem was proposed by Yuan and Lin (2006) with the *Group LASSO* penalty. Another version of LASSO is the two-step procedure *Relaxed LASSO* proposed by Meinshausen (2007) which essentially performs standard LASSO in the first step and then debiases the non-zero estimated coefficients by fitting the model without any penalty.

9.4.2 Sparse LTS Regression

Neither ridge regression nor the LASSO takes care of outliers which might be present in the data—neither vertical outliers contaminating the response variable nor leverage points contaminating the predictors. A number of robust versions have been proposed, most of them being penalized M-estimators. A least absolute deviations (LAD) estimator called LAD-LASSO is proposed by Wang et al. (2007). Alfons et al. (2013) have shown that both the classical LASSO and the robust LAD-LASSO have breakdown points of $1/n$. One of the first estimators with high breakdown point which can produce sparse solutions was the sparse LTS introduced by Alfons et al. (2013) which adds an L_1 penalty term with penalty parameter λ to the LTS estimation equation given in (3.48)

$$\hat{\beta}_{\text{sparseLTS}} = \arg\min_{\beta} \left\{ \sum_{i=1}^{h} \left(r_{[i]}^2(\beta) \right) + h\lambda \sum_{j=1}^{p} |\beta_j| \right\}, \tag{9.14}$$

where h is the trimming parameter. Alfons et al. (2013) prove that the so-defined sparse LTS with subset size $h \leq n$ has the high breakdown point of $(n - h + 1)/n$; the smaller the value of h, the higher the breakdown point. However, they recommend taking $h = 0.75n$, which results in *bdp* 25%, which should guarantee a sufficiently high statistical efficiency. The efficiency might be improved by taking a reweighted version of the estimates. However, Chap. 5 shows the deleterious effect reweighting can have on the size and power of tests. It is important to note that the *bdp* of sparse LTS does not depend on the dimension p, as is the case of standard LTS estimates which can have the maximal possible breakdown of $([(n - p)/2] + 1)/n$ for $h = [n/2] + [(p + 1)/2]$ (see Rousseeuw and Leroy 1987, p. 132). An algorithm, similar to the Fast-LTS described in Sect. 3.11.2, has been proposed and is implemented as function `sparseLTS()` in the R package **robustHD**. We use this function to continue our example from the previous section. Since we expect outliers in the NCI60 dataset, we will try to fit a sparse robust model. We run the function `sparseLTS()` with all default settings (e.g. *bdp* = 0.25) on a penalty grid with 50 values of λ. Figure 9.5 shows the coefficient path plotted versus the penalty parameter λ—with increasing λ the coefficients are shrunk to 0 so that, first, it is possible to estimate the model although $p > n$ and second, a more tractable model with fewer non-zero coefficients is obtained. Although, in Fig. 9.5 there are no coefficients which change sign, for different values of $\lambda < 0.18$ the models can be quite distinct.

The optimal penalty parameter λ can be selected in a similar way as for the classical LASSO, using cross validation and estimating a measure of the prediction accuracy.

Fig. 9.5 NCI60 data: coefficient path for the sparse LTS, plotted versus the penalty parameter λ

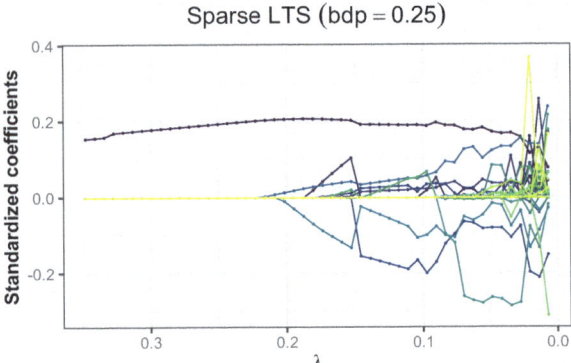

However, the computational cost of the robust method is much higher than that of the classical LASSO and to solve this issue Alfons et al. (2013) proposed to optimize the Bayesian Information Criterion (BIC), which for a model with shrinkage parameter λ is given by

$$\text{BIC}(\lambda) = \log(\hat{\sigma}^2) + df(\lambda)\frac{\log(n)}{n}, \qquad (9.15)$$

where $\hat{\sigma}$ is the residual scale estimate (of the raw or reweighted model) and $df(\lambda)$ is the degrees of freedom of the model given by the number of non-zero estimated parameters in β (Zou et al. 2007). This is equivalent to the form of BIC introduced in (4.11) but with the sign changed and the expression divided by n. It is the standard form used in the literature on the LASSO. In our example the BIC criterion selected 29 non-zero parameters which is not satisfactory for modelling 59 observations. The rule of thumb is that n/p should be greater than 5. We therefore consider applying cross validation. The mean squared error could be used as an estimate of the prediction error as in the classical case but this would allow the potential outliers to influence the selection of the parameter λ. Alternatively, the *root trimmed mean squared prediction error* (RTMSPE) can be used, trimming the same proportion as that used in computing the sparse LTS:

$$\text{RTMSPE}(\lambda) = \sqrt{\frac{1}{h}\sum_{i=1}^{h} e_{[i]}^2}, \qquad (9.16)$$

where $e^2 = (e_1^2, \ldots, e_n^2)^T$ is the vector of prediction errors, $e_i^2 = (y_i - \hat{y}_i)^2$. Here $\hat{y}_i$ is the predicted response value from a cross-validation fold and y_i is the respective response value in the test set. We search for a λ which minimizes $\text{BIC}(\lambda)$ or $\text{RTMSPE}(\lambda)$ over a grid of values in the interval $[0\ \lambda_0]$ where λ_0 is a value of the shrinkage parameter that would shrink all parameters to zero. An estimate $\hat{\lambda}_0$ for the parameter λ_0 can be found as proposed by Efron et al. (2004),

9.4 Regularization Methods

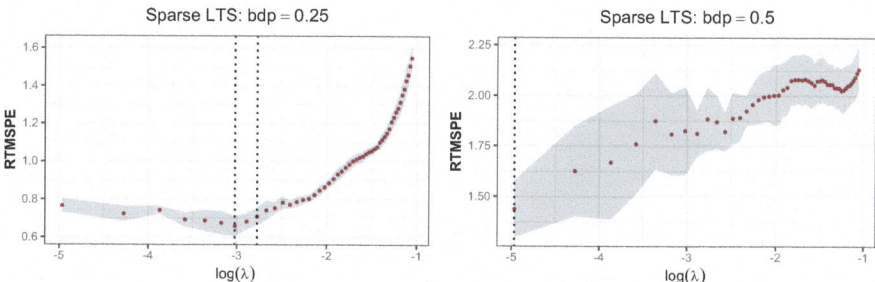

Fig. 9.6 NCI60 data: a cross-validated estimate of the robust root-mean-squared prediction error for sparse LTS together with upper and lower standard deviation curves, as a function of the penalty parameter λ (on a log scale). The value of λ which gives the minimal RTMSPE and the largest value of λ such that the error is within one standard error of the minimum are indicated by vertical dotted lines. Left-hand panel: $bdp = 0.25$ and right-hand panel: $bdp = 0.5$. Note that the variation is so large that the upper dotted vertical line in the right-hand panel cannot be included in the plot

$$\hat{\lambda}_0 = \frac{2}{n} \max_{j \in \{1,\ldots,p\}} \mathrm{cor}\left(y, x_j\right), \tag{9.17}$$

where $\mathrm{cor}(y, x_j)$ is the Pearson correlation between the response variable y and each of the columns of the design matrix X. If $p > n$, then 0 is excluded from the candidate set (because otherwise, if $\lambda = 0$, all $p > n$ regressors will be selected (no regularization) which is not possible for LTS). To obtain a robust estimate of λ_0 the Pearson correlation can be replaced by Winsorization as suggested by Khan et al. (2007). Applying this procedure to our example results in the plot shown in the left-hand panel of Fig. 9.6 of the robust prediction error as a function of the parameter λ. The λ for which the best prediction error of 0.6598 is obtained is 0.0626 and the corresponding model has 11 non-zero coefficients. This is already much more tractable than the model estimated by the classical LASSO with 24 non-zero parameters. Note that the selected model is not the one which minimizes the robust measure, but the one selected by the "one-standard-error" rule mentioned in Sect. 9.4.1 and by Hastie et al. (2009, p. 61).

The right-hand panel of Fig. 9.6 shows the penalty parameter selection with the maximum $bdp = 0.5$ (in both cases we are looking at the reweighted version of the sparse LTS). The estimated prediction errors are significantly higher and the minimal value is obtained for 24 non-zero coefficients—exactly as many as estimated by the classical LASSO but, importantly, only four of them overlap. Clearly, such estimates should be treated with great caution. In particular, some form of monitoring the value of bdp is recommendable. Is there a value of bdp less than 0.5, but greater than 0.25, for which the plot becomes like the left-hand panel of Fig. 9.6? Is this a robust, or non-robust, fit?

Once we have selected a model (with 11 non-zero coefficients for $bdp = 0.25$) we can look at the robust regression diagnostic plots, two of which are presented in Fig. 9.7. The left-hand panel shows a normal QQ-plot of the standardized residuals

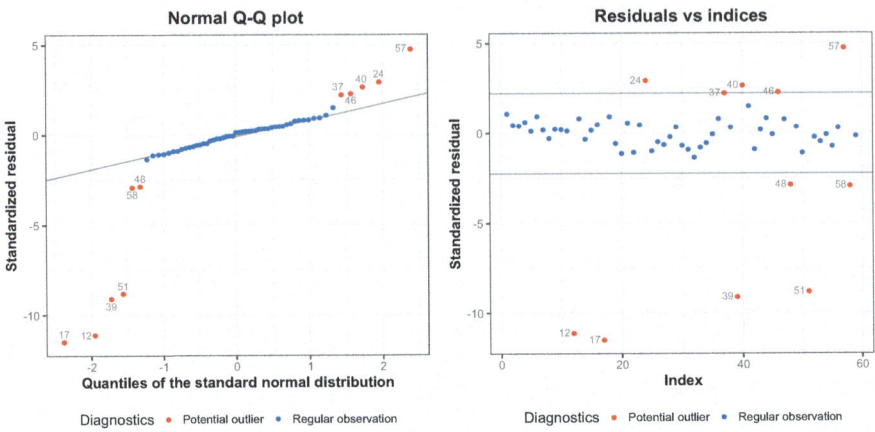

Fig. 9.7 NCI60 data: robust regression diagnostic plots with 11 non-zero coefficients for $bdp = 0.25$. Left-hand panel: normal QQ-plot of the standardized residuals against the quantiles of the standard normal distribution. Right-hand panel: standardized residuals against the index of the observations. Note that these "standardized" residuals are the scaled residuals r_i, see Sect. 3.2.2

against the quantiles of the standard normal distribution. 11 observations do not follow the diagonal line and are marked as outliers. The right-hand panel shows the residual plot presenting the standardized residuals against the index of the observations. Again the 11 observations are identified as outliers.

Remark: note that these "standardized" residuals are the scaled residuals r_i, see Sect. 3.2.2)

Maronna (2011) introduced S-Ridge and MM-Ridge estimators, i.e. L_2-penalized S- and MM-estimators. Although L_2-penalized regression estimators do not produce sparse models, this was the first attempt to make S- and MM-estimators available for $p > n$ while remaining robust. Smucler and Yohai (2017) proposed a penalized MM-LASSO estimator and studied the robust and asymptotic properties of MM-LASSO and adaptive MM-LASSO estimators: that is L_1-penalized MM-estimators and MM-estimators with an adaptive L_1 penalty. The most recent addition to the set of robust penalized estimators is the PENSE (penalized elastic net S-estimators) introduced by Freue et al. (2019) and Kepplinger (2023). The PENSE estimates and their adaptive version is implemented in the R package **pense** (Kepplinger et al. 2024).

9.4.3 Monitoring Sparse LTS

The main improvement of the robust LASSO, as implemented in the sparse LTS algorithm, over the classical LASSO is that it asymptotically guarantees a high breakdown

9.4 Regularization Methods

point which does not depend on the number of variables p. This is also an improvement over the LTS algorithm of Rousseeuw and Van Driessen (2006) which can attain maximal bdp =50% but this is only possible if $p < n/2$. However, the authors of sparse LTS (Alfons et al. 2013) recommend using bdp = 25% in order to base the estimate on a sufficiently large number of observations and thus to increase the statistical efficiency of the estimates. We saw the example in the previous section that the results significantly depend on the chosen bdp. Figure 9.6 shows two completely different pictures in the two panels representing the selection of the penalty parameter with 25% (left-hand) and 50% (right-hand) bdp; the estimated prediction errors with bdp = 50% are significantly higher and the minimal value is obtained for 24 non-zero coefficients (compared to 11 non-zero coefficients if bdp = 25% is used). This example once again shows that the arbitrary choice of values of bdp may well not lead to a clear understanding of the data. As mentioned many times in this book, it is preferable to investigate the results for a set of values of bdp and to choose that which provides the robust analysis with lowest bdp.

We illustrate the procedure of monitoring in the case of sparse LTS continuing to use the NCI60 cancer cell panel. We vary the bdp from maximal robustness with bdp = 50% down to bdp = 0% with a step of 1%. The complete procedure is run for each value: selecting the optimal penalty parameter λ by ten-fold cross validation, repeated 10 times to reduce variability, and then estimating the sparse LTS model with the chosen parameter λ. In many runs a different number of non-zero coefficients was estimated, from less than 10 up to more than 30. Figure 9.8 shows the number

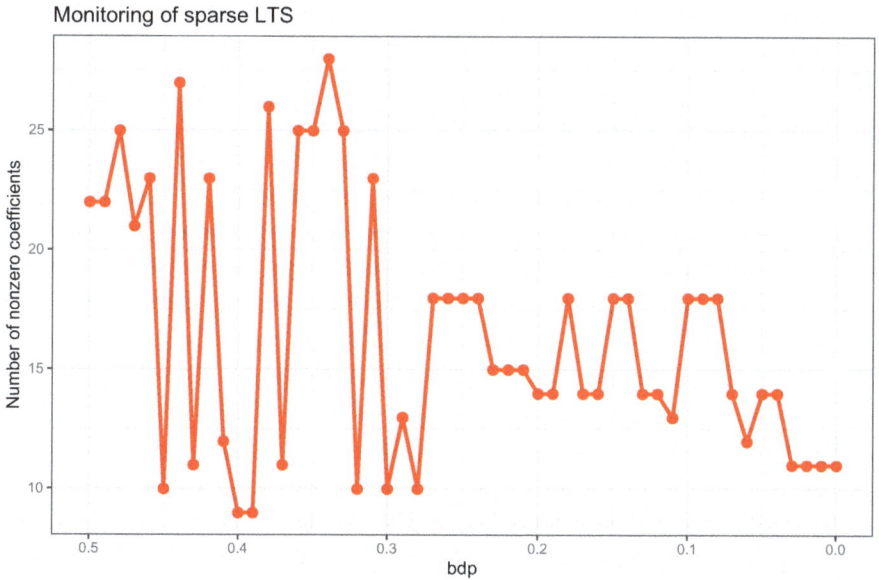

Fig. 9.8 NCI60 data: monitoring of sparse LTS presented by the number of non-zero coefficients estimated for each value of bdp

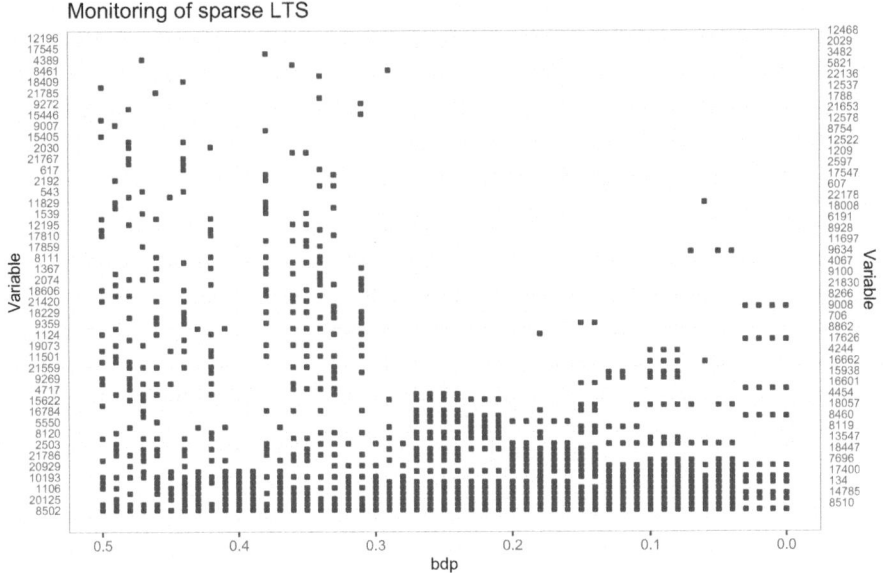

Fig. 9.9 NCI60 data: monitoring of sparse LTS presented by the frequency of selection of the different variables for each value of *bdp*. The left- and right-hand *x* axes give labels for alternating rows of the figure

of variables selected for each value of *bdp*. We see that for higher values of *bdp* (say, greater than 30%), there is a huge instability in the numbers of variables which are selected.

Figure 9.9 presents the monitoring of sparse LTS as a map of the (selected) variables against the values of *bdp*. Each blue square shows a variable which was selected. The variables are ordered by the frequency with which they are selected across the different *bdp* values—the most often selected are at the bottom. There is one variable which is selected always (x8502) and six which are selected for more than 30 values of *bdp*. These most frequently selected variables are extracted in a dataset which is analysed separately in Sect. 10.6, where it is simply called Cancer data. The codes for the seven selected variables are given in Table 10.9.

Problems

9.1 Analysis of the reduced logged ozone data using RAVAS
Apply RAVAS to the logged reduced ozone using as regressors Time plus variables 4, 5, and 6. Find the best model and produce the ANOVA table in the transformed scale.

Table 9.3 Air Pollution and Mortality data for 60 U.S. metropolitan areas, 1959–1961: variable names and description of the explanatory variables

Variable	Description
`Precip`	Mean annual precipitation (in inches)
`Humidity`	Percent relative humidity (annual average at 1:00pm)
`JanTemp`	Mean January temperature (degrees Fahrenheit)
`JulyTemp`	Mean July temperature (degrees Fahrenheit)
`Over65`	Percentage of the population aged 65 years or over
`House`	Population per household
`Educ`	Median number of school years completed for persons 25 years or older
`Sound`	Percentage of housing that is sound with all facilities
`Density`	Population density (in persons per square mile of urbanized area)
`NonWhite`	Percentage of population that is nonwhite
`WhiteCol`	Percentage of employment in white-collar occupations
`Poor`	Percentage of households with annual income under 3,000 USD in 1960
`HC`	Relative pollution potential of hydrocarbons
`NOX`	Relative pollution potential of oxides of nitrogen
`SO2`	Relative pollution potential of sulphur dioxide

9.2 Analysis of air pollution and mortality data

A dataset on air pollution and mortality in 60 metropolitan statistical areas in the United States is available in the R package **Sleuth3** (Ramsey et al. 2024) which accompanies the book Ramsey and Schafer (2013). The original reference is McDonald and Schwing (1978). The name of the dataset inside **FSDA** is `air_pollution`. The response variable is the total age-adjusted mortality from all causes, counted as deaths per 100,000 population. The explanatory variables are listed in Table 9.3 and include four climate variables, eight demographic variables, and three pollution related variables. "Relative pollution potential" is the product of the tonnes emitted per day per square kilometre and a factor correcting for the area dimension and exposure. The three pollution variables are skewed and for simplicity we follow standard practice in transforming them by taking logarithms (see Sect. 6.5.1 for a statistical approach to selecting such transformations, which should be chosen robustly).

1. Perform some preliminary data analysis by visualizing the correlations using either R package **corrplot** (Taiyun and Simko 2024) or the function `spmplot` in **FSDA**. Provide comments.
2. Perform classical ridge regression and LASSO and compare the models.
3. Perform sparse LTS using $bdp = 0.25$ and $bdp = 0.5$. Check if these two values of bdp provide confidence intervals for λ of similar width and whether the number of non-zero coefficients is the same. Compare this analysis to that with the classical LASSO

4. Are there outliers in the data? If so what is their effect? Show the plot of scaled residuals (called standardized residuals in R package **robustHD**!). Comment on these plots remembering that the observations are ordered by Mortality, the response variable.

9.3 Analysis of air pollution and mortality data II

Conduct the analysis of the data from the previous problem using the robust PENSE method with $\alpha = 1$ and with $bdp = 0.25$ and $bdp = 0.4$. Note that PENSE does not perform correct cross validation for the maximal breakdown point 0.5. Therefore, it is suggested to use $bdp = 0.4$. Compare the models obtained with those from sparse LTS with $bdp = 0.25$ and $bdp = 0.50$. Create a table which shows the significant variables using the 4 different procedures mentioned above and comment on the stability of the selection of variables.

Open Access This chapter is licensed under the terms of the Creative Commons Attribution 4.0 International License (http://creativecommons.org/licenses/by/4.0/), which permits use, sharing, adaptation, distribution and reproduction in any medium or format, as long as you give appropriate credit to the original author(s) and the source, provide a link to the Creative Commons license and indicate if changes were made.

The images or other third party material in this chapter are included in the chapter's Creative Commons license, unless indicated otherwise in a credit line to the material. If material is not included in the chapter's Creative Commons license and your intended use is not permitted by statutory regulation or exceeds the permitted use, you will need to obtain permission directly from the copyright holder.

Chapter 10
Some Robust Data Analyses

Abstract In this last chapter, we exhibit the power of the methods described earlier, by analysing five datasets. We start in Sect. 10.2 with the two sets of income data from Sect. 1.4. Without explanatory variables, we found the log transformation for the former, while the analysis of the latter remained inconclusive. When explanatory variables are included, and outliers deleted, the square-root transformation is indicated for both. In Sect. 10.4 we analyse 1711 responses to a survey on customer loyalty, in which there are six explanatory variables. Parametric methods lead to $\sqrt{y}$ as the response, the identification of 41 outliers, and a skewed distribution of residuals. RAVAS followed by the FS provides a good approximation to normally distributed errors, when only nine observations are deleted. This analysis is summarized in tabular form in Sect. 10.4.6 to provide a template for the modern robust analysis of regression data. Despite transformation and outlier detection, the t-statistics for the significance of the variables in the customer loyalty data hardly change. Accordingly, in Sect. 10.5 we modify 25 observations: monitoring plots reveal the outliers and the results of the RAVAS analysis are close to those for the uncontaminated data. Finally, we analyse the NCI-60 cancer cell data (Chap. 9). With only seven explanatory variables, we monitor LS diagnostics, detect outliers, and apply RAVAS, which gives the best-fitting model. The generalized candlestick plot provides further model selection.

10.1 Introduction

In the previous chapters of this book we have used the analysis of carefully selected datasets to illustrate particular points in the robust, or other, analyses of data. The focus in this chapter is rather more on analysing five sets of data and seeing where the analysis takes us. A general conclusion is that robust analysis combined with the monitoring approach is necessary to extract all the information about the data, its quirks and quiddities. But the moral seems to be that for some datasets parametric analysis is to be preferred, but for others the non-parametric approach is more fruitful.

© The Author(s) 2025
A. C. Atkinson et al., *Robust Statistics Through the Monitoring Approach*,
Springer Series in Statistics, https://doi.org/10.1007/978-3-031-88365-1_10

In Chap. 1, in order to emphasize the importance of response transformation, we analysed two sets of income data. In those analyses we ignored the explanatory variables and obtained the customary log transformation for income data for the first dataset. In Sects. 10.2 and 10.3 we repeat these analyses including the explanatory variables. These analyses are more complicated than those in Chap. 1 and do not suggest the log transformation.

The example in Sect. 10.4 concerns 1711 observations from an online survey of the respondents' assessment of the importance of six explanatory variables in maintaining customer loyalty. Given the subjective nature of these numbers, there is no reason to expect that a simple linear regression will be satisfactory. We find that RAVAS provides the best model, identifying only 16 outliers. The inferences about the importance of the variables are strangely stable to transformations and outlier detection. In Sect. 10.5 we slightly modify 25 observations, a change that has an appreciable effect on the data analyses. Although LS diagnostics suggests there is something untoward with the data, it requires the monitoring of robust procedures to find the introduced outliers. RAVAS identifies 35 outliers, including the 25 introduced ones.

The chapter concludes with the analysis of the NCI-60 cancer cell panel data introduced in Chap. 9. We take the seven explanatory variables which the earlier analysis had indicated as important. With this more compact data we are able to monitor LS diagnostics, detect outliers, and apply RAVAS, which provides the best-fitting model. The generalized candlestick plot provides a method for comparing models when some explanatory variables are omitted.

10.2 Income Data 1: United States Census Bureau

In Sect. 1.4 we studied the analysis of two sets of income data. These were presented as univariate data and were analysed as such, with regard both to robustness and transformation. In fact, both are sets of regression data, albeit with different explanatory variables. For the data from the United States Census Bureau the explanatory variables are:

x_1: Number of persons in household (HNUMPER)
x_2: All other types of household income (HOTHVAL)
x_3: Household income—social security (HSSVAL)

Figure 10.1 is the yX plot of the 200 observations selected from the census. The diagonal boundaries to the lower values of x_2 and x_3 are a function of the way the sample was taken. The fan plot of Fig. 1.5 indicates the logarithmic transformation ($\lambda = 0$) when the explanatory variables are ignored. Interestingly and importantly, this is the standard transformation for income data in economic and econometric analyses.

10.2 Income Data 1: United States Census Bureau

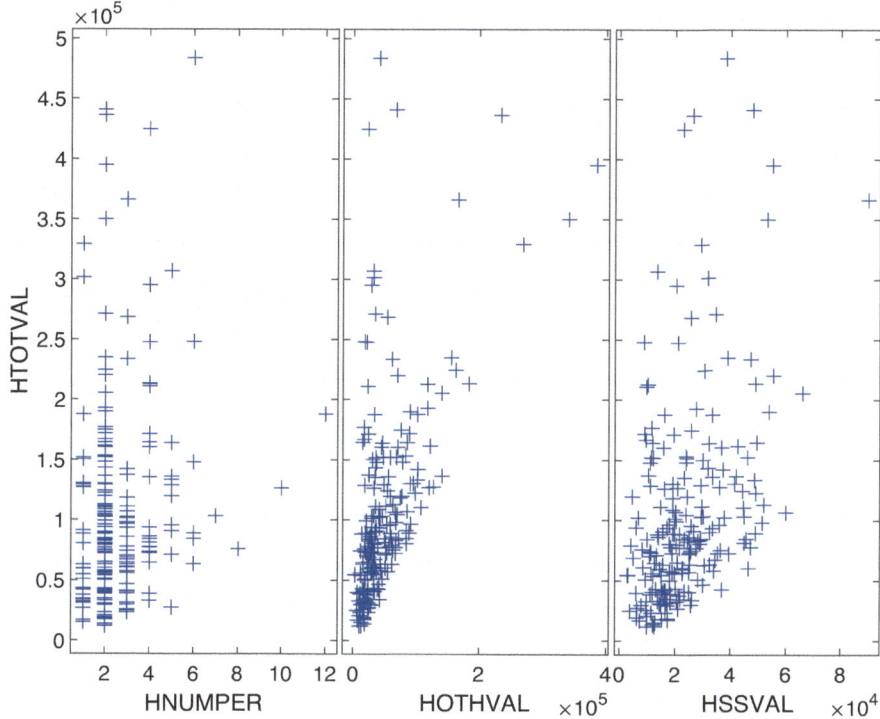

Fig. 10.1 Income data 1: yX plot; y has a positive skewed distribution. The data are likely to need transformation

We now consider the transformation of the response in these regression data. In Fig. 1.5 the grid of values for λ_0 is the coarse grid with values $-1, -0.5, 0, 0.5,$ and 1. Repeating the analysis on a finer grid for values of λ_0, but including the explanatory variables, shows that, at the end of the search, $\lambda_0 = 0.25$ is almost as good a value as zero. The left-hand panel of Fig. 10.2 is a fan plot using the finer grid for λ, which elucidates this finding. It seems that $\lambda = 0$ and 0.25 may be almost as good as each other at $n = 200$. But the plot shows that the trajectories leave the 99% confidence region before the end of the search and then both decrease at the end to return very close to the boundaries of the confidence interval. All trajectories give a clear indication of the presence of outliers or other influential observations on all scales in the plot, with the effect increasing as λ goes from zero to one. It seems that, if these observations are excluded, $\lambda = 0.5$ provides the best transformation. The three-panel plot from the automatic method of Sect. 6.3.4 in the right-hand panel of Fig. 10.2 shows that 0.5 is, just, the preferred transformation. However, the white rectangle atop the red bar for this value of λ shows that there are some observations which do not affect the transformation but are still outlying. The FS on the square-root scale identifies 9 outliers, which are shown as circles in the yX plot of Fig. 10.3.

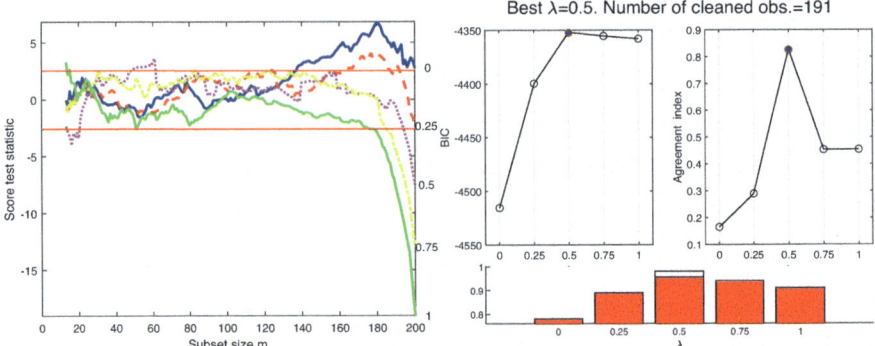

Fig. 10.2 Income data 1: left-hand panel: fan plot, monitoring plot of score statistics for five values of λ_0 for the Box-Cox transformation. Right-hand panels, automatic analysis of Box-Cox transformation: extended BIC, agreement index AGI and $h(\lambda)/n$ and $m^*(\lambda)/n$, the proportion of observations used in fitting the model. Calculations for five values of λ

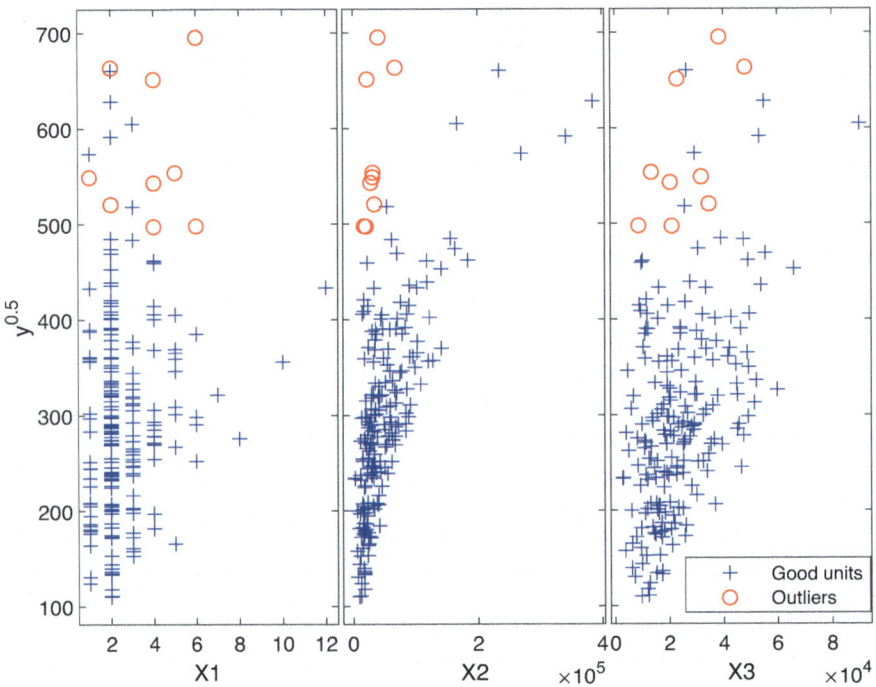

Fig. 10.3 Transformed Income Data 1 ($\lambda = 0.5$): yX plot. The 9 outliers are shown as red circles

10.2 Income Data 1: United States Census Bureau

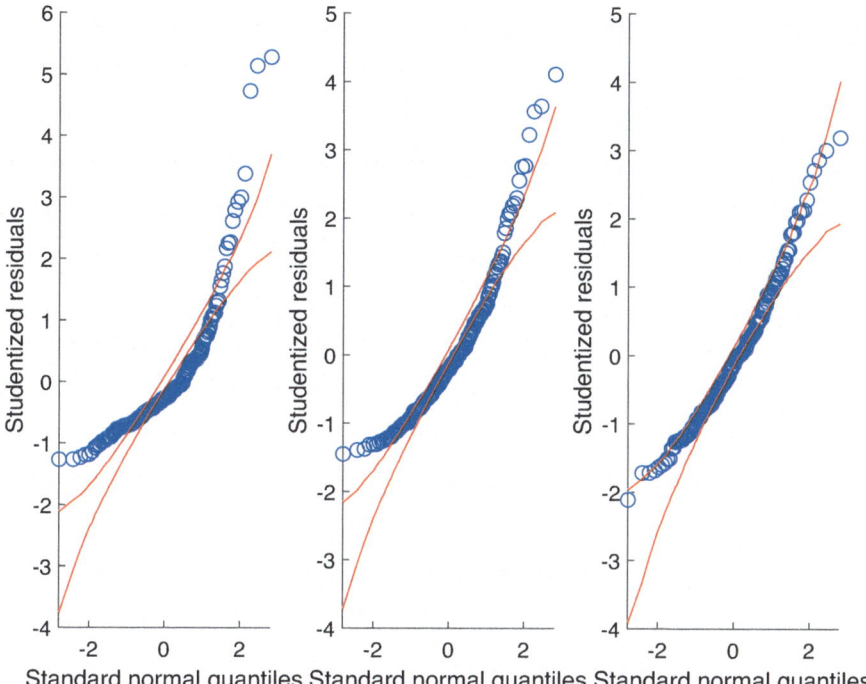

Fig. 10.4 Income Data 1: QQ-plots of studentized residuals for three fitted models: left-hand panel, original data; centre panel, square-root transformed data and, right-hand panel, square-root transformed data with nine outliers deleted

As a result we have three analyses based on LS: the original 200 observations; the same on the square-root scale and the 191 observations on the square-root scale. We now compare the results of these analyses. The three panels of Fig. 10.4 show the QQ-plots of the studentized residuals with simulation envelopes. The left-hand panel, for the original data, is the most curved. The square-root transformation reduces the curvature of the plot, as does the deletion of outliers in the right-hand plot. However, the plot is still smoothly curved; there may still be some systematic departure from the model.

Table 10.1 summarizes the analysis of variance tables for these three fitted models. The first three lines give the values of the t-statistics for the three explanatory variables, with the value of R^2_{adj} in the fourth line. It is clear from the last line that the values of R^2_{adj} increase due to the response transformation and deletion of outliers, an increase of 43%. The corresponding t-statistic for β_3 has a significance level of 0.0101, very much on the boundary of the 1% significance level. Transformation increases the value of all t-statistics (centre column), while deletion of the nine outliers increases the significance of regression on x_2, but decreases that on x_1 and x_3. The explanation of this behaviour is in Fig. 10.5, which shows the monitoring plot of

Table 10.1 Income Data 1: summary of t-statistics from ANOVA tables for three different fitted models

Property	Original	Transformed	Deletion[a]
t_1	3.93	4.59	3.83
t_2	7.56	7.40	11.10
t_3	2.60	3.23	2.69
R^2_{adj}	0.39	0.42	0.56

[a] 9 outliers deleted on the transformed (square root) scale

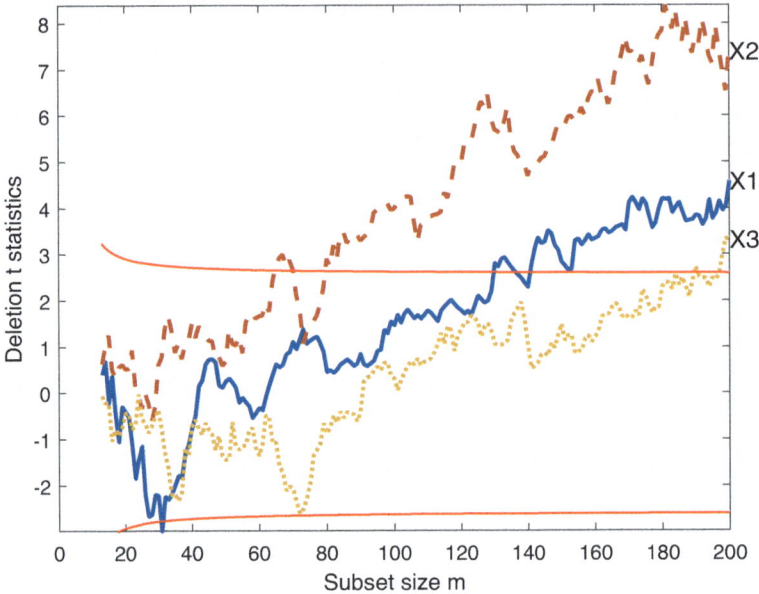

Fig. 10.5 Transformed Income Data 1 ($\lambda = 0.5$): monitoring plot of the three added variable t-statistics, with 99% confidence interval

the three added variable t-statistics. In the latter half of the search those for x_1 and x_3 trend upwards, with fluctuations. Although the same is true for the statistics for x_2, the values being more significant than those for the other two, there is a downward trend at the end of the search, particularly as the outliers enter. We conclude the LS analysis of the data with the monitoring plot of residuals in Fig. 10.6 which is not as uneventful as might have been hoped.

There are two observations, 107 and 200 that, around $m = 150$ until unit 107 enters the subset at $m = 186$, have egregiously large negative residuals. When these two units are brushed, the effect is to reveal these are the two "leverage" points in the panels of Fig. 10.8. These panels suggest that these observations may be far from the majority of the other observations. A monitoring plot of the leverages h_i is not especially revealing. However, rotating point clouds is helpful. Figure 10.7 is the

10.2 Income Data 1: United States Census Bureau

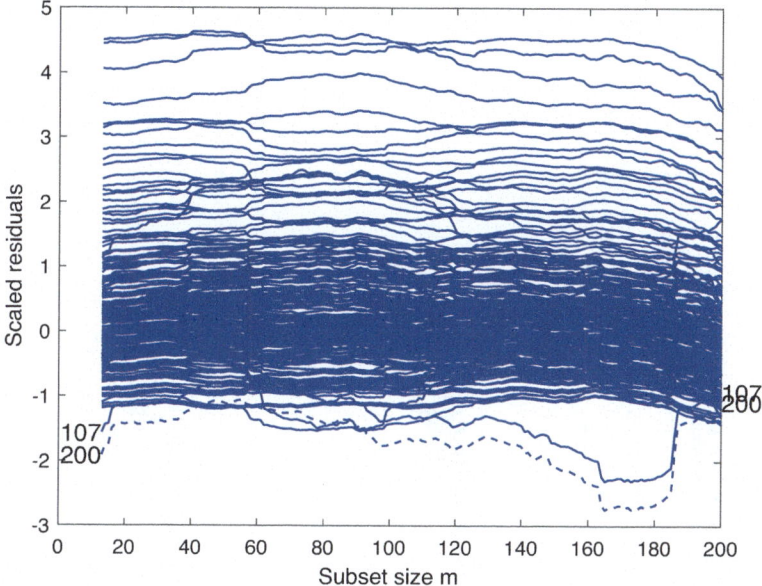

Fig. 10.6 Transformed Income Data 1 ($\lambda = 0.5$): monitoring residuals. Two observations, 107 and 200, around $m = 170$ have large negative residuals

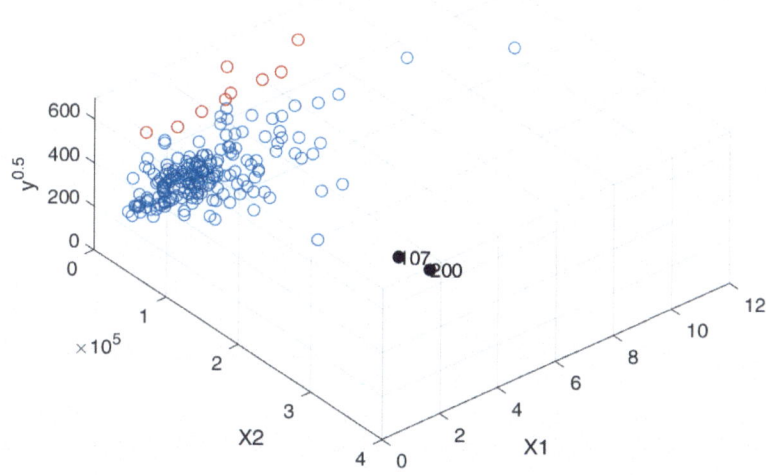

Fig. 10.7 Transformed Income Data 1 ($\lambda = 0.5$): 3-dimensional scatter plot of $\sqrt{y}$, x_1 and x_2 with rotation. The outliers are shown as red circles, the leverage points as black dots

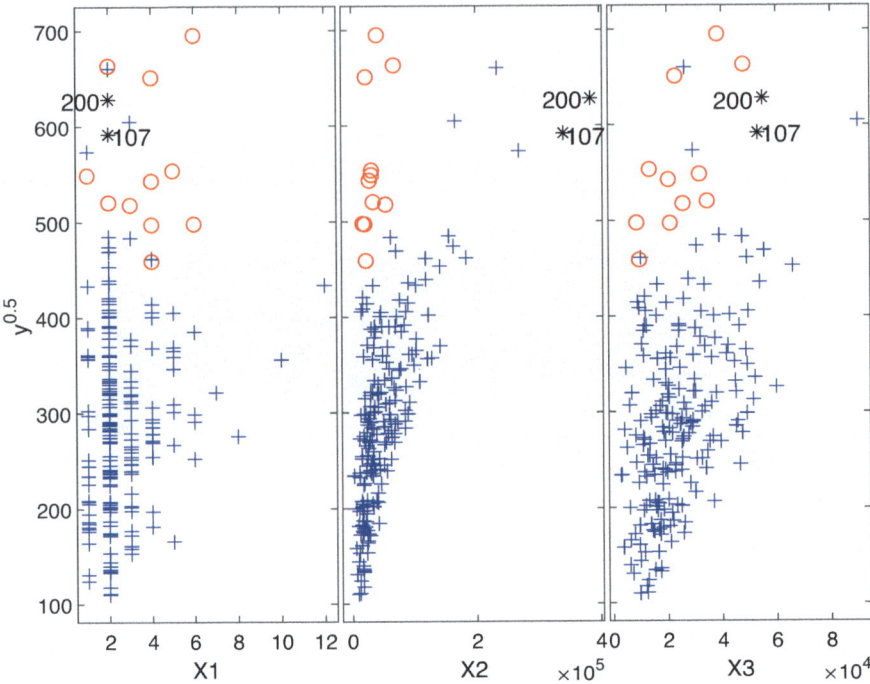

Fig. 10.8 Transformed Income Data 1 ($\lambda = 0.5$): outliers from heteroskedastic model. The two units declared as high leverage from the homoskedastic model are shown by * and labelled with their row number

point cloud formed by $\sqrt{y}$, x_1 and x_2. Since the regression on x_3 is much weaker, it is the natural candidate to be excluded from a three-dimensional plot. The figure suggests that the outliers form a weak cluster on one side of the regression plane, with the two leverage points on the opposite side, but in such a position that they do not lie far from the plane.

The yX plot in the square-root scale shown in Fig. 10.3 suggests that, even after the deletion of outliers, there may be some residual heteroskedasticity. We accordingly try heteroskedastic regression, Sect. 8.4. Since the weights for the heteroskedastic regression are re-estimated at each step of the search, we can again use the FS to establish whether any outliers are present. Figure 10.8 shows the 13 outliers detected on fitting the heteroskedastic model. These include the nine outliers and two leverage points shown in Fig. 10.7. A disappointing feature of this analysis is that the observations identified as outliers by the heteroskedastic regression consist mostly of those with the largest values of y.

Heteroskedastic regression does however provide some interesting insights into the data. We start with analysis of the untransformed data. Figure 10.9 presents monitoring plots, from $m = 150$, of the parameter estimates together with 95% and 99% confidence intervals. To start at the bottom, the estimate of σ^2 gently increases with

10.2 Income Data 1: United States Census Bureau

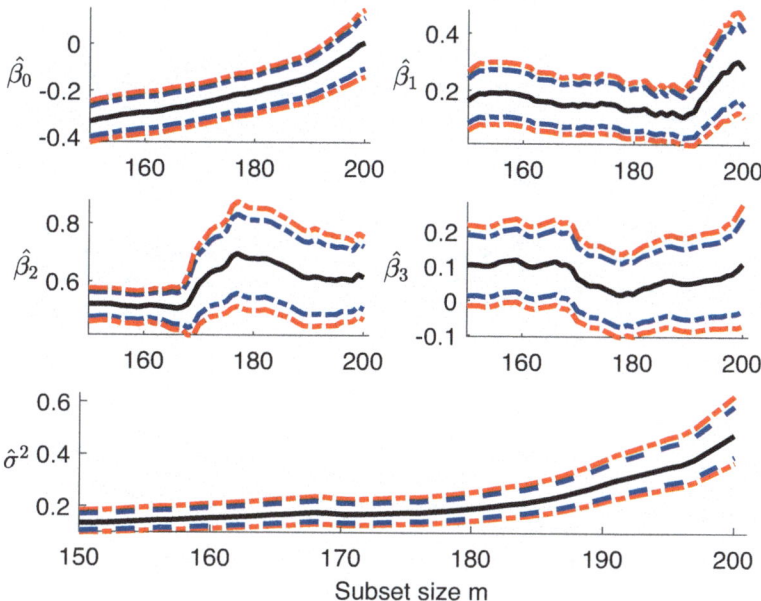

Fig. 10.9 Original Income Data 1: heteroskedastic regression; monitoring plots of estimates of β and of σ^2

m. However, in the calculation of the confidence intervals we use a consistency correction. Thus, in the panel for $\hat{\beta}_0$, the width of the confidence interval is virtually constant. The parameter estimate $\hat{\beta}_1$ is relatively steady until the last 11 observations start to enter the subset, when, until the very end, the parameter estimate increases steadily. The most surprising plot is that for $\hat{\beta}_2$ which has a minimum close to $m = 170$ and a maximum around $m = 180$, accompanied by a confidence interval, the width of which more than doubles. The estimate $\hat{\beta}_3$ has a peak when that of $\hat{\beta}_2$ has a minimum before decreasing slightly and then increasing again. The width of the confidence interval steadily increases.

The rather busy structure of the plots for the untransformed data is suggestive of an inadequate model. The plots, in Fig. 10.10, with response $\sqrt{y}$, are the more stable of the two sets; all parameter estimates, apart from that for $\hat{\beta}_3$, decrease towards the end of the search ($\hat{\beta}_3$ increases slightly). In comparison with Fig. 10.9, the trajectories of the confidence intervals have a more nearly constant width; this is especially marked in those for $\hat{\beta}_2$ and $\hat{\beta}_3$.

The effect of outliers is more clearly seen on the transformed scale rather than on the original scale, including an effect on the estimation of σ^2. The combination of transformation augmented with the estimation of heteroskedasticity leads to more stable behaviour and clearer inference about the effects of outliers.

Finally, we provide, in Fig. 10.11, an analysis using RAVAS. Only five outliers are found. The upper left-hand panel shows that the plot of $g(y)$ against y is close

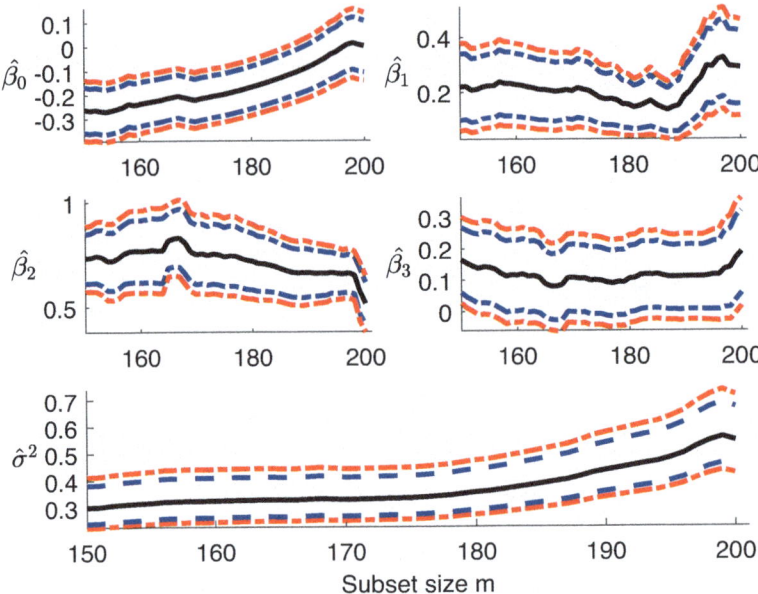

Fig. 10.10 Transformed Income Data 1 ($\lambda = 0.5$): heteroskedastic regression; monitoring plots of estimates of β and of σ^2

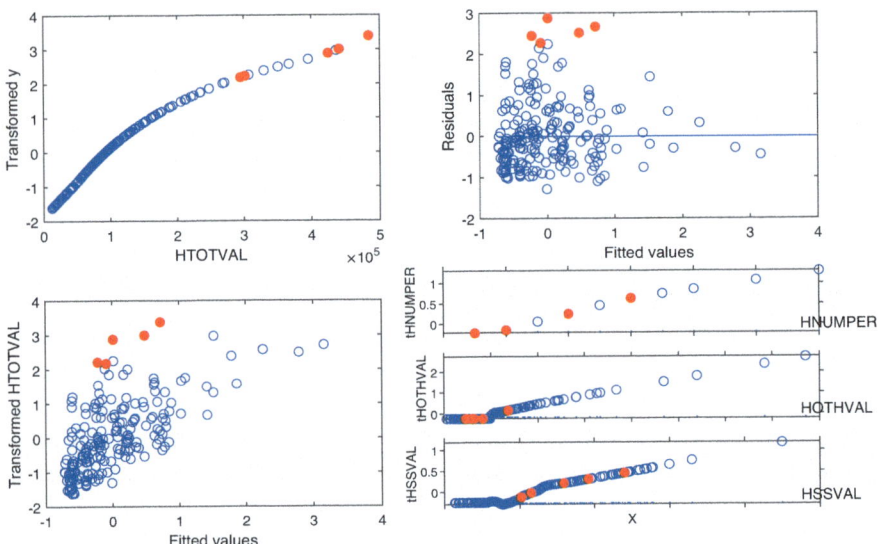

Fig. 10.11 Income Data 1: RAVAS. Upper left-hand panel, $g(y)$ against y; upper right-hand panel, residuals against fitted values; lower left-hand panel, $g(y)$ against fitted values; lower right-hand panel transformations $f_j(x_j)$ of the explanatory variables against original x_j

to the square-root transformation. The plot also shows that three of the five outliers found correspond to the largest values of y and lie slightly away from the estimated $g(y)$. The upper right-hand panel and the lower left-hand panel show that the outliers are properly separated from the other observations. In both panels the two leverage points have fitted values around three, with small residuals. The transformation of x_1 is shown in the top of the lower right-hand panel. Since there are only 10 levels of x_1, with outliers at four of them, the plot gives a visual impression of an excessive number of outliers. In the other two panels, the transformations of x_2 and x_3 show that the two transformations are similar, consisting of a horizontal part and a straight line for large x_j with hard to interpret structures for the intermediate values. The value of R^2_{adj} for the regression using these transformations is 0.458 (with the outliers, of course, deleted). Table 10.1 shows that this is appreciably worse than the result from regression with the square-root transformation when nine outliers are deleted.

10.3 Income Data 2: An Italian Municipality

We now turn to the regression analysis of the second example in Chap. 1 of income data. Like the analysis of the first example, the study of just the response suggested that income should be transformed. But the fan plot of Fig. 1.6 indicated the presence of outliers. We now explore whether and how these suspected outliers influence the need for transformation of the response in the regression model.

The data are again a sample of 200 observations. The variables are:

y: Income;
x_1: Age (Age);
x_2: Number of years of eduction (5–17), (Educational-num);
x_3: Gender (1 = male), (Gender);
x_4: Return from capital investments, (ExtraGain);
x_5: Hours worked per week, (Hours).

Our sample of 200 subjects is of those with income $\leq$ 50000 Euros.

Figure 10.12 is the yX plot for the five variables. Two quick conclusions are that there appears to be a relationship between income and age and that the sample contains many more men than women. Whether the men have greater individual incomes is certainly a question of interest. A third impression is that the long-tailed distribution of income may suggest that a response transformation will be advantageous. However, these are regression data. The apparent need for a transformation may be a function of the values of the explanatory variables and of the regression coefficients.

The starting point for the analysis is a regression on the five variables. The ANOVA table for this regression (Table 10.2) gives a value of 0.0167 for R^2_{adj}, with no variables significant at the 1% level. Perhaps there are too many variables in the model. Standard automatic stepwise variable selection leads to regression on only x_1, which

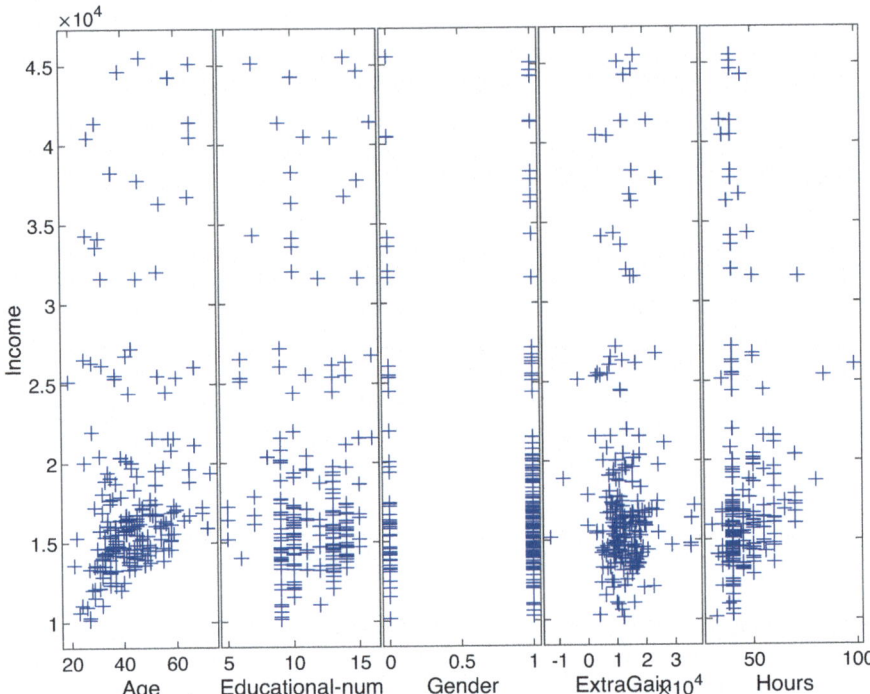

Fig. 10.12 Income Data 2: yX plot; y shows a positive skewed distribution. The data are likely to need transformation

is not quite significant at the 1% level and a value of 0.0267 for R^2_{adj}. Instead, we now consider the Box-Cox transformation of the data. The fan plot is in Fig. 10.13, which, although it is over a different set of values of λ_0 from the fan plot of Fig. 1.6, has a similar structure. It is clear that, for the greater part of the search, there is no evidence for any transformation of the data. Around $m = 140$, the value of -2 is rejected and, by the time $m = 160$, there is evidence against all negative values for λ_0. The best value seems to be $\lambda_0 = 0.5$ with $\lambda_0 = 0$ and 1 having slightly less good performance. In all three cases many outliers are indicated. Values of the statistics for these values of λ_0 lie within the envelope until they leave with negative values of the statistic and do not return.

The fan plot indicates the potential for transformation but leaves the choice of transformation slightly open. The same is true of the three-panel plot of the automatic procedure in Fig. 10.14. The BIC indicates a value of $\lambda = 1$ or greater, whereas the values of the agreement index have a peak at $\lambda_0 = 0.5$. In such a situation, where the value of the score statistic changes abruptly, we follow the guidance of the AGI and explore the properties of the square-root transformation.

The FS analysis with $\sqrt{y}$ as the response leads to the identification of 36 outliers a number identical to that suggested by the automatic procedure using AGI. Inspection

10.3 Income Data 2: An Italian Municipality

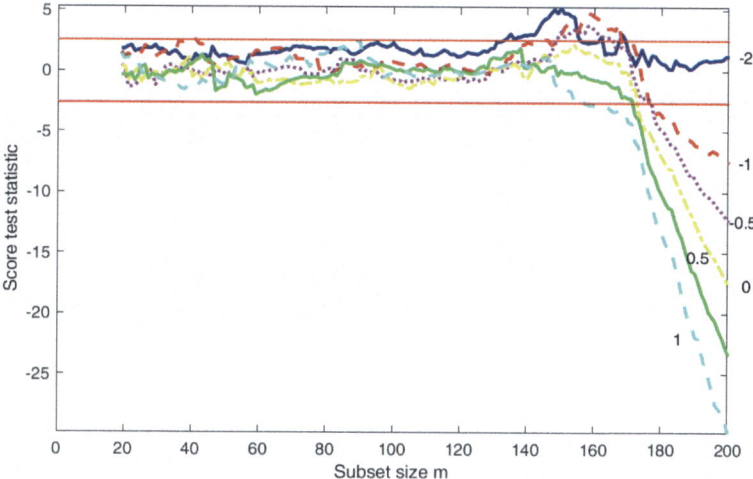

Fig. 10.13 Income Data 2: fan plot; monitoring plot of score statistics for the Box-Cox transformation with values of $\lambda_0 = -2, -1, -0.5, 0, 0.5$ and 1

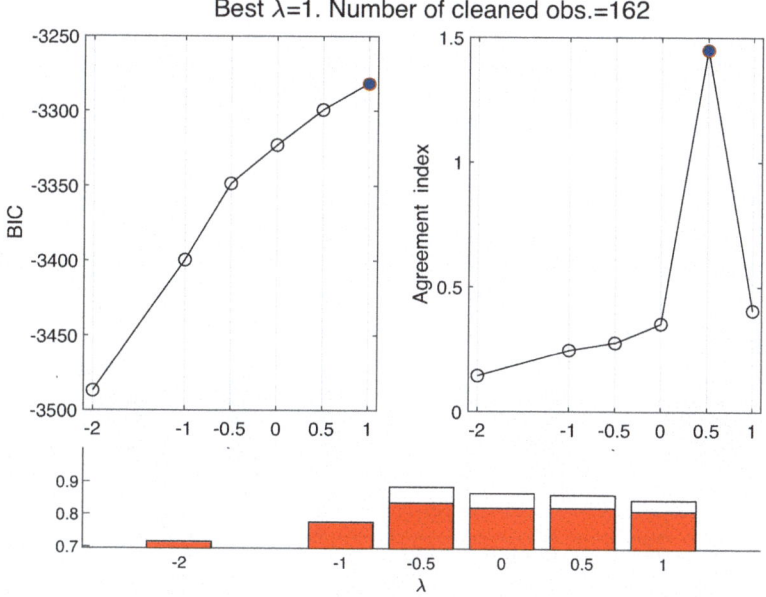

Fig. 10.14 Income Data 2: automatic analysis of Box-Cox transformation: extended BIC, agreement index AGI and $h(\lambda)/n$ and $m^*(\lambda)/n$, the proportion of observations used in fitting the model

Table 10.2 Income Data 2: ANOVA in original scale for *y*

	Estimate	SE	tStat	pValue
(Intercept)	12327	3825.6	3.2223	0.0014914
x1	111.34	44.673	2.4924	0.013527
x2	112.29	224.55	0.50005	0.6176
x3	-1454.1	1228.9	-1.1833	0.23815
x4	-0.024045	0.082962	-0.28983	0.77226
x5	31.439	51.42	0.61141	0.54165

```
Number of observations: 200, Error degrees of freedom: 194
Root Mean Squared Error: 7.19e+03
R-squared: 0.0414,  Adjusted R-Squared: 0.0167
F-statistic vs. constant model: 1.67, p-value = 0.142
```

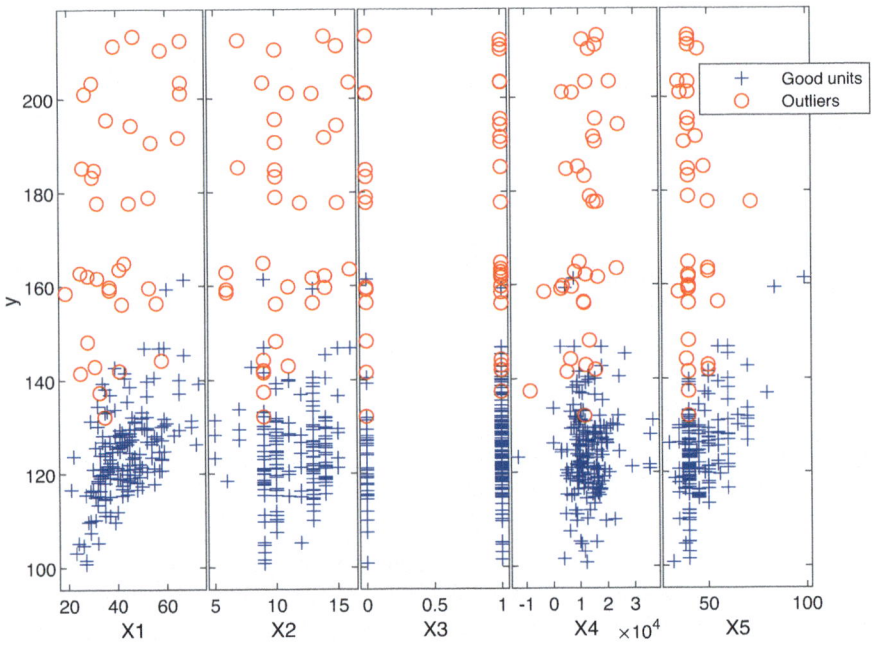

Fig. 10.15 Income Data 2: square-root transformation: yX plot with the 36 outliers highlighted

10.3 Income Data 2: An Italian Municipality

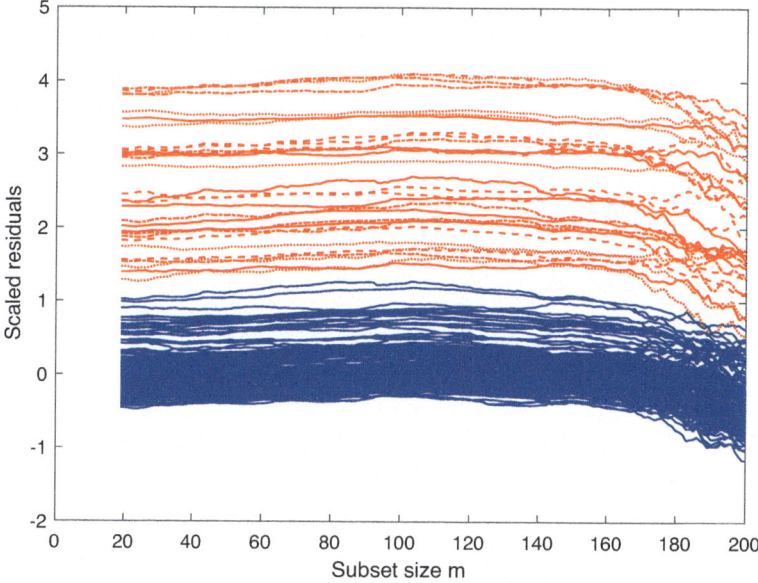

Fig. 10.16 Income Data 2: square-root transformation; monitoring plot of scaled residuals

of Fig. 10.13 also indicates that this is a plausible value for λ and we use it for the next few analyses, starting with Fig. 10.15. This figure shows that the majority of the observations identified as outliers have the highest values of y.

The monitoring plot of scaled residuals in Fig. 10.16 confirms that this is not an overestimate of the number of outliers; initially the plot is stable. But, from about 40 units from the end, turbulence sets in, that is the trajectories start to intersect. The plot also shows, during the whole search, that there is a set of around 40 observations with large positive residuals. Brushing these units for $m < 160$ gives a set of units with complete overlap with the units declared as outliers.

The ANOVA in Table 10.3, that is with response $\sqrt{y}$ and the outliers removed, gives a value of 0.659 for R^2_{adj}, compared with the value of 0.0167 for all data on the original scale. A statistically significant regression model has certainly been established. One statistical effect of deleting the outliers is clear in Fig. 10.17, showing QQ-plots of studentized residuals with and without the outliers. Deletion of the 36 outliers greatly improves the normality of the residuals. A further comparison is in Fig. 10.18. The left-hand panel shows the monitoring plot of the five added variable t-statistics over the whole search; the right-hand panel truncates the same plot, stopping at $m = 164$, that is before any of the outliers are included in the subset. The comparison of the two panels of the figure shows not only the dramatic effect of the outliers on the model, but also that x_1 and x_5 are highly significant with x_2 just significant at the 1% level, agreeing with the results of Table 10.2.

Table 10.3 Income Data 2: square-root scale; ANOVA after removing 36 outliers

	Estimate	SE	tStat	pValue
(Intercept)	69.456	3.6291	19.139	5.3309e-43
x1	0.5316	0.04117	12.912	1.6896e-26
x2	0.53244	0.20183	2.6381	0.0091717
x3	2.2693	1.136	1.9976	0.047473
x4	-3.7771e-05	7.1116e-05	-0.53111	0.59609
x5	0.53283	0.043091	12.365	5.3678e-25

```
Number of observations: 164, Error degrees of freedom: 158
Root Mean Squared Error: 5.72
R-squared: 0.669,  Adjusted R-Squared: 0.659
F-statistic vs. constant model: 63.9, p-value = 3.47e-36
```

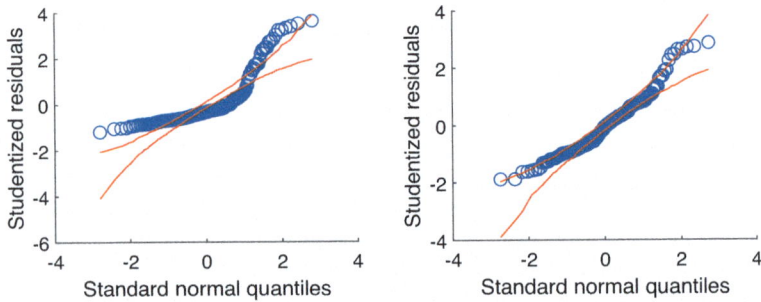

Fig. 10.17 Income Data 2: square-root transformation; QQ-plots of studentized residuals with 99% simulation envelope using all the data (left-hand panel) or just the units not declared as outliers (right-hand panel)

One check of the number of outliers found is to try a different robust analysis. For this we used very robust LTS, that is with a *bdp* of 50%. The index plot of residuals showed that 41 observations had residuals lying outside the 97.5% confidence band and 33 outside the 99% confidence band. If we use a confidence band comparable with the one based on FS (that is 1% simultaneous size) we find just 29 outliers. This result is in agreement with the comparative sizes and powers of FS and LTS procedures for outlier detection presented in Sects. 5.2 and 5.3. If LTS were the preferred robust method, monitoring could be used to search for a fit with lower *bdp*. This might well indicate a slightly reduced number of outliers.

10.3 Income Data 2: An Italian Municipality

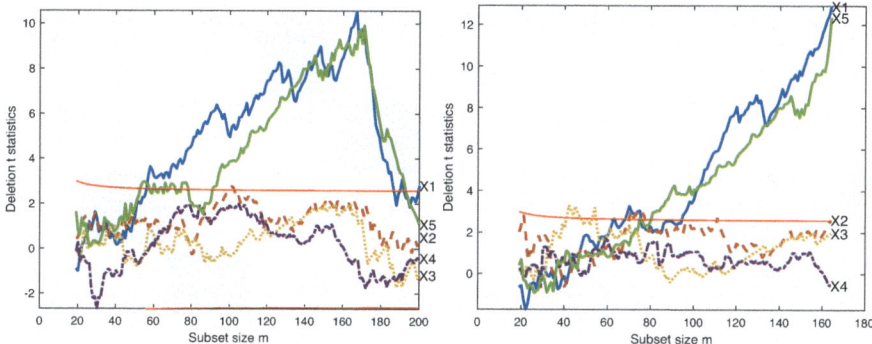

Fig. 10.18 Income Data 2: square-root transformation; monitoring of the five added variable t-statistics using all observations (left-hand panel) and after removing the outliers (right-hand panel)

Since not all variables are significant, we again explored automatic model selection, this time with $\sqrt{y}$ as response and with the 36 outliers deleted. Working with only first-order terms, x_4 (return from capital investment) is dropped. This change increases the value of R^2_{adj} in Table 10.3 from 0.659 to 0.660. This is not much of an improvement and leaves x_3 (gender) with a t value of 1.98. It is natural also to consider model selection starting from the model with all two-factor interactions, that is a model with 15 terms, plus the intercept. This leads to the same four-variable model as before, but with the inclusion of the interaction $x_2 x_5$, that is of number of years of education and of hours worked. Any non-cynical theory of economics would predict that this virtuous combination would lead to a higher income. Unfortunately, it here has a negative coefficient, with a t-statistic of -2.22. One interpretation is that the combined effect of the two variables is a little less than additive. Inclusion of the interaction also has the effect of reducing the t-statistic for x_5 from 12.36 in the first-order model (see Table 10.3) to 4.72 when the interaction is included. For this model R^2_{adj} increases from 0.660 for the model excluding x_4 to 0.668. Given the slight improvement in fit, the extra complexity of the model, and the not very high significance of all t-statistics, except that for x_1 (age) = 13.27, we would not recommend this model.

In the ANOVA table (Table 10.3) both x_1 and x_5 were highly significant, with x_2 marginally significant at the 1% level. We now describe a RAVAS analysis with the intention of obtaining more information about the potential importance of x_2.

The analysis is on the original scale. Figure 10.19 shows the plots for the choice of options in RAVAS. The augmented star plot in the left-hand panel shows that robustness is important in the four best solutions. The heat map in the right-hand panel shows that the four solutions are equivalent. Plots of the statistical results are in Figs. 10.20 and 10.21. In interpreting these plots, we keep in mind the plotting method. The "good" observations are plotted first (here in blue), followed by the

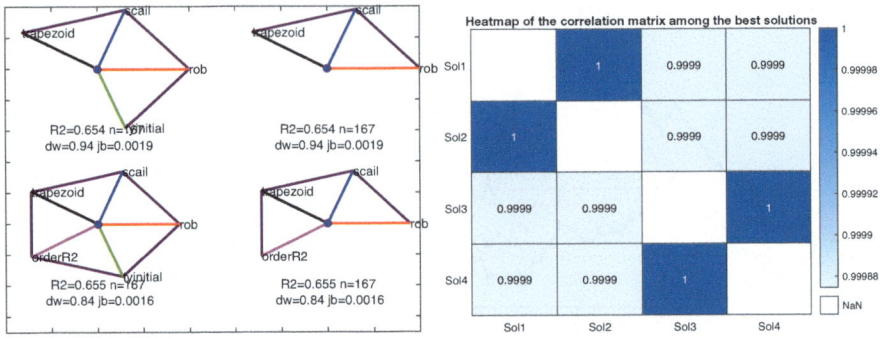

Fig. 10.19 Income Data 2: augmented star plot from RAVAS (left-hand panel) and correlations among the solutions (right-hand panel)

outliers in red, which overprint the blue symbols. The presence of outliers can be over-emphasized.

The plots of residuals against fitted values and of $g(y)$ against fitted values in Fig. 10.20 confirm the finding of the outlying nature of the largest observations, here 33. Care is needed in interpreting the upper left-hand panel of Fig. 10.20 as the apparent linearity of the transformation is an optical effect of the visual dominance of the outliers. In the analyses so far, the most important variables have been x_1 and x_5. In the plots of transformations $f_j(x_j)$ against x_j in Fig. 10.21 both variables have small horizontal initial sections followed by increasing almost straight lines. The transformations of x_3 and x_4 are unimportant, given the low significance of these variables in the fitted regression model of Table 10.3. However, the transformation of x_3 is a reminder that a discrete variable with two levels cannot be transformed into any other kind of variable. Here the constraints that the transformed variable have mean zero and variance one give the more numerous males a smaller transformed value than the less numerous females. The transformation of x_2 suggests that no education is harmful to earnings and that values near the maximum are helpful. However, especially once the outliers have been removed, we are dealing with small incomes.

The final analysis is summarized in the ANOVA of Table 10.4, based on the transformations from RAVAS with 33 observations deleted. The results of this analysis are close to those of the ANOVA with the square-root transformed response and 36 outliers deleted, summarized in Table 10.3. The use of the RAVAS variables slightly decreases the value of R^2_{adj} from 0.659 to 0.644 while increasing the t-statistic for x_2 from 2.638 to 2.922. The local conclusions are that there is no evidence for the significance of x_3 and x_4, but that x_2 cannot be discarded. The final model will thus include, in order of significance in their effect on income, the variables age, hours worked and years of education.

A wider significance is that the regression analysis, involving decisions about transformations and the significance of variables, and the semi-automatic procedure

10.4 Customer Loyalty

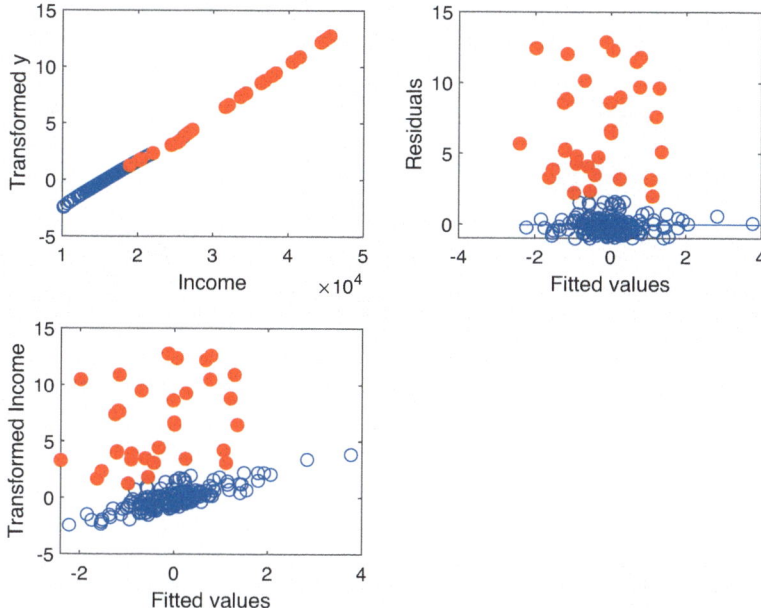

Fig. 10.20 Income Data 2, RAVAS: transformed response $g(y)$. Upper left-hand panel, $g(y)$ against y; upper right-hand panel, residuals against fitted values; lower panel, $g(y)$ against fitted values. 33 outliers highlighted. After outlier deletion the plot of the remaining observations in the upper left-hand panel is curved

of RAVAS gave very similar and so mutually supportive results. Further support came from brief analyses using LTS and regression variable selection. The common truncation of Tukey's suggestion (see Sect. 1.5) to become that a robust and a non-robust analysis be compared should, at least, be replaced with the suggestion of the comparison of dissimilar robust methods. If the results disagree, further analysis is needed.

10.4 Customer Loyalty

10.4.1 Data and Regression

Loyalty in marketing is highly desirable for both brands and retailers. The attempt to build loyalty uses many marketing strategies such as the development of discount programmes for customers, for example through loyalty cards; free goods; prizes, and special services. Developing customer loyalty is seen as a form of protection from competition and gives more control over marketing planning. However, the variables

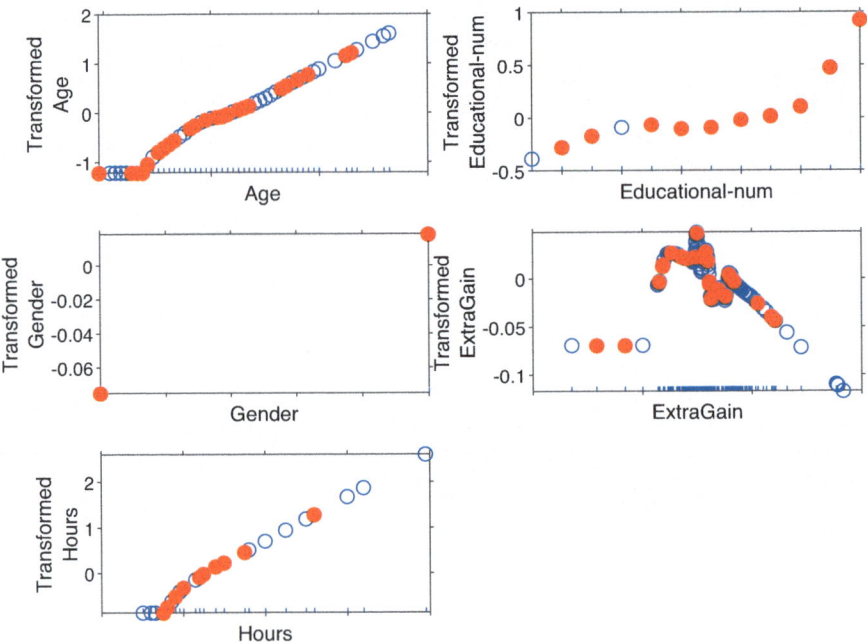

Fig. 10.21 Income Data 2, RAVAS: transformations $f_j(x_j)$ of the explanatory variables against original x_j

that affect loyalty are not well understood. Often the variables are expressed as ratings of customer perceptions.

The data analysed here, the results of a consumer-loyalty questionnaire, come from the website https://data.world/cesarpolo/consumer-loyalty-in-retail. These data were chosen because the information about them is limited solely to the names of the variables. On the web there is no description of the scale of the variables, of any data pre-processing nor any other information. The purpose of our blind analysis is to see whether the methods of our book lead to conclusions that reflect or criticize existing standard analyses of data of this type, such as linear regression.

If regression data come from measuring a physical system, the data are often generated by a smooth underlying model. Taylor series expansion of this unknown model should lead to useful polynomial models. However, here all variables are taken from the subjective responses to the questionnaire. Consequently, the variables may be highly non-linear and may benefit from transformations to produce a simple regression model. The purpose of the analysis is to determine which of the variables above most affect loyalty and what is the sign and the structure of the effect. As the analysis progresses, we explore non-linear transformations of both the response and the explanatory variables in order to better understand which levels of these variables cause a change in the level of loyalty.

10.4 Customer Loyalty

Table 10.4 Income Data 2: ANOVA in transformed RAVAS scale for y after removing the outliers

	Estimate	SE	tStat	pValue
(Intercept)	-2.8103e-16	0.046313	-6.0681e-15	1
x1	0.99871	0.07776	12.843	1.9213e-26
x2	0.91271	0.31232	2.9223	0.0039741
x3	1.3457	1.2601	1.0679	0.28715
x4	1.2	1.578	0.76046	0.44809
x5	1.0012	0.085572	11.701	2.8165e-23

Number of observations: 167, Error degrees of freedom: 161
Root Mean Squared Error: 0.598
R-squared: 0.655, Adjusted R-Squared: 0.644
F-statistic vs. constant model: 61, p-value = 2e-35

There are 1711 observations, all complete. The variables for a regression model are:

y: (response) Loyalty
x_1: Price
x_2: Quality
x_3: Community Outreach
x_4: Trust
x_5: Customer Satisfaction
x_6: Negative Publicity.

Figure 10.22 is a yX plot of the data, the panels of which all indicate a relationship between customer loyalty and the individual regressors, that for negative publicity being the most obviously non-linear. However, we start with fitting a linear regression model.

The analysis of variance for regression on all variables is in Table 10.5, with a summary in Table 10.6. All variables are highly significant, with $R^2_{adj} = 0.741$. Quality, price, and trust are the most significant variables. It is natural to ask whether, and if so how, we might do better. One possibility, given the curved nature of the relationships in Fig. 10.22, is to try a model with all interactions. After using a stepwise procedure for model selection, a model is obtained with 17 terms and a value of 0.783 for R^2_{adj}. This is a small improvement in fit for such a cumbersome model. Instead we look at plots of residuals and the fitted model to assess how well the data and fitted model agree.

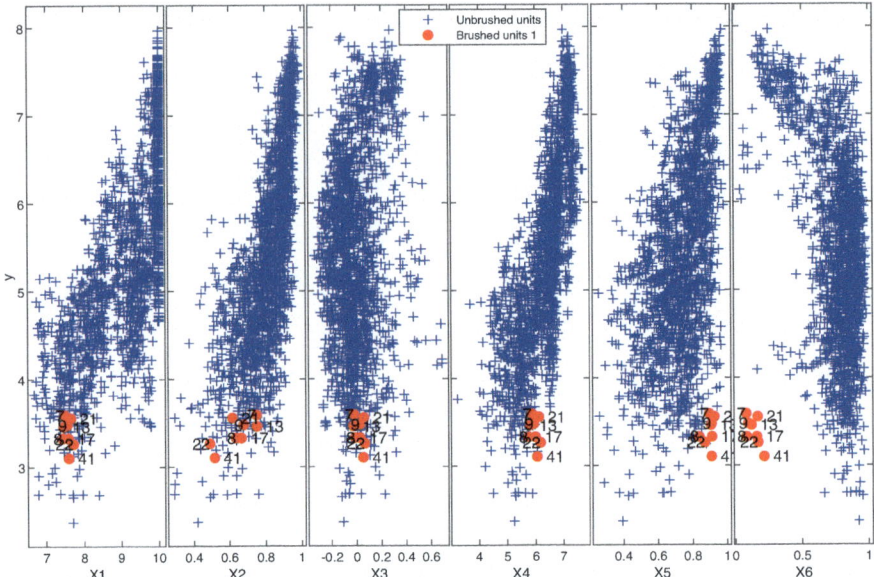

Fig. 10.22 Customer Loyalty data with the 8 brushed units from Fig. 10.24 highlighted

The plot of residuals against fitted values is uninformative about any lack of fit of the first-order model. But the QQ-plot of studentized residuals in the upper panel of Fig. 10.23 shows that, although the extreme residuals are not extreme, smaller positive and negative residuals lie outside the 95% simulation envelope. In particular, there are too many negative residuals that are too large to have come from a normal distribution. The lower panel of the figure is a symmetry plot of residuals about their median. For n even, let the ordered residuals be $e_{[i]}$ with the median residual $e_m = e_{[n/2]}$. Working out from the median, the points plot $e_{[m-i]}$ against $e_{[m+i]}$. (The procedure is the same for n odd, but the notation is, unhelpfully, more complicated). The upper right-hand part of the plot exhibits, in an accentuated form, the large negative residuals lying outside the envelope in the plot of the upper panel.

The histogram of the residuals is slightly asymmetric. Superimposition of a fitted normal distribution shows that there are too few residuals with values around minus one for the distribution to be normal. Fuller information about the structure of the data can be found by looking at the monitoring plot of all residuals. Figure 10.24 shows the plot of scaled residuals from the FS. There are several interesting features in this plot. One is that there is a change in the fitted model around $m = 1000$, leading to a change in the structure of the residuals. For $m > 1000$ the pattern is remarkably constant; the distribution of the upper tail is not far from normal, but the lower tail is too thin.

10.4 Customer Loyalty

Table 10.5 Customer Loyalty data: ANOVA

```
                         Estimate       SE        tStat       pValue

    (Intercept)          -2.0312     0.18383     -11.05     1.8399e-27
    Price                 0.32801    0.031277     10.487    5.5983e-25
    Quality               2.5894     0.16721      15.486    1.043e-50
    CommunityOutreach     0.75773    0.095933      7.8986   5.016e-15
    Trust                 0.37445    0.035882     10.436    9.338e-25
    CustomerSatifaction   0.99442    0.12458       7.9822   2.6177e-15
    NegativePublicity    -0.95476    0.089834    -10.628    1.3692e-25
Number of observations: 1711, Error degrees of freedom: 1704
Root Mean Squared Error: 0.578
R-squared: 0.742,  Adjusted R-Squared: 0.741
F-statistic vs. constant model: 815, p-value = 0
```

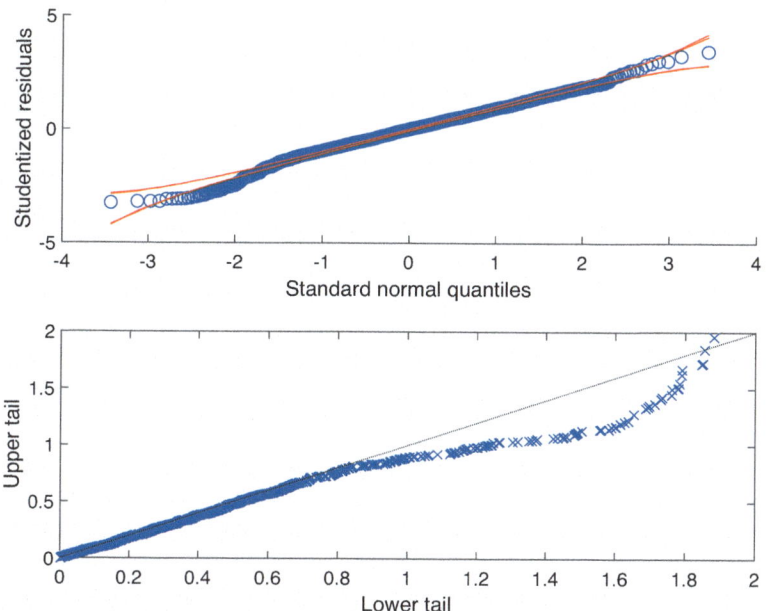

Fig. 10.23 Customer Loyalty data: cumulative plots of residuals. Upper panel; QQ-plots of studentized residuals with 95% simulation envelope: lower panel; symmetry plot of studentized residuals

A second feature is the band of large negative residuals for m around 600. The yellow rectangle in the bottom left-hand corner of Fig. 10.24 highlights, in red, the trajectories of the 8 most negative residuals (units 7, 8, 9, 13, 17, 21, 22, and 41). Only four of these (7, 9, 13, and 17) are among the most negative by the end of the search. The location of these observations is shown in red in the yX plot of Fig. 10.22. All

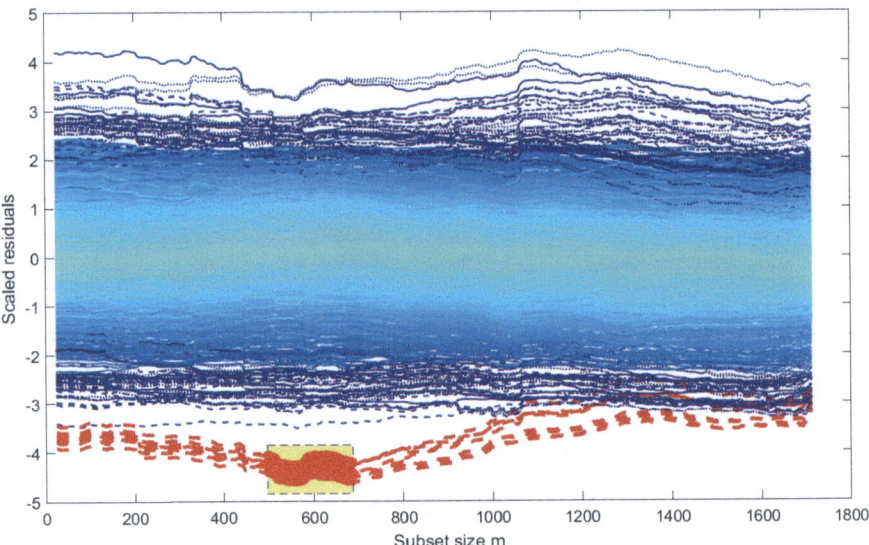

Fig. 10.24 Customer Loyalty data: monitoring residuals from the FS with brushing. Eight observations are highlighted

these observations have low values of y. The plot of y against x_6, in particular, shows that, in this view, they form a distinct outlying cluster. Since these observations are close together in index number, it would be interesting to explore these observations further, if such fine detail of the data were available.

It is clear, especially from the residual plots of Fig. 10.23, that there is some unexplained structure in the data, perhaps caused by outliers. In addition, the fitted model may be incorrect. The FS procedure for outlier detection on the original data detects 28 outliers.

To see whether outlying observations affect the fitted model, Fig. 10.25 shows the monitoring plots of all six t-statistics from the six forward searches. There is a steady increase in the significance of all variables as the search progresses. If the outliers were important in changing the values of the tests, we would expect an appreciable change in the plotted values, especially when the outliers start to enter the subset. The entry of the 28 outliers found by the overall FS into each individual search is shown in red in the figure. The behaviour for x_1, x_3, x_4 and x_5 is similar. The outliers enter near the end of the search and cause a noticeable, but not practically significant, change in the values of the t-statistics. The behaviour for x_2 and x_6 is smoother. If all outliers enter at the end, they will have no effect until $m = 1684$. Inspection of a zoom of the last part of the FS shows that a few of the outliers in each search enter well before this value. The search with x_6 as the added variable is most extreme in this way, with the first outlier entering at $m = 1595$.

10.4 Customer Loyalty

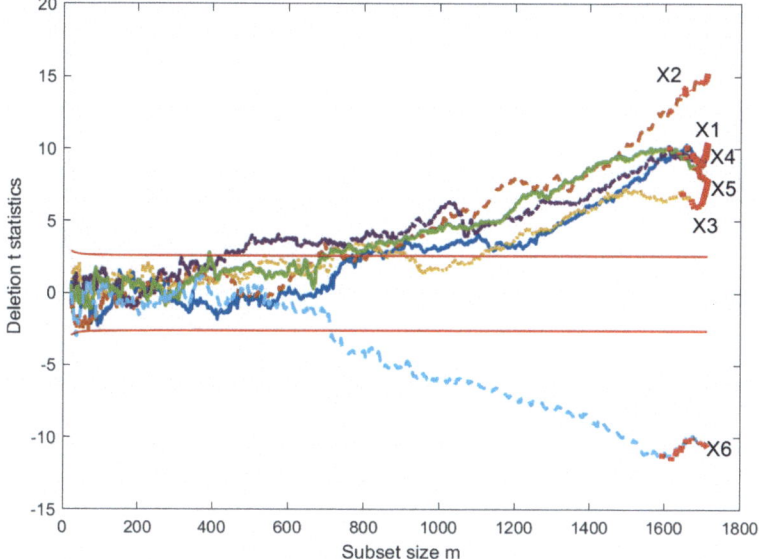

Fig. 10.25 Customer Loyalty data: monitoring of the six added variable t-statistics. Highlighted units are the 28 outliers found from a single FS

10.4.2 Robust Box-Cox Transformation

The plot of t-statistics in Fig. 10.25 fails to explain the presence of an appreciable number of outliers in the FS analysis. We now consider the alternative of a systematic shortcoming of the model which may be removed by response transformation. Figure 10.26 is the fan plot for the approximate score statistics for five standard values of the Box-Cox parameter λ, together with 99% intervals for testing the value of the parameter. The outliers from the individual searches are highlighted. The lowest trajectory is for $\lambda = 1$, that is: no transformation. Although this transformation is acceptable at the end of the FS, the score statistics lie steadily outside the envelope until $m = 1682$. Working upwards in the plot, the curve for $\lambda = 0.5$ is within the envelope until $m = 1615$. Thereafter the trajectory rises rapidly. The trajectories for the more negative values of $\lambda = 0, -0.5$ and -1 have a similar shape but move outside the envelope for smaller values of m. The indication is that these values of λ will not provide a satisfactory transformation for all the data. Although this plot indicates taking $\lambda = 0.5$, there is no obvious discontinuity in the slope of any of the curves to provide a clear indication of the presence of outliers and so to provide a guide to an unambiguous selection of the transformation. The pattern of the outliers for $\lambda = 1$ is different but again does not show any sharp change as outliers enter the subset for the FS. These trajectories are rather more an indication of a systematic failing in the fitted model.

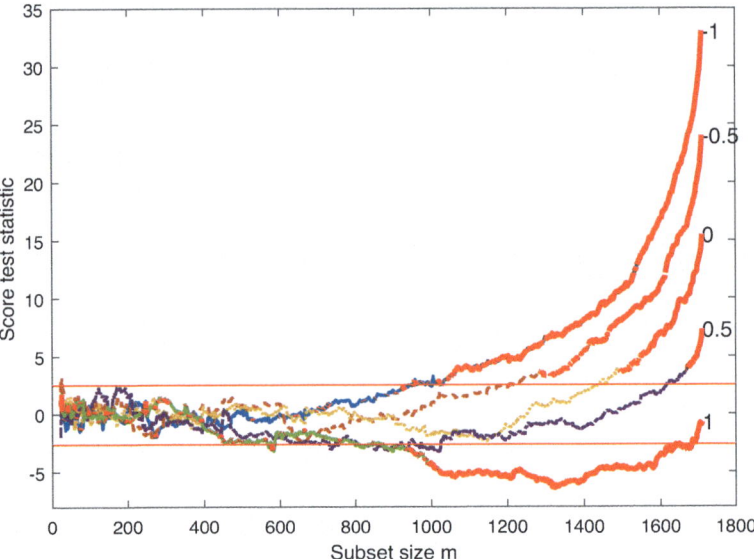

Fig. 10.26 Customer Loyalty data: fan plot. Monitoring plot of score statistics for five values of λ_0 for the Box-Cox transformation. Highlighted observations are the outliers from the individual searches

The conclusions from the fan plot agree with those from the automatic analysis which faithfully extracts all features of Fig. 10.26. In addition, the plot of extended BIC for the five values of λ (Exercise 10.2) has a clear peak at $\lambda = 0.5$; the square-root transformation is again indicated. There is also strong agreement of the index AGI with the plot of the BIC values. The proportion of observations included in the regression analysis of the square-root transformed data is close to one (0.976).

The FS on the first-order model with $\sqrt{y}$ as response identifies 41 outliers. Table 10.6 includes a summary of the analysis of variance for the 1670 observations remaining after outlier deletion. The value of adjusted R^2 has increased to 0.779, compared to 0.741 for the original analysis, also given in Table 10.6. In the new table four of the variables (price, quality, customer satisfaction, and negative publicity) all have higher levels of significance, with the remaining two levels of significance being slightly reduced. To check the fit of this new model, we show plots of studentized residuals in Fig. 10.27 which show only a slight reduction in the patterns seen in Fig. 10.23. In particular, the symmetry plot in the lower panel of Fig. 10.27 is close in form to the lower panel of Fig. 10.23. The change in the shape of the simulation envelopes of the QQ-plots in going from Figs. 10.23, 10.24, 10.25, 10.26 and 10.27 is an effect of the deletion of the 41 observations indicated as outlying for this transformation.

It is clear that there is still some unexplained structure in the data that is not illuminated by this analysis. The monitoring plot of FS residuals from the square-

10.4 Customer Loyalty

Table 10.6 Customer Loyalty data: summary ANOVA table for four models, with indicated outliers deleted; t-statistics t_j for variable j, number of observations m^* and adjusted R^2

j	Variable	Original	Square root	AVAS	RAVAS
0	Intercept	−11.1	18.6		
1	Price	10.5	11.7	9.7	18.3
2	Quality	15.5	16.6	16.6	16.9
3	Community outreach	7.9	6.7	10.5	11.0
4	Trust	10.4	10.3	16.0	14.5
5	Customer satisfaction	8.0	11.0	6.1	7.6
6	Negative publicity	−10.6	−11.1	12.3	13.4
	Observations m^*	1711	1670	1711	1695
	Adjusted R^2	0.741	0.779	0.790	0.806

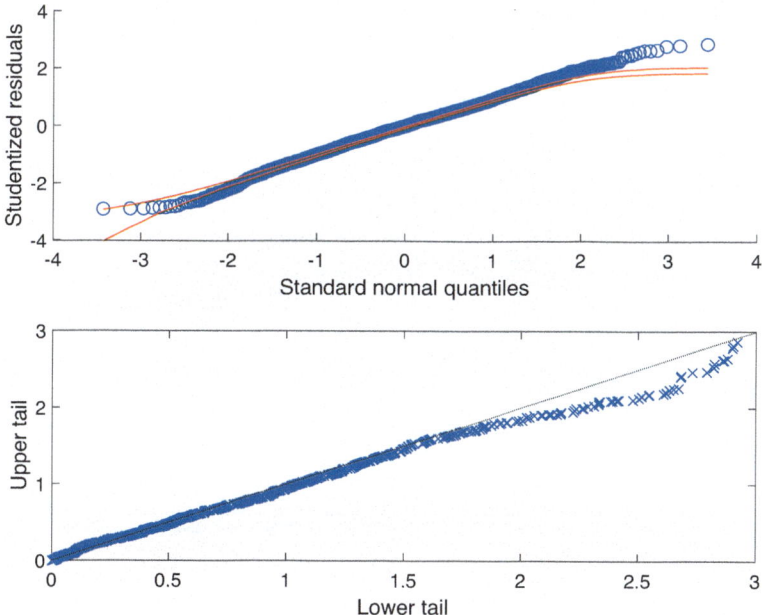

Fig. 10.27 Customer Loyalty data: square-root transformation; cumulative plots of residuals. Upper panel; QQ-plots of residuals with 95% simulation envelope: lower panel; symmetry plot of studentized residuals

root fit in Fig. 10.28 is helpful. Comparison with the monitoring plot of FS residuals for the untransformed data (Fig. 10.24) shows that the change in structure around $m = 1000$ has virtually vanished. A second feature is that the pattern of residuals is very stable over the whole range of m. The trajectories of the 41 outliers are highlighted in red in the figure. These show that the effect of the transformation has been to symmetrize the distribution of residuals. There is appreciably more identification of negative residuals as coming from outlying observations than there is of positive residuals. The outliers are shown with red circles in the yX plot of Fig. 10.29. The group of 8 outliers forming a clear low cluster in the panel for x_6 in Fig. 10.22 is still identified as outlying, as are several observations with the lowest value of y. Some intermediate values of y also show as outlying in some plots, for example y against x_2.

It seems that the parametric Box-Cox transformation cannot provide a sufficiently flexible family of transformations to explain these data. Accordingly, in the next sections, we explore the use of non-parametric transformations of both the response and of the explanatory variables.

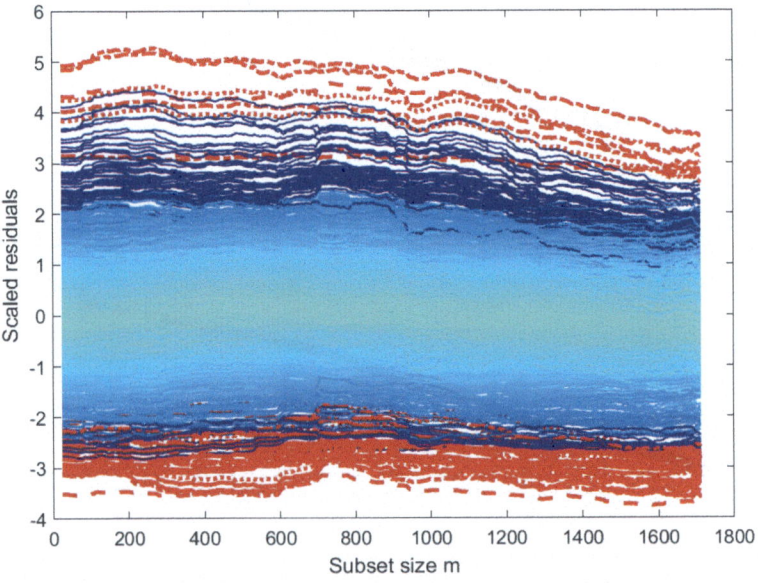

Fig. 10.28 Customer Loyalty data: square-root transformation; monitoring residuals with 41 outliers highlighted

10.4 Customer Loyalty

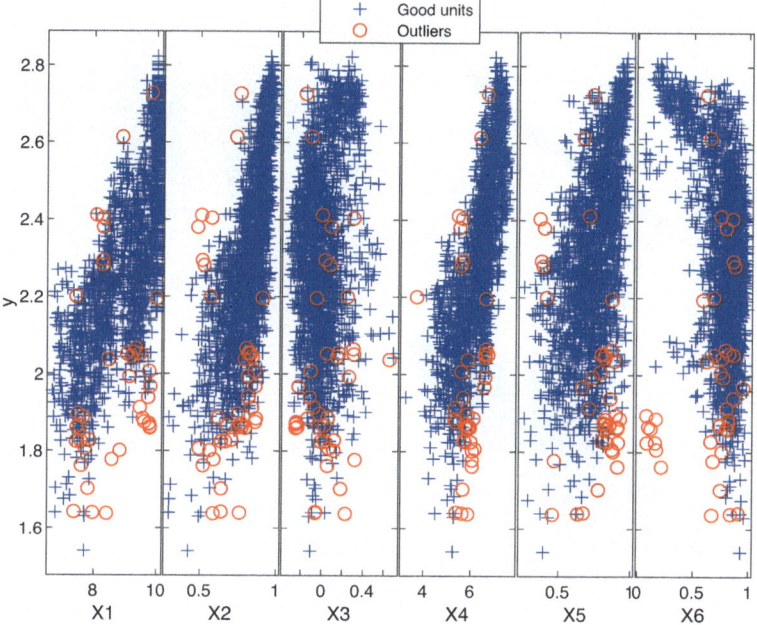

Fig. 10.29 Customer Loyalty data: square-root transformation; yX plot with the 41 outliers shown as red circles

10.4.3 Non-robust Analysis of the Loyalty Data with AVAS

Figure 10.30 shows the transformed response from AVAS. Unlike the estimated Box-Cox transformation, here the transformation of the response is slight. The plot of residuals versus fitted values indicates that there may be several outliers as does the plot of loyalty against fit. There is no trend of the residuals against fitted values. The more interesting plot is Fig. 10.31 which shows the highly non-linear transformations of the explanatory variables found by AVAS. These are similar to those found after the deletion of outliers using RAVAS, so we discuss them following our robust analysis. Table 10.6 shows that this non-robust analysis produces a value of $R^2_{adj} = 0.790$, the highest of any analysis so far without any deletion of outliers. The two most significant variables are now quality and price. The value of the estimated intercept from AVAS and RAVAS is zero, as expected from the definition of the GAM.

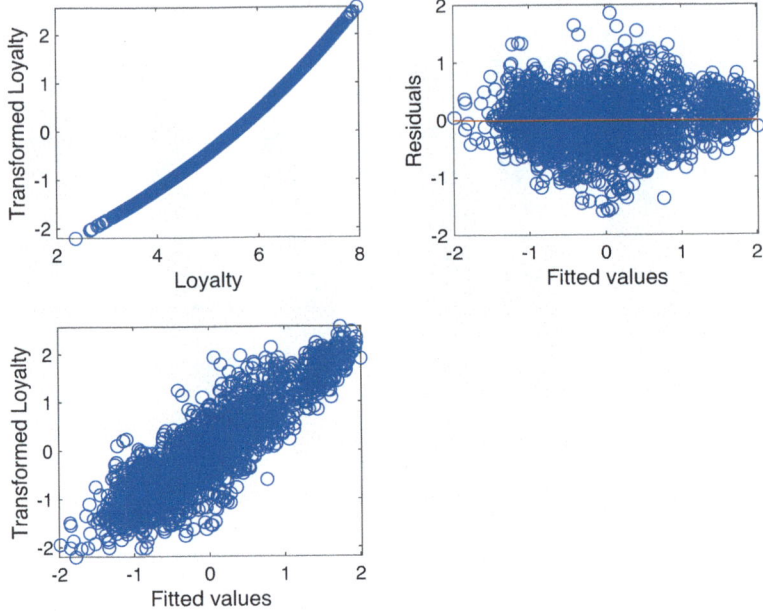

Fig. 10.30 Customer Loyalty data: AVAS; transformed response $g(y)$. Upper left-hand panel, $g(y)$ against y; upper right-hand panel, residuals against fitted values; lower panel, $g(y)$ against fitted values

10.4.4 Robust Non-parametric Regression with Response and Explanatory Variable Transformations: RAVAS

We conclude with the results of the robust analysis using RAVAS with all automatic options. With the robustness of RAVAS we are able to detect outliers—our procedure finds 16 which are highlighted in Fig. 10.32. The plots of residuals against fitted values and of loyalty against fitted values in Fig. 10.32 are pretty much what would be expected from the AVAS (non-robust) analysis except that, with the robust analysis, the residuals have moved further from the main cloud of points. The plot of transformed response against the original values shows that, although many of the outliers occur among the higher values of y, nothing like all do so. The plots of the transformed explanatory variables, in Fig. 10.33, are similar to those of the non-robust transformations in Fig. 10.31 except for the transformation of trust. We leave the discussion of these curves until we have considered the statistical properties of this fitted model.

The summary analysis of variance table, Table 10.6, gives a value of 0.806 for R^2_{adj}, the highest value achieved so far, despite the modest number of outliers detected. As a check on this model we look at the plots that, for the untransformed data, gave the strongest warning of nonstandard behaviour of the data. The first were the two

10.4 Customer Loyalty

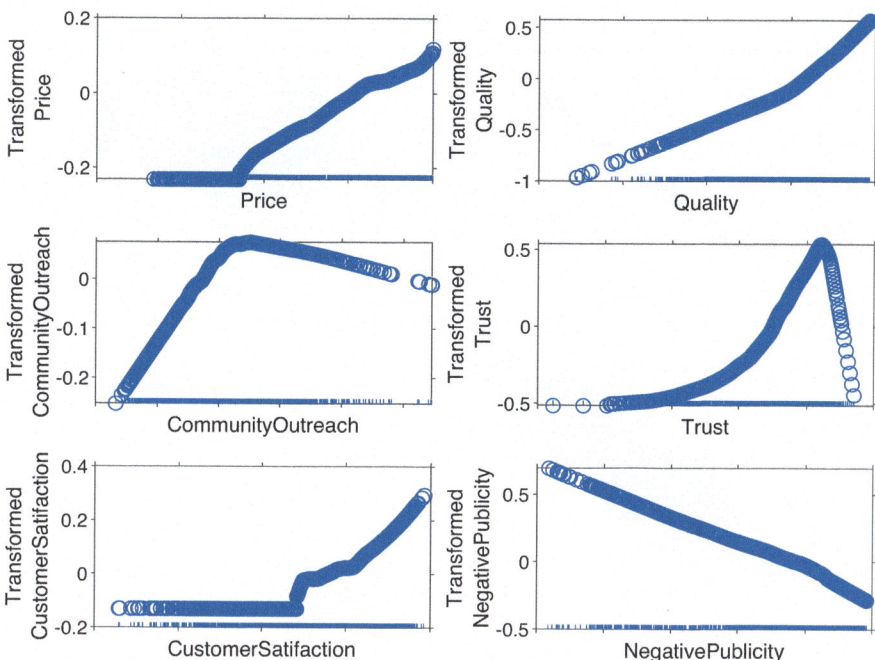

Fig. 10.31 Customer Loyalty data: AVAS; transformations $f_j(x_j)$ of the explanatory variables against original x_j

QQ-plots of residuals given in Fig. 10.23 for the untransformed data. They are in Fig. 10.34 for the RAVAS analysis. The comparison between the two figures shows that the use of RAVAS and the deletion of 16 observations (less than 1% of the total) has led to an acceptable pattern of residuals, especially in comparison with that of Fig. 10.23. The envelope for the larger residuals in the upper panel of the plot shows that the residuals are much closer to the envelope than they are in Fig. 10.27 from the analysis on the square-root scale.

Figure 10.35 shows histograms of the residuals from the analysis with RAVAS with 16 outliers detected. The left-hand panel has a superimposed normal curve and the right-hand panel a superimposed kernel density estimate. The similarity of the two superimposed curves indicates that a close approximation to normality has been achieved. This is very different from the pair of plots from the analysis of the data on the original scale which gave a clear indication of skewness.

The histograms in Fig. 10.35 show the distribution of the residuals at the stopping point of the FS, that is when the 16 outliers are excluded. These are the outliers found in the process of estimating the transformations. As a final analysis, we take the transformation of the response in Fig. 10.32 and of the explanatory variables in Fig. 10.33 and apply the outlier detection procedure of the FS. We do not show the resulting version of Fig. 10.34 when only 9 outliers are deleted, but both panels

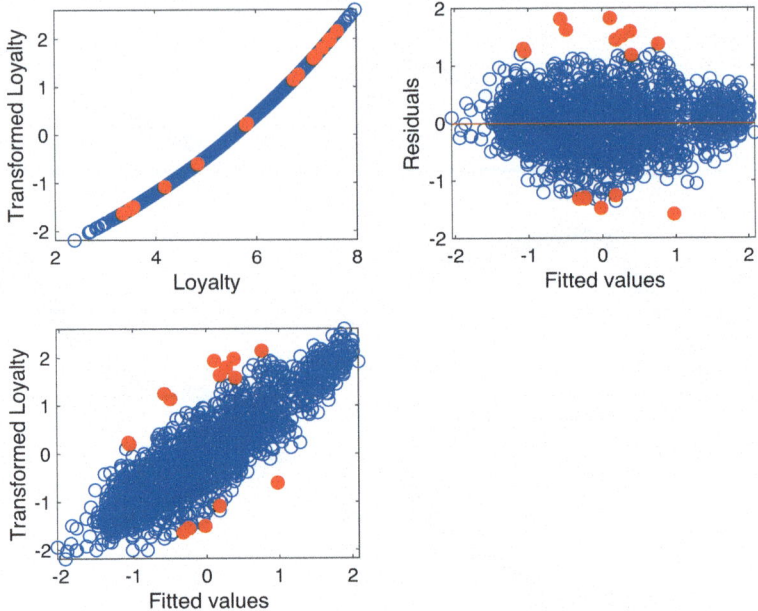

Fig. 10.32 Customer Loyalty data: RAVAS; transformed response $g(y)$. Upper left-hand panel, $g(y)$ against y; upper right-hand panel, residuals against fitted values; lower panel, $g(y)$ against fitted values. 16 outliers highlighted

are a slight improvement on Fig. 10.34; in particular, the large positive residuals in the upper panel lie between the simulation bands. The histograms of normality are virtually unchanged. The monitoring plot of the residuals in Fig. 10.36 shows that the pattern of residuals, including the nine identified as outliers, is virtually constant over the whole search. This is an improvement, both in constancy and symmetry, on the monitoring plot of residuals from the square-root transformation in Fig. 10.28, which was obtained after 41 observations had been identified as outlying. It is interesting that the majority of the observations deleted for the square-root transformation were positive. In this example transformation of the response alone, as judged by these monitoring plots, is not sufficient to provide a good model; the joint transformation of the response and the explanatory variables is needed to produce a well-behaved model.

10.4.5 Interpretation

The summary ANOVA in Table 10.6 shows that the most important explanatory variable is price, followed by quality and trust. At least the first two of these are straightforward and objective properties, the general levels of which providers of

10.4 Customer Loyalty

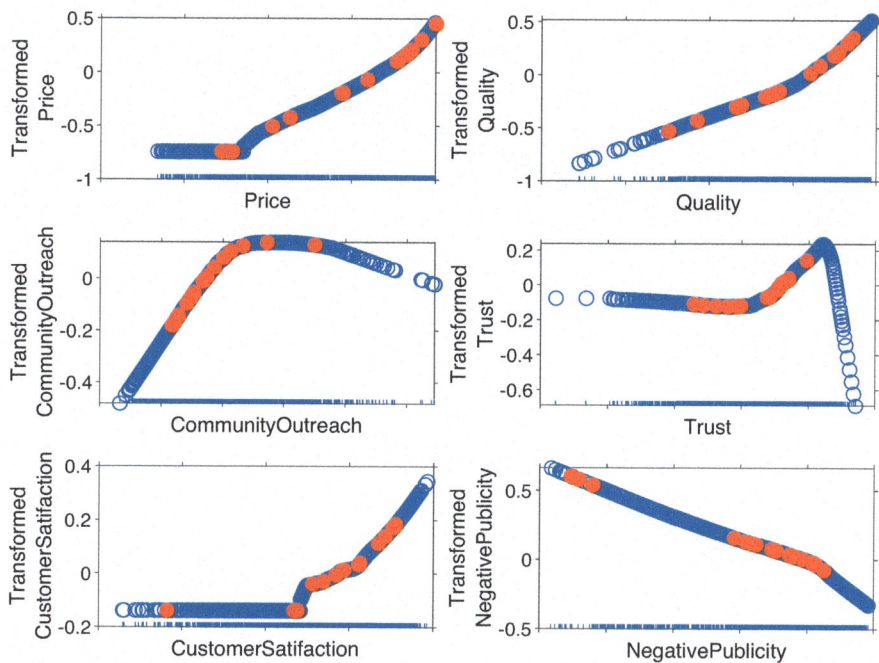

Fig. 10.33 Customer Loyalty data: RAVAS; transformations $f_j(x_j)$ of the explanatory variables against original x_j

goods and services should be able to choose and maintain. Some of the others provide less straightforward targets. Since all variables are highly significant, we conclude with brief comments on the transformations found by RAVAS.

The plots of the transformations $f_j(x_j)$ against x_j in Fig. 10.33 show some interesting and unexpected shapes. We suggest some interpretations.

1. x_1—Price. The first part of the curve shows low loyalty for cheap items. Perhaps for these products, maybe without any particular characteristics, such as household essentials, price is important, but the brand is not. But, increasingly, for more expensive items loyalty becomes higher as specific characteristics are felt to be important—including showing off (conspicuous consumption).
2. x_2—Quality. These are data on what people believe they are doing. Here loyalty increases with perceived quality, in a surprisingly linear way, although with proportionally increased loyalty for products of high perceived quality.
3. x_3—Community Outreach. Here the relationship is roughly quadratic, although the decreasing upper part is relatively sparse. Community outreach promotes loyalty up to a certain level. But too much may make customers feel that the firm is more interested in its image than in serving customers.
4. x_4—Trust. This is unimportant for loyalty at low levels, perhaps corresponding to low priced goods. Then loyalty increases with trust, but then flattens and starts to

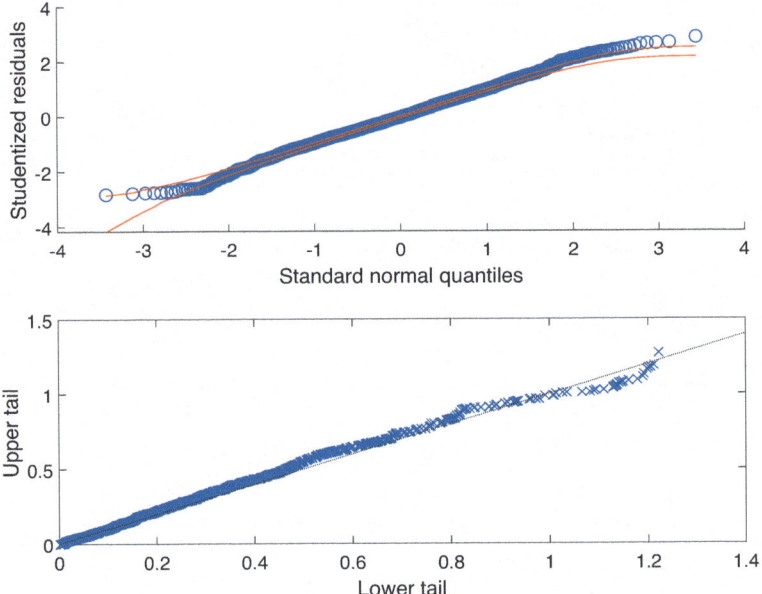

Fig. 10.34 Customer Loyalty data: RAVAS (16 observations deleted); cumulative plots of residuals. Upper panel; QQ-plots of residuals with 95% simulation envelope: lower panel; symmetry plot of residuals

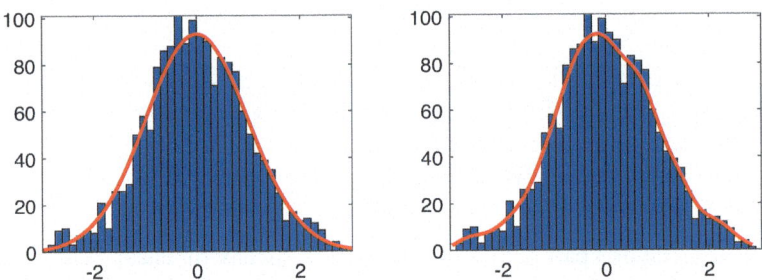

Fig. 10.35 Customer Loyalty data: RAVAS (16 observations deleted); histogram of residuals with superimposed curves. Left-hand panel; superimposed normal curve; right-hand panel; superimposed kernel density

decline. The decline could be caused by trustworthy firms being seen to produce less interesting products.

5. x_5—Customer Satisfaction. This is like price and trust in that, at low levels, loyalty is low. It then increases steadily with customer satisfaction.
6. x_6—Negative Publicity. Here loyalty decreases in a surprisingly linear way with negative publicity. However, it may be that, since this is a survey of attitudes,

10.4 Customer Loyalty

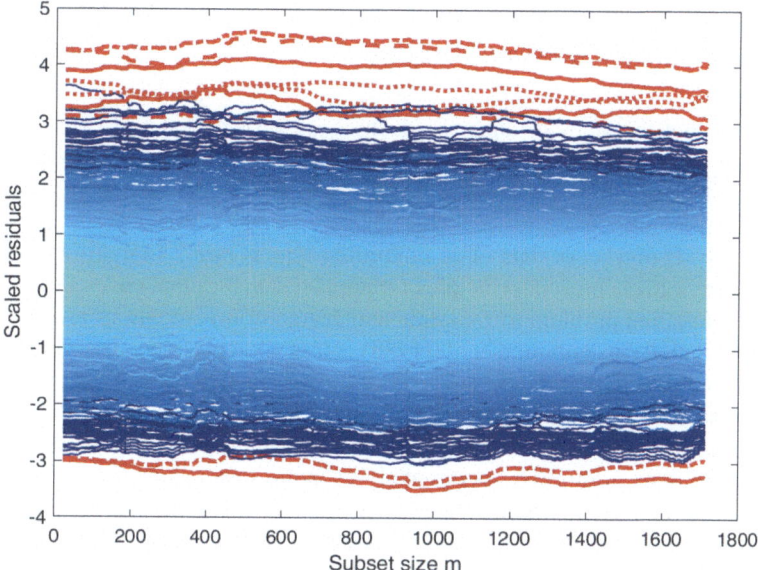

Fig. 10.36 Customer Loyalty data: RAVAS; the very stable monitoring plot of FS residuals using transformed variables from RAVAS; the trajectories of the resulting nine outliers are highlighted

consumers believe they behave in such a rational way. In some market segments it may indeed be true that "there is no such thing as bad publicity".

In this analysis we have investigated a seemingly innocuous set of marketing data about the subjective determinants of customer loyalty. Despite the significance of the variables in the analysis of variance table of the original observations (Table 10.5), our further analysis shows that the data do not satisfy the conditions for a regression analysis to be efficient. Our final robust analysis, with non-parametric transformations of the response and explanatory variables, yields a fitted generalized additive model for which statistical assumptions are satisfied and for which the proportion of variation in the data explained by the model is appreciably increased in comparison with the original regression model. More important for the discussion and understanding of consumer behaviour are the transformations of the explanatory variables shown in Fig. 10.33. These provide appreciably more insight into human behaviour than do models using the first-order terms of the initial regression analysis. These transformations should also be helpful in the planning of marketing campaigns.

One of the YouTube videos accompanying Chap. 10 provides a conversation around the analysis of the customer loyalty data presented in this section. Riani et al. (2025b) extend this analysis to include a short discussion of the particular importance of our approach to robust data analysis in interrogating the large data sets used in machine learning, including the clustered data arising in attempts to establish causality (Bühlmann 2020).

10.4.6 A Procedure for a Structured Approach to Modern Robust Regression Analysis

We believe that the steps of analysis given above provide a set of checks on any analysis of regression data. In this section we present a structured approach to the building and checking of regression models using the tools of dynamic regression modelling.

The analysis of a set of data of an unfamiliar type is rarely straightforward. Cox and Donnelly (2011) present principles for applied statistics that cover a wide range of topics, including problem formulation and design of investigations as well as the analysis of data. Wolstenholme et al. (1988) is less wide in scope, describing a programme for the semi-automatic analysis of data of many types. These references are not concerned about single or multiple sets of observations that may seriously affect the fitted model and its interpretation. If robustness is included, the choices during analysis are multiplied. As our analyses in this paper show, robust regression methods do not follow a simple path. However, we find that there is sufficient structure to indicate the helpfulness of the procedures summarized in Table 10.7, which suggests five major steps. We comment briefly on them, particularly to provide both bibliographic references and references to sections of this book.

10.4.6.1 Step 1: Variable Transformation

The fan plot for transformation of the response is introduced in Sect. 6.1.3; the robust non-parametric transformation RAVAS is in Sect. 7.4.

10.4.6.2 Step 2: Robust Variable Selection

The added variable plot for t-statistics is introduced in Sect. 4.8. In the later stages of the FS, the candlestick plot of Sect. 9.3.1 provides a structured plot of the values of the C_p criterion for model choice (Mallows 1973). Monitoring the robust LASSO of Freue et al. (2019) and of Kepplinger (2023) is in Sect. 9.4.3. Since the outliers are removed and the appropriate scale estimated, robust variable selection enables us to address the potential problem of (approximate) collinearity among the variables, since we restrict attention to a subset of variables.

10.4.6.3 Step 3: Monitoring of Scaled Residuals

Monitoring of the forward plot of residuals has been used frequently. It was introduced in Sect. 4.9.2 for S-estimation. It is possible to monitor these or other residuals in order to determine an optimal value of the empirical *bdp* or efficiency to use in the automatic outlier detection procedure based on a particular estimator.

10.4 Customer Loyalty

Table 10.7 Modern procedure for robust and efficient data analysis

	Description	Tools to use
STEP 1	Variable transformation	Parametric (fan plot + automatic procedure for finding best value of λ)
		Non-parametric (RAVAS + automatic option selection)

Output: best value of transformation parameter for the response. Find observations influential for transformation. Find best transformation to work with. Compare parametric and non-parametric approach

	Description	Tools to use
STEP 2	Robust variable selection	Monitoring of added value t-statistics, candle stick plot or if the number of variables is very large, robust LASSO

Output: find the effect of influential subsets of units on t-statistics. Find a set of relevant explanatory variables

	Description	Tools to use
STEP 3	Monitoring of scaled residuals. Analysis of FS, S and MM residuals	Brushing in the monitoring plots to understand the position of outlying residuals and the eventual presence of subgroups of units

Output: analysis of correctness of the model and detection of the optimal value of *bdp* or efficiency to use

	Description	Tools to use
STEP 4	Outlier detection and removal of outlying observations	Routines for automatic outlier detection. Analysis of the position of the outlying units in the yX plot

Output: find a subset of clean units

	Description	Tools to use
STEP 5	Check residuals on the subset of clean units: heteroskedasticity, normality and serial correlation. Comparison of parametric and non-parametric approach. Find whether the linear approach is reasonable	QQ-plots with envelopes, normality plots, autocorrelation tests

Output: if some test fails go to step 2 and restart with another model (e.g. heteroskedastic approach)

10.4.6.4 Step 4: Outlier Detection and Removal of Outlying Observations

The FS routine for automatic outlier detection, which we prefer, is described in detail in Sect. 4.3. Analysis of the position of the outlying units in the yX plot is demonstrated in Fig. 4.14. Spinning a point cloud to obtain a sense of the three-dimensional distribution of the units is in Fig. 4.13.

10.4.6.5 Step 5: Check Residuals on the Subset of Clean Units: Heteroskedasticity, Normality and Serial Correlation

QQ-plots with envelopes are introduced in Sect. 3.5. Histograms for assessing normality are first used in Fig. 10.35. We use the procedure of Durbin and Watson (1950) to test for the correlations of regression residuals.

10.5 Modified Customer Loyalty Data

10.5.1 Least Squares Analysis

A strange feature of the customer loyalty data analysed in Sect. 10.4 is that, although the robust analysis provides an improved model, from the outset all explanatory variables are highly significant; there are no dramatic changes in the importance of variables as the data are transformed or outliers removed. In this section we slightly modify the data (25 observations) to produce a few outliers that have a surprisingly powerful effect on the data analysis, but are, for example, hard to detect using LS or AVAS. They are fully revealed by RAVAS.

Figure 10.39 shows the yX plot for the modified data, which is to be compared with Fig. 10.22. The differences are slight and are not immediately evident. However, the resulting analysis of variance table for the first-order model, given in Table 10.8 is very different from the analysis of variance in Table 10.5 in that the t-statistic for x_6 (negative publicity) is now a non-significant -0.92, as opposed to a highly significant -10.63. The value of R^2_{adj} has decreased from 0.741 to 0.646.

There is however an appreciable change in the plots of residuals, which now clearly suggest the presence of outliers. The left-hand panel of Fig. 10.37 is the QQ-plot of the residuals, with simulation envelopes. This exhibits the presence of an excessive number of negative residuals. The symmetry plot in the right-hand panel of the figure curves away from the line of equality, again showing the effect of too many negative residuals. The smooth shape of this plot, distinct from the upper panel of Fig. 10.23, suggests that there may be some systematic failure of the model.

The scatterplot matrix does not provide any useful information about the structure, if any, leading to the generation of outliers. This is amply provided by brushing the

10.5 Modified Customer Loyalty Data

Table 10.8 Modified customer loyalty data: ANOVA

```
Estimated Coefficients:
                       Estimate         SE         tStat        pValue

   (Intercept)          -2.7179      0.21348     -12.731      1.5574e-35
   Price                0.32155      0.036321      8.853      2.0886e-18
   Quality              2.482        0.19418      12.782      8.5701e-36
   CommunityOutreach    0.8503       0.1114        7.6325     3.8096e-14
   Trust                0.37205      0.041669      8.9288     1.0886e-18
   CustomerSatifaction  1.2095       0.14467       8.3601     1.2842e-16
   NegativePublicity   -0.096345     0.10432      -0.92353    0.35586


Number of observations: 1711, Error degrees of freedom: 1704
Root Mean Squared Error: 0.672
R-squared: 0.647,  Adjusted R-Squared: 0.646
F-statistic vs. constant model: 521, p-value = 0
```

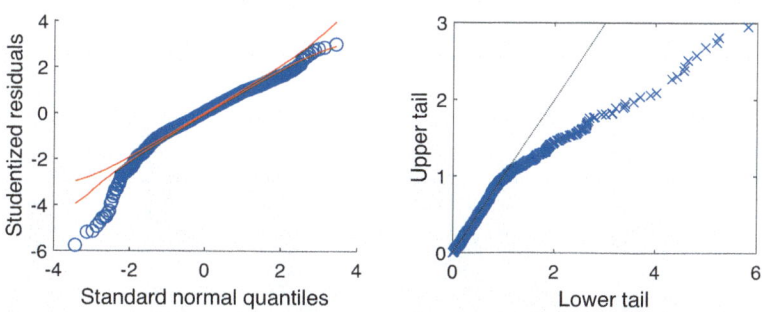

Fig. 10.37 Modified Customer Loyalty data: cumulative plot of residuals. Left-hand panel; QQ-plot of residuals with 95% simulation envelope: right-hand panel; symmetry plot of studentized residuals

monitoring plot of residuals in Fig. 10.38. This plot, for most of its trajectory, has a very stable structure, with a clearly visible set of 25 large negative residuals. The result of brushing these negative residuals is shown in the yX plot of Fig. 10.39. The highlighted observations are those that were generated by the modification; for all observations with price $x_1 = 10$ and negative publicity $x_6 < 0.2$, the response was reduced by three. The panel for y against x_1 in Fig. 10.39 shows a strip of brushed observations for $x_1 = 10$, that all have egregiously low responses. The same is true in the panel for x_6 where the brushed units are seen to have a low value of x_6 and a low value of y. At the end of the search, the modified units will start to join the subset used in fitting. The effect of these observations is clearly shown at the end of the monitoring plot of residuals repeated in Fig. 10.40. From m close to 1675 onwards, there is an upward turn, that is a decrease in the absolute values of the large negative

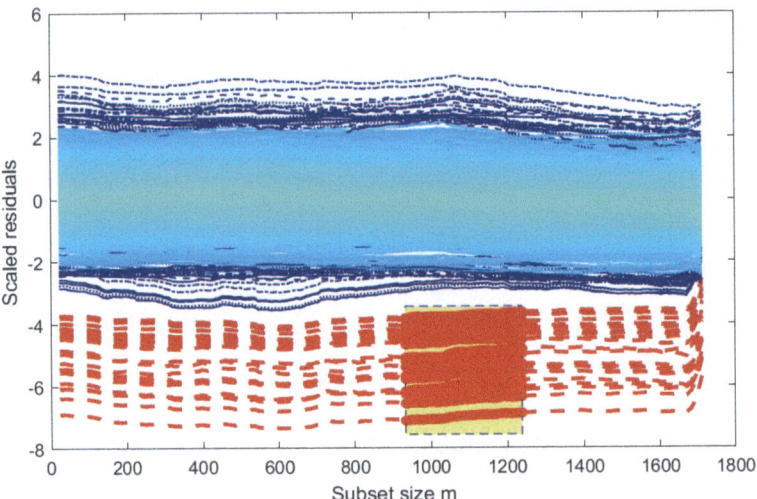

Fig. 10.38 Modified Customer Loyalty data: monitoring residuals from the FS with brushing. 25 observations are highlighted

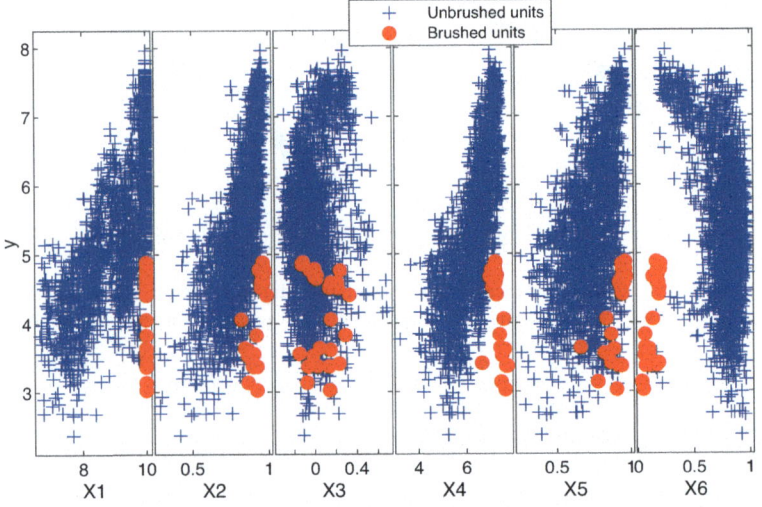

Fig. 10.39 Modified Customer Loyalty data: yX plot with 25 brushed units

residuals. This is a repeat of the monitoring plot of Fig. 10.38 since the representation of brushing makes it hard to see this not very large effect.

The FS analysis of the data identifies 51 outliers, including the modified observations and some others. The monitoring plot of the six added variable t-statistics in Fig. 10.41 shows the effect of the outliers on inference about the values of the parameters. Until the outliers start to enter, the trajectories are virtually identical to those of

10.5 Modified Customer Loyalty Data

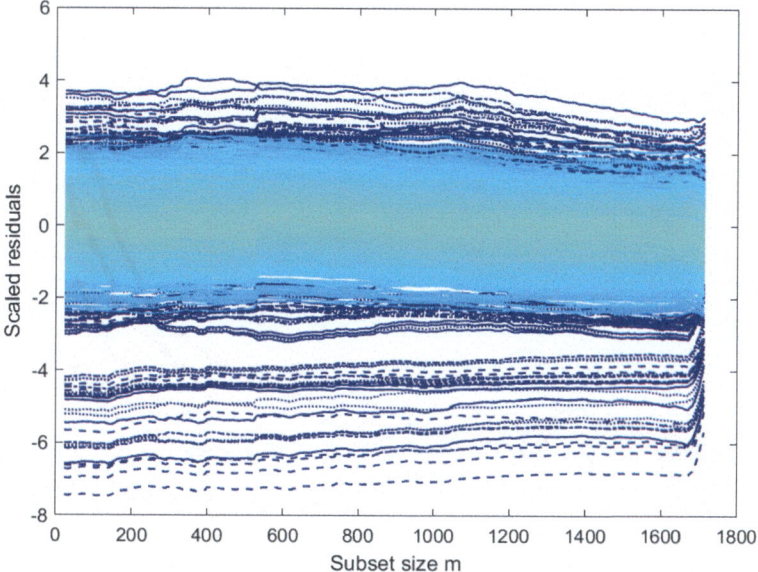

Fig. 10.40 Modified Customer Loyalty data: monitoring residuals from the FS. Note the change of residuals in the last steps. This is a repeat of the monitoring plot of Fig. 10.38 without brushing

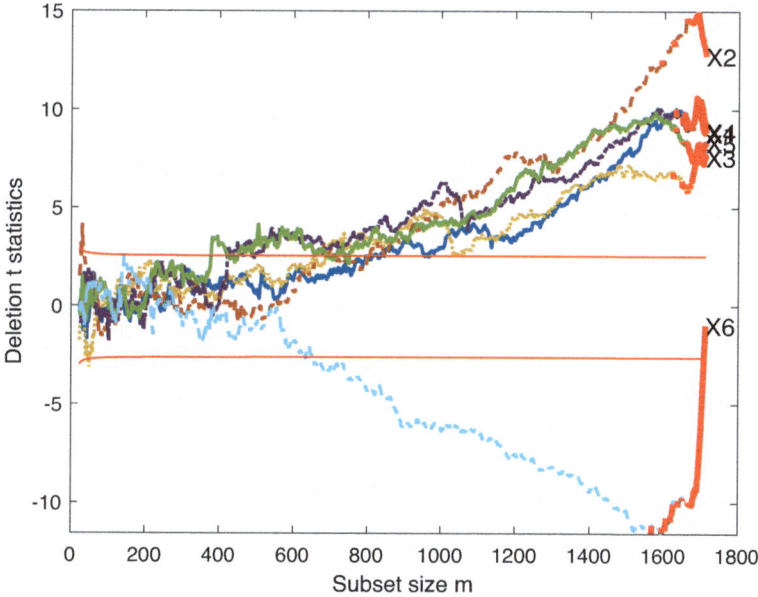

Fig. 10.41 Modified Customer Loyalty data: monitoring of the six added variable t-statistics. Highlighted units are the 51 outliers found from a single FS

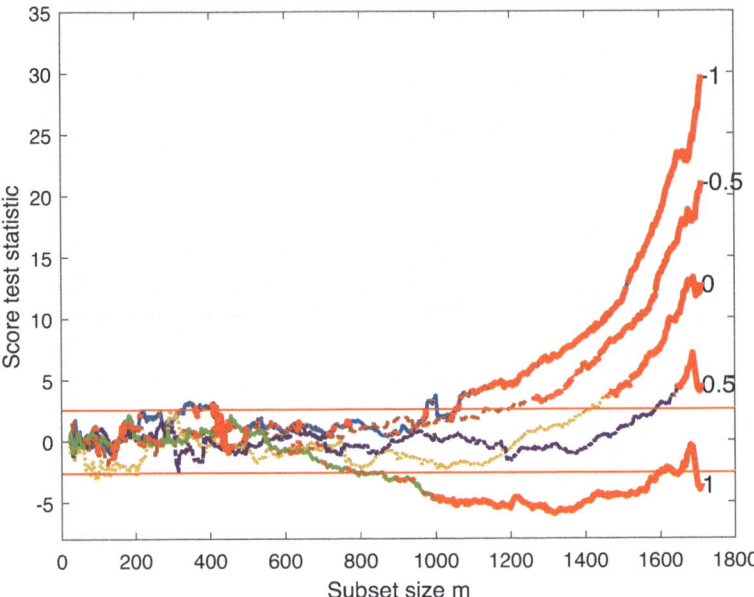

Fig. 10.42 Modified Customer Loyalty data: fan plot. Monitoring plot of score statistics for five values of λ_0 for the Box-Cox transformation. Highlighted observations are the outliers from the individual searches

Fig. 10.25. The effect of the outliers, particularly on the coefficient of x_6, is dramatic. The t-statistics for x_6 go from a value of around -10 to being non-significant. The trajectory of t_2 is a little different from that in the earlier figure, since the outliers now cause the value to decrease slightly. The other four trajectories are not much changed.

Although the outliers also affect the trajectories of the score tests for the Box-Cox transformations, the fan plot of Fig. 10.42 shows that the effects are slight, although increasing with the value of λ. In particular, for the original data, the hypothesis of no transformation is acceptable at the end of the search (Fig. 10.26). However, now, although the trajectory returns within the confidence limit towards the end of the search, at the very end it leaves again. The outliers make the value of 0.5 for λ more plausible towards the end of the search. The automatic Box-Cox procedure again suggests the square-root transformation. The analysis in the square-root scale is explored in Exercise 10.3.

10.5 Modified Customer Loyalty Data

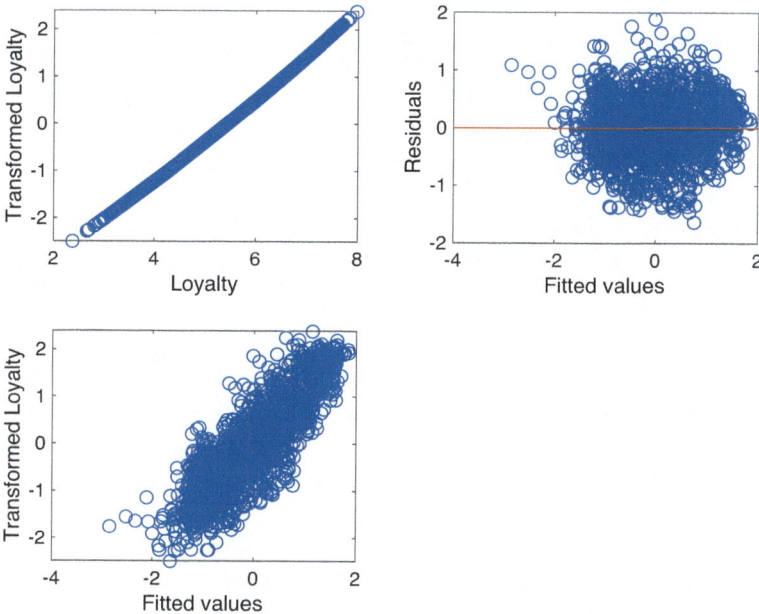

Fig. 10.43 Modified Customer Loyalty data: non-robust AVAS, transformed response $g(y)$. Upper left-hand panel, $g(y)$ against y; upper right-hand panel, residuals against fitted values; lower panel, $g(y)$ against fitted values

10.5.2 AVAS and RAVAS

In an attempt to find a model which leads to the indication of fewer outliers, we first try (non-robust) AVAS. Figure 10.43 shows the three panels of plots involving residuals, the equivalent of Fig. 10.30 for the unmodified data. The two sets of plots are similar, although in Fig. 10.43 there is virtually no transformation of the response. A surprising feature is that there is now even less indication of outliers than there was in Fig. 10.30. Figure 10.44 shows the estimated transformations $g_j(x_j)$ of the explanatory variables. Some are very much like those for the unmodified data Fig. 10.31 and two are very much not. The most different transformation is that of price, x_1, which is not monotonic. It is, at least to us, hard to think of any situations in which such a relationship could exist between loyalty and price. The second difference from Fig. 10.31 is the transformation of negative publicity, x_6. Since the estimate of β_6 is not significant in Table 10.8, interpretation of this variable has no meaning, as it does not affect the fitted model. It may be that the AVAS algorithm has over-responded to the presence of outliers.

Use of RAVAS, with its accompanying methods of outlier detection, on the other hand, leads to the expected solution, that is one close to that for the uncontaminated data when analysed with RAVAS. Selection of the five options shows that the best

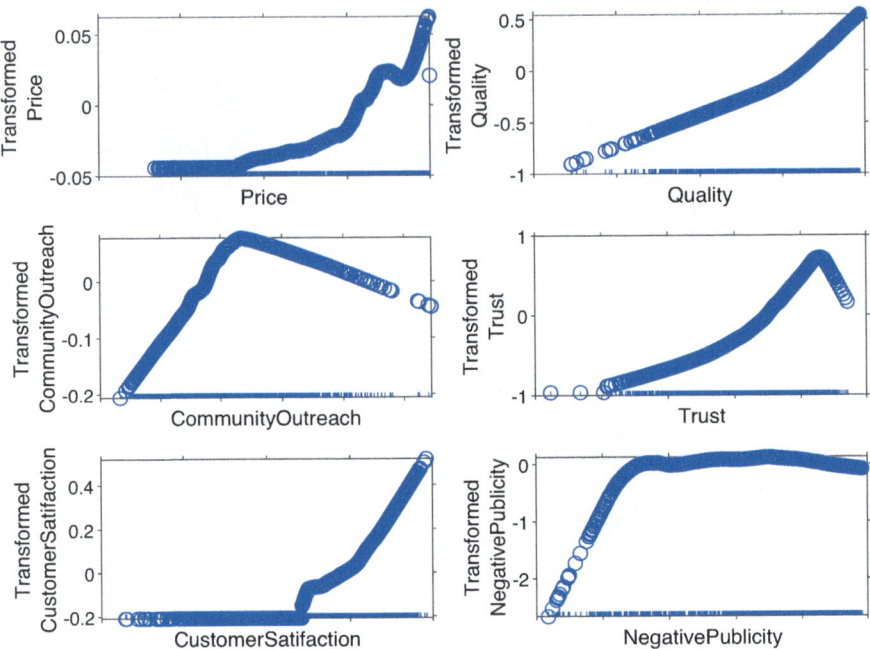

Fig. 10.44 Modified Customer Loyalty data: non-robust AVAS; transformations $f_j(x_j)$ of the explanatory variables against original x_j

solution is found when just robustness and the initial transformation of the response (tyinitial) are used. This leads to the identification of 35 outliers and to the transformation and residual plots of Fig. 10.45. The top left-hand panel of the plot shows the smooth transformation of the response as in Fig. 10.30. The upper right-hand panel of residuals against fitted values very nicely not only shows (highlighted) the residuals found in the analysis of the unmodified data, but also the separate cluster of outliers generated by the modification. The lower panel shows the transformed response against fitted values, again with the two clusters of outliers.

Figure 10.46 shows the yX plot of the data in the trasformed RAVAS scale, with highlighting of the 35 outliers. An interesting feature of the plot is that the panel for x_6 is reversed. Estimating this negative transformation does not affect the fitted model, as the sign of the parameter estimate is also reversed. Once the robust transformations had been established by RAVAS, we finished the analysis by using the more sophisticated outlier detection procedure from the FS. This identified 34 outliers. The two yX plots with highlighted outliers are virtually indistinguishable.

The conclusion from this analysis of the Modified Customer Loyalty data is that the use of robust methods, together with monitoring, is vital for the detection and understanding of outlying observations and incorrect structure. In the least squares analysis the residual plots indicated some kind of departure; the nature of this departure was made manifest by the monitoring plots of Figs. 10.38 and 10.40. The robust fea-

10.5 Modified Customer Loyalty Data

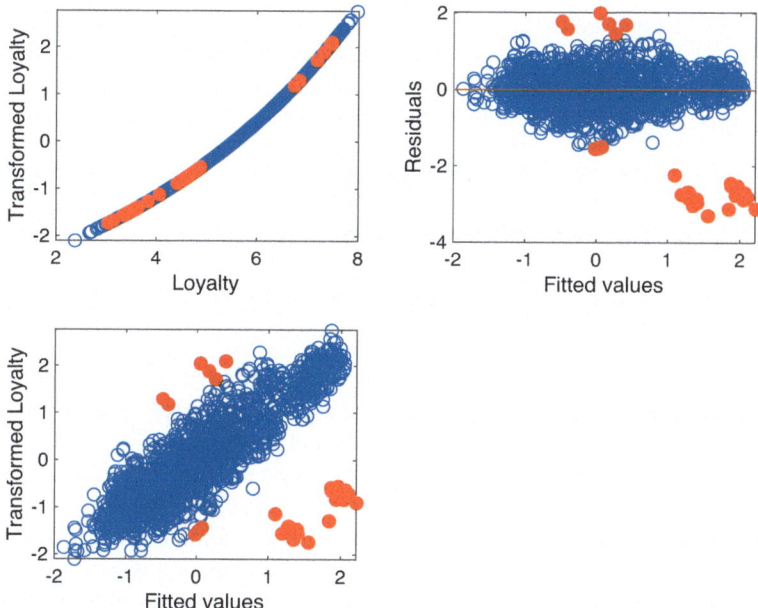

Fig. 10.45 Modified Customer Loyalty data: RAVAS, 35 outliers deleted, transformed response $g(y)$. Upper left-hand panel, $g(y)$ against y; upper right-hand panel, residuals against fitted values; lower panel, $g(y)$ against fitted values

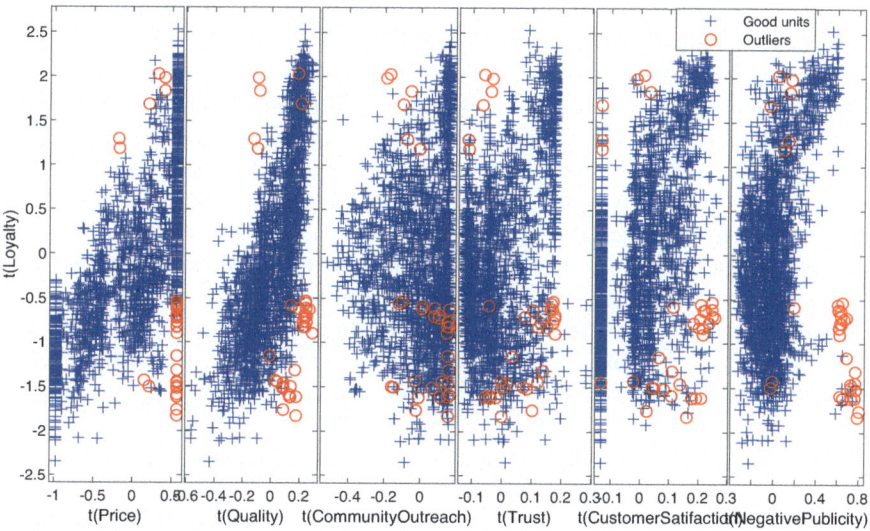

Fig. 10.46 Modified Customer Loyalty data in the transformed RAVAS scale: yX plot with 35 outliers from RAVAS highlighted

tures of RAVAS detected outliers, removed them, and led to an interpretable model. The failure of AVAS is concerning, since it shows how a non-robust non-parametric method can be so flexible as to adjust to any outliers and failures of structure. These are no longer evident when the fitted non-robust model is scrutinized.

10.6 NCI60 Cancer Cell Panel Data

10.6.1 Monitoring and Stability

In the monitoring plots introduced in Chap. 4 and frequently exemplified thereafter, there is a structure which is often stable, but changes, perhaps to a new form of stability, often after the introduction of outliers into the observations used in fitting. Figure 4.18 is an example from monitoring the residuals from S-estimation for the AR regression data. For high *bdp* down to 0.26 there is a stable structure of residuals including outliers; below this value there is a sharp transition to a stable pattern of LS residuals. An example of a related phenomenon in a fan plot is given in Fig. 6.2. Here, for the Loyalty Card data, there is a set of diverging trajectories until, about 20 units from the end of the FS, the slopes of many of the curves change abruptly. This is most clearly seen for $\lambda = 0.4$ or 0.3. However, the monitoring plot for sparse LTS in Fig. 9.9 is not of this form. The indicator of the significance of the majority of the variables changes in an erratic way as *bdp* decreases. There are seven variables which were selected most often over the range of values of *bdp*. In general, the strategy is to select over a range of values of *bdp* (say 0.5–0.1).

In Fig. 9.9 the variables with the highest frequency of significance are plotted at the bottom of the figure. The seven lowest trajectories (ignoring that for the intercept term) are not of the form of the plots mentioned in the previous paragraph. As $bdp \to 0$, there are initially some non-significant results until $bdp = 0.3$. Thereafter, the seven variables are mostly significant until $bdp = 0.03$, when some of the seven lose significance, whereas others do not. This irregular behaviour is more pronounced for all the less frequently significant variables. In order to try to understand the structure, we focus our modelling on the seven most frequently occurring variables and start with an LS analysis. The NCI designations of these samples and the regression labelling we use are given in Table 10.9. To avoid confusion we refer to this set of data with 7 explanatory variables as the Cancer data.

10.6 NCI60 Cancer Cell Panel Data

Table 10.9 Cancer data: NCI identification of seven most frequently selected variables (old names and new names)

Original NCI name	Regression variable
x134	x1
x10193	x2
x1106	x3
x14785	x4
x20125	x5
x8510	x6
x8502	x7

10.6.2 Least Squares

In the data for analysis there are 59 observations and 7 explanatory variables. Figure 10.47 shows the yX plot of the data with five outliers identified by the FS. Four of these are particularly clear in the panel for x_7. Table 10.10 is the analysis of

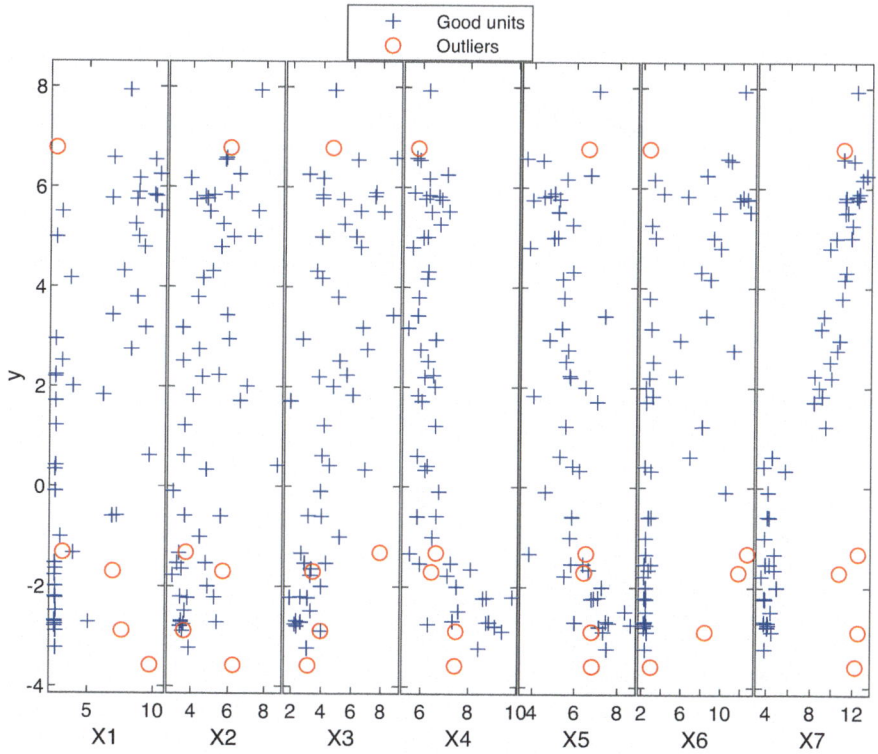

Fig. 10.47 Cancer data: yX plot of the data with five outliers identified by the FS

Table 10.10 Cancer data: ANOVA

```
Estimated Coefficients:
                   Estimate         SE         tStat       pValue
                   _____       _____      _____     _____

    (Intercept)     1.5508        2.8039        0.5531      0.58261
    x1              0.037889      0.1065        0.35575     0.7235
    x2              0.57128       0.20805       2.7458      0.0083164
    x3              0.17336       0.17621       0.98384     0.32984
    x4             -0.42891       0.31051      -1.3813      0.1732
    x5             -0.76755       0.28288      -2.7134      0.009061
    x6              0.017995      0.086735      0.20747     0.83647
    x7              0.43813       0.11084       3.9527      0.00023871

Number of observations: 59, Error degrees of freedom: 51
Root Mean Squared Error: 1.87
R-squared: 0.747,  Adjusted R-Squared: 0.712
F-statistic vs. constant model: 21.5, p-value = 3.48e-13
```

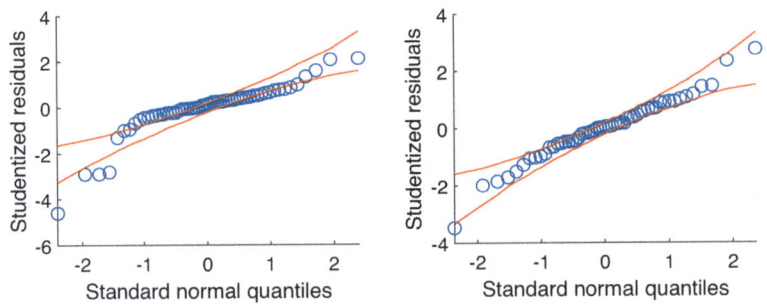

Fig. 10.48 Cancer data: QQ-plots of studentized residuals with 95% simulation envelope using all the data (left-hand panel) or just the 54 units not declared as outliers (right-hand panel)

the variance table for all observations. This gives a value of 0.712 for R^2_{adj}. However, in line with the results of Fig. 9.9, only three of the variables are significant: $x2$, $x5$ and $x7$. These are also the three significant variables found by stepwise regression, which gives the same value of R^2_{adj}.

We next consider the effect of the five outliers detected by the FS. The left-hand panel of Fig. 10.48 is the QQ-plot with 95 per cent simulation envelopes of the Studentized residuals. Not only is the effect of four outlying observations clear, but the distribution of the remaining residuals does not lie in the envelopes. The right-hand panel of the Figure repeats the analysis, but with the five outliers removed. Now the residuals are well-behaved with only one point (out of 54) lying marginally outside the envelopes. The structure of the residuals is clearly shown in the monitoring

10.6 NCI60 Cancer Cell Panel Data

plot of residuals in Fig. 10.49. There are four large outliers (units 12, 17, 39, and 51), one minor outlier (57), and a swamping effect (unit 40) at the end of the FS.

The effect of the outliers on the significance of the LS tests of the parameters is less transparent. Figure 10.50 is the monitoring plot of added variable t-statistics for all observations. This shows that x_7 is highly significant, but that the last observations do appreciably change its significance, as they do, less strongly, change the significance of other variables. In line with the results for $n = 59$ in the analysis of variance table of Table 10.10, only three variables are significant. Figure 10.51 is the monitoring plot of added variable t-statistics with the five outliers removed. The trajectories, not only that for x_7, are now well-behaved. It seems that of the seven variables, x_3 is now not significant.

10.6.3 Response Transformation

The indication of the QQ-plot of the left-hand panel of Fig. 10.48 and of the effect of the five outliers might be interpreted as indicating a need for response transformation. To examine the effect of the outliers on the analysis of transformations, we return to the analysis of all the data (59 observations).

Since the responses can be positive or negative, power transformation of the response requires the Yeo-Johnson transformation or its extended form (see

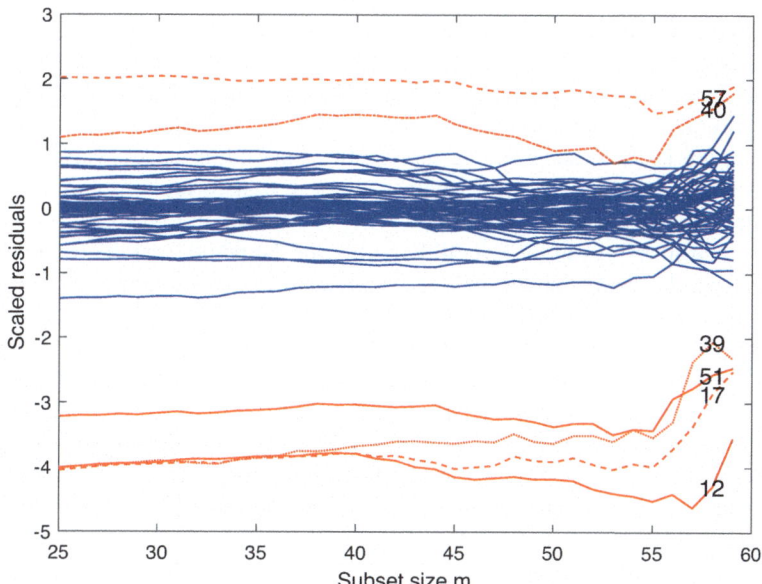

Fig. 10.49 Cancer data: monitoring residuals from the FS

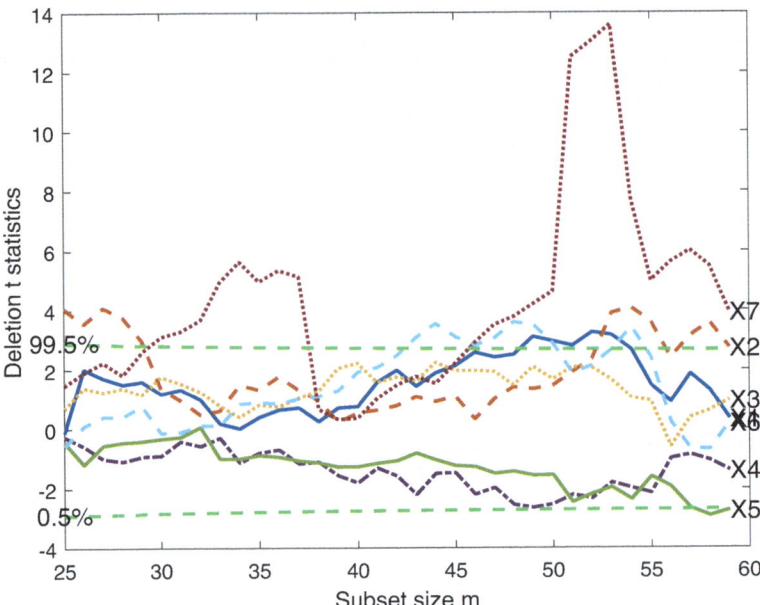

Fig. 10.50 Cancer data: monitoring of the added variable *t*-statistics

Sect. 6.2.3). We start with estimating a common transformation for the positive and negative transformations. The fan plot of Fig. 10.52 shows trajectories on a fine grid of values of λ_0 from 0.5 in steps of 0.1 to 1. It is clear from the figure that this finer grid is needed—the interval between 0.5 and 1 containing all trajectories that might indicate a suitable transformation. The main feature of the figure is that the outliers, entering in the last five steps, cause appreciable, but similar changes in the trajectories of the score statistic. To test whether different transformations are needed we look at the three test statistics for $\lambda_0 = 0.9$, a value which, in Fig. 10.52, produces a trajectory that just stays in the overall bounds over the whole search. The result, in Fig. 10.53 clearly shows that any evidence of a need for two values of λ is caused by the last observations to enter the search.

The automatic analysis leads to the same conclusions. As we did in Fig. 10.52, we start with the Yeo-Johnson transformation. Figure 10.54 shows the output from the automatic analysis using the coarse grid $\mathcal{G} = 0, 0.25, 0.5, 0.75, 1, 1.25$, and 1.5. The upper left-hand panel plots the extended BIC, and the upper right-hand panel the agreement index AGI. Both suggest a transformation of $\lambda = 1$, more firmly in the case of the AGI than of the extended BIC, where the value for 0.75 is close to that for $\lambda = 1$. The lower panel of the figure shows that fewer observations are deleted when $\lambda = 0.75$ rather than one.

The heatmaps of Fig. 10.55 confirm that the data do not require transformation. The left-hand panel shows the values of the extended BIC for the grids $\lambda_N = 1$,

10.6 NCI60 Cancer Cell Panel Data

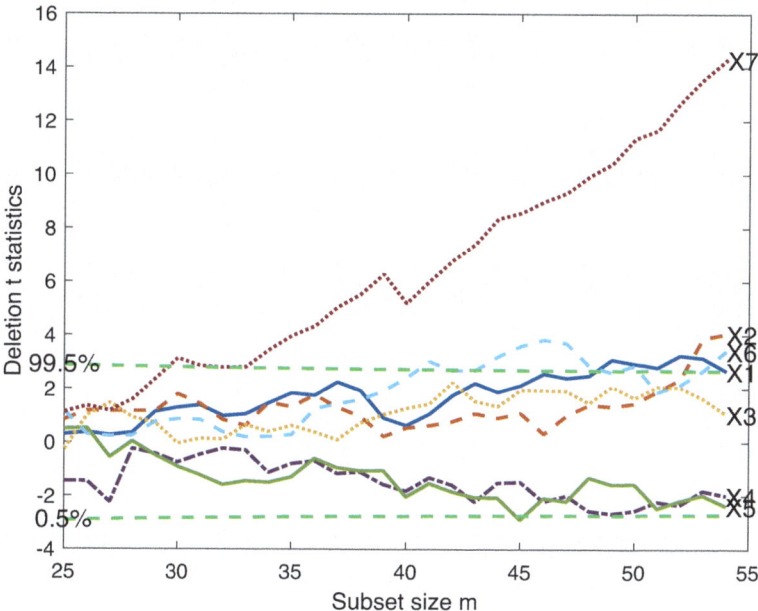

Fig. 10.51 Cancer data: monitoring of the added variable t-statistics with the 5 outliers removed

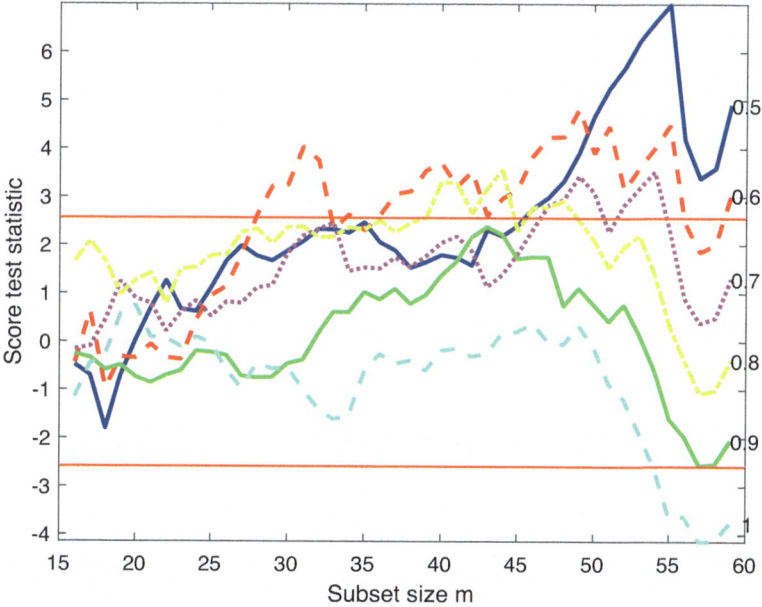

Fig. 10.52 Cancer data: fan plot; monitoring plot of score statistics for the Yeo-Johnson transformation with values of $\lambda_0 = 0.5, (0.1), 1$

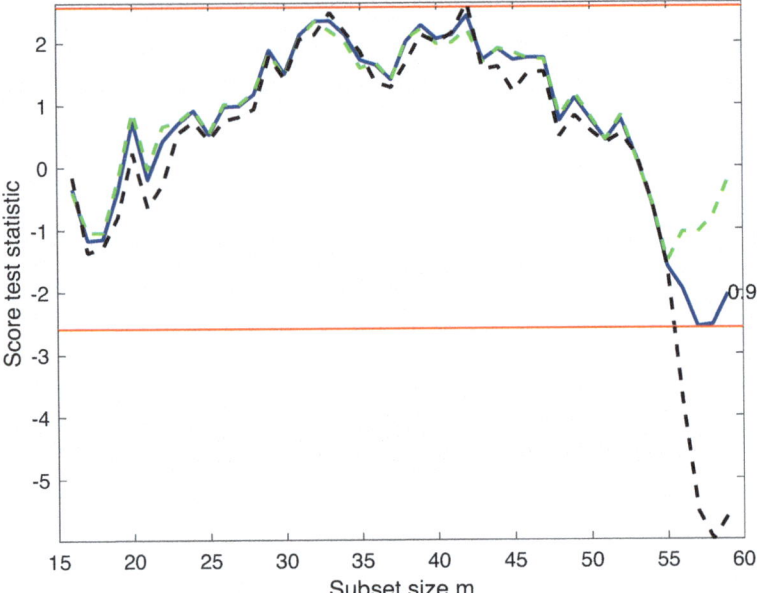

Fig. 10.53 Cancer data, $\lambda_0 = 0.9$: checking the two transformation parameters for the extended Yeo-Johnson transformation. Extended fan plots for $\lambda_0 = 0.9$. Upper (green) trajectory, positive observations, lower (black) trajectory, negative observations

1.25, 1.5 and 1.75. For λ_P the grid has values 0.25, 0.5, 0.75, and 1. For both quantities the maxima are at $\lambda_N = \lambda_P = 1$, that is at the edge of the plot. However, the adjacent panels, not plotted here, show a decline in both measures. The analysis of untransformed data is indicated.

The conclusions so far are that:

1. There are four large outliers and one minor outlier (Fig. 10.49);
2. There is no reason to transform the response (Fig. 10.55);
3. The least significant variable seems to be x_3 (Fig. 10.51) and
4. After removing x_3 and the five outliers, x_7 is by far the most important variable (Table 10.11).

The analysis of the variance table when the five outliers are deleted and x_3 excluded from the model is shown as Table 10.11. For this fitted model R^2_{adj} has the extremely high value of 0.96, compared to the appreciably smaller value of 0.712 in Table 10.10.

10.6 NCI60 Cancer Cell Panel Data

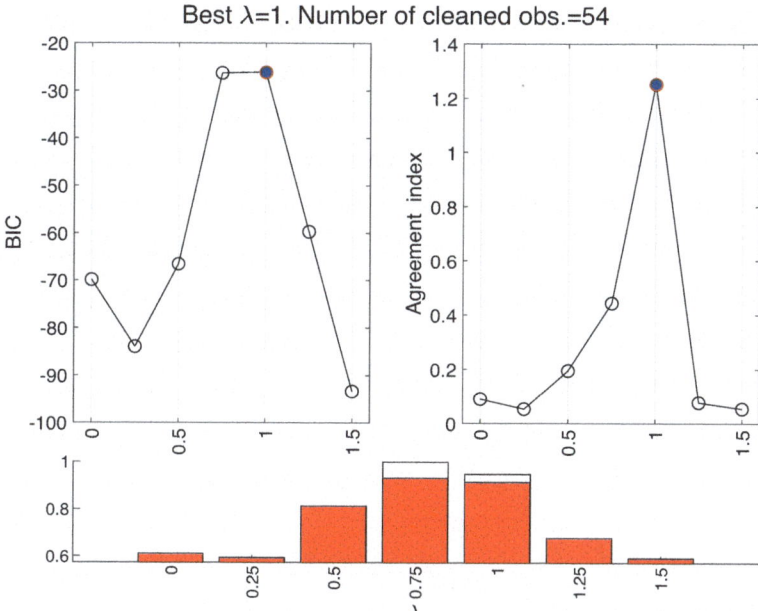

Fig. 10.54 Cancer data, Yeo-Johnson transformation: output from automatic analysis with coarse grid; $\mathcal{G} = 0, 0.25, 0.5, 0.75, 1, 1.25,$ and 1.5. Upper left-hand panel, extended BIC; upper right-hand panel, agreement index AGI. Lower panel, proportions $h(\lambda)/n$ and $m^*(\lambda)/n$

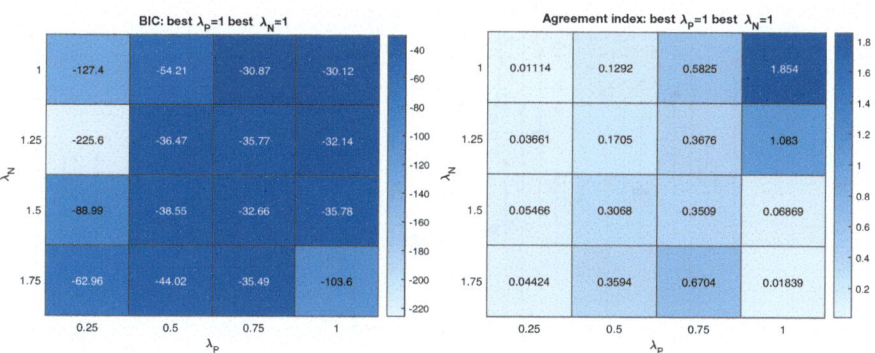

Fig. 10.55 Cancer data, extended Yeo-Johnson transformation: heat maps as functions of λ_P and λ_N. Left-hand panel extended BIC; right-hand panel agreement index AGI. Cell left blank if $h(\lambda) \leq 0.6n$

Table 10.11 Cancer data: ANOVA after removing x3 and the 5 outliers

```
Estimated Coefficients:
                  Estimate        SE          tStat       pValue
                  _____       _____     _____     _____

   (Intercept)    -2.8575        0.96807      -2.9517     0.0049181
   x1              0.11873       0.041589      2.8548     0.0063887
   x2              0.33727       0.079784      4.2273     0.00010812
   x4             -0.25229       0.11271      -2.2384     0.029971
   x5             -0.27454       0.10816      -2.5384     0.01451
   x6              0.1253        0.03409       3.6755     0.00060798
   x7              0.63635       0.044133     14.419      7.08e-19

Number of observations: 54, Error degrees of freedom: 47
Root Mean Squared Error: 0.678
R-squared: 0.965,  Adjusted R-Squared: 0.96
F-statistic vs. constant model: 214, p-value = 2.16e-32
```

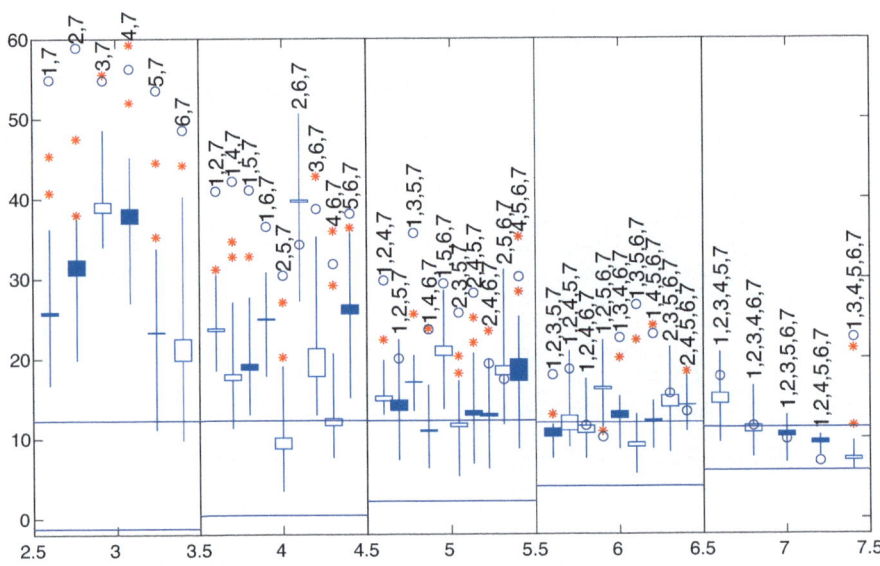

Fig. 10.56 Cancer data: generalized candlestick plot (5 outliers removed)

10.6.4 The Generalized Candlestick Plot

Figure 10.56 shows the candlestick plot for the Cancer data with five outliers removed. Since the intercept is included in all models, $p^+ = 8$. The plot shows information on models with p going from 3 to 7. These plots are given in adjacent panels, with horizontal lines marking the 2.5% and 97.5% points of the distribution of $C_p(n)$ given in (9.7).

The interpretation of the plot illuminates the challenges in the analysis of these data with highly correlated explanatory variables. The model with the lowest central box is 134567, that is with x_2 omitted. However, this model has three high final values of $C_p(m)$, which are not included in the calculation for the central box. These observations are included in the monitoring plot of the added variable t-statistics in Fig. 10.51, even though 5 outliers have been removed. This plot indicates the removal of x_3, that is the model 124567. Figure 10.56 shows that this is the next best model for $p = 7$; there are no values of $C_p(m)$ lying outside the central box and so good agreement with Fig. 10.51. The candlestick plot shows that there is also a good model with $p = 6$; 13567. That is, dropping terms in both x_2 and x_4, although there are then two values of $C_p(m)$ lying outside the confidence band.

10.6.5 AVAS

In the last sections we investigate the use of non-parametric transformations starting with AVAS. We use all seven explanatory variables.

The analysis with AVAS can be briefly described. The upper left-hand panel of Fig. 10.57 shows the transformed response $g(y)$ against y, which is similar in shape to using $\lambda = 0.65$ with the Yeo and Johnson transformation (note that this is not $y^{0.65}$, because the y are negative as well as positive). The residuals against fitted values in the upper right-hand panel suggest the four outliers we have already encountered. Since AVAS is not robust, these observations are not deleted. The lower panel shows $g(y)$ against the fitted values, a relationship in which there is considerable scatter. Regression with the transformed variables from this fit gives a value for R^2_{adj} of 0.781, appreciably less than the value of 0.96 obtained by LS in the absence of transformation, but with the deletion of 5 outlying observations.

10.6.6 RAVAS

The analysis using RAVAS after imposing the monotonicity constraint, presents a much clearer picture. The results for the transformation of y are in Fig. 10.58. The automatic combination of options which was used is `rob, scail` and `tyinitial`. Option `rob` was included in all the models which were found reasonable. All three plots show the four outliers deleted by RAVAS, which are well-separated from the residuals and fitted values. The response function is similar to that of AVAS, indicating a value of λ close to 0.7.

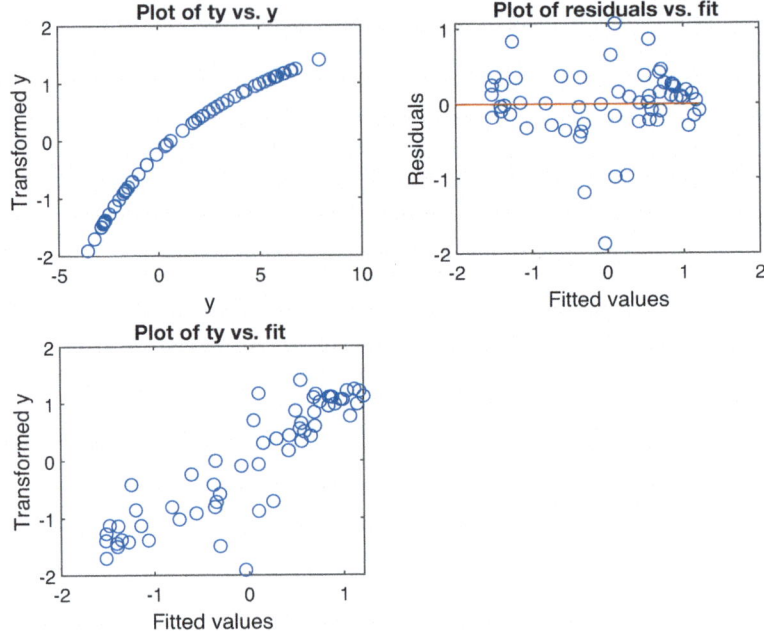

Fig. 10.57 Cancer data: non-robust AVAS, transformed response $g(y)$. Upper left-hand panel, $g(y)$ against y; upper right-hand panel, residuals against fitted values; lower panel, $g(y)$ against fitted values

The monotonic transformations of the explanatory variables are shown in Fig. 10.59. They are similar to those obtained in the absence of the monotonicity constraint. Regression on these transformations with the four outliers deleted gives a value of 0.968 for R^2_{adj}, higher than that of 0.96 for the first-order model with five deleted outliers.

Figure 10.60 shows the monitoring plot of the added variable t-statistics from a regression on the RAVAS transformed data with the 4 outliers removed. This is very close to Fig. 10.51 for the untransformed data with 5 outliers removed. Here the (arbitrary) signs of x_4 and x_5 have been changed and x_1 replaces x_3 as the non-significant variable; it is also excluded by variable selection.

The candlestick plot in Fig. 10.61 indeed shows that the preferred model is 234567, that is with x_1 omitted. Two other good models are 24567 and 23457, that is with x_3 or x_6 also excluded. Unlike the models suggested by Fig. 10.56, none of these include x_1. It is noticeable in the candlestick plot that all models displayed include x_7. However, it needs to be in a model with other variables.

Figure 10.62 shows the yX plot for transformed y and X using the transformations from RAVAS when the monotonicity constraint is imposed. The most striking feature is the panel for x_7. When the four outliers are deleted there is a clear linear relation with the transformed response. This plot explains the ubiquitous presence of x_7 in the

10.6 NCI60 Cancer Cell Panel Data

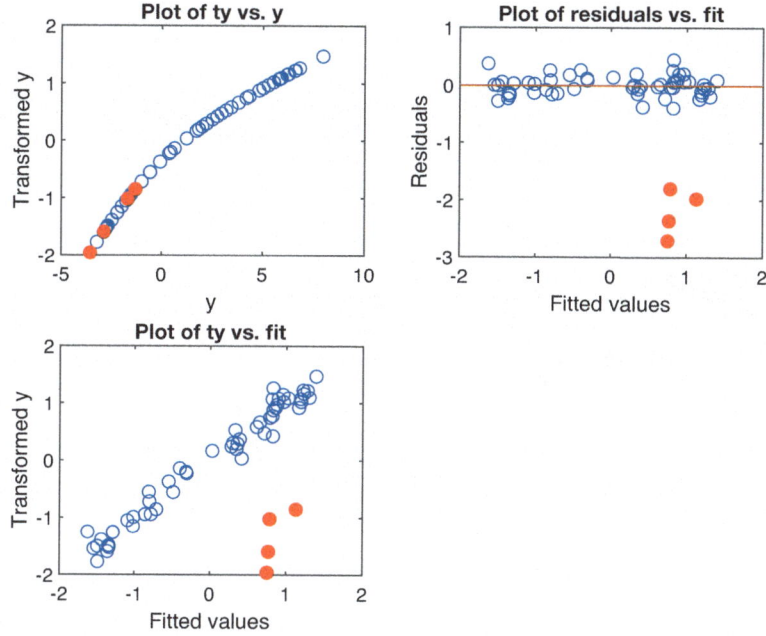

Fig. 10.58 Cancer data: RAVAS, transformed response $g(y)$. Upper left-hand panel, $g(y)$ against y; upper right-hand panel, residuals against fitted values; lower panel, $g(y)$ against fitted values

candlestick plot of Fig. 10.61 and the large values of the statistics for x_7 in Fig. 10.60, the monitoring plot of the added variable t-statistics.

As with Hald's cement data, a challenge in analysing these data is the high correlations between the explanatory variables. The analysis of this section provides a fine example of the effectiveness of the methods for monitoring robust procedures that form the basic message of our book.

The monitoring of LTS regression shown in Fig. 9.9 showed that there were just seven explanatory variables that occurred in at least 30 of the selected models. The problem with 7 variables and 59 observations was amenable to the monitoring approach to robust regression. Fitting all variables gave the analysis of variance in Table 10.10 in which only three variables were significant. However, the monitoring plot of residuals, Fig. 10.49, showed that there were outliers in the data. Deletion of these and of one explanatory variable led to a model that fitted the data well (Table 10.11). The generalized candlestick plot of Fig. 10.56 suggested three satisfactory models, all with five observations deleted.

A parallel analysis with RAVAS led to the identification of four outliers and to a model which gave the best fit of any investigated, with a value of 0.968 for R^2_{adj}, higher than that of 0.96 for the first-order model with five deleted outliers. The generalized candlestick plot of Fig. 10.61 indicates three satisfactory models.

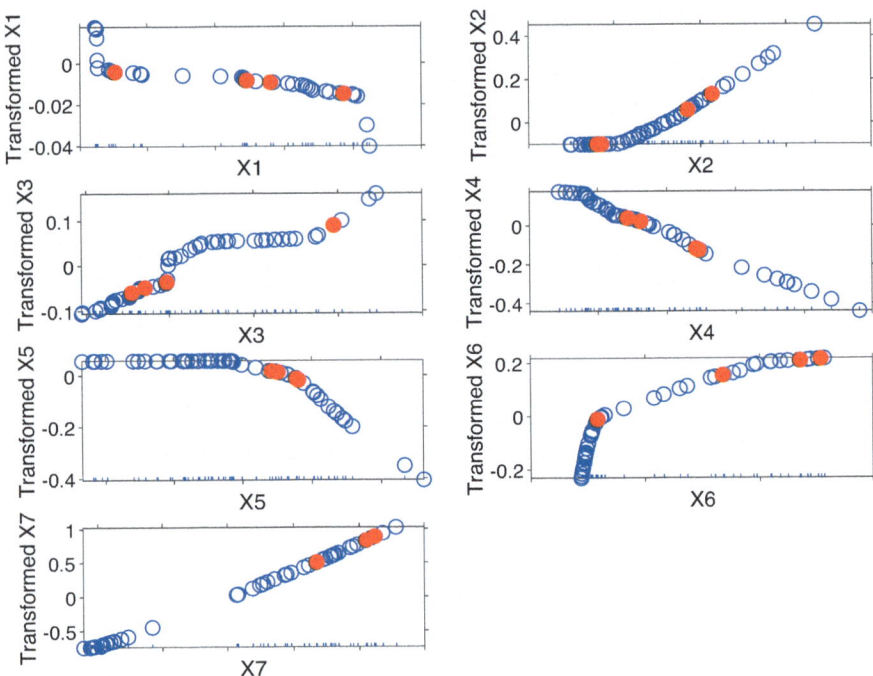

Fig. 10.59 Cancer data: RAVAS imposing the monotonicity constraint; transformations $f_j(x_j)$ of the explanatory variables against original x_j

In these two analyses, the greatest breakthrough seems to be the ability to detect outliers and to respond constructively to this information in the building of further models of greatly improved informativeness.

Problems

10.1 Analysis of the heart rate data

Students in an introductory statistics class at the University of Queensland participated in a simple experiment. They took their own pulse rate. They were then randomized to run in place for one minute or to sit for that minute. Then everyone measured their pulse rate again. There are 109 complete observations, nine explanatory variables covering physiological and lifestyle data and the two pulse rates.

One research question was how does the difference in pulse rate before and after the minute depend on lifestyle and physiological measurements? It is expected to depend heavily on whether the students ran or not. The data are available inside

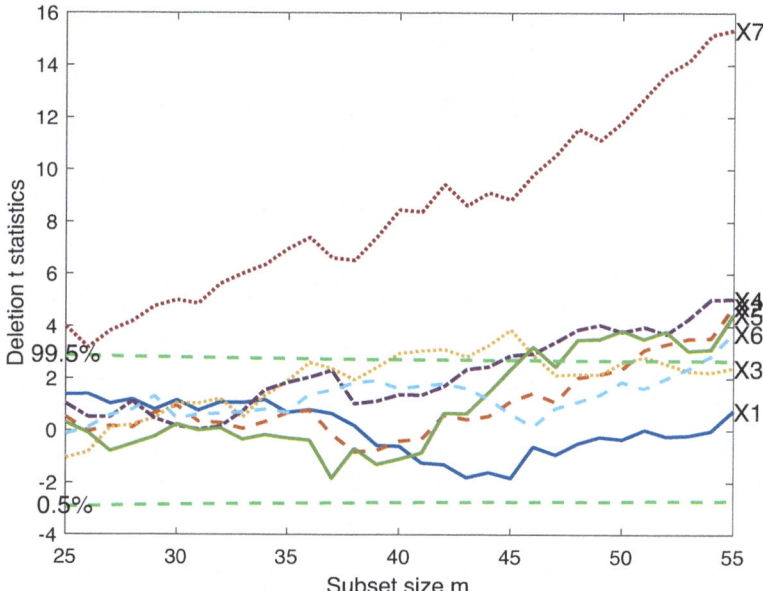

Fig. 10.60 Cancer data: monitoring of the added variable t-statistics on RAVAS transformed data imposing the monotonicity constraint with the 4 outliers removed

FSDA with name `ms212`. For a detailed description see http://www.statsci.org/data/oz/ms212.html.

1. Analyse the distribution of the response and comment on the possibility of treating the integer valued observations as continuous. Is a long right-hand tail present?
2. Analyse the difference in pulse rates with a regression analysis (ignoring the explanatory variable year). Run the variable selection tools and find the most important variables. Check the possibility of the Yeo-Johnson transformation of the response using a single value of λ, the difference in pulse rates.
3. Find if there are outliers causing abrupt changes in the indicated transformation.
4. In the transformed space, run the different tools of variable selection and find the most important variables.
5. Check if two values of λ are needed to transform the response.
6. Compare the QQ plots of the studentized residuals in the original and transformed scale and comment on the validity of the assumption of homogeneity of variance.
7. Comment on the effect of transformation on the F test.
8. Analyse the original and transformed data using FS- and MM-estimators.

10.2 Additional analysis of the customer loyalty data

Show the plot of extended BIC for the five values of λ, the AGI index and the proportion of observations included in the final transformed analyses. Discuss the best value of the transformation parameter and relate this plot to the fan plot shown in Fig. 10.26.

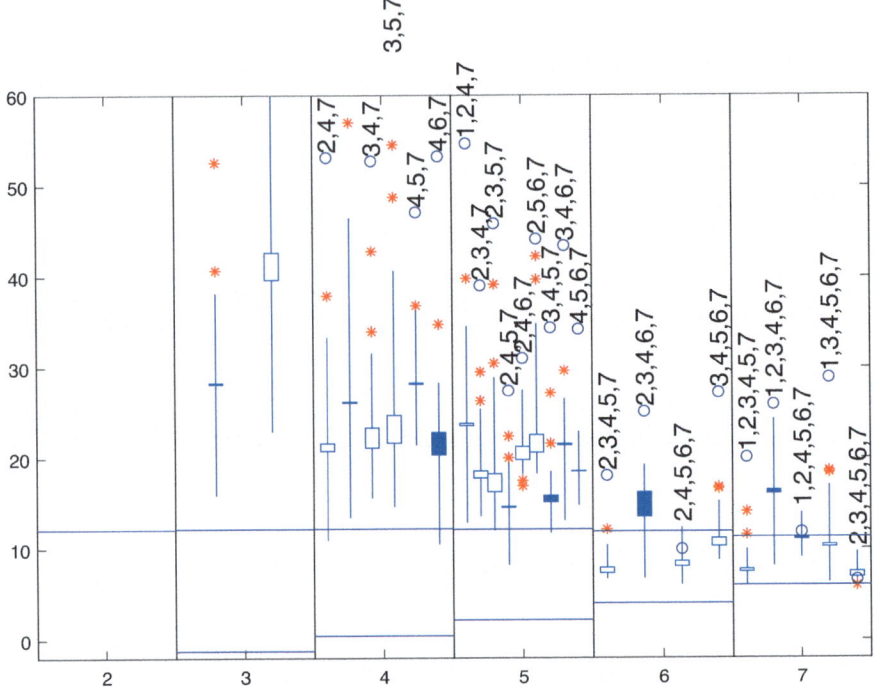

Fig. 10.61 Cancer data: generalized candlestick plot on RAVAS transformed data imposing the monotonicity constraint with the 4 outliers removed

10.3 Analysis of the modified customer loyalty data in the square-root scale

Analyse the Modified Customer Loyalty data in the square-root scale. Compute and comment on the monitoring plot of residuals, the cumulative plot of residuals, the QQ plot of studentized residuals, and the monitoring plot of the six added variable t-statistics. Comment on the effect of the outliers on inference about the values of the t-statistics. How many observations need to be deleted to remove the effect of the outliers using the automatic FS procedure?

10.4 Analysis of the auto mpg data

The auto mpg dataset has 398 rows and 9 columns and provides mileage, horsepower, model year, and further technical specifications for cars. The number of rows without missing values is 392. The website from which the dataset has been downloaded https://code.datasciencedojo.com/datasciencedojo/datasets/ states that "this dataset is recommended for learning and practicing your skills in exploratory data analysis, data visualization, and regression modelling techniques". The name of the dataset inside **FSDA** is autompg. The variables are:

Problems

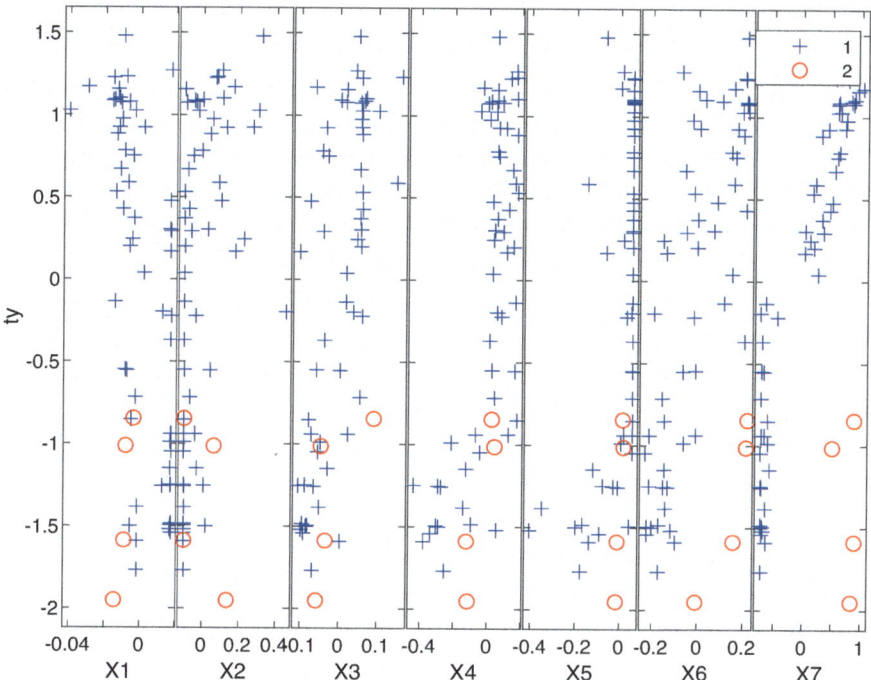

Fig. 10.62 Cancer data: RAVAS transformation with the monotonicity constraint. yX plot with the 4 units detected as outliers

mpg (y):	fuel efficiency measured in miles per gallon (mpg);
cylinders (x_1):	number of cylinders in the engine;
displacement (x_2):	engine displacement (in cubic inches);
horsepower (x_3):	engine horsepower;
weight (x_4):	vehicle weight (in pounds);
acceleration (x_5):	time to accelerate from 0 to 60 mph (in seconds);
modelYear (x_6):	model year;
origin (x_7):	origin of car (1: American, 2: European, 3: Japanese).

The goal is to predict mpg (the fuel efficiency of a car).

1. Show the yX plot and comment on the linearity of the relationship between y and the predictors. Comment on the univariate relationship between cylinders and y. Compute the ANOVA on the original scale.
2. Check whether the response has to be transformed and for the presence of outliers.
3. In the transformed scale for y, run the different tools of variable selection and find the most important variables. Comment on the parsimonious model which is obtained and on the percentage of variance explained and show the yX plot in the transformed space for y.

4. Analyse the residuals and find the characteristics of the units which show positive and negative large residuals.
5. Compare the above solutions with RAVAS and find the best solution using RAVAS. Check on the presence of outliers. Comment on the transformation of the continuous explanatory variables. In the transformed RAVAS space build the ANOVA using the transformed variables which are significant. Check whether the significance of the explanatory variables changes when the outliers detected by RAVAS are included in the analysis based on the non-parametrically transformed data.
6. Compute a 3D plot using the two most significant explanatory variables and the response. Compare the original view with a view which best highlights the relationship between the two explanatory variables and the response.

Open Access This chapter is licensed under the terms of the Creative Commons Attribution 4.0 International License (http://creativecommons.org/licenses/by/4.0/), which permits use, sharing, adaptation, distribution and reproduction in any medium or format, as long as you give appropriate credit to the original author(s) and the source, provide a link to the Creative Commons license and indicate if changes were made.

The images or other third party material in this chapter are included in the chapter's Creative Commons license, unless indicated otherwise in a credit line to the material. If material is not included in the chapter's Creative Commons license and your intended use is not permitted by statutory regulation or exceeds the permitted use, you will need to obtain permission directly from the copyright holder.

Appendix
Software and Datasets

The first appendix lists the software routines we have used; the second lists our datasets in order of first occurrence.

A.1 Software Details

A.1.1 The Systematic Naming of Functions

All functions like $\rho(x)$, $\psi(x)$, $w = \psi(x)/x$, $\psi'(x)$, $\psi(x)x$ described in this book have been implemented in the toolbox **FSDA**.

The names of the **MATLAB** functions are formed by a prefix specifying the type of ρ function which is one of `"HU"`, `"TB"`, `"OPT"`, `"HA"`, `"HYP"`, `"PD"` and `"AS"` followed by the name of the function specifying the different ingredients. It can be one of the following: `"rho"`, `"psi"`, `"wei"`, `"psider"` and `"psix"`. Each function is stored in an .m file and a **HTML** documentation file is related to it. For example, the function which computes ρ for `"OPT"` is called `OPTrho` and is stored in the **MATLAB** file `OPTrho.m`; the function which computes ψ for the Tukey biweight is called `TBpsi` and is stored in the **MATLAB** file `TBpsi.m`. To see the corresponding documentation for the ρ function for the hyperbolic, type in your browser http://rosa.unipr.it/FSDA/HYPrho.html; to see the corresponding documentation of the derivative of the ψ function of Hampel, type http://rosa.unipr.it/FSDA/HApsider.html.

The functions to find the constant c associated with a particular value of breakdown point (see Sect. 3.9) end with the suffix `bdp`. For example, the function which computes the constant c associated with the Tukey biweight given a prefixed value of bdp is `TBbdp`, is stored in file `TBbdp.m`, and the corresponding documentation can be found at http://rosa.unipr.it/FSDA/TBbdp.html. The functions to find the constant c associated with a particular value of the efficiency, see (3.37), end with the suffix `eff`. Finally, the functions which, given a particular value of c, compute bdp and *eff* end with suffix c. For example, the function to compute *bdp* and *eff* for the power divergence estimator given c is PDc and the associated documentation is at http://rosa.unipr.it/FSDA/PDc.html.

Many of these functions are also implemented in the R package **fsdaR** which provides an interface to the MATLAB toolbox **FSDA**. Table A.1 for each topic (first column) shows the name of the function inside **FSDA** (second column) and its availability in **fsdaR** (third column).

The function which computes S-estimators in regression is called Sreg. The function MMreg computes the MM-estimators, whereas the function taureg computes τ estimators. The option inside these functions which lets the user choose among the different ρ functions is called rhofunc.

The function which computes LMS or LTS estimators is called LXS. The option inside named lms enables choice between the two.

The function which computes the forward search with automatic outlier detection is called FSR; the function which stores the information about the statistics generated during the search (required for further exploratory data analysis) is called FSReda; the routine which computes the monitoring of the minimum deletion residuals is called FSRmdr and the function which simply returns the units belonging to the subset in each step is called FSRbsb. Heteroskedastic and Bayesian routines follow the same rule but they respectively add the letters "B" and "H" to their names. For example, functions FSRH, FSRHeda, FSRHmdr, and FSRHbsb generalize the routines for linear regression to heteroskedastic regression. Inside each of them, the option typeH enables the user to choose between ART (default) and HAR types of heteroskedasticity.

For plots of residuals the output of functions FSReda, Sregeda, MMregeda, FSRHeda, FSRBeda can be shown by function resfwdplot (see for example Figs. 4.12 and 4.18). Option corres in resfwdplot enables the display of the three measures of correlation between adjacent residuals (see for example Figs. 4.10 and 4.16). Finally, options standand, fground and bground control the style of, respectively, normal units (which by default are shown in dark blue, see for example Fig. 4.35), foreground units which have to be highlighted (see for example the red lines in Fig. 10.16) and background units (which by default are shown in pale blue; see for example the trajectories associated with small residuals in Figs. 4.28 and 4.29).

The output for properties related to explanatory variables (for example t statistics, score tests and $\hat{\beta}$), that are monitored as a function of *bdp*, *eff* or subset size, can be shown using function fanplotFS in a single panel (see for example Figs. 4.22, 4.33, 6.1) or separate panels (see for example Figs. 4.39 and 4.40). Option highlight inside fanplotFS selects a set of units whose entry order has to be highlighted (see for example Figs. 10.25 and 10.26).

In the case of transformations depending on a single λ, the function fanplotBIC may be used to show the automatic choice of λ (by taking as input the output produced by FSRfan) (see for example Figs. 6.15 and 6.16). In the case of two values of λ, function fanplotpnBIC may be used (see for example Figs. B.28 and B.29).

Table A.1 First column: topic. Second column: name inside **FSDA** or **fsdaR**. Third column: availability in **fsdaR** ($\checkmark$)

Linear regression with soft-trimming estimators

M-estimator of scale	Mscale	
S-estimator	Sreg	$\checkmark$
Monitoring of S-estimators for different values of *bdp*	Sregeda	$\checkmark$
MM-estimator	MMreg	$\checkmark$
MM-estimator starting from the output of S-estimator	MMregcore	
Monitoring of MM-estimators for different values of *eff*	MMregeda	$\checkmark$
τ-estimator	Taureg	
Covariance matrix of robust regression coefficients	RobCov	
Proper threshold for robust estimators to obtain an empirical size close to nominal size	RobRegrSize	

Linear regression with hard trimming estimators

LMS and LTS estimator	LXS	$\checkmark$
Forward search with automatic outlier detection procedure	FSR	$\checkmark$
Forward search storing statistical information for exploratory data analysis	FSReda	$\checkmark$
Monitoring of minimum deletion residuals	FSRmdr	$\checkmark$
Random start monitoring of minimum deletion residuals	FSRmdrrs	$\checkmark$
Monitoring of the units belonging to subset	FSRbsb	
LTS estimator extended to time series	LTSts	
Variable selection in the robust time series model LTSts	LTStsVarSel	
Forward search using in each step LTSts based on *m* observations	FSRts	

Parametric transformations

Score test for Box-Cox,	Score	$\checkmark$
Score test for Yeo-Johnson	ScoreYJ	$\checkmark$
Score test for extended Yeo-Johnson	ScoreYJpn	$\checkmark$
Monitoring of score test	FSRfan	$\checkmark$
Automatic choice of the transformation parameter in linear regression using Box-Cox or Yeo-Johnson	fanBIC	
Automatic choice of the transformation parameter in linear regression using extended Yeo-Johnson transformation	fanBICpn	
Compute (normalized) transformed values according to Box-Cox, Yeo-Johnson and extended Yeo-Johnson	normBoxCox, normYJ, normYJpn	$\checkmark$
Transform both sides of a linear regression model	tBothSides	

(continued)

Table A.1 (continued)

Non-parametric transformations	
Alternative conditional expectation	`ace`
Additivity and variance stabilization for regression	`avas`
Automatically find the best options to use in `avas`	`avasms`
Find updated transformed values of response using or not using option `trapezoid`	`ctsub`
Friedman's supersmoother algorithm	`supsmu`

Bayesian regression	
Bayesian estimates of regression parameters	`regressB`
Automatic outlier detection based on the FS in Bayesian regression	`FSRB`
Forward search in Bayesian regression storing statistical information for exploratory data analysis	`FSRBeda`
Monitoring of minimum deletion residuals in Bayesian regression	`FSRBmdr`
Monitoring of the units belonging to subset in each step of the Bayesian forward search	`FSRBbsb`

Heteroskedastic regression	
Heteroskedastic estimates of regression parameters	`regressH`
Automatic outlier detection based on the FS in heteroskedastic regression	`FSRH`
Forward search in heteroskedastic regression storing statistical information for exploratory data analysis	`FSRHeda`
Monitoring of minimum deletion residuals in heteroskedastic regression	`FSRHmdr`
Monitoring of the units belonging to subset in each step of the heteroskedastic forward search	`FSRHbsb`

Regression clustering	
Trimmed clusterwise linear regression and trimmed cluster-weighted model (TCWM)	`tclustreg`
Monitoring trimmed clusterwise linear regression and trimmed cluster-weighted model (TCWM) storing statistical information for exploratory data analysis	`tclustregeda`
Compute `tclustreg` for different number of groups k and restriction factor c in order to automatically select the best combination (for a fixed trimming level α)	`tclustregIC`
Takes as input the output of function `tclustregIC` and extracts the first best solutions (plausible partitions) and shows them in the yX plot	`tclustICsol`

Censored regression	
Estimates of parameters in censored regression	`regressCens`
Estimates of transformation parameter and signed square-root likelihood ratio test in censored regression	`regressCensTra`

(continued)

Appendix: Software and Datasets

Table A.1 (continued)

Forward search in censored regression storing statistical information for exploratory data analysis	`FSRedaCens`	
Monitoring of signed square-root likelihood ratio test in censored regression	`FSRfanCens`	
Compositional data analysis		
Transformation to pivot coordinates	`pivotCoord`	
Inverse of pivot coordinates transformation	`pivotCoordInv`	
Variable selection		
Added variable test	`addt`	
Monitoring of deletion t tests for each explanatory variable (boolean option `ilr` specifies whether the matrix X contains compositional data when the test which is monitored is the first coordinate of the *ilr* transformation (see for example Fig. 8.34))	`FSRaddt`	
Robust model selection with flexible trimming based on FS	`FSRms`	
Graphics		
Index plot of residuals from `LXS`, `LTSts`, `Sreg`, `MMreg` and `taureg`	`resindexplot`	✓
Monitoring plot of residuals as function of subset size, breakdown point or efficiency from `Sregeda`, `MMregeda`, `FSReda`, `FSRHeda` and `FSRBeda`	`resfwdplot`	✓
Monitoring plot of minimum deletion residuals from `FSRmdr`, `FSRHmdr` and `FSRBmdr`	`mdrplot`	✓
Monitoring plot of minimum deletion residuals from random starts computed with function `FSRmdrrs`	`mdrrsplot`	✓
Monitoring plot of t deletion or score tests from `FSRaddt` or `FSRfan` (i.e. show the fan plot)	`fanplot`	✓
Multiple panel output of ACE or AVAS from `ace` or `avas`	`aceplot`	
Plot the output of `avasms` (i.e. show the augmented star plot and the heatmaps)	`avasmsplot`	
Show the output of monitoring of C_p from `FSRms` (i.e. show the candlestick plot)	`cdsplot`	
Scatterplot matrix or yX plot with clickable legends, grouping variables, and brushing, with background colour, number or circles or squares depending on the correlation. In the scatter plots it is possible to add ellipses, bivariate boxplots, or non-parametric density contours	`spmplot, yXplot`	✓
Plot the output of `tclustregIC`	`tclustICplot`	
Plot the output of `tclustICsol` (i.e. show the car-bike plot)	`carbikeplot`	✓
Create the wedge plot taking the output of `LTSts`	`wedgeplot`	

Note that all the graphical routines above inside **FSDA** *, with the exception of* `wedgeplot`, `aceplot` *and* `carbikeplot`, *allow interactive brushing and linking*

The output of AVAS can be shown by function `aceplot` using all the 3 plots (as in Figs. B.57 and 10.58) or a subset of them (e.g. Figs. 7.6 and 7.14). Using option `oneplot` inside `aceplot` it is possible to specify whether the plot of the transformation of the explanatory variables should be embedded inside the previous one as an extra panel included in the figure (e.g. Figs. 7.5 and 7.11) or presented as a separate figure (e.g. Figs. 7.15 and 10.59).

The output of automatic model selection using RAVAS (function `avasms`) can be shown by function `avasmsplot` together with the details of the best solutions shown using either text (see for example Figs. 7.10 and 7.22) or bars (see for example Figs. 7.13 and 7.16), with or without the heatmap of the correlations among the solutions (see for example Figs. 7.17 and 10.19).

The output of variable selection (function `FSRms`) can be shown by the function `cdsplot` (see for example Figs. 9.1 and 9.2).

The output of random start monitoring (e.g. function `FSRmdrrs`) can be shown by the function `mdrrsplot` (see for example Figs. 8.20 and 8.23).

Function `LTSts`, which is used in Sect. 8.6 to analyse the Airline time series requires using the `model` option to set the parameters of equation 8.33. This option provides a rich set of further options to control the underlying LTS estimator and the Alternating Least Squares algorithm. Its output can be directly passed to the `wedgeplot` function, to produce the representation of residuals in Fig. 8.32.

A.1.2 Interactive Brushing

All the above plots are brushable using option `databrush`. The output of brushing depends on the particular plot:

- brushing from `resfwdplot` enables the user to see the position of the units in the yX plot and (if it is already open) also in the monitoring plot of minimum deletion residuals;
- brushing from the augmented star plot automatically calls function `aceplot` for the selected solution;
- brushing the trajectory of the score test in the fan plot shows the units which entered the search in the brushed area highlighted in the yX plot based on transformed y;
- brushing the candles in the Candlestick plot shows the corresponding trajectories of C_p.

If `databrush` is a scalar equal to 1 it is possible to brush only once. On the other hand, if `databrush` is defined as a `struct` (in MATLAB language), it is possible to have repeated brushing in a cumulative way (in this case `databrush.persist='on'`) (see for example Figs. 4.24, 4.25, 4.26) or in a non-cumulative way (`databrush.persist='off'`).

In the R package **fsdaR** `databrush` can be TRUE for brushing only once or a `list` instead of `struct` for repeated brushing. However, not all options available in MATLAB are supported.

A.1.3 Personalized Data Tips

In each plot when the user clicks on a particular point (or line) a yellow rectangle appears with information about the point which has been clicked. The output of the click depends on the particular plot:

- clicking on a point in the graph produced by resfwdplot, which takes as input the output of Sregeda, automatically highlights in red the trajectory of the residual and shows, in a yellow rectangle, the *bdp* associated with the selected step, together with the weight of the selected unit and the unit number (or label of the unit);
- clicking on the graph produced by resfwdplot, which takes as input the output of FSReda, highlights in red, the trajectory of the residual and in a yellow rectangle shows the unit number (or label of the unit), the value of *m* (subset size) and the step(s) in which the unit entered the subset;
- clicking on the plot of minimum deletion residuals provides information about the unit(s) entering in the selected steps.

Finally, as in the documentation for databrush, if datatooltip is a struct, it is possible to further personalize the action of the mouse click on a single point. For example, once the user clicks on a point of the *x* axis, datatooltip.SubsetLinesColor=[1 0 0] highlights in red the units belonging to the subset (see for example Fig. 4.12).

These details are up to date at the time of writing and need MATLAB at least version 2020b. We expect to develop further software tools which will be documented in the release notes of **FSDA** at the page http://rosa.unipr.it/FSDA/release_notes.html

In the R package **fsdaR** datatooltip can be TRUE to activate the interactive data tips or a list instead of struct for further personalization. However, not all options available in MATLAB are supported.

A.2 Details of the Datasets That We Have Used

In this section we list, in order of first appearance, all the datasets that have been used. Further appearances are also listed. All the datasets are contained inside **FSDA** and can be loaded with the instruction load DataSetName (Table A.2).

Table A.2 First column: conventional name of the dataset inside **FSDA**. Second column: brief description. Third column: statistical techniques which have been used for analysis. Fourth column: Sections and/or Exercises where the dataset is used

Name	Brief description	Topic	Section or exercise
Income1	Survey of the United States Census Bureau, available in their annual social and economic supplements	Robust regression and transformations	Sections 1.4, 1.5, 2.1.1, 2.4.5, 10.2—Exercise 1.1
Income2	Income data from a municipality in the north of Italy	Robust regression and transformations	Sections 1.4, 1.5, 10.3—Exercise 1.4
multiple_ regression	Data from AR (2000)	Robust regression	Sections 3.5, 3.13, 4.9.3—Exercise 4.1
stars	Hertzsprung–Russell diagram of star cluster CYG OB1	Robust regression	Sections 4.5.1, 4.9.2
wool	Data from Box-Cox	Robust regression and transformation	Section 4.5.2
hawkins	Simulated data from Hawkins	Robust regression and variable selection	Section 4.9.4—Exercises 4.2, 4.5
hospitalFS	Logged survival time of patients undergoing liver surgery	Robust regression and variable selection	Sections 4.9.5, 6.1.4—Exercises 4.3, 4.4
bank_data	Amount of money made from personal banking customers over a year with 13 potential explanatory variables	Robust regression and transformation	Section 4.10
illness07	Observational data on the assessment of mental illness of 53 patients	Robust regression and transformation	Sections 4.11.5, 6.1.6
inttradedata	International trade data with a mixture of two regression models	Robust regression, robust regression clustering	Section 5.8
loyalty	509 observations on the behaviour of customers with loyalty cards from a supermarket chain in Northern Italy	Robust regression and transformation	Sections 6.1.5, 6.3.5
fondi_large	Medium term performance of 309 investment funds	Robust regression and transformation	Section 6.2.5
BalanceSheets	1,405 observations on profitability, calculated as return over sales and a set of 5 potential explanatory variables	Robust regression and transformation	Section 6.2.6—Exercises 6.3 6.4

(continued)

Appendix: Software and Datasets

Table A.2 (continued)

Name	Brief description	Topic	Section or Exercise
fetus	167 observations on the mandible length in foetuses	Transformation of both sides of a regression model	Section 6.5.2
leafpine	70 observations, taken from Bruce and Schumacher (1935), on the volume in cubic feet of shortleaf pine	Transformation of Both Sides of a Regression Model	Section 6.5.2.3
D1	Simulated dataset	Robust regression and transformation	Exercise 6.5
D2	Simulated dataset	Robust regression and transformation	Exercise 6.6
D3	Simulated dataset	Robust regression and transformation	Exercise 6.7
WangSimData	200 observations and 4 explanatory variables all independently uniformly distributed on $[-1, 1]$	Non-parametric regression	Section 7.6
fish	Data on the weight of 159 fish caught in a lake	Non-parametric regression	Section 7.7
Marketing_Data	200 results of an experiment on internet marketing	Non-parametric regression	Section 7.8
InvFundExt	Augmented Investment Funds data	Non-parametric regression	Exercise 7.2
hprice	Sales prices of 546 houses in the city of Windsor, Ontario, Canada, during July, August, and September, 1987	Robust Bayesian regression	Section 8.3.1
Fishery2002 Fishery2003 Fishery2004	Three years importation of a particular seafood into the European Union from the American continent	Robust Bayesian regression	Sections 8.3.3, 8.3.4 8.3.5
inttrade1	International Trade Data 1: 3867 import flows of product with TARIC code 4801000000 from Switzerland into Italy (POD_4801000000_CH_IT)	Heteroskedastic regression	Section 8.4.3 Exercise 8.2
inttrade2	International Trade Data 2: 1302 import flows of product with TARIC code 0307491800	Heteroskedastic regression	Section 8.4.4

(continued)

Table A.2 (continued)

Name	Brief description	Topic	Section or Exercise
`inttrade3`	International Trade Data 3: 389 import flows of product with TARIC code 0307591000 from Serbia into Italy (POD_0307591000_SN_IT)	Heteroskedastic regression	Exercise 8.3
`facemasks`	352 import flows from one of the newly defined codes specific for FFP2 and FFP3 face masks	Robust regression clustering	Sections 8.5, 8.5.6
`valueadded`	Data on value added by industry from the UNIDO INDSTAT database	Regression with compositional data	Section 8.7.1.1
`FishMorphology`	26 morphometric measurements, in millimetres, on a sample of 75 fish	Regression with compositional data	Section 8.7.3.1
`fat`	Physical measurements of 252 males	Regression with compositional data	Exercise 8.5
`Data_Airline`	Airline data introduced in Box and Jenkins (1976)	Robust time series monitoring	Section 8.6— Exercise 8.4
`cement`	13 observations on the relationship between heat generated in the setting of cement and the concentration of four chemicals	Variable selection	Section 9.3
`ozone_330_obs`	330 daily measurements, from the beginning of the year, of ozone concentration and eight meteorological variables	Variable selection	Section 9.3.2
`ozone`	Subset of 80 daily measurements from the `ozone_330_obs` dataset	Variable selection	Section 9.3.2— Exercise 9.1
`nci60`	Data on 60 human cancer cell lines	Variable selection	Sections 9.4, 10.6
`air_pollution`	Air pollution and mortality in 60 metropolitan statistical areas in the United States	Variable selection	Exercise 9.2

(continued)

Table A.2 (continued)

Name	Brief description	Topic	Section or Exercise
ConsLoyaltyRet	1711 observations from an online survey of the respondents' assessment of the importance of six explanatory variables in maintaining customer loyalty	Robust regression, variable selection and transformation	Sections 10.4, 10.5—Exercise 10.2
ms212	Heart rate data on 109 people and 9 explanatory variables	Robust regression, variable selection and transformation	Exercise 10.1
autompg	Characteristics of 398 cars. There are 9 explanatory variables	Robust regression, variable selection and transformation	Exercise 10.4

Solutions

Problems: Chap. 1

1.1 Trimmed Means
The first 15 observations of variable HTOTVAL in dataset `Income1` are
51112 84682 135960 26999 63228 160783 93940
164837 130712 60393 161923 12422 112027 40988 83325.
When $\alpha = 0.05$ and $n = 15$, $m = [14 \times 0.05] = [0.7] = 0$ and therefore the truncated mean is equal to the simple arithmetic mean $\bar{y}_{0.05} = \hat{\mu} = \bar{y}_{15} = 92222.07$.
When $\alpha = 0.10$ $m = [14 \times 0.1] = [1.4] = 1$ and the truncated mean is

$$\bar{y}_{0.10} = \frac{1}{13} \sum_{i=2}^{14} y_{[i]} = \frac{26999 + 40988 + 51112 + \cdots + 161923}{13} = 92774.77.$$

1.2 Consistency factor for IQR
Under the normal model $\mathcal{N}(\mu, \sigma^2)$, the order statistics $y_{[n-[n/4]+1]}$ and $y_{[n/4]}$ tend to the first and third quartiles of the normal distribution. Therefore

$$\begin{aligned} \text{E(IQR)} &= [\sigma \Phi^{-1}(0.75) + \mu - \{\sigma \Phi^{-1}(0.25) + \mu\}] \\ &= \sigma\{0.6745 - (-0.6745)\} = 2\sigma \times 0.6745 = 1.3490\sigma. \end{aligned}$$

1.3 Consistency factor for MAD
Asymptotically MAD covers 50% (between 1/4 and 3/4) of the standard normal cumulative distribution function, i.e.

$$\frac{1}{2} = \Pr(|X - \mu| \leq \text{MAD}) = \Pr\left(\left|\frac{X-\mu}{\sigma}\right| \leq \frac{\text{MAD}}{\sigma}\right) = \Pr\left(|Z| \leq \frac{\text{MAD}}{\sigma}\right),$$

whence

$$\Phi(\text{MAD}/\sigma) - \Phi(-\text{MAD}/\sigma) = 1/2.$$

© The Editor(s) (if applicable) and The Author(s) 2025
A. C. Atkinson et al., *Robust Statistics Through the Monitoring Approach*,
Springer Series in Statistics, https://doi.org/10.1007/978-3-031-88365-1

Given that
$$\Phi(-\text{MAD}/\sigma) = 1 - \Phi(\text{MAD}/\sigma),$$

$\text{MAD}/\sigma = \Phi^{-1}(3/4) = 0.67449$, from which we obtain the consistency factor $1/\Phi^{-1}(3/4) = 1.4826$.

Another way of establishing the relationship is by noting that MAD equals the half-normal distribution median. In other words, from

$$\int_0^{\text{MAD}} \frac{1}{\sigma}\sqrt{\frac{2}{\pi}} \exp\left(-\frac{y^2}{2\sigma^2}\right) dy = 0.5,$$

using the change-of-variables $z = y/(\sqrt{2}\sigma)$, so that $y = z\sqrt{2}\sigma$ and $dy = dz\sqrt{2}\sigma$, yields

$$\int_0^{\text{MAD}/(\sqrt{2}\sigma)} \frac{1}{\sigma}\sqrt{\frac{2}{\pi}} \exp\left(-z^2\right) dz\sqrt{2}\sigma = 0.5,$$

which leads to

$$0.5 = \frac{2}{\sqrt{\pi}} \int_0^{\text{MAD}/(\sqrt{2}\sigma)} \exp\left(-z^2\right) dz = \text{erf}\left(\frac{\text{MAD}}{\sqrt{2}\sigma}\right),$$

where the error function $\text{erf}(z) = \frac{2}{\sqrt{\pi}} \int_0^z \exp(-t^2) dt$. Then

$$\text{MAD} = \sigma\sqrt{2}\,\text{erf}^{-1}(1/2) \approx 0.6745\sigma.$$

1.4 Univariate analysis
(a) Figure B.1 shows the values of the trimmed mean for different values of the trimming proportion α. In this figure the x axis is reversed so that $\alpha = 0$ is on the right. When α is in the range $(0.25\ 0.5]$, the value of the truncated mean is slightly smaller than that of the median. For values of α in the range $[0\ 0.25]$ the value of the truncated mean increases steadily to reach 18395 due to the effect of the inclusion of the observations which belong to the right tail of the distribution.

(b) Figure B.2 shows boxplots of the distribution of income after six different Box-Cox transformations. The six values of λ used in Fig. B.2 are $1, 0, -0.5, -1, -1.5$ and -2. From this figure it is clear that the variable has to be transformed, but it is not clear which is the best value of λ to use; any value of λ in the range $[-2\ 0]$ seems reasonable.

1.5 Likelihood concepts
(a) Figure B.3 shows the typical loglikelihood curve in the normal model. The three likelihood-based tests, namely Wald, likelihood ratio and score tests, are shown on the curve. The tests use different information about the function.

Solutions

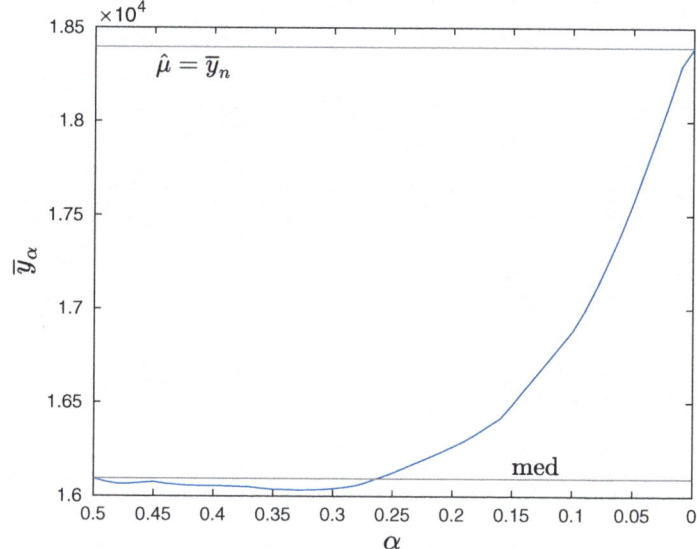

Fig. B.1 Income data from a municipality in the north of Italy: monitoring the truncated mean as a function of the trimming proportion α

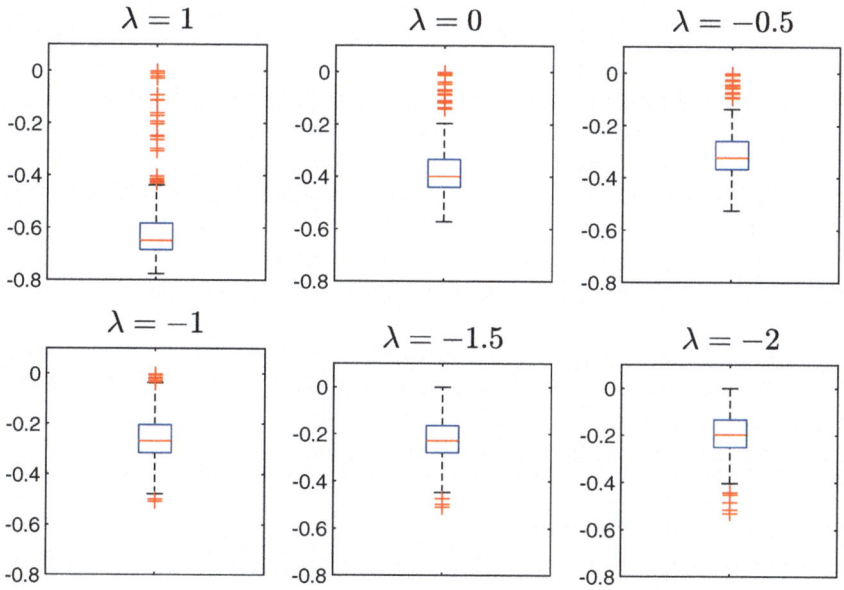

Fig. B.2 Income data from a municipality in north of Italy: boxplots for six values of λ using the normalized Box-Cox power transformation after preliminary rescaling of the data to a maximum value of one

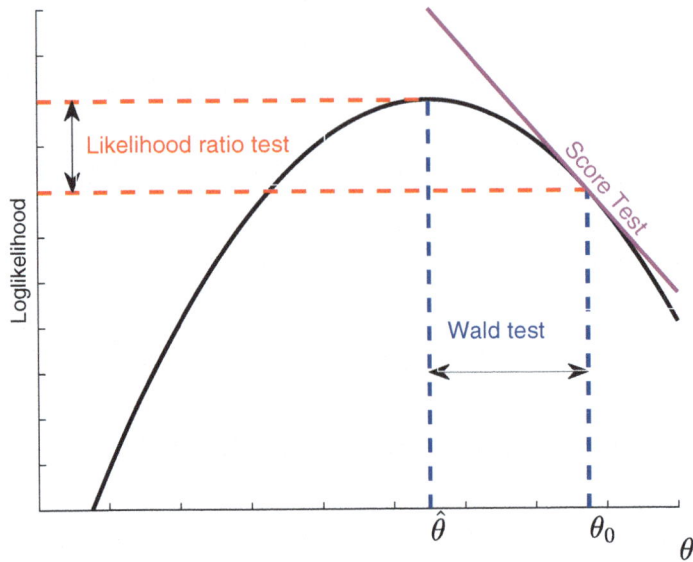

Fig. B.3 Loglikelihood curve: the three likelihood-based tests, namely Wald, likelihood ratio, and score tests, are shown on the curve. The tests use different information about the function

(b) The value of T_{SC} when $\theta_0 = \hat{\theta}$ is 0. In other words, the slope of the tangent to the loglikelihood function when $\theta = \hat{\theta}$ is 0.

(c) For the normal distribution and a single observation $\mathcal{L}(\theta) = -\log(\sigma\sqrt{2\pi}) - (y-\theta)^2/(2\sigma^2)$. Since differences in loglikelihood are important, the constant can be ignored and we take $\mathcal{L}(\theta) = -a(y-\theta)^2$. For a sample of n observations $\mathcal{L}(\theta) - \mathcal{L}(\hat{\theta}) = -a\sum(y_i - \theta)^2 - a\sum(y_i - \hat{\theta})^2$. For a simple sample $\hat{\theta} = \bar{y}$. The difference in loglikelihoods is then $-na(\hat{\theta} - \theta)^2$. The likelihood ratio test is minus twice this difference $= 2na(\hat{\theta} - \theta)^2$ which is distributed χ_1^2 when θ is the true value of the parameter. Differentiation gives the score test as $2na(\hat{\theta} - \theta)$, which is distributed as $\mathcal{N}(0, 1)$. The Wald test $= \theta - \hat{\theta}$, so needs rescaling.

For other distributions the curve will not be parabolic, so the three statistics will not be related in this way. However, under simple conditions, many parameter estimates converge to normality as $n \to \infty$. Then the three statistics will converge to obeying the relationship given here for the normal distribution.

Problems: Chap. 2
2.1 SC for the arithmetic mean

$$\begin{aligned}
\text{SC}(y, \hat{T}_n, Y_{n-1}) &= n\left(\frac{y_1 + \cdots + y_{n-1} + y}{n} - \frac{y_1 + \cdots + y_{n-1}}{n-1}\right) \\
&= \frac{(n-1)(y_1 + \cdots + y_{n-1} + y) - n(y_1 + \cdots + y_{n-1})}{n-1} \\
&= \frac{-y_1 - \cdots - y_{n-1} + (n-1)y}{n-1} \\
&= y - \bar{y}_{n-1}.
\end{aligned}$$

2.2 SC for the median
When $n = 2m + 1$ is odd and the estimator is the median, the two median observations of the sample $(y_1, \ldots, y_{n-1})$ are $y_{[m]}$ and $y_{[m+1]}$,

$$\hat{T}_{n-1}(y_1, \ldots, y_{n-1}) = \text{med}(y_1, \ldots, y_{n-1}) = \frac{1}{2}(y_{[m]} + y_{[m+1]})$$

$$\hat{T}_n(y_1, \ldots, y_{n-1}, y) = \text{med}(y_1, \ldots, y_{n-1}, y) = \begin{cases} y_{[m]} & \text{if } y < y_{[m]} \\ y & \text{if } y_{[m]} \leq y \leq y_{[m+1]} \\ y_{[m+1]} & \text{if } y > y_{[m]} \end{cases}$$

Therefore

$$\text{SC}(y, \hat{T}_n, Y_{n-1}) = \begin{cases} n\{y_{[m]} - \frac{1}{2}(y_{[m]} + y_{[m+1]})\} = \frac{n}{2}(y_{[m]} - y_{[m+1]}) & \text{if } y < y_{[m]} \\ n\{y - \frac{1}{2}(y_{[m]} + y_{[m+1]})\} & \text{if } y_{[m]} \leq y \leq y_{[m+1]} \\ n\{y_{[m+1]} - \frac{1}{2}(y_{[m]} + y_{[m+1]})\} = \frac{n}{2}(y_{[m+1]} - y_{[m]}) & \text{if } y > y_{[m+1]} \end{cases}$$

An observation y has one of the constant effects on the sample median unless y lies between the two median observations of the original sample. These constant effects are proportional to the distance between the two median observations $|y_{[m]} - y_{[m+1]}|$.

2.3 IF for the mean

$$\begin{aligned} T(F_\varepsilon) &= \mathrm{E}_{F_\varepsilon}(Y) \\ &= (1-\varepsilon)\mathrm{E}_F(Y) + \varepsilon \mathrm{E}(\delta_y) \\ &= (1-\varepsilon)T(F) + \varepsilon y \end{aligned}$$

$$\text{IF}(y; T, F) = \lim_{\varepsilon \to 0} \frac{(1-\varepsilon)T(F) + \varepsilon y - T(F)}{\varepsilon} = y - T(F).$$

2.4 IF for the median
(a) If the distribution is well-behaved, when $n \to \infty$ the distance $y_{[m+1]} - y_{[m]}$ tends to zero and the sample points $y_{[m+1]}, y_{[m]}$ tend to $F^{-1}(0.5)$. In general

$$\frac{\text{Number of points } y | y_0 < y < y_0 + \delta_y}{n} \approx \Pr(y_0 < y < y_0 + \delta_y) \approx f(y_0)\delta_y.$$

The ratio between the number of points in a short interval δ_y containing the median of the population $F^{-1}(0.5)$, and the length of the small interval δ_y is approximately equal to $f\{F^{-1}(0.5)\}n$. Alternatively we can say that the number of sample points in the interval $[y_{[m]}\ y_{[m+1]}]$, which is one, is approximately equal to $nf\{F^{-1}(0.5)\}\delta_y$. Therefore asymptotically

$$1 = nf\{F^{-1}(0.5)\}\delta_y,$$

$$\delta_y = \frac{1}{nf\{F^{-1}(0.5)\}}.$$

Thus in the sensitivity curve the expression $y_{[m+1]} - y_{[m]} \to 1/[nf\{F^{-1}(0.5)\}]$ and

$$\frac{1}{2}n\left(y_{[m+1]} - y_{[m]}\right) \to \frac{1}{2}n\frac{1}{nf\{F^{-1}(0.5)\}} = \frac{1}{2}\frac{1}{f\{F^{-1}(0.5)\}}.$$

The change in the median is positive or negative depending on whether the additional point is greater than or less than the median of the uncontaminated distribution. Thus the tail sections of the sensitivity curve for the median are negative and positive constants, $\pm 1/[2f\{F^{-1}(0.5)\}]$. As the sample size increases the central part of the sensitivity curve becomes steeper and the sensitivity curve tends to the sign function multiplied by $1/[2f\{F^{-1}(0.5)\}]$.

(b) In part (a) we saw that the average distance between points tends to $1/[nf\{F^{-1}(0.5)\}]$. The addition of a point greater than the old median changes the value of the median by $\frac{1}{2}f\{F^{-1}(0.5)\}$. The addition of a fraction ε of points to the right of the median therefore changes the value of the median by $\varepsilon\frac{1}{2}f\{F^{-1}(0.5)\}$. In conclusion

$$\hat{T}_\infty\left\{(1-\varepsilon)F_\theta + \varepsilon\delta_{y_0}\right\} - \hat{T}_\infty(F_\theta) = \text{sgn}(y-\mu)\varepsilon\frac{1}{2}f\{F^{-1}(0.5)\}$$

when dividing by ε yields

$$\text{IF}_{\text{med}}(y, F_\theta) = \frac{\text{sgn}(y-\mu)}{2f\{F^{-1}(0.5)\}}. \tag{B.1}$$

2.5 IF for the quantile $y_s = F^{-1}(s)$
(a) The function of the s-th quantile is

$$y_s = F^{-1}(s) = \inf\{y|F(y) \geq s\}, \quad 0 < s < 1.$$

(b) Given,
$$F_\varepsilon = (1-\varepsilon)F + \varepsilon G$$

we want to find $\frac{d}{d\varepsilon}F_\varepsilon^{-1}(s)$. We start from the identity

$$F_\varepsilon\{F_\varepsilon^{-1}(s)\} = s$$

and differentiate this identity with respect to ε. Then

$$\frac{d}{d\varepsilon}[(1-\varepsilon)F\{F_\varepsilon^{-1}(s)\} + \varepsilon G\{F_\varepsilon^{-1}(s)\}] = \frac{d}{d\varepsilon}s$$

$$-F\{F_\varepsilon^{-1}(s)\} + (1-\varepsilon)f\{F_\varepsilon^{-1}(s)\}\frac{d}{d\varepsilon}F_\varepsilon^{-1}(s) + G\{F_\varepsilon^{-1}(s)\} = 0$$

$$(1-\varepsilon)f\{F_\varepsilon^{-1}(s)\}\frac{d}{d\varepsilon}F_\varepsilon^{-1}(s) = F\{F_\varepsilon^{-1}(s)\} - G\{F_\varepsilon^{-1}(s)\}.$$

When $\varepsilon \to 0$, $F_\varepsilon^{-1}(s) \to F^{-1}(s)$ and we obtain

$$\frac{d}{d\varepsilon}F_\varepsilon^{-1}(s) = \frac{s - G\{F^{-1}(s)\}}{f\{F^{-1}(s)\}}.$$

When $G = \delta_y$ is the pointmass 1 at y, the influence function of T_s is

$$\text{IF}_{y_s}(y, F_\theta) = \begin{cases} \frac{s-1}{f\{F^{-1}(s)\}} & \text{for } y < F^{-1}(s) \\ \frac{s}{f\{F^{-1}(s)\}} & \text{for } y > F^{-1}(s). \end{cases} \quad (B.2)$$

When $s = 1/2$ we obtain the IF for the median of Eq. (B.1).

2.6 IF for the truncated mean
(a) The function for the truncated mean is

$$T(F) = \frac{1}{1-2\alpha}\int_\alpha^{1-\alpha} F^{-1}(s)ds \quad 0 < s < 1 \quad (B.3)$$

$$= \frac{1}{1-2\alpha}\int_{F^{-1}(\alpha)}^{F^{-1}(1-\alpha)} y\,dFy. \quad (B.4)$$

(b) Using result (B.2), the IF of the truncated mean can be written as

$$\text{IF}_{\bar{y}_\alpha}(y, F_\theta) = \frac{1}{1-2\alpha}\int_\alpha^{1-\alpha} \text{IF}_{y_s}(y, F_\theta)\,ds = \frac{1}{1-2\alpha}\int_\alpha^{1-\alpha} \frac{s - 1_{y<F^{-1}(s)}}{f\{F^{-1}(s)\}}ds.$$

where $1_{(\cdot)}$ is an indicator function (5.9).

In this case it is necessary to distinguish the three cases $F(y) < \alpha, \alpha \le F(y) \le 1-\alpha$ and $F(y) > 1-\alpha$. When $F(y) < \alpha$ we must solve the following integral:

$$\int_\alpha^{1-\alpha} \frac{s-1}{f\{F^{-1}(s)\}}ds.$$

After the change of variable $s = F(y)$, $ds = f(y)dy$, $y = F^{-1}(s)$, $ds = f\{F^{-1}(s)\}dy$. The extremes of integration change from α and $1-\alpha$ to $F^{-1}(\alpha)$ and $F^{-1}(1-\alpha)$. The previous integral can then be rewritten as:

$$\int_{F^{-1}(\alpha)}^{F^{-1}(1-\alpha)} (s-1) dF^{-1}(s)$$

or

$$\int_{F^{-1}(\alpha)}^{F^{-1}(1-\alpha)} s\, dF^{-1}(s) - \{F^{-1}(1-\alpha) - F^{-1}(\alpha)\}. \tag{B.5}$$

Now, using the formula for integration by parts $\int u\, dv = uv - \int v\, du$, where $u = s$ and $dv = dF^{-1}(s)$,

$$sF^{-1}(s)\Big|_{F^{-1}(\alpha)}^{F^{-1}(1-\alpha)} - \int_{\alpha}^{1-\alpha} F^{-1}(s) ds - \{F^{-1}(1-\alpha) - F^{-1}(\alpha)\}$$

$$= \{F^{-1}(1-\alpha)\}(1-\alpha) - \{F^{-1}(\alpha)\}\alpha - \int_{\alpha}^{1-\alpha} F^{-1}(s) ds + F^{-1}(\alpha) - F^{-1}(1-\alpha).$$

In summary, when $F(y) < \alpha$, the IF for the truncated mean can be written as:

$$\text{IF}_{\bar{y}_\alpha}(y, F_\theta) = \frac{1}{1-2\alpha}\{F^{-1}(\alpha) - W(F)\} \quad \text{for} \quad y < F^{-1}(\alpha)$$

where

$$W(F) = \int_{\alpha}^{1-\alpha} F^{-1}(s) ds + \alpha F^{-1}(\alpha) + \alpha F^{-1}(1-\alpha)$$
$$= (1-2\alpha) T(F) + \alpha F^{-1}(\alpha) + \alpha F^{-1}(1-\alpha).$$

When $y > F^{-1}(1-\alpha)$ the integral which has to be solved is

$$\int_{F^{-1}(\alpha)}^{F^{-1}(1-\alpha)} s\, dF^{-1}(s),$$

that is Eq. (B.5) without the final term in braces. It is easy to see that in this case

$$\text{IF}_{\bar{y}_\alpha}(y, F_\theta) = \frac{1}{1-2\alpha}\{F^{-1}(1-\alpha) - W(F)\}.$$

When $F^{-1}(\alpha) < F^{-1}(y) < F^{-1}(1-\alpha)$ we have to solve the sum of the following two integrals

Solutions

$$\int_{F^{-1}(\alpha)}^{F^{-1}(y)} s dF^{-1}(s) + \int_{F^{-1}(y)}^{F^{-1}(1-\alpha)} (s-1) dF^{-1}(s).$$

We obtain

$$sF^{-1}(s)\Big|_{F^{-1}(\alpha)}^{F^{-1}(y)} - \int_{\alpha}^{F(y)} F^{-1}(s) ds +$$
$$sF^{-1}(s)\Big|_{F^{-1}(y)}^{F^{-1}(1-\alpha)} - \int_{F(y)}^{1-\alpha} F^{-1}(s) ds - \{F^{-1}(1-\alpha) - y\}$$
$$= \{F^{-1}(y)\}y - \{F^{-1}(\alpha)\}\alpha + \{F^{-1}(1-\alpha)\}(1-\alpha) - \{F^{-1}(y)\}y$$
$$- \int_{\alpha}^{1-\alpha} F^{-1}(s) ds - \{F^{-1}(1-\alpha) - y\} =$$
$$y - \left(\int_{\alpha}^{1-\alpha} F^{-1}(s) ds + \alpha F^{-1}(\alpha) + \alpha F^{-1}(1-\alpha) \right) = y - W(F).$$

As expected, the slope of the IF of the truncated mean in the interval $F^{-1}(\alpha) \le y \le F^{-1}(1-\alpha)$ is the same as that of the arithmetic mean. In summary, the IF for the truncated mean can be written as

$$\text{IF}_{\bar{y}_\alpha}(y, F_\theta) = \begin{cases} \frac{1}{1-2\alpha}\{F^{-1}(\alpha) - W(F)\} & y < F^{-1}(\alpha) \\ \frac{1}{1-2\alpha}\{y - W(F)\} & F^{-1}(\alpha) \le y \le F^{-1}(1-\alpha). \\ \frac{1}{1-2\alpha}\{F^{-1}(1-\alpha) - W(F)\} & y > F^{-1}(1-\alpha) \end{cases}$$

If F is a continuous unimodal density symmetric around 0, $W(F) = 0$.
(c) If F is a continuous unimodal density symmetric around μ, $W(F) = \mu$. This proves Eq. (2.8).

2.7 ARE when the underlying distribution is normal or Student's t
(a) When the data come from the normal distribution $\mathcal{N}(\mu, \sigma^2)$

$$\begin{aligned}
\text{var(med)} &= \frac{1}{4\left[f\{F^{-1}(0.5)\}\right]^2} \\
&= \frac{1}{4\{f(\mu)\}^2} \\
&= \frac{1}{4\left(\frac{1}{\sigma\sqrt{2\pi}}\right)^2} \\
&= \frac{2\pi\sigma^2}{4} \\
&= \frac{\pi}{2}\sigma^2.
\end{aligned}$$

Therefore

$$\text{ARE}(\text{med}; \bar{y}) = \frac{\sigma^2}{\frac{\pi}{2}\sigma^2}$$
$$= \frac{2}{\pi} \approx 0.64.$$

When the data come from the Student's t distribution with ν degrees of freedom:

$$\text{var}(t_\nu) = \frac{\nu}{\nu - 2} \qquad \nu > 2.$$

When $\nu = 2$ $\quad$ ARE(med; $\bar{y}$) $= \infty$.

$$\text{var}(\text{med}) = \frac{1}{4\left[f\{F^{-1}(0.5)\}\right]^2}$$
$$= \frac{1}{4\{f(\mu)\}^2}$$
$$= \frac{1}{4\left[\frac{\Gamma\{(\nu+1)/2\}}{\sqrt{\nu\pi}\,\Gamma(\nu/2)}\right]^2},$$

where Γ is the Gamma function. Remembering that $\Gamma(1) = 1$, $\Gamma(1/2) = \sqrt{\pi}$, $\Gamma(n) = (n-1)!$ when n is a positive integer and that $\Gamma(n) = (n-1)\Gamma(n-1)$, we obtain

$$\frac{\Gamma\{(\nu+1)/2\}}{\sqrt{\nu\pi}\,\Gamma\left(\frac{\nu}{2}\right)} = \frac{8}{3\sqrt{5\pi}} = 0.3796 \quad \text{if} \quad \nu = 5,$$
$$= \frac{3}{8} \quad \text{if} \quad \nu = 4,$$
$$= \frac{2}{\sqrt{3\pi}} \quad \text{if} \quad \nu = 3.$$

$$\text{var}(\text{med}) = \frac{45\pi^2}{256} = 1.7349 \quad \text{if} \quad \nu = 5,$$
$$= \frac{64}{36} = 1.7778 \quad \text{if} \quad \nu = 4,$$
$$= \frac{3\pi^2}{16} = 1.8506 \quad \text{if} \quad \nu = 3.$$

Solutions

$$\text{var}(\bar{y}) = \frac{5}{3} = 1.6667 \quad \text{if} \quad \nu = 5,$$
$$= 2 \quad \text{if} \quad \nu = 4,$$
$$= 3 \quad \text{if} \quad \nu = 3.$$

$$\text{ARE(med; } \bar{y}) = 0.9607 \quad \text{if} \quad \nu = 5,$$
$$= 1.1250 \quad \text{if} \quad \nu = 4,$$
$$= 1.6211 \quad \text{if} \quad \nu = 3.$$

2.8 ARE when not all measurements are equally precise
(a) When the data come from $N(\mu, \sigma^2)$, $\text{var}(\bar{y}) = \sigma^2/n$ (in a fraction $1 - \varepsilon$ of cases). Similarly when the data come from $N\{\mu, (\tau\sigma)^2\}$, $\text{var}(\bar{y}) = \tau^2\sigma^2/n$ (in a fraction ε of cases). The overall asymptotic variance is

$$n \, \text{var}(\bar{y}) = \sigma^2\{(1-\varepsilon) + \varepsilon\tau^2\}.$$

(b) As we have seen in Eq. (2.13) for normal random samples the asymptotic variance of the median is

$$n \, \text{var(med)} = \frac{1}{4\left[f\{F^{-1}(0.5)\}\right]^2} = \frac{1}{[4\{f(m)\}]^2},$$

where $f(m) = 1/(\sigma\sqrt{2\pi})$. For the contaminated normal distribution

$$f(m, \varepsilon) = (1-\varepsilon)\frac{1}{\sigma\sqrt{2\pi}} + \varepsilon\frac{1}{\sigma\tau\sqrt{2\pi}}$$

$$\frac{1}{4}\{1/f(m,\varepsilon)\}^2 = \sigma^2\frac{\pi}{2}\left(\frac{1}{1-\varepsilon+\varepsilon/\tau}\right)^2.$$

Therefore, the asymptotic variance of the median is

$$n \, \text{var(med)} = \sigma^2 \frac{\pi}{2(1-\varepsilon+\varepsilon/\tau)^2}.$$

2.9 Rate of convergence
If Y denotes a random variable with mean μ and variance σ^2, then using the Chebyshev inequality we have that

$$\Pr(|Y - \mu| > \sigma \cdot m) \leq \frac{1}{m^2} \quad \text{for all} \quad m > 0.$$

Therefore, if $\hat{\theta}_n$ is an unbiased estimator of an unknown parameter θ satisfying $\text{var}\left(\hat{\theta}_n\right) = Cn^{-1}$ for some $0 < C < \infty$, applying the Chebyshev inequality we have

$$\Pr\left(\left|\hat{\theta}_n - \theta\right| > n^{-1/2}\sqrt{C} \cdot \frac{1}{\sqrt{\epsilon}}\right) \leq \epsilon \text{ for all } \epsilon > 0.$$

The sample mean $\bar{y} = \bar{y}_n$ is an unbiased estimator of μ with variance $\text{var}(\bar{y}) = \sigma^2/n$. For large n we have by the central limit theorem that, approximately, $\sqrt{n}(\bar{y} - \mu) \sim N(0, \sigma^2)$. Therefore, for example:

- with $\epsilon = 0.05$ we obtain

$$\Pr\left(|\bar{y}_n - \mu| \geq 1.96\sigma \cdot n^{-1/2}\right) = 0.05,$$

- with $\epsilon = 0.01$ we obtain

$$\Pr\left(|\bar{y}_n - \mu| \geq 2.58\sigma \cdot n^{-1/2}\right) = 0.01.$$

Generalizing this argument for all possible $\epsilon > 0$, we can conclude that $\bar{Y} - \mu = O_P\left(n^{-1/2}\right)$. On the other hand for any $r > 1/2$ we have $n^{-r}/n^{-1/2} \to 0$ as $n \to \infty$. Hence, for any constant $c > 0$

$$\Pr\left(|\bar{Y}_n - \mu| \geq c\sigma \cdot n^{-r}\right) =$$
$$= \Pr\left\{|\bar{Y}_n - \mu| \geq \left(c\sigma \cdot n^{-1/2}\right) \cdot \frac{n^{-r}}{n^{-1/2}}\right\} \to 1 \quad \text{as} \quad n \to \infty.$$

Therefore $n^{-1/2}$ is the rate of convergence of $\bar{y}$.

Problems: Chap. 3

3.1 **Relationship between $\hat{\beta}_{(i)}$ and $\hat{\beta}$**

We start from the Sherman–Morrison–Woodbury formula. Let A be a square $p \times p$ matrix and let U and V be matrices of dimension $p \times m$. Then it is easy to verify that (see for example Atkinson & Riani, 2000)

$$(A - UV^T)^{-1} = A^{-1} + A^{-1}U(I_m - V^T A^{-1} U)^{-1} V^T A^{-1}, \tag{B.6}$$

where it is assumed that all necessary inverses exist. For regression we let $A = X^T X$. The ith row of X is x_i^T. Deletion of this row leaves the matrix $X_{(i)}$, where the subscripted i in parentheses is to be read as "with observation i deleted". With this definition

$$X_{(i)}^T X_{(i)} = (X^T X - x_i x_i^T).$$

Solutions

It then follows from (B.6) that

$$(X_{(i)}^T X_{(i)})^{-1} = (X^T X)^{-1} + (X^T X)^{-1} x_i x_i^T (X^T X)^{-1}/(1 - h_i). \quad \text{(B.7)}$$

The new inverse thus depends on the old inverse $(X^T X)^{-1}$, on x_i and on the leverage measure h_i. The vector of parameter estimates after deletion of observation i is $\hat{\beta}_{(i)}$ defined by

$$\hat{\beta}_{(i)} = (X_{(i)}^T X_{(i)})^{-1}(X^T y - x_i y_i).$$

Using Eq. (B.7),

$$\begin{aligned}
\hat{\beta}_{(i)} &= \left\{(X^T X)^{-1} + (X^T X)^{-1} x_i x_i^T (X^T X)^{-1}/(1 - h_i)\right\} (X^T y - x_i y_i) \\
&= \hat{\beta} + (X^T X)^{-1} x_i \frac{\hat{y}_i - (1 - h_i) y_i - h_i y_i}{1 - h_i} \\
&= \hat{\beta} + (X^T X)^{-1} x_i \frac{\hat{y}_i - y_i + h_i y_i - h_i y_i}{1 - h_i} \\
&= \hat{\beta} - (X^T X)^{-1} x_i \frac{e_i}{1 - h_i}.
\end{aligned}$$

3.2 Relationship between $s_{(i)}^2$ and s^2

(a) We start from the following equations

$$(n - p)s^2 = y^T y - \hat{\beta}^T X^T y,$$

$$(n - p - 1)s_{(i)}^2 = y^T y - y_i^2 - \hat{\beta}_{(i)}^T (X^T y - x_i y_i).$$

Using Eq. (3.52),

$$\begin{aligned}
(n - p - 1)s_{(i)}^2 &= y^T y - y_i^2 - \left\{\hat{\beta} - (X^T X)^{-1} x_i \frac{e_i}{1 - h_i}\right\}^T (X^T y - x_i y_i) \\
&= y^T y - y_i^2 - \hat{\beta}^T X^T y + \frac{\hat{y}_i e_i}{1 - h_i} + \hat{\beta}^T x_i y_i - \frac{h_i e_i y_i}{1 - h_i} \\
&= (n - p)s^2 + \frac{-y_i^2(1 - h_i) + \hat{y}_i e_i + \hat{\beta}^T x_i y_i (1 - h_i) - h_i e_i y_i}{1 - h_i} \\
&= (n - p)s^2 + \frac{-y_i^2 + y_i^2 h_i + \hat{y}_i y_i - \hat{y}_i^2 + \hat{y}_i y_i - h_i \hat{y}_i y_i - h_i y_i^2 + h_i y_i \hat{y}_i}{1 - h_i} \\
&= (n - p)s^2 - \frac{y_i^2 + \hat{y}_i^2 - 2\hat{y}_i y_i}{1 - h_i} \\
&= (n - p)s^2 - \frac{e_i^2}{1 - h_i}.
\end{aligned}$$

(b) The quantities $\hat{\beta}_{(i)}$ and $s^2_{(i)}$, by definition, do not contain any information about y_i, because they are computed excluding observation i and therefore are independent of y_i.

3.3 Relationship between deletion residual r_i^* and studentized residual $\tilde{r}_i$

From Eq. (3.52) we have

$$y_i - x_i^T \hat{\beta}_{(i)} = y_i - x_i^T \hat{\beta} + \frac{x_i^T (X^T X)^{-1} x_i e_i}{1 - h_i}$$

$$= e_i + \frac{h_i}{1 - h_i} e_i$$

$$= \frac{e_i}{1 - h_i}.$$

Using Eq. (B.7),

$$1 + x_i^T (X_{(i)}^T X_{(i)})^{-1} x_i = 1 + h_i + \frac{h_i^2}{1 - h_i}$$

$$= \frac{1}{1 - h_i}.$$

Using these results we can write

$$\frac{y_i - x_i^T \hat{\beta}_{(i)}}{\sqrt{1 + x_i^T (X_{(i)}^T X_{(i)})^{-1} x_i}} = \frac{e_i/(1 - h_i)}{\sqrt{1/(1 - h_i)}}$$

$$= \frac{y_i - \hat{y}_i}{\sqrt{1 - h_i}}.$$

Then $r_i^* s_{(i)} = \tilde{r}_i s$.

3.4 Interpretation of $h_i/(1 - h_i)$

Using Eq. (3.52) we can write:

$$\hat{y}_i = x_i^T \hat{\beta} = x_i^T \left\{ \frac{e_i}{1 - h_i} (X^T X)^{-1} x_i + \hat{\beta}_{(i)} \right\}$$

$$= h_i \frac{e_i}{1 - h_i} + x_i^T \hat{\beta}_{(i)}$$

$$= h_i (y_i - x_i^T \hat{\beta}_{(i)}) + x_i^T \hat{\beta}_{(i)}$$

$$= (1 - h_i) x_i^T \hat{\beta}_{(i)} + h_i y_i.$$

3.5 r_i^* and mean shift outlier model

Let $H_0 : E(Y) = X\beta$ and $H_1 : E(Y) = X\beta + d\theta$. Under the normality assumption the F statistic for testing H_0 versus H_1 is

Solutions

$$F = \frac{\{SS(e_0) - SS(e_1)\}/1}{SS(e_1)/(n-p-1)}, \tag{B.8}$$

where $SS(e_j)$ is the residual sum of squares under the hypothesis H_j $j = \{0, 1\}$. Using the identity in Eq. (3.53) we find that

$$F = \frac{e_i^2/(1-h_i)}{s_{(i)}^2} = r_i^{*2}. \tag{B.9}$$

Note that since $F \sim F_{1,n-p-1}$, r_i^* has a t distribution on $n - p - 1$ degrees of freedom, that is, t_{n-p-1}.

3.6 Find s_w^2 in the added variable test

The normal equations of the partitioned model (3.19) are

$$X^T X \hat{\beta} + X^T w \hat{\gamma} = X^T y \tag{B.10}$$

and

$$w^T X \hat{\beta} + w^T w \hat{\gamma} = w^T y. \tag{B.11}$$

If the model without γ can be fitted, $(X^T X)^{-1}$ exists and

$$\hat{\beta} = (X^T X)^{-1} X^T y - (X^T X)^{-1} X^T w \hat{\gamma}. \tag{B.12}$$

Now $(n - p - 1)s_w^2$ is given by

$$\begin{aligned}(n-p-1)s_w^2 &= (y - X\hat{\beta} - w\hat{\gamma})^T (y - X\hat{\beta} - w\hat{\gamma}) & \text{(B.13)}\\ &= y^T y - y^T X\hat{\beta} - y^T w\hat{\gamma} & \text{(B.14)}\\ &\quad -\hat{\beta}^T X^T y + \hat{\beta}^T X^T X\hat{\beta} + w^T X\hat{\beta}\hat{\gamma} & \text{(B.15)}\\ &\quad + w^T X\hat{\beta}\hat{\gamma} - w^T y\hat{\gamma} + \hat{\gamma}^2 w^T w & \text{(B.16)}\\ &= y^T y - y^T X\hat{\beta} - y^T w\hat{\gamma} & \text{(B.17)}\\ &\quad + \hat{\beta}^T \underbrace{\left(-X^T y + X^T X\hat{\beta} + X^T w\hat{\gamma}\right)}_{=0,\ \text{equation (A.10)}} & \text{(B.18)}\\ &\quad + \hat{\gamma}\underbrace{\left(w^T X\hat{\beta} - w^T y + \hat{\gamma}w^T w\right)}_{=0,\ \text{Eq. (A.11)}} & \text{(B.19)}\\ &= y^T y - \hat{\beta}^T X^T y - \hat{\gamma}w^T y. & \text{(B.20)}\end{aligned}$$

Now, using expressions (B.12) and (3.20),

$$(n - p - 1)s_w^2 = y^T y - \{y^T X(X^T X)^{-1} - \hat{\gamma} w^T X(X^T X)^{-1}\} X^T y \quad (B.21)$$

$$- \frac{w^T A y}{w^T A w} w^T y \quad (B.22)$$

$$= y^T (I_n - H) y + \frac{w^T A y w^T H y - w^T A y w^T y}{w^T A w} \quad (B.23)$$

$$= y^T A y - \frac{w^T A y w^T (I_n - H) y}{w^T A w} \quad (B.24)$$

$$= y^T A y - \frac{(w^T A y)^2}{w^T A w}. \quad (B.25)$$

3.7 Comparison of the distributions of squared deletion residuals and squared studentized residuals

When $n = 50$ and $p = 5$ the density of the squared deletion residual is $\mathcal{F}(1, 44)$ while the density of the squared studentized residual is $\mathcal{B}(1/2, 22)$. Figure B.4 compares these two densities when the x axis coordinates lie in the interval $(0\ 0.6]$ and the y axis coordinates are in the interval $[0\ 20]$. Both of the densities are unbounded in the point $x = 0$. The figure shows that the difference between the two densities decreases as $x \to \infty$.

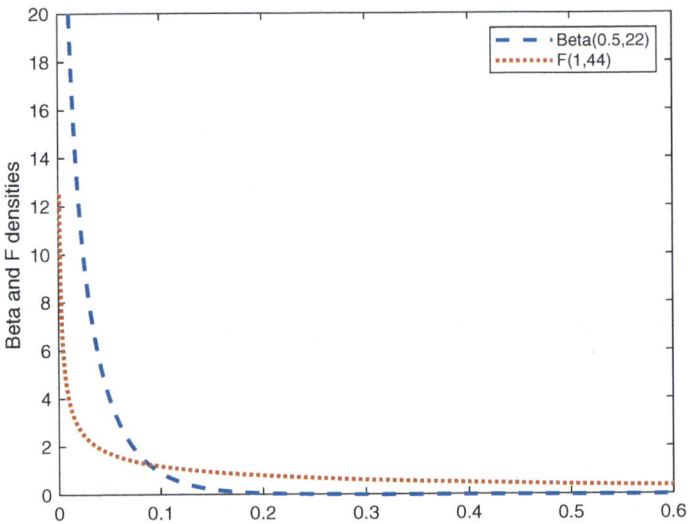

Fig. B.4 Comparison between the distribution of the squared deletion residual (dotted line) and of the squared studentized residual (dashed line) when $n = 50$ and $p = 5$. The x axis coordinates lie in the interval $(0\ 0.6]$ and the y axis coordinates are in the interval $[0\ 20]$

The only difference between $\tilde{r}_i$ and r_i^* is the estimate of σ. From Eq. (3.12)

$$s_{(i)}^2 = \{(n-p)s^2 - e_i^2/(1-h_i)\}/(n-p-1),$$

so that it follows that the difference between $s_{(i)}^2$ and s^2 gets larger as $h_i \to 1$.

3.8 Value of γ in Sect. 3.7.2 when the data come from the normal distribution

When the data come from the standard normal distribution $\rho(y) = y^2/2$ so that $\psi(y) = y$ and $\psi'(y) = 1$. Therefore when $n \to \infty$, $E\psi(r_i^2) = E(e_i/\hat{\sigma})^2 \to 1$ and $\{E\psi'(r_i)\}^2 = 1$.

3.9 Computation of the tuning constant c for the Tukey biweight

Conditions (3.35) and (3.32) imply that c for the Tukey biweight must satisfy the following equation:

$$\int_{-c}^{c}\left(\frac{x^2}{2} - \frac{x^4}{2c^2} + \frac{x^6}{6c^4}\right) d\Phi_{0,1}(x) + \frac{c^2}{6}\Pr(|X| > c) = bdp\frac{c^2}{6}. \tag{B.26}$$

Using the expression for the expectation of the k-th order truncated central moment given in (3.40), Eq. (B.26) can be rewritten as

$$6\left[\frac{\Pr(\chi_3^2 < c^2)}{2} - 3\frac{\Pr(\chi_5^2 < c^2)}{2c^2} + 15\frac{\Pr(\chi_7^2 < c^2)}{6c^4} + \frac{c^2}{3}\{1 - \Phi(c)\}\right] = bdp\,c^2. \tag{B.27}$$

Simplifying, the result follows immediately.

The constant c associated with a nominal efficiency eff must satisfy Eq. (3.37). Given the expressions for $\psi'(x)$ (3.38) and $\{\psi(x)\}^2$ (3.39) we obtain

$$\int_{-c}^{c}\psi'(x)d\Phi(x) = 15\frac{\Pr(\chi_5^2 < c^2)}{c^4} - 6\frac{\Pr(\chi_3^2 < c^2)}{c^2} + \Pr(\chi_1^2 < c^2). \tag{B.28}$$

Similarly,

$$\int_{-c}^{c}\{\psi(x)\}^2 d\Phi(x) = 9!!\frac{\Pr(\chi_{11}^2 < c^2)}{c^8} - 4 \times 7!!\frac{\Pr(\chi_9^2 < c^2)}{c^6} + 6 \times 5!!\frac{\Pr(\chi_7^2 < c^2)}{c^4}$$
$$-4 \times 3!!\frac{\Pr(\chi_5^2 < c^2)}{c^2} + \Pr(\chi_3^2 < c^2). \tag{B.29}$$

Taking the square of Eq. (B.28) and dividing by Eq. (B.29), the result follows.

3.10 First truncated moment in the folded normal distribution

When $k = 1$ Eq. (3.41) reduces to

$$\int_{a<|x|<b}|x|d\Phi(x) = \sqrt{\frac{2}{\pi}}\left\{F_{\chi_2^2}(b^2) - F_{\chi_2^2}(a^2)\right\}.$$

Now, remembering that
$$F_{\chi_2^2}(x) = 1 - e^{-x/2}$$
we obtain
$$\int_{a<|x|<b} |x|d\Phi(x) = 2\frac{1}{\sqrt{2\pi}}\left(e^{-a^2/2} - e^{-b^2/2}\right) = 2\{\phi(a) - \phi(b)\}. \quad \text{(B.30)}$$

If $a = 0$ and $b = \infty$ we get $E(|X|) = \sqrt{2/\pi}$, the usual formula for the expectation of the half-normal distribution.

3.11 Equality of the truncated mean in a folded normal distribution to the MAD in a standard normal distribution

The MAD in a $\mathcal{N}(0, 1)$ distributon (as we have seen in Exercise 1.3) is the median of the folded normal distribution and equals $\Phi^{-1}(3/4)$. The tuning constant c, which imposes the constraint that the truncated mean must be equal to the MAD in a folded normal sample, must satisfy

$$\int_{0<|x|<c} |x|d\Phi(x) = \Phi^{-1}(3/4).$$

Using Eq. (B.30) we obtain

$$2\{\phi(0) - \phi(c)\} = \Phi^{-1}(3/4).$$

Tedious algebra leads to

$$c = \sqrt{-2\log\left\{\frac{2 - \sqrt{2\pi}\Phi^{-1}(3/4)}{2}\right\}}.$$

3.12 Test whether the current solution in Sect. 3.10 is an admissible candidate to find the minimum scale

Given that the function ρ is monotonic, if $\hat{\sigma}_j < \hat{\sigma}_{\min}$, we must have

$$K = \overline{\rho}\{r_{rw}(j)/(c\hat{\sigma}_j)\} \geq \overline{\rho}\{r_{rw}(j)/(c\hat{\sigma}_{\min})\}. \quad \text{(B.31)}$$

Thus, if
$$\overline{\rho}\{r_{rw}(j)/(c\hat{\sigma}_{\min})\} > K \quad \text{(B.32)}$$

we may discard $\hat{\beta}_{rw}(j)$ because in this case $\hat{\sigma}_j > \hat{\sigma}_{\min}$.

Solutions   463

Problems: Chap. 4

4.1 AR data: standard static way of data analysis (non-robust and robust)

(a) The ANOVA based on all the units is shown in Table B.1, while that after removing observation 43 is shown in Table B.2. After deleting observation 43 variable x_1 becomes almost significant. From this standard way of analysing the data it is not possible to understand the presence of subgroups in the data or of hidden structure. Moreover, given that robust methods are not used, the results above might (as they do in this case) suffer from masking and swamping problems.

(b) Figure B.5 shows the typical static output of robust data analysis based on MM-regression. In this case the hyperbolic ρ function and a nominal efficiency of 0.95 are used. The top panel shows the weights, while the bottom panel shows the scaled residuals. This static analysis detects (using a confidence level of 0.975) three out-

Table B.1 AR regression data: ANOVA table based on all the observations

	Estimate	SE	tStat	pValue
(Intercept)	11.174	0.67501	16.553	3.1288e-23
x1	-0.21796	0.17244	-1.264	0.21146
x2	1.4981	0.15534	9.6439	1.6733e-13
x3	2.2596	0.13668	16.531	3.3265e-23

Number of observations: 60, Error degrees of freedom: 56
Root Mean Squared Error: 1.09
R-squared: 0.965, Adjusted R-Squared: 0.963
F-statistic vs. constant model: 510, p-value = 1.33e-40

Table B.2 AR regression data: ANOVA table after deleting observation 43

	Estimate	SE	tStat	pValue
(Intercept)	11.724	0.66472	17.637	2.6658e-24
x1	-0.32034	0.1664	-1.9251	0.059389
x2	1.441	0.14773	9.7541	1.3511e-13
x3	2.3627	0.13378	17.661	2.5023e-24

Number of observations: 59, Error degrees of freedom: 55
Root Mean Squared Error: 1.03
R-squared: 0.969, Adjusted R-Squared: 0.967
F-statistic vs. constant model: 570, p-value = 2.19e-41

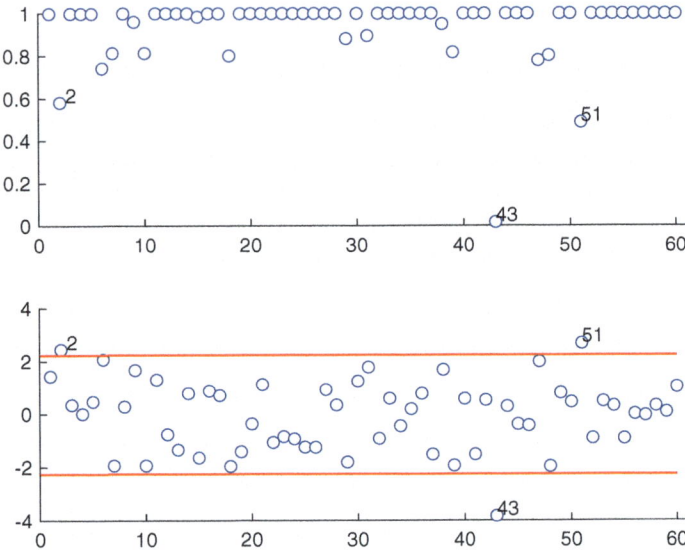

Fig. B.5 Typical output from traditional static robust statistics. AR regression data, MM-estimation: weights and residuals for hyperbolic ρ function and a nominal efficiency of 0.95. Using a 0.975 confidence band, three units are declared as outliers

Table B.3 AR regression data: ANOVA table after deleting observation 2, 43 and 51

```
               Estimate       SE         tStat        pValue
               _____    _____    _____    _____

(Intercept)     12.075      0.63013     19.162      1.7287e-25
x1             -0.44075     0.16022     -2.751       0.0081134
x2              1.2608      0.1493       8.4448      2.1939e-11
x3              2.4626      0.12873     19.131       1.8654e-25


Number of observations: 57, Error degrees of freedom: 53
Root Mean Squared Error: 0.961
R-squared: 0.972,   Adjusted R-Squared: 0.971
F-statistic vs. constant model: 615, p-value = 3.76e-41
```

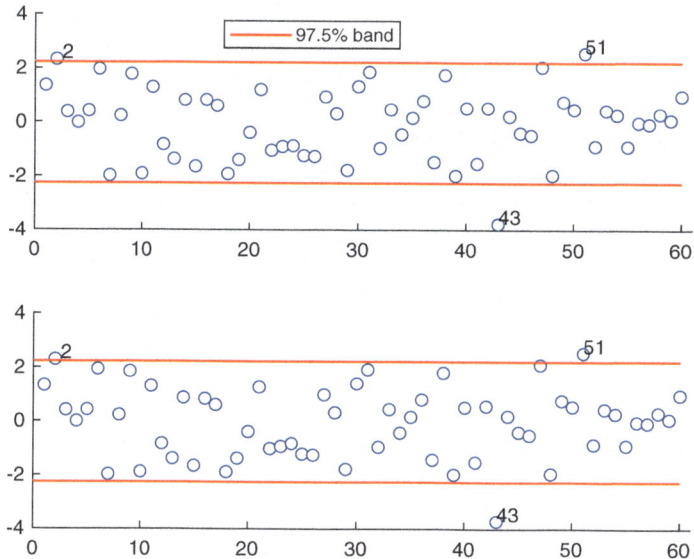

Fig. B.6 AR regression data: scaled residuals for Andrews' sine (upper panel) and power divergence (lower panel) using a nominal efficiency of 0.95 for MM-estimation

liers. While observation 43 is far from the confidence bands, the residuals for units 2 and 51 are close to the confidence band. In this case the swamping effect is even more evident. Note that, from these traditional index plots of weights and residuals, there is no indication of such an effect and also that the residuals which are obtained from the use of the hyperbolic ρ function, are virtually indistinguishable from those based on the Tukey biweight shown in the bottom panel of Fig. 3.9.

The ANOVA table after removing observations 2, 3 and 43 is shown in Table B.3. The coefficient of the first variable remains negative and becomes strongly significant.

(c) Figure B.6 shows the scaled residuals coming from MM-estimation based on a nominal efficiency of 0.95 and Andrews' sine ρ function (top panel) and power divergence (bottom panel). Once again, the residuals which are obtained are virtually identical to those coming from the hyperbolic ρ (Fig. B.5) and the Tukey biweight (bottom panel of Fig. 3.9).

4.2 Hawkins data: brushing residuals from monitoring S residuals

(a) The left-hand panel of Fig. B.7 shows the brushing of the units in pale blue in the left-hand panel of Fig. 4.29. These are the units entering the subset in the first 86 steps (see right-hand panel of Fig. B.7). The brushed units are also shown in the yX plot of Fig. B.8. The brushed units correspond to those shown in Fig. 4.27 as "units not declared as outliers".

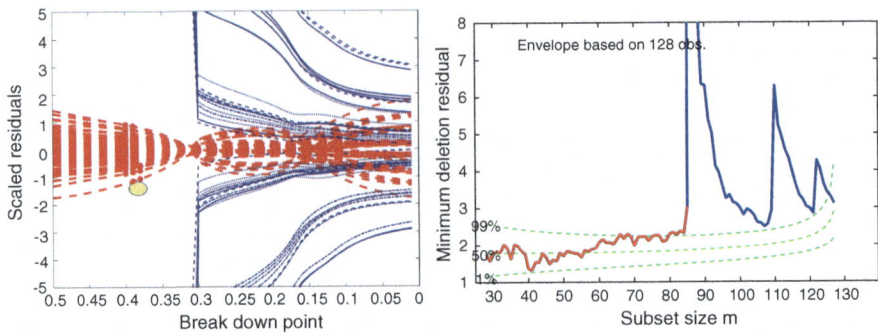

Fig. B.7 Hawkins data: left-hand panel, brushing of the units in pale blue in the left-hand panel of Fig. 4.29. Right-hand panel, order of entry of these units in the plot of minimum deletion residuals

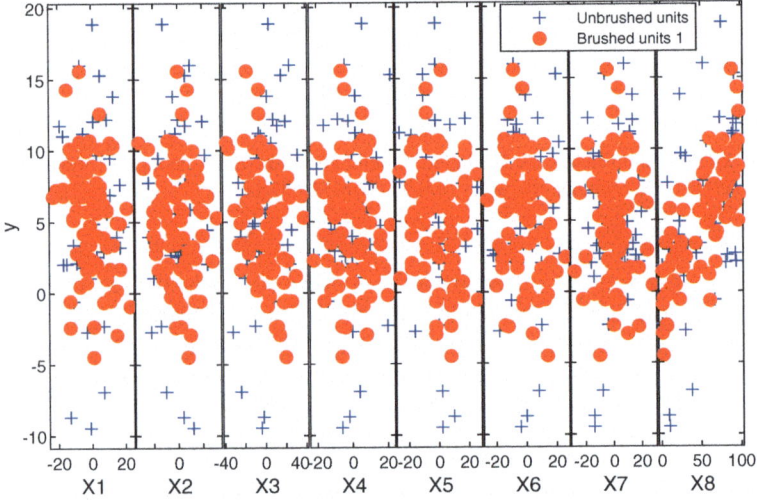

Fig. B.8 Hawkins data: yX plot with units in pale blue in the left-hand panel of Fig. 4.29 highlighted

(b) The left-hand panel of Fig. B.9 shows the brushing of the units with the largest 6 absolute residuals. These are the units entering the subset in the last 6 steps (see right-hand panel of Fig. B.9).

Remark: in order to produce these plots you need both the FS (minimum deletion residuals) and the monitoring of S residuals.

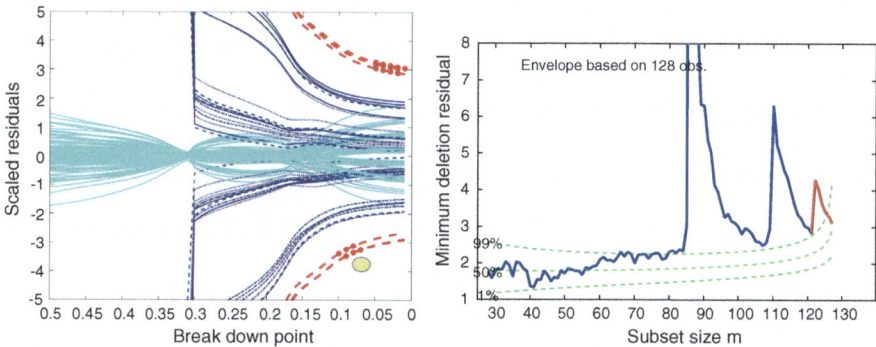

Fig. B.9 Hawkins data: left-hand panel, brushing of the units with the 6 largest absolute residuals in Fig. 4.29; right-hand panel, order of entry of these units in the plot of minimum deletion residuals

4.3 Surgical Unit data: analysis to test the difference between the two groups

Table B.4 shows the ANOVA table for the Surgical Unit data after adding a dummy variable for Unit 1 (variable *dum*). This variable is highly significant and therefore the conclusion is that the origin of the data matters in the prediction of the response. However, from this standard approach, it is not possible to understand if this difference is due to the presence of particular observations, and whether there are subgroups of data.

(b) Figure B.10 shows the monitoring of the five added variable t statistics. It seems that the significance of the dummy variable does not depend on the presence of particular observations.

Table B.4 Surgical Unit data: ANOVA table after adding a dummy variable for Unit 1

	Estimate	SE	tStat	pValue
(Intercept)	0.37855	0.037893	9.9899	8.5715e-17
x1	0.068493	0.0037849	18.097	1.2955e-33
x2	0.0097408	0.00030352	32.093	4.0681e-55
x3	0.0096583	0.0002738	35.275	5.821e-59
x4	-0.0008379	0.0068242	-0.12278	0.90252
dum	0.072997	0.0093189	7.8333	4.685e-12

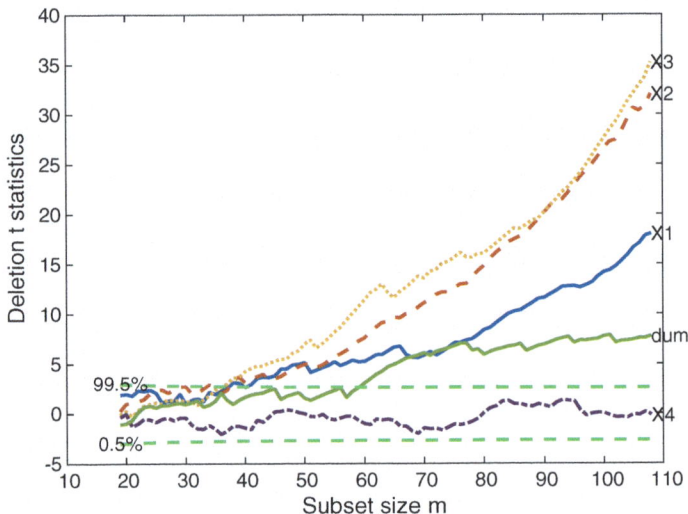

Fig. B.10 Surgical unit data: monitoring plot of the five added variable t-statistics showing the growth of significance of t-statistics for the parameters. The trajectory for the dummy variable is labelled dum

4.4 Surgical Unit data: traditional robust analysis

Figure B.11 shows the index plot of residuals coming from S-estimation and the Tukey biweight for two values of breakdown point. The units declared as outliers strongly depend on the value of *bdp* which is used. When *bdp* is 0.25 (top panel) no unit has a residual outside the 99% confidence band. On the other hand, when *bdp* is 0.5 (bottom panel) there are several units whose residuals are outside the band.

Figure B.12 has the same information as Fig. B.11, but it comes from the use of the power divergence ρ function. These two figures are virtually indistinguishable. This is in agreement with what has been seen so far; the most important parameter is *bdp* and not the choice of the ρ function. When $bdp = 0.5$ it is possible to see that the residuals for the observations in the first Unit are much more concentrated around 0 than those from the second Unit.

4.5 Hawkins data: BIC monitoring

The left-hand panel of Fig. B.13 shows the plot of BICW for LTS estimation and the right-hand panel for FS. Both show the almost linear increase as more observations are included in the fit until a peak for $bdp = 0.33$ which corresponds to 86 observations, and to $h = 86$ for FS, after which there is a sharp decline.

Solutions

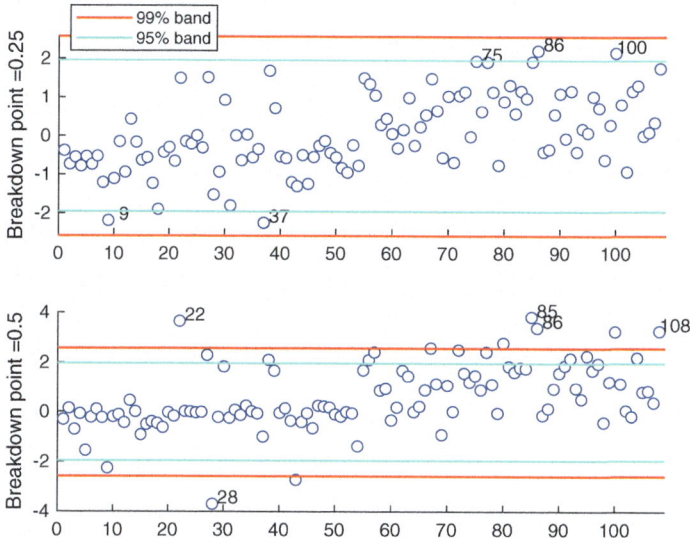

Fig. B.11 Surgical unit data: S-estimation with TB ρ function. Index plot of residuals for two values of *bdp*. Upper panel, $bdp = 0.25$; lower panel, $bdp = 0.5$

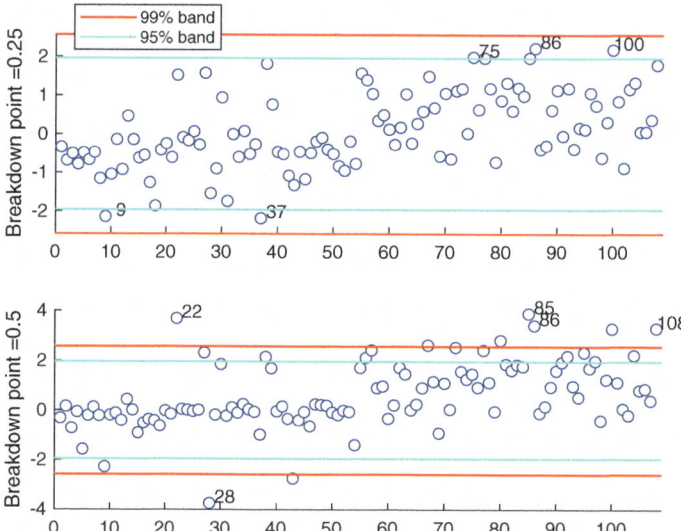

Fig. B.12 Surgical unit data: S-estimation with PD ρ function. Index plot of residuals for two values of *bdp*. Upper panel, $bdp = 0.25$; lower panel, $bdp = 0.5$

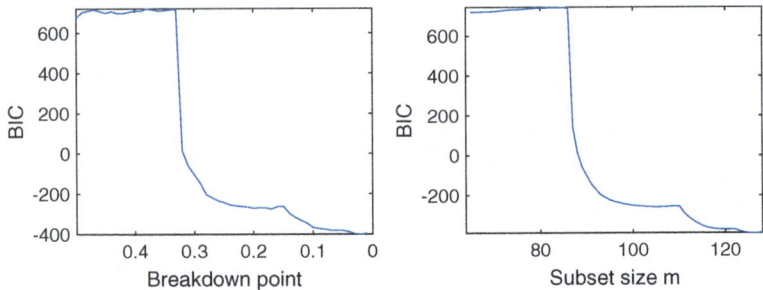

Fig. B.13 Hawkins data: monitoring plots of BICW. Left-hand panel, Least trimmed squares (LTS); right-hand panel, Forward Search (FS). The stepping for LTS arises from a search over 50 values of *bdp*

Problems: Chap. 5

5.1 Multiple tests and the Bonferroni correction
(a)

$$\begin{aligned} \text{Pr(at least one significant result)} &= 1 - \text{Pr(no significant result)} \\ &= 1 - (1 - 0.05)^{20} \\ &\approx 0.6415. \end{aligned}$$

With 20 tests being considered, we have a probability under H_0 of 0.6415 of observing at least one significant result even if all the tests are actually not significant.
(b) Figure B.14 shows how the probability of finding at least one significant result varies as a function of the number of tests and the significance level α. The probability of getting a significant result simply due to chance keeps going up. While for $\alpha = 0.05$ to reach 0.99 it is enough to have at least 90 tests, when $\alpha = 0.01$ and the number of tests is 200 the probability of at least one significant test is 0.866.
(c) The Bonferroni correction sets the significance cut-off at α/n. For example, in the settings of (a) above, with 20 tests and $\alpha = 0.05$, one rejects a null hypothesis if the *p*-value is less than 0.0025.
(d) The probability of observing at least one significant result when using the Bonferroni correction described in point (c) is

$$\begin{aligned} \text{Pr(at least one significant result)} &= 1 - \text{Pr(no significant result)} \\ &= 1 - (1 - 0.0025)^{20} \\ &\approx 0.0488. \end{aligned}$$

The value which is obtained is a bit smaller than the nominal one. In this case we benefitted from the assumption that all tests are mutually independent. In practical applications, this is often not the case. Depending on the correlation structure of the

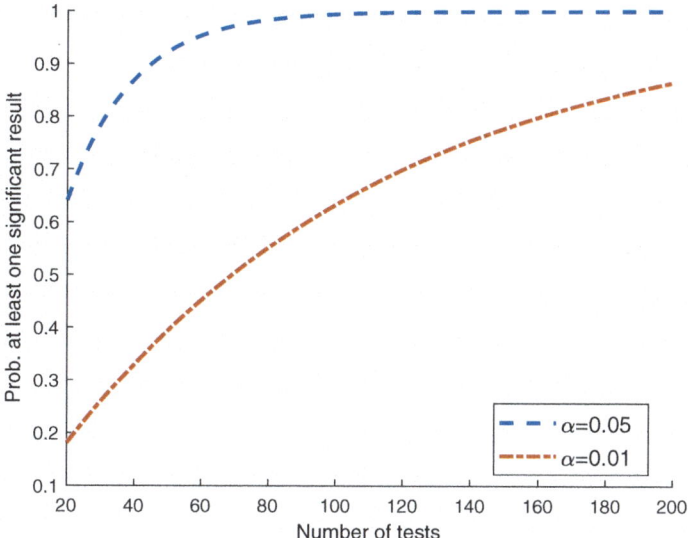

Fig. B.14 Probability under H_0 of finding at least one significant result as function of the number of tests and the significance level α

tests, the Bonferroni correction could be extremely conservative, leading to a high rate of false negatives.

5.2 Bonferroni correction for confidence intervals
The right-hand side of Eq. (5.17) is one minus the sum of the probabilities of each of the intervals missing their true values. Therefore, if simultaneous multiple interval estimates are desired with an overall confidence coefficient $1 - \alpha$, one can construct each interval with confidence coefficient $(1 - \alpha/g)$, and the Bonferroni inequality ensures that the overall confidence coefficient is at least $1 - \alpha$. That is, the Bonferroni correction states that the confidence coefficient is at least $1 - \alpha$ that simultaneously all the g confidence intervals are "correct" (or capture their respective true values).

5.3 Theoretical individual and simultaneous size
Figure B.15 shows the plot of the theoretical individual and simultaneous sizes—based on the χ^2 approximation calculated from (5.7)—for n in the interval [30 5000] in percentages multiplied by 10 when $p = 2$. Figure B.16 is similar to Fig. B.15 but is for $p = 10$. In both plots the horizontal line associated with the nominal size is also given. The distance of the approximate size from the nominal one is much greater for the simultaneous size. Note that the simultaneous size does not tend to 10. The distance between the nominal and the theoretical size increases with p.

5.4 Probability of overlapping
It is assumed that the outlying observations, including the response, have a multivariate normal distribution $W \sim \mathcal{N}(\mu, \Sigma)$. To find the probability of the proportion of outliers in a particular strip around the regression plane, we need the marginal dis-

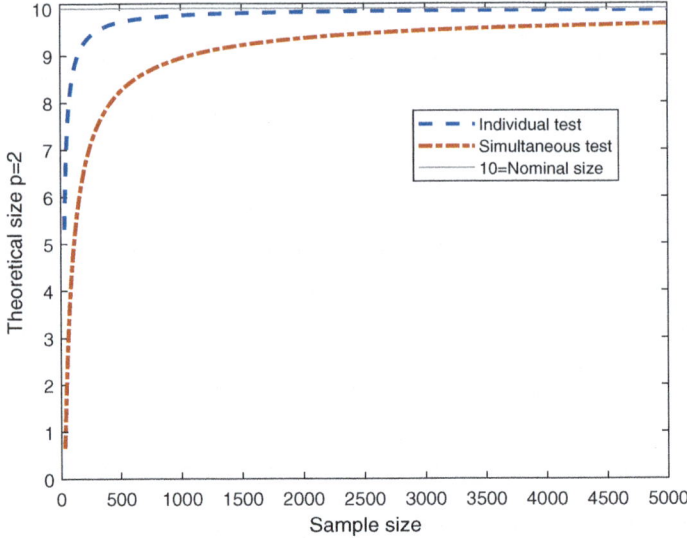

Fig. B.15 Theoretical individual and simultaneous size, percentages multiplied by 10, based on the χ^2 approximation, calculated from (5.7) for n in the interval [30 5000] when $p = 2$

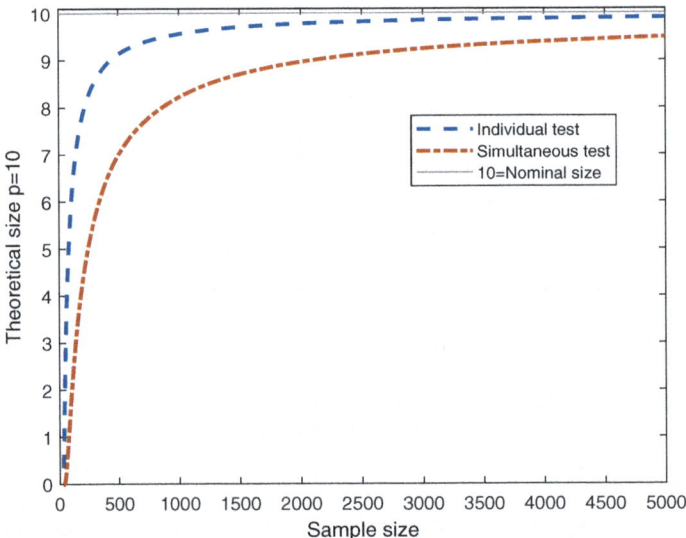

Fig. B.16 Theoretical individual and simultaneous size, percentages multiplied by 10, based on the χ^2 approximation, calculated from (5.7) for n in the interval [30 5000] when $p = 10$

Solutions

tribution of W parallel to the normal to the plane, given by (5.13). The specific value of w_0 is irrelevant; all that is required is the direction. From the particular properties of the multivariate normal distribution, the marginal distribution is given by omitting all remaining variables from the original multivariate distribution. With the required direction given by z_1 (5.13), the marginal distribution is $Z_1 \sim \mathcal{N}(b^T \mu, b^T \Sigma b)$. Note that the calculations may be more complicated if other continuous multivariate distributions are specified.

5.5 Theoretical and empirical overlapping indexes with point mass contamination

Point mass contamination means that the outliers are at a single point. Then both overlapping indexes are one if this point is within the band around the plane and zero otherwise. The results in the two cases may differ, because the band in the empirical index only covers a specified range given by $\mathcal{X}$.

Problems: Chap. 6

6.1 Variance-stabilizing transformation
The first-order Taylor expansion about μ is

$$g(Y_i) \approx g(\mu) + g'(\mu)(Y_i - \mu).$$

Consequently

$$\text{var}[g(Y_i)] \approx \{g'(\mu)\}^2 \text{var}(Y_i) \qquad (B.33)$$
$$\approx \{g'(\mu)\}^2 \mu^{2\alpha}. \qquad (B.34)$$

Now for $\text{var}\{g(Y_i)\}$ to be approximately constant, $g(Y_i)$ must be chosen so that

$$\{g'(\mu)\}^2 \mu^{2\alpha} = \text{const}_1$$

$$g'(\mu) = \text{const}_2 \times \mu^{-\alpha}.$$

So that, on integration,

$$g(\mu) = \begin{cases} \mu^{-\alpha+1} & \text{if } \alpha \neq 1 \\ \log \mu & \text{if } \alpha = 1, \end{cases}$$

since the constant does not matter. For example, if the standard deviation of a variable is proportional to the mean ($\alpha = 1$), a logarithmic transformation (the base is irrelevant) will give a constant variance. If the variance is proportional to the mean ($\alpha = 1/2$), the square-root transformation will give a constant variance and so on. Table B.5 reports the transformation required to stabilize the variance for different values of α.

6.2 Confidence interval for transformation parameter λ
Figure B.17 shows the profile loglikelihood together with the 95% confidence interval for λ based on the change in the profile loglikelihood. The top panel is for the

Table B.5 Transformations to constant variance when the variance depends on the mean

α	$\mathrm{var}(Y) = k\mu^{2\alpha}$	Transformation
0	k	Y
1/2	$k\mu$	$\sqrt{Y}$
1	$k\mu^2$	$\log Y$
3/2	$k\mu^3$	$1/\sqrt{Y}$
2	$k\mu^4$	$1/Y$

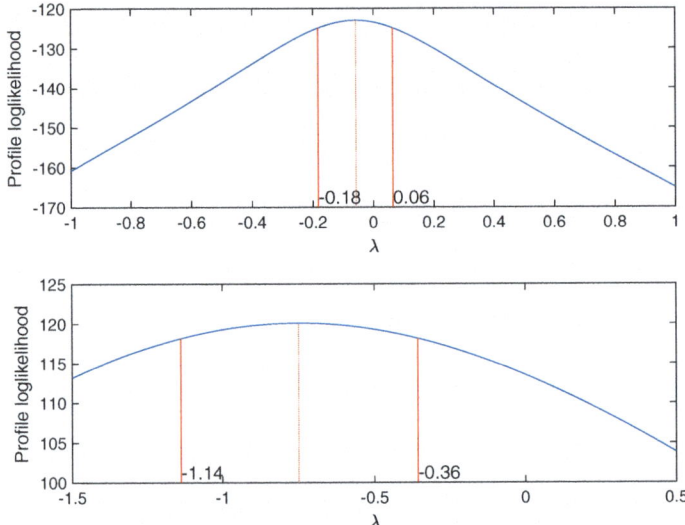

Fig. B.17 Profile loglikelihood together with $\hat{\lambda}$ and the 95% confidence interval for λ. Top panel: wool data, bottom panel: poison data

wool data while the bottom panel is for the poison data. These maximum likelihood confidence intervals are unlikely to remain unchanged in the presence of outliers.

6.3 Further analysis of the Balance Sheet data

Figure B.18 shows the yX plot with normal units (1), 19 brushed units from Fig. 6.14 (2) and the remaining units declared as outliers (3). This plot shows that the remaining units declared as outliers are mostly for extreme values of the response with a very few for extreme values of individual explanatory variables.

6.4 Transformed Balance Sheet data and F test

Figure B.19 shows the monitoring of the F test for untransformed and transformed balance sheet data. Note that the F test has been rescaled using the correction factor based on the variance of the truncated normal distribution. The trajectory of the F test based on transformed data is (as expected) uniformly above that based on

Solutions

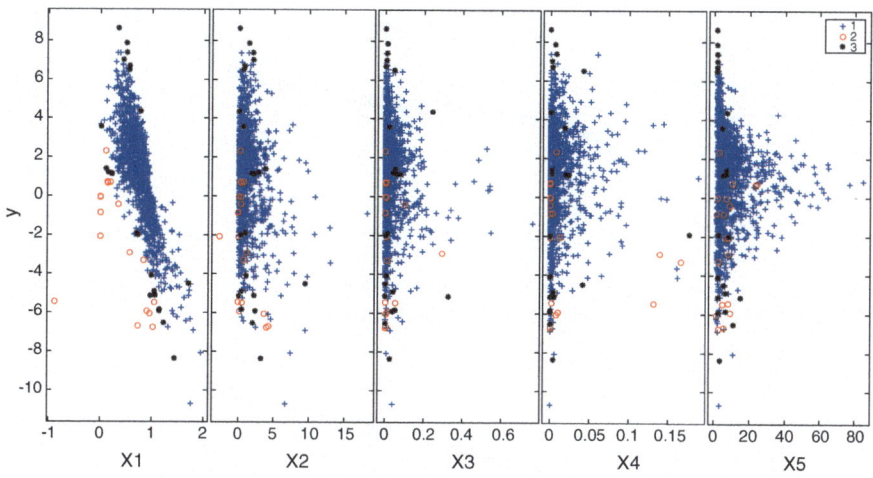

Fig. B.18 Balance sheet data: yX plot with normal units (1), 19 brushed units from Fig. 6.14 (2) and the remaining units declared as outliers (3)

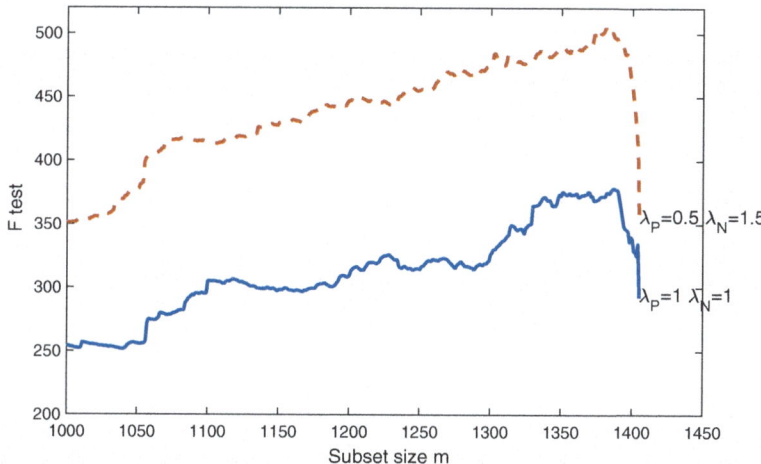

Fig. B.19 Balance sheet data: monitoring of F test (corrected for truncation) for untransformed and transformed data

untransformed data. Note the effect of the outliers which makes the two values much closer as we move towards considering all observations.

The third property sought by the Box-Cox transformation is normality. Figure B.20 compares the QQ-plots of studentized residuals for untransformed data and for transformed data after removing the 42 outliers. The residuals based on transformed data almost all lie inside the 99% confidence envelope.

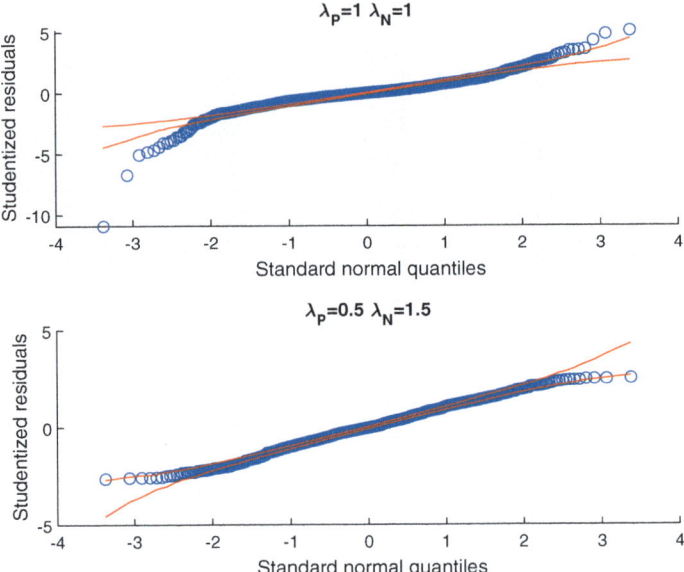

Fig. B.20 Balance sheet data: QQ-plots of studentized residuals for untransformed and transformed data after removing the 42 outliers

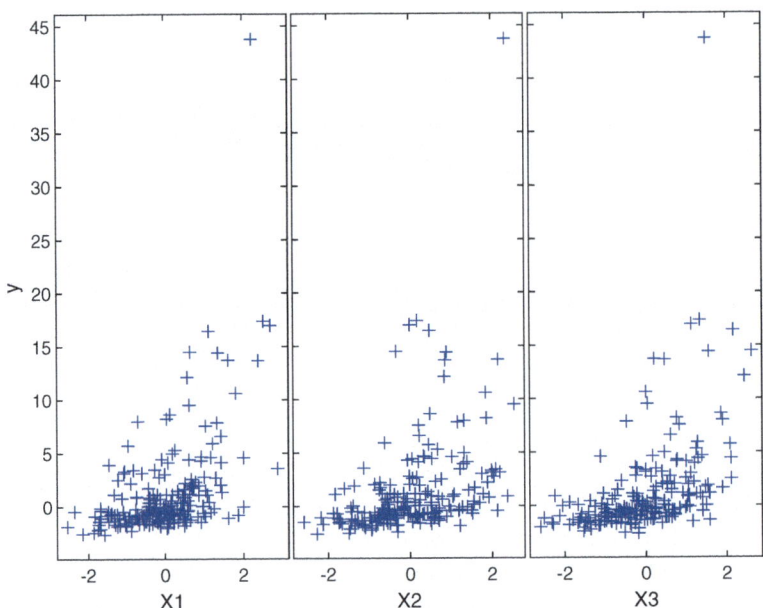

Fig. B.21 D1: yX plot. Since y shows a positively skewed distribution, the data are likely to need transformation

Solutions

Table B.6 D1: ANOVA in the original scale for y

	Estimate	SE	tStat	pValue
(Intercept)	1.474	0.21501	6.8556	9.0202e-11
x1	2.2605	0.21645	10.443	1.3969e-20
x2	1.895	0.20203	9.3799	1.673e-17
x3	2.002	0.20459	9.7853	1.1552e-18

```
Number of observations: 200, Error degrees of freedom: 196
Root Mean Squared Error: 3.04
R-squared: 0.604, Adjusted R-Squared: 0.598
F-statistic vs. constant model: 99.5, p-value = 3.57e-39
```

Table B.7 D1: Yeo-Johnson transformation; value of λ and score test

λ	Score test
−1.0000	57.8295
−0.7500	46.7749
−0.5000	34.7105
−0.2500	22.1881
0	10.3900
0.2500	−0.8932
0.5000	−14.6884
0.7500	−33.2317
1.0000	−51.8021

6.5 Score test and fan plot 1

Figure B.21 shows the yX plot of the data D1: y has a positive skewed distribution. The minimum value of y is smaller than 0 and therefore a Yeo-Johnson transformation should be considered, rather than that of Box and Cox. It is not clear whether the largest value of y can be considered as an outlier. The ANOVA table based on all the observations and untransformed data is given in Table B.6. In this scale the percentage of variance which is explained is around 60%. Table B.7 shows that using all the observations $\lambda = 0.25$ is clearly the best transformation.

The left-hand panel of Fig. B.22 gives the fan plot for the Yeo-Johnson transformation of these data. The value of 0.25 is accepted throughout the search, with the neighbouring values of 0 and 0.5 rejected at the 1% level from around $m = 150$ and 160. The right-hand panel of Fig. B.22 presents the extension of this plot for testing whether negative and positive responses require distinct transformations. These two constructed variables each give rise to a score test and so to an additional line on the extended fan plot, the upper green line of the three corresponding to testing positive y. For the indicated Yeo-Johnson transformation, $\lambda = 0.25$, the two test statistics are close to that for the overall transformation; the transformation has achieved approx-

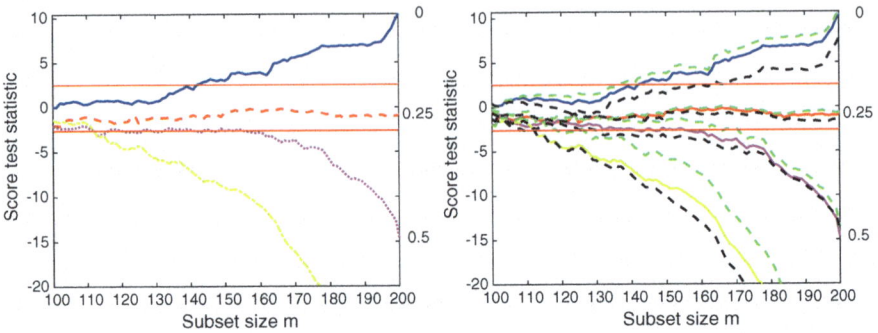

Fig. B.22 D1: left-hand panel, fan plots for, reading down at the right, $\lambda_0 = 0, 0.25, 0.5$ and 1. Right-hand panel, extended fan plots for the homogeneity of transformation for the same four values of λ; the highest trajectory in green for each λ comes from testing the transformation of the positive observations

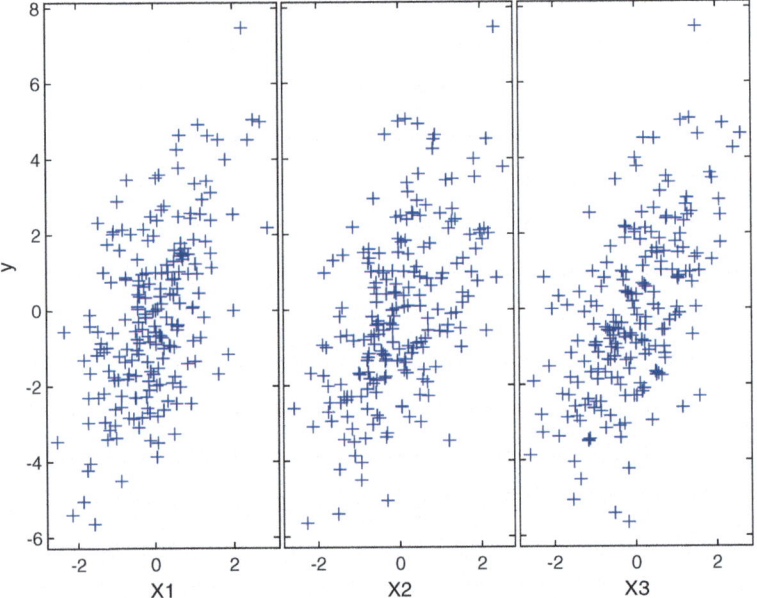

Fig. B.23 D1: yX plot in the transformed scale for y

imate symmetry. For the rejected transformation values of 1 and 0, and to a lesser extent for 0.5, the spread between the values of the positive and negative statistics is greater and decreasingly symmetrical around the value for the overall transformation. The automatic procedure which searches for two values of λ briefly described in Sect. 6.3.6 confirms that 0.25 is the best value both for positive and negative observations.

Table B.8 D1: ANOVA in the transformed scale for y ($\lambda = 0.25$)

	Estimate	SE	tStat	pValue
(Intercept)	-0.0067019	0.04161	-0.16107	0.87221
x1	1.1927	0.041887	28.474	1.4341e-71
x2	1.1779	0.039097	30.128	1.7303e-75
x3	1.2068	0.039593	30.479	2.6612e-76

```
Number of observations: 200, Error degrees of freedom: 196
Root Mean Squared Error: 0.588
R-squared: 0.932,  Adjusted R-Squared: 0.931
F-statistic vs. constant model: 897, p-value = 3.66e-114
```

Figure B.23 shows the yX plot after transforming the response with $\lambda = 0.25$. Approximate normality has been achieved and no outliers seem to be present in the dataset. The ANOVA table based on all the observations and transformed data using $\lambda = 0.25$ is given in Table B.8. In this scale the percentage of variance which is explained is above 90%. The analysis of the p-values of the t-statistics shows that the explanatory variables are now much more significant.

6.6 Score test and fan plot 2

Figure B.24 shows the yX plot of the data D2: y again has a positive skewed distribution. The minimum value of y is smaller than 0 and therefore a Yeo-Johnson transformation should again be considered. It is not clear whether the larger value of y can be considered as outliers or if they can be reconciled with the rest of the data after transformation. The ANOVA table based on all the observations and untransformed data is given in Table B.9. In this scale the percentage of variance which is explained is below 28%. Table B.10 shows that, using all the observations, the best Yeo-Johnson transformation is clearly $\lambda = 0$.

The left-hand panel of Fig. B.25 gives the fan plot for the Yeo-Johnson transformation of these data. For $\lambda = 0.25$ the statistic lies within the band up to $m = 180$. Thereafter, as the remaining observations enter, the value of 0.25 is increasingly rejected. The value of $\lambda = 0$ is accepted when all the observations are included but lies well outside the bands starting from $m = 130$. The trajectory for $\lambda = 0.5$ shows the more complicated form that arises when, because of the incorrect transformation, some influential observations enter the subset more than 20 steps from the end of the search. The extended fan plot is shown in the right-hand panel of the figure. It is interesting that, in the searches with $\lambda = 0$ and $\lambda = 0.25$, the green dashed lines which come from testing the transformation of the positive observations are (as expected, see Sect. 6.2.4) above the black dashed lines which are associated with the negative observations up to $m = 180$. In the last 20 steps the order is reversed. Once the last 20 units start to enter for $\lambda = 0.25$, the upper test is close to the overall test while the other separates. The jumps in the curves due to the outliers are noticeable in the

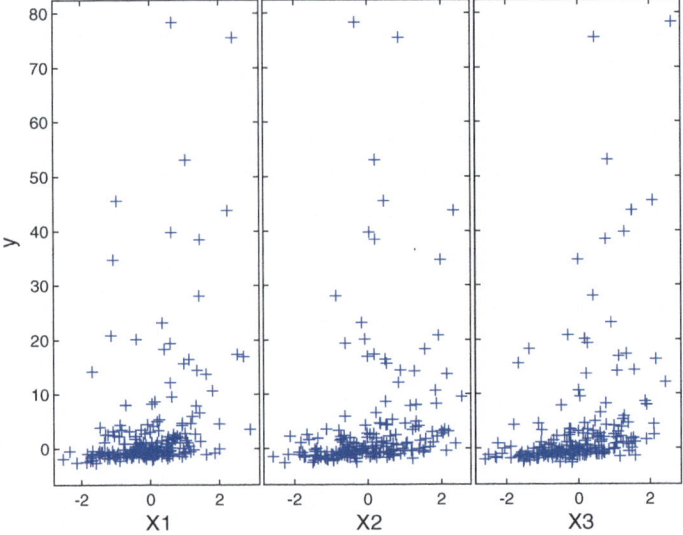

Fig. B.24 D2: yX plot. Again y shows a positive skewed distribution. The data are likely to need transformation

Table B.9 D2: ANOVA in the original scale for y

	Estimate	SE	tStat	pValue
(Intercept)	4.0076	0.69215	5.79	2.7595e-08
x1	3.4516	0.69676	4.9538	1.5678e-06
x2	2.6805	0.65036	4.1215	5.5519e-05
x3	3.6795	0.65861	5.5867	7.6626e-08

```
Number of observations: 200, Error degrees of freedom: 196
Root Mean Squared Error: 9.78
R-squared: 0.276,  Adjusted R-Squared: 0.265
F-statistic vs. constant model: 24.9, p-value = 1.08e-13
```

curves for 0.5 and 1. The curves for $\lambda = 0$ show appreciable variability in the centre of the plot which decreases towards the end of the search; this value is accepted once most of the outliers are included, illustrating the need of robust monitoring for the correct estimation of the transformation parameter. The three-panel plot from the automatic method of Sect. 6.3.4 in Fig. B.26 shows that 0.25 is the preferred transformation. However, the white rectangle atop the red bar for this value of λ shows that there are some observations which do not affect the transformation but are still outlying. The FS on the fourth root scale identifies 20 outliers, which are shown as

Solutions

Table B.10 D2: Yeo-Johnson transformation; value of λ and score test

λ	Score test
−1.0000	50.9914
−0.7500	36.0461
−0.5000	21.9813
−0.2500	10.0863
0	0.1375
0.2500	−10.5116
0.5000	−25.7420
0.7500	−46.3590
1.0000	−68.2085

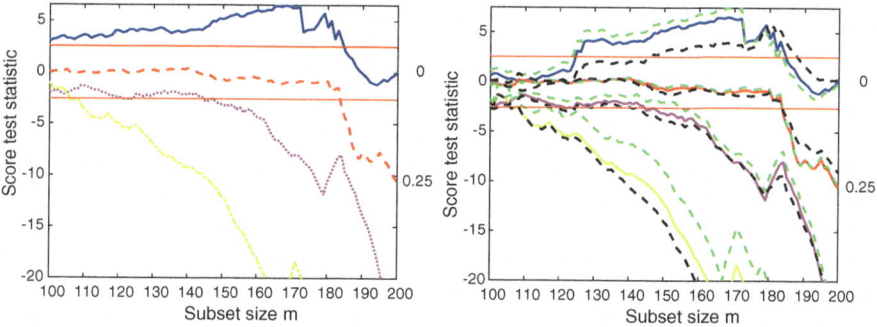

Fig. B.25 D2: left-hand panel, fan plots for, reading down at the right, $\lambda_0 = 0, 0.25, 0.5$ and 1. Right-hand panel, extended fan plots for the homogeneity of transformation for the same four values of λ; the green trajectory for each λ comes from testing the transformation of the positive observations. Some trajectories in both plots show the effect of influential units on the value of λ entering before the end of the search

circles in the yX plot of Fig. B.27. The automatic procedure which searches for two values of λ, described in Sect. 6.3.6, confirms that 0.25 is the best value both for positive and negative observations (see the heatmaps of BIC and AGI in Fig. B.28). The two heatmaps in Fig. B.29, which respectively give $h(\lambda)$ (left-hand panel) and the extended coefficient of determination R^2_{EXT} (right-hand panel) show that this transformation is supported by 180 units and that this is the combination of values of λ with the highest percentage of variance explained (corrected for truncation).

The ANOVA table based on the transformed data using $\lambda = 0.25$ after removing the outliers is given in Table B.11. In this scale the percentage of variance which is explained is above 90%. The analysis of the p-values and of the t-statistics shows that the explanatory variables are now much more significant. This example is yet another case where, from the simple examination of the scatterplot matrix, it is impossible to detect the presence of outliers or influential observations.

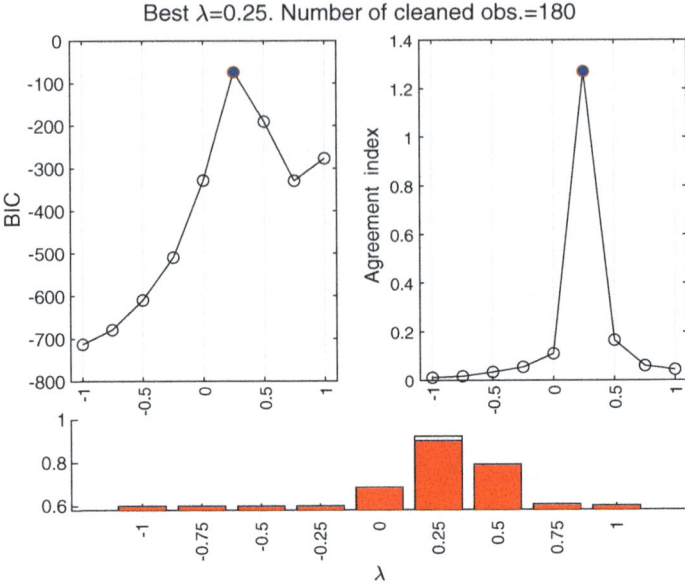

Fig. B.26 D2: automatic analysis of Yeo-Johnson transformation. Upper left-hand panel, extended BIC; upper right-hand panel, agreement index AGI; lower panel $h(\lambda)/n$ and $m^*(\lambda)/n$, the proportion of observations used in fitting the model

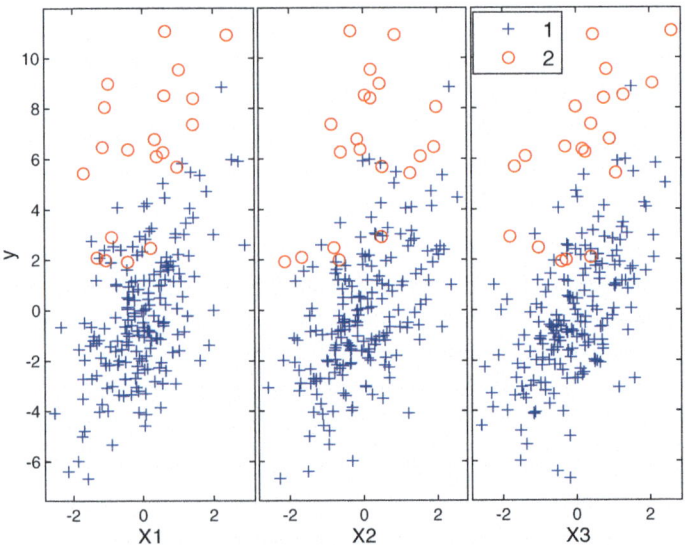

Fig. B.27 D2: yX plot in the transformed scale for y ($\lambda = 0.25$) with the outliers shown

Solutions

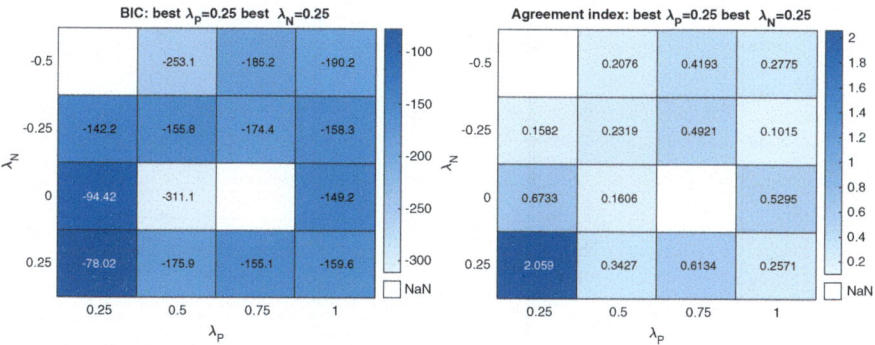

Fig. B.28 D2: heat maps as functions of λ_P and λ_N. Left-hand panel, extended BIC; right-hand panel, agreement index AGI. Cell left blank if $h(\lambda) \leq 0.6n$

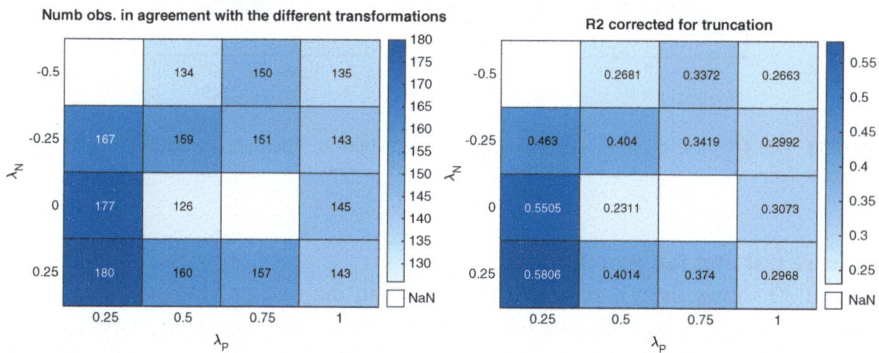

Fig. B.29 D2: heat maps as functions of λ_P and λ_N. Left-hand panel, $h(\lambda)$; right-hand panel, extended coefficient of determination R^2_{EXT}. Cell left blank if $h(\lambda) \leq 0.6n$

Table B.11 D2: ANOVA in the transformed scale for y ($\lambda = 0.25$)

	Estimate	SE	tStat	pValue
(Intercept)	-0.038855	0.051365	-0.75645	0.45039
x1	1.4066	0.052119	26.989	1.8656e-64
x2	1.3872	0.048197	28.782	1.8091e-68
x3	1.3997	0.049318	28.381	1.382e-67

```
Number of observations: 180, Error degrees of freedom: 176
Root Mean Squared Error: 0.687
R-squared: 0.932,  Adjusted R-Squared: 0.931
F-statistic vs. constant model: 804, p-value = 1.94e-102
```

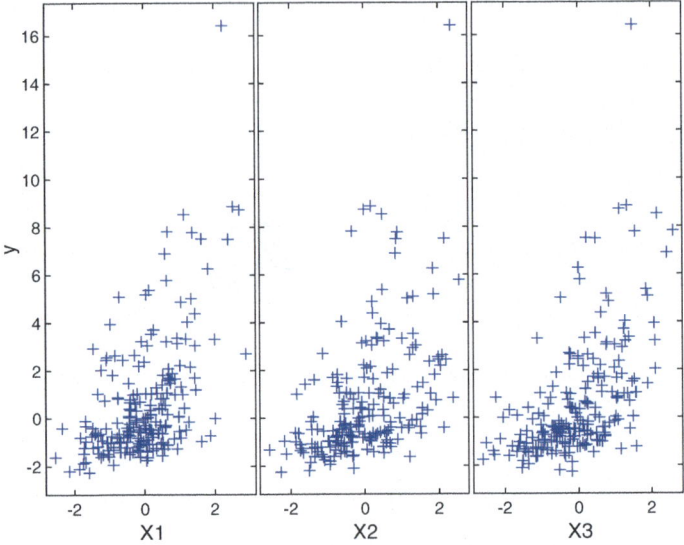

Fig. B.30 D3: yX plot. y shows a positive skewed distribution. The data are likely to need transformation

6.7 Score test and fan plot 3

Figure B.30 shows the yX plot of data D3. As in the two previous examples, the minimum value of y is smaller than 0, the distribution of y seems right skewed and it is not clear whether or how many outliers might be present in the data. The ANOVA table based on all the observations and untransformed data is given in Table B.12. In this scale the adjusted R^2 is around 78%. Table B.13 shows that using all the observations $\lambda = 0.25$ is the best Yeo-Johnson transformation among those from the grid, even if its value is at the boundary of significance. The traditional way of proceeding would be to try the third root transformation.

The left-hand panel of Fig. B.31 gives the fan plot for the Yeo-Johnson transformation of these data. For $\lambda = 0.25$ the statistic lies within the band but steadily around the value of 2. The values of λ, 0 and 0.5, start to be rejected when m is smaller than $[n \times 0.7]$. The extended fan plot is shown in the right-hand panel of the figure. The striking thing in this plot is the large difference in the values of the statistics for the overall transformation and for testing positive and negative values. This plot provides strong evidence that a single transformation is not satisfactory for these data. It is interesting also to notice that this difference gets larger as λ departs from the interval [0 0.5]. The results from the automatic analysis (not given here) show that using a single value of λ, the best transformation is $\lambda = 0.25$. The automatic procedure which searches for two values of λ described in Sect. 6.3.6 produces Figs. B.32 and B.33. The heatmaps of the BIC and AGI index agree (Fig. B.32) and show that the data need the following two values of λ: $\lambda_P = 0.5$ and $\lambda_N = 0$. The left-hand panel of Fig. B.33 shows that this transformation is supported by all the data, while the

Solutions

Table B.12 D3: ANOVA in the original scale for y

	Estimate	SE	tStat	pValue
(Intercept)	0.7694	0.08502	9.0496	1.4325e-16
x1	1.3053	0.085587	15.251	3.8544e-35
x2	1.1736	0.079887	14.691	1.9653e-33
x3	1.2566	0.0809	15.533	5.3606e-36

```
Number of observations: 200, Error degrees of freedom: 196
Root Mean Squared Error: 1.2
R-squared: 0.782,  Adjusted R-Squared: 0.779
F-statistic vs. constant model: 234, p-value = 1.53e-64
```

Table B.13 D3: value of λ in Yeo-Johnson transformation and score test

λ	Score test
−1.0000	46.4657
−0.7500	37.4030
−0.5000	28.1622
−0.2500	19.0683
0	10.4780
0.2500	2.2486
0.5000	−6.6266
0.7500	−17.4951
1.0000	−30.4731

right-hand panel of the same figure shows that the percentage of variance explained in this transformed scale is 0.932.

Figure B.34 shows the yX plot after transforming the response with these two values of λ. Approximate normality has been achieved. The ANOVA table based on the transformed data is given in Table B.14.

Problems: Chap. 7
7.1 New updated transformed values in AVAS
(a) The initial values of $\widehat{ty}_i$ (namely $\widehat{ty}_i^{(0)}$), with $i = 1, 2, \ldots, 10$, are just the standardized values of y using the mle of σ and are given in the first column of Table B.15.
(b) As Fig. B.35 shows, the value of $\widehat{ty}_1^{(0)}$ is smaller than $\widehat{ty}_{[1]}^{(1)}(10) = \widehat{ty}_{[1]}^{(1)}$. If option trapezoid is false, the updated value of $\widehat{ty}_1$ is given by (see Eq. 7.7):

$$\widehat{ty}_1^{(1)} = (\widehat{ty}_1^{(0)} - \widehat{ty}_{[1]}^{(1)})v_1 = \{-1.5802 - (-1.2401)\}2.1869 = -0.744. \quad (B.35)$$

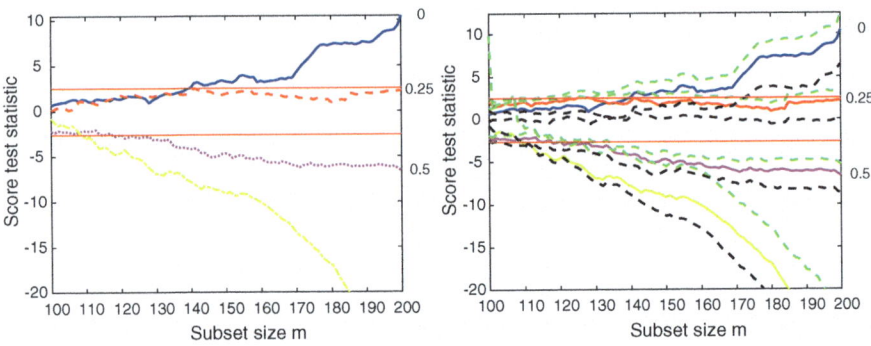

Fig. B.31 D3: left-hand panel, fan plots for, reading down at the right, $\lambda_0 = 0, 0.25, 0.5$ and 1. Right-hand panel, extended fan plots for the homogeneity of transformation for the same four values of λ; the higher green trajectory for each λ comes from testing the transformation of the positive observations. There is a large difference in the values of the statistics for the overall transformation and for testing positive and negative values

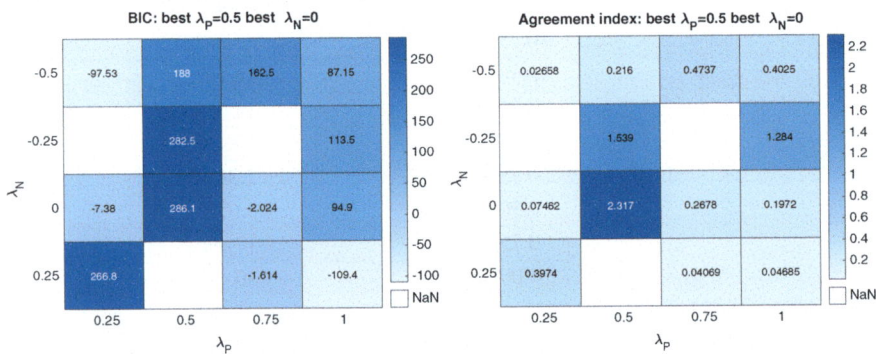

Fig. B.32 D3: heat maps as functions of λ_P and λ_N. Left-hand panel extended BIC; right-hand panel agreement index AGI. Cell left blank if $h(\lambda) \leq 0.6n$

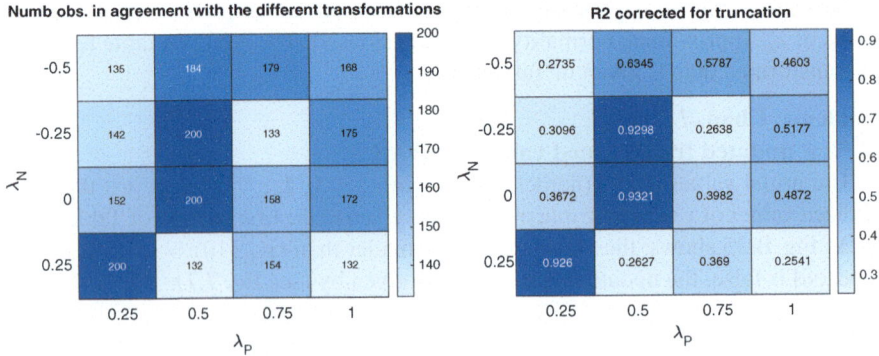

Fig. B.33 D3: heat maps as functions of λ_P and λ_N. Left-hand panel, $h(\lambda)$; right-hand panel, extended coefficient of determination R^2_{EXT}

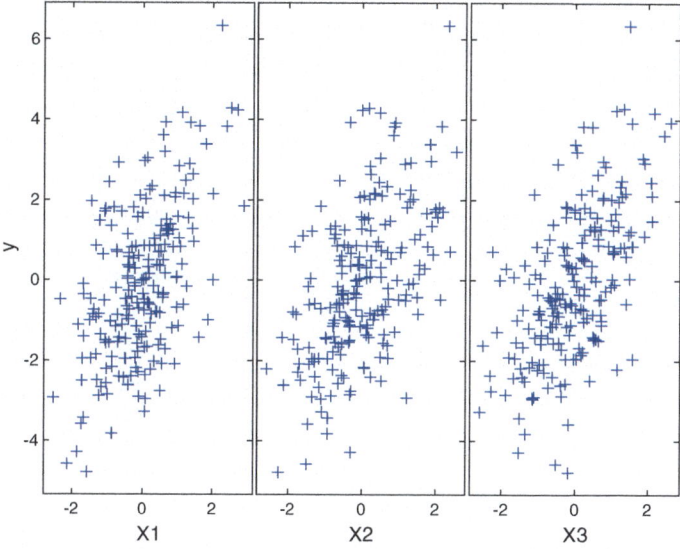

Fig. B.34 D3: yX plot in the transformed scale for y ($\lambda_P = 0.5$ and $\lambda_N = 0$). Approximate normality is achieved

Table B.14 D3: ANOVA in the transformed scale for y ($\lambda_P = 0.5$ and $\lambda_N = 0$)

	Estimate	SE	tStat	pValue
(Intercept)	-0.0056834	0.035286	-0.16107	0.87221
x1	1.0114	0.035522	28.474	1.4341e-71
x2	0.99893	0.033156	30.128	1.7303e-75
x3	1.0234	0.033576	30.479	2.6612e-76

```
Number of observations: 200, Error degrees of freedom: 196
Root Mean Squared Error: 0.499
R-squared: 0.932,  Adjusted R-Squared: 0.931
F-statistic vs. constant model: 897, p-value = 3.66e-114
```

On the other hand, if option trapezoid is true, the updated value of $\widehat{ty}_1$ is given by (see Eq. 7.6)

$$\widehat{ty}_1^{(1)} = (\widehat{ty}_1^{(0)} - \widehat{ty}_{[1]}^{(1)})(v_1 + \bar{v})/2 = \{-1.5802 - (-1.2401)\}(2.1869 + 5.2588)/2 = -1.266. \tag{B.36}$$

(c), (d) The results of the calculations of all updated values using trapezoid=false and trapezoid=true are given in columns 2 and 3 of Table B.15. These results can be automatically obtained calling **FSDA** function ctsub(x,y,z,trapezoid) where x and y are the first two columns of Table 7.1

Table B.15 Old values of $\widehat{ty}_i$, updated value of $\widehat{ty}_i$ using trapezoid=false and trapezoid=true. The updated values will be standardized again

Old values $\widehat{ty}_i^{(0)}$	Updated values $\widehat{ty}_i^{(1)}$ using trapezoid=false	Updated values $\widehat{ty}_i^{(1)}$ using trapezoid=true
−1.5802	−0.7436	−1.2659
−0.9256	0.7608	0.7608
−0.1329	4.0208	4.0208
0.6547	11.6434	11.6434
1.2119	12.9856	13.5337
1.3358	13.2717	14.0027
0.9770	12.4435	12.6451
0.2745	8.6367	8.6367
−0.5402	1.8804	1.8804
−1.2749	−0.0760	−0.1294

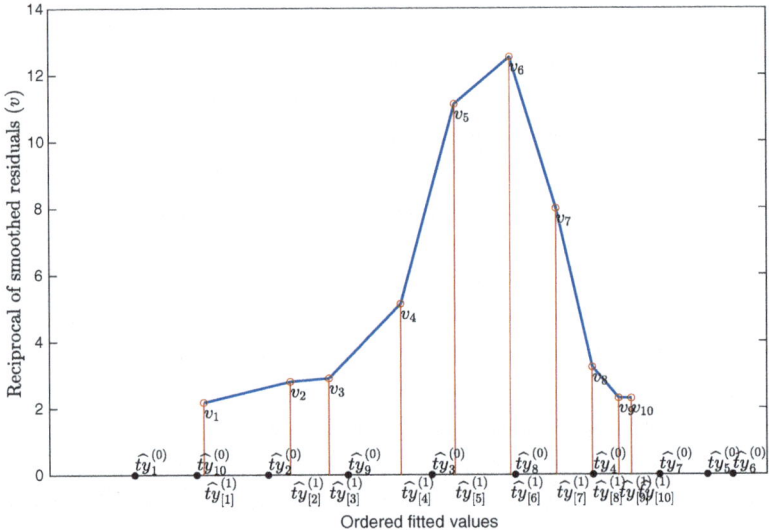

Fig. B.35 Reciprocal of smoothed residuals versus ordered fitted values in the first iteration of the AVAS algorithm. The black dots are associated with the initially transformed values $\widehat{ty}_i^{(0)}$

and z is the vector containing $\widehat{ty}_i^{(0)}$ given in the first column of Table B.15 and trapezoid is a Boolean value.

7.2 Augmented Investment Funds data

1. The left-hand panel of Fig. B.36 gives the fan plot for the Yeo-Johnson transformation of these data. $\lambda = 0.75$ is surely the best value, but the score statistic fluctuates above and below the bands. The extended fan plot is shown in the right-

Solutions

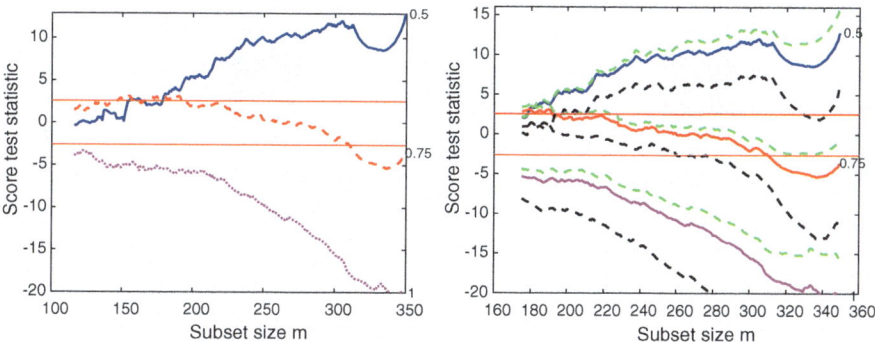

Fig. B.36 Augmented investment funds data: left-hand panel, fan plot for, reading down at the right, $\lambda_0 = 0.5, 0.75$ and 1. Right-hand panel, extended fan plot for the homogeneity of transformation for the same three values of λ; the higher green trajectory for each λ comes from testing the transformation of the positive observations. There is a large difference in the values of the statistics for the overall transformation and for testing positive and negative values

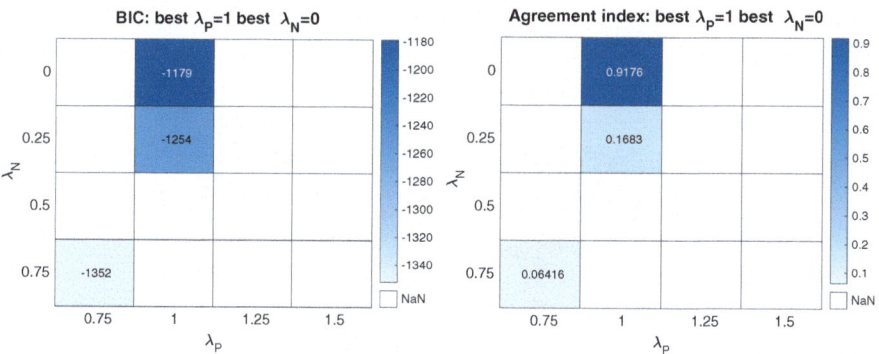

Fig. B.37 Augmented investment funds data: heat maps as functions of λ_P and λ_N. Left-hand panel extended BIC; right-hand panel agreement index AGI. Cell left blank if $h(\lambda) \leq 0.6n$

hand panel of the figure. The striking thing in this plot is the large difference in the values of the statistics for the overall transformation and for testing positive and negative values. This plot provides strong evidence that a single transformation parameter is not satisfactory for these data. It is also interesting that this difference gets larger in the last part of the search. The results from the automatic analysis (not given here) show that, using a single value of λ, the best transformation is $\lambda = 0.75$.

The automatic procedure which searches for two values of λ, described in Sect. 6.3.6, produces Figs. B.37 and B.38. Both the heatmaps of the BIC and AGI (Fig. B.37) show that the data need the following two values of λ: $\lambda_P = 1$ and $\lambda_N = 0$. The left-hand panel of Fig. B.38 shows that this transformation is supported just by 314 observations, while the right-hand panel of the same figure

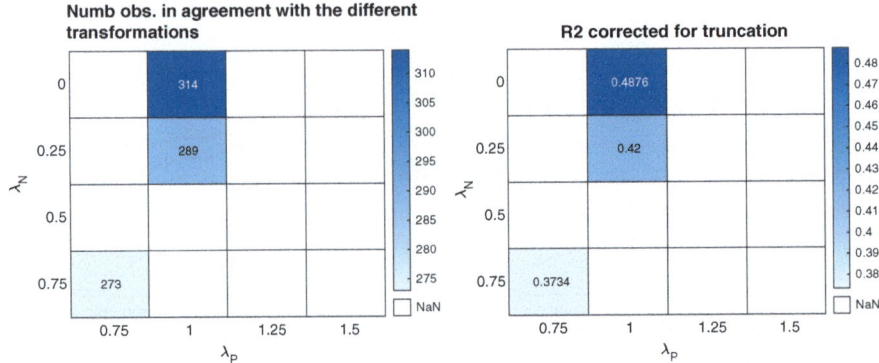

Fig. B.38 Augmented investment funds data: heat maps as functions of λ_P and λ_N. Left-hand panel, $h(\lambda)$; right-hand panel, extended coefficient of determination R^2_{EXT}

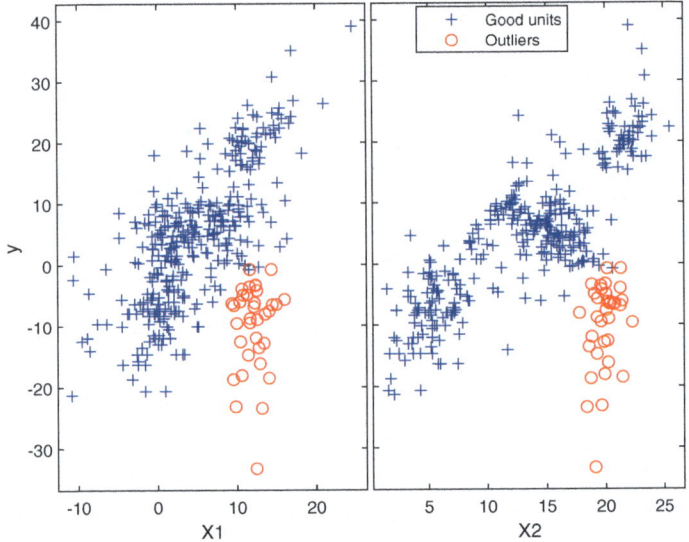

Fig. B.39 Augmented investment funds data: yX plot in the transformed scale for y ($\lambda_P = 1$, $\lambda_N = 0$) with the outliers shown as red circles. Approximate normality is achieved

shows that the percentage of variance explained in this transformed scale is below 0.5. It is interesting to notice that all the cells, apart from 3, are left blank. This indicates that just three combinations provide reasonable values and that the best one is by far the one which uses the greatest number of observations and has the largest value of R^2 (corrected for truncation).

2. The automatic procedure for outlier detection in the transformed scale detects 35 outliers. Figure B.39 shows the yX plot after transforming the response with

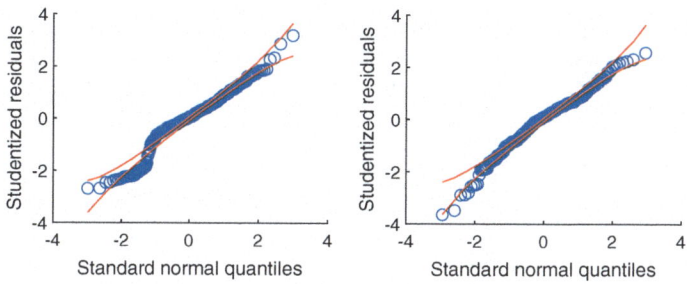

Fig. B.40 Augmented investment funds data: comparison of normal QQ-plots of residuals. Left-hand panel, untransformed data. Right-hand panel, transformed cleaned data from EYJ ($\lambda_P = 1, \lambda_N = 0$)

the two values of λ ($\lambda_P = 1, \lambda_N = 0$) and the detected outliers. Approximate normality has been achieved.

3. The left-hand panel of Fig. B.40 shows the QQ-plot of the residuals of all 349 observations from regression with the original response. The sigmoid shape of this plot indicates that the observations are not normally distributed. The right-hand panel is the QQ-plot for residuals of the transformed cleaned data. The distribution of residuals is much closer to normality, although the centre of the curve indicates that many small residuals are slightly too large in absolute value.

4. The value of R^2 for regression on the untransformed augmented data is 0.399. For the cleaned transformed data it is 0.783 and for the original data 0.816. The left-hand column of Table B.16 lists the values of R^2 achieved by regression on parametric and non-parametric transformations of the augmented data. The largest value is 0.421 for unconstrained ACE. The monotonicity constraint on ACE comes from isotonic regression on the unconstrained transformation and yields a slightly reduced value of 0.417. AVAS produces a value of 0.241, less than that for EYJ.

5. Figure B.41 provides plots of transformed against untransformed response for these four transformations. The top left-hand panel of the figure shows that the EYJ transformation for $y > 0$ is linear (no transformation) whereas for negative y the transformation is concave, transforming the more negative observations to be more extreme. AVAS, in the top right-hand panel, provides a more smooth concave curve, which not only makes the more negative values more extreme, but makes the more positive values less extreme. Unconstrained ACE is virtually linear for $y > 12$, but shrinks the most negative observations, some of which are outliers. Constrained ACE is formed by isotonic regression on the unconstrained version and, as the figure shows, is similar in structure to ACE. Both transformations show several points of inflection for $y < 12$, especially just above zero.

6. If the errors are approximately normally distributed and the model is correct, the plot of residuals against fitted values should be without any features, apart from those from the distribution of fitted values. The left-hand panel of Fig. B.42 shows the plot of residuals from constrained ACE. The plot is wedge shaped,

Table B.16 Augmented Investment Funds data: summary properties of regression for parametric and non-parametric transformations of the original augmented data and cleaned transformed data

	Augmented	Cleaned and transformed
Untransformed	0.399	–
EYJ	0.360	0.783
AVAS	0.241	0.778
ACE	0.421	0.806
ACE (monotonic)	0.417	0.805

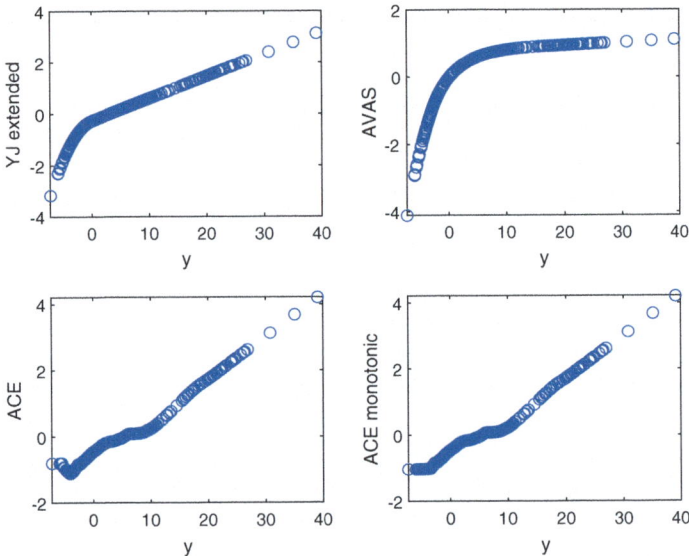

Fig. B.41 Augmented investment funds data: transformed responses against untransformed responses. Top row, EYJ, and AVAS. Bottom row, ACE, unconstrained and constrained

with a sharp lower diagonal bound. The other panel, for AVAS, also has some structure, in this case a cloud of large negative residuals for fitted values around 0.5; the non-parametric transformations indicate faults in the model or data.

7. Figure B.43 shows the output from the best model which comes from the application of the RAVAS procedure. The necessity of transforming positive and negative observations in a different way is evident in the top left-hand panel. The outliers found with RAVAS are those found with the other procedures. The bottom right-hand panel shows that the transformation of the explanatory variables is monotonic.

The ANOVA table after removing the outliers in the transformed RAVAS space is shown in Table B.17. The percentage of variance explained is much greater than those with the models shown in Table B.16.

Solutions

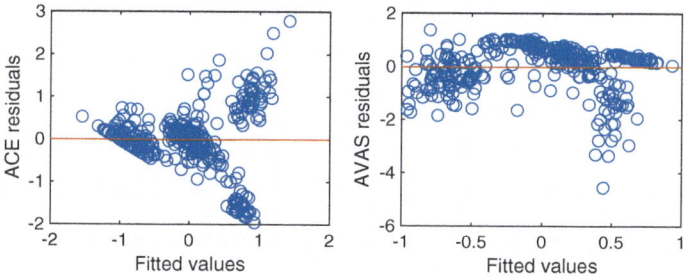

Fig. B.42 Augmented investment funds data: residuals against fitted values. Left-hand panel, constrained ACE. Right-hand panel AVAS (note the scales of these residuals)

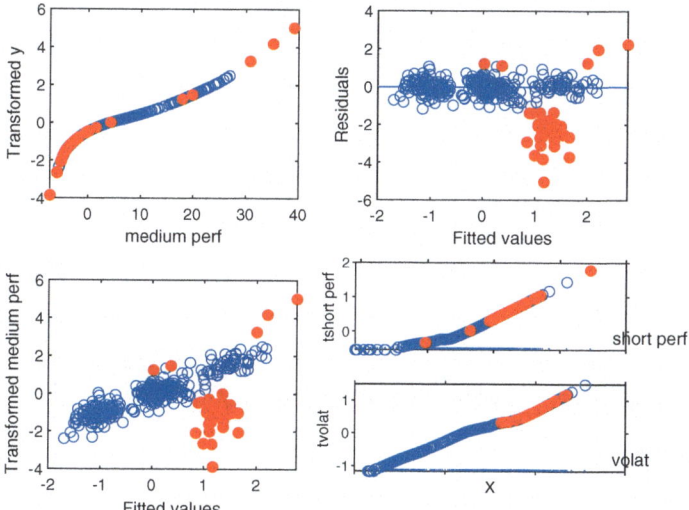

Fig. B.43 Augmented investment funds data: output from RAVAS. Top left-hand panel, transformed y against y; top right-hand panel, residuals against fitted values; bottom left-hand panel transformed y against fitted values; bottom right-hand panel, transformations of the two explanatory variables

Problems: Chap. 8

8.1 Frequentist analysis of the Windsor House Price data

Figure B.44 shows the forward plot of minimum deletion residuals in the absence of any prior information. The automatic outlier detection procedure finds 17 of the 18 outliers found in the Bayesian search; the 18th outlier is given a special symbol in Fig. 8.4. The data trajectory of minimum deletion residuals is similar to that for weak prior information in Fig. 8.1. The main difference is the, perhaps, slightly more jagged appearance of the trajectory for least squares; the prior information serves to smooth the curve. This effect is appreciably more evident in Fig. 8.6 where n_0 was

Table B.17 Augmented Investment Funds data: ANOVA in the transformed RAVAS scale for y

	Estimate	SE	tStat	pValue
(Intercept)	-4.8704e-16	0.022248	-2.1892e-14	1
short perf	0.98845	0.067077	14.736	2.3742e-37
volat	1.0306	0.043482	23.701	1.0403e-70

Number of observations: 303, Error degrees of freedom: 300
Root Mean Squared Error: 0.387
R-squared: 0.852, Adjusted R-Squared: 0.851
F-statistic vs. constant model: 860, p-value = 5.67e-125

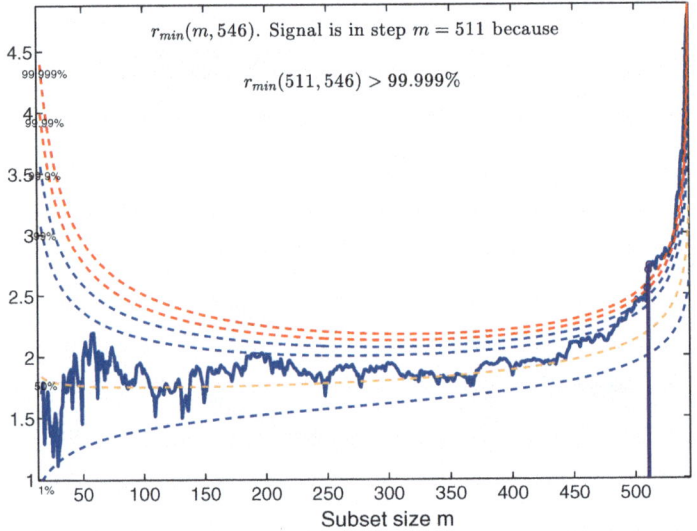

Fig. B.44 Windsor house price data: frequentist analysis. Forward plot of absolute minimum deletion residuals. Ultimately 17 outliers are identified

taken as 250 and is another aspect of the effect of appreciable prior information on the analysis of data.

8.2 Differences and similarities between HAR and ART heteroskedasticity

Figure B.45 shows the outliers detected using HAR heteroskedasticity for dataset inttrade1, while Fig. B.46 shows the fitted regression line with 99% Bonferronized confidence bands. In the example ART and HAR forms of heteroskedasticity provide very similar results for the region involved with large quantities. The main difference consists in the region close to the origin. While the width of the confidence band in the case of HAR heteroskedasticity tends to zero and detects outliers associated with

Solutions

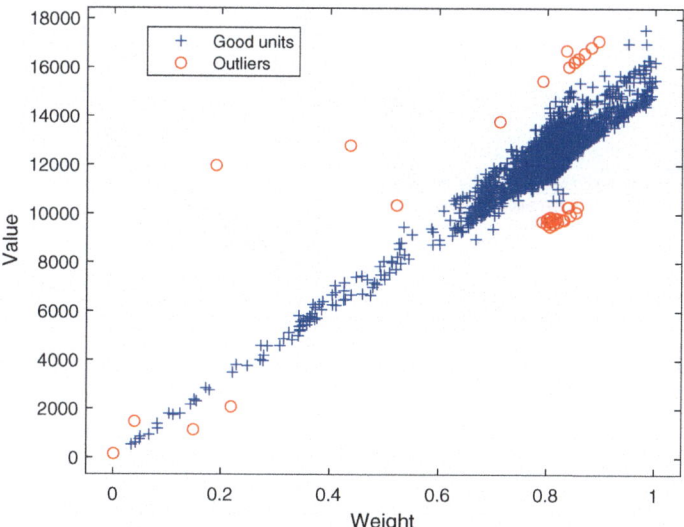

Fig. B.45 Dataset inttrade1: scatter with outliers using HAR heteroskedasticity

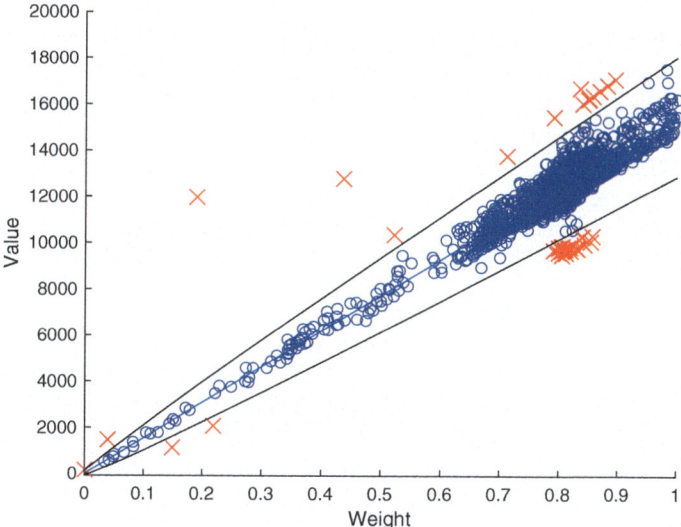

Fig. B.46 Dataset inttrade1: fitted regression line with 99% Bonferronized confidence bands using HAR heteroskedasticity

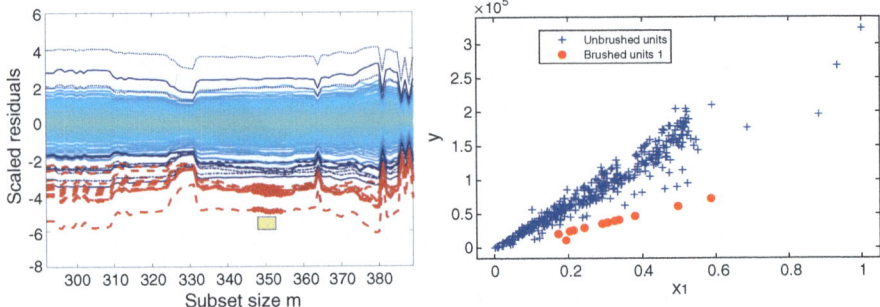

Fig. B.47 Dataset inttrade3: ART heteroskedasticity. Brushing the units with the largest negative residuals (left-hand panel) together with their position in the original scatter plot (right-hand panel)

very small quantities which slightly deviate from the fitted regression line, the one based on ART does not suffer from this drawback.

8.3 Analysis of an additional international trade dataset: inttrade3

Figure B.47 shows the monitoring of residuals (based on ART heteroskedasticity) and the brushing of the units with large negative residuals. These units correspond to a set of transactions which lie in a different group (are characterized by a smaller price). The 99% Bonferronized confidence bands of predicted values, together with the units detected as outliers, are shown in Fig. B.48. Note that the confidence bands based on ART heteroskedasticity do not go to zero with x.

The homoskedastic analysis is shown in Fig. B.49. A large number of outliers is detected and the confidence bands of the plot do not reflect in any way the increasing spread of the data as the quantity increases.

8.4 Analysis of transformed Airline data

We consider the original (uncontaminated) Airline data. There is obviously no need to search for a level shift, so we set $\delta_1 = 0$. We keep the other model parameters as in Chap. 8, that is, a linear trend ($A = 1$) and two harmonics ($B = 2$) with amplitude varying linearly ($G = 1$), which we hope are sufficient to address the increasing variance of the seasonal fluctuations. With these model choices, LTSts finds some outliers, visible in Fig. B.50. Given that the outliers are located on top of the seasonal peaks, we interpret this as a sign that the amplitude of the variance grows more than linearly. On the other hand, an amplitude increasing quadratically ($G = 2$) is excessive, as Fig. B.51 reveals: now several outliers appear.

Figure B.52 clearly shows that the logarithmic transformation of the data is more appropriate. In fact the left-hand panel, obtained using a simple linearly varying amplitude, exhibits a very good fit, with only three borderline outliers. The fit is even better with a fixed amplitude, that is, by setting $G = 0$ in the model. We can conclude that the logarithmic transformation is appropriate for these data.

The logarithmic transformation is only one of the power transforms in the Box-Cox family. We could therefore apply this more general framework to the non-linear

Solutions

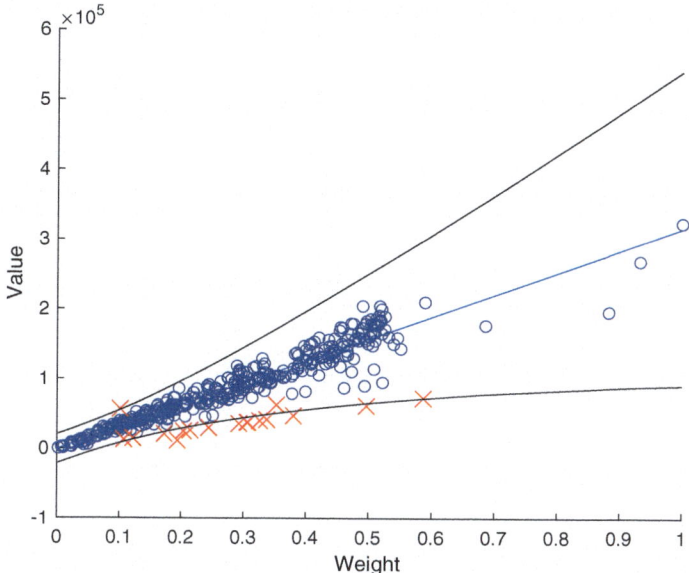

Fig. B.48 Dataset inttrade3: ART heteroskedasticity. Fitted regression line with 99% Bonferronized confidence bands

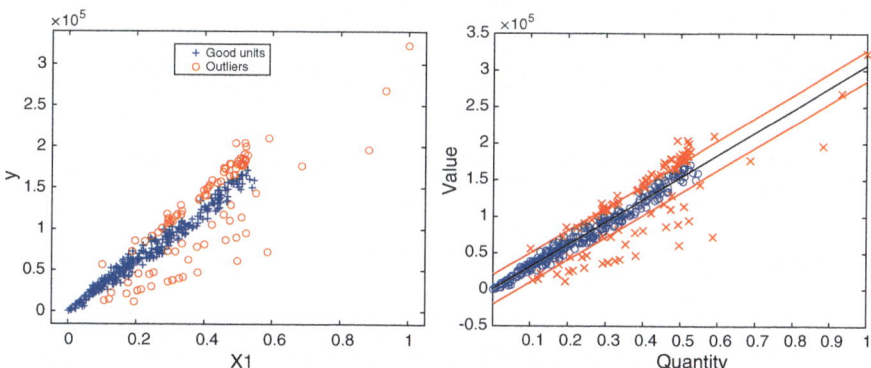

Fig. B.49 Dataset inttrade3: homoskedastic analysis. Left-hand panel: scatter plot of the original data with detected outliers. Right-hand panel: 99% Bonferronized confidence bands with detected outliers

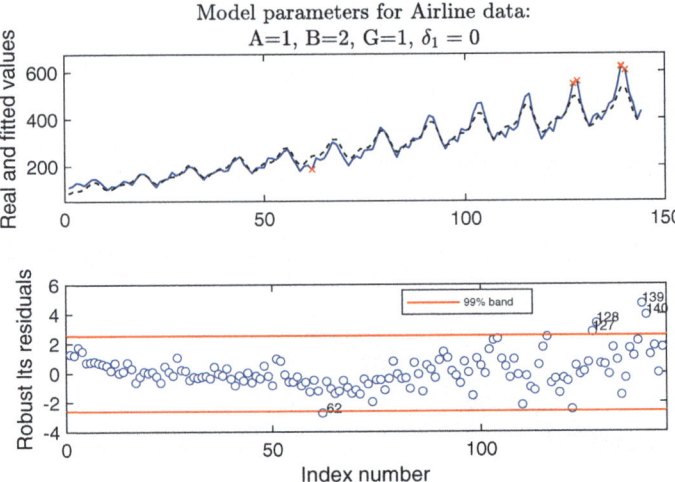

Fig. B.50 Airline data in the original scale: seasonal amplitude growing linearly. Upper panel, fitted values: the intensity of the symbols x increases with the outlyingness of the observations. Lower panel, residuals

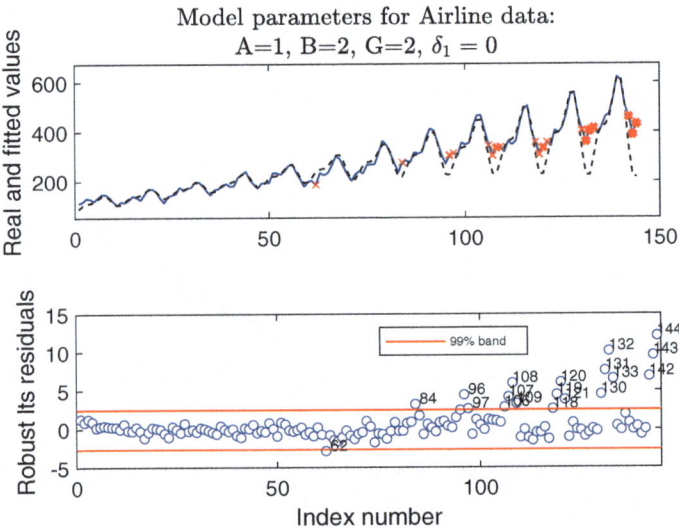

Fig. B.51 Airline data in the original scale: seasonal amplitude growing quadratically. Upper panel, fitted values: the intensity of the symbols x increases with the outlyingness of the observations. Lower panel, residuals

Solutions

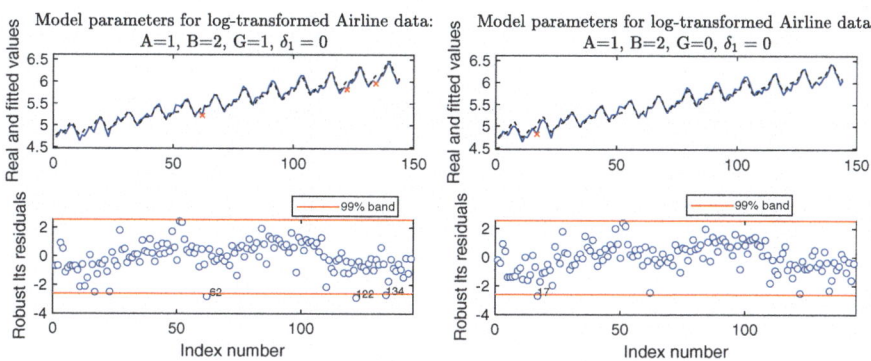

Fig. B.52 Airline data in the logarithmic scale: left-hand panels: amplitude growing linearly. Right-hand panels: no varying amplitude. Upper panels, fitted values: the intensity of the symbols x increases with the outlyingness of the observations. Lower panels, residuals

time series model used to fit the Airline data, using a fan plot to investigate the dependence of estimated transformations on the outliers.

8.5 Analysis of body measurements to predict percentage of body fat

We start by loading the data and selecting the specified variables to form the predictor matrix X and put the transformed variable body.fat into the response variable y,

$$y = \log\{\text{body.fat}/(100 - \text{body.fat})\},$$

the extension of the logit transformation (6.42) to the analysis of percentages.

Observation 182 is problematic, with a value of 0 in the body.fat variable; therefore we ignore it. The selected predictors which are all body measurements can be regarded as a composition because the size of the body is not relevant. The important information is contained in the ratios between the different body measurements. Then we transform the predictors using the *ilr*-transformation as implemented in the function pivotCoord(). This function is implemented both in the R package **robCompositions** (Filzmoser et al., 2018) and inside **FSDA**.

Using S-estimation with *bdp* equal to 0.25, a value sometimes recommended in the literature, produces the plot of residuals shown in the upper panel of Fig. B.53. Observations 171, 172, and 223 are outside the 99% bands and observations 9 and 26 are close to the border line. This is quite similar to the results that we obtain with LS regression. The lower panel of Fig. B.53 shows the results of regression with the maximal breakdown point, $bdp = 0.5$. Here additionally to observations 171, 172, and 223, three other outliers are clearly visible: 31, 86, and 153.

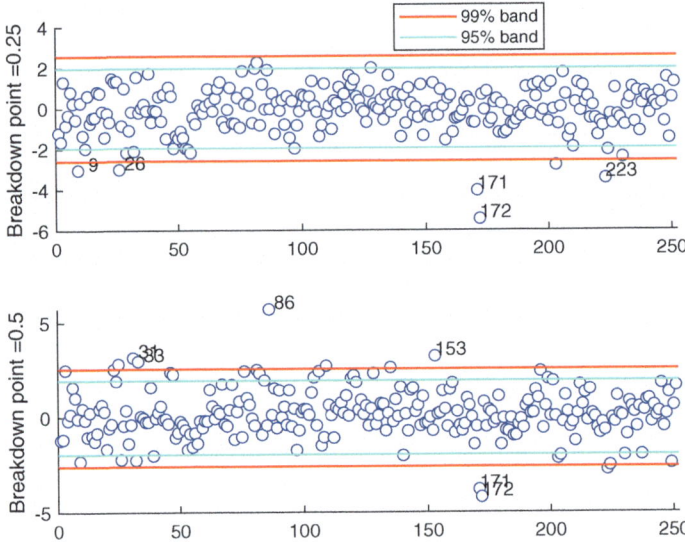

Fig. B.53 Body measurements data: *ilr*-transformed, S-estimation. Index plots of residuals from Tukey's biweight ρ function; two values of *bdp*. Upper panel, *bdp* = 0.25; lower panel, *bdp* = 0.5

Figure B.54 shows the residual plots for MM-estimation, again with Tukey's biweight ρ function. The two panels are quite similar. In the upper panel, with *eff* = 0.80, two more outliers are identified, 9 and 203, but both are very close to the border line.

Figure B.55 shows the monitoring plot of scaled residuals for S-estimation. This is generated by a series of robust fits on the *ilr*-transformed data, starting from *bdp* = 0.5 and decrementing the value by 0.01 down to 0.01. There are therefore 50 robust fits leading to the plot of scaled residuals in the figure. The abrupt change happens at *bdp* = 0.48. The monitoring of the scaled residuals from MM-estimation shown in Fig. B.56 does not show specific structure and all correlations in the right-hand panel are close to one.

Problems: Chap. 9

9.1 Analysis of the reduced logged ozone data using RAVAS
For the reduced ozone data with response logged ozone concentration and regressors Time plus variables 4, 5, and 6, the best model from analysis with RAVAS produces the output shown in Figs. B.57 and B.58. The results in Fig. B.57 show a smooth transformation of the response and a linear relationship between the transformed response and fitted values, once the 3 outliers are ignored. The transformed explanatory variables all have smoothly changing shapes (Fig. B.58) and all are significant; the least significant *t*-statistic has a value of 4.8 (see Table B.18).

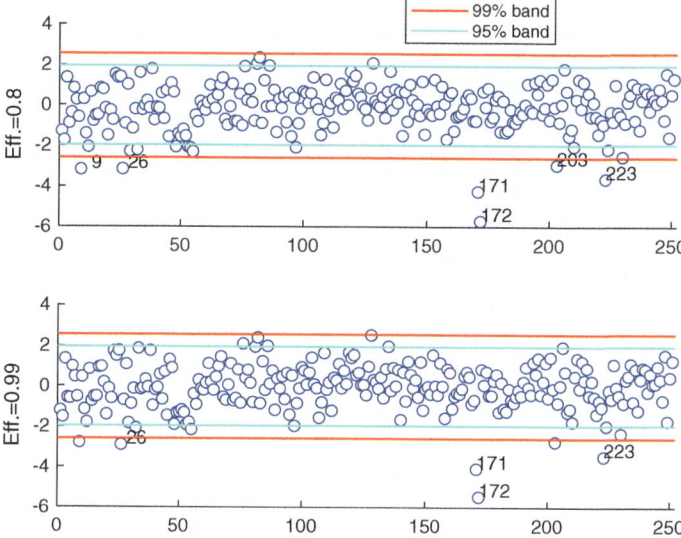

Fig. B.54 Body measurements data: *ilr*-transformed, MM-estimation. Index plots of residuals from Tukey's biweight ρ function; two values of *eff*. Upper panel, *eff* = 0.80; lower panel, *eff* = 0.99

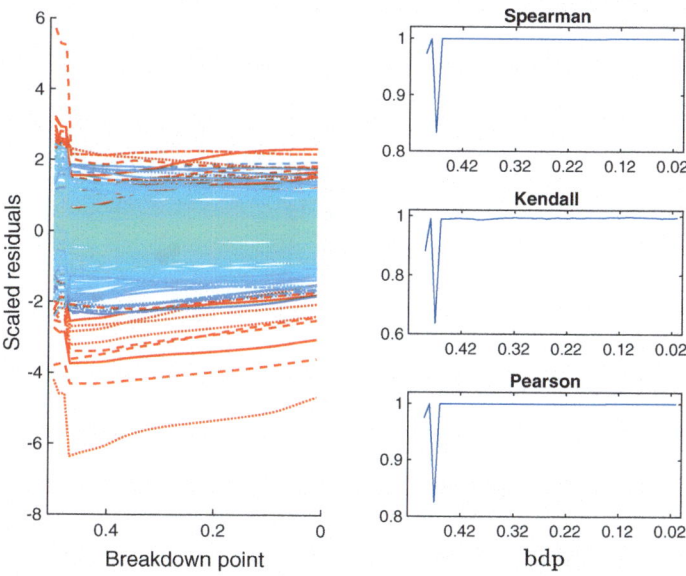

Fig. B.55 Body measurements data: *ilr*-transformed, S-estimation. Left-hand panel, plot of scaled residuals. Right-hand panel, three measures of the correlations of adjacent residuals

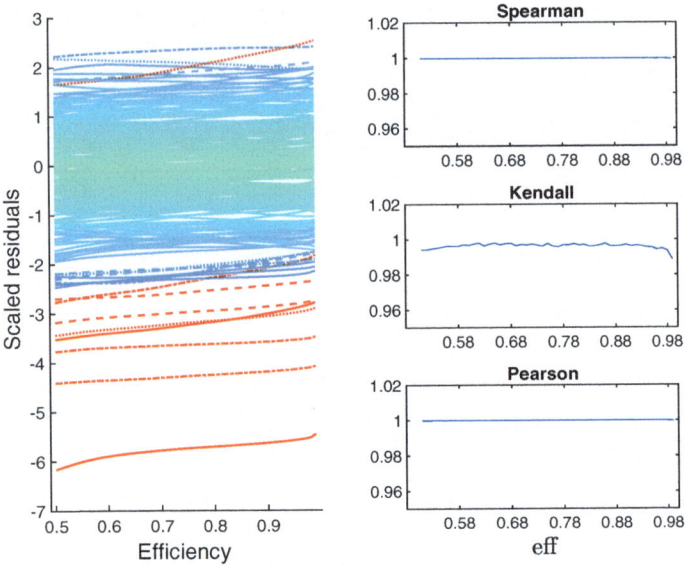

Fig. B.56 Body measurements data: *ilr*-transformed, MM-estimation. Left-hand panel, plot of scaled residuals. Right-hand panel, three measures of the correlations of adjacent residuals—all correlations are close to one

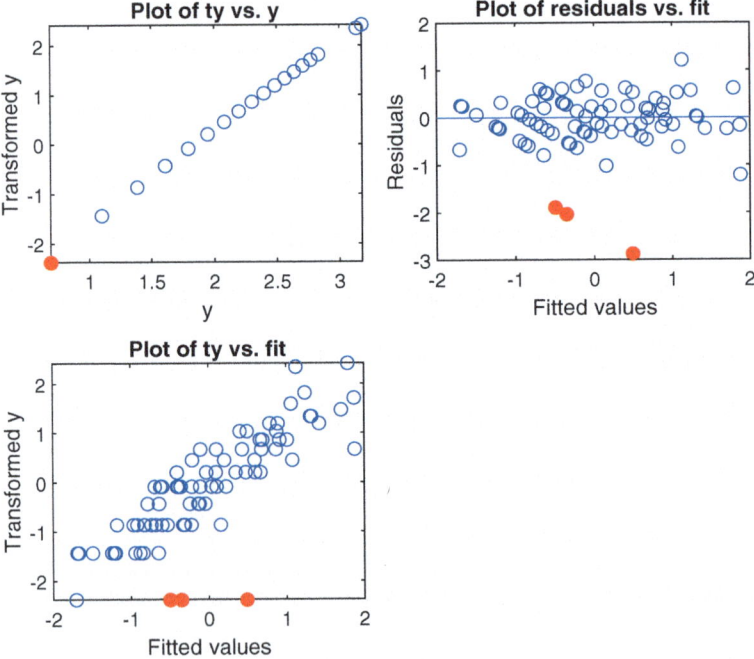

Fig. B.57 Reduced logged ozone data: RAVAS: transformed response $g(y)$. Upper left-hand panel, $g(y)$ against y; upper right-hand panel, residuals against fitted values; lower panel, $g(y)$ against fitted values. 3 outliers highlighted

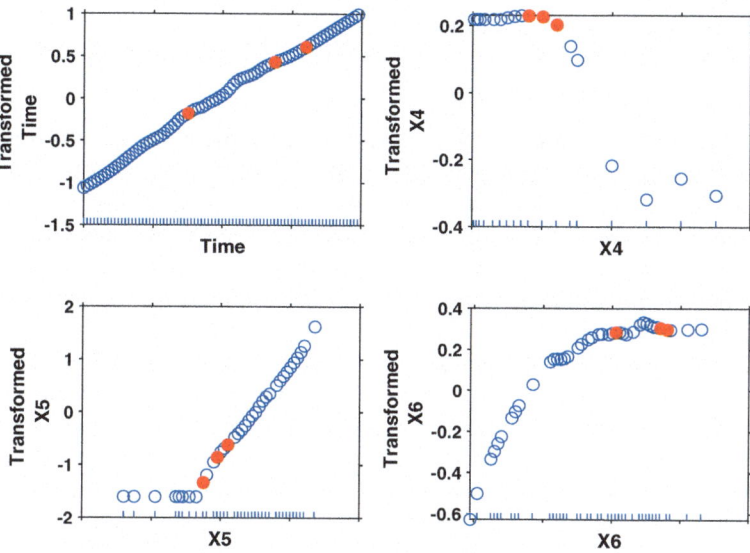

Fig. B.58 Reduced logged ozone data: RAVAS: transformations $f_j(x_j)$ of the explanatory variables against original x_j

Table B.18 Reduced Logged Ozone data: RAVAS: ANOVA in the transformed scale after deleting the 3 outliers

```
Estimated Coefficients:
                 Estimate        SE            tStat         pValue

   (Intercept)   -4.6083e-18     0.051029      -9.0308e-17        1
   Time           0.99852        0.088477      11.286        1.3881e-17
   X4             1.1771         0.24475        4.8095       8.0623e-06
   X5             0.99857        0.080376      12.424        1.3934e-19
   X6             1.0206         0.16383        6.23         2.8049e-08


Number of observations: 77, Error degrees of freedom: 72
Root Mean Squared Error: 0.448
R-squared: 0.813,   Adjusted R-Squared: 0.802
F-statistic vs. constant model: 78, p-value = 2.03e-25
```

9.2 Analysis of air pollution and mortality data

We begin by loading the data either from the R package **Sleuth3** or from **FSDA**. The dependent variable is Mortality.

1. **Perform some preliminary data analysis by visualizing the correlations.**
   After taking logs of the three pollution variables we do some exploratory analysis, looking at the correlations. For this purpose we can use the R package **corrplot** (Taiyun and Simko, 2024) or the function spmplot in **FSDA**. Figure B.59 shows the correlation plot obtained with the latter, which we prefer because it uses the main diagonal to show a boxplot for each data variable rather than an (uninformative) series of unitary circles. A second reason is that spmplot also has options for making the plot interactive, in order to summarize and visualize data by group more effectively. We see that two variables, namely, HC and NOX are very strongly positively correlated. Also, there are pairs of variables with high negative correlation: Poor and Sound and NonWhite and Over65.

2. **Perform classical ridge regression and LASSO and compare the models.**
   We start by estimating the LASSO for a range of the penalty parameter λ (using the function glmnet() from the R package **glmnet**). The result is shown in the left-hand panel of Fig. B.60. The penalty parameter λ varies from 0 on the

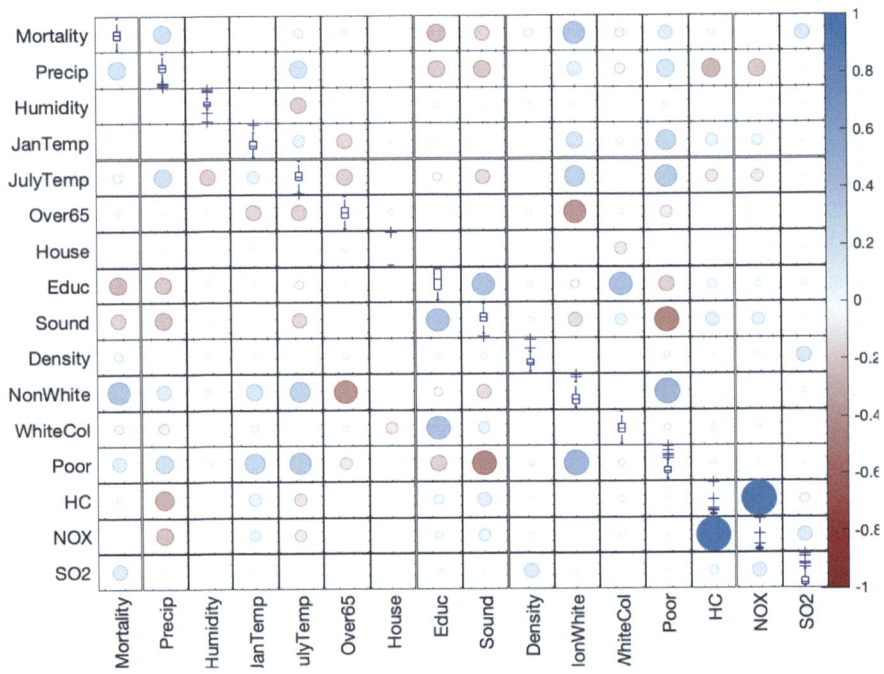

Fig. B.59 Air pollution and mortality data: correlation plot. The radius of the circle is proportional to the absolute value of the correlation. The colours are those of the correlation colour bar on the right

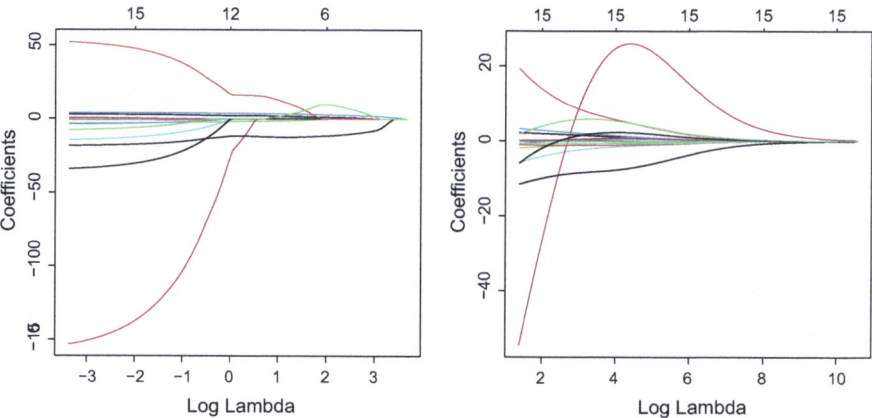

Fig. B.60 Air pollution and mortality data: coefficient paths plotted against the penalty parameter λ (on a log scale). Upper x axes, number of non-zero coefficients. Left-hand panel: LASSO; right-hand panel: ridge regression

left, where it does not have any effect, to some large value on the right where all coefficients β_j become 0 (λ is shown on a log scale). The upper x-axis shows the degrees of freedom (or the number of effective predictors in the model). Many of the coefficients become exactly zero early for large values of λ. The right-hand panel of Fig. B.60 shows related coefficient paths for ridge regression (also computed with the function glmnet() from the R package **glmnet**). We see that while all coefficients are shrunk towards 0 (as λ moves from 0 to some large value) none of them becomes exactly 0. Ridge regression provides a solution, even if the number of predictors $p > n$. However, it does not allow variable selection, i.e. does not produce a sparse solution.

The results for the LASSO in the left-hand panel of Fig. B.60 show the dependence of the number of non-zero parameter estimates on the value of the penalty parameter λ, but do not provide information as to which model is the most accurate for predicting new independent data from the same population, which is also a deficiency of ridge regression. Conducting cross validation and estimating the MSE (or some other measure) we can choose the penalty parameter for both algorithms. We can perform the selection of the penalty parameter with the function cv.glmnet() which reports the results of the cross validation in the plots shown in Fig. B.61 for the LASSO (left-hand panel) and ridge regression (right-hand panel). The dotted vertical lines show the value of λ which gives the minimal MSE and the largest value of λ such that error is within one standard error of the minimum. The minimal MSE for LASSO is 1590 obtained for $\lambda = 2.674$ which selects a model with 8 effective predictors. The next (within 1 SD) MSE is 1957 for $\lambda = 13.002$ which selects a model with 4 non-zero parameters. For ridge regression the minimal MSE is 1765 obtained for $\lambda = 19.31$. The next (within 1 SD) MSE is 2101 for $\lambda = 124.11$.

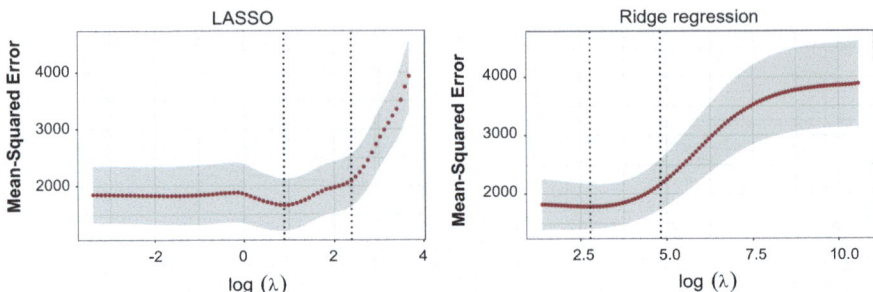

Fig. B.61 Air pollution and mortality data: cross-validated estimate of mean squared prediction error together with upper and lower standard deviation curves, as a function of the penalty parameter λ (on a log scale). Left-hand panel: LASSO, right-hand panel: ridge regression. The value of λ which gives the minimal MSE and the largest value of λ such that the error is within one standard error of the minimum are indicated by vertical dotted lines

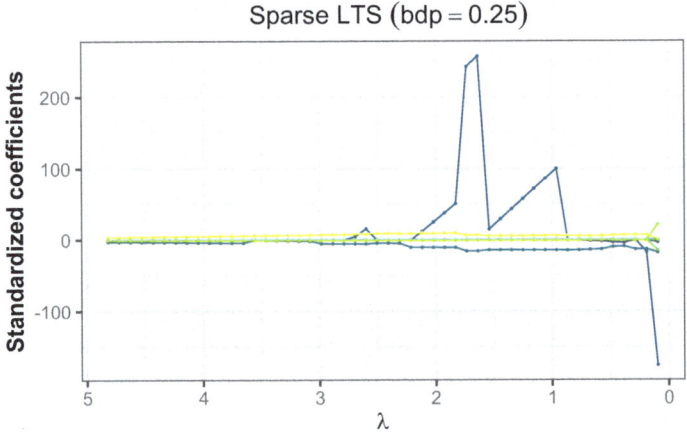

Fig. B.62 Air pollution and mortality data: coefficient path for the sparse LTS, plotted versus the penalty parameter λ

3. **Perform sparse LTS and compare to the classical LASSO.**
   We expect outliers in the air pollution and mortality dataset, we fit a sparse robust regression model. We use the function `sparseLTS()` from the R package **robustHD** with all default settings (e.g. $bdp = 0.25$) on a penalty grid with 50 values of λ. Figure B.62 shows the coefficient path plotted versus the penalty parameter λ—with increasing λ the coefficients are shrunk towards 0 so that, first it is possible to estimate the model although $p > n$ and second, a more tractable model with fewer non-zero coefficients is obtained.
   The optimal penalty parameter λ can be selected in a similar way as for the classical LASSO, using cross validation and estimating a measure of the prediction accuracy. However, instead of using the MSE, it is preferable to use a robust

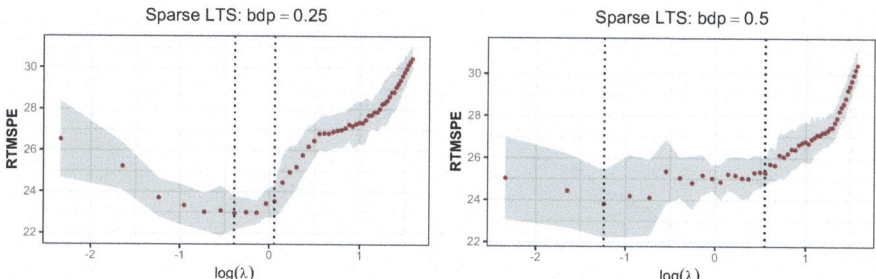

Fig. B.63 Air pollution and mortality data: cross-validated estimate of the robust root-mean squared prediction error for sparse LTS together with upper and lower standard deviation curves, as a function of the penalty parameter λ (on a log scale). The value of λ which gives the minimal RTMSPE and the largest value of λ such that the error is within one standard error of the minimum are indicated by vertical dotted lines. Left-hand panel: $bdp = 0.25$ and right-hand panel: $bdp = 0.5$. Note that the width of the interval depends strongly on the value of bdp which is used

measure such as the *root trimmed mean squared prediction error* (RTMSPE) proposed by Alfons et al. (2013). The results of cross validation for the sparse LTS with different breakdown points are shown in Fig. B.63. When $bdp = 0.25$ the λ for which the best prediction error of 23.540 is obtained is 1.0595 and the corresponding model has 6 non-zero parameters. The right-hand panel of Fig. B.63 shows the penalty parameter selection with the maximum $bdp = 0.5$ (in both cases we are looking at the reweighted version of the sparse LTS). Note that the width of the interval of λ depends strongly on the value of bdp which is used. When $bdp = 0.5$ the minimal value of RTMSPE 25.317 is obtained for $\lambda = 1.7337$ with 8 non-zero coefficients. Although in both cases ($bdp = 0.25$ and $bdp = 0.50$) only one of the three pollution related variables was selected: SO2, the number of non-zero coefficients is different and the confidence interval for λ is much wider for $bdp = 0.5$. If we add one of the other two (highly correlated) pollution variables (NOX or HC) to the model the results do not change (the estimated coefficient of the newly added variable is not significant, R-squared and adjusted R-squared slightly decrease).

4. **Are there outliers in the data? What is their effect?**

Robust sparse LTS provides diagnostic plots for identifying outliers. With the selected model (with six non-zero coefficients for $bdp = 0.25$) we can look at the robust regression diagnostic plots which are presented in Fig. B.64. The top left-hand panel shows a normal QQ-plot of the standardized residuals (that is scaled residuals see Sect. 3.2.2) against the quantiles of the standard normal distribution. Four observations (4, 7, 20, and 60) lie away from the diagonal line and are marked as outliers. The top right-hand panel shows the regression diagnostic plot presenting the standardized residuals against robust Mahalanobis distances. Again the four observations are identified as outliers. However, this traditional static analysis fails to reveal the effect of these outlying units on the fitted model; a dynamic analysis is required to answer this question. The bottom left-hand panel

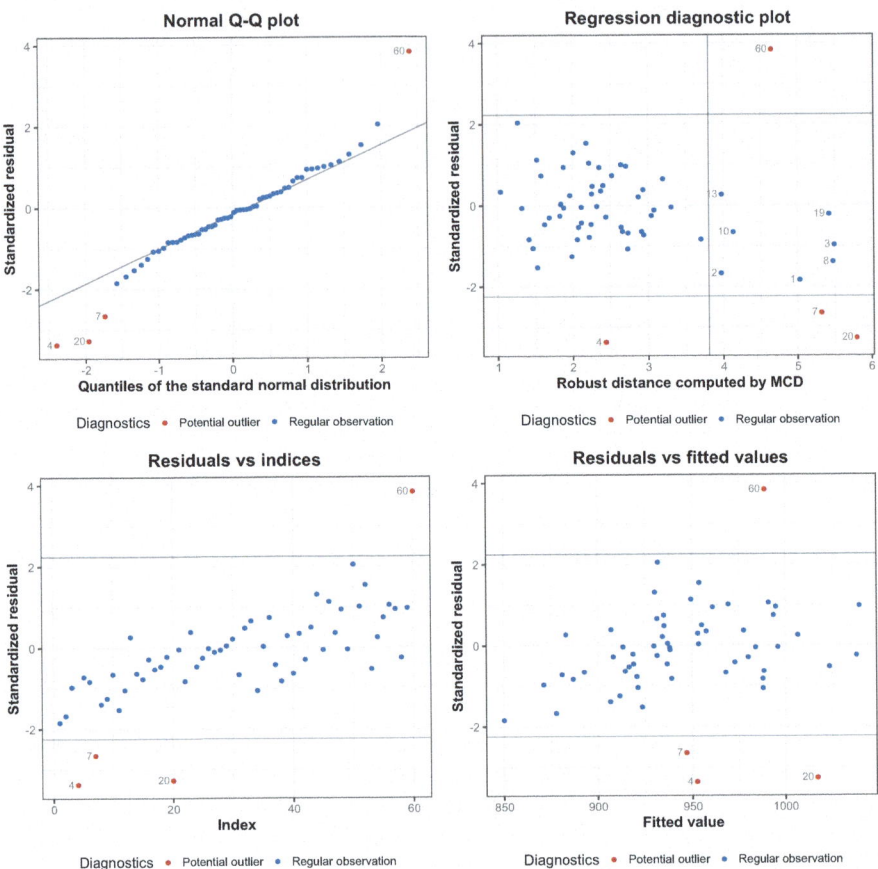

Fig. B.64 Air pollution and mortality data: robust regression diagnostic plots with 6 non-zero coefficients for $bdp = 0.25$. Top left-hand panel: normal QQ-plot of the standardized residuals against the quantiles of the standard normal distribution; Top right-hand panel: standardized residuals against robust Mahalanobis distances; Bottom left-hand panel: Index plot of the residuals; Bottom right-hand panel: plot of the residuals versus fitted values

shows the standardized residuals plotted against the index of the observations. Note that the observations are ordered by Mortality, the response variable, and the pattern of residuals suggests that we may have missed some structure. The effect of missing structure could be investigated by the monitoring of the added variable t-statistics.

9.3 Analysis of air pollution and mortality data II

We use the function pense_cv() from the package **pense** (Kepplinger et al., 2024) with all default settings (e.g. $bdp = 0.25$) to find the optimal penalty parameter λ. Since this function implements the *elastic net* as given in Eq. (9.13), the parameter $\alpha \in [0, 1]$ controls the compromise between the two penalties. With $\alpha = 1$ we have

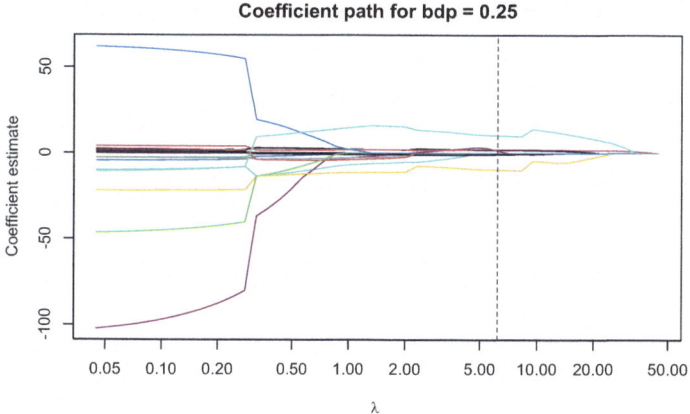

Fig. B.65 Air pollution and mortality data: coefficient path for the PENSE method, plotted versus the penalty parameter λ

the LASSO penalty and, with $\alpha = 0$, the ridge penalty. As indicated in the text of the exercise we set $\alpha = 1$. Figure B.65 shows the coefficient path plotted versus the penalty parameter λ—with increasing λ the coefficients are shrunk towards 0 so that, first it is possible to estimate the model although p is close to n and second, a more tractable model with fewer non-zero coefficients is obtained.

The optimal penalty parameter λ can be selected in a similar way as for the classical LASSO or sparse LTS, using cross validation and estimating a measure of the prediction accuracy. However, instead of using the mean squared prediction error, a robust version is obtained by using the τ-scale estimate (Maronna and Zamar, 2002) of the prediction errors. The results of cross validation with different breakdown points are shown in Fig. B.66. The λ for which the best prediction error of 35.240 is obtained is 4.675 and the corresponding model has 8 non-zero parameters. In this case, PENSE does not perform correct cross validation for the maximal breakdown point 0.5. Even for *bdp* as low as 0.4, it selects a model with all 15 predictors. The penalty selection with *bdp* = 0.4 is shown in the right-hand panel of Fig. B.66. The variables which are selected by LASSO, PENSE (robust LASSO in this case) and sparse LTS with *bdp* = 0.25 and *bdp* = 0.5 are shown in Table B.19. Each procedure selects a different set of variables. Those selected by PENSE with *bdp* = 0.4 contain all variables selected by all the other methods.

Problems: Chap. 10

10.1 Analysis of the heart rate data

1. The integer valued observations of the difference in pulse rates range from -12 to 94; 69 observations are greater than zero, 8 are zero and the rest negative. With such a great range the data may be treated as continuous. The histogram of the data shows high concentration around zero; the highest frequencies are 9, 10, 8, and

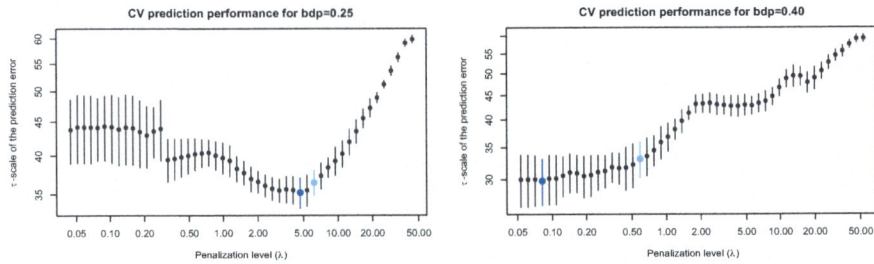

Fig. B.66 Air pollution and mortality data: cross-validated estimate of the prediction error for PENSE as a function of the penalty parameter λ. The value of λ which gives the minimal prediction error is shown in dark blue. The largest value of λ such that the error is within one standard error of the minimum is marked in light blue. Left-hand panel: $bdp = 0.25$ and right-hand panel: $bdp = 0.4$

Table B.19 Air pollution and mortality data: variables selected by the different procedures

	LASSO	LTS $bdp=$ 0.25	LTS $bdp=$ 0.50	PENSE $bdp=$ 0.25	PENSE $bdp=$ 0.40
Precip	X	X	X	X	X
Humidity					X
JanTemp			X		X
JulyTemp					X
Over65					X
House		X	X	X	X
Educ	X	X	X	X	X
Sound			X		X
Density		X		X	X
NonWhite	X	X	X	X	X
WhiteCol			X	X	X
Poor					X
HC					X
NOX					X
SO2	X	X	X	X	X
	4	6	8	7	15

9 for $y = -4, -2, 0$, and 2. There is a long right-hand tail mostly with 0 or unit frequencies.

2. The explanatory variables, including the indicators for a three-level factor, account for nine degrees of freedom with x_9 the indicator for running or sitting. The only significant variable from the regression (forward, backward, or full-model t-tests) is x_9. We now consider Yeo-Johnson transformation of the data. The left-hand panel of Fig. B.67 shows the fan plot for the transformation when all the variables are included in the model. It is clear that a transformation with $\lambda = 0.7$ is indicated; values of 0.6 and 0.8 are rejected when all the observations are fitted.

Solutions

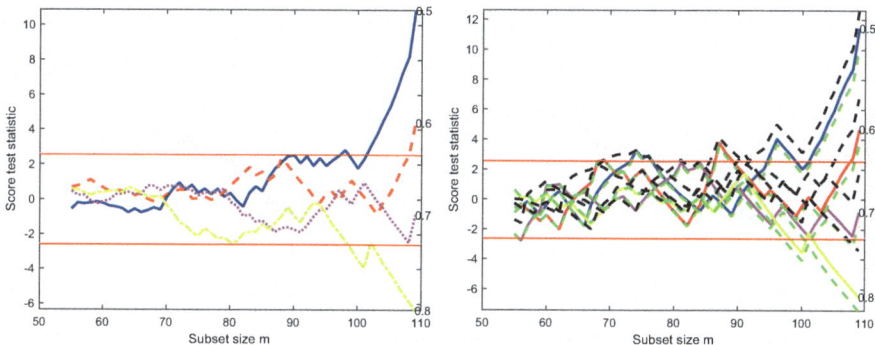

Fig. B.67 Heart rate data, Yeo-Johnson transformations: left-hand panel, fan plot for four values of λ, nine explanatory variables; right-hand panel, extended fan plot, only x_9. A value of 0.7 is indicated for λ

3. What is also clear is that there are no outliers causing abrupt changes in the indicated transformation. The plot for $\lambda = 0.7$ lies smoothly within the bounds throughout.

4. Model selection with the transformed response leads again to the model with one explanatory variable, x_9.

5. The second part of the analysis of transformations is to use the constructed variable for the two halves of the transformation to see whether negative and positive responses require distinct transformations. The extended fan plot in the right-hand panel of Fig. B.67 is for one fitted explanatory variable. For the three values $\lambda = 0.6, 0.7$ and 0.8 the dotted lines of the score tests for positive and negative observations lie close to the full lines for the overall transformation. We infer that both parts of the distribution require the same transformation.

6. We now exhibit some advantages of transformation. For the Box and Cox transformation these include approximate normality, homogeneity of variance, and a simple linear model. Here we already have a simple linear model so we concentrate on quality of inferences and the distribution of residuals, throughout for the linear model with one explanatory variable. QQ-plots of the studentized residuals are in Fig. B.68. That for the untransformed data in the left-hand panel shows symmetrical long tails which are much reduced in the plot in the right-hand panel for the transformed data. Normality has been better approximated by the transformation. In particular, the upper tail of the distribution for the transformed data is now close to normal. A minor feature of both plots is the occurrence of horizontal parts corresponding to the repeated response values listed above. However, it is more informative to look at the boxplots of the studentized residuals against the dichotomous variable x_9. The plots for the untransformed observations are in the left-hand panel of Fig. B.69. The boxplot for the lower value of x_9 is appreciably smaller in scale than that for the upper value; the assumption of homogeneity of variance in the data is clearly violated. After transformation, the right-hand panel of Fig. B.69 shows two groups with boxplots of much more similar size.

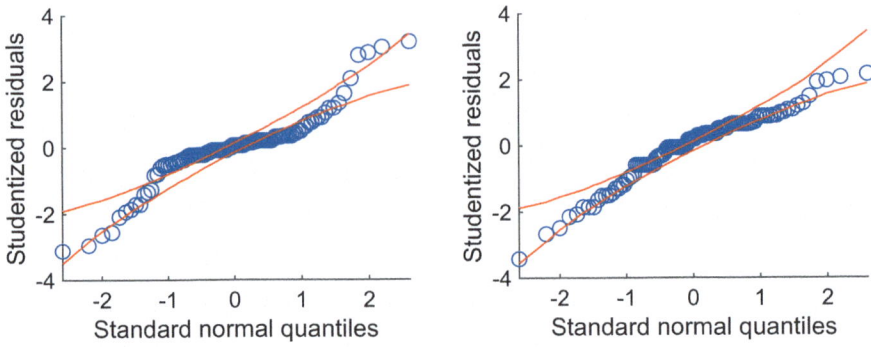

Fig. B.68 Heart rate data: normal QQ-plots of studentized residuals. Left-hand panel, untransformed data, right-hand panel, Yeo-Johnson transformation with $\lambda = 0.7$, showing improvement from transformation

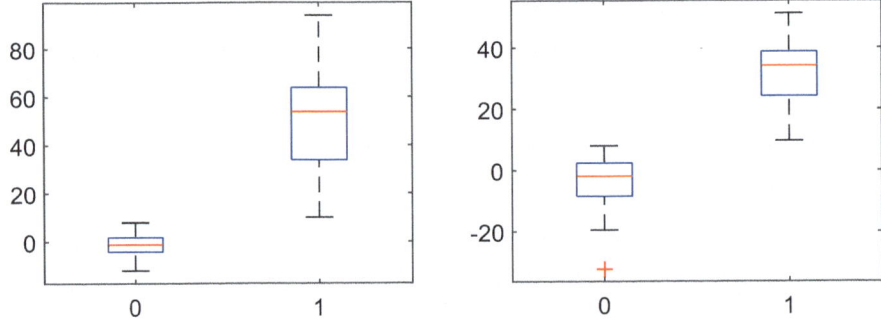

Fig. B.69 Heart rate data: boxplots of the response against x_9. Left-hand panel, untransformed data; right-hand panel, after transformation with $\lambda = 0.7$

7. The effect of transformation on the F-test for the effect of x_9 is appreciable. For the untransformed data the residual mean squared error is 14 and the F-statistic on 1 and 107 °C of freedom has the large value of 372. After transformation the residual mean squared error reduces to 8.91 and the F-statistic rises to 428. In either case there is no doubt about the significance of x_9. In less clear cut cases, the increase in power of the test due to transformation will be an important aspect of inference.

8. Application of the forward search to the untransformed data reveals 16 outliers, virtually all coming from the lower values of y from the upper value of x_9. Removing such observations improves the homogeneity of variance of the remaining observations. No outliers are detected when the search is run on the transformed data. The last observation to enter, which is close to being outlying, is number 40. An analysis on the original data with MM-estimation indicates 13 outliers when they are tested, as shown in the upper panel of Fig. B.70, using a 99% Bonferroni limit, that is 99.9908% pointwise. The lower panel of the figure, for the transformed data, shows that observation 40 is now just outlying. The point here is that, in this case, outlier

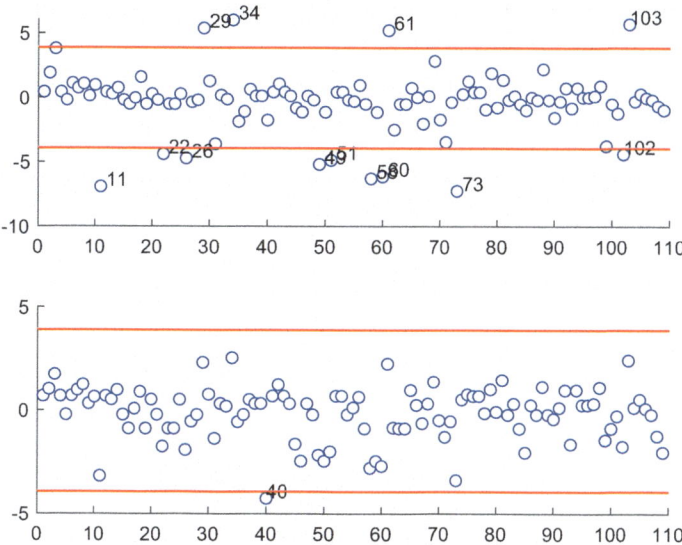

Fig. B.70 Heart rate data: index plot of standardized residuals from MM-regression. Upper panel untransformed data, lower panel $\lambda = 0.7$. 99% Bonferroni limits

detection does not depend critically on the robust method chosen. However, robust estimation of the transformation parameter requires the identification of influential observations in the fan plot coming from the ordering of observations provided by the forward search.

10.2 Additional analysis of the customer loyalty data

The plot of extended BIC for the five values of λ (Fig. B.71) has a clear peak at $\lambda = 0.5$; the square-root transformation is indicated. There is also strong agreement of the index AGI with the plot of the BIC values. The proportion of observations included in the regression analysis of the square-root transformed data is close to one (0.976). Figure 10.26, which contains the fan plot for the approximate score statistics for the five standard values of the Box-Cox parameter λ, together with 99% intervals for testing, shows that the last units to enter the search in the scale $\lambda = 0.5$ bring the value of the score statistic much outside the boundaries.

10.3 Analysis of the modified customer loyalty data in the square-root scale

The effect of the modified observations is clearly shown at the end of the monitoring plot of residuals given in Fig. B.72. From m close to 1675 onwards, there is an upward turn, that is a decrease in the absolute values of the large negative residuals. This figure is very similar to Fig. 10.40 which shows the monitoring plot in the original scale. The QQ-plot of studentized residuals in the upper panel of Fig. B.73 shows a strong deviation from normality and the presence of fat tails. The lower panel of the figure is a symmetry plot of residuals about their median. This panel confirms the interpretation of the upper panel. The FS procedure for outlier detection in the

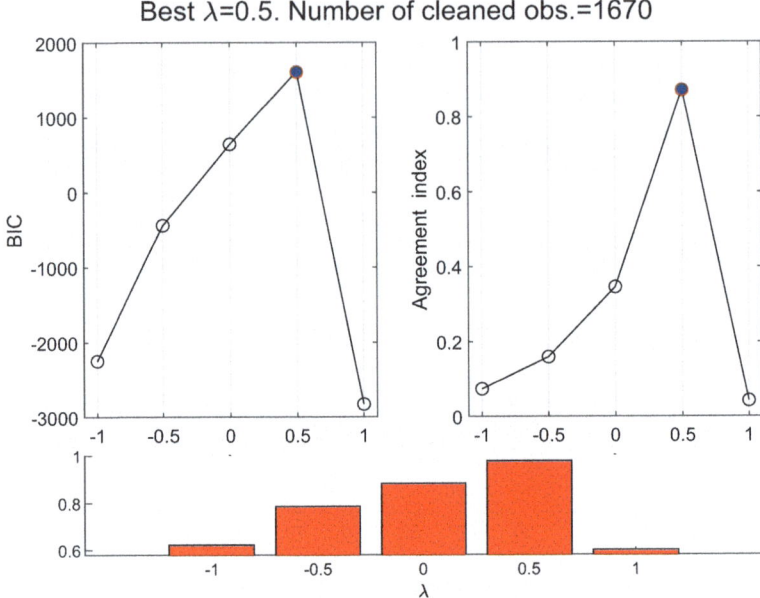

Fig. B.71 Customer loyalty data, Box-Cox transformation: output from automatic analysis with grid; $\mathcal{G} = -1, -0.5, 0, 0.5, 1$. Upper left-hand panel, extended BIC; upper right-hand panel, agreement index AGI. Lower panel, proportions $h(\lambda)/n$

square-root scale detects 65 outliers. The monitoring plot of the six added variable t-statistics in Fig. B.74 shows the effect of the outliers on inference about the values of the parameters. Until the outliers start to enter, the trajectories are virtually identical to those of Fig. 10.25. The effect of the outliers, particularly on the coefficient of x_6, is dramatic. The t-statistic goes from a value of around -10 to being non-significant.

10.4 Analysis of the auto mpg data

Figure B.75 shows the yX plot of these data: the relationship between y (mpg) and some of the explanatory variables seems non-linear. The univariate relationship between cylinders and y seems strong and inverse, as is to be expected. The ANOVA table is given in Table B.20. In this scale the percentage of variance which is explained is around 82%. Note that not all the variables are significant.

Figure B.76 gives the fan plot for the Box-Cox transformation. The value of –0.5 is accepted throughout the search. The three-panel plot from the automatic method of Sect. 6.3.4 in Fig. B.77 confirms that –0.5 is the preferred transformation and shows that there are no outliers. The ANOVA table based on all the observations and transformed data using $\lambda = -0.5$ (not given here) shows that the percentage of variance which is explained is around 89.1%. The automatic variable selection procedure implemented in the **MATLAB** function `stepwiselm` leads to identification of the

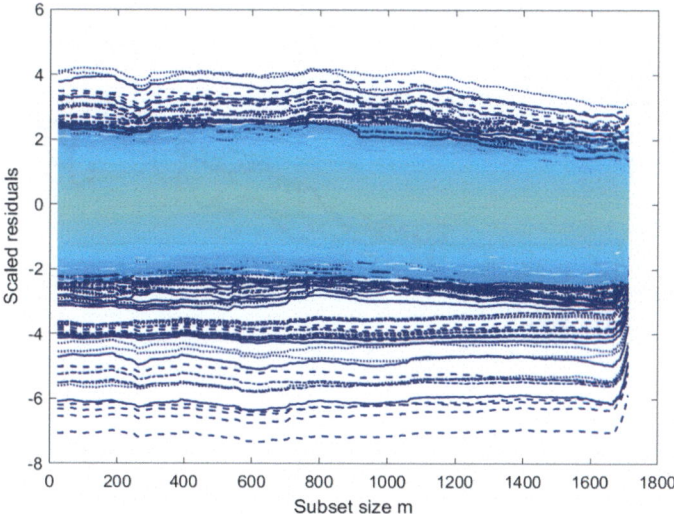

Fig. B.72 Modified customer loyalty data: square-root transformation; monitoring residuals from the FS. Note the change of residuals in the last steps

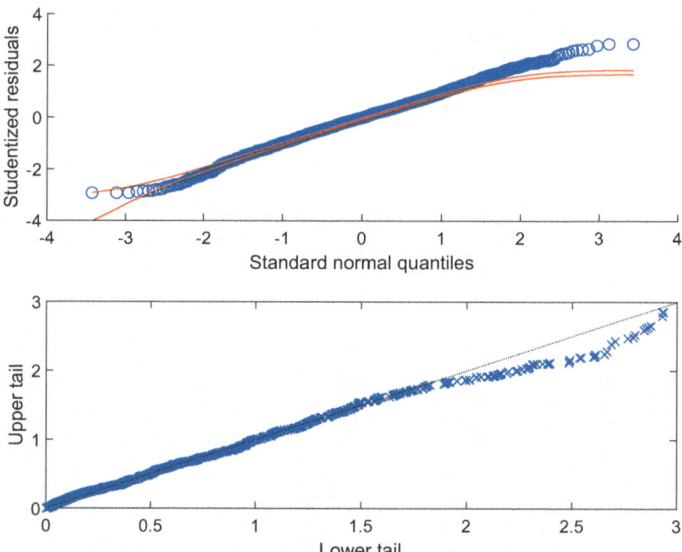

Fig. B.73 Modified customer loyalty data: square-root transformation; cumulative plots of residuals. Upper panel; QQ-plots of residuals with 95% simulation envelope: lower panel; symmetry plot of studentized residuals

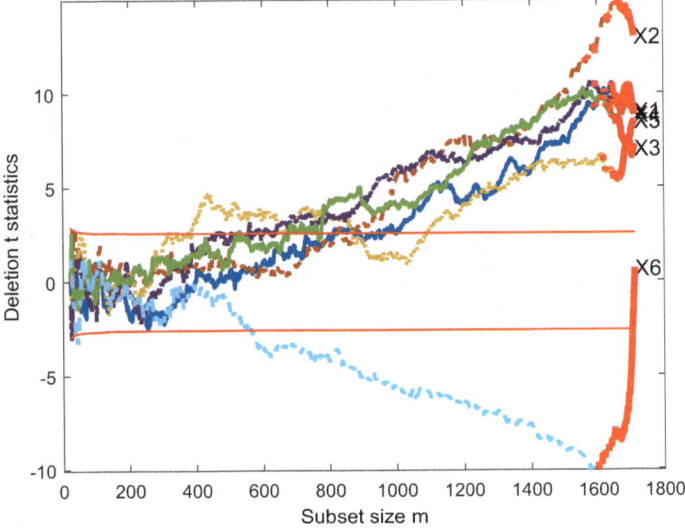

Fig. B.74 Modified customer loyalty data: monitoring of the 6 added variable t-statistics. Highlighted units are those declared as outliers by a single FS

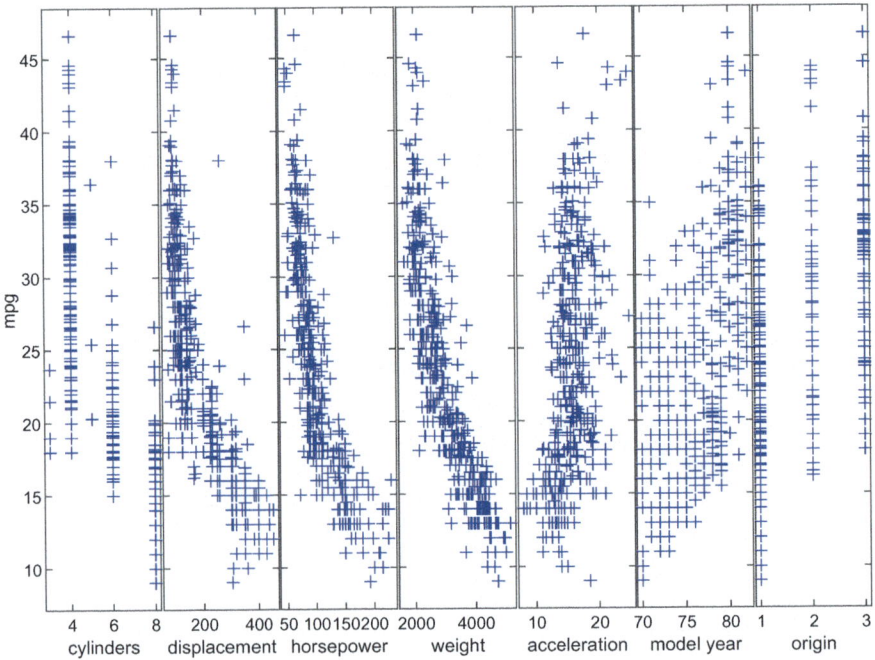

Fig. B.75 MPG: yX plot. The relationship between y and some of the explanatory variables seems non-linear

Table B.20 MPG: ANOVA in the original scale for y

	Estimate	SE	tStat	pValue
(Intercept)	-15.101	4.6807	-3.2263	0.001362
cylinders	-0.48971	0.32123	-1.5245	0.12821
displacement	0.023979	0.0076533	3.1331	0.0018627
horsepower	-0.018183	0.013709	-1.3264	0.18549
weight	-0.0067104	0.00065513	-10.243	6.3756e-22
acceleration	0.079103	0.098218	0.80538	0.4211
modelYear	0.77703	0.051784	15.005	2.3329e-40
ori1	-2.8532	0.55274	-5.162	3.9332e-07
ori2	-0.22323	0.56608	-0.39434	0.69355

Number of observations: 392, Error degrees of freedom: 383
Root Mean Squared Error: 3.31
R-squared: 0.824, Adjusted R-Squared: 0.821
F-statistic vs. constant model: 224, p-value = 1.79e-139

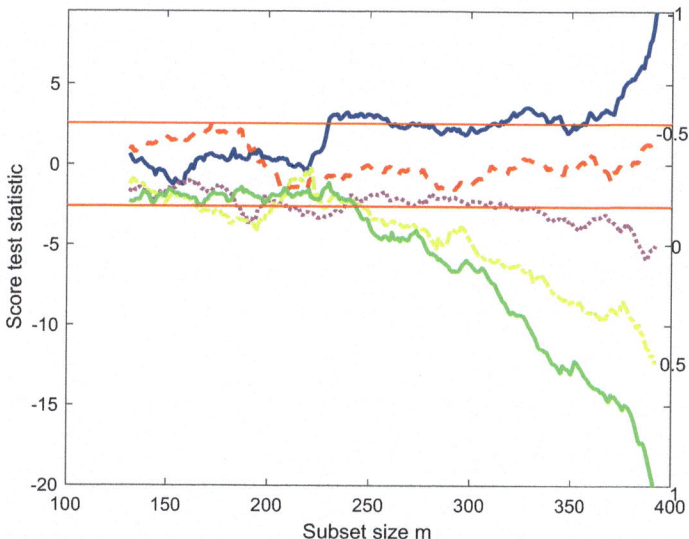

Fig. B.76 MPG, Box-Cox transformation: fan plot using the 5 most common values of λ

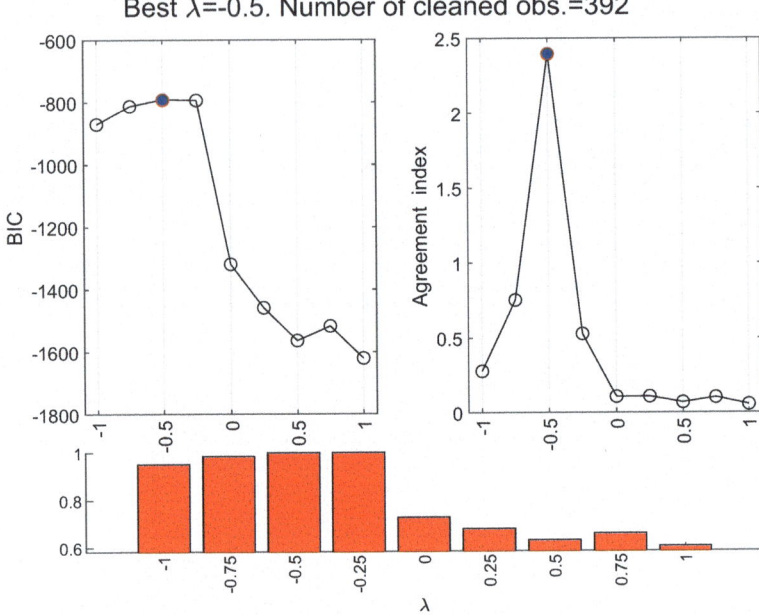

Fig. B.77 MPG: automatic analysis of Box-Cox transformation. Upper left-hand panel, extended BIC: upper right-hand panel, agreement index AGI; lower panel $m^*(\lambda)/n$, the proportion of observations used in fitting the model

Table B.21 MPG: ANOVA in the transformed scale for y after the variable selection procedure

	Estimate	SE	tStat	pValue
(Intercept)	-0.35847	0.015651	-22.904	6.1568e-74
horsepower	-0.00014031	3.4877e-05	-4.023	6.9148e-05
weight	-2.7049e-05	1.6456e-06	-16.438	2.1015e-46
modelYear	0.0031064	0.00019249	16.138	3.8328e-45
origin_2	0.0063097	0.0019361	3.259	0.0012173
origin_3	0.00507	0.0019573	2.5904	0.0099507

```
Number of observations: 392, Error degrees of freedom: 386
Root Mean Squared Error: 0.0125
R-squared: 0.888,  Adjusted R-Squared: 0.887
F-statistic vs. constant model: 614, p-value = 2.79e-181
```

importance of just four variables: horsepower, weight, modelYear and origin. The same result is obtained using the R package **olsrr** (Hebbali, 2024) and the BIC criterion. The ANOVA table after variable selection is given in Table B.21.

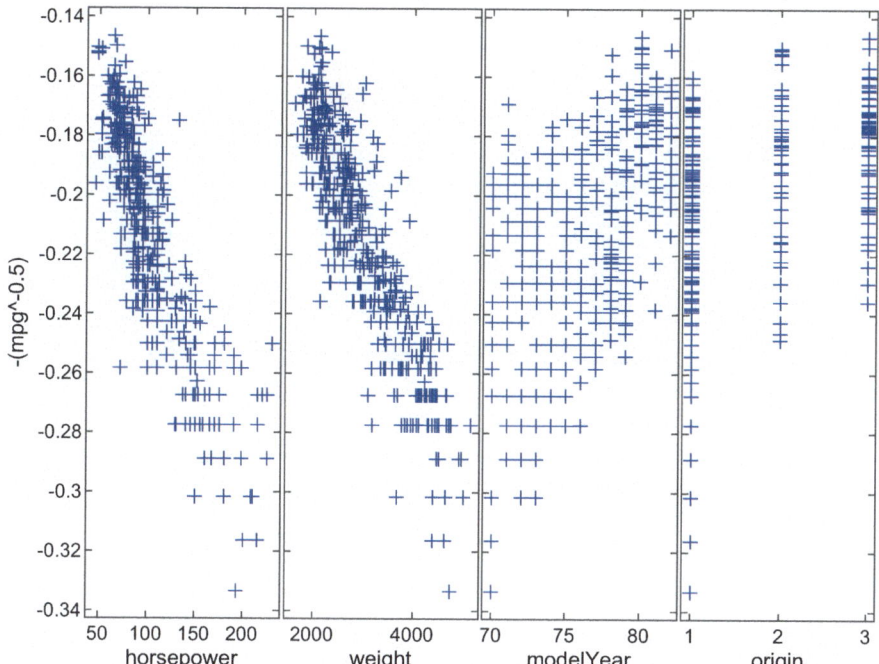

Fig. B.78 MPG, $\lambda = -0.5$: yX plot in the transformed scale for y versus the significant explanatory variables after the variable selection procedure

It is interesting to notice that cylinders do not appear after the variable selection procedure. This is because the correlation between this variable and those already present is very high. Figure B.78 shows the yX plot for the important variables after the variable selection procedure with $\lambda = -0.5$. Figures B.79 and B.80 which contain the monitoring of the scaled residuals, show that the brushing of the units with the largest positive and largest negative residuals leads to the identification of a series of cars with very high or low values of y given their weight.

Now we move to the non-parametric approach provided by RAVAS. Given that there is high correlation among the explanatory variables, options `scail` and `orderR2` should help.

The left-hand panel of Fig. B.81 shows that just two solutions have p-values of DW and normality greater than the threshold of 0.05. In both these solutions options `scail` and `orderR2` are used. The right-hand panel shows that these two solutions are virtually indistinguishable.

The details of the first solution of RAVAS are in Fig. B.82. There are 6 outliers. The transformation of all the explanatory variables is monotonic except for acceleration.

The final analysis is summarized in the ANOVA of Table B.22, based on the transformation from RAVAS with 6 observations deleted. All the original variables are significant and the value of the adjusted R^2 is greater than 0.9.

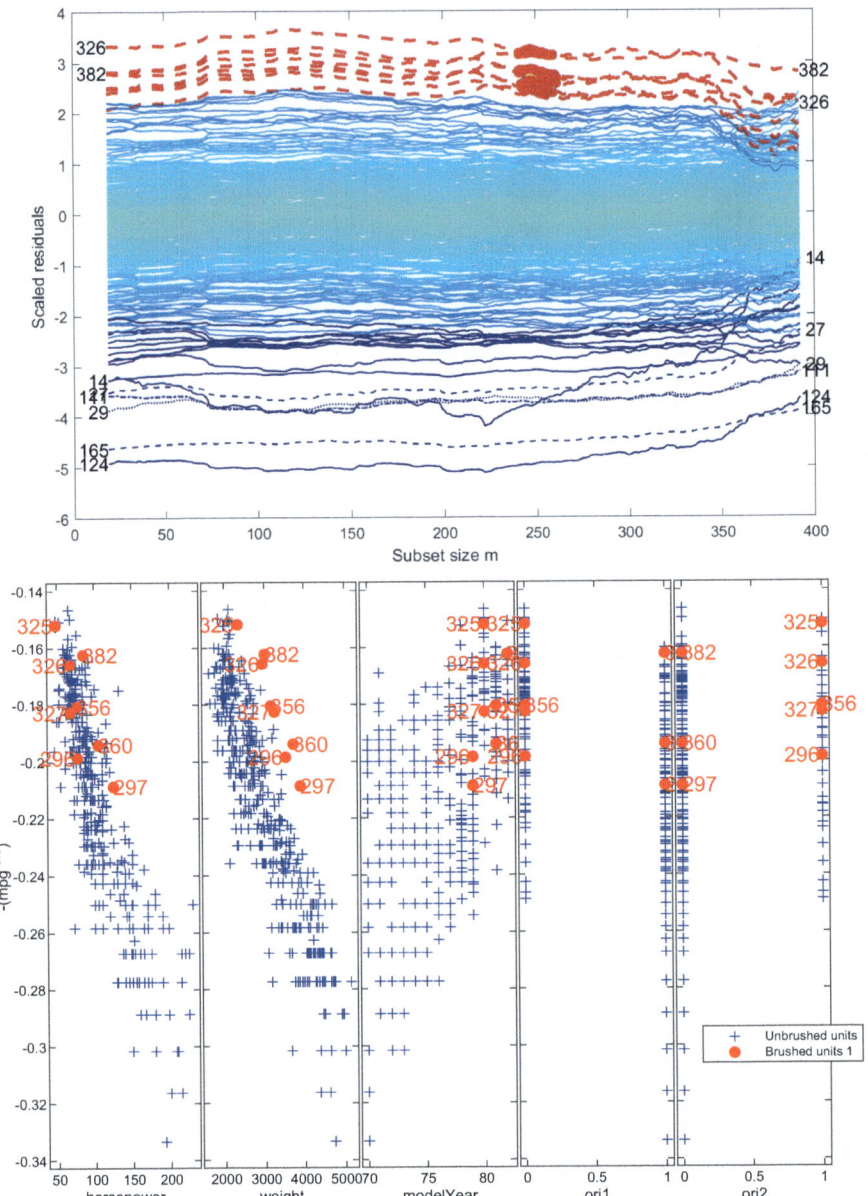

Fig. B.79 MPG, $\lambda = -0.5$. Upper panel: monitoring of scaled residuals with brushing of the trajectories with large residuals. Lower panel: yX plot in the transformed scale for y vs the significant explanatory variables after the variable selection procedure with the units brushed shown with red circles. Units with large positive residuals are associated with cars with very large mpg given their weight

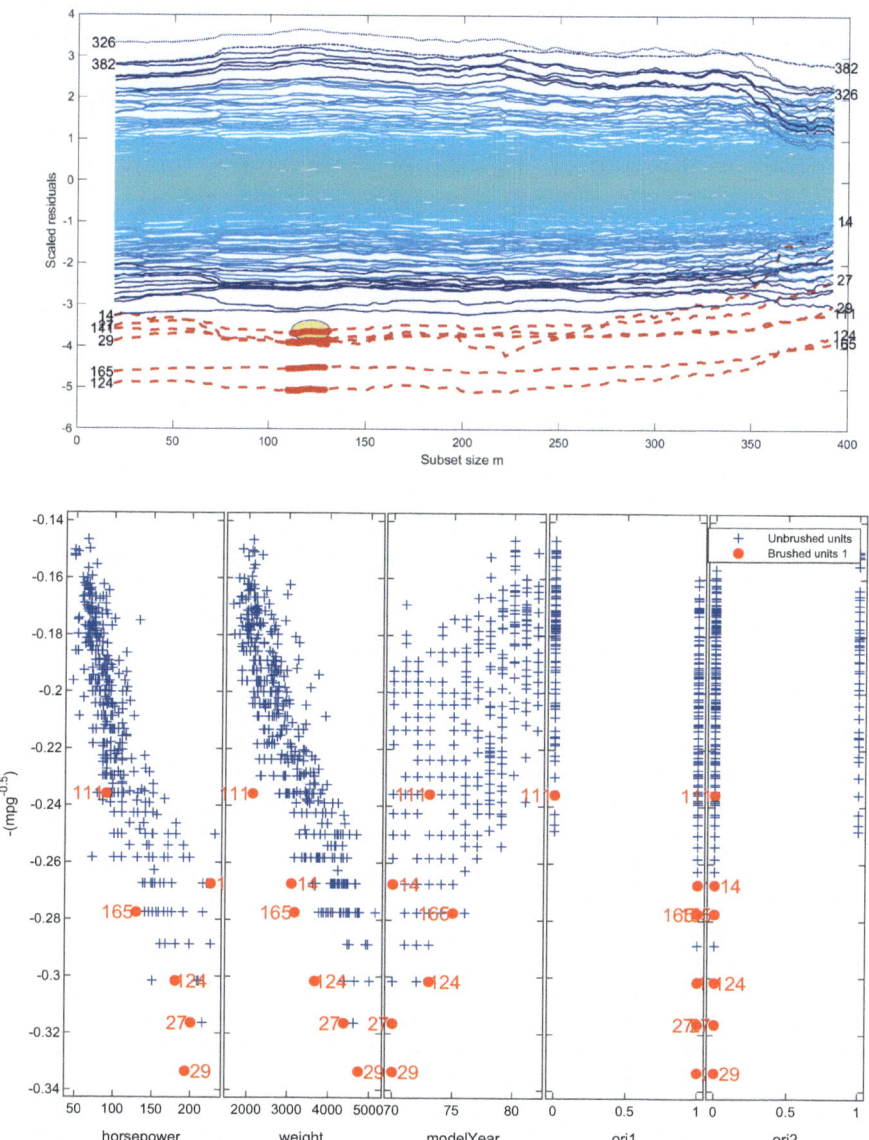

Fig. B.80 MPG, $\lambda = -0.5$. Upper panel: monitoring of scaled residuals with brushing of the trajectories with large negative residuals. Lower panel: yX plot in the transformed scale for y vs the significant explanatory variables after the variable selection procedure with the units brushed shown with red circles. Units with large negative residuals are associated with cars with very small mpg given their weight

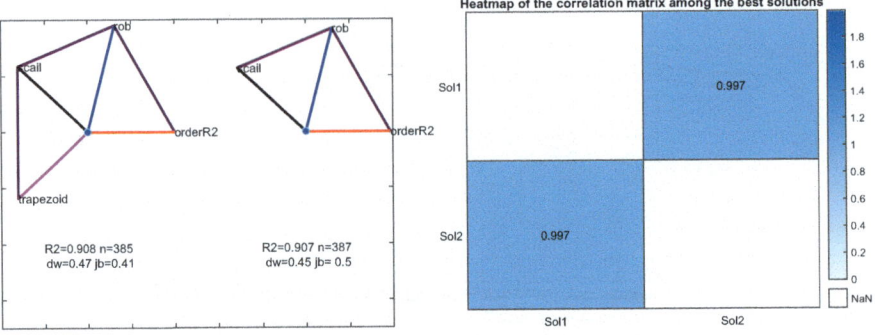

Fig. B.81 MPG: left-hand panel, augmented star plot of options in RAVAS. The best solution has four options and the statistically indistinguishable second-best solution does not include the trapezoid option. Right-hand panel, correlations among the solutions

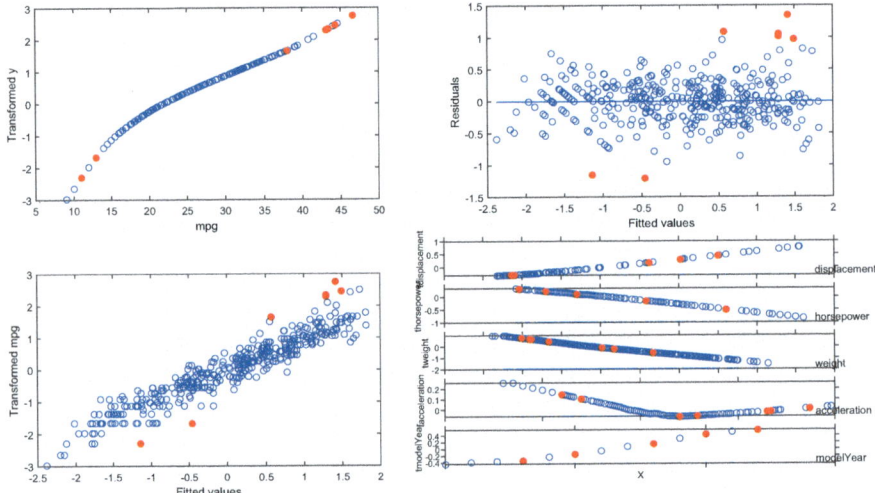

Fig. B.82 MPG, RAVAS: upper left-hand panel, $g(y)$ against y; upper right-hand panel, residuals against fitted values; lower left-hand panel, $g(y)$ against fitted values; lower right-hand panel transformations $f_j(x_j)$ of the explanatory variables against original x_j (all explanatory variables except cylinders and origin are shown)

The monitoring of the added variable t-statistics in the transformed RAVAS space (Fig. B.83) shows that the significance of the explanatory variables does not depend on the outliers which have been found. However, the t-statistics cease to grow with m when the outliers start to be included.

The three-dimensional scatter plots of $g(y)$ in the transformed RAVAS space, against the two most significant variables (weight and modelYear), before and after rotation, are given in Fig. B.84. Note that after appropriate rotation the outliers are far from the bulk of the data.

Solutions

Table B.22 ANOVA in transformed RAVAS scale for y after removing the outliers

	Estimate	SE	tStat	pValue
(Intercept)	-6.5829e-16	0.015451	-4.2604e-14	1
cylinders	0.86561	0.15469	5.5957	4.2393e-08
displacement	0.5338	0.19423	2.7482	0.0062804
horsepower	1.1045	0.18472	5.9793	5.2155e-09
weight	0.85468	0.076392	11.188	2.8138e-25
acceleration	1.759	0.39443	4.4596	1.0859e-05
modelYear	0.98379	0.051538	19.088	2.7915e-57
ori1	0.71431	0.18533	3.8544	0.00013643
ori2	0.76336	0.63725	1.1979	0.23171

```
Number of observations: 385, Error degrees of freedom: 376
Root Mean Squared Error: 0.303
R-squared: 0.91,  Adjusted R-Squared: 0.908
F-statistic vs. constant model: 477, p-value = 1.33e-191
```

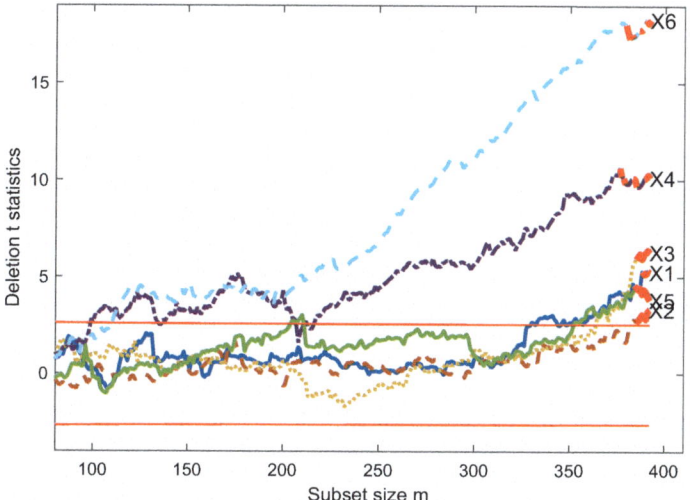

Fig. B.83 MPG: monitoring of the added variable t-statistics in the transformed RAVAS space for the first 6 continuous variables. Highlighted units are the 6 outliers found from RAVAS

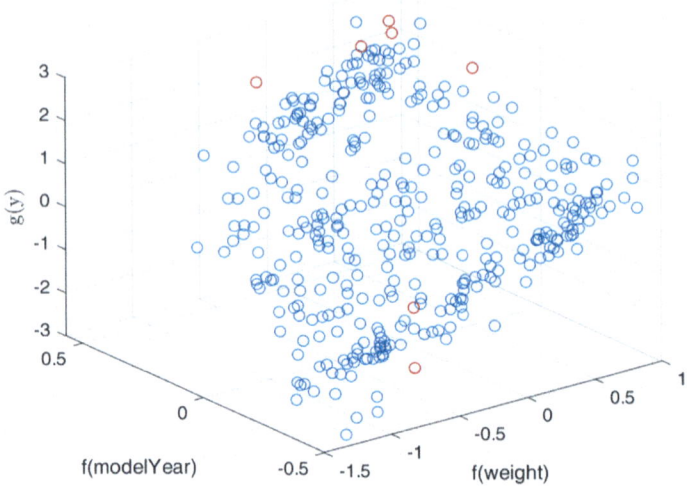
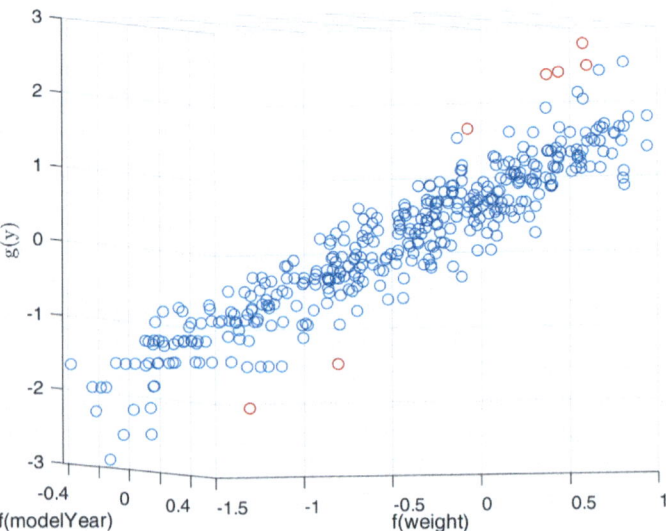

Fig. B.84 MPG transformed using RAVAS: three-dimensional scatter plot of $g(y)$, f(weight) and f(modelYear) without (top panel) and with rotation. The outliers are shown as red circles

References

Agulló, J. J. E. (1997). Exact algorithms for computing the least median of squares estimate in multiple linear regression. *IMS Lecture Notes-Monograph Series, 31*, 133–146.

Aitchison, J. (1986). *The statistical analysis of compositional data*. Monographs on statistics and applied probabilityLondon, UK: Chapman & Hall.

Aitchison, J. (2005). *A concise guide to compositional data analysis*. Available online at CoDaWeb: CoDaWeb. Lecture Notes.

Akaike, H. (1974). A new look at the statistical model identification. *IEEE Transactions on Automatic Control, 19*, 716–723.

Alfons, A. (2021). RobustHD: An R package for robust regression with high-dimensional data. *Journal of Open Source Software, 6*(67), 3786.

Alfons, A., Croux, C., & Gelper, S. (2013). Sparse least trimmed squares regression for analyzing high-dimensional large data sets. *The Annals of Applied Statistics, 7*(1), 226–248.

Alimadad, A., & Salibian-Barrera, M. (2011). An outlier-robust fit for generalized additive models with applications to disease outbreak detection. *Journal of the American Statistical Association, 106*, 719–731.

Amemiya, T. (1984). Tobit models: A survey. *Journal of Econometrics, 24*, 3–61.

Andrews, D. F. (1974). A robust method for linear regression. *Technometrics, 16*, 523–531.

Andrews, D. F., Bickel, P. J., Hampel, F. R., Tukey, J. W., & Huber, P. J. (1972). *Robust estimates of location: survey and advances*. Princeton, NJ: Princeton University Press.

Anglin, P., & Gençay, R. (1996). Semiparametric estimation of a hedonic price function. *Journal of Applied Econometrics, 11*, 633–648.

Anscombe, F. J., & Tukey, J. W. (1963). The examination and analysis of residuals. *Technometrics, 5*(2), 141–160.

Aranda-Ordaz, F. J. (1981). On two families of transformations to additivity for binary response models. *Biometrika, 68*, 357–363.

Argote-Espino, D., Lopez-García, P., & Facevicova, K. (2018). Statistical processing of compositional data. The case of ceramic samples from the archaeological site of Xalasco, Tlaxcala. Mexico. *Journal of Archaeological Science: Reports, 19*, 100–114.

Atkinson, A. C. (1982). Regression diagnostics, transformations and constructed variables (with discussion). *Journal of the Royal Statistical Society, Series B, 44*, 1–36.

Atkinson, A. C. (1985). *Plots, transformations, and regression*. Oxford: Oxford University Press.

Atkinson, A. C. (1994). Fast very robust methods for the detection of multiple outliers. *Journal of the American Statistical Association, 89*, 1329–1339.

Atkinson, A. C. & Cox, D. R. (1988). Transformations I. *Wiley Statsref.*

Atkinson, A. C., & Lawrance, A. J. (1989). A comparison of asymptotically equivalent tests of regression transformation. *Biometrika, 76,* 223–229.

Atkinson, A. C., & Riani, M. (2000). *Robust diagnostic regression analysis.* New York: Springer.

Atkinson, A. C., & Riani, M. (2002). Forward search added-variable *t* tests and the effect of masked outliers on model selection. *Biometrika, 89,* 939–946.

Atkinson, A. C., & Riani, M. (2002). Tests in the fan plot for robust, diagnostic transformations in regression. *Chemometrics and Intelligent Laboratory Systems, 60,* 87–100.

Atkinson, A. C., & Riani, M. (2006). Distribution theory and simulations for tests of outliers in regression. *Journal of Computational and Graphical Statistics, 15,* 460–476.

Atkinson, A. C., Pericchi, L. R., & Smith, R. L. (1991). Grouped likelihood for the shifted power transformation. *Journal of the Royal Statistical Society, Series B, 53,* 473–482.

Atkinson, A. C., Riani, M., & Cerioli, A. (2004). *Exploring multivariate data with the forward search.* New York: Springer.

Atkinson, A. C., Donev, A. N., & Tobias, R. D. (2007). *Optimum experimental designs, with SAS.* Oxford: Oxford University Press.

Atkinson, A. C., Riani, M., & Cerioli, A. (2010). The forward search: theory and data analysis (with discussion). *Journal of the Korean Statistical Society, 39,* 117–134. https://doi.org/10.1016/j.jkss.2010.02.007

Atkinson, A. C., Riani, M., & Torti, F. (2016). Robust methods for heteroskedastic regression. *Computational Statistics and Data Analysis, 104,* 209–222.

Atkinson, A. C., Corbellini, A., & Riani, M. (2018). Robust Bayesian regression with the forward search: Theory and data analysis. *International Statistical Review.* https://doi.org/10.1111/insr.12247

Atkinson, A. C., Riani, M., & Corbellini, A. (2020). The analysis of transformations for profit-and-loss data. *Applied Statistics, 69,* 251–275. https://doi.org/10.1111/rssc.12389

Atkinson, A. C., Riani, M., & Corbellini, A. (2021). The Box-Cox transformation: Review and extensions. *Statistical Science, 36,* 239–255. https://doi.org/10.1214/20-STS778

Azzini, I., Perrotta, D., and Torti, F. (2023). A practically efficient fixed-pivot selection algorithm and its extensible Matlab suite. arXiv, stat.ME. https://doi.org/10.48550/arXiv.2302.05705

Barlow, R. E., Bartholomew, D. J., Bremner, J. M., & Brunk, H. D. (1972). *Statistical inference under order restrictions.* Chichester: Wiley.

Barndorff-Nielsen, O. E., & Cox, D. R. (1989). *Asymptotic techniques for use in statistics.* London: Chapman and Hall.

Barndorff-Nielsen, O. E., & Cox, D. R. (1994). *Inference and asymptotics.* London: Chapman and Hall.

Bartlett, M. S. (1947). The use of transformations. *Biometrics, 3,* 39–52.

Bartoloni, E. (2013). Profitability and innovation: New empirical findings based on Italian data 1996–2003. *Rivista Internazionale di Scienze Sociali, 2*(2013), 137–170.

Beaton, A. E., & Tukey, J. W. (1974). The fitting of power series, meaning polynomials, illustrated on band-spectroscopic data. *Technometrics, 16,* 147–185.

Belsley, D. A., Kuh, E., & Welsch, R. E. (1980). *Regression diagnostics.* New York: Wiley.

Berrendero, J. R., & Zamar, R. H. (2001). Maximum bias curves for robust regression with non-elliptical regressors. *Annals of Statistics, 29,* 224–251.

Berrendero, J. R., Mendes, B. V. M., & Tyler, D. E. (2007). On the maximum bias functions of MM-estimates and constrained M-estimates of regression. *Annals of Statistics, 35,* 13–40.

Bhat, H. S., & Kumar, N. (2010). On the derivation of the Bayesian Information Criterion. Technical report, University of California, Merced CA 95343.

Biernacki, C., Celeux, G., & Govaert, G. (2000). Assessing a mixture model for clustering with the integrated completed likelihood. *IEEE Transactions on Pattern Analysis and Machine Intelligence, 22,* 719–725.

Bliss, C. I. (1934). The method of probits. *Science, 79*(2037), 38–39.

Boente, G., Martínez, A., & Salibian-Barrera, M. (2017). Robust estimators for additive models using backfitting. *Journal of Nonparametric Statistics, 29*, 744–767.

Box, G. E. P. (1953). Non-normality and tests on variances. *Biometrika, 40*, 318–335.

Box, G. E. P. (1980). Sampling and Bayes' inference in scientific modelling and robustness (with discussion). *Journal of the Royal Statistical Society, Series A, 143*, 383–430.

Box, G. E. P., & Cox, D. R. (1964). An analysis of transformations (with discussion). *Journal of the Royal Statistical Society, Series B, 26*, 211–252.

Box, G. E. P., & Jenkins, G. M. (1976). *Time series analysis: forecasting and control*. San Francisco: Holden-Day.

Box, G. E. P., & Tidwell, P. W. (1962). Transformations of the independent variables. *Technometrics, 4*, 531–550.

Breiman, L. (1988). Comment on "Monotone regression splines in action" (Ramsey, 1988). *Statistical Science, 3*, 442–445.

Breiman, L., & Friedman, J. H. (1985). Estimating optimal transformations for multiple regression and transformation (with discussion). *Journal of the American Statistical Association, 80*, 580–619.

Brown, B. W., & Hollander, M. (1977). *Statistics: A biomedical introduction*. New York: Wiley.

Bruce, D., & Schumacher, F. X. (1935). *Forest mensuration*. New York: Mc-Graw Hill.

Bühlmann, P. (2020). Invariance, causality and robustness. *Statistical Science, 35*, 404–426.

Buja, A., & Kass, R. E. (1985). Comment on "Estimating optimal transformations for multiple regression and transformation" by Breiman and Friedman. *Journal of the American Statistical Association, 80*, 602–607.

Buja, A., & Rolke, W. (2003). Calibration for simultaneity: (re)sampling methods for simultaneous inference with applications to function estimation and functional data. Technical report, The Wharton School, University of Pennsylvania.

Buja, A., Hastie, T., & Tibshirani, R. (1989). Linear smoothers and additive models. *Annals of Statistics, 17*, 453–510.

Burnham, K. P., & Anderson, D. R. (2002). *Model selection and multi-model inference*. A practical information-theoretic approachNew York: Springer.

Cantoni, E., & Ronchetti, E. M. (2001). Robust inference for generalized linear models. *Journal of the American Statistical Association, 96*, 1022–1030.

Carroll, R. J. (1982). Prediction and power transformations when the choice of power is restricted to a finite set. *Journal of the American Statistical Association, 77*, 908–915.

Carroll, R. J., & Ruppert, D. (1988). *Transformation and weighting in regression*. London: Chapman and Hall.

Cerioli, A., Farcomeni, A., & Riani, M. (2014). Strong consistency and robustness of the forward search estimator of multivariate location and scatter. *Journal of Multivariate Analysis, 126*, 167–183.

Cerioli, A., Garcia-Escudero, L. A., Mayo-Iscar, A., & Riani, M. (2018a). Finding the number of normal groups in model-based clustering via constrained likelihoods. *Journal of Computational and Graphical Statistics, 27*(2), 404–416.

Cerioli, A., Riani, M., Atkinson, A. C., & Corbellini, A. (2018b). The power of monitoring: How to make the most of a contaminated multivariate sample (with discussion). *Statistical Methods and Applications, 27*, 559–666. https://doi.org/10.1007/s10260-017-0409-8

Cerioli, A., Farcomeni, A., & Riani, M. (2019). Wild adaptive trimming for robust estimation and cluster analysis. *Scandinavian Journal of Statistics, 46*(1), 235–256.

Chaloner, K., & Brant, R. (1988). A Bayesian approach to outlier detection and residual analysis. *Biometrika, 75*, 651–659.

Cheng, T.-C. (2011). Robust diagnostics for the heteroscedastic regression model. *Computational Statistics and Data Analysis, 55*, 1845–1866.

Claeskens, G., & Hjort, N. L. (2008). *Model selection and model averaging*. Cambridge: Cambridge University Press.

Cook, R. D., & Hawkins, D. M. (1990). Outliers everywhere: Comment on Rousseeuw and Van Zomeren (1990). *Journal of the American Statistical Association, 85*, 640–4.
Cook, R. D., & Weisberg, S. (1982). *Residuals and influence in regression*. London: Chapman and Hall.
Cook, R. D., & Weisberg, S. (1994). *An introduction to regression graphics*. New York: Wiley.
Cox, D. R. (1977). Nonlinear models, residuals and transformations. *Mathematische Operationsforschung und Statistik, Serie Statistik, 8*, 3–22.
Cox, D. R., & Donnelly, C. A. (2011). *Principles of applied statistics*. Cambridge: Cambridge University Press.
Cox, D. R., & Hinkley, D. V. (1974). *Theoretical statistics*. London: Chapman and Hall.
Cox, D. R., & Reid, N. (1987). Parameter orthogonality and approximate conditional inference (with discussion). *Journal of the Royal Statistical Society, Series B, 49*, 1–39.
Cox, D. R., & Snell, E. J. (1981). *Applied statistics: Principles and examples*. London: Chapman and Hall.
Croux, C., & Rousseeuw, P. J. (1992). A class of high-breakdown scale estimators based on subranges. *Communications in Statistics - Theory and Methods, 21*, 1935–1951.
Croux, C., Dhaene, G., & Hoorelbeke, D. (2004). Robust standard errors for robust estimators. CES - Discussion paper series OR 0367, Department of applied economics, KU Leuven.
Croux, C., Haesbroeck, G., & Ruwet, C. (2013). Robust estimation for ordinal regression. *Journal of Statistical Planning and Inference, 143*, 1486–1499.
Davies, L. (1990). The asymptotics of S-estimators in the linear regression model. *The Annals of Statistics, 18*(4), 1651–1675.
Davison, A. C. (2003). *Statistical models*. Cambridge: Cambridge University Press.
De Leeuw, J. (1968). Nonmetric discriminant analysis. Technical report, Department of data theory, University of Leiden, Netherlands. Research Note 06–68.
DeSarbo, W., & Cron, W. (1988). A maximum likelihood methodology for clusterwise linear regression. *Journal of Classification, 5*(2), 249–282.
Durbin, J., & Watson, G. S. (1950). Testing for serial correlation in least squares regression: I. *Biometrika, 37*, 409–428.
Economist. (2014). Uncontained. *Economist, UK Edn, 411*(8885), 59–60.
Efron, B., Hastie, T., Johnstone, I., & Tibshirani, R. (2004). Least angle regression. *Annals of Statistics, 32*, 407–499.
Egozcue, J. J., & Pawlowsky-Glahn, V. (2006). Simplicial geometry for compositional data. In *Compositional data analysis in the geosciences: From theory to practice. Special publications*, vol. 264 (pp. 145–160). London: Geological Society.
Egozcue, J. J., & Pawlowsky-Glahn, V. (2019). Compositional data: the sample space and its structure. *TEST: An Official Journal of the Spanish Society of Statistics and Operations Research, 28*(3), 599–638.
Egozcue, J. J., Pawlowsky-Glahn, V., Mateu-Figueras, G., & Barceló-Vidal, C. (2003). Isometric logratio transformations for compositional data analysis. *Mathematical Geology, 35*, 279–300.
Egozcue, J. J., Barceló-Vidal, C., Martín-Fernández, J. A., Jarauta-Bragulat, E., Díaz-Barrero, J. L., & Mateu-Figueras, G. (2011). Elements of simplicial linear algebra and geometry. In V. Pawlowsky-Glahn & A. Buccianti (Eds.), *Compositional data analysis: Theory and applications* (pp. 141–157). Chichester, UK: Wiley.
European Union. (2022). Judgment in case C-213/19. *Court of Justice of the European Union, Press Release No 42/22*.
Farcomeni, A., & Greco, L. (2015). *Robust methods for data reduction*. Boca Raton, FL: Chapman & Hall/CRC Press.
Fedorov, V. V., & Leonov, S. L. (2014). *Optimal design for nonlinear response models*. Boca Raton, FL: Chapman and Hall/ CRC Press.
Field, C. A., & Hampel, F. R. (1982). Small-sample asymptotic distributions of M-estimators of location. *Biometrika, 69*, 29–46.

References

Field, C. A., & Ronchetti, E. M. (1990). *Small sample asymptotics* (Vol. 13). Hayward, CA: Institute of Mathematical Statistics.

Filzmoser, P., & Hron, K. (2008). Outlier detection for compositional data using robust methods. *Mathematical Geosciences, 40*, 233–248.

Filzmoser, P., Hron, K., & Reimann, C. (2012). Interpretation of multivariate outliers for compositional data. *Computational Geosciences, 39*, 77–85.

Filzmoser, P., Hron, K., & Templ, M. (2018). *Applied compositional data analysis: with worked examples in R*. Cham, Switzerland: Springer. in statistics.

Firth, D. (1988). Multiplicative errors: Lognormal or gamma? *Journal of the Royal Statistical Society, Series B, 50*, 266–268.

Fisher, R. A. (2015). Frequency distribution of the values of the correlation coefficient in samples from an indefinitely large population. *Biometrika, 10*, 507–521.

Foi, A. (2008). Direct optimization of nonparametric variance-stabilizing transformations. In *Meeting on mathematical statistics*. C.I.R.M. Marseille

Forbes, J. D. (1857). Further experiments and remarks on the measurement of heights by the boiling point of water. *Transactions of the Royal Society of Edinburgh, 21*, 235–243.

Fowlkes, E. B., & Kettenring, J. R. (1985). Comment on "Estimating optimal transformations for multiple regression and transformation" by Breiman and Friedman. *Journal of the American Statistical Association, 80*, 607–613.

Fraley, C., & Raftery, A. E. (2002). Model-based clustering, discriminant analysis, and density estimation. *Journal of the American Statistical Association, 97*, 611–631.

Freue, G. V. C., Kepplinger, D., Salibian-Barrera, M., & Smucler, E. (2019). Robust elastic net estimators for variable selection and identification of proteomic biomarkers. *The Annals of Applied Statistics, 13*(4), 2065–2090.

Friedman, J. H., & Stuetzle, W. (1982). Smoothing of scatterplots. Technical report, Department of Statistics, Stanford University, Technical Report ORION 003.

Friedman, J. H., Hastie, T., & Tibshirani, R. (2010). Regularization paths for generalized linear models via coordinate descent. *Journal of Statistical Software, 33*(1), 1–22.

Fuller, W. A. (1987). *Measurement error models*. New York: Wiley.

Garcia-Escudero, L. A., Gordaliza, A., Mayo-Iscar, A., & San Martin, R. (2010). Robust clusterwise linear regression through trimming. *Computational Statistics and Data Analysis, 54*, 3057–3069. https://doi.org/10.1016/j.csda.2009.07.002

Garcia-Escudero, L. A., Gordaliza, A., Greselin, F., Ingrassia, S., & Mayo-Iscar, A. (2017). Robust estimation of mixtures of regressions with random covariates, via trimming and constraints. *Statistics and Computing, 27*(2), 377–402.

Gershenfeld, N. (1997). Nonlinear inference and cluster-weighted modeling. *Annals of the New York Academy of Sciences, 808*(1), 18–24.

Gershenfeld, N., Schoner, B., & Metois, E. (1999). Cluster-weighted modelling for time-series analysis. *Nature, 397*(6717), 329–332.

Gilmour, S. G. (1996). The interpretation of Mallows's C_p-statistic. *The Statistician, 45*, 49–56.

Giudici, P., Raffinetti, E., & Riani, M. (2024). Robust machine learning models: Linear and nonlinear. *International Journal of Data Science and Analytics*. https://doi.org/10.1007/s41060-024-00512-1

Gnedenko, B. V. (1962). *The theory of probability*. New York: Chelsea.

Greenacre, M. (2019). *Compositional data analysis in practice*. Boca Raton, FL: Chapman & Hall/CRC Press.

Greene, W. H. (2012). *Econometric analysis* (7th ed.). Upper Saddle River, NJ: Prentice Hall.

Gruen, B., & Leisch, F. (2007). Fitting finite mixtures of generalized linear regressions in R. *Computational Statistics and Data Analysis, 51*(11), 5247–5252.

Guenther, W. C. (1977). An easy method for obtaining percentage points of order statistics. *Technometrics, 19*, 319–321.

Guerrero, V. M., & Johnson, R. A. (1982). Use of the Box-Cox transformation with binary response models. *Biometrika, 69*, 309–314.

Hadi, A. S. (1992). Identifying multiple outliers in multivariate data. *Journal of the Royal Statistical Society, Series B, 54*, 761–771.

Hadi, A. S., & Simonoff, J. S. (1993). Procedures for the identification of multiple outliers in linear models. *Journal of the American Statistical Association, 88*, 1264–1272.

Hald, A. (1952). *Statistical theory with engineering applications*. New York: Wiley.

Hampel, F. R. (1974). The influence curve and its role in robust estimation. *Journal of the American Statistical Association, 69*(346), 383–393.

Hampel, F. R. (1975). Beyond location parameters: Robust concepts and methods. *Bulletin of the International Statistical Institute, 46*, 375–382.

Hampel, F. R., Rousseeuw, P. J., & Ronchetti, E. M. (1985). The change-of-variance curve and optimal redescending M-estimators. *Journal of the American Statistical Association, 76*, 643–648.

Hampel, F. R., Ronchetti, E. M., Rousseeuw, P. J., & Stahel, W. A. (1986). *Robust statistics*. New York: Wiley.

Harrell Jr, F. E. (2024). *Hmisc*. R package version 5.2-1.

Harvey, A. C. (1976). Estimating regression models with multiplicative heteroscedasticity. *Econometrica, 44*, 461–465.

Hastie, T., & Tibshirani, R. (1986). Generalized additive models. *Statistical Science, 1*, 297–318.

Hastie, T., & Tibshirani, R. (1988). Comment on "Monotone regression splines in action" (Ramsey, 1988). *Statistical Science, 3*, 451–456.

Hastie, T., & Tibshirani, R. (1990). *Generalized additive models*. London: Chapman and Hall.

Hastie, T., Tibshirani, R., & Friedman, J. H. (2009). *The elements of statistical learning: data mining, inference, and prediction*. New York: Springer. in Statistics.

Hastie, T., Tibshirani, R., & Wainwright, M. (2015). *Statistical learning with sparsity: The lasso and generalizations*. Monographs on statistics and applied probabilityBoca Raton, FL: Chapman Hall/CRC.

Hastie, T., Tibshirani, R., & Tibshirani, R. (2020). Best subset, forward stepwise or lasso? Analysis and recommendations based on extensive comparisons. *Statistical Science, 35*(4), 579–592.

Hawkins, D. M. (1980). *Identification of outliers*. London: Chapman and Hall.

Hawkins, D. M., & Olive, D. J. (2002). Inconsistency of resampling algorithms for high-breakdown regression estimators and a new algorithm (with discussion). *Journal of the American Statistical Association, 97*, 136–159.

Hebbali, A. (2024). *OLSRR: Tools for building OLS regression models*. R package version 0.6.1.

Hettmansperger, T. P., & McKean, J. W. (2011). *Robust nonparametric statistical methods* (2nd ed.). Boca Raton, FL: Chapman Hall/CRC.

Hinkley, D. V. (1975). On power transformations to symmetry. *Biometrika, 62*, 101–111.

Hoaglin, D. C., Mosteller, F., & Tukey, J. W. (Eds.). (1983). *Understanding robust and exploratory data analysis*. New York: Wiley.

Hoerl, A. E., & Kennard, R. W. (1970). Ridge regression: Biased estimation for nonorthogonal problems. *Technometrics, 12*(1), 55–67.

Hoffmann, I., Serneels, S., Filzmoser, P., & Croux, C. (2015). Sparse partial robust M regression. *Chemometrics and Intelligent Laboratory Systems, 149, Part A*, 50–59.

Hössjer, O. (1992). On the optimality of S-estimators. *Statistics and Probability Letters, 14*, 413–419.

Hoyle, M. H. (1973). Transformations—An introduction and a bibliography. *International Statistical Review, 41*, 203–223.

Hron, K., & Fišerová, E. (2011). On the interpretation of orthonormal coordinates for compositional data. *Mathematical Geosciences, 43*, 455–468.

Hron, K., Filzmoser, P., & Thompson, K. (2012). Linear regression with compositional explanatory variables. *Journal of Applied Statistics, 39*, 1115–1128.

Hrusova, K., Todorov, V., Hron, K., & Filzmoser, P. (2016). Classical and robust orthogonal regression between parts of compositional data. *Statistics, 50*, 1261–1275.

References

Huber, P. J. (1964). Robust estimation of a location parameter. *Annals of Mathematical Statistics, 35*, 73–101.

Huber, P. J. (1973). Robust regression: Asymptotics, conjectures and Monte Carlo. *Annals of Statistics, 1*, 799–821.

Huber, P. J. (1981). *Robust statistics*. New York: Wiley.

Huber, P. J., & Ronchetti, E. M. (2009). *Robust statistics* (2nd ed.). New York: Wiley.

Insolia, L., Kenney, A., Chiaromonte, F., & Felici, G. (2021). Simultaneous feature selection and outlier detection with optimality guarantees. *Biometrics*.

Jarque, C. M., & Bera, A. K. (1987). A test for normality of observations and regression residuals. *International Statistical Review, 52*, 163–172.

Johansen, S., & Nielsen, B. (2016). Analysis of the forward search using some new results for martingales and empirical processes. *Bernoulli, 21*, 1131–1183.

Jurečková, J., Picek, J., & Schindler, M. (2019). *Robust statistical methods with R*. Boca Raton, FL: Chapman Hall/CRC.

Kassambara, A. (2019). *Datarium: Data bank for statistical analysis*. R package version 0.1.0.

Kendall, M. G. (1938). A new measure of rank correlation. *Biometrika, 30*, 81–89.

Kepplinger, D. (2019). *Complmrob: Robust linear regression with compositional data as covariates*. R package version 0.7.0.

Kepplinger, D. (2023). Robust variable selection and estimation via adaptive elastic net S-estimators for linear regression. *Computational Statistics & Data Analysis, 183*, 107730.

Kepplinger, D., Salibian-Barrera, M., & Cohen Freue, G. (2024). *Pense: Penalized elastic Net S/MM-estimator of regression*. R package version 2.2.2.

Khan, J. A., Van Aelst, S., & Zamar, R. H. (2007). Robust linear model selection based on least angle regression. *Journal of the American Statistical Association, 102*(480), 1289–1299.

Kiefer, J. (1959). Optimum experimental designs (with discussion). *The Journal of the Royal Statistical Society, Series B, 21*, 272–319.

Kleiber, C., & Zeileis, A. (2008). *Applied econometrics with R*. Springer Science & Business Media.

Kleinbaum, D. G., & Kupper, L. (1978). *Applied regression analysis and other multivariable methods*. Boston, Mass: Duxbury.

Koop, G. (2003). *Bayesian econometrics*. Chichester: Wiley.

Kruskal, J. B. (1965). Analysis of factorial experiments by estimating monotone transformations of the data. *Journal of the Royal Statistical Society. Series B (Methodological), 27*(2), 251–263.

Lawrance, A. J. (1987). The score statistic for regression transformation. *Biometrika, 74*, 275–279.

Lehmann, E. (1991). *Point estimation*. New York: Wiley.

Maechler, M., Rousseeuw, P. J., Croux, C., Todorov, V., Ruckstuhl, A., Salibian-Barrera, M., Verbeke, T., Koller, M., Conceição, E. L. T., & di Palma, M. A. (2024). *Robustbase: Basic robust statistics*. R package version 0.99-4-1.

Mallows, C. L. (1973). Some comments on C_p. *Technometrics, 15*, 661–675.

Marazzi, A., Villar, A. J., & Yohai, V. J. (2009). Robust response transformations based on optimal prediction. *Journal of the American Statistical Association, 104*, 360–370. https://doi.org/10.1198/jasa.2009.0109

Maronna, R. A. (2011). Robust ridge regression for high-dimensional data. *Technometrics, 53*(1), 44–53.

Maronna, R. A., & Yohai, V. J. (2010). Correcting MM estimates for "fat" data sets. *Computational Statistics & Data Analysis, 54*(12), 3168–3173.

Maronna, R. A., & Zamar, R. H. (2002). Robust estimates of location and dispersion for high-dimensional datasets. *Technometrics, 44*(4), 307–317.

Maronna, R. A., Martin, R. D., Yohai, V. J., & Salibian-Barrera, M. (2019). *Robust statistics: theory and methods (with R)* (2nd ed.). Chichester: Wiley.

Mazza, A., Ingrassia, S., & Punzo, A. (2018). A flexible framework for cluster-weighted models. *Journal of Statistical Software, 86*(2), 1–30.

McCullagh, P., & Nelder, J. A. (1989). *Generalized linear models* (2nd ed.). London: Chapman and Hall.

McDonald, G. C., & Schwing, R. C. (1978). Instabilities of regression estimates relating air pollution to mortality. *Technometrics, 15*, 463–481.

Meinshausen, N. (2007). Relaxed lasso. *Computational Statistics & Data Analysis, 52*(1), 374–393.

Michaelis, L., & Menten, M. (1913). Die Kinetik der Invertinwirkung. *Biochemische Zeitschrift, 49*, 333–369.

Mosteller, F., & Tukey, J. W. (1977). *Data analysis and regression*. Reading, Mass: Addison-Wesley.

Neter, J., Kutner, M. H., Nachtsheim, C. J., & Wasserman, W. (1996). *Applied linear statistical models* (4th ed.). New York: McGraw-Hill.

Neykov, N. M., Filzmoser, P., & Neytchev, P. N. (2012). Robust joint modeling of mean and dispersion through trimming. *Computational Statistics and Data Analysis, 56*, 34–48.

Olive, D. J. (2020). *Robust statistics*. On-line. http://parker.ad.siu.edu/Olive/robbook.htm

Pawlowsky-Glahn, V., & Egozcue, J. J. (2006). Compositional data and their analysis: An introduction. In A. Buccianti, G. Mateu-Figueras, & V. Pawlowsky-Glahn (Eds.), *Compositional data analysis in the geosciences: From theory to practice* (pp. 1–10). London: Geological Society.

Pawlowsky-Glahn, V., Egozcue, J. J., & Tolosana-Delgado, R. (2008). Lecture notes on compositional data analysis. Report, Universitat de Girona, Girona.

Pawlowsky-Glahn, V., Egozcue, J. J., & Tolosana-Delgado, R. (2015). *Modeling and analysis of compositional data*. New York: Wiley.

Pawlowsky-Glahn, V., Egozcue, J. J., & Lovell, D. (2015). Tools for compositional data with a total. *Statistical Modelling, 2*(15), 175–190.

Pearson, E. S. (1929). Statistics in biological research. *Nature, 123*, 866–867.

Pearson, E. S., & Please, N. W. (1975). Relation between the shape of population distribution and the robustness of four simple test statistics. *Biometrika, 62*, 223–241.

Pearson, K. (1895). Notes on regression and inheritance in the case of two parents. *Proceedings of the Royal Society of London, 58*, 240–242.

Perrotta, D., Riani, M., & Torti, F. (2009). New robust dynamic plots for regression mixture detection. *Advances in Data Analysis and Classification, 3*, 263–279. https://doi.org/10.1007/s11634-009-0050-y

Perrotta, D., Checchi, E., Torti, F., Cerasa, A., & Arnés Novau, X. (2020). Addressing price and weight heterogeneity and extreme outliers in surveillance data—The case of face masks. Technical report JRC122315, European commission, Joint Research Centre, Publications Office of the European Union, Luxembourg. ISBN 978-92-76-24707-4. https://doi.org/10.2760/817681, EUR 30431 EN.

Piepel, G., & Redgate, T. (1998). A mixture experiment analysis of the Hald cement data. *American Statistician, 452*, 23–30.

Pison, G., Van Aelst, S., & Willems, G. (2002). Small sample corrections for LTS and MCD. *Metrika, 55*, 111–123. https://doi.org/10.1007/s001840200191

Pregibon, D. (1980). Goodness of link tests for generalized linear models. *Applied Statistics, 29*(15–23), 14.

Pronzato, L., & Pázman, A. (2013). *Design of experiments in nonlinear models*. New York: Springer.

Raftery, A. E., Madigan, D., & Hoeting, J. A. (1997). Bayesian model averaging for linear regression models. *Journal of the American Statistical Association, 92*, 179–191.

Ramsay, J. O. (1988). Monotone regression splines in action. *Statistical Science, 3*, 425–461.

Ramsey, F., Schafer, D., Sifneos, J., & Turlach, B. A. (2024). *Sleuth3: Data sets from Ramsey and Schafer's, statistical sleuth* (3rd ed.). R package version 1.0-6.

Ramsey, F. L., & Schafer, D. W. (2013). *The statistical sleuth: A course in methods of data analysis* (3rd ed.). London, UK: Cengage Learning.

Rao, C. R. (1973). *Linear statistical inference and its applications* (2nd ed.). New York: Wiley.

Raymaekers, J., & Rousseeuw, P. J. (2021). Transforming variables to central normality. *Machine Learning*.

Riani, M., & Atkinson, A. C. (2007). Fast calibrations of the forward search for testing multiple outliers in regression. *Advances in Data Analysis and Classification, 1*, 123–141. https://doi.org/10.1007/s11634-007-0007-y

References

Riani, M., & Atkinson, A. C. (2010). Robust model selection with flexible trimming. *Computational Statistics and Data Analysis, 54,* 3300–3312. https://doi.org/10.1016/j.csda.2010.03.007

Riani, M., Atkinson, A. C., & Cerioli, A. (2009). Finding an unknown number of multivariate outliers. *Journal of the Royal Statistical Society, Series B, 71,* 447–466.

Riani, M., Perrotta, D., & Torti, F. (2012). FSDA: a MATLAB toolbox for robust analysis and interactive data exploration. *Chemometrics and Intelligent Laboratory Systems, 116,* 17–32. https://doi.org/10.1016/j.chemolab.2012.03.017

Riani, M., Cerioli, A., Atkinson, A. C., & Perrotta, D. (2014a). Monitoring robust regression. *Electronic Journal of Statistics, 8,* 642–673.

Riani, M., Cerioli, A., & Torti, F. (2014b). On consistency factors and efficiency of robust S-estimators. *TEST, 23,* 356–387.

Riani, M., Atkinson, A. C., & Perrotta, D. (2014c). On-line supplement to "A parametric framework for the comparison of methods of very robust regression". *Statistical Science.*

Riani, M., Atkinson, A. C., & Perrotta, D. (2014). A parametric framework for the comparison of methods of very robust regression. *Statistical Science, 29,* 128–143.

Riani, M., Perrotta, D., & Cerioli, A. (2015). The forward search for very large datasets. *Journal of Statistical Software, 67*(1), 1–20.

Riani, M., Atkinson, A. C., Corbellini, A., & Perrotta, D. (2020). Robust regression with density power divergence: Theory, comparisons and data analysis. *Entropy, 22*(399). https://doi.org/10.3390/e22040399

Riani, M., Atkinson, A. C., & Corbellini, A. (2022). Automatic robust Box-Cox and extended Yeo-Johnson transformations in regression. *Statistical Methods and Applications.* https://doi.org/10.1007/s10260-022-00640-7

Riani, M., Atkinson, A. C., Corbellini, A., Farcomeni, A., & Laurini, F. (2022). Information criteria for outlier detection avoiding arbitrary significance levels. *Econometrics and Statistics.* https://doi.org/10.1016/j.ecosta.2022.02.002

Riani, M., Atkinson, A. C., Torti, F., & Corbellini, A. (2022). Robust correspondence analysis. *Journal of the Royal Statistical Society Series C: Applied Statistics, 71*(5), 1381–1401.

Riani, M., Atkinson, A. C., and Corbellini, A. (2023a). Robust response transformations for generalized additive models via additivity and variance stabilization. In L. Grilli, M. Lupparelli, C. Rampichini, E. Rocco, & M. Vichi (Eds.), *Statistical models and methods for data science.* Studies in classification, data analysis, and knowledge. Springer, Heidelberg.

Riani, M., Atkinson, A. C., & Corbellini, A. (2023b). Robust transformations for multiple regression via additivity and variance stabilization. *Journal of Computational and Graphical Statistics, 32.*

Riani, M., Atkinson, A. C., Corbellini, A., & Morelli, G. (2025a). Robust tobit regression for censored observations using extended Box-Cox transformations. *Statistical Methods and Applications (in press).*

Riani, M., Atkinson, A. C., Morelli, G., & Corbellini, A. (2025b). The use of modern robust regression analysis with graphics: An example from marketing. *STATS, 8*(6). https://doi.org/10.3390/stats8010006

Rocke, D. M., & Woodruff, D. L. (1996). Identification of outliers in multivariate data. *Journal of the American Statistical Association, 91*(435), 1047–1061.

Rousseeuw, P. J. (1984). Least median of squares regression. *Journal of the American Statistical Association, 79,* 871–880.

Rousseeuw, P. J., & Croux, C. (1993). Alternatives to the median absolute deviation. *Journal of the American Statistical Association, 88*(424), 1273–1283.

Rousseeuw, P. J., & Leroy, A. M. (1987). *Robust regression and outlier detection.* New York: Wiley.

Rousseeuw, P. J., & Van Driessen, K. (1999). A fast algorithm for the minimum covariance determinant estimator. *Technometrics, 41,* 212–223.

Rousseeuw, P. J., & Van Driessen, K. (2006). Computing LTS regression for large data sets. *Data Mining and Knowledge Discovery, 12,* 29–45.

Rousseeuw, P. J., & Van Zomeren, B. C. (1990). Unmasking multivariate outliers and leverage points. *Journal of the American Statistical Association, 85,* 633–9.

Rousseeuw, P. J., & Yohai, V. J. (1984). Robust regression by means of S-estimators. In J. Franke, W. Härdle, & R. D. Martin (Eds.), *Robust and nonlinear time series analysis: Lecture notes in statistics 26* (pp. 256–272). New York: Springer.

Rousseeuw, P. J., Perrotta, D., Riani, M., & Hubert, M. (2019). Robust monitoring of time series with application to fraud detection. *Econometrics and Statistics, 9*, 108–121.

Royston, P. J., & Altman, D. G. (1994). Regression using fractional polynomials of continuous covariates: Parsimonious parametric modelling (with discussion). *Applied Statistics, 43*, 429–467.

Sakia, R. M. (1992). The Box-Cox transformation technique: A review. *The Statistician, 41*, 169–178.

Salibian-Barrera, M. (2023). Robust nonparametric regression: Review and practical considerations. *Econometrics and Statistics*. https://doi.org/10.1016/j.ecosta.2023.04.004

Salibian-Barrera, M., & Yohai, V. J. (2006). A fast algorithm for S-regression estimates. *Journal of Computational and Graphical Statistics, 15*, 414–427.

Salini, S., Cerioli, A., Laurini, F., & Riani, M. (2015). Reliable robust regression diagnostics. *International Statistical Review, 84*, 99–127.

Salini, S., Laurini, F., Morelli, G., Riani, M., & Cerioli, A. (2022). Covariance matrices of S robust regression estimators. *Journal of Statistical Computation and Simulation, 92*(4), 724–747.

Scheffé, H. (1958). Experiments with mixtures. *The Journal of the Royal Statistical Society, Series B, 20*, 344–360.

Schimek, M. G., & Turlach, B. A. (2006). Additive and generalized additive models. Technical report, Sonderforschungsbereich 373, Humboldt University, Berlin.

Schlesselman, J. (1971). Power families: A note on the Box and Cox transformation. *Journal of the Royal Statistical Society, Series B, 33*, 307–311.

Schwarz, G. (1978). Estimating the dimension of a model. *Annals of Statistics, 6*, 461–464.

Shertzer, K., & Prager, M. (2002). Least median of squares: A suitable objective function for stock assessment models? *Canadian Journal of Fisheries and Aquatic Sciences, 59*, 1474–1481.

Shiller, R. J. (1999). The ET interview: Professor James Tobin. *Econometric Theory, 15*, 867–900.

Smith, R. L. (1985). Maximum likelihood estimation in a class of non-regular cases. *Biometrika, 72*, 67–92.

Smucler, E., & Yohai, V. J. (2017). Robust and sparse estimators for linear regression models. *Computational Statistics & Data Analysis, 111*, 116–130.

Spearman, C. (1904). The proof and measurement of association between two things. *American Journal of Psychology, 15*, 72–101.

Spector, P., Friedman, J. H., Tibshirani, R., Lumley, T., Garbett, S., & Baron, J. (2024). *Acepack: ACE and AVAS for selecting multiple regression transformations*. R package version 1.4.2.

Spiegel, E., Kneib1, T., & Otto-Sobotka, F. (2019). Generalized additive models with flexible response functions. *Statistics and Computing, 29*, 123–138.

Stigler, S. M. (2010). The changing history of robustness. *The American Statistician, 64*, 277–281.

Stromberg, A. J. (1993). Computation of high breakdown nonlinear regression parameters. *Journal of the American Statistical Association, 88*, 237–244.

Stromberg, A. J. (1993). Computing the exact least median of squares estimate and stability diagnostics in multiple linear regression. *SIAM Journal on Scientific Computing, 14*(6), 1289–1299.

Taiyun, W., & Simko, V. (2024). *R package 'corrplot': Visualization of a correlation matrix*. (Version 0.95).

Tallis, G. M. (1963). Elliptical and radial truncation in normal samples. *Annals of Mathematical Statistics, 34*, 940–944.

Taylor, J. M. G. (2004). Transformations II. *Wiley Statsref*.

Tibshirani, R. (1988). Estimating transformations for regression via additivity and variance stabilization. *Journal of the American Statistical Association, 83*, 394–405.

Tibshirani, R. (1996). Regression shrinkage and selection via the lasso. *Journal of the Royal Statistical Society. Series B (Methodological), 58*, 267–288.

Tobin, J. (1958). Estimation of relationships for limited dependent variables. *Econometrica, 26*, 24–36.
Todorov, V. (2021). Monitoring robust estimates for compositional data. *Austrian Journal of Statistics, 50*, 16–37.
Todorov, V., & Filzmoser, P. (2009). An object oriented framework for robust multivariate analysis. *Journal of Statistical Software, 32*(3), 1–47.
Todorov, V., & Sordini, E. (2023). *fsdaR: Robust data analysis through monitoring and dynamic visualization*. R package version 0.9-0.
Tong, H. (2001). A personal journey through time series in Biometrika. *Biometrika, 88*, 195–218.
Torti, F. (2011). *Advances in the forward search: Methodological and applied contributions*. Padova, Italy: Italian Statistical Society.
Torti, F., Perrotta, D., Atkinson, A. C., & Riani, M. (2012). Benchmark testing of algorithms for very robust regression: FS, LMS and LTS. *Computational Statistics and Data Analysis, 56*, 2501–2512. https://doi.org/10.1016/j.csda.2012.02.003
Torti, F., Perrotta, D., Riani, M., & Cerioli, A. (2018). Assessing trimming methodologies for clustering linear regression data. *Advances in Data Analysis and Classification, 13*, 227–257.
Torti, F., Corbellini, A., & Atkinson, A. C. (2021). fsdaSAS: A package for robust regression for very large datasets including the batch forward search. *Stats, 4*, 327–347.
Torti, F., Riani, M., & Morelli, G. (2021). Semiautomatic robust regression clustering of international trade data. *Statistical Methods & Applications, 30*, 863–894.
Tukey, J. W., et al. (1960). A survey of sampling from contaminated distributions. In I. Olkin (Ed.), *Contributions to probability and statistics: essays in honor of Harold Hotelling* (pp. 448–485). Palo Alto, CA: Stanford University Press.
Tukey, J. W. (1979). Robust techniques for the user. In R. L. Launer & G. N. Wilkinson (Eds.), *Robustness in statistics* (pp. 103–106). London: Academic Press.
UN. (2002). International standard industrial classification of all economic activities (ISIC) Rev. 3.1. Statistical papers, United Nations, New York.
UN. (2015). Transforming our world: the 2030 agenda for sustainable development. Resolution adopted by the general assembly on 25th September 2015, United Nations, New York.
UNIDO. (2010). *Industrial statistics: guidelines and methodology*. United Nations, Vienna: Report.
Verzani, J. (2022). *UsingR: Data sets, etc. for the text "Using R for introductory statistics"* (2nd ed.). R package version 2.0-7.
Wang, D., & Murphy, M. (2005). Identifying nonlinear relationships in regression using the ACE algorithm. *Journal of Applied Statistics, 32*, 243–258.
Wang, H., Li, G., & Jiang, G. (2007). Robust regression shrinkage and consistent variable selection through the LAD-Lasso. *Journal of Business & Economic Statistics, 25*(3), 347–355.
Wang, M., Lyu, B., & Yu, G. (2021). ConvexVST: A convex optimization approach to variance-stabilizing transformation. In *Proceedings of the 38th international conference on machine learning*, vol. 139, pp. 10839–10848.
Weisberg, S. (2005a). *Applied linear regression* (3rd ed.). New York: Wiley.
Weisberg, S. (2005b). Yeo-Johnson power transformations. Online. https://www.stat.umn.edu/arc/yjpower.pdf.
Welsh, A. H., Carroll, R. J., & Ruppert, D. (1994). Fitting heteroscedastic regression models. *Journal of the American Statistical Association, 89*, 100–116.
White, H. (1980). A heteroskedasticity-consistent covariance matrix estimator and a direct test for heteroskedasticity. *Econometrica, 48*, 817–838.
Wiens, B. L. (1999). When log-normal and gamma models give different results: A case study. *The American Statistician, 53*, 89–93.
Wolstenholme, D. E., O'Brien, C. M., & Nelder, J. A. (1988). GLIMPSE: A knowledge-based front end for statistical analysis. *Knowledge-Based Systems, 1*, 173–178.
Yeo, I.-K., & Johnson, R. A. (2000). A new family of power transformations to improve normality or symmetry. *Biometrika, 87*, 954–959.

Yohai, V. J. (1987). High breakdown-point and high efficiency estimates for regression. *The Annals of Statistics, 15*, 642–656.

Yohai, V. J., & Zamar, R. H. (1988). High breakdown-point estimates of regression by means of the minimization of an efficient scale. *Journal of the American Statistical Association, 83*, 406–413.

Yohai, V. J., & Zamar, R. H. (1997). Optimal locally robust M-estimates of regression. *Journal of Statistical Planning and Inference, 64*(2), 309–323.

Yuan, M., & Lin, Y. (2006). Model selection and estimation in regression with grouped variables. *Journal of the Royal Statistical Society: Series B (Statistical Methodology), 68*(1), 49–67.

Zou, H., & Hastie, T. (2005). Regularization and variable selection via the elastic net. *Journal of the Royal Statistical Society. Series B (Statistical Methodology), 67*(2), 301–320.

Zou, H., Hastie, T., & Tibshirani, R. (2007). On the "degrees of freedom" of the lasso. *The Annals of Statistics, 35*(5), 2173–2192.

Author Index

A
Agulló, J. J. E., 88
Aitchison, J., 330–332
Akaike, H., 149, 353
Alimadad, A., 277
Altman, D. G., 234
Amemiya, T., 343
Anderson, D. R., 349, 350
Andrews, D. F., 2, 15, 43
Anglin, P., 286
Anscombe, F. J., 249
Aranda-Ordaz, F. J., 232
Argote-Espino, D., 331
Arnés Novau, X., 307
Atkinson, A. C., ix, x, 15, 16, 40, 68, 71–73, 77, 80, 102, 105, 110, 111, 129, 138, 143, 147, 164, 165, 179, 180, 184, 196, 199–203, 206–209, 211, 213, 214, 217, 218, 221, 227–233, 236, 237, 239, 242, 246, 259, 263, 297–300, 303, 316, 343–345, 355, 356, 456
Azzini, I., ix

B
Barceló-Vidal, C., 331, 333
Barlow, R. E., 244
Barndorff-Nielsen, O. E., 62
Baron, J., 263, 276
Bartholomew, D. J., 244
Bartlett, M. S., 196
Bartoloni, E., 223
Beaton, A. E., 38
Belsley, D. A., 68

Bera, A. K., 261
Berrendero, J. R., 171
Bhat, H. S., 145
Bickel, P. J., 2
Biernacki, C., 316
Bliss, C. I., 239
Boente, G., 277
Box, G. E. P., 1, 104, 196, 199, 201, 202, 206, 208, 209, 233, 235, 239, 286, 328, 442
Brant, R., 283
Breiman, L., 242, 244, 246, 275, 277, 355, 357
Bremner, J. M., 244
Brown, B. W., 207
Bruce, D., 441
Brunk, H. D., 244
Buja, A., 71, 242–244, 276
Burnham, K. P., 349, 350

C
Cantoni, E., 277
Carroll, R. J., 199, 203, 209, 226, 234, 238, 239, 298
Celeux, G., 316
Cerasa, A., 307
Cerioli, A., ix, x, 16, 48, 77, 78, 80, 83, 85, 98, 99, 101, 138, 145, 152, 153, 158, 310, 315–317
Chaloner, K., 283
Checchi, E., 307
Cheng, T.-C., 298
Chiaromonte, F., 70, 88

Claeskens, G., 143, 145, 349, 350
Conceição, E. L. T., viii, 337
Cook, R. D., 15, 67, 68, 70, 72, 151, 160, 169
Corbellini, A., ix, x, 40, 107, 143, 146–148, 199, 204, 207, 211, 213, 214, 217, 221, 227, 229, 239, 242, 246, 259, 263, 297, 315, 343–345
Cox, D. R., 61, 105, 165, 196, 199, 201, 202, 206, 208, 235, 239, 249
Cron, W., 313
Croux, C., viii, 48, 80, 89, 282, 337

D
Davies, L., 89
Davison, A. C., 352, 353, 355
De Leeuw, J., 325
DeSarbo, W., 313
Dhaene, G., 80
Díaz-Barrero, J. L., 331
di Palma, M. A., viii, 337
Donev, A. N., 80
Durbin, J., 258, 408

E
Economist, 292
Egozcue, J. J., 331–333
European Union, 292

F
Facevicova, K., 331
Farcomeni, A., 77, 142, 145, 148, 310
Fedorov, V. V., 299
Felici, G., 70, 88
Field, C. A., 2, 61
Filzmoser, P., viii, 282, 298, 331, 335, 336, 339, 347, 499
Firth, D., 207
Fisher, R. A., 239
Fišerová, E., 333, 334
Foi, A., 276
Forbes, J. D., 15
Fowlkes, E. B., 242
Fraley, C., 171, 189, 305, 316
Friedman, J. H., 242–244, 263, 275, 276, 356, 358
Fuller, W. A., 332

G
Garbett, S., 263, 276

Garcia-Escudero, L. A., 171, 189, 190, 315–317
Gençay, R., 286
Gershenfeld, N., 313
Giudici, P., 112
Gnedenko, B. V., 61
Gordaliza, A., 171, 189, 190, 315
Govaert, 316
Greco, L., 77
Greenacre, M., 340
Greene, W. H., 298, 300
Greselin, F., 189, 315
Gruen, B., 321
Guenther, W. C., 99
Guerrero, V. M., 231

H
Hössjer, O., 47, 83
Hadi, A. S., 16
Haesbroeck, G., 282
Hald, A., 352
Hampel, F. R., 2, 16, 24, 36, 38, 39, 61, 77
Harrell Jr, F. E., 274, 277
Harvey, A. C., 298, 300, 303
Hastie, T., 242–245, 276
Hawkins, D. M., 15, 92, 151, 169
Hettmansperger, T. P., 16
Hinkley, D. V., 61, 165, 201
Hjort, N. L., 143, 145, 349, 350
Hoffmann, I., 282
Hollander, M., 207
Hoorelbeke, D., 80
Hoyle, M. H., 239
Hron, K., 331, 333, 335, 336, 339, 347, 499
Hrusova, K., 332
Huber, P. J., 2, 15, 25, 37, 45, 78, 80, 109, 170
Hubert, M., ix, 325, 326, 329

I
Ingrassia, S., 189, 315, 321
Insolia, L., 70, 88

J
Jarauta-Bragulat, E., 331
Jarque, C. M., 261
Jenkins, G. M., 328, 442
Johansen, S., 146
Johnson, R. A., 210, 211, 231
Jurečková, J., 16

Author Index

K
Kassambara, A., 270
Kass, R. E., 242, 276
Kendall, M. G., 112
Kenney, A., 70, 88
Kepplinger, D., 337
Kettenring, J. R., 242
Kiefer, J., 80
Kleiber, C., 287
Kleinbaum, D. G., 147, 207
Kneib1, T., 277
Koller, M., viii, 337
Koop, G., 286, 287
Kruskal, J. B., 325
Kuh, E., 68
Kumar, N., 145
Kupper, L., 147, 207
Kutner, M. H., 134, 202

L
Laurini, F., 48, 80, 143, 146, 148, 152, 153, 157
Lawrance, A. J., 200
Lehmann, E., 99
Leisch, F., 321
Leonov, S. L., 299
Leroy, A. M., 16, 82, 88, 89, 102, 106, 125, 363
Lopez-Garcá, P., 331
Lovell, D., 330
Lumley, T., 263, 276
Lyu, B., 276

M
Maechler, M., viii, 337
Marazzi, A., 9, 184, 228
Maronna, R. A., 16, 32, 62, 79, 81, 91, 149, 157, 170, 171
Martínez, A., 277
Martín-Fernández, J. A., 331
Martin, R. D., 16, 32, 62, 70, 79, 83, 91, 149, 170, 171
Mateu-Figueras, G., 331, 333
Mayo-Iscar, A., 171, 189, 190, 315, 317
Mazza, A., 321
McCullagh, P., 207, 208
McKean, J. W., 16
Mendes, B. V. M., 171
Menten, M., 234
Metois, E., 313
Michaelis, L., 234
Morelli, G., 80, 189, 316, 343–346

Mosteller, F., 230
Murphy, M., 263, 264

N
Nachtsheim, C. J., 134, 202
Nelder, J. A., 207, 208
Neter, J., 134, 202
Neykov, N. M., 298
Neytchev, P. N., 298
Nielsen, B., 146

O
Olive, D. J., 16, 92, 189
Otto-Sobotka, F., 277

P
Pawlowsky-Glahn, V., 330–332
Pázman, A., 89
Pearson, E. S., 2, 112
Pearson, K., 112
Pericchi, L. R., 209
Perrotta, D., ix, 40, 78, 85, 98, 102, 112, 138, 164, 179, 180, 184, 189, 307, 315, 325–327, 329
Picek, J., 16
Pison, G., 90, 157
Please, N. W., 2
Prager, M., 89
Pregibon, D., 208
Pronzato, L., 89
Punzo, A., 321

R
Raffinetti, E., 112
Ramsay, J. O., 246, 277
Rao, C. R., 284
Raymaekers, J., viii
Reid, N., 199
Reimann, C., 331
Riani, M., ix, x, 16, 40, 48, 68, 71–72, 77, 78, 80, 83, 85, 98, 99, 101, 102, 105, 110, 111, 129, 138, 142, 146–148, 152, 153, 158, 164, 179, 180, 184, 189, 196, 199, 201, 203, 207, 208, 211, 213, 214, 218, 221, 227, 229, 236–239, 242, 246, 259, 263, 297, 298, 300, 303, 310, 315–317, 325–327, 329, 343–346, 355, 356, 456

Rocke, D. M., 55
Rolke, W., 71

Ronchetti, E. M., 15, 36–39, 61, 77, 80, 109, 277
Rousseeuw, P. J., viii, ix, 16, 36, 38, 39, 48, 77, 80, 82, 88, 89, 102, 106, 125, 147, 151, 186, 325–327, 329, 335, 337, 363, 367
Royston, P. J., 234
Ruckstuhl, A., viii, 337
Ruppert, D., 209, 234, 238, 239, 298
Ruwet, C., 282

S
Sakia, R. M., 239
Salibian-Barrera, M., viii, 16, 32, 62, 70, 79, 81, 86, 91, 149, 170, 171, 277, 337
Salini, S., 48, 80, 152, 153, 157
San Martin, R., 171, 189, 315
Schimek, M. G., 243
Schindler, M., 16
Schlesselman, J., 231
Schoner, B., 313
Schumacher, F. X., 236, 441
Schwarz, G., 70, 142, 145
Serneels, S., 282
Shertzer, K., 89
Shiller, R. J., 343
Simonoff, J. S., 16
Smith, R. L., 209
Snell, E. J., 249
Sordini, E., xi, 2
Spearman, C., 112
Spector, P., 263, 276
Spiegel, E., 277
Stahel, W. A., 16, 36, 38, 77
Stigler, S. M., 1
Stromberg, A. J., 88
Stuetzle, W., 244

T
Tallis, G. M., 89
Taylor, J. M. G., 239
Templ, M., 331, 347, 499
Thompson, K., 331, 336, 339
Tibshirani, R., 242–245, 263, 276
Tidwell, P. W., 233
Tobias, R. D., 80

Tobin, J., 343
Todorov, V., viii, xi, 2, 332, 335, 337
Tolosana-Delgado, R., 330, 331
Torti, F., ix, 83, 107, 164, 189, 204, 206, 298–300, 303, 307, 310, 315
Tukey, J. W., 2, 15, 38, 230, 249
Turlach, B. A., 243
Tyler, D. E., 171

U
UN, 336
UNIDO, 334

V
Van Aelst, S., 90, 157
Van Driessen, K., 89, 186, 335, 367
Van Zomeren, B. C., 151
Verbeke, T., viii, 337
Verzani, J., 346
Villar, A. J., 9, 184, 228

W
Wang, D., 263, 264
Wang, M., 276
Wasserman, W., 134, 202
Watson, G. S., 258, 408
Weisberg, S., 15, 67, 68, 70, 72, 160, 211
Welsch, R. E., 68
Welsh, A. H., 298
White, H., 298
Wiens, B. L., 207
Willems, G., 90, 157
Woodruff, D. L., 55

Y
Yeo, I.-K., 210, 211
Yohai, V. J., 9, 16, 32, 39, 58, 62, 70, 79, 81, 86, 90, 149, 157, 170, 171, 184, 228
Yu, G., 276

Z
Zamar, R. H., 39, 91, 171
Zeileis, A., 287

Subject Index

Symbols
ψ-function, 35–39
  redescending, 36, 38, 48, 59, 60, 86
  weight function $\psi(u)/u$, 35–42
ρ-function family, 35–44
  Andrews' sine function, 43, 149
  bounded vs. unbounded, 34–35
  comparisons between, 35–36
  Hampel three-part ρ, 36–38
  Huber ρ, 35–36
  hyperbolic tangent ρ, 39–40
  optimal ρ (Yohai–Zamar), 38–39
  power-divergence ρ, 41–43
  Tukey biweight (bisquare), 38
  tuning constant c, 36–37

A
Added variable, 72–74, 76, 93, 95, 110–112, 135, 137, 140, 142, 144, 149, 200, 337–339, 350, 355, 376, 385, 387, 394, 395, 406, 410, 411, 419–421, 425–427, 429, 430, 437, 459, 467, 468, 507, 508, 514, 516, 522, 523
  plot, 110–112, 142
  t-statistics, 352
Additive model, 15, 241, 242, 405
Additivity and Variance Stabilization (AVAS), 241–246, 252–254, 397–408, 413–416
  backfitting algorithm, 243–245
Agreement Index (AGI), 224–225
Akaike Information Criterion (AIC), 353–354
Aranda–Ordaz transformation, 232

AR regression data, 116–128
Assumptions in regression, 66
Asymptotics, 5–6
  asymptotic properties, 25–30
  asymptotic relative efficiency, 25–26
  asymptotic value, 7, 23, 91, 117, 151, 234
  asymptotic variance, 25–27
  rate of convergence, 6
  simulation studies, 5
Augmented star plot, 261–265
Autocorrelation, 326, 407
Automatic procedure for choosing transformations, 223–228

B
Balance-Sheet data, 219–223
Bank data, 137–142
Bayesian analysis, 281–286, 296–297
  conjugate prior, 283
  forward search, 284–286, 295–296
  highest posterior density (HPD) intervals, 285
  normal inverse–gamma prior, 282–283
  outlier detection, 285–286
  parameter estimation, 284
  posterior distributions, 284
  prior distribution from fictitious observations, 283
Bayesian Information Criterion (BIC), 144–149
  extended, 143, 145–148, 151, 152, 184–189, 223–229, 396, 420, 423, 429, 482, 483, 489, 513, 514, 518

Bias, 23–25, 39, 40, 55–58, 80, 151, 152, 157, 169–172, 174–176, 178–180, 189, 275, 358
Binary data, 231
Binomial distribution, 158
Boundary, 4, 29, 186, 209, 375, 484
Box-Cox transformation, 18, 195, 197, 199–203, 206–208, 210–212, 218, 224, 225, 227–231, 233–235, 237, 239, 240
   link, 207
   power transformation, 8–10
   score tests for λ, 9–11
   traditional vs. robust analysis, 10–12
Box–Tidwell transformation (explanatory variables), 233
Breakdown point (bdp), 29, 34–36, 45–48
   definition, 4
   explosion vs. implosion, 34
   finite-sample vs. asymptotic, 4–5
   influence of tuning constant, 37–38
   inverse relation with eff, 6
   nominal vs. achieved values, 41–43
   relation with efficiency, 25–26
Brushing and linking, 106–109, 117–120

C

Candlestick plot, generalized, 424–428
Change-of-Value Curve (CVC), 51–52
Change-of-Value Sensitivity (CVS), 51–52
Chi-squared approximation, 152–162
Clusterwise linear regression, 312–316
   basic concepts, 312–313
   information criterion, 316
   monitoring, 316
   number of groups k, 311, 316
   trimming level α, 316
Coefficient of determination (extended), 224–225
Collinearity (near), 353–358
Consistency, 23, 61–62
   consistency factor, 6–7
   Fisher consistency, 23, 34
   small-sample correction, 6
   strong vs. weak, 61–62
Constructed variable, 9, 72, 200, 201, 211, 213, 214, 228–231, 233, 236, 237, 345, 477, 511
   both sides, 236
Contaminated normal, 23, 455
Contaminated normal model, 23–24
   point-mass contamination, 23, 55–58
   shift contamination, 28, 55–59

Contamination models, 152–168
Convergence (stochastic), 60–62
   almost-sure convergence, 60
   convergence in probability, 60
   quadratic-mean convergence, 60
Correlation, 112–115, 122–125
Cross-validation, 361–364
Customer Loyalty data, 393–405

D

Deletion diagnostics, 68–71
Deletion estimates, 69
Deletion residuals, 69–75, 94, 95, 98, 99, 102–108, 129–135, 137, 138, 149, 161, 162, 164, 204, 220–222, 226, 285–288, 290–294, 296, 307–312, 330, 346, 434–439, 458, 460, 466, 493, 494

E

Effect of transformation, 218, 239, 429, 512
Eigenvalues, 314
Elastic net, 362–363
Elemental set, 151
Empirical breakdown point *see* Breakdown point
Envelopes, 73, 75, 99–107, 130–132, 138, 159, 204, 217, 220, 222, 278, 286, 287, 291, 375, 382, 386, 392–397, 401, 404, 407–409, 418, 466, 475, 515
Error distribution, 80, 137, 196
Error rate, 101
Estimates of location, 3–6
   mean (classical), 3
   median (robust), 3
   trimmed mean, 4, 12
   breakdown point of estimators, 4–5
Estimates of scale, 6–7
   inter-quartile range (IQR), 6
   median absolute deviation (MAD), 7
   normalized versions (IQRN, MADN), 6–7
Exploratory analysis, 504
Exponential distribution, 30
Extended Yeo–Johnson transformation, 210–219

F

Face-masks trade dataset, 307–312, 321–323
   random-start forward searches, 308–312
   TCLUSTREG classification, 323
False signals (oversized tests), 151–159

Subject Index

Fan plot, 14, 195, 201–203, 208, 214–221, 223–226, 228, 229, 236–238, 240, 247, 278, 345, 372–374, 381–383, 395, 396, 406, 407, 412, 416, 420–422, 429, 437, 438, 477–479, 481, 484, 486, 488, 489, 499, 510, 511, 513, 514, 517
Fisher consistency, 23, 34
Fisher information, 32, 34
Folded power transformation, 230, 231
  proportions, 230–231
Forward plot, 14, 102–108, 110, 120, 121, 126, 129–135, 137, 138, 142, 143, 205, 206, 220–222, 287–289, 291, 292, 296, 297, 310, 312, 330, 346, 406, 493, 494
Forward Search (FS)
  Bayesian forward search, 295–296
  fan plots, 13–14
  minimum deletion residuals, 286, 291–296
Forward selection, 352–353

G

Gamma distribution, 160, 207, 282
Gamma GLM & log link, 207–208
Gamma models, 207
Gasoline data, 231
Generalized Additive Model (GAM), 242–245
Generalized linear models, 206, 207, 277
Geometric mean, 198, 208, 230, 231, 332, 333
Grand plan (history of robust statistics), 1–3
  early expectations (Stigler), 2
  reasons for partial failure, 2–3
Gross Error Sensitivity (GES), 24–25, 48–49
Group LASSO, 362
Guerrero–Johnson transformation, 231

H

Hald's cement data, 352–353
Hat matrix, 66–67
Hawkins data, 128–133
Heteroskedastic regression, 378–381
  ART skedastic equation, 299–300
  HAR skedastic equation, 300
  International Trade Data 1, 300–303
  International Trade Data 2, 303–304
  robust heteroskedastic regression, 298
  White heteroskedastic-robust errors, 298
Homoskedasticity, 298

I

Income Data 1 (United States Census), 372–381
Income Data 2 (Italian Municipality), 381–388
Indicator function, 451
Influence function, 21–24, 51–53
  examples: mean, median, trimmed mean, 22–24
  gross error sensitivity, 24–25
Information criterion, 316
Initial subset, 186, 188, 311, 327
Interchange, 309
International Trade data, 290, 300–304
  background and importance, 290
  Data set 1, 300–303
  Data set 2, 303–304
Internet-marketing data, 270–273
Interquartile range, 6
  normalized (IQRN), 6
Inverse Gamma distribution, 282–283
Investment Funds data, 214–218
Iteratively reweighted least squares, 31

J

Jacobian of transformation, 197–198, 210–212, 230

L

Large datasets, 405
LASSO, 358–363
  LASSO estimator, 366
Least median of squares, 88–90
  least median of squares estimator, 85, 95, 97, 98
Least squares, 21, 31, 65–67, 70, 72, 74, 77, 78, 84, 86, 87, 89, 91, 93, 97, 98, 104, 106, 109, 112–116, 122–124, 144, 148, 154, 186, 197, 198, 201, 207, 224, 236, 238, 244, 277, 282, 284, 285, 287, 289, 291–293, 298, 299, 325, 332, 336, 358, 360, 408, 414, 417, 438, 493
  least squares estimator, 154, 299
  weighted least squares, 299
Least trimmed squares, 65, 77, 87, 89, 90, 92, 98, 108, 113, 116, 124–127, 142, 145–149, 151–157, 163–169, 171, 175–177, 179, 180, 182–191, 242, 246, 325, 335, 349, 360, 363–370, 386, 389, 416, 427, 434, 435, 438, 468, 470, 506, 507, 509, 510

least trimmed squares estimator, 65, 89, 92, 154, 434, 435, 438
Leave-one-out, 361
Level shifts, 281, 326, 328–330, 496
Leverage, 67, 69, 72, 80, 95, 98, 110, 148, 152, 160, 173, 180, 181, 187–189, 284, 286, 295, 304, 322, 363, 376–378, 381, 457
Likelihood, 5, 6, 9–11, 18, 21, 29, 30, 32, 66, 98, 143, 145, 195, 197, 198, 200, 201, 207–209, 232, 245, 298, 299, 305, 314–317, 322, 344, 345, 354, 436, 437, 446, 448, 474
  likelihood ratio test, 9, 18, 145, 198, 200, 201, 345, 436, 437, 448
  maximum likelihood estimator, 5, 6, 18, 30, 66, 198, 354
Linear model
  linearized model, 233
  linear predictor, 207, 236, 349
Link function, 207, 208, 243, 277
Lobster data, 293–297
  least-squares analysis (2002), 293–295
  Bayesian forward search (2003), 295–296
  Bayesian forward search (2004), 296–297
  timeliness & power, 297
Local minima, 86, 209
Location estimators, 21–22, 29–34
  mean (classical), 22
  median (robust), 21–23
  M-estimator of location, 30–32
  trimmed mean, 21–22
Location model, 10, 19, 30
Location parameter, 2, 3, 5, 28, 31, 59
Loglink, 206–208
L'Hôpital's rule, 200
Loyalty data, 371, 392–405, 408–415, 429, 513–516
Loyalty Cards data, 191–196, 214–215

M

Mahalanobis distance, 99, 174, 176, 335, 507, 508
Mallows' C_p, 353–355
Mandible-length data (foetal growth), 234–235
Masking, 68, 98, 100, 104, 106, 117, 129, 166, 168, 204, 216, 463
Maximum likelihood, 5, 6, 9–11, 18, 21, 29, 30, 32, 66, 98, 143, 195, 197, 198, 207, 208, 245, 344, 354, 474
  maximum likelihood estimator, 18

Mean shift outlier model, 70, 95, 97, 143, 145, 224, 458
Mean squared error, 5, 180, 182, 183, 274, 275, 354, 360, 364, 512
Median absolute deviation, 7, 177
M-estimation, 29–44
  downweighting / soft trimming, 29
  M-estimation (regression), 78–80
  monotone vs. redescending ψ, 31, 36
  properties of M-estimators, 30–36
M-estimator of location, 30, 61
M-estimator of scale, 34, 81, 435
  iterative algorithm, 33–36
  explosion / implosion breakdown, 29
  robust scale (MAD, MADN), 24, 31, 50
Minimum Covariance Determinant (MCD), 335
MM, 45, 58, 65, 77, 90, 91, 93, 94, 107, 109, 113–116, 123, 124, 148, 149, 151–158, 163–169, 171, 175–186, 189–191, 246, 318, 337, 338, 340–343, 366, 407, 429, 434, 435, 464, 465, 500–502, 512, 513
MM-estimator, 45, 58, 90, 91, 93, 94, 114–116, 123, 124, 148, 149, 153, 156, 157, 164, 171, 184, 185, 337, 338, 341–343, 464, 465, 500–502, 512
Model-based clustering, 314
Model failure, 237
  systematic, 237
Monitoring
  adaptive subset size m, 13–14
  percentage of data in agreement, 14
  monitoring graphics, 16
  monitoring size & power, 163–169
Multicollinearity, *see* Collinearity
Multiple regression, 43, 65, 78, 97, 221–223, 246, 267, 275, 281, 332, 336, 354
  extensions of, 281
Multivariate normal, 169, 173, 196, 313, 471, 473

N

NCI60 cancer cell panel data, 416–425
  cancer data, 416–425
Non-constant variance, 299–300
Non linear least squares, 236
Normal equations, 66, 72, 244, 459
  partitioned, 72
Normal inverse–gamma prior distribution, 282
Numerical variance-stabilising transformation, 249–255

Subject Index

O

Order statistics, 3, 6, 73, 88, 99, 100, 204, 282, 445
Outlier deletion, 222, 223, 389, 396
Outliers
  dispersed vs. grouped, 1–2
  outlier detection, 205–208, 217–221
  outlier detection tests, 151–168, 184–189
  robustness to moderate outliers, 5
  outlyingness, 98, 127, 498, 499
Overlap index, 172–175
Ozone data, 500, 502, 503

P

Parameter estimates
  least squares, 66
  parameter estimation, 282, 284, 287
  tests, 73
  t-statistics, 73, 80
Parameterised departures, 171–175
Parametric model, 3, 16, 170, 241, 261
Percentage of variance explained, 481, 485, 490, 492
Plot of residuals, 71, 75, 93, 94, 125, 207, 258–260, 264, 267, 270, 273, 303, 305, 330, 376, 386, 392, 399, 402, 404, 409, 419, 427, 430, 437, 468, 469, 491, 499, 513
Poison data, 201, 236, 239, 474
Poisson, 276
Posterior distribution, 283–284
Power of tests, 164–169
Predictor, 207, 236, 247, 349, 357, 358, 360–363, 431, 499, 505, 509
Principal components, 358
Prior distribution, 282–284
  fictitious observations, 283–284
  prior information (general), 281, 286
  strong prior information (effect), 281
Profile loglikelihood, 198

Q

QQ-plot, 278, 429, 430
Quasi-likelihood, 298

R

Rank, 88, 112, 319, 333
Rate of convergence, 6, 28, 63, 79, 125, 455, 456
Reciprocal link, 207, 208

Regression estimators, 87, 151, 169, 304, 366
Regression parameters, 58, 81, 173, 179, 184, 186, 277, 313, 336, 436
Regularization, 349, 357–360, 365
  methods, 357–364
Residual plots, 102–107, 117–149
Residuals
  deletion, 73, 99
  least squares, 72
  scaled, 71
  studentized, 73
Residual sum of squares, 66, 68, 89, 94, 110, 111, 144, 146, 159, 198, 199, 209, 224, 236, 238, 244, 283, 299, 353–355, 459

Ridge regression, 349, 359–363, 369, 504–506
Robust AVAS (RAVAS), 379–381, 400–406, 414–422
  algorithm, 245–248
  options, 245–48
  trapezoid option (integration), 249–255
Robust regression clustering, 304–325
  face-masks trade dataset, 307–325
  monitoring approach, 316–317
  traditional information criterion, 317
Robust regression estimators, 87, 151, 169–171, 304
Root Trimmed Mean-Squared Prediction Error (RTMSPE), 364–365

S

Sample mean, 3, 4, 17, 23, 25–27, 63, 170, 335, 456
Sample median, 4, 5, 23, 25–27, 31, 63, 449
Scale family, 32
Scatterplot matrix, 408, 437, 481
Score test
  transformation, 8–10, 12–14
  income datasets, 8–12
  monitoring plot of score test, 13–14
Selection of variables, 370
Sensitivity Curve (SC), 21–25
  link to IF, 23–24
S-estimator, 81, 82, 88, 107, 124, 140, 154, 156, 186, 189, 366, 434, 435
Shifted power transformation, 208–210
Shortleaf Pine, 236, 238, 269, 441
Sign function, 450
Simple power transformation, 197

Simulation envelope, 73, 75, 375, 386, 392, 393, 396, 397, 404, 408, 409, 418, 515

Simulation studies, 256–262, 273–277
Simultaneous estimation of μ and σ, 44–45
Size of tests, 151–161
Small-sample comparisons, 53–59
   contamination scenarios, 54–58
   choice of bdp vs. variance, 56–58
Sparse LTS, 363–367
   monitoring, 366–368
Square-root transformation, 371–381, 384–387, 397–398
Stars data, 102–106, 113–116
Starting values, 44, 185
Stepwise regression, 339–340
Structured procedure for modern robust analysis (Table 10.7), 406–407
Student's t, 26, 63, 67, 110, 111, 171, 201, 428, 453, 454
   t statistics, 110–112, 121–122, 134–137
Subsampling, 96
Surgical Unit data, 134–137
Swamping, 68, 204, 419, 463, 465

T

Three classes of estimators, 77
Timeliness and power, 297
Traditional robust approach, 8–12
   comparison with monitoring, 12–14
Transformations
   transformation to normality, 196, 235, 239

   transform both sides, 234–235
Trimmed mean *see* estimates of location

Tuning constants, 9, 36–38, 40, 42, 45, 65, 79, 82, 85, 90, 91, 95, 96, 461, 462
   tuning constant c, 85

U

Univariate income data, 8–14
   Italian municipality dataset, 10–12, 14
   U.S. Census dataset, 8–13

V

Variable selection, 358–360, 362, 381, 389, 406, 407, 426, 429, 431, 435, 437, 438, 440, 442, 443, 505, 514, 518–521
Variance-stabilising transformation (general), 195–199, 208

W

Wald test, 18, 79, 448
Weighted least squares, 31, 78, 86, 282, 284, 285, 299
Weighted mean, 31, 33
Weight function, 35–42
Weight-of-fish data set, 265–271
Windsor House Price data, 286–291
   prior–data mismatch, 289–293
   strong prior information, 291
   minimum deletion residuals, 288, 279
Wool data, 104, 106–108, 199, 201, 202, 474

Y

Yeo–Johnson transformation, 210–217
   extended, 419–422

The manufacturer's authorised representative in the EU is Springer Nature Customer Service Centre GmbH, Europaplatz 3, 69115 Heidelberg, Germany. If you have any concerns regarding our products, please contact ProductSafety@springernature.com

Printed and bound by CPI Group (UK) Ltd, Croydon, CR0 4YY
26/03/2026
02078986-0001